Advanced Structured Materials

Volume 206

Common engineering materials are reaching their limits in many applications, and new developments are required to meet the increasing demands on engineering materials. The performance of materials can be improved by combining different materials to achieve better properties than with a single constituent, or by shaping the material or constituents into a specific structure. The interaction between material and structure can occur at different length scales, such as the micro, meso, or macro scale, and offers potential applications in very different fields.

This book series addresses the fundamental relationships between materials and their structure on overall properties (e.g., mechanical, thermal, chemical, electrical, or magnetic properties, etc.). Experimental data and procedures are presented, as well as methods for modeling structures and materials using numerical and analytical approaches. In addition, the series shows how these materials engineering and design processes are implemented and how new technologies can be used to optimize materials and processes.

Advanced Structured Materials is indexed in Google Scholar and Scopus.

Dambarudhar Mohanta · Purushottam Chakraborty
Editors

Nanoscale Matter and Principles for Sensing and Labeling Applications

Springer

Editors
Dambarudhar Mohanta
Department of Physics
Tezpur University
Sonitpur, Assam, India

Purushottam Chakraborty
Surface Physics and Materials Science
Saha Institute of Nuclear Physics
Kolkata, West Bengal, India

ISSN 1869-8433 ISSN 1869-8441 (electronic)
Advanced Structured Materials
ISBN 978-981-99-7847-2 ISBN 978-981-99-7848-9 (eBook)
https://doi.org/10.1007/978-981-99-7848-9

This Springer imprint is published by the registered company Springer Nature Singapore Pte Ltd.
The registered company address is: 152 Beach Road, #21-01/04 Gateway East, Singapore 189721, Singapore

Paper in this product is recyclable.

Foreword

"This unique book brings together a wide range of important topics on nanoscale materials for sensing. With a focus on presenting some of the latest research results, it will be of interest to graduate students and researchers who are seeking to understand the way that the chemical, physical, electrical, magnetic, and structural properties of materials and surfaces influence their performance as sensors."

Andrew Kirk
Professor, Department of Electrical and Computer Engineering
McGill University, Canada

"Sensing and labeling using the materials and ideas at nanoscale is an important and versatile modern area to which this volume brings a rather unique approach. It will be valuable for the scientists at different stages in their career, from graduate students to senior and practising scientists. The chapters are diversified but contemporary which will allow the readers to comprehend both the underlying ideas and technology aspects of the fast-growing field of nanoscale in sensing research."

Ashutosh Sharma
Institute Chair Professor, IIT Kanpur
President, INSA, New Delhi, India
Former Secretary to Government of India and Head
DST New Delhi, India

"I am glad that Springer has taken the responsibility to publish the collection of chapters entitled "Nanoscale Matter and Principles for Sensing and Labelling Applications" edited by Prof. D. Mohanta and Prof. P. Chakraborty. It is enough to give the flavour of sensing and labelling applications. I congratulate the authors."

C. N. R. Rao, FRS
Linus Pauling Professor
JNCSR, Bengaluru, India

Preface

Ever since the dawn of human civilization, there has been an intrinsic human instinct to know about nature and its surroundings through sensory organs, such as eyes for sight, ears for sound, nose for smell, tongue for taste, and skin for touch. All fundamentally express the sensing actions; form and intensity of which largely depend on the genetic and epigenetic factors of individuals that are essentially governed by biophysical and physicochemical courses of nature. On the other hand, men, through creative thinking and analytical skill, can make testbeds and components for detecting and sensing species of interest. Remarkable advancement in novel materials and, at the same time, innovative scientific approaches have helped solve the existing problems and to confront obstreperous challenges. Although energy, environment, health, and hygiene are the main issues of common interest, interdisciplinary pathways may appear to bridge the gaps making science and society go hand in hand in the pursuit of happy and harmonious living.

The growth of biological cells through self-regulated gradual mass accumulation has essentially triggered the desire of scientists to replicate the phenomenon by the manipulation of individual atoms and molecules in laboratories. The bottom-up approach adopted in physical or chemical methods essentially deals with the atom-by-atom self-assembled growth of nanostructured matters of various size, shape, and morphology. Because of large surface-to-volume ratios, the atoms and molecules in nanoscale materials behave entirely in a different manner exhibiting unusual properties, resulting in appreciably lower consumption of both materials and energy in devices. Concomitantly, innovative experimental methods are being developed to design, develop, and characterize new nanomaterials with improved quality and performances. This volume is merely a handpicked collection of articles unfolding our knowledge and existing challenges in nanoscience and nanotechnology, primarily focusing on principles and scope of sensing and labeling phenomena in various technologically important fields.

Obviously, a pertinent question may arise: what is the need of such a book when such a topic is not new and for being regarded as a prime topic of most advanced research? To answer this in one sentence, this volume is an exclusive assortment of new articles describing the current developments and challenges in the frontline areas of nanosystem-based sensing. Additionally, two more relevant reasons are invoked: first, the edited volume can help the beginners, graduate students, and researchers who look for a comprehensive guidebook on basic and state-of-the art research related to sensing-based nanoscale materials and phenomena. Second, it can be a reference manual for the practicing scientists and experts who are already into the sensing business. Despite numerous books available on sensing based on the principles of nanotechnology, this book is unique, and also because it covers a wide range of key materials and applications with proper weightage to the necessary theoretical concepts. The chapters are carefully chosen from the contributions of the experts in their domain areas and compiled in the most appropriate manner.

Out of 28 chapters, one can easily find that sensing and detection employing optical, chemical, optoelectronic, plasmonic, and secondary ion mass-spectrometry (SIMS) methods form the basis of the entire volume. To name a few: quantum-dot or nanoparticle-based sensing fluorescence, humidity, organic medium, heavy metals, explosives, etc., form the core of the book. Carbon dots, rare-earth oxides and compounds, metal-organic framework, reinforced polymer, guest-host interaction as well as physics of bioconjugation and bio-functionality have strengthened the scope of sensing phenomena to a greater extent. Applications in flexible and wearable chemical sensing, as well as sensing of foodborne pathogen, macrolide, antibiotic residue, etc., are certainly noteworthy. A perfect balance between diversity in content and insightful depth is intended for readers, enabling them to refresh their knowledge in sensing and related phenomena. Both nanoscale matter and nanoscale principles are at the verge of sensing at large which, in fact, justify the title of the book. Each chapter, unique in itself, is prepared and edited with extreme care and dedication keeping the interest of the readers of prime importance.

It was never an easy task to bring a good number of expert authors, reviewers, and colleagues into the same platform and, especially, when it comes to the documentation of their specialized knowledge. Although words of the book have limitations, but the scope is not. Undoubtedly, starting from inviting manuscripts to reviewing and editing is a Herculean task that involves the participation and support of many. Nevertheless, unfailing effort and goodwill among academia play a big role in making differences even in the time of adversity. One of the purposes was to utilize the Corona pandemic period, and this book is a thriving effort in this context. At the same time, our fellow-feeling will remain forever with those colleagues and family friends who came strong and proved fit to counter the tragedy that mankind has witnessed recently.

We do express our heartfelt thanks to all the reviewers who scrupulously reviewed and potentially enriched the quality of chapters by their constructive criticisms and thoughtful comments. Last but not the least, we express our earnest thankfulness and utmost gratitude to those distinguished experts who have generously reviewed the overall volume with their enlightening words.

We sincerely hope, the book will create enthusiasm among young researchers and aspirants, particularly those pursuing research in materials science and nanobiotechnology. Keeping aside the application strategies, the overlapping of biotechnology and nanotechnology has also been highlighted in view of generating fundamental interest. We, the editors, are thankful to the publishers and concerned parties for giving permission to the authors to reproduce the figures and textual facts in the relevant chapters. While every effort has been made to make the book simple, clear and flawless as much as possible, some errors may remain inadvertently. The suggestions in this regard will be appreciated.

Tezpur, Sonitpur, India Dambarudhar Mohanta
Kolkata, India Purushottam Chakraborty

Contents

About the Editors

Dambarudhar Mohanta is Professor at the Department of Physics, Tezpur University, Assam, India, which he joined as Lecturer in 1998. Earlier, he worked as UGC/CSIR-NET JRF at the Department of Physics, Utkal University, Bhubaneswar, Odisha, for about a year. He obtained his Master's degree in Physics from Maharaja Purna Chandra College, Utkal University, and Ph.D. from Tezpur University, respectively, in 1994 and 2004. After his Ph.D., he availed the Summer Teacher Fellowship and worked at Physical Research Laboratory, Ahmedabad, and at S. N. Bose National Centre for Basic Sciences, Kolkata, respectively, in 2003 and 2006. A long term association with Inter University Accelerator Centre (IUAC), New Delhi has also helped him pursue distinct and diverse research avenues. He received the BOYSCAST Fellowship from DST-SERB, New Delhi, during 2006–2007 and worked at the Department of Computer and Electrical Engineering, University of Wisconsin-Madison, USA.

He has visited several countries on academic pursuits and presented conference papers at National Institute for Nuclear Physics, Italy (2003), National University of Singapore (2009), and Aristotle University, Thessaloniki, Greece (2011). He was awarded Indo-US Research Fellowship during 2011–2012 to work in the School of Engineering and Applied Sciences, Harvard University, USA. With a student, one of his articles published in the Bulletin of Materials Science bagged the Material Research Society of India (MRSI) best paper award for the year 2012. He is a life member of MRSI, Indian Physics Association (IPA), Indian

Physical Society (IPS), Orissa Physical Society (OPS) and a regular member of American Physical Society (APS), MRS, Optical Society of America (OSA), and Optical Society of India (OSI). With more than 140 papers, including review articles, papers in proceedings, book chapters, and technical reports, he has supervised fifteen PhD theses and many Masters' projects. His current research interest includes optoelectronic features of nanoscale semiconductors and rare-earth oxides, 2D materials, radiation-induced phenomena, nano-bio interface, electrochemical and biosensing, biophotonics, and soft matter physics.

Purushottam Chakraborty is a former Senior Professor of Physics at the Surface Physics and Materials Science Division, Saha Institute of Nuclear Physics, Kolkata, India, and a former Adjunct Professor of Physics, University of Pretoria, South Africa. With more than 150 research articles including reviews, monographs, and invited book chapters, he is Editor of the books, *Ion Beam Analysis of Surfaces and Interfaces of Condensed Matter Systems, Photonic Materials, Encyclopaedia of Materials: Electronics*, and *Low-dimensional Materials, Systems and Applications*. He also is a regular contributor to the *Journal of Physics*. A leading expert on materials analysis using ion beams, Prof. Chakraborty was awarded the "Most Eminent Mass Spectrometrist of India" and conferred with the gold medal by the Department of Atomic Energy, the Government of India, for his outstanding contributions in secondary ion mass spectrometry (SIMS). He received the "Premchand Roychand Scholarship" and the "Mouat Medal" of the University of Calcutta. His research areas include atomic collisions in solids, ion-beam modifications and analysis, low-dimensional materials, X-UV optics, nonlinear optics, photonics, etc.

He is an elected Fellow of the West Bengal Academy of Science and Technology and the Indian Chemical Society. He indigenously fabricated an RF-quadrupole mass spectrometer for the first time in India. His "MCs*n*+ molecular-ion based SIMS" is considered to be innovative for compositional analysis of nanostructured materials. His work on the fabrication of "layered synthetic microstructures (LSM)" is recognized as a pioneering contribution in the realization of optical devices for the extreme ultraviolet to soft X-rays. His works on "ion beam synthesis of metal-glass nanocomposites" have led to the remarkable achievements in the development of novel photonic materials.

Professor Chakraborty worked at FOM-Institute for Atomic and Molecular Physics, the Netherlands; International Centre for Theoretical Physics (ICTP) and Padova University, in Italy; Laval University, Canada; Osaka Electro-Communication University, Japan; University of Pretoria, South Africa; Pontifical Catholic University of Rio de Janeiro, Brazil; etc. He delivered lectures at Imperial College, London, UK; Maria Curie-Skłodowska University and Polish Academy of Science in Poland; Vanderbilt University, IBM T. J. Watson Research Centre, Furman University, Yale University, and Rutgers University, in the USA; Bielefeld University, Friedrich Schiller University, and Kaiserslautern University, in Germany; University of Western Australia; Newcastle University, UK; Chinese Academy of Science, Kyoto University, and Spring-8, in Japan; Witwatersrand University, iThemba Labs for Accelerator Science, Nelson Mandela Metropolitan University in South Africa; Asian Institute of Technology in Thailand, Tata Institute of Fundamental Research (TIFR), several Indian Institute of Technology, Tezpur University, Lucknow University, University of Delhi, International School of Photonics, etc. in India.

Chapter 1
Alkali Containing Molecular Ions in SIMS: A Cutting-Edge Ion-Beam Technique for Materials Quantification in Nanoscale Systems

Purushottam Chakraborty

Abstract Secondary Ion Mass Spectrometry (SIMS) is an extremely powerful and sensitive ion-beam technique for detection and quantification in materials. Extent of secondary ions of a specific species in a sample sturdily depends on instantaneous local surface chemistry of that sample causing a strong variation in ionization cross section of the emitted species. This is so-called "Matrix Effect", which makes the SIMS technique challenging for quantitative analysis of materials, even though the technique has highest detection sensitivity (<ppb) and exceptional depth resolution (<1 nm). Therefore, the compensation of "matrix effect" is necessary for direct quantification without calibration standards. If an alkali atom (lithium, rubidium, potassium, sodium, caesium, ... etc.), referred to as "A", comes to the vicinity of a matrix element (M) on sample surface, this alkali ion can attach to a sputtered neutral atom (M^0) and forms a quasi-molecular $(MA)^+$ ion. Such formation of $(MA)^+$ molecular ion is strongly correlated to atomic polarizability of element M. Since M^0 emission is completely independent of MA^+ ion formation, the "matrix effect" drastically decreases. Although these $(MA)^+$ molecular ions have great applications in material analysis without the need for "calibration standards", they are generally very low in yields. On the contrary, $(MA_2)^+$ molecular ions have yields much higher even by orders of magnitude compared to $(MA)^+$ ions and can magnificently meet the requirements for quantification of low-concentration species. Caesium is mostly preferred among all alkali metals because of strongest reactivity and electropositive nature of caesium. Present chapter deals with ion emission phenomena and matrix effect in conventional SIMS, and state-of-the-art "MCs_n^+ (n = 1, 2, ...)-based SIMS" method for precise quantification of materials, in general and nanoscale systems, in particular.

Keywords SIMS · Alkali ion · Matrix · ZnO · Nano · Heterostructure

P. Chakraborty (✉)
Saha Institute of Nuclear Physics, Kolkata 700064, India
e-mail: purushottam.chakraborty@gmail.com

D. Mohanta and P. Chakraborty (eds.), *Nanoscale Matter and Principles for Sensing and Labeling Applications*, Advanced Structured Materials 206,
https://doi.org/10.1007/978-981-99-7848-9_1

1.1 Introduction

Nanoscale materials have low-dimensional structures. Controlled fabrication of these materials crucially relies on how precise the chemical and structural analyses of these materials are. Plentiful methods have been in practice, out of which the methods based on sputtering involve the sectioning of atomic layers over a period of time and subsequent detection of individual elements present in the layers. Sputtering is, however, a complex ion–surface interaction process, introducing some distortional effects into surface morphology and sample composition. Secondary Ion Mass Spectrometry (SIMS) is unique among all sputter-based analytical methods, in which a minute fraction of sputtered particles is ionized positively and negatively. These ions are called "secondary ions". Effective thickness of a sizzled layer is of few atomic dimensions and can be estimated from the expression of sputter yield that involves "information depth", "average energy of sputtered particles", and "nuclear stopping power" [1]. Thus, secondary ions provide information about surface and near-surface compositions.

Sputter products constitute electrons, atoms, molecules, clusters, molecular fragments, and ions. Minute fraction of sputtered neutral species is excited above ground states and subsequently undergoes processes, such as (1) de-excitation through "radiative electronic transition" leading to photon emission in ultra-violet to visible optical range, (2) "non-radiative electronic transition" leading to emission of "Auger electrons", and (3) "ionization via tunneling of electrons" between valence band of the solid and energy levels of sputtered species in resonant energy states. These dynamic charge-transfer processes can be even more complex due to ion-induced electronic and structural perturbations. Exact charge-state formation mechanism can be understood in terms of "symmetric and asymmetric collisional effects" on single- and double-charged ion formation in sputtering process [2]. Kinetic energy distributions of secondary ions emitted from Si surface suggested that Si^+ ions are largely formed in the upper surface layers and Si^{2+} ions are produced owing to the Si 2p shell vacancies created under target–target symmetric collisions, followed by Auger-electron emission [2]. Sputtered ions have large kinetic energy distributions that are mostly peaked at around 2–5 eV. The energy-selected ions are mass analyzed and detected using a combination of a suitable mass spectrometer and an appropriate ion-collection optics. As ion bombardment goes on, the material is continuously eroded layer by layer, providing in-depth quantitative information about material constituents through dynamic SIMS. As a material is irrevocably consumed in this mode of SIMS analysis, it is a consumptive process. However, this problem is principally compromised at the expense of highest "detection sensitivity", allowing the technique to provide maximum information from infinitesimal volumes of materials. Figure 1.1 shows schematic diagram of an instrument used in surface and depth-profile analysis.

As secondary ion intensity strongly depends on "ionization efficiency" of a sputtered atom or molecule of that element, the instantaneous local surface chemistry has a dominant role on secondary ion emission. Such ion-yield variation in SIMS

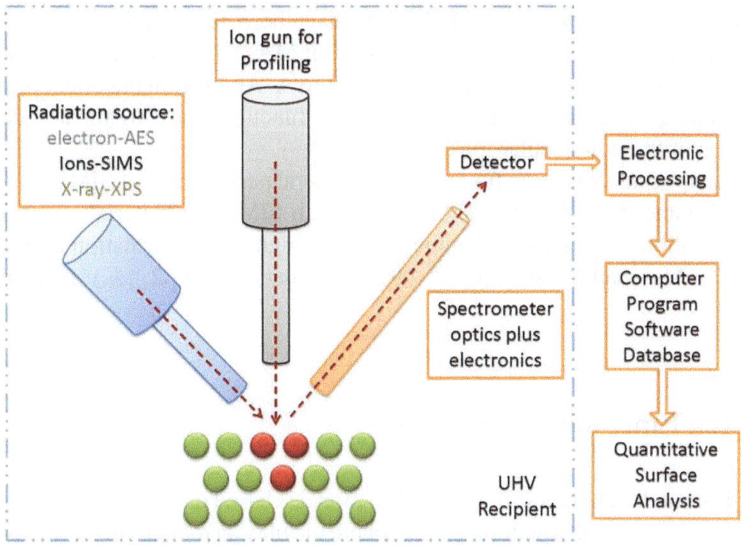

Fig. 1.1 Schematic diagram of the principal arrangement of a surface-analysis instrument used for sputter depth-profile analysis

for a spccics, known as "matrix cffcct", rcsults in non-lincar variation of ion signals with variation in concentration of that element in the matrix. This matrix effect is, however, intentionally employed in detecting elements that extremely low ion yields. Obviously, "matrix effect" requires appropriate compensation for estimating exact concentration of a certain species in the matrix. Matrix effect is compensated by using "calibration standards" or through estimation of "relative sensitivity factor" (RSF). In context of calibration standards, "ion-implanted standards" allow the precision measurements while quantifying through SIMS [3]. The drive toward reduction of matrix effect has led to discovery of another technique called "secondary neutral mass spectrometry" (SNMS), in which secondary neutral yields seem to be unaffected by changes in local surface chemistry, making this technique directly quantitative without the need of calibration standards [4].

A great success in matrix effect compensation has been achieved using "alkali-ion-based molecular SIMS" approach [5]. If an alkali metal, referred to as "A", is present at proximity of probing element (M), a quasi-molecular ion like $(MA)^+$ is formed by the attachment of this alkali ion to the probing atom (M). Formation of such $(MA)^+$ molecular ion is strongly correlated to atomic polarizability of element M. Emission process for M^0 is independent of MA^+ ion formation process, resulting in a dramatic decrease in "matrix effect", a process like SNMS. Caesium, compared to other alkali elements, is strongest in electropositivity and hence the MCs^+ ions are highly preferred in SIMS analysis. Although MA^+ molecular ions have found applicability in direct material quantification, these ions usually suffer from low yields. On the contrary, $(MA)_2{}^+$ ions provide a detection sensitivity better than several orders of magnitude, as these $(MA)_2^+$ ions have been found to be of much higher yields. "Recombination coefficients" for MA^+ or MA_2^+ ions depend on whether M is electropositive or electronegative with respect to A. Apart from the effect of surface binding energy, changes in "local surface work function" have significant effects in formation of these molecular ions.

The chapter addresses the fundamentals, general perspectives of SIMS, and prodigious advantages of "alkali-based molecular-ion SIMS" in material quantification without "calibration standards", giving proper weightage on formation mechanisms of MA_n^+ (M—matrix element, A—alkali element; n = 1, 2, ...) molecular-ion complexes.

1.2 Ion Emission Mechanisms

Local Thermodynamic Equilibrium (LTE) model is based on the formation of a high-temperature plasma at "sputtering site", where the plasma remains in thermodynamic equilibrium with relative ion populations. Considering the validity of LTE model, positive ionization cross section is governed by an Equation [6].

$$\frac{n^+}{n^0} \propto \exp\left(\frac{-I}{kT}\right) \tag{1.1}$$

where "I" is ionization potential of an element, T is plasma temperature, K is Boltzmann constant, and n^+ and n^0 are densities of positive ions and neutral atoms, respectively, in the plasma. Fraction of positive ions (n^+/n^0) denotes "positive ion emission probability". Energy distributions of ions, electrons, and neutral atoms follow the Maxwellian distribution of velocities. Andersen and Hinthorne [6] used "Saha–Langmuir equation" of thermal ionization for calculating "dissociation constant", expressed as.

$$\frac{n^+ n_e}{n^0} = \frac{u^+(T)}{u^0(T)} 2 \frac{(2\pi mkT)^{3/2}}{h^3} e^{\frac{-(E-\Delta E)}{kT}} \tag{1.2}$$

where n_e is density of electrons in the plasma; u^+, u^o, and 2 are partition functions for ions, atoms, and electrons, respectively; and (ΔI) is reduction of ionization potential due to plasma. The product (kT) indicates average energy of emitted particles.

Using Eqs. (1.1) and (1.2), the modified positive ionization probability $\beta^+ (= n^+/n^0)$ is expressed as

$$\beta^+ \propto n_e^{-1} \exp\left(\frac{-I}{kT}\right) \tag{1.3}$$

Equation (1.3) explains "matrix effect in SIMS", i.e., the increase in β^+ with the decrease of n_e in the plasma. This effect is equivalent to decrease in "surface work function" due to oxygen present on the surface or in an oxide surface, thereby decreasing the probability of neutralization of the emitted positive ions by electron capture. On the contrary, the increase of n_e or increase in work function for caesium-coated surfaces causes an increase in the negative ion emission. Therefore, LTE model explains the matrix effect effectively, although it dauntingly fails to explain the varying "kinetic energy distributions" of sputtered atoms, molecules, and cluster ions, in contrast to Boltzmann distributions.

"Bond-breaking Model" [7] also explains the matrix effect observed during sputtering of ionic solids or oxides [8], where a metal atom, sputtered due to breaking of an oxide bond, creates an "oxygen-surrounded vacancy" on surface. This oxygen-surrounded cation vacancy having strong electron affinity attracts an electron from ejected atom, causing increase in positive ion emission. This phenomenon is qualitatively like the "Landau–Zener treatment" of "ion-pair dissociation of polar molecules" [9]. The model, however, restricted to the emission of positive ions only, fails to explain the emission of negative ions, molecular ions, and multiple-charged ions.

"Electron-tunneling model" [10, 11] is essentially based on survival probability of an evading ion above the surface, thus providing the emission probability for both positive and negative ions. Fermi energy E_F has an important role on the survival of an emitted ion. Depending on the location of an ion relative to surface, the ionization energy changes. At the crossing distance, where the ionization energy equals to Fermi energy, electrons in metal will tunnel out to occupy the atomic levels of emitted ion through "resonant tunneling", leading to the neutralization of the emitted positive ion. Since the position of Fermi level is a measure of surface work function (ϕ), the positive and negative ionization probabilities are expressed as [3].

$$\beta^+ \propto \exp\left[-\frac{I-\phi}{v_\perp}\right] \tag{1.4}$$

$$\beta^- \propto \exp\left[-\frac{\phi-A}{v_\perp}\right] \tag{1.5}$$

Electron-tunneling model has been found to be remarkably effective in explaining the dependency of ionization probabilities on surface work function, emission velocity and emission angle for atoms sputtered from metallic, non-metallic, and oxide surfaces.

1.3 Matrix Effect: A Bottleneck in Quantitative Analysis

Irrespective of ion formation mechanisms, the quantification in SIMS is, in principle, achieved simply through conversion of secondary ion current (I_x) of a given species (x) into its concentration C_x in a matrix (m). Concentration versus depth is done by calibrating time scale into depth scale, provided the erosion time is linearly proportional to the eroded depth in the sample. Considering that a steady situation exists between surface composition of the matrix and sputtering conditions, I_x can be expressed [3] as

$$I_x^{\pm} = I_p Y_{tot} \beta_{x,\,m}^{\pm} f(\Delta E)\, f(\Delta\Omega)\, T_{x,m} \eta_{x,m} C_x \tag{1.6}$$

where $I_x^{\pm}$ (positive or negative) is secondary ion current and I_p is primary ion current. Both are expressed as ions/second. The combined subscript (x, m) for a certain parameter in Eq. (1.6) indicates a possible dependence of that parameter on both sputtered species and target matrix. Partial sputtering is ignored as the surface composition gets rearranged under steady-state condition making sputtered flux composition identical to that of bulk. Y_{tot} is total sputter yield of target matrix, $\beta_{x,\,m}^{\pm}$ is positive or negative ionization probability of species (x) sputtered from matrix (m), T is transmission factor of mass spectrometer, f(ΔE) and f(Ω) are fractions of ionized particles that enter through energy- and solid-angle acceptance windows of mass spectrometer, η is detector efficiency, and C_x is "fractional concentration" of element x. Parameter $\beta_{x,\,m}^{\pm}$ depends strongly on local surface chemistry, making the ion yield vary dramatically. This is known as "matrix effect", as evidenced by significant rise in positive ion yield under the presence of an electronegative species (oxygen or fluorine) on surface [12]. Ion enhancement is more as the electronegativity is stronger [13]. An enhancement is similarly observed in negative ion emission when an electropositive element, like Cs, is present on surface. The matrix effect being unpredictable and not so well characterized results in a serious bottleneck on quantification, and thus needs to be appropriately compensated.

1.3.1 Matrix Effect Compensation

Matrix effect is inclusive in SIMS measurements and cannot be ignored. Therefore, several approaches to minimize matrix effect are adopted in quantification. Using "calibration standards" involves the preparation of closely equivalent standards. The "Ion implantation standard" is a widely used technique for preparing fairly accurate standards [3]. Other methods are "relative sensitivity factor" (RSF) [3], "infinite velocity approach [14]", "highly charged ion-based SIMS" [15], etc. The strong decoupling of secondary ion production from ionization probability in highly charged ion-based SIMS has made the technique free from matrix effect. Each of the above techniques has its advantages and limitations over the others. RSF approach is the most popular among all, although the accurate determination of RSF values is increasingly critical in converting signal intensities to concentrations in ppm range [16].

MCs_n^+ (N = 1, 2, …..) Molecular Ions

Although quantification of a material is possible by using "standards" having composition almost equivalent to that of the sample material, it is difficult to quantify or interpret data across the interfaces of different materials. In this context, "alkali element (A)-based MA^+-SIMS" approach has been found remarkable in reduction of matrix effect [17]. As caesium has highest reactivity among all alkali elements, Cs^+ ions are preferred as primary ions and MCs^+ ions are monitored in MA^+-SIMS approach. The method of analysis is known as MCs^+-SIMS technique [3]. M is referred to as "matrix element". Under steady-state condition, MCs^+ ion flux reflects the atomic concentration of element M in the matrix. For a variety of semiconductors, MCs^+ yield quadratically depends on "atomic polarizability" (α) of element M (ion yield $\propto \alpha^n$), where $n \approx 2$ [18], establishing the conjecture that polarizability of M principally governs the creation of MCs^+ ion. In this case, ion and ion-induced dipole moment interact each other. MCs^+-SIMS technique has been found to be extremely useful while analyzing a metal/semiconductor interface (Fig. 1.2). The absence of "matrix effect" was established in MCs^+-SIMS analysis of a thin (0.3 μm) gold film on GaAs substrate [17]. The appearance of a large hump of Au^+ signal across the interface signifies a characteristic interface peak due to native oxygen or other electronegative species present at interface. This interface peak, essentially appeared due to "matrix effect", was found to be successively reduced under different conditions and completely disappeared while $AuCs^+$ signals were monitored (Fig. 1.4e).

Formation mechanisms of MCs^+ ions, followed by diagnostic capability of MCs^+-SIMS methodology, were systematically examined [19]. Based on all experimental observations, MCs^+ ions have been proposed to form via recombination of independently sputtered neutral M^0 atoms with Cs^+ ions. For such formation mechanism to be valid, intensity ratio $I(MCs^+)/I(Cs^+)$ is expected to vary linearly with sputtering yield of M and inversely with mean velocity of M [20].

$$I(MCs^+)/I(Cs^+) \propto Y_M/ < V_M > \tag{1.7}$$

where $<v_M>$ is mean velocity of sputtered M atoms that reflects the density of M atoms in effective emission volume. Normalized MCs^+ intensity has been found to exponentially vary with surface work function [21]. A close agreement between the variation of $AgCs^+$ and Cs^+ intensities with variation in surface work function directly indicates a correlation between Cs^+ and MCs^+ intensities.

From Eq. (1.4), the relationship between the probability of ionization (P^+), ionization energy (I) and surface work function (ϕ) can be extracted as

$$P+ \propto \exp\left[-(I-\phi)/\varepsilon_0\right] \tag{1.8}$$

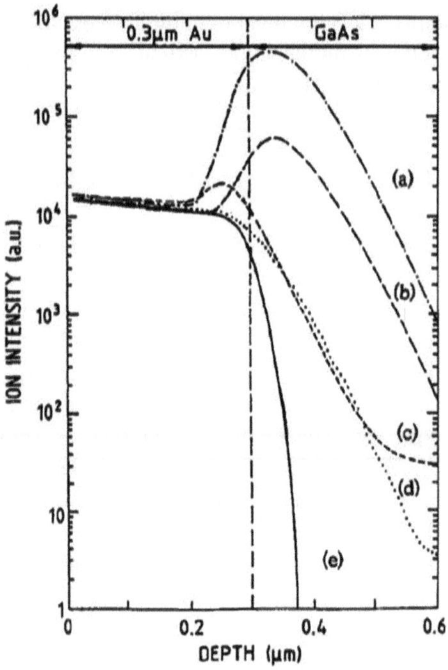

Fig. 1.2 Au profiles under different experimental conditions of a gold layer (0.3 μm thick) deposited on a GaAs substrate. **a** Au^+ secondary ions under oxygen bombardment, **b** Au^- ions with an offset of 80 eV under oxygen bombardment, **c** Au^- under Cs bombardment, **d** Au^0 atoms (SNMS) under O_2 bombardment, **e** $(AuCs)^+$ ions under Cs bombardment [17]

where ε_0 is related to normal component of emission velocity of a secondary ion. Variation of "normalized Cs^+ counts" with surface work function changes ($\Delta\phi$) has shown a good agreement with theoretical prediction for $\varepsilon_0 = 0.29$ eV [5]. The leading parts of energy distributions of secondary ions are reflected by instantaneous surface chemistry, and the trailing parts are reflected by kinematics of secondary ions. Therefore, the varying slopes of leading edges of energy distributions give measures of relative changes in surface work function [19]. Figure 1.3 gives the estimation of surface work function changes extracted from the tangents drawn at the leading edges of Cs^+ energy distributions under varying oxygen environments.

Under combined influence of caesium and oxygen, $AgCs^+$ intensity was found to linearly increase with Cs^+ intensity for all oxygen exposures in the sputter region [5], supporting the notion that MCs^+ formation occurs via recombination of sputtered neutral M^0 atom with re-sputtered Cs^+ ion. Recombination coefficient for MA^+ molecular ion (M—matrix element; A—alkali element) is seen to be greater for higher electropositivity of element M. The fact that MCs^+-SIMS approach explains the nonappearance of matrix effect and can be effective for direct quantification has been established from the analysis of a Zn-diffused InGaAsP-bilayer of varying compositions [22]. As seen in Fig. 1.4, the compositional variation of the matrix has almost insignificant effect on ion yields of the major elements. In fact, the signals for major elements represent the actual densities of the elements. Thus, MCs^+-SIMS method can be directly employed in quantification without the need of "calibration standards". However, as MCs^+ ions have extremely low yields, detection of

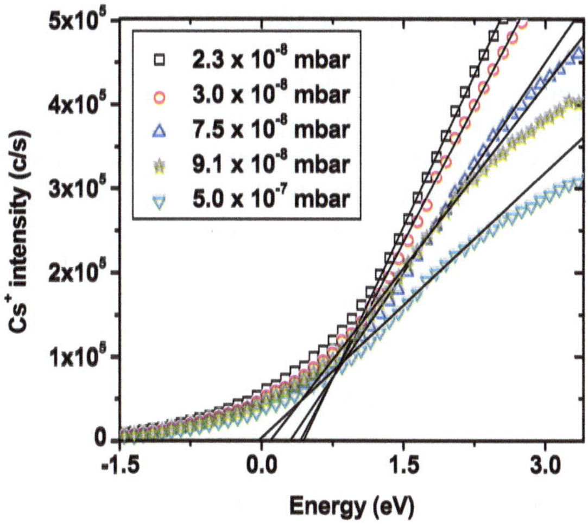

Fig. 1.3 Work function estimation from Cs^+ energy distributions. Relative changes in the intercepts on the energy axis give an estimate of relative changes in surface work function [5]

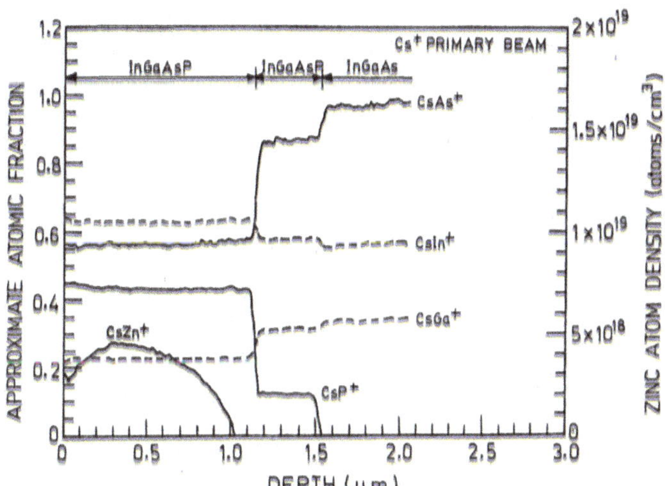

Fig. 1.4 MCs^+-SIMS profiles of an InGaAsP bi-layered structure of varying composition, grown on InGaAs substrate. Intensities of $InCs^+$, $GaCs^+$, $AsCs^+$, and PCs^+ straightway represent In, Ga, As, and P contents in the layers [22]

MCs_2^+ molecular ions shows a better sensitivity. This becomes more prominent when matrix elements have relatively stronger electronegativities with respect to caesium. A comparison of detection sensitivities for MCs^+ and MCs_2^+ ions from a fluorine-implanted Si wafer has evidently established the phenomenon, where the intensity of FCs_2^+ ions was found to be higher compared to FCs^+ ions by orders of magnitude [17].

Energy distributions of MCs_2^+ ions help understand the formation mechanisms of these ions [23]. From the variation of energy distributions for MCs_2^+ ions and their constituents with respective surface binding energies, it is evident that the sputtered neutral atoms (M^0) are responsible in MCs_2^+ formation [24]. Out of two possible formation mechanisms [24], $MCs^0 + Cs^+ \rightarrow MCs_2^+$ mechanism appears to be more feasible, especially for electronegative elements (M) [24]. In this case, MCs^+ ion gets neutralized capturing an electron resulting in the formation of a neutral MCs^0 [24] molecule. Intensity of MCs_2^+ molecular ions can be expressed [19] as

$$I_{MCs_2^+} \propto Y c_M Y^2 (c_{Cs})^2 P^+ f_{MCs^+} f_{MCs_2^+} \tag{1.9}$$

where Y is "total sputter yield", C_{Cs} is caesium concentration in the matrix, C_M is concentration of M, P^+ denotes "positive ionization probability" for Cs, and f_{MCs+} and f_{MCs2+} are formation probabilities of MCs^+ and MCs_2^+ molecular ions, respectively. Considering M, c_M, f_{MCs+}, and f_{MCs2+} as constant for a particular matrix element, $I^+_{MCs2} \propto P^+$ [25], indicating that MCs_2^+ ion yield is directly proportional to positive ionization cross section of caesium. This was further evidenced by exponential dependence of $AgCs_2^+$ intensity on change in surface work function ($\Delta\phi$) for silver, similar to AgCs + emission. These experimental observations obviously verify the perception that MCs_2^+ molecular ions form via combination of MCs^0 molecules with Cs^+ ions in near-surface regions [19]. Formation probabilities for MCs^+ are MCs^{2+} ions are proportional to the spatial and temporal correlation factors of these molecular ions which are seen to remain constant for any change in surface environment [19]. The proportionality constant is related to the square of the atomic polarizability of M, as expressed in the following [24].

$$f_{MCs_n^+} = \gamma_{M^0}^{pol} \mathrm{T}_{MCs_n^+} \tag{1.10}$$

where $\mathrm{T}_{MCs_n^+}$ is special and temporal correlation factor and $\gamma_{M^o}^{pol}$ is constant of proportionality, related to square of atomic polarizability.

1.4 Composition Analysis Using MCs⁺-SIMS Approach

1.4.1 Alloy and Superlattice Structures

MCs^+-SIMS was found effective in quantifying Ge in SiGe alloys up to Ge concentration of 23.5% [26], whereas for higher Ge concentrations, quantification was possible with lower sputtering energies [27–29]. However, no reliable data of MCs^+-SIMS analysis were available for Ge contents higher than 50 at% in SiGe alloy films. For accurate quantification of Ge in $Si_{1-x}Ge_x$ alloys ($0 < x < 0.72$), MCs^+-SIMS has been found innovative, where a complete (100%) compensation in "matrix effect" was achieved over an entire range of Ge concentrations, irrespective of Cs^+ ion energy [30].

Following Eq. (1.9), MCs^+ intensity is expressed as [31].

$$I_{MCs^+} \propto Y c_M Y c_{Cs} P^+ f_{MCs^+} \tag{1.11}$$

where the parameters Y and P^+ carry same meanings as discussed in case of MCs_2^+ ions. The parameters c_M and c_{Cs} are fractional concentrations of M (matrix) and Cs, respectively. As already discussed, f_{MCs^+} is constant both for germanium and silicon. Therefore, Eq. (1.12) can be rewritten as

$$\frac{I_{GeCs^+}}{I_{SiCs^+}} = K\frac{c_{Ge}}{c_{Si}} \tag{1.12}$$

where the constant "K" corresponds to "relative sensitivity factor" (RSF) in composition analysis of $Si_{1-x}Ge_x$ using MCs^+-SIMS. Figure 1.5 shows the plots of intensity ratios (I_{GeCs+}/I_{SiCs+}) obtained from $GeCs^+$-SIMS and $SiCs^+$-SIMS data against concentration ratios (c_{Ge}/c_{Si}) obtained from Energy-Dispersive X-ray Analysis (EDX) for various ion-impact energies [30]. Obviously, the slopes of the lines indicate RSF values for different impact-ion energies. The linear relationship between these two ratios indicates the "complete absence of matrix effect" for all possible Si and Ge concentrations in Si_xGe_{1-x} alloys under any impact-ion energy. Such a "full compensation" of matrix effect is possibly due to low steady-state surface concentration of caesium under higher ion-impact angles. This experiment has obviously validated the exquisiteness of "MCs_n^+-SIMS (n = 1, 2)" procedure in absolute quantification of materials.

With monitoring of MCs_2^+ ions in the "MCs_2^+-SIMS" mode of analysis, slope of the linearity of I_{GeCs+}/I_{SiCs+} versus c_{Ge}/c_{Si} increased to 45^0, signifying a one-to-one

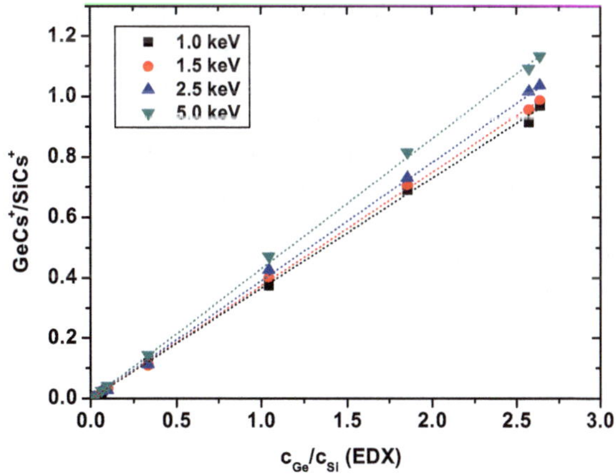

Fig. 1.5 $GeCs^+/SiCs^+$ intensity ratio versus germanium concentration ratio [30]

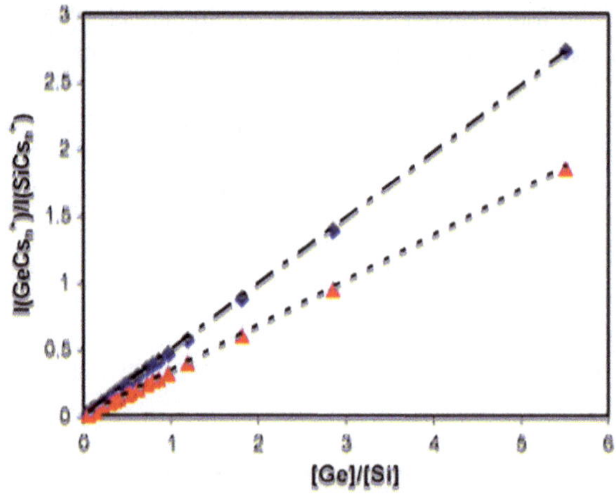

Fig. 1.6 I_{GeCs+}/I_{SiCs+} as a function of c_{Ge}/c_{Si} for a set of Ge_xSi_{1-x} samples. The red solid triangles represent MCS^+-SIMS and black solid squares represent MCs_2^+-SIMS. Slope of the linearity in case of MCs_2^+-SIMS is 45°, indicating a one-to-one match between intensity ratios ($GeCs^+/SiCs^+$) and concentration ratios (c_{Ge}/c_{Si}) [27]

match between intensity ratios ($GeCs^+/SiCs^+$) and concentration ratios (c_{Ge}/c_{Si}), as evident in Fig. 1.6. This experiment has once again established the superiority of analytical capability of MCs_2^+-SIMS over MCs^+-SIMS [27].

MCs^+-SIMS methodology was found to be highly successful in quantifying a multilayer structure [30]. Study was undertaken on an MBE-grown Si/Ge multilayer with a nominal thickness ~10 nm for each Si/Ge bilayer [30]. Using Eq. (1.12), Relative Sensitivity Factor (RSF)s for Ge were calculated from the slopes of I_{GeCs+}/I_{SiCs+} (MCs^+-SIMS) versus c_{Si}/c_{Ge} (EDX). From the concentration profiles obtained through MCs^+-SIMS, "electron density (ρ_e) versus thickness (z) profile" for multilayer is extracted through expression [30],

$$\rho_e(z) = N_A \sum_{M=Ge,Si} \frac{Z_M c_M(z)}{A_M} \tag{1.13}$$

where N_A is Avogadro number, Z_M is atomic number of matrix element M, $c_M(z)$ is concentration profile for M (Ge, Si), and A_M is mass number of M (Ge, Si).

Figure 1.7 represents the electron density profile (EDP) (red circles) extracted through MCs^+-SIMS data [30] and EDP (blue lines) obtained independently from X-ray reflectivity (XRR) [32]. As evident from Fig. 1.7, interface positions obtained from XRR and MCs^+-SIMS depth profiles match perfectly well. Discrepancy between electron densities obtained separately by these two measurements is due to possible inter-diffusion between Si and Ge across interfaces of the multilayer occurred in sputtering.

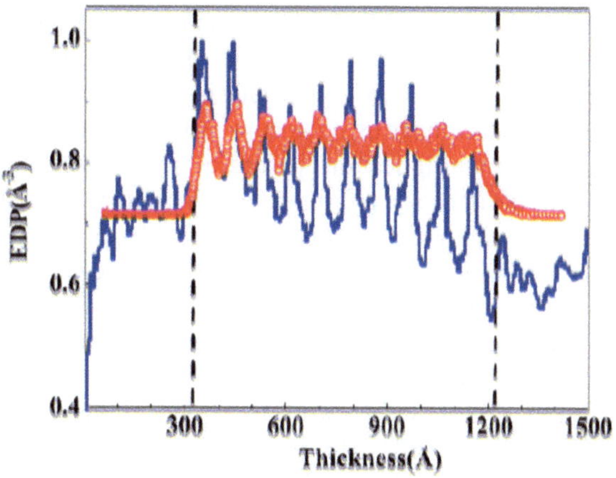

Fig. 1.7 Electron-density profiles of a Si/Ge multilayer, as obtained from XRR (blue solid line) and MCs^+-SIMS (red circles). Two vertical dashed lines indicate the depth positions for expected buffer and cap layers [30]

1.4.2 *ZnO and ZnS/ZnO Heterostructured Nanowalls*

ZnO nanostructures exhibit a range of beneficial optoelectronic properties. In bare ZnO nanostructures, "oxygen-surrounded vacancies" form on surfaces following sputtering of Zn atoms through breaking of oxide bonds of ZnO. Obviously, the enhanced Zn^+ emission follows the "bond-breaking mechanism". X-ray diffraction of "self-assembled ZnO nanowalls", grown on Al substrates [33], warranted hexagonal wurtzite crystal structure of ZnO and formation of Al_2O_3 layer at interfaces between Al substrate and ZnO nanowalls. SIMS spectra of unannealed and vacuum-annealed ZnO nanowalls, grown on Zn substrates (Fig. 1.8), have shown no appreciable change in Zn^+ emission under thermal treatment [33] and also ascertained the presence of O_2^- and OH^- surface adsorbates that essentially

govern the luminescence features of ZnO nanowalls. Large deficiency of lattice oxygen in vacuum-annealed nanowalls, compared to unannealed nanowalls, was established by SIMS analysis [33]. In another work, a modified light emission was detected in SnO-coated ZnO needle-like nanostructures [34]. EDX and SIMS analysis exhibited the formation of SnO phase on ZnO surface. Luminescence response

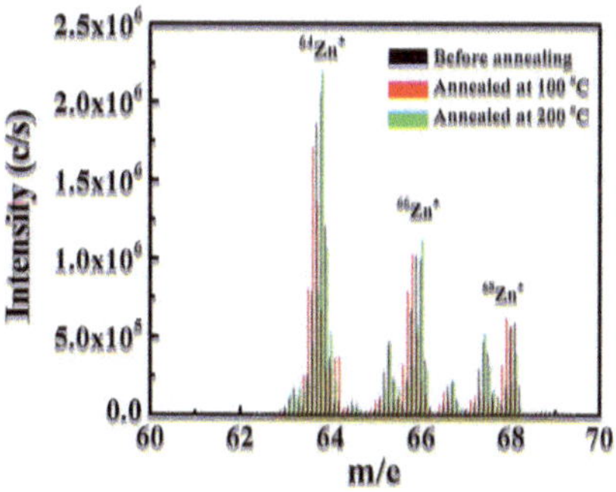

Fig. 1.8 SIMS spectra for Zn^+ ion emission from unannealed and vacuum-annealed nanoscale ZnO walls [36]

of hetero-structured SnS/ZnO system revealed quenching of "defect-related emission" through surface passivation of ZnO. Lower Sn:Zn molar ratios in hetero-structured material exhibited higher photoconductive responses compared to bare ZnO. Enhancement of photocurrent was explained in terms of "carrier separation and carrier multiplication" processes occurring at SnO-ZnO heterojunctions [34].

Matrix effect was prominent in SIMS analysis of bare ZnO- and hetero-structured ZnS/ZnO nanowalls, indicating ZnS phases present over ZnO surface. Use of "MCs^+-SIMS" was found effective in quantitative estimation of Zn and O surface–atom concentrations in these two systems [35]. Surface work function in the heterostructure, as estimated through MCs^+-SIMS, was found to decrease, thereby supporting the presence of sulfur on top of ZnO nanowalls. Higher electron density on sulfide surfaces lowers surface work function leading to declining of Zn^+ ion intensity in ZnS/ZnO nanowalls. Thus, "Matrix effect-compensated MCs^+-SIMS" provided a comprehensive understanding of the overall surface chemistry of ZnS nanoparticle-decorated ZnO nanowalls (ZnS/ZnO hetero-structured systems) [35]. Luminescence responses of ZnS/ZnO heterostructures were found to be strongly dependent on extent of ZnS phases over ZnO. Higher luminescence responses in such heterostructures fabricated with smaller ZnS nanoparticles were explained in terms of charge-carrier transfer from ZnS to ZnO [35].

In a study of a ZnS/ZnO hetero-structured film, presence of ZnS nanostructure formed over the top of ZnO nanowalls was clearly visible in SEM image and EDX spectra (Fig. 1.9). In a measurement of $ZnCs^+$-to-OCs^+ intensity ratios through MCs^+-SIMS analysis, depth distributions of $ZnCs^+$ and OCs^+

ions were seen to remain steady indicating homogeneity of ZnO film (Fig. 1.10). On contrary, SCs^+ intensity variation indicated transitory existence of ZnS in the heterostructure network. For MCs^+ molecular ions, as $\frac{I_{ZnCs^+}}{I_{OCs^+}} \propto \frac{c_{Zn}}{c_O}$, higher intensity ratios (I_{ZnCs+}/I_{OCs+}) reflect higher concentration ratios (c_{Zn}/c_O). This obviously indicates the increase in Zn concentration compared to O due to ZnS decoration on top of ZnO surface.

1.5 Conclusion and Outlook

Major limitation of SIMS lies in material quantification owing to its strong "matrix effect". Although quantification can be achieved using "calibration standards", it is difficult to quantify the interfaces between matrices of different materials. Continuous attempts toward correction of "matrix effect" for accurate quantification without calibration standards eventually led to "MCs_n^+-SIMS ($n = 1, 2, \ldots\ldots$)" method which has proven to be extremely valuable in precise quantitative analysis. Concurrently, considerable attempts have been made to conceptually understand the formation mechanisms of MCs_n^+ molecular ions.

Interface analysis of multilayers is of continued interest in materials science. In particular, problems in semiconductor technology have been a driving force.

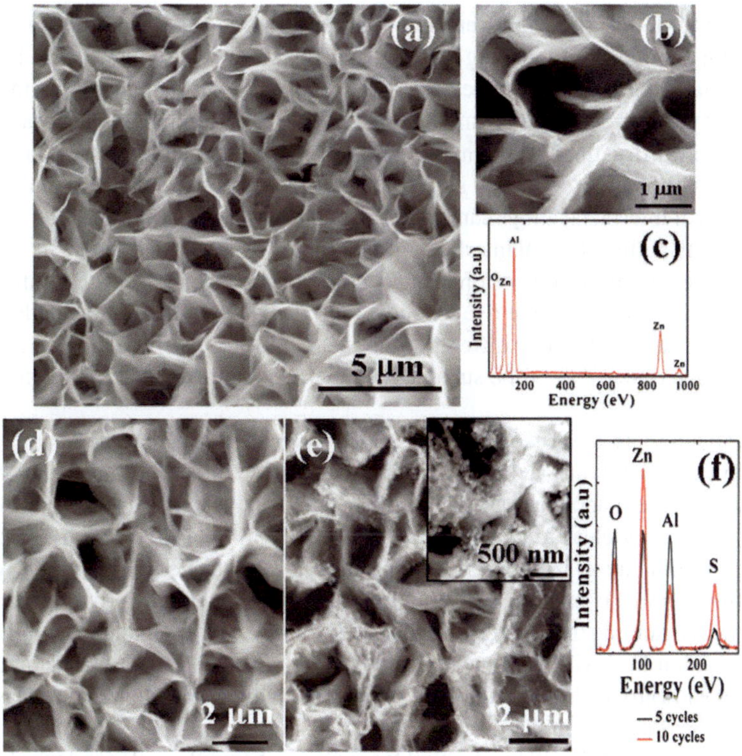

Fig. 1.9 **a** SEM image of ZnO nanowall film, **b** Magnified image of nanowalls, and **c** EDX spectra of nanowalls, **d**, **e** SEM images of nanowalls after ZnS growth in 5 and 10 cycles. Inset of (**e**) depicts a magnified view of ZnS nanostructures formed on top of ZnO nanowalls. **f** EDX spectra of ZnS-decorated ZnO nanowalls [35]

Fig. 1.10 Intensity ratios of $ZnCs^+$-to-OCs^+ molecular ions as a function of sputter time (depth) [35]

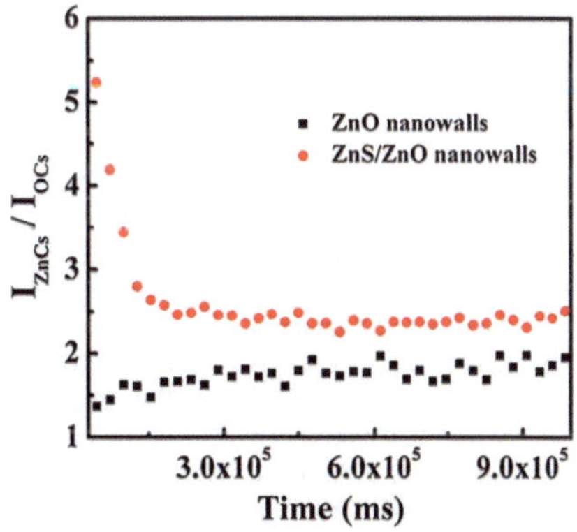

Dynamic SIMS has shown its tremendous applications in various multilayer systems including semiconductor Bragg mirrors and quantum wells. Interfacial diffusion across interfaces is a natural process causing compositional changes across the interfaces in layered structures. In combination with other measurements like X-ray reflectivity, X-ray diffraction, etc., dynamic SIMS has been found to be useful for compositional analysis of interfaces in semiconductor multilayers. In metallic multilayers, formation of interfacial alloys can exhibit various crystalline structures. The structure of a particular alloy phase at interfaces is determined by many parameters like interfacial energies, composition and ordering condition. Moreover, the strong confinement effect at interfaces may give rise to a novel interfacial alloy phase that could have a crystal structure completely different from that in the bulk. SIMS and XRD analyses together can effectively probe such novel "interfacial alloys" in metallic multilayer systems.

Although there have been enormous instrumental and methodological developments in SIMS for surface and interface analysis in condensed matter systems in general, futuristic application range of this sputter-induced surface analytical technique is not very clear at this moment. An in-situ SIMS analysis with an extreme level of optimization in depth resolution during a controlled film growth will be a useful attempt to make a real-time analysis of an interface, providing a much better understanding of the interfacial behavior. Furthermore, adequate flexibility and improvisation in SIMS instrumentation could possibly lead to an extension of this technique toward another direction, for example, analysis of instantaneous "layering of liquid" or "surface melting" states. Multi-technique approaches involving Time-of-Flight (TOF) SIMS will be essential for analysis of complex soft condensed matter systems such as colloids, functional materials, supra-molecular organic assemblies, liquid crystals, crystalline polymers, granular media, various sensing materials, and photonic metamaterials of negative refractive indices.

References

1. Brundle, C.R., Evans, C.A., Wilson, S.: In: Brundle, C.R., Evans, C.A., Wilson, S. (eds.) Encyclopaedia of Materials Characterization. Butterworth-Heinemann, MA, USA (1992)
2. Mondal, S., Gnaser, H., Chakraborty, P.: Eur. Phys. J. D **66**, 197–203 (2012)
3. Chakraborty, P.: Dynamic secondary ion mass spectrometry. In: Chakraborty, P. (ed.) Ion Beam Analysis of Surfaces and Interfaces of Condensed Matter Systems, pp. 217–265. Chap 7 Nova Science Inc., New York (USA) (2002)
4. Oechsner, H., Gerhard, W.: Phys. Lett. A **40**, 211–212 (1972)
5. Saha, B., Chakraborty, P.: Energy Procedia **41**, 80–109 (2013)
6. Andersen, C.A., Hinthorne, J.R.: Anal. Chem. **45**, 1421–1438 (1973)
7. Slodzian, G.: Phys. Scripta **T6**, 54–66 (1983)
8. Williams, P.: Surf. Sci. **90**, 588–634 (1979)
9. Landau, L.: Physics of the Soviet Union 2: 46–51; Zener C (1932). Proc. R. Soc. Lond. A **137**, 696–702 (1932)
10. Yu, M.L.: Charged and excited states of sputtered atoms. In: Behrisch, R., Wittmaack, K. (eds.) Sputtering by Particle Bombardment III, pp. 91–160. Springer, Berlin (1991)
11. Van der Weg, W.F., Rol, P.K.: Nucl. Instrum. Meth. **38**, 274–276 (1965)

12. Yu, M.L.: In: Benninghoven, A. (ed.) Proceedings of SIMS VI, p. 41. Wiley, NY (1987)
13. Canteri, R., Moro, L., Anderle, M.: In: Benninghoven, A. (ed) Proceedings of SIMS VIII, p. 135. Wiley, NY, USA (1991)
14. Wittmaack, K.: Surf. Sci. **429**, 84–101 (1999)
15. Schenkel, T.: Surface analysis of slow, highly-charged ions like Au^{69+}: TOF-SIMS and the probing of nano-environments. In: Purushottam, C (ed.) Ion Beam Analysis of Surfaces and Interfaces of Condensed Matter Systems, pp. 267–310. Chap 8 Nova Science Inc., New York (USA) (2002)
16. Angle, J.W., Palczewski, A.D., Reece, C.E., Stevie, F.A., Kelley, M.J.: J. Vac. Sci. Tech. B **39**, 024004 (2021)
17. Gao, Y., Marie, Y., Saldi, F., Migeon, H.N.: Int. J. Mass Spectrom Ion Proc. **143**, 11–18 (1995)
18. Gnaser, H., Oechsner, H.: Surf. Sci. **302**, L289–L292 (1994)
19. Saha, B., Sarkar, S., Chakraborty, P., Gnaser, H.: Surf. Sci. **602**, 1061–1065 (2008)
20. Gnaser, H., Oechsner, H.: Surf. Interface Anal. **21**, 257–260 (1994)
21. Gnaser, H.: Low-Energy Ion Irradiation of Solid Surfaces (Springer Tracts in Modern Physics), vol. 146. Springer, Berlin (1999)
22. Wilson, R.G., Stevie, F.A., Magee, C.W.: SIMS—A Practical Handbook for Depth Profiling and Bulk Impurity Analysis. Wiley, NY, USA (1989)
23. Sarkar, S., Chakraborty, P., Gnaser, H.: Phys. Rev. B **70**(2004), 195427–195431 (2004)
24. Mootz, T., Adriaens, A., Adams, F.: Int. J. Mass Spectrom. Ion Process **156**, 1–10 (1996)
25. Saha, B., Chakraborty, P.: Nucl. Instrum. Methods Phys. Res.. Instrum. Methods Phys. Res. **B258**, 218–225 (2007)
26. Prudon, G., Gautier, B., Dupuy, J.C., Dubois, C., Bonneau, M., Delmas, J., Vallard, J.P., Bremond, G., Brenier, R.: Thin Solid Films **294**, 54–59 (1997)
27. Marseilhan, D., Barnes, J.P., Fillot, F., Hartmann, J.M., Holliger, P.: Appl. Surf. Sci. **255**, 1412–1414 (2008)
28. Holliger, P., Laugier, F., Dupuy, J.C.: Surf. Interface Anal. **34**, 472–476 (2002)
29. Gavelle, M., Scheid, E., Cristiano, F., Armand, C., Hartmann, J.M., Campidelli, Y., Halimaoui, A., Fazzini, P.F., Marcelot, O.: J. Appl. Phys. **102**, 074904 (2007)
30. Saha, B., Chakraborty, P., Gnaser, H., Sharma, M., Sanyal, M.K.: Appl. Phys. A **108**, 671–677 (2012)
31. Gnaser, H.: Surf. Sci. **342**, 319–326 (1995)
32. Sharma, M., Sanyal, M.K., Mukhopadhyay, M., Bera, M., Saha, B., Chakraborty, P.: J. Appl. Phys. **110**, 102204–102208 (2011)
33. Bayan, S., Chakraborty, P.: Appl. Surf. Sci. **303**, 233–240 (2014)
34. Bayan, S., Mishra, S.K., Satpati, B., Srivastava, R.K., Shukla, R.K., Chakraborty, P.: J. Vac. Sci. Technol. B **34**, 061201–061208 (2016)
35. Bayan, S., Satpati, B., Chakraborty, P.: Surf. Interface Anal. **47**, 37–44 (2014)

Chapter 2
Quantum-Dot-Based Fluorescence Sensing

T. K. Nideep, M. Ramya, and M. Kailasnath

Abstract Quantum dots are artificial atom clusters that can be used in a variety of photonic devices due to their size-dependent absorption and emission characteristics. Compared to the quantum dots, many of the organic dyes suffer from self-quenching, photobleaching, small Stokes shift, short-term aqueous stability, and short excited state fluorescence lifetime [3, 166]. Quantum dots show tunable optical properties, simultaneous excitation of multiple fluorescence colors, large surface areas, and excellent photostability. The intense emission of the quantum dots makes them especially useful in various optical sensing applications. The surface passivation techniques help to improve the electro-optic properties of quantum dots and reduce their toxicity. Quantum dots have found application in a wide variety of areas like displays, photovoltaics, lasers, sensing, etc. In particular, the application of quantum dots in sensing has been discussed in detail in this chapter. Because of their unique characteristics like photostability, single molecule detection, NIR emission, and long-term sensitivity, quantum dots are increasingly replacing organic dyes in sensing applications. Today, the detection of more biomolecules and DNA tracking are both made possible by quantum dot sensing. It promises pH sensing, ratiometric fluorescence sensing, and a multi-color probing method. According to recent advancements in quantum-dot-based sensing, the management of cancer may soon be profoundly impacted by quantum dots [46, 117].

Keywords Quantum dot · Fluorescence · Sensing · Photostability · Color

---

T. K. Nideep (✉) · M. Ramya · M. Kailasnath
International School of Photonics, Cochin University of Science and Technology, Kochi, India
e-mail: nideeptk@gmail.com

M. Kailasnath
e-mail: kailas@cusat.ac.in

D. Mohanta and P. Chakraborty (eds.), *Nanoscale Matter and Principles for Sensing and Labeling Applications*, Advanced Structured Materials 206,
https://doi.org/10.1007/978-981-99-7848-9_2

2.1 Quantum Dots and Confinement Effect

Today, quantum dots are the most promising fluorescent nanomaterials used in a variety of commercial applications including photovoltaics [89, 138, 147], light concentrators [71, 170], displays [152, 213], sensors [104, 228], and optical probes in biological applications [5, 104]. Quantum dots are semiconductor materials made up of ten to few hundreds of atoms and have a size regime of less than 10 nm. Due to their reduced size, quantum dots usually behave like a single atom, and their allowed energy levels are discrete. Figure 2.1 shows the distinction between the band structure of the atom, the bulk semiconductor, and quantum nanostructures. The material properties of QDs lie between those of bulk material and a single atom. They are distinguished from other materials by their size-dependent optical and electronic properties. With the decrease in size, the discreteness in electronic energy levels as well as the energy level separation increases [9, 40, 60, 84, 86, 165, 175, 186, 216, 244, 246].

An increase in the energy gap between the upper conduction band and the lower valence band leads to a change in the absorption and emission spectra. A semiconducting material absorbs only that amount of energy that is equal to or greater than the bandgap energy of the semiconductor. Due to this, absorption and emission shift with respect to the size of the quantum dot. So, quantum dots exhibit highly tunable absorption and emission properties compared to other nanomaterials. The narrow and sharp emission spectra of quantum dots with a low full width at half maximum (FWHM) value give rise to very high quantum yields (QY). Fluorescence quantum yield is one of the special characteristics of quantum dots. It is the ratio of photons absorbed to photons emitted through fluorescence [91]. Because of this highly tunable

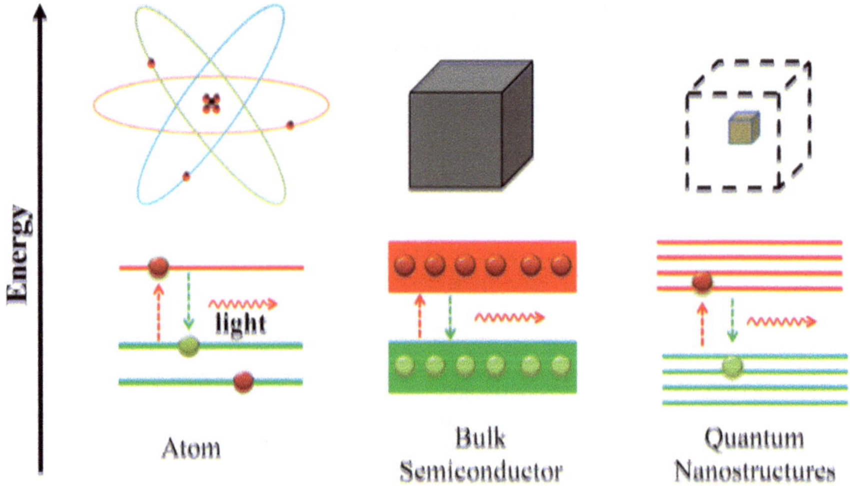

Fig. 2.1 The distinction between band structures of the atom, bulk semiconductors, and quantum nanostructures

optical property, they are widely used in a variety of applications such as solar cells, transistors, LEDs, lasers, quantum computing, medical, and bioimaging [5, 41, 123, 133, 235, 236].

When discussing nanomaterials, the individual size of the materials is very important because of the observed drastic changes in the bandgap of the material even under a minute variation in size. When the size approaches the exciton Bohr radius of the atom, changes in size control the material characteristics. In this context, the materials can be divided into three regimes, viz., strong confinement, weak confinement, and intermediate confinement, by comparing the exciton radius of an atom with the radius of the quantum dots. In the strong confinement regime ($R \ll a_B$), where R is the radius of the quantum dot and a_B is the exciton Bohr radius, the Coulomb attraction between electrons and holes will be very weak, and hence electrons and holes can be considered as separate entities. In the weak confinement regime ($R \gg a_B$), the particle diameter will be much greater than the exciton Bohr radius of the atom. In the intermediate confinement regime ($R \sim a_B$), the electronic properties of the material are affected both by the confinement effect and the Coulomb interaction [54, 67].

Upon illumination with light of energy equal to or greater than the energy bandgap of the quantum-dot material, it gets absorbed by the electrons in the material, and the electrons in the valance band are excited into the conduction band, leaving holes in the valance band. An exciton is a single system made up of an electron and a hole. Energy is emitted in the form of a photon when the electron and hole in an exciton recombine due to the deexcitation of the electron in the conduction band. This process is known as emission or fluorescence. The energy of this emitted photon can be described as the sum of bandgap energy, the bound energy of the exciton, and the confinement energy of the hole and the excited electron. As the confinement energy is size dependent, both the absorption onset and the fluorescence emission can be tuned by changing the size of the quantum dot [102, 141, 204]. A schematic representation of the size-dependent emission mechanism and the band structure in a quantum dot are shown in Fig. 2.2.

In organic dyes, optical properties originate from electronic transitions. Most organic dyes, such as Rhodamines, Cyanine, and Fluorescein, have high molecular absorption coefficients, narrow emission and absorption bands, and moderate-to-high quantum yields. But in the case of quantum dots, optical properties are controlled by constituent material, particle size, dispersity, and surface functional groups. Size-tunable optical properties, simultaneous excitation of multiple fluorescence colors, large surface area, and excellent chemical and photostability are the unique properties of quantum dots compared with organic dyes. These properties make QDs useful as fluorescence probes for real-time imaging of biological cells. The photoluminescence spectra of monodispersed QDs are narrower than those of organic dyes. Meanwhile, in the case of heterogenous QDs, broadband emission is observed due to different particle sizes [66, 77]. In multiplexed bioimaging applications, narrowband emission is the key point where the overlap between different emission spectra must be at a minimum.

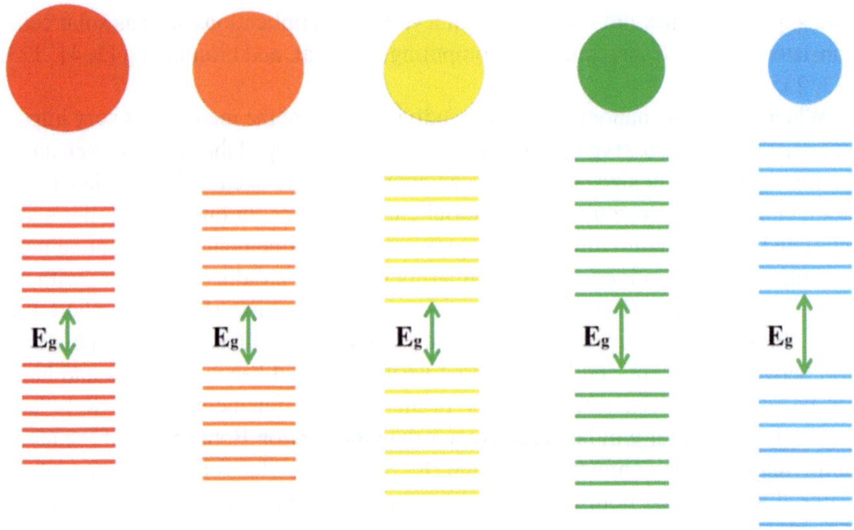

Fig. 2.2 Size-dependent shift in emission as a consequence of quantum confinement effect

The surface of the QDs must be suitably passivated to achieve a high quantum yield. Surface states have a sufficient effect on electro-optical parameters such as absorbance, fluorescence intensity, quantum efficiency, and photostability due to the high surface-to-volume ratio of the QD. In general, these surface defects are trap states that enhance the redox reaction, which simultaneously affects the overall properties of the QDs. This trap state reduces the quantum yield as well as the radiative recombination rate. Hence, the surface passivation or capping over the QDs confines the charge carriers inside the core and improves the photostability and quantum yield. The perfectly capped QDs possess saturated dangling bonds and internally confine the motion of charge carriers. Thus, surface functionalization is very important in quantum-dot-based applications, and, in general, organic and inorganic materials are coated over the QDs. Monodispersed quantum dots with organic molecule capping are synthesized using different methods [38, 143]. The organic layers of phosphenes or mercaptans groups are most commonly used to increase colloidal stability and bioconjugation ability. Wang [212] studied the luminescence properties of bovine serum albumin (BSA)-capped CdTe QDs to show improvements in chemical stability and PL quantum yield. The CdS QDs passivated using DNA molecules demonstrated their suitability for bioimaging [126]. In the second approach, quantum dots are functionalized using a wide bandgap material wherein a passivating shell is formed over the quantum-dot surface. The uniform shell coating increases the fluorescence quantum yield and stability. The thickness of the shell and lattice parameter mismatch between the core and shell affect the absorption and emission properties. Chen [31] studied the coherency strain effect on the CdSe/CdS core–shell heteronanostructure. The absorption and emission spectra were red-shifted when the CdS shell was formed on the CdS core. A thicker CdS layer causes lattice mismatch, which creates nonradiative recombination sites, lowering the quantum yield.

The physicochemical properties of individual quantum dots are widely varied. The toxicity of quantum dots is related to quantum-dot size, concentration, charge, type of coating material, oxidative nature, mechanical stability, and photolytic nature. Quantum dots made up of a heavy metal core, such as cadmium or lead, contain toxic components. Since cadmium or lead ions released from QDs are harmful to living organisms, they should be well coated with organic molecules/polymers/inorganic molecular layers. Quantum dots encapsulated with silica molecules show less toxicity than QDs coated with mercaptoundecanoic acid, 2-aminoethanethiol, mercaptopropionic acid, and mercaptoacetic acid [21, 205]. The silica layer has a micrometer thickness, which reduces the leakage of heavy metals. When PEG-coated CdSe–ZnS quantum dots were reattached with silica molecules, no toxic effect was observed [97]. Table 2.1 shows the comparison of properties between the organic dye and quantum dots.

Table 2.1 Comparison of properties of organic dyes and quantum dots

Property	Dye	Quantum dots
Absorption spectra	Narrow, discrete bands	Broad, steady increase toward UV region
Molar absorption spectra	2.5×10^4–2.5×10^5 M^{-1} cm^{-1}	10^5–10^6 M^{-1} cm^{-1}
Emission spectra	Asymmetric	Symmetric, Gaussian profile
Quantum yield (QY)	0.5–1.0 (visible) 0.05–0.25 (NIR)	0.1–0.8 (visible) 0.2–0.7 (NIR)
Stokes shift	Normally <50 nm Up to >150 nm	Typically, <50 nm For QDs with visible emission
Size	~0.5 nm molecule	1–10 nm colloids
Fluorescence lifetime	1–10 ns, monoexponential decay	10–100 ns, multiexponential decay
Two-photon cross section	1×10^{-52}–5×10^{-48} $cm^4 s photon^{-1}$	2×10^{-47}–4.7×10^{-46} $cm^4 s photon^{-1}$
Photochemical stability	Low photobleaching threshold	High photobleaching threshold
Dispersibility	Controlled by substitution pattern	Control via surface chemistry (ligand molecules)
Toxicity	From very low to high—depends on dye	Heavy metal leakage causes toxicity
Thermal stability	Low, depends on dye	High, depends on ligands
Spectral multiplexing	Not ideal due to the spectral overlap	Ideal for multi-color experiment

2.2 Classification of Quantum Dots

In recent years, fluorescent nanodots have drawn much attention and are considered a versatile tool for a wide range of applications, including bioimaging, photothermal therapy, drug delivery, and chemical sensing. Quantum dots are generally classified as semiconductor or carbon-based quantum dots. Semiconductor quantum dots (SQDs) are commonly classified into three groups, III–V semiconductor quantum dots (GaAs, GaP, AlSb), II–VI semiconductor quantum dots (ZnO, ZnS, ZnSe, CdS, CdTe, CdSe), and IV–VI semiconductor quantum dots (PbS, PbSe). Carbon-based quantum dots are classified as graphene quantum dots, carbon quantum dots, and carbon nanodots.

The emission mechanisms observed in semiconductors are named "band-edge electron–hole recombination/intrinsic bandgap fluorescence/HOMO–LUMO transition," where no intermediate states are present in the lattice. This mechanism is observed in defect-free and impurity-free semiconductor quantum dots. Here, the PL properties entirely arise from quantum confinement effects and are independent of exciton emission. Photoluminescence emission spectra have a fixed maximum position, irrespective of excitation wavelength. In general, the emission spectra observed in semiconductor quantum dots are very narrow, with an FWHM value of less than 40 nm. In carbon quantum dots and graphene quantum dots, the emission mechanisms are different. Their lattice contains different types of trap states due to the presence of impurities, functional groups, surface defects, and adsorbed molecules. Here the excited photoelectrons or holes are trapped, and their recombination leads to the emission of higher wavelengths. These types of emission mechanisms are due to the intrinsic quantum size effect and also to the properties of surface functional groups and surface defects. These types of emission mechanisms are exciton wavelength dependent, and the spectrum is shifted toward higher wavelength regions. These types of mechanisms are commonly observed in carbon quantum dots and graphene quantum dots. In general, fluorophore molecules or functional groups are present over the quantum-dot surface, and the emission mechanism results from the superposition of assembled individual emitters. Here, the fluorescence behavior is very similar to that observed in metal nanoclusters, and quantum confinement or exciton effects are not involved. The carbon nanodots prepared at low temperatures show a very broad emission band, which is the consequence of the superposition of several emissions.

Another classification is based on chemical composition, with core-type quantum dots, core–shell quantum dots, and alloyed quantum dots being the most common. Core-type quantum dots can be single component materials with a uniform composition such as sulfide, selenide, or telluride of metals such as lead, cadmium, and zinc. These quantum dots possess excellent electroluminescence and optical properties that are widely used in fluorescence imaging and biosensing. These properties can be tuned by varying the quantum size effects. Although cadmium-based core-type quantum dots (CdS, CdSe, and CdTe) have received the most attention, the toxic cadmium ions that leak from them limit their further practical applications. The demand for more biocompatible, heavy metal-free quantum dots for practical applications introduced IV quantum dots, such as carbon-based (carbon/graphene quantum dots) and silicon/germanium quantum dots (Si/Ge quantum dots). Carbon or graphene quantum dots are considered the most prominent quantum dots for bioimaging and sensing applications. This is due to their particular characteristics such as higher solubility, lower toxicity, strong chemical inertness, high surface area, versatility, and a large number of edge sites for functionalization [159, 196, 224]. Silicon (Si) is an indirect bandgap semiconductor with weak optical properties. The confinement effect makes silicon quantum dots a potential material for optoelectronic devices and biological applications, including biosensors and bioimaging [140]. Germanium quantum dots are less toxic compared to other metal oxides and their doping decreases the toxicity of other quantum dots [90, 124].

Reduced nonradiative decay of excitons is one methodology used to improve the efficiency and brightness of semiconductor nanocrystals. This is accomplished by enclosing them in shells of a larger bandgap semiconducting material. Core–shell quantum dots or core–shell semiconducting nanocrystals are particles with tiny areas of one material embedded in another with a wider bandgap. These structures exhibit improved stability, optical properties, and quantum efficiency due to the formation of shells around the quantum-dot core structure. The quantum yield of quantum dots decreases when excitons decay through the nonradiative path. One of the common methods used by researchers to enhance the quantum yield is to grow a shell layer over the quantum-dot surface. To construct a core–shell structure, both layers should have a different bandgap. For example, the CdSe–ZnS core–shell exhibits a quantum yield greater than 50% [201]. Most commonly, core–shell is typically compared to type III–V, IV–VI, and II–VI semiconductors such as (CdS)ZnS, (CdSe)ZnS, (CdSe)CdS, and (InAs)CdSe [201]. Depending upon the position of conduction and the valence band energy level, core–shell occurs in different categories. In the Type-1 configuration, the bandgap of the core semiconductor is less than that of the shell semiconductor, as the core bandgap falls within the bandgap of the shell. In the Type-II configuration, the valence band edge of the core lies within the bandgap of the shell. The formation of a shell over the core structure protects the core against oxidation and prevents the formation of metal ions, thereby reducing the toxicity of quantum dots. As a result, the core–shell structure not only improves the quantum yield but also makes QDs less harmful, which is desirable for fluorescence bioimaging.

Alloyed semiconductor quantum dots, which are formed by alloying two semiconductors with distinct bandgap energies, exhibit intriguing properties that distinguish them not only from their bulk counterparts, but also from their parent semiconductors. The formation of alloyed quantum dots consists of two nanocrystal semiconductors having different properties and bandgaps. Alloyed quantum dots exhibit clear quantum confinement effects, and their properties are changed by changing the composition, not the color. Several studies have shown that the internal structure and composition of alloyed quantum dots are the parameters used to tune their optical and electrical properties [7, 76]. Alloyed quantum dots both have homogenous and gradient internal structures. In a homogenous structure, the alloyed quantum dots have uniform internal structure and composition everywhere. In gradient quantum dots, the alloyed composition varies radially [112]. Unlike the core–shell structure, the alloyed quantum dots have no abrupt boundary change.

2.3 Synthesis of Quantum Dots

The choice of the synthesis method is a key factor in the determination of quantum-dot properties. Synthesis routes are mainly classified into two categories, (1) top-down and (2) bottom-up. In top-down synthesis processes, a bulk semiconductor is thinned to form the ~30 nm-sized quantum dots. Electron beam lithography, etching techniques, focused ion beam techniques, laser ablation, and electrochemical methods are commonly used top-down methods. The major advantages of these methods are their simplicity of operation and abundance of raw materials. Non-controllability in size and shape, low yield, and large defect state presence are some of the disadvantages of these methods. It is very difficult to prepare small-sized uniform quantum dots. In electron beam lithography, a focused electron beam is used to draw custom shapes on the electron-beam-sensitive resist. The resist consists of long-chain polymers and the electron beam changes the solubility. This is a popular and versatile method for patterning micro- and nanostructures [118, 155].

The focused ion beam (FIB) is a commonly used technique in the semiconductor manufacturing process that offers the formation of high-precision quantum dots. The technique of FIB closely resembles that of the scanning electron microscope (SEM) and makes use of ion beams to form the sample. The shape and size of the quantum dots depend on the size of the ion beam [22]. Another method to prepare quantum dots is laser ablation. This method is effectively used for the synthesis of various nanostructures, including core–shell quantum dots [167], semiconductor quantum dots [17], and metal nanoparticles [136]. This method is a very fast single-step method for generating quantum dots with a size less than 10 nm and produces fewer harmful by-products [96]. Laser pulse width, wavelength, power, focused spot size, and composition of precursors affect the shape and size distribution of the resultant nanoparticles [18, 162].

Bottom-up methods are broadly classified into two types: (1) the vapor phase method and (2) the wet chemical method. The vapor phase deposition method grows quantum-dot layers atom by atom without any patterning [105, 109, 189]. Molecular beam epitaxy [10, 108, 179, 222], physical vapor deposition [15, 47, 103], and chemical vapor deposition [106, 190, 230] are the different vapor phase techniques for the deposition of quantum-dot layers. The wet chemical method is a conventional synthesis method using a single or mixture of solutions with precise control of external parameters. Quantum dots of different sizes and shapes can be synthesized by tuning temperature, time, pH of the reaction, type of surfactant, varying concentration, and solvent of the reaction. Nucleation and Ostwald ripening are the major mechanisms to be considered in the growth of nanoparticles. Sol–gel process [8, 13, 100, 129, 156], microemulsion [38, 42, 56, 107, 119, 151, 225, 226], hot-solution method [10, 144], microwave irradiation [56, 249], hydrothermal method [150, 174, 214], and solvothermal method [34, 49, 221] are common wet chemical methods.

2.4 Quantum-Dot-Based Sensing

Because of their unique optical characteristics, QDs are exceptional fluorophores that can be employed to fabricate highly sensitive and selective sensors. In addition, as compared to organic dyes, QD emission profiles are symmetric and have relatively small bandwidths. So they find potential applications in a variety of sensors, including biosensors, chemical sensors, pH sensors, temperature sensors, humidity sensors, etc. To create sensitive and selective sensors even at the ppm level and occasionally even at the ppb level, changes in both photoluminescent lifetimes and spectral response are crucial [59, 236]. The majority of QD applications are based on their exceptional optical characteristics and their functions in light emission, conversion, and detection. Quantum dots show higher brightness, chemical stability, efficiency, narrow bandwidth, and surface area. These advantages make quantum dots a good candidate for sensing applications. In fluorescence-based sensors, the binding or unbinding of analyte species changes the fluorescence intensity of the sensing agent. The fluorescence property of the quantum dot is influenced by the nature of the surface. The presence of ligands or surface charges creates trap sites and affects the quantum efficiency. Hence, the presence of metallic ions in a solution affects the quantum yield in different ways. This is the basis for ion detection and optical sensing applications of quantum dots. Traditionally, chemiluminescence, voltammetry, plasma mass spectroscopy, atomic absorption spectroscopy, and organic dyes are used for ion detection [39]. Quantum dots overcome most of the problems of organic dyes, such as low photostability, narrow excitation bands, low signal intensity, and broad emission spectra. Therefore, quantum dots can be a better alternative for simple and efficient optical sensing. Since quantum dots are used for the detection of heavy metal ions in an aqueous solution, they have great relevance in environmental applications. The presence of heavy metal ions in drinking water resources, even at very low concentrations, creates severe adverse effects on human health and marine ecosystems. Even though some literature demonstrates an enhancement in the quantum yield of quantum dots due to the presence of external ions [32, 187], fluorescence quenching is the most common sensing mechanism. The reduction in emission is attributed to several mechanisms, namely, nonradiative recombination, ion-binding interactions, inner filter effect, and electron transfer processes [12, 58, 81, 116]. Chen et al. [32] studied the CdS quantum-dot luminescence properties in the presence of Cu^{2+}, Na^{+}, Ca^{2+}, and Mg^{2+} ions. In the presence of cu^{2+} ions, the quantum-dot emission is quenched, but the emission is unresponsive to other cations. Jin wang [211] fabricated a CdTe@SiO_2@CdSe dual emitting probe for ratiometric fluorescence detection of Cd^{2+} ions via turn-on and ion-imprinted techniques. The green fluorescence of the probe is quenched during the EDTA etching on the CdSe

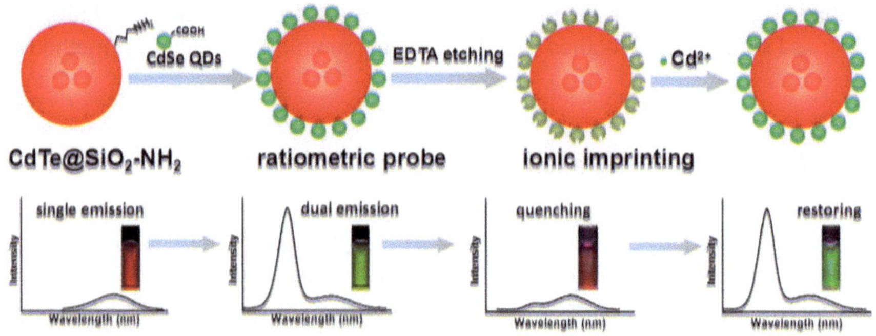

Fig. 2.3 Schematic illustration for the synthesis of the ratiometric probe and the sensing mechanism for fluorescence turn-on detection of Cd^{2+} [211]

quantum-dot surface, which increases the surface defects. Ionic imprinting repairs the CdSe surface, recovering the green emission as illustrated in Fig. 2.3.

In quantum dots, another type of dipole–dipole interaction that causes energy transfer between the excited states of the smaller donor quantum dots and the bigger acceptor quantum dots is known as the Forster Resonance Energy Transfer (FRET) [24]. FRET is the transfer of energy from a donor particle to an acceptor particle when the distance between them is smaller than the Forster radius (distance between the donor and acceptor particle for 50% of FRET efficiency) [83]. This transfer decreases the donor's excited state lifetime and increases the acceptor's emission intensity. Quantum-dot-mediated FRET has several advantages compared with conventional FRET due to its unique properties such as size tunability, broad excitation spectra, narrow emission spectra, high photostability, strong signal intensity, and good biocompatibility. These superior properties of the quantum-dot-based FRET phenomena are used in many fields including cell imaging, virus detection, drug testing, molecular biology, electron transfer detection, medical diagnosis, etc. In quantum-dot FRET sensing, quantum dots are used as either donors or acceptors [29, 35, 36, 44, 64, 87, 110, 149, 192, 234]. In most cases, fluorescent proteins (FPs), fluorescent nanoparticles (NPs), lanthanides, and organic dyes are used as FRET donors. In a few particular cases, quantum dots are used as FRET donors due to their special properties that can improve the energy transfer efficiency: (i) quantum dots have greater flexibility in the selection of donor excitation wavelengths due to their large absorption cross-section and broad absorption spectra; (ii) quantum dots exhibit narrower emission spectra than organic dye molecules and the emission does not tail-off to the red, reducing spectral overlap and cross-talk with the acceptor molecules. (iii) Size-dependent emission allows the donor emission spectra of quantum dots to be tuned into resonance with the acceptor absorption spectrum of any visible absorbing chromophore. (iv) Because of their high photostability, brightness, and sustained fluorescence emission, quantum dots are a more efficient FRET donor in FRET-based bioassays [3, 4, 37, 72, 98, 137, 166, 171]. Doris et al. [163] reported the design of a new FRET-based ratiometric method for arginine detection.

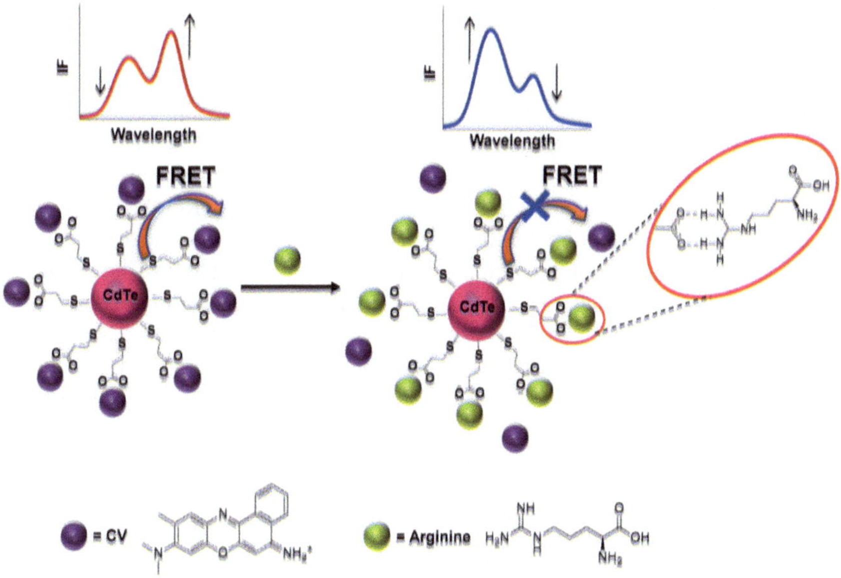

Fig. 2.4 Schematic illustration of FRET between QD and CV dye and the mechanism for the arginine detection [163]

The mechanism is found to be based on the inhibition of FRET formed between MPA-capped CdTe quantum dot (donor) and Crystal Violet (CV) dye (acceptor) in the presence of arginine at pH 8. A schematic illustration of FRET mechanism is shown in Fig. 2.4.

In general, two types of acceptor molecules can be employed in FRET: fluorophores or quenchers [37, 176]. Typically, fluorescent proteins or organic dyes act as fluorophores, while organic molecules and metal nanoparticles play the role of quenchers. The merits of the quantum dot as an acceptor are that (i) large extinction coefficients permit very large spectral overlap and Forster distance and (ii) large surface area increases the probability of FRET sensitization. Sebastian et al. [48] reported a flexible and modular design for a lanthanide (Tb)-quantum-dot time-resolved FRET sensor, in which lanthanide acts as a donor and quantum dot acts as an acceptor. This time-resolved FRET bioassay was used as an ADP sensor working at nM concentrations. The findings of this study revealed two factors that must be considered for an efficient quantum-dot acceptor FRET sensor: (1) the size of the quantum-dot acceptor must be less than the Forster radius and (2) a longer lifetime and higher brightness of the quantum dot result in better sensing.

Quantum dots act both as acceptors and donors in the quantum dot-to-quantum dot FRET devices. The advantages of quantum dots in this type of FRET compared to dyes or fluorescent protein sensing are extraordinary brightness, large absorption cross-section, and extreme photostability, potentially lowering the limit of detection.

Controlling the donor–acceptor ratio is a difficult task while engineering quantum–quantum FRET systems because of stringent conditions on particle size for tailoring absorption and emission. The FRET mechanism has been reported to be used for detecting toxic elements like melanine [192, 241], arsenic [191], Cu^{2+} [180], Zn^{2+} [215], and protein and biomolecules [33]. Tang et al. [191] demonstrated the arsenic (III) detection based on the FRET mechanism between mercaptopropionic acid-capped CdTe quantum dots (MPA-CdTe) and Rhodamine 6G (R6G). In this FRET system, MPA-CdTE quantum dots act as energy donors while R6G acts as an energy acceptor. The attachment of glutathione (GSH) on the surface of MPA-CdTe quantum dot destroys the chelation between GSH and the MPA ligand, leading to a decrease in the FRET efficiency. The addition of arsenic (III) (As(III)) removes the GSH capping layer from the surface of the CdTe quantum dot due to the binding of GSH and As (III). Removal of GSH increases the chance of recombination of R6G with CdTe quantum dot, recovering the FRET phenomena. Upon addition of As(III), the emission intensity of CdTe (donor) decreased while that of R6G (acceptor) increased, and this spectral change of R6G led to the determination of arsenic (III). Figure 2.5 shows the schematics of the various synthesis strategies of biologically nonfunctionalized quantum-dot probes for sensing applications [185]. This means that the quantum-dot probes can be prepared in multiple ways, viz., by capping, coating, or conjugation with other molecules, leading to different functionalities for the prepared quantum dots. For example, quantum dots coated with hydrophilic thiol are stabilized ionically in solution [20]. A quantum-dot capped with a cross-linked silica shell may be easily modified with a variety of organic functionalities [3]. Quantum dots modified by amphiphilic polymers form highly stable micelle structure [50, 217]. Quantum-dot-capped PEG reduces surface charges and improves colloidal stability [184]. Quantum dots are conjugated with streptavidin, antibody, ligands, and peptides to make them bioaffinity probes [25, 185, 194, 223]. Furthermore, quantum dots can also be used to detect biomolecules that act as donors or acceptors in energy transfer processes with the quantum dot. For example, quantum dots are used to sense the existence of sugar maltose by conjugating the maltose binding protein to the surface of the quantum dot [134]. Figure 2.6 [27] shows the two-step FRET mechanism used to identify the DNA degradation process by introducing an intermediate acceptor of nucleic acid.

An immunoassay is a chemical test for detecting the presence of biomolecules. In the first stage of an immunoassay, the target molecule is probed with a specific antibody. In the second stage, called visualization, detection techniques such as chemiluminescence, colorimetric, and fluorescence are employed. Surface-treated quantum dots are powerful fluorescent probes for sensing a wide variety of chemical and biological analytes. The main advantage of a quantum dots in immunoassays is that even a lower concentration of target molecules can be detected due to the higher quantum yield of quantum dots. In immunofluorescence assays, quantum dot–avidin conjugates [69, 70] are excellent alternatives to organic dyes due to their high quantum yield, photostability, and versatility for sensing biomolecules. They are used for multi-protein detection due to their narrow emission band and wide absorption band, size-controlled color tunability, and their surface engineering. This type of multi-target detection is provided using modified ELISA techniques

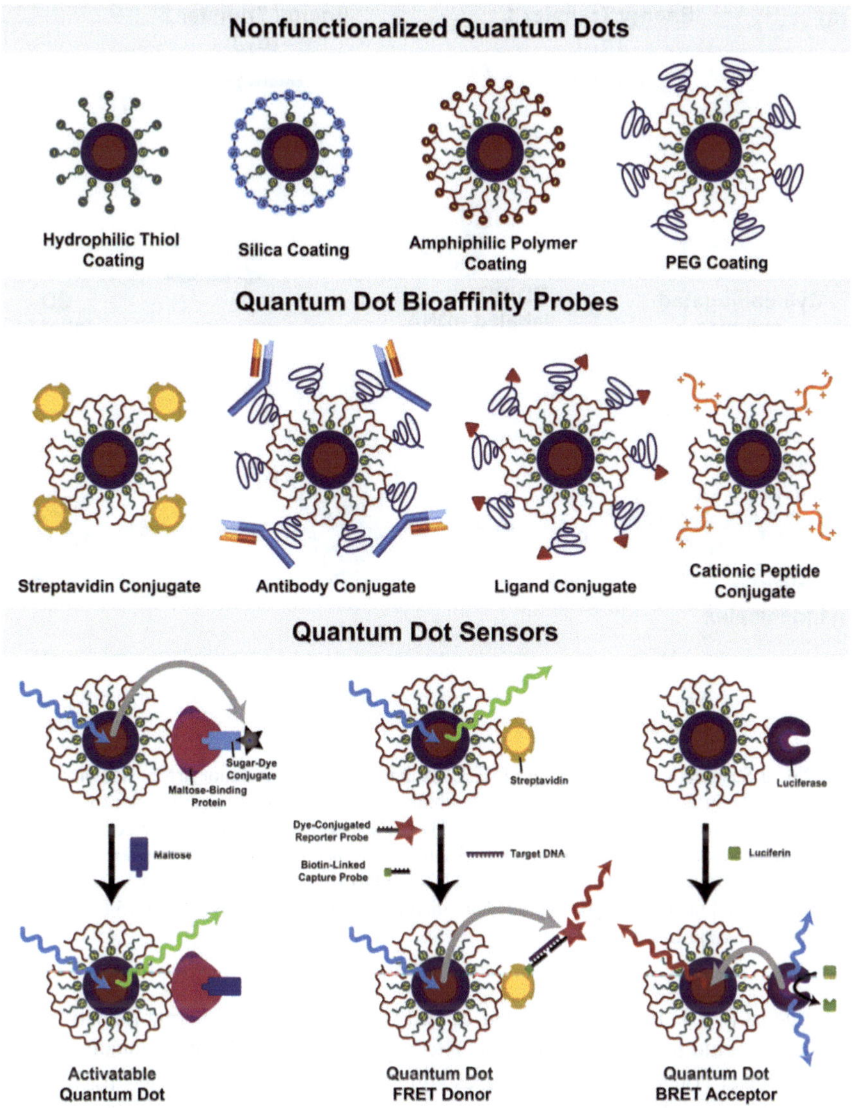

Fig. 2.5 Schematics of various quantum-dot probes for sensing applications [185]

[78, 128]. The development of quantum-dot-based immunoassay has been beneficial for microfluidics and electrochemical detection techniques [135]. The quantum-dot-based ultrasensitive Western Blot (WB) technique is currently used in the diagnosis of HIV, lyme diseases, and bovine spongiform encephalopathy [177]. Rapid and sensitive response, high sensitivity, and multiplexed detection are the advantages of the quantum-dot-based immunoassay. The multichannel detection of drug residues, toxins, or cancer biomarkers has been confirmed using multicolored antibody–QD

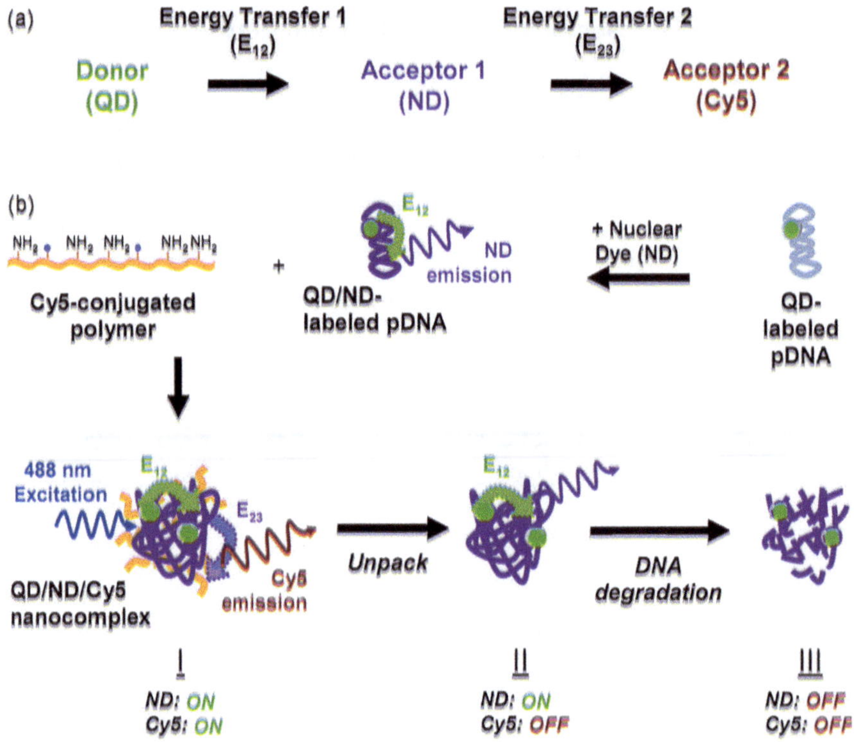

Fig. 2.6 **a** Schematic of two-step QD-FRET. Excitation of the QD donor drives stepwise energy transfer (E12) through the nuclear dye (ND) which serves as the first acceptor/donor (or relay) for energy transfer to the second acceptor Cy5 (E23). **b** Plasmid DNA is double-labeled with QD and nuclear dyes before complexation with a Cy5-labeled cationic polymer to form nanocomplexes by self-assembly. Three distinct states of plasmid DNA (pDNA): (I) condensed within a nanocomplex, (II) released and intact, and (III) degraded are distinguished by relative ND and Cy5 emission and calculated E12 and E23 efficiencies [27]

conjugates [115]. In 2006, Agrawal et al. [1] reported the real-time detection of a single molecule using quantum dots. The high quantum yield of the quantum dots, even at lower concentrations, produces significant emission and allows for tracking of a single molecule [19, 61, 145]. This capacity of the quantum dot to detect a single molecule gives information about gene expression and regulation [6, 101, 206].

Irrespective of material types, quantum dots exhibit temperature-dependent variation in their emission intensity, which naturally lends quantum dots to optical temperature sensing. According to the Varshni relation [200], the semiconductor bandgap is temperature dependent, and hence the photoluminescence emission is affected by temperature. CdSe [11, 229], CdTe [139], CdHgTe [82], ZnSe/ZnS [219], InGaN [240], HgTe [92], and CdSe-ZnS/ZnS [158] quantum-dot structures are reported to show temperature-dependent fluorescence. The surrounding environment [154, 207], surface ligands [85, 198], and dopants [75, 154] are the other factors that affect the temperature-dependent emission property. In temperature-sensing applications, toluene-dispersed CdS_xSe_{1-x}/ZnS quantum dots are mixed with VGE-7021 varnish to produce a quantum-dot paint which can be used for real-time temperature sensing [199]. Since the illumination of quantum dots with NIR light causes localized heating and destruction of cancerous cells, photothermal therapy has been developed [14, 68]. Figure 2.7 [121] shows the fluorescence image of the surface temperature of resistors with different resistance values using ZnCuInS/ZnSe/ZnS quantum-dot film. Different surface temperatures caused by the differences in resistance value lead to various emission intensities for the quantum dots covering the surface.

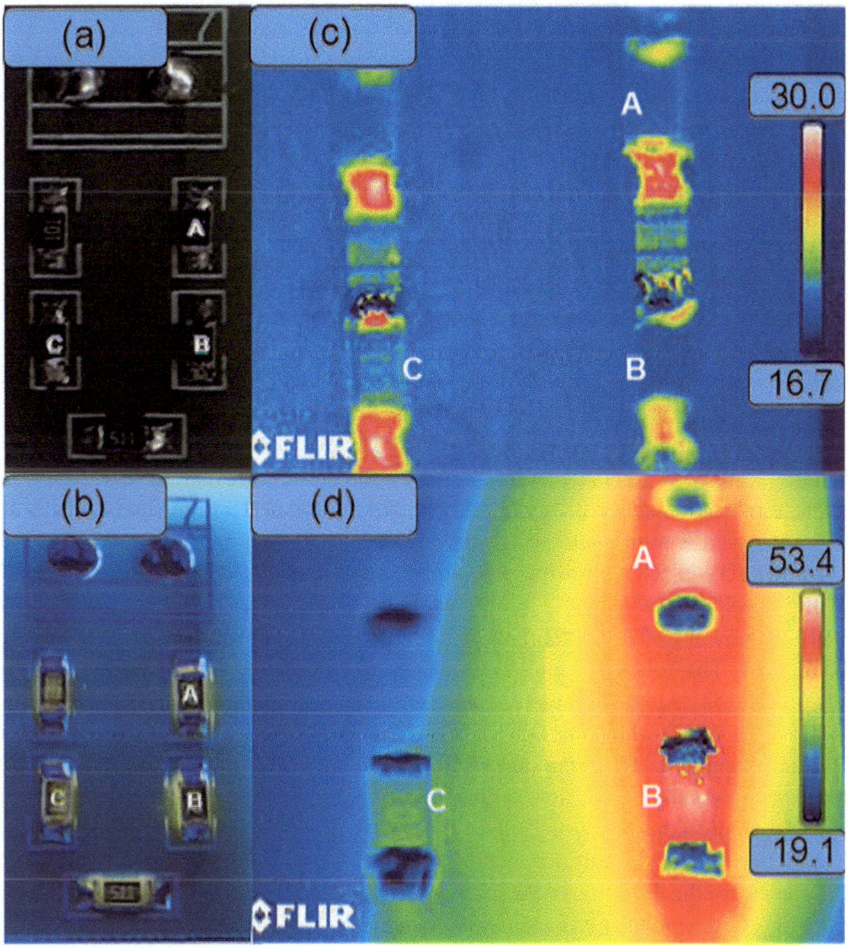

Fig. 2.7 Images of PCB without QDs (**a**) and with QDs under UV lamp (**b**). Thermal infrared images of PCB with QDs operated at 0 mA (**c**) and 7.9 mA (**d**). A, B, and C are three resistors in a series circuit [121]

In general, quantum-dot photoluminescence emission intensity changes with aggregation, concentration, and environment, and such quantum dots can exhibit low sensitivity to the target analyte. To compensate for this inherent sensitivity to solution environment conditions, the ratiometric fluorescence measurement method can be used. Dual-color quantum-dot sensors are prepared for ratiometric measurement, which monitors the change in the ratio of PL emission intensity at two different wavelengths. So, dual-color quantum-dot sensors are less sensitive to environmental factors. Dual-color quantum-dot emitters are synthesized by (i) growing core/shell/shell quantum dots; (ii) including dopants during nucleation; and (iii) growing seeded quantum rods tipped with the second material. A double-well PbS/CdS/CdSe heterostructure dual-color sensor was developed by Razgoniaeva et al. [164]. In the presence of methyl viologen and 3-mercaptopropionic acid quantum dots, PL emission profile changes are studied. In the presence of methyl viologen, PbS emission was found to be quenched more effectively than CdSe emission. But in the presence of MPA, CdSe emission is more quenched. In both cases, one color is more efficiently quenched than the other, and the analyte concentration was quantified by calculating the ratio of emission colors. Ratiometric fluorescence temperature sensors were developed using doped quantum dots, where the presence of dopant in the quantum-dot lattice introduces an additional energy level. Depending upon the position of these levels, quantum dots exhibit dual emission, and the ratio of emission peak intensity is sensitive to the temperature [203].

Many other interesting instances of ratiometric sensing of quantum dots have been reported in the literature. Water-soluble CdTe QDs were employed by Lue et al. [120] to differentiate 29 green teas based on their emission properties. Using ZnCdSe-CdTe double QDs, Ou Hu's group [79] later discovered a way to identify 53 teas. Feng et al. [57] researched polymer-coated QDs, revealing their ability to detect tetrabromobisphenol-A, which is primarily found in electronic waste. Hao et al. [73] employed a combination of dual-wavelength overlapping resonance Rayleigh scattering (DWO-RRS) and fluorescence characteristics to determine the presence of diminazene aceturate (DA) in milk samples. Quinet et al. [160] reported a TNT explosive sensor based on a fluorescence ratiometric sensor using L-cysteine-capped CdTe QDs. The folic acid and methotrexate were simultaneously determined by Ensafi et al. [55] using dual-color CdTe quantum dots. An immunosensor for prostate-specific antigen detection was fabricated by Zhao et al. [242] using CdTe@SiO_2. Uric acid and dopamine detection studies were reported by Yu [232] using CdTe QD-decorated graphene composite. A glucose oxidase-induced fluorescence quenching of mercaptopropionic-acid-modified CdTe QDs was used by Yingingwu et al. [218] for the detection of Hepatitis B.

A probe for the detection of the uranyl ion, which causes adverse effects on human kidneys, was developed by Hu et al. [80]. Chen et al. [23] developed a portable device for Ag sensing with CdTe QDs as fluorescence probes using an electron transfer process. Yang et al. [227] developed a method for the detection of benzo pyrene-β-cytotoxic functionalized CdTe QDs. Zhu et al. [247] used CdTe QDs for Hg detection. The histamine in tuna fish was detected by using thioglycolic-acid-modified CdTe QDs by Khan et al. [94]. Silver ion detection was carried out by Wang et al. [209]

using L-glutathione-capped CdTe QDs. Ebrahim et al. [53] performed the sensing of ochratoxin-A using CdTe QDs with a combined coating of thioglycolic acid and L-cysteine. Ascorbic acid detection based on CdTe QD emission wavelength shift was demonstrated by Zhu et al. [248]. Glutathione-capped CdTe QD was used by Yuan et al. [233] for the detection of glucose. The pH-sensitive CdTe QDs capped with MPA were developed by Maule et al. [131]. Shi et al. [182] found a method for the detection of lead ions using the sensitization effect of CdTe QDs on MoS_2–CdS nanocomposite. Zheng et al. [243] fabricated CdTe–CdS core–shell QDs that can detect perfluorooctanoic acid, which is a water contaminant that causes immunological and neurological toxicity. Colorimetric detection of the colorless compounds was done by Seidi et al. [178] using CdTe QDs. The energy transfer between CdTe QDs and cyanine dyes was used for the detection of ochratoxin-A by Gao et al. [62]. Tao et al. [193] found a method for the detection of Hg^{2+} using CdTe QDs. Work on Hg (II) detection in aqueous media was done by Saikia et al. [172] using CdTe/ZnS core–shell QD by the mechanism of photoelectron transfer. A method for the visual detection of cadmium ions was found by QIian et al. [161] using a CdTe QD-based hybrid probe. Chen et al. [28] developed an optical humidity sensor using MPA-capped CdTe QDs. Peng et al. [157] found a method for the detection of curcumin using a CdTe-based composite. A gaseous formaldehyde sensor was developed by Ma et al. [127] using CdTe QD film. Chen et al. [26] developed a method for the detection of 32 Chinese vinegars by the method of fluorescence on–off technique. The MSA-coated CdTe QD thermometers were developed by Jiang et al. [85]. A method for sulfadimidine detection by CdTe QDs was found by Zheng et al. [245]. The CdTe DNA sensor was developed by Shamsipus [181] for the detection of human papilloma viruses. In addition to the abovementioned studies about CdTe QDs as different sensors, there are a number of reports which discuss the limitless sensing property of the CdTe QDs. Quantum dots such as CdSe are 20 times brighter and 100 times more stable than the commonly used organic dye, Rhodamine 6G and they exhibit size-dependent absorption and emission properties. In addition, their emission spectra span a wide visible region, and single-wavelength excitation of different particle sizes leads to multiple-wavelength emission. When compared to their organic counterparts, these prominent emission properties result in a wide range of applications such as cell labeling, DNA probing, immunity labeling, and binding assay [30, 45, 65, 113, 122, 148, 153, 173, 183, 208, 210, 220, 231, 237, 238].

pH sensing is another important property of quantum dots. It is well known that most biological and chemical processes depend on pH. So, pH is an important parameter to be measured and controlled. Since most chemical and biological reactions are dependent on pH, its measurement is important in almost every piece of laboratory work. In chemical reactions, the speed of the reaction process, solubility of chemicals, etc. depend on the pH of the environment. So, controlling the pH is very essential in these types of processes in order to prevent the occurrence of unwanted reactions. The detection of pH is very important in water purification because it helps to identify the presence of dangerous chemicals in the water. Most industries, such as food, drugs, textiles, semiconductors, cement, and biomedicine, utilize pH measurements for monitoring and controlling various processes.

pH is the negative logarithm of the hydrogen-ion concentration present in a solution, and it measures the acidity or basicity of the material at a given temperature. It can be mathematically expressed as $pH = -\log [H^+]$. pH test strips, glass electrodes, ISFET, pH images, optical fiber, magnetoelastic, conductometric, cantilever, potentiometry, etc. are the commonly used experimental techniques [93]. According to the literature, the intensity of emission of water-soluble QD is pH dependent [132, 169, 197, 237]. Depending upon the pH of the medium, the emission intensity either increases or decreases. Most often, an enhancement in PL intensity occurs with pH. The first such pH-dependent studies started in 1987 by Lubbmir and his group [187]. In this study, the pH-dependent properties of CdS colloids were investigated. In 2000, Mattoussi et al. [130] developed a novel and direct method for bioactive fluorescence probe sensing using a CdSe–ZnS core–shell structure.

High water stability of the QD is a necessary condition for biological applications. To ensure stability in aqueous media, the QD surface is modified using hydrophilic capping ligands. Surface-modified QDs are used for the detection of metallic and non-metallic ions, explosive gases, and molecules. Most of the reported studies focused on the pH-dependent emission properties of surface-capped QDs. This pH-dependent emission property is most probably a function of surface functionalization. For QDs, surface modification and defect states are the most important characteristics involved in pH probing applications. Surface passivation helps to increase the chemical stability via the thick organic coating layer, which decreases the diffusion of hydrogen ions. Many reports show that surfactant-capped QDs exhibit excellent photoluminescence quantum yield [51, 131, 169].

Although most studies show an increase in PL intensity with a decrease in pH value, there have been few reports of the opposite effect. For example, Kharlampieva [95] and Gao [63] studied the pH-dependent emission properties of the mercaptoacetic-acid-capped CdTe QDs. In both cases, PL intensity was found to increase at lower pH values. So, the pH-dependent emission properties of QDs are highly variable. The type of semiconductor material and organic coatings are two major factors that affect pH sensing and the range of sensing. Zhang et al. [74] investigated the effect of a capping agent on QD emission. In that work, they capped CdTe QDs with different capping molecules and identified the presence of polyacrylic acid. Besides PL intensity variation, a shift in photoluminescence peak with respect to the change in pH of the system is also commonly used for quantification of analytes. Several studies have been conducted in this direction using various types of QDs [88, 114, 195]. There are other factors like the presence of organic coating, and the history of QDs, which contribute to the PL shift [16, 43, 52, 125, 239]. The interaction of carboxylic acid and thiol groups with the QD surface causes an internal electric field, leading to the shift in PL. Moreover, reversibility is very important in sensing mechanisms, which determines how efficient the sensing is. Any pH-sensing methodology based on QDs must consider the following important aspects:

(1) The effect of organic coating on the pH-dependent emission.
(2) The reversible or irreversible nature of pH change.
(3) The spectral shift in the PL spectra.
(4) The reproducibility of the pH sensor.

The change in PL emission also coincides with the fluorescence lifetime (τ) measurements. According to Spanhel et al. [188], fluorescence lifetime decreases with a decrease in pH value, whereas Durisic et al. [51] found an increase in fluorescence lifetime with a decrease in pH value. The fluorescence lifetime is the inverse of the total recombination rate which is the sum of the radiative and nonradiative recombination rates. The nonradiative recombination itself is a sum of various processes, like inter-system crossing, internal conversion, vibrational relaxation, etc. CdSe/ZnS quantum dots capped with mercaptopropionic acid (MPA) are reported to be used as lifetime-based pH nanosensors [146, 147, 168]. Nideep et al. [146, 147] synthesized MSA-capped CdTe quantum dots and studied their pH sensitivity by measuring the intensity of emission. As shown in Fig. 2.8, the emission intensity was found to decrease with a decrease in pH value. The protonation and deprotonation of the carboxylic acid group present in the MSA molecule make the CdTe quantum dots pH sensitive.

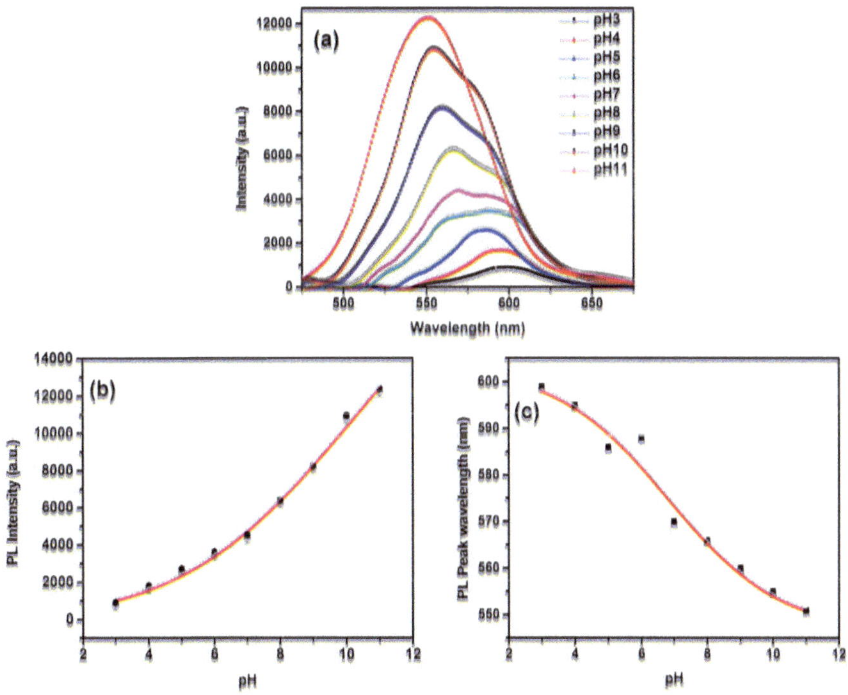

Fig. 2.8 **a** Emission spectra of CdTe QDs with a variation of pH, **b** variation of peak emission intensity with change in pH, and **c** shift in peak emission wavelength with change in pH [146, 147]

The facts presented above indicate that quantum dots are interesting candidates for nanotechnology and nanomedicine. Massive efforts are currently being made to synthesize biocompatible quantum dots for biological sensing and the development of quantum devices. Changes in photoluminescence intensity are often used to monitor sensing in quantum dots. Controlling the surface chemistry is another form of optical sensing that is still challenging. In addition to photoluminescence, quantum dots facilitate optical interferometry, a unique detection mechanism in the field of biosensing. In optical interferometry, the adsorption of analytes induces a change in refractive index, which leads to sensing at the femtomolar level [111, 142, 202]. Although medical imaging can detect disorders, it cannot provide visual guidance during surgery. This difficulty has been solved to an extent after the introduction of magnetic quantum-dot probes. Magnetic nanoparticles combined with highly luminescent quantum dots have been developed for multimodal imaging by positron emission tomography (PET) imaging, computed tomography (CT) imaging, and magnetic resonance imaging (MRI) [2, 99]. Quantum dots have been utilized for the treatment of diseased tissue as well as for real-time imaging. More detailed research on this topic is required to validate additional feasible multifunctional applications of quantum dots. The positive results obtained demonstrate the success of quantum dots in several application areas. In essence, applied nanomaterial research on QDs continues to be an active and significant topic of exploration, and the recent contributions point toward many future technological leaps in this field.

2.5 Conclusion and Perspectives

The fundamental characteristics, synthesis, categorization, and sensing capabilities of quantum dots are the main topics of this chapter. In applied science, zero-dimensional nanostructures provide countless benefits and outperform all other dimensional nanostructures. Compared to their bulk counterparts, quantum dots show size-dependent physical, electrical, and optical properties. The quantum confinement effect is thought to be the primary cause of the unique characteristics of quantum dots and their suitability for diverse applications in photonics. As a luminous material with superior photostability, size tunability, higher quantum yield and brightness, and narrow and symmetric emission, quantum dots have replaced fluorophores in optical and biological sensing as well as in imaging. The capacity to examine biomolecules such as antibodies, genes, peptides, proteins, cells, and nucleic acids is another possible use of quantum dots in sensing. As a result, quantum dots can create bioactive fluorescent probes for sensing applications. The quantum dots are also employed for pH sensing, DNA tracking, and protein identification.

The toxicity of quantum dots is still unpredictable. Studies on the toxicity evaluation of quantum dots, particularly the degradation of quantum dots in non-human primates, are still ongoing. Prior to their use in clinical settings, the toxicity and biocompatibility of quantum dots in vivo and in vitro should be thoroughly and properly assessed. There is still a long way to go before quantum dots are used in the human body. Biocompatible quantum dots will be developed for biomedical applications in the future thanks to the rapid advancement of nanotechnology and biotechnology. Future concerns in the area of quantum-dot applications include multiple functional group attachments on the quantum-dot surface, an increase in quantum-dot efficiency, and sensitivity for better targeting and drug delivery.

Acknowledgements The authors would like to thank Science and Engineering Research Board (SERB) India grant numbers EMR/2016/003614 and EEQ/2018/ 000468.

References

1. Agrawal, A., et al.: Counting single native biomolecules and intact viruses with color-coded nanoparticles. Anal. Chem. **78**(4), 1061–1070 (2006). https://doi.org/10.1021/ac051801t
2. Saha, A.K., Sharma, P., Sohn, H.B., Ghosh, S., Das, R.K., Hebard, A.F., Zeng, H., Baligand, C., Walter, G.A., Moudgil, B.M.: Fe Doped CdTeS magnetic quantum dots for bioimaging. J. Mater. Chem. B (207890), 1–9 (2013). https://doi.org/10.1039/b000000x
3. Alivisatos, A.P., Gu, W., Larabell, C.: Quantum dots as cellular probes. Annu. Rev. Biomed. Eng. **7**, 55–76 (2005). https://doi.org/10.1146/annurev.bioeng.7.060804.100432
4. Alivisatos, P.: The use of nanocrystals in biological detection. Nat. Biotechnol.Biotechnol. **22**(1), 47–52 (2004). https://doi.org/10.1038/nbt927
5. Wagner, A.M., Knipe, J.M., Orive, G., Peppas, N.A.: Quantum dots in biomedical applications. Acta Biomater **94**(1), 44–63 (2019). https://doi.org/10.1016/j.actbio.2019.05.022.Quantum
6. Baba, K., Nishida, K.: Single-molecule tracking in living cells using single quantum dot applications. Theranostics **2**(7), 655–667 (2012). https://doi.org/10.7150/thno.3890
7. Bailey, R.E., Nie, S.: Alloyed semiconductor quantum dots: Tuning the optical properties without changing the particle size. J. Am. Chem. Soc. **125**(23), 7100–7106 (2003). https://doi.org/10.1021/ja035000o
8. Bang, J., Yang, H., Holloway, P.H.: Enhanced and stable green emission of ZnO nanoparticles by surface segregation of Mg. Nanotechnology **17**(4), 973–978 (2006). https://doi.org/10.1088/0957-4484/17/4/022
9. Banin, U., et al.: Identification of atomic-like electronic states in indium arsenide nanocrystal quantum dots. Nature **400**(August), 542–544 (1999)
10. Bera, D., et al.: Quantum dots and their multimodal applications: a review. Materials **3**, 2260–2345 (2010). https://doi.org/10.3390/ma3042260
11. Biju, V., et al.: Temperature-sensitive photoluminescence of CdSe quantum dot clusters. J. Phys. Chem. B **109**(29), 13899–13905 (2005). https://doi.org/10.1021/jp050424l
12. Bo, C., Ping, Z.: A new determining method of copper(II) ions at ng ml-1 levels based on quenching of the water-soluble nanocrystals fluorescence. Anal. Bioanal. Chem.Bioanal. Chem. **381**(4), 986–992 (2005). https://doi.org/10.1007/s00216-004-2963-9
13. Bokov, D., et al.: Nanomaterial by sol-gel method: synthesis and application. Adv. Mater. Sci. Eng. **2021**, 1–21 (2021). https://doi.org/10.1155/2021/5102014
14. Brunetti, J., et al.: Near-infrared quantum dots labelled with a tumor selective tetrabranched peptide for in vivo imaging. J. Nanobiotechnol.Nanobiotechnol. **16**(1), 1–10 (2018). https://doi.org/10.1186/s12951-018-0346-1
15. Burda, C., et al.: Chemistry and Properties of Nanocrystals of Different Shapes (2005)
16. Byrne, S.J., et al.: Optimisation of the synthesis and modification of CdTe quantum dots for enhanced live cell imaging. J. Mater. Chem. **16**, 2896–2902 (2006). https://doi.org/10.1039/b605333e

17. Calabro, R.L., Yang, D., Kim, D.Y.: Liquid-phase laser ablation synthesis of graphene quantum dots from carbon nano-onions: Comparison with chemical oxidation. J. Colloid Interface Sci. **527**, 132–140 (2018). https://doi.org/10.1016/j.jcis.2018.04.113
18. Calabro, R.L., Yang, D.S., Kim, D.Y.: Controlled nitrogen doping of graphene quantum dots through laser ablation in aqueous solutions for photoluminescence and electrocatalytic applications. ACS Appl. Nano Mater. **2**(11), 6948–6959 (2019). https://doi.org/10.1021/acsanm.9b01433
19. Cappello, G., Dahan, M.: Tracking individual kinesin motors in living cells using single quantum-dot imaging. Nano Lett. **6**, 1–5 (2006)
20. Chan, W.C.W., Nie, S.: Quantum dot bioconjugates for ultrasensitive nonisotopic detection. Science **281**(5385), 2016–2018 (1998). https://doi.org/10.1126/science.281.5385.2016
21. Chang, E., et al.: Quantifying the influence of surface coatings on quantum dot uptake in cells. J. Biomed. Nanotechnol.Nanotechnol. **1**(4), 397–401 (2006). https://doi.org/10.1166/jbn.2005.053
22. Chason, E., et al.: Ion beams in silicon processing and characterization. J. Appl. Phys. **81**(10), 6513–6561 (1997). https://doi.org/10.1063/1.365193
23. Chen, B., et al.: Development of a portable device for Ag+ sensing using CdTe QDs as fluorescence probe via an electron transfer process. Talanta **191**, 357–363 (2019). https://doi.org/10.1016/j.talanta.2018.08.088
24. Chen, C.Y., et al.: Potassium ion recognition by 15-crown-5 functionalized CdSe/ZnS quantum dots in H_2O. Chem. Commun.Commun. **3**(3), 263–265 (2006). https://doi.org/10.1039/b512677k
25. Chen, F., et al.: Fluorescent CdSe/ZnS nanocrystal–peptide conjugates for long-term , nontoxic imaging and nuclear targeting in living cells. Nano Lett. **4**, 1827–1832 (2004)
26. Chen, H., et al.: A colorimetric sensor array for recognition of 32 Chinese traditional cereal vinegars based on "turn-off/on" fluorescence of acid-sensitive quantum dots. Spectrochim. Acta-Part A: Mol. Biomol. Spectrosc. **227**, 117683 (2020). https://doi.org/10.1016/j.saa.2019.117683
27. Chen, H.H., et al.: Simultaneous non-invasive analysis of DNA condensation and stability by two-step QD-FRET. NanoToday **4**, 125–134 (2009). https://doi.org/10.1016/j.nantod.2009.02.008
28. Chen, M., et al.: A highly stable optical humidity sensors based on nano-composite film. Sens. Actuators, B Chem. **287**, 329–337 (2019). https://doi.org/10.1016/j.snb.2019.02.051
29. Chen, N.T., et al.: Recent advances in nanoparticle-based főrster resonance energy transfer for biosensing, molecular imaging and drug release profiling. Int. J. Mol. Sci. **13**(12), 16598–16623 (2012). https://doi.org/10.3390/ijms131216598
30. Chen, S., et al.: Room-temperature synthesis of fluorescent carbon-based nanoparticles and their application in multidimensional sensing. Sens. Actuators, B Chem. **288**, 749–756 (2019). https://doi.org/10.1016/j.snb.2019.03.067
31. Chen, X., et al.: Coherency strain effects on the optical response of core/shell heteronanostructures. Nano Lett. **3**(6), 799–803 (2003). https://doi.org/10.1021/nl034243b
32. Chen, Y., Rosenzweig, Z.: Luminescent CdS quantum dots as selective ion probes. Anal. Chem. **74**(19), 5132–5138 (2002). https://doi.org/10.1021/ac0258251
33. Cheng, W., et al.: Effective cell capture with tetrapeptide-functionalized carbon nanotubes and dual signal amplification for cytosensing and evaluation of cell surface carbohydrate. Anal. Chem. **80**(10), 3867–3872 (2008)
34. Choucair, M., Thordarson, P., Stride, J.A.: Gram-scale production of graphene based on solvothermal synthesis and sonication. Nat. Nanotechnol.Nanotechnol. **4**(1), 30–33 (2009). https://doi.org/10.1038/nnano.2008.365
35. Christiaans, M.P.T., et al.: Photoinduced electron transfer from conjugated polymers onto TiO_2. Synth. Met. **101**(1), 265–266 (1999). https://doi.org/10.1016/S0379-6779(98)01437-4
36. Clapp, A.R., et al.: Fluorescence resonance energy transfer between quantum dot donors and dye-labeled protein acceptors. J. Am. Chem. Soc. **126**(1), 301–310 (2004). https://doi.org/10.1021/ja037088b

37. Clapp, A.R., et al.: Can luminescent quantum dots be efficient energy acceptors with organic dye donors? J. Am. Chem. Soc. **127**(4), 1242–1250 (2005). https://doi.org/10.1021/ja045676z
38. Colvin, V.L., Goldstein, A.N., Alivisatos, A.P.: Semiconductor nanocrystals covalently bound to metal surfaces with self-assembled monolayers. J. Am. Chem. Soc. **114**(13), 5221–5230 (1992). https://doi.org/10.1021/ja00039a038
39. Costa-Fernández, J.M., Pereiro, R., Sanz-Medel, A.: The use of luminescent quantum dots for optical sensing. TrAC—Trends Anal. Chem. **25**(3), 207–218 (2006). https://doi.org/10.1016/j.trac.2005.07.008
40. Cui, L., et al.: Advances in the integration of quantum dots with various nanomaterials for biomedical and environmental applications. Analyst **143**(11), 2469–2478 (2018). https://doi.org/10.1039/c8an00222c
41. Huffaker, D.L., Park, G., Zou, Z., Shchekin, O.B., Deppe, D.G.: 1.3 μm room-temperature GaAs-based quantum-dot laser. Appl. Phys. Lett. **73**, 2564–2566 (1998)
42. Darbandi, M., Thomann, R., Nann, T.: Single quantum dots in silica spheres by microemulsion synthesis. Chem. Mater. **17**(23), 5720–5725 (2005). https://doi.org/10.1021/cm051467h
43. Debruyne, D., Deschaume, O., Coutiño-gonzalez, E.: The pH-dependent photoluminescence of colloidal CdSe/ZnS quantum dots with different organic coatings. Nanotechnology **26**, 1–9 (2015). https://doi.org/10.1088/0957-4484/26/25/255703
44. Dennis, A.M., et al.: Quantum dot-fluorescent protein fret probes for sensing intracellular pH. ACS Nano **6**(4), 2917–2924 (2012). https://doi.org/10.1021/nn2038077
45. Desai, M.L., et al.: Influence of doping ion, capping agent and pH on the fl uorescence properties of zinc sul fi de quantum dots: Sensing of Cu_{2+} and Hg_{2+} ions and their biocompatibility with cancer and fungal cells. Spectrochim. Acta Part A Mol. Biomol. Spectrosc.. Acta Part A Mol. Biomol. Spectrosc. **210**, 212–221 (2019). https://doi.org/10.1016/j.saa.2018.11.027
46. Devi, S., et al.: Quantum dots: an emerging approach for cancer therapy. Front. Mater. **8**, 1–18 (2022). https://doi.org/10.3389/fmats.2021.798440
47. Dhawan, S., Dhawan, T., Vedeshwar, A.G.: Growth of Nb_2O_5 quantum dots by physical vapor deposition. Mater. Lett. **126**, 32–35 (2014). https://doi.org/10.1016/j.matlet.2014.03.107
48. Díaz, S.A., et al.: Quantum dots as Förster resonance energy transfer acceptors of lanthanides in time-resolved bioassays. ACS Appl. Nano Mater. **1**(6), 3006–3014 (2018). https://doi.org/10.1021/acsanm.8b00613
49. Du, C.F., et al.: Controllable synthesis of ultrasmall $CuInSe_2$ quantum dots for photovoltaic application. RSC Adv. **4**(64), 33855–33860 (2014). https://doi.org/10.1039/c4ra04727c
50. Dubertret, B., et al.: In vivo imaging of quantum dots encapsulated in phospholipid micelles. Science **1759**, 1–5 (2013). https://doi.org/10.1126/science.1077194
51. Durisic, N., et al.: A common mechanism underlies the dark fraction formation and fluorescence blinking of quantum dots. ACS Nano **3**(5), 1167–1175 (2009). https://doi.org/10.1021/nn800684z
52. Durisic, N., et al.: Probing the "Dark" fraction of core À shell quantum dots by ensemble and single particle pH-dependent spectroscopy. ACS Nano **5**(11), 9062–9073 (2011). https://doi.org/10.1021/nn203272p
53. Ebrahim, S., et al.: CdTe quantum dots capped with different stabilizing agents for sensing of ochratoxin A. J. Lumin. **182**, 154–159 (2017). https://doi.org/10.1016/j.jlumin.2016.09.038
54. Einevoll, G.T.: Confinement of excitons in quantum dots. Phys. Rev. B **45**(7), 3410–3417 (1992)
55. Ensafi, A.A., Nasr-Esfahani, P., Rezaei, B.: Simultaneous detection of folic acid and methotrexate by an optical sensor based on molecularly imprinted polymers on dual-color CdTe quantum dots. Anal. Chim. Acta **996**, 64–73 (2017). https://doi.org/10.1016/j.aca.2017.10.011
56. Entezari, M.H., Ghows, N.: Micro-emulsion under ultrasound facilitates the fast synthesis of quantum dots of CdS at low temperature. Ultrason. Sonochem.. Sonochem. **18**(1), 127–134 (2011). https://doi.org/10.1016/j.ultsonch.2010.04.001
57. Feng, J., et al.: Highly sensitive and selective fluorescent sensor for tetrabromobisphenol-a in electronic waste samples using molecularly imprinted polymer coated quantum dots. Microchem. J.. J. **144**, 93–101 (2019). https://doi.org/10.1016/j.microc.2018.08.041

58. Fernández-Argüelles, M.T., et al.: Surface-modified CdSe quantum dots for the sensitive and selective determination of Cu(II) in aqueous solutions by luminescent measurements. Anal. Chim. Acta **549**(1–2), 20–25 (2005). https://doi.org/10.1016/j.aca.2005.06.013
59. Foerster, A., Besley, N.A.: Quantum chemical characterization and design of quantum dots for sensing applications. J. Phys. Chem. A **126**, 2899–2908 (2022). https://doi.org/10.1021/acs.jpca.2c00947
60. Franceschetti, A., Zunger, A.: Direct pseudopotential calculation of exciton coulomb and exchange energies in semiconductor quantum dots. Phys. Rev. Lett. **78**(5), 915–918 (1997)
61. Gambhir, S.S., Weiss, S.: Quantum dots for live cells, in vivo imaging, and diagnostics. Science **307**(January), 538–545 (2005)
62. Gao, J., et al.: Electrochemiluminescent aptasensor based on resonance energy transfer system between CdTe quantum dots and cyanine dyes for the sensitive detection of Ochratoxin A. Talanta **199**, 178–183 (2019). https://doi.org/10.1016/j.talanta.2019.02.044
63. Gao, M., et al.: Strongly photoluminescent CdTe nanocrystals by proper surface modification. J. Phys. Chem. B **5**, 8360–8363 (1998). https://doi.org/10.1021/jp9823603
64. Gao, X., et al.: In vivo cancer targeting and imaging with semiconductor quantum dots. Nat. Biotechnol.Biotechnol. **22**(8), 969–976 (2004). https://doi.org/10.1038/nbt994
65. Gao, X., Chan, W.C.W.: Quantum-dot nanocrystals for ultrasensitive biological labeling and multicolor optical encoding. J. Biomed. Opt. **7**(4), 532–537 (2002). https://doi.org/10.1117/1.1506706
66. Gerion, D., et al.: Synthesis and properties of biocompatible water-soluble silica-coated CdSe/ZnS semiconductor quantum dots. J. Phys. Chem. B **105**(37), 8861–8871 (2001). https://doi.org/10.1021/jp0105488
67. Gieben, H., et al.: Quantum dots in the strong confinement regime: a model system for gain in quasi zero-dimensional semiconductors. SPIE Proc. **2693**, 233–237 (2016)
68. Gil, H.M. et al.: NIR-quantum dots in biomedical imaging and their future. iScience **24**(3), 1–26 (2021). https://doi.org/10.1016/j.isci.2021.102189.
69. Goldman, E.R., et al.: Avidin: a natural bridge for quantum dot-antibody conjugates. J. Am. Chem. Soc. **124**(22), 6378–6382 (2002). https://doi.org/10.1021/ja0125570
70. Goldman, E.R. et al.: Luminescent quantum dot-adaptor protein-antibody conjugates for use in fluoroimmunoassays. Phys. Status Solidi (B) Basic Res. **229**(1), 407–414 (2002). https://doi.org/10.1002/1521-3951(200201)229:1<407::AID-PSSB407>3.0.CO;2-S.
71. Purcell-Milton, F., Gun'ko, Y.K.: Quantum dots for luminescent solar concentrators. J. Mater. Chem. **22**(33), 16687–16697 (2012). https://doi.org/10.1039/c2jm32366d
72. Han, M., et al.: Quantum-dot-tagged microbeads for multiplexed optical coding of biomolecules. Nat. Biotechnol.Biotechnol. **19**(7), 631–635 (2001). https://doi.org/10.1038/90228
73. Hao, C., et al.: Dual-channel optical sensing platform for detection of diminazene aceturate based on thioglycolic acid-wrapped cadmium telluride/cadmium sulfide quantum dots. J. Colloid Interface Sci. **472**, 76–83 (2016). https://doi.org/10.1016/j.jcis.2016.03.017
74. Zhang, H., Zhou, Z., Bai Yang, M.G.: The influence of carboxyl groups on the photoluminescence of mercaptocarboxylic acid-stabilized CdTe nanoparticles. J. Phys. Chem. B **107**, 8–13 (2003). https://doi.org/10.1021/jp025910c
75. Harbord, E., et al.: Influence of p-doping on the temperature dependence of InAs/GaAs quantum dot excited state radiative lifetime. Appl. Phys. Lett. **101**(18), 1–5 (2012). https://doi.org/10.1063/1.4765349
76. He, J., et al.: Composition dependence of structure and optical properties of Cu 2ZnSn(S, Se)4 solid solutions: an experimental study. J. Alloy. Compd. **511**(1), 129–132 (2012). https://doi.org/10.1016/j.jallcom.2011.08.099
77. Hines, M.A., Guyot-Sionnest, P.: Synthesis and characterization of strongly luminescing ZnS-capped CdSe nanocrystals. J. Phys. Chem. **100**(2), 468–471 (1996). https://doi.org/10.1021/jp9530562
78. Hoshino, A., et al.: Simultaneous multicolor detection system of the single-molecular microbial antigen with total internal reflection fluorescence microscopy. Microbiol. Immunol.. Immunol. **49**(5), 461–470 (2005). https://doi.org/10.1111/j.1348-0421.2005.tb03750.x

79. Hu, O., et al.: "Turn-off" fluorescent sensor based on double quantum dots coupled with chemometrics for highly sensitive and specific recognition of 53 famous green teas. Anal. Chim. Acta **1008**, 103–110 (2018). https://doi.org/10.1016/j.aca.2017.12.042
80. Hua, M., et al.: Highly selective and sensitive determination of uranyl ion by the probe of CdTe quantum dot with a specific size. Talanta **190**, 278–283 (2018). https://doi.org/10.1016/j.talanta.2018.08.012
81. Isarov, A.V., Chrysochoos, J.: Optical and photochemical properties of nonstoichiometric cadmium sulfide nanoparticles: surface modification with copper(II) ions. Langmuir **13**(12), 3142–3149 (1997). https://doi.org/10.1021/la960985r
82. Jagtap, A.M. et al.: Size and temperature dependence of the photoluminescence properties of NIR emitting ternary alloyed mercury cadmium telluride quantum dots. J. Phys. D: Appl. Phys. **49**(13) (2016). https://doi.org/10.1088/0022-3727/49/13/135302
83. Jamieson, T., et al.: Biological applications of quantum dots. Biomaterials **28**(31), 4717–4732 (2007). https://doi.org/10.1016/j.biomaterials.2007.07.014
84. Jdira, L., et al.: Size-dependent single-particle energy levels and interparticle Coulomb interactions in CdSe quantum dots measured by scanning tunneling spectroscopy. Phys. Rev. B-Condens. Matter **73**(11), 1–6 (2006). https://doi.org/10.1103/PhysRevB.73.115305
85. Jiang, X., Shao, J., Li, B.Q.: Ceiling temperature and photothermalsensitivity of aqueous MSA-CdTe quantum dots thermometers. Appl. Surf. Sci. **394**, 554–561 (2017). https://doi.org/10.1016/j.apsusc.2016.10.149
86. Jin, S., et al.: Application of quantum dots in biological imaging. J. Nanomater. 1–13 (2011). https://doi.org/10.1155/2011/834139
87. Jin, T., et al.: A quantum dot-based ratiometric pH sensor. Chem. Commun.Commun. **46**(14), 2408–2410 (2010). https://doi.org/10.1039/b921602b
88. Jin, W.J., Ferna, T., Sanz-Medel, A.: Photoactivated luminescent CdSe quantum dots as sensitive cyanide probes in aqueous solutions. Chem. Commun. 883–885 (2005). https://doi.org/10.1039/b414858d.
89. Jun, H.K., Careem, M.A., Arof, A.K.: Quantum dot-sensitized solar cells-perspective and recent developments: a review of Cd chalcogenide quantum dots as sensitizers. Renew. Sustain. Energy Rev. **22**, 148–167 (2013). https://doi.org/10.1016/j.rser.2013.01.030
90. Karatutlu, A., et al.: Synthesis and structure of free-standing germanium quantum dots and their application in live cell imaging. RSC Adv. **5**(26), 20566–20573 (2015). https://doi.org/10.1039/c5ra01529d
91. Lawson-Wood, K., Upstone, S., Evans, K.: Determination of Relative Fluorescence Quantum Yields using the FL6500 Fluorescence Spectrometer. Fluoresc. Spectrosc. 1–5 (2018).
92. Keuleyan, S., Kohler, J., Guyot-Sionnest, P.: Photoluminescence of mid-infrared HgTe colloidal quantum dots. J. Phys. Chem. C **118**(5), 2749–2753 (2014). https://doi.org/10.1021/jp409061g
93. Khan, M.I., et al.: PH sensing materials for MEMS sensors and detection techniques. Int. Conf. Solid State Integr. Circuit **32**, 18–22 (2012)
94. Khan, S., et al.: Determination of histamine in tuna fish by photoluminescence sensing using thioglycolic acid modified CdTe quantum dots and cationic solid phase extraction. J. Lumin. **182**, 71–78 (2017). https://doi.org/10.1016/j.jlumin.2016.09.041
95. Kharlampieva, E., et al.: PH-responsive photoluminescent LbL hydrogels with confined quantum dots. Soft Matter **6**, 800–807 (2010). https://doi.org/10.1039/b917845g
96. Kim, M., et al.: Synthesis of nanoparticles by laser ablation: a review. Kona Powder Part. J. **34**, 80–90 (2017). https://doi.org/10.14356/kona.2017009
97. Kirchner, C., et al.: Cytotoxicity of colloidal CdSe and CdSe/ZnS nanoparticles. Nano Lett. **5**(2), 331–338 (2005). https://doi.org/10.1021/nl047996m
98. Klostranec, J.M., Chan, W.C.W.: Quantum dots in biological and biomedical research: recent progress and present challenges. Adv. Mater. **18**(15), 1953–1964 (2006). https://doi.org/10.1002/adma.200500786
99. Koole, R., et al.: Magnetic quantum dots for multimodal imaging. Adv. Rev. **1**, 475–491 (2009). https://doi.org/10.1002/wnan.014.

100. Korala, L., et al.: Uniform thin films of CdSe and CdSe(ZnS) core(shell) quantum dots by sol-gel assembly: enabling photoelectrochemical characterization and electronic applications. ACS Nano **7**(2), 1215–1223 (2013). https://doi.org/10.1021/nn304563j
101. Korn, K.: Gene expression analysis using single molecule detection. Nucleic Acids Res. **31**(16), 89e–889 (2003). https://doi.org/10.1093/nar/gng089
102. Kumar, D.S., Kumar, B.J., Mahesh, H.M.: Quantum nanostructures (QDs): an overview. Elsevier Ltd., Synthesis of Inorganic Nanomaterials (2018). https://doi.org/10.1016/B978-0-08-101975-7.00003-8
103. Kumar, M.M.D., Devadason, S., Rajesh, S.: Formation of CdSe/CdTe quantum dots in multilayer thin films using PVD method. AIP Conf. Proc. **178**, 176–178 (2012). https://doi.org/10.1063/1.4732406
104. Kumar, Y.R., et al.: Graphene quantum dot based materials for sensing, bio-imaging and energy storage applications: a review. RSC Adv. **10**(40), 23861–23898 (2020). https://doi.org/10.1039/d0ra03938a
105. Kurtz, E., et al.: Formation and properties of self-organized II-VI quantum islands. Thin Solid Films **367**(1–2), 68–74 (2000). https://doi.org/10.1016/S0040-6090(00)00665-9
106. Lee, J., et al.: Uniform graphene quantum dots patterned from self-assembled silica nanodots. Nano Lett. **12**, 6078–6083 (2012)
107. Lemke, K., Koetz, J.: Polycation-capped CdS quantum dots synthesized in reverse microemulsions. J. Nanomater.Nanomater. **2012**, 1–11 (2012). https://doi.org/10.1155/2012/478153
108. Leonard, D., et al.: Direct formation of quantum-sized dots from uniform coherent islands of InGaAs on GaAs surfaces. Appl. Phys. Lett. **63**(23), 3203–3205 (1993). https://doi.org/10.1063/1.110199
109. Leonardi, K., et al.: Formation of self-assembling II-VI semiconductor nanostructures during migration enhanced epitaxy. J. Cryst. GrowthCryst. Growth **184–185**, 259–263 (1998). https://doi.org/10.1016/S0022-0248(98)80055-9
110. Leschkies, K.S., et al.: Photosensitization of ZnO nanowires with CdSe quantum dots for photovoltaic devices. Nano Lett. **7**, 1793–1798 (2007)
111. Leuermann, J., et al.: Optimizing the limit of detection of waveguide-based interferometric biosensor devices. Sensors **19**, 1–13 (2019). https://doi.org/10.3390/s19173671
112. Li, H., et al.: A review on the synthesis methods of CdSeS-based nanostructures. J. Nanomater.Nanomater. **2015**, 1–16 (2015). https://doi.org/10.1155/2015/519385
113. Li, L., et al.: Excitation-independent hollow orange-fluorescent carbon nanoparticles for pH sensing in aqueous solution and living cells. Talanta **196**, 109–116 (2019). https://doi.org/10.1016/j.talanta.2018.12.034
114. Li, Y., Li, B., Zhang, J.: H_2O_2- and pH-sensitive CdTe quantum dots as fl uorescence probes for the detection of glucose. Luminescence **28**, 667–672 (2012). https://doi.org/10.1002/bio.2413
115. Li, Z., et al.: Simultaneous determination of nine types of phthalate residues in commercial milk products using HPLC-ESI-MS-MS. J. Chromatogr. Sci.Chromatogr. Sci. **49**(4), 338–343 (2011). https://doi.org/10.1093/chrsci/49.4.337
116. Liang, J.G., et al.: Functionalized CdSe quantum clots as selective silver ion chemodosimeter. Analyst **129**(7), 619–622 (2004). https://doi.org/10.1039/b317044f
117. Liang, Z., et al.: Bio-conjugated quantum dots for cancer research: detection and Imaging. Front. Oncol. **11**(October), 1–14 (2021). https://doi.org/10.3389/fonc.2021.749970
118. Lin, Y.P., et al.: Design and fabrication of densely integrated silicon quantum dots using a VLSI compatible hydrogen silsesquioxane electron beam lithography process. Microelectron. Eng. **98**, 386–390 (2012). https://doi.org/10.1016/j.mee.2012.07.011
119. Linehan, K., Doyle, H.: Size controlled synthesis of carbon quantum dots using hydride reducing agents. J. Mater. Chem. C **2**(30), 6025–6031 (2014). https://doi.org/10.1039/c4tc00826j
120. Liu, L., et al.: Turn-off" fluorescent sensor for highly sensitive and specific simultaneous recognition of 29 famous green teas based on quantum dots combined with chemometrics. Anal. Chim. Acta. Elsevier B.V. (2017). https://doi.org/10.1016/j.aca.2017.01.032.

121. Liu, W., et al.: Planar temperature sensing using heavy-metal-free quantum dots with micrometer resolution. Nanotechnology **25**(28), 1–9 (2014). https://doi.org/10.1088/0957-4484/25/28/285501
122. Liu, Y., et al.: PH-sensitive photoluminescence of CdSe/ZnSe/ZnS quantum dots in human ovarian cancer cells. J. Phys. Chem. C **111**, 2872–2878 (2007)
123. Liu, Z., et al.: Micro-light-emitting diodes with quantum dots in display technology. Light: Sci. Appl. **9**(83), 1–23 (2020). https://doi.org/10.1038/s41377-020-0268-1.
124. Luo, J., et al.: Materials science in semiconductor processing one-pot aqueous synthesis of germanium-doped cadmium sulfide quantum dots as fluorescent probes for cell imaging. Mater. Sci. Semicond. Process. **34**, 1–7 (2015). https://doi.org/10.1016/j.mssp.2015.01.024
125. Franke, M., Leubner, S., Dubavik, A., George, A., Savchenko, T., Pini, C., Frank, P., Melnikau, D., Rakovich, Y., Gaponik, N., Eychmüller, A.: Immobilization of pH-sensitive CdTe quantum dots in a poly (acrylate) hydrogel for microfluidic applications. Nanoscale Res. Lett. **12**, 1–8 (2017). https://doi.org/10.1186/s11671-017-2069-x
126. Ma, N., et al.: DNA-passivated CdS nanocrystals: Luminescence, bioimaging, and toxicity profiles. Langmuir **23**(26), 12783–12787 (2007). https://doi.org/10.1021/la7017727
127. Ma, Q., Cui, H., Su, X.: Highly sensitive gaseous formaldehyde sensor with CdTe quantum dots multilayer films. Biosens. Bioelectron.. Bioelectron. **25**(4), 839–844 (2009). https://doi.org/10.1016/j.bios.2009.08.038
128. Makrides, S.C., Gasbarro, C., Bello, J.M.: Bioconjugation of quantum dot luminescent probes for Western blot analysis. Biotechniques **39**(4), 501–505 (2005). https://doi.org/10.2144/000112004
129. Mashford, B., et al.: Synthesis of quantum dot doped chalcogenide glasses via sol-gel processing. J. Appl. Phys. **109**(9), 1–8 (2011). https://doi.org/10.1063/1.3579442
130. Mattoussi, H., et al.: Self-assembly of CdSe-ZnS quantum dot bioconjugates using an engineered recombinant protein. J. Am. Soc. **49**(8), 12142–12150 (2000). https://doi.org/10.1021/ja002535y
131. Maule, C., Gonçalves, H., et al.: Wavelength encoded analytical imaging and fiber optic sensing with pH sensitive CdTe quantum dots. Talanta **80**(5), 1932–1938 (2010). https://doi.org/10.1016/j.talanta.2009.10.048
132. Maule, C., Gonc, H., et al.: Wavelength encoded analytical imaging and fiber optic sensing with pH sensitive CdTe quantum dots. Talanta **80**, 1932–1938 (2010). https://doi.org/10.1016/j.talanta.2009.10.048
133. McDaniel, H., et al.: An integrated approach to realizing high-performance liquid-junction quantum dot sensitized solar cells. Nat. Commun.Commun. **4**, 1–10 (2013). https://doi.org/10.1038/ncomms3887
134. Medintz, I.L., et al.: Self-assembled nanoscale biosensors based on quantum dot FRET donors. Nature **2**, 1–9 (2003). https://doi.org/10.1038/nmat961
135. Mehrzad-Samarin, M., et al.:A novel metronidazole fluorescent nanosensor based on graphene quantum dots embedded silica molecularly imprinted polymer. Biosens. Bioelectron. **92**, 618–623 (2017). https://doi.org/10.1016/j.bios.2016.10.047
136. Menazea, A.A.: Femtosecond laser ablation-assisted synthesis of silver nanoparticles in organic and inorganic liquids medium and their antibacterial efficiency. Radiat. Phys. Chem.. Phys. Chem. **168**, 108616 (2019). https://doi.org/10.1016/j.radphyschem.2019.108616
137. Michalet, X., et al.: Properties of fluorescent semiconductor nanocrystals and their application to biological labeling. Single Mol. **2**(4), 261–276 (2001). https://doi.org/10.1002/1438-5171(200112)2:4%3c261::AID-SIMO261%3e3.0.CO;2-P
138. Mohammad Mominur Rahman, M.R.K. et al.: Cadmium selenide quantum dots for solar cell applications: a review. Chem.–Asian J. **16**(8), 902–921 (2021). https://doi.org/10.1002/asia.202001369
139. Morello, G., et al.: Temperature and size dependence of nonradiative relaxation and exciton-phonon coupling in colloidal CdTe quantum dots. J. Phys. Chem. C **111**(16), 5846–5849 (2007). https://doi.org/10.1021/jp068307t

140. Morozova, S., et al.: Silicon quantum dots: synthesis, encapsulation, and application in light-emitting diodes. Front. Chem. **8**, 1–8 (2020). https://doi.org/10.3389/fchem.2020.00191
141. Mousavi, S.H., et al.: Light-emitting devices–luminescence from low-dimensional nanostructures. Adv. Opt. Commun. (2014). https://doi.org/10.5772/59103
142. Mun, K., et al.: A stable, label-free optical interferometric biosensor based on TiO_2 nanotube arrays. ACS Nano **4**(4), 2070–2076 (2010)
143. Murray, C.B., Kagan, C.R., Bawendi, M.G.: Self-organization of CdSe nanocrystallites into three-dimensional quantum dot superlattices. Science **270**(5240), 1335–1338 (1995). https://doi.org/10.1126/science.270.5240.1335
144. Murray, C.B., Norris, D.J., Bawendi, M.G.: Synthesis and characterization of nearly monodisperse CdE (E = S, Se, Te) semiconductor nanocrystallites. J. Am. Chem. Soc. **115**(19), 8706–8715 (1993). https://doi.org/10.1021/ja00072a025
145. Neuhauser, R.G., et al.: Correlation between fluorescence intermittency and spectral diffusion. Phys. Rev. Lett. **85**, 3301–3304 (2000)
146. Nideep, T.K., et al.: MSA capped CdTe quantum dots for pH sensing application. Mater. Res. Express **6**(10), 1–9 (2019). https://doi.org/10.1088/2053-1591/ab35a0
147. Nideep, T.K., Ramya, M., Kailasnath, M.: An investigation on the photovoltaic performance of quantum dot solar cells sensitized by CdTe, CdSe and CdS having comparable size. Superlattices Microstruct. **141**, 106477 (2020). https://doi.org/10.1016/j.spmi.2020.106477
148. Nie, H. et al.: Carbon dots with continuously tunable full-color emission and their application in ratiometric pH sensing. Chem. Mater. **26**, 3104–3112 (2014). https://doi.org/10.1021/cm5003669
149. Niu, X., et al.: Green and economical synthesis of nitrogen-doped carbon dots from vegetables for sensing and imaging applications. RSC Adv. **5**(115), 95223–95229 (2015). https://doi.org/10.1039/c5ra17439b
150. O'Connor, É., et al.: Near-infrared electroluminescent devices based on colloidal HgTe quantum dot arrays. Appl. Phys. Lett. **86**(20), 1–3 (2005). https://doi.org/10.1063/1.1928321
151. Ogawa, S., et al.: Photoelectrochemistry of films of quantum size lead sulfide particles incorporated in self-assembled monolayers on gold. J. Phys. Chem. B **101**(29), 5707–5711 (1997). https://doi.org/10.1021/jp970737j
152. Osypiw, A.R.C., et al.: Solution-processed colloidal quantum dots for light emission. Mater. Adv. **3**, 6773–6790 (2022). https://doi.org/10.1039/d2ma00375a
153. Ou, J., Tan, H., Chen, Z.: FRET-based semiconducting polymer dots for pH sensing. Sensors **19**(1455), 1–11 (2019). https://doi.org/10.3390/s19061455
154. Pal, U., et al.: Near band gap photoreflectance studies in CdTe, CdTe: V and CdTe: Ge crystals. Mater. Sci. Eng. B **42**(1–3), 297–301 (1996). https://doi.org/10.1016/S0921-5107(96)01725-4
155. Palankar, R., et al.: Fabrication of quantum dot microarrays using electron beam lithography for applications in analyte sensing and cellular dynamics. ACS Nano **7**(5), 4617–4628 (2013). https://doi.org/10.1021/nn401424y
156. Park, H.Y., et al.: PbS quantum dot solar cells integrated with sol-gel-derived ZnO as an n-type charge-selective layer. J. Phys. Chem. C **118**(31), 17374–17382 (2014). https://doi.org/10.1021/jp504156c
157. Peng, J. et al.: Photoelectrochemical sensor based on composite of CdTe and nickel tetra-amined phthalocyanine covalently linked with graphene oxide for ultrasensitive detection of curcumin. Sens. Actuators, B: Chem. **294**, 157–165 (2019). https://doi.org/10.1016/j.snb.2019.05.047
158. Phadnis, C., et al.: The implications of grading on the emission line width of core-shell nanocrystals. J. Phys. D: Appl. Phys. **50**(14) (2017). https://doi.org/10.1088/1361-6463/aa5f25
159. Ponomarenko, L.A., et al.: Chaotic dirac billiard in graphene quantum dots. Science **320**(5874), 356–358 (2008). https://doi.org/10.1126/science.1154663
160. Qian, J., et al.: Fabrication of L-cysteine-capped CdTe quantum dots based ratiometric fluorescence nanosensor for onsite visual determination of trace TNT explosive. Anal. Chim. Acta **946**, 80–87 (2016). https://doi.org/10.1016/j.aca.2016.10.007

161. Qian, J., et al.: Ratiometric fluorescence nanosensor for selective and visual detection of cadmium ions using quencher displacement-induced fluorescence recovery of CdTe quantum dots-based hybrid probe. Sens. Actuators, B Chem. **241**, 1153–1160 (2017). https://doi.org/10.1016/j.snb.2016.10.020
162. Qin, W.J., et al.: Control of Cu-doping and optical properties of ZnO quantum dots by laser ablation of composite targets. Mater. Chem. Phys. **130**(1–2), 425–430 (2011). https://doi.org/10.1016/j.matchemphys.2011.07.001
163. Ramírez-Herrera, D.E., et al.: Ratiometric arginine assay based on FRET between CdTe quantum dots and Cresyl violet. Microchim. Acta. **184**(7), 1997–2005 (2017). https://doi.org/10.1007/s00604-017-2205-4
164. Razgoniaeva, N., et al.: Double-well colloidal nanocrystals featuring two-color photoluminescence. Chem. Mater. **29**(18), 7852–7858 (2017). https://doi.org/10.1021/acs.chemmater.7b02585
165. Reed, M.A., et al.: Observation of discrete electronic states in a zero-dimensional semiconductor nanostructure. Phys. Rev. Lett. **60**(6), 535–537 (1988)
166. Resch-Genger, U., et al.: Quantum dots versus organic dyes as fluorescent labels. Nat. Methods **5**(9), 763–775 (2008). https://doi.org/10.1038/nmeth.1248
167. Riedel, R., et al.: Synthesis of gold-silica core-shell nanoparticles by pulsed laser ablation in liquid and their physico-chemical properties towards photothermal cancer therapy. Nanoscale **12**(5), 3007–3018 (2020). https://doi.org/10.1039/c9nr07129f
168. Ruedas-rama, M.J., Orte, A., Hall, E.A.H., Alvarez-pez, J.M., et al.: Quantum dot photoluminescence lifetime-based pH nanosensor. Chem. Commun.Commun. **47**(1), 2898–2900 (2011). https://doi.org/10.1039/c0cc05252c
169. Ruedas-Rama, M.J., Orte, A., Hall, E.A.H., Alvarez, J.M.: Quantum dot photoluminescence lifetime-based pH nanosensor. Chem. Commun. (2011)
170. Gallagher, S. J., Rowan, B. C., Doran, J., Norton, B.: Quantum dot solar concentrator: device optimisation using spectroscopic techniques. Solar Energy **81**(4), 540–547 (2007). https://doi.org/10.1016/j.solener.2006.07.006
171. Saha, J., et al.: Role of quantum dot in designing FRET based sensors. Mater. Today: Proc. **5**(1), 2306–2313 (2018). https://doi.org/10.1016/j.matpr.2017.09.234
172. Saikia, D., et al.: CdTe/ZnS core/shell quantum dot-based ultrasensitive PET sensor for selective detection of Hg (II) in aqueous media. Sens. Actuators, B Chem. **230**, 149–156 (2016). https://doi.org/10.1016/j.snb.2016.02.035
173. Salvo, P. et al.: Graphene-based devices for measuring pH. Sens. Actuators: B. Chem. **256**, 976–991 (2018). https://doi.org/10.1016/j.snb.2017.10.037
174. Sankaran, V., et al.: Small PbS clusters prepared via ROMP block copolymer technology. J. Am. Chem. Soc. **112**(19), 6858–6859 (1990). https://doi.org/10.1021/ja00175a019
175. Santra, S.: The potential clinical impact of quantum dots. Nanomedicine **7**(5), 623–626 (2012)
176. Sapsford, K.E., Berti, L., Medintz, I.L.: Materials for fluorescence resonance energy transfer analysis: beyond traditional donor-acceptor combinations. Angewandte Chemie-International Edition **45**(28), 4562–4589 (2006). https://doi.org/10.1002/anie.200503873
177. Scholl, B., et al.: Single particle quantum dot imaging achieves ultrasensitive detection capabilities for Western immunoblot analysis. ACS Nano **3**(6), 1318–1328 (2009). https://doi.org/10.1021/nn9000353
178. Seidi, S., et al.: A promising design of microfluidic electromembrane extraction coupled with sensitive colorimetric detection for colorless compounds based on quantum dots fluorescence. Talanta **194**, 298–307 (2019). https://doi.org/10.1016/j.talanta.2018.10.046
179. Seravalli, L., Trevisi, G., Frigeri, P.: Design and growth of metamorphic InAs/InGaAs quantum dots for single photon emission in the telecom window. CrystEngComm **14**(100), 6833–6838 (2012). https://doi.org/10.1039/c2ce25860a
180. Sha, J., et al.: Dyes and Pigments CdTe QDs functionalized mesoporous silica nanoparticles loaded with conjugated polymers : a facile sensing platform for cupric (II) ion detection in water through FRET. Dyes Pigm.Pigm. **113**, 102–109 (2015). https://doi.org/10.1016/j.dyepig.2014.07.040

181. Shamsipur, M., et al.: A highly sensitive quantum dots-DNA nanobiosensor based on fluorescence resonance energy transfer for rapid detection of nanomolar amounts of human papillomavirus 18. J. Pharm. Biomed. Anal. **136**, 140–147 (2017). https://doi.org/10.1016/j.jpba.2017.01.002
182. Shi, J.J., et al.: Ultrasensitive photoelectrochemical aptasensor for lead ion detection based on sensitization effect of CdTe QDs on MoS_2-CdS: Mn nanocomposites by the formation of G-quadruplex structure. Talanta **183**, 237–244 (2018). https://doi.org/10.1016/j.talanta.2018.02.087
183. Slewa, L.H., Abbas, T.A., Ahmed, N.M.: Synthesis of quantum dot porous silicon as extended gate fi eld e ff ect transistor (EGFET) for a pH sensor application. Mater. Sci. Semicond. Process. **100**(April), 167–174 (2019). https://doi.org/10.1016/j.mssp.2019.04.045
184. Smith, A.M., et al.: A systematic examination of surface coatings on the optical and chemical properties of semiconductor quantum dots. Phys. Chem. Chem. Phys. **8**, 3895–3903 (2006). https://doi.org/10.1039/b606572b
185. Smith, A.M., et al.: Bioconjugated quantum dots for in vivo molecular and cellular imaging. Adv. Drug Deliv. Rev. **60**(11), 1226–1240 (2008). https://doi.org/10.1016/j.addr.2008.03.015
186. Smyder, J.A., Krauss, T.D.: Coming attractions for applications of colloidal semiconductor quantum dots (QDs) have recently. Mater. Today **14**(9), 382–387 (2011). https://doi.org/10.1016/S1369-7021(11)70182-1
187. Spanhel, L. et al.: 'Photochemistry of colloidal semiconductors. 20. Surface modification and stability of strong luminescing CdS particles. J. Am. Chem. Soc. (13), 5649–5655 (1987). https://doi.org/10.1021/ja00253a015
188. Spanhel, Y., et al.: Photochemistry of colloidal semiconductors. 20. surface modification and stability of strong luminescing CdS particles. J. Am. Chem. Soc. **109**(19), 5649–5655 (1987). https://doi.org/10.1021/ja00253a015
189. Swihart, M.T.: Vapor-phase synthesis of nanoparticles. Curr. Opin. Colloids Interface Sci. **8**, 127–133 (2003). https://doi.org/10.1016/S1359-0294
190. Tandon, R.M.Æ.R.P.: Growth of InAs quantum dots on germanium substrate using metal organic chemical vapor deposition technique. Nano Express **5**, 31–37 (2010). https://doi.org/10.1007/s11671-009-9439-y
191. Tang, G., et al.: Determination of arsenic(iii) based on the fluorescence resonance energy transfer between CdTe QDs and Rhodamine 6G. RSC Adv. **5**(23), 17519–17525 (2015). https://doi.org/10.1039/c4ra16789a
192. Tang, G., Du, L., Su, X.: Detection of melamine based on the fluorescence resonance energy transfer between CdTe QDs and Rhodamine B. Food Chem. **141**(4), 4060–4065 (2013). https://doi.org/10.1016/j.foodchem.2013.06.135
193. Tao, H., et al.: Determination of trace Hg2+ ions based on the fluorescence resonance energy transfer between fluorescent brightener and CdTe quantum dots. J. Lumin. **146**, 376–381 (2014). https://doi.org/10.1016/j.jlumin.2013.10.005
194. Taylor, P., et al.: Bioconjugated luminescent nanoparticles for biological applications bioconjugated luminescent nanoparticles for biological applicationsJ. Dispers. Sci. Technol. 37–41 (2007). https://doi.org/10.1081/DIS-120021801
195. Thuy, U.T.D., et al.: CdTe quantum dots for an application in the life sciences. Adv. Nat. Sci.: Nanosci. Nanotechnol. **1**(4) (2010). https://doi.org/10.1088/2043-6262/1/4/045009
196. Tian, P., et al.: Graphene quantum dots from chemistry to applications. Mater. Today Chem. **10**, 221–258 (2018). https://doi.org/10.1016/j.mtchem.2018.09.007
197. Oymak, T., Nusret Ertaş, U.T.: Use of water soluble and phosphorescent MPA-capped CdTe quantum dots for the detection of urea. Turk. J. Pharm. Sci. **15**(1), 44–49 (2018). https://doi.org/10.4274/tjps.02986
198. Turyanska, L., et al.: Temperature dependence of the photoluminescence emission from thiol-capped PbS quantum dots. Appl. Phys. Lett. **90**(10), 1–4 (2007). https://doi.org/10.1063/1.2711529
199. Tycko, R.: Remote sensing of sample temperatures in nuclear magnetic resonance using photoluminescence of semiconductor quantum dots. J. Magn. Reson.Reson. **244**, 64–67 (2014). https://doi.org/10.1016/j.jmr.2014.04.021

200. Varshni, Y.P.: Temperature dependence of the energy gap in semiconductors. Physica **34**(1), 149–154 (1967). https://doi.org/10.1016/0031-8914(67)90062-6
201. Vasudevan, D., et al.: Core-shell quantum dots: properties and applications. J. Alloy. Compd. **636**, 395–404 (2015). https://doi.org/10.1016/j.jallcom.2015.02.102
202. Lin, V.S.-Y., et al.: 3 Lin1997, A porous silicon-based optical interferometric biosensor. **278**(5339), 840–843 (1973)
203. Vlaskin, V.A., et al.: Tunable dual emission in doped semiconductor nanocrystals. Nano Lett. **10**(9), 3670–3674 (2010). https://doi.org/10.1021/nl102135k
204. Vokhmintcev, K.V., Samokhvalov, P.S., Nabiev, I.: Charge transfer and separation in photoexcited quantum dot-based systems. Nano Today **11**(2), 189–211 (2016). https://doi.org/10.1016/j.nantod.2016.04.005
205. Voura, E.B., et al.: Tracking metastatic tumor cell extravasation with quantum dot nanocrystals and fluorescence emission-scanning microscopy. Nat. Med. **10**(9), 993–998 (2004). https://doi.org/10.1038/nm1096
206. Vu, T.Q., et al.: Quantum dots for quantitative imaging: from single molecules to tissue. HHS Public Access **360**(1), 71–86 (2015). https://doi.org/10.1007/s00441-014-2087-2
207. Walker, G.W., et al.: Quantum-dot optical temperature probes. Appl. Phys. Lett. **83**(17), 3555–3557 (2003). https://doi.org/10.1063/1.1620686
208. Wang, C., et al.: A hydrothermal route to water-stable luminescent carbon dots as nanosensors for pH and temperature. Carbon **82**, 1–9 (2014). https://doi.org/10.1016/j.carbon.2014.10.035
209. Wang, H., et al.: Turn-on fluorescence probe for high sensitive and selective detection of Ag+ by L-glutathione capped CdTe quantum dots in aqueous medium. J. Lumin. **180**, 14–19 (2016). https://doi.org/10.1016/j.jlumin.2016.07.049
210. Wang, H., et al.: A rapid microwave synthesis of green-emissive carbon dots with solid-state fluorescence and pH-sensitive properties. R. Soc. Open Sci. **5**, 1–10 (2018). https://doi.org/10.1098/rsos.180245
211. Wang, J., et al.: Fabrication of an "ion-imprinting" dual-emission quantum dot nanohybrid for selective fluorescence turn-on and ratiometric detection of cadmium ions. Analyst **141**(20), 5886–5892 (2016). https://doi.org/10.1039/c6an00868b
212. Wang, Q., et al.: Luminescent properties of water-soluble denatured bovine serum albumin-coated CdTe quantum dots. J. Phys. Chem. B **110**(34), 16860–16866 (2006). https://doi.org/10.1021/jp062279x
213. Wei, S., et al.: Thin-shell CdSe/ZnCdS core/shell quantum dots and their electroluminescent device application. J. Mater. Chem. C **6**(41), 11104–11110 (2018). https://doi.org/10.1039/C8TC03906B
214. Winiarz, J.G., et al.: Inorganic: organic hybrid nanocomposites for photorefractivity at communication wavelengths. J. Phys. Chem. B **106**(5), 967–970 (2002). https://doi.org/10.1021/jp013805h
215. Wu, J., et al.: Quantum dot optoelectronic devices: lasers, photodetectors and solar cells. J. Phys. D Appl. Phys. **48**(36), 1–28 (2015). https://doi.org/10.1088/0022-3727/48/36/363001
216. Wu, L., et al.: Fluorescence resonance energy transfer-based ratiometric fluorescent probe for detection of Zn2+ using a dual-emission silica-coated quantum dots mixture. Anal. Chem. **87**(10), 5318–5323 (2015). https://doi.org/10.1021/acs.analchem.5b00514
217. Wu, X., et al.: Immunofluorescent labeling of cancer marker Her2 and other cellular targets with semiconductor quantum dots. Nature **21**, 41–46 (2003). https://doi.org/10.1038/nbt764
218. Wu, Y., et al.: Fluorescence ELISA based on glucose oxidase-mediated fluorescence quenching of quantum dots for highly sensitive detection of Hepatitis B. Talanta **181**, 258–264 (2018). https://doi.org/10.1016/j.talanta.2018.01.026
219. Wu, Y.H., Arai, K., Yao, T.: Temperature dependence of the photoluminescence of ZnSe/ZnS quantum-dot structures. Phys. Rev. B-Condens. Matter Mater. Phys. **53**(16), R10485–R10488 (1996). https://doi.org/10.1103/PhysRevB.53.R10485
220. Wu, Z.L., et al.: A general quantitative pH sensor developed with dicyandiamide N-doped high quantum yield graphene quantum dots. Nanoscale **6**, 3868–3874 (2014). https://doi.org/10.1039/c3nr06353d

221. Xie, R.C., Shang, J.K.: Morphological control in solvothermal synthesis of titanium oxide. J. Mater. Sci. **42**(16), 6583–6589 (2007). https://doi.org/10.1007/s10853-007-1506-0
222. Xin, S.H., et al.: Formation of selfassembling CdSe quantum dots on ZnSe by molecular beam epitaxy. Appl. Phys. Lett. **3884**, 1994–1997 (1996). https://doi.org/10.1063/1.117558
223. Xing, Y., Smith, A.M.: Molecular pro fi ling of single cancer cells and clinical tissue specimens with semiconductor quantum dots. Int. J. Nanomediciinene **1**(4), 473–481 (2006)
224. Xu, X., et al.: Electrophoretic analysis and purification of fluorescent single-walled carbon nanotube fragments. J. Am. Chem. Soc. **126**(40), 12736–12737 (2004). https://doi.org/10.1021/ja040082h
225. Yang, H., et al.: CdS: Mn nanocrystals passivated by ZnS: Synthesis and luminescent properties. J. Chem. Phys. **121**(20), 10233–10240 (2004). https://doi.org/10.1063/1.1808418
226. Yang, H., Holloway, P.H.: Enhanced photoluminescence from CdS:Mn/ZnS core/shell quantum dots. Appl. Phys. Lett. **82**(12), 1965–1967 (2003). https://doi.org/10.1063/1.1563305
227. Yang, M., Wang, Y., Wang, H.: β-cyclodextrin functionalized CdTe quantum dots for electro-chemiluminescent detection of benzo[a]pyrene. Electrochim. Acta. Acta **169**, 7–12 (2015). https://doi.org/10.1016/j.electacta.2015.04.057
228. Yoo, D., et al.: Carbon dots as an effective fluorescent sensing platform for metal ion detection. Nanoscale Res. Lett. **14**(1), 1–13 (2019). https://doi.org/10.1186/s11671-019-3088-6
229. You, A., Be, M.A.Y., In, I.: Temperature-dependent micro-photoluminescence of individual CdSe self- assembled quantum dots. Appl. Phys. Lett. **214**, 1997–2000 (2003)
230. You, A., Be, M.A.Y., In, I.: Temperature-dependent photoluminescence of ZnSe / ZnS quantum dots fabricated under the Stranski–Krastanov mode. Appl. Phys. Lett. **2656** (2008). https://doi.org/10.1063/1.1612898
231. Yu, D., et al.: Quantum dot-based pH probe for quick study of enzyme reaction kinetics. Enzyme Microb. Technol.Microb. Technol. **41**, 127–132 (2007). https://doi.org/10.1016/j.enzmictec.2006.12.012
232. Yu, H.W., et al.: Preparation of quantum dots CdTe decorated graphene composite for sensitive detection of uric acid and dopamine. Anal. Biochem. **519**, 92–99 (2017). https://doi.org/10.1016/j.ab.2016.12.001
233. Yuan, J., et al.: Glutathione-capped CdTe quantum dots for the sensitive detection of glucose. Talanta **77**(5), 1858–1863 (2009). https://doi.org/10.1016/j.talanta.2008.10.032
234. Zhang, C.Y., et al.: Single-quantum-dot-based DNA nanosensor. Nat. Mater. **4**(11), 826–831 (2005). https://doi.org/10.1038/nmat1508
235. Zhang, H., et al.: Amplified spontaneous emission and random lasing using $CsPbBr_3$ quantum dot glass through controlling crystallization. Chem. Commun.Commun. **56**(19), 2853–2856 (2020). https://doi.org/10.1039/c9cc07676j
236. Zhang, J. et al.: Quantum dots-based hydrogels for sensing applications. Chem. Eng. J. **408**, 127351 (2021). https://doi.org/10.1016/j.cej.2020.127351
237. Zhang, M., et al.: Red/orange dual-emissive carbon dots for pH sensing and cell imaging. Nano Res. **12**(4), 815–821 (2019). https://doi.org/10.1007/s12274-019-2293-z
238. Zhang, X., Zhang, W., Tang, S.: 'L-Tryptophan functionalized graphene quantum dots as a fluorescence indicator for pH detection in real water'. J. Photochem. Photobiol. A: Chem. **372**, 71–77 (2019). https://doi.org/10.1016/j.jphotochem.2018.12.014
239. Zhao, C., et al.: Green and microwave-assisted synthesis of carbon dots and application for visual detection of cobalt (II) ions and pH sensing. Microchem. J.. J. **147**, 183–190 (2019). https://doi.org/10.1016/j.microc.2019.03.029
240. Zhao, J., et al.: Label-free fluorescence turn-on sensing for melamine based on fluorescence resonance energy transfer between CdTe/CdS quantum dots and gold nanoparticles. RSC Adv. **4**(106), 61667–61672 (2014). https://doi.org/10.1039/c4ra08776c
241. Zhao, W.R., et al.: Temperature dependence of emission properties of self-assembled InGaN quantum dots. Chin. Phys. Lett. **31**(11), 1–5 (2014). https://doi.org/10.1088/0256-307X/31/11/114205

242. Zhao, Y., et al.: CdTe@SiO_2 signal reporters-based fluorescent immunosensor for quantitative detection of prostate specific antigen. Anal. Chim. Acta **1057**, 44–50 (2019). https://doi.org/10.1016/j.aca.2019.01.019
243. Zheng, J., Cheng, K., Yu, P.: Simple preparation of fluorescence probe based on CdTe quantum dots combined with MIP for selective detection of sulfadimidine. J. Mater. Sci. Mater. Electron. **30**(24), 21177–21184 (2019). https://doi.org/10.1007/s10854-019-02491-7
244. Zheng, J., Zhang, C., Dickson, R.M.: Highly fluorescent, water-soluble, size-tunable gold quantum dots. Phys. Rev. Lett. **93**(7), 1–4 (2004). https://doi.org/10.1103/PhysRevLett.93.077402
245. Zheng, L., et al.: Core-shell quantum dots coated with molecularly imprinted polymer for selective photoluminescence sensing of perfluorooctanoic acid. Talanta **194**, 1–6 (2019). https://doi.org/10.1016/j.talanta.2018.09.106
246. Zhu, H., et al.: A simple hydrothermal route for synthesizing SnO_2 quantum dots. Nanotechnology **17**(2015), 2386–2389 (2006). https://doi.org/10.1088/0957-4484/17/9/052
247. Zhu, J., et al.: CdTe quantum dot-based fluorescent probes for selective detection of Hg (II): the effect of particle size. Spectrochim. Acta—Part A: Mol. Biomol. Spectrosc. **177**, 140–146 (2017). https://doi.org/10.1016/j.saa.2017.01.043
248. Zhu, J., et al.: Fluorescent detection of ascorbic acid based on the emission wavelength shift of CdTe quantum dots. J. Lumin. **192**, 47–55 (2017). https://doi.org/10.1016/j.jlumin.2017.06.015
249. Zhu, J., Koltypin, Y., Gedanken, A.: General sonochemical method for the preparation of nanophasic selenides: synthesis of ZnSe nanoparticles. Chem. Mater. **12**(1), 73–78 (2000). https://doi.org/10.1021/cm990380r

Chapter 3
Nanomaterial for Humidity Sensor Applications

Y. T. Ravikiran, CH. V. V. Ramana, S. K. Alla, M. Prashantkumar, B. Arundhati, D. K. Mishra, and Sabu Thomas

Abstract Humidity sensors are developed to detect humidity in the surrounding and used to assess the metabolic and physiological conditions of people. As humidity sensors are used in so many different applications, the materials utilized to make them could be adaptable. Nanomaterials are an excellent candidate for these components in this context. The many features of nanomaterial-based humidity sensor production have been explored in this mini-review chapter. The scope and challenges involved in humidity-based sensing are also highlighted.

Keywords Nanomaterial · Humidity · Sensing · Polymer

Y. T. Ravikiran
Department of PG Studies and Research in Physics, Government Science College, Chitradurga, Karnataka, India

CH. V. V. Ramana (✉)
Department of Physics, Vignan's Institute of Information Technology (A), Duvvada, Visakhapatnam 530049, Andhra Pradesh, India
e-mail: ramana6@gmail.com

Department of CEER, Vignan's Institute of Information Technology (A), Duvvada, Visakhapatnam 530049, Andhra Pradesh, India

S. K. Alla (✉)
Department of Physics, Aditya Institute of Technology and Management, 532201, Andhra Pradesh Tekkali, India
e-mail: santhosh.it.bhu@gmail.com

M. Prashantkumar
Department of Physics, Government College (Autonomous), Kalaburagi, Karnataka, India

B. Arundhati
Department of EEE, Vignan's Institute of Engineering for Women, Kapujaggaraju Peta,VSEZ Post, Andhra Pradesh 530 046 Visakhapatnam, India

D. K. Mishra
Department of Physics, Faculty of Engineering and Technology (ITER), Siksha 'O' Anusandhan (Deemed to Be University), Bhubaneswar 751030, Odisha, India

S. Thomas
School of Energy Materials, Mahatma Gandhi University, Kottayam, Kerala, India

D. Mohanta and P. Chakraborty (eds.), *Nanoscale Matter and Principles for Sensing and Labeling Applications*, Advanced Structured Materials 206,
https://doi.org/10.1007/978-981-99-7848-9_3

3.1 Introduction

Manufacturing and manipulation of materials on the scale of atoms or groups of atoms have been explored in the last two decades with care, rigor, and employing these materials to fabricate electronics on the same scale is known as nanotechnology. The production of devices on the same scale is a challenge. However, researchers primarily focused on the present area of research on nanotechnology and the remarkable work that is being done with nanomaterials. Nanomaterials are used in low-density high-strength materials, smart structures, sensors, highly efficient solar cells, electronic, medical, and nano-scale devices due to their small size, large surface area (high surface-to-volume ratio), high reactivity, and recyclability like these properties have made nanomaterials for the development of cost-effective devices. Invisible sunscreens (made with ZnO nanomaterials), stain-resistant clothing (made with nano-whiskers on fibers), and protective bandages (made with silver nanomaterials) are few examples of nanomaterials in consumer products [1]. Nanomaterials have a large commercial influence, which is expected to grow in future [2].

Many industries focus on the manufacturing of devices that employ nanomaterials in the current context as nanomaterials have superior optical, electrical, electronic, catalytic, thermal, and mechanical capabilities. These materials have opened a world of possibilities for devices like sensors, solar cells, memory devices, negative differential resistance, switches, light-emitting diodes (LEDs), field-effect transistors (FETs), and so on. Researchers have been working on the production of composites employing nanomaterials for decades due to the advantages of the combination of two or more different elements which could not be combined any other way. However, dissimilar materials could combine using solely the nanocomposites approach [1].

Nanocomposites developed by combining two or more materials (chemical substances); one of the chemical substances (materials) should be nano-sized (1–100 nm) [1]. Nanomaterials and nanocomposites play an important role in fabrication of sensors due to their improved properties. Sensor innovations are significant because of their numerous uses in environmental monitoring, medical, pharmaceutical, food processing, storage, agricultural, construction, chemical processing, and electronics.

In this chapter, the introduction section aims to provide information about fundamental ideas, developments, and trends in nanomaterials by discussing synthesis techniques, properties, and potential discoveries in the vast and fascinating field of nanomaterials, as depicted in Scheme 3.1 [3]. The introduction part also covers the basic physical, chemical, and biological properties of nanomaterials [4].

- Size and shape of nanomaterials [one-dimensional (1D), two-dimensional (2D), and three-dimensional (3D)], surface effects, quantum confinement effects,

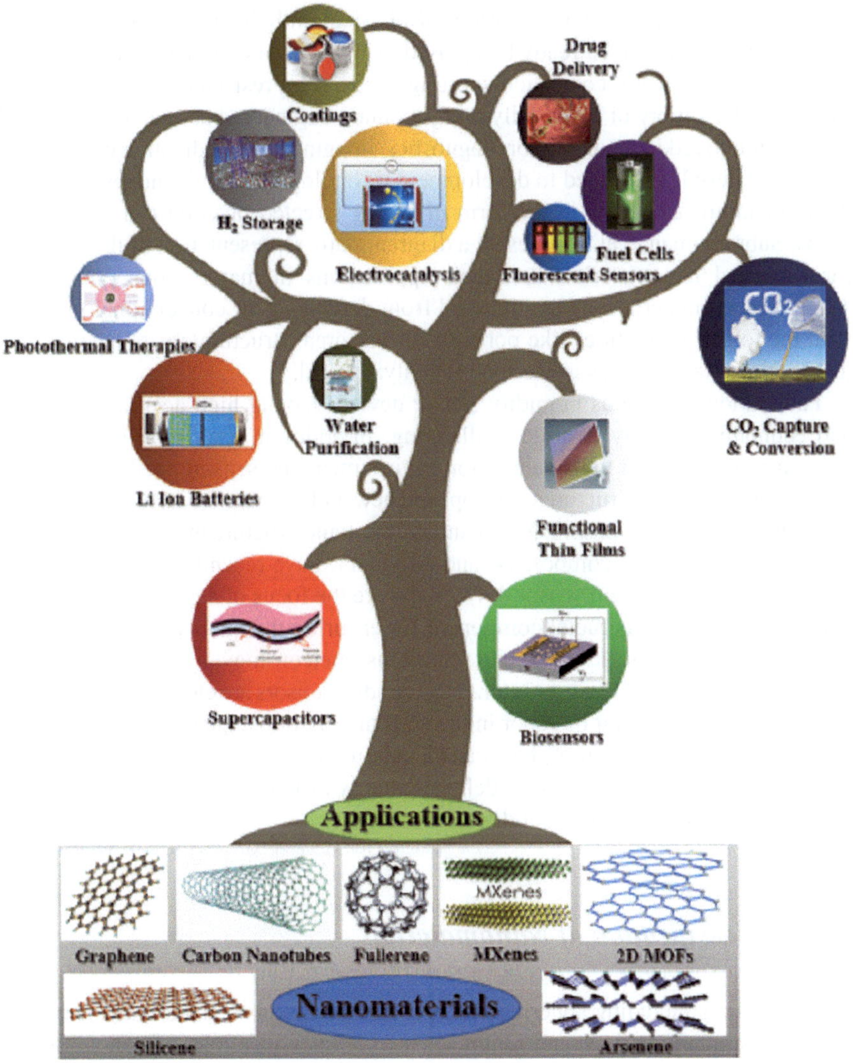

Scheme 3.1 Schematic representation of nanomaterials and their applications [3]

surface charge and stability, and other factors all influence the physical properties of nanomaterials.

- Chemical properties of nanomaterials are primarily determined by their chemical structure and composition, catalytic reactivity, optical, magnetic, and electrical properties.

Some nanomaterials can be found naturally, but they are of particular interest to researchers since they are manufactured as by-products of combustion reactions or engineered to fulfill a specific purpose. Due to their bulk form, the engineering materials generated can have a variety of physical and chemical properties. Engineered nanoparticles are already being used in a variety of commercial products and operations. Sunscreens, cosmetics, sporting goods, stain-resistant clothes, tires, electronics, and a variety of other daily things contain engineered nanomaterials, which are also employed in medicine for diagnostics, imaging, and medication delivery [5].

Nanomaterials are linked to development technologies with the hopes of gaining direct or indirect benefits in the form of highly targeted outcomes in a variety of fields. Subramanian et al. [6] present a diagrammatic representation of the approach for advanced nanomaterials and their applications in major problematic areas. The advanced materials are synthesized from derivatives of conventional materials with the following features like porosity/surface area, structure/crystallization, size distribution, size, shape, surface functionality, etc. [6].

Fabrication of various humidity sensor devices having high accuracy, stability, and economic viability is quite challenging. In addition, each sensor's degree of efficiency with respect to its calibration environment and sensing method is needed to be improved. Nanostructures have opened new and exciting options for addressing the needs of humidity sensors. Novel and unique nanostructure materials (e.g., metal oxide semiconductors, composites, and polymers) have recently played a key role in the creation of efficient sensor devices and are the focus of a global investigation. Humidity sensing is a crucial component for ensuring the quality of the product in industries like nuclear power plants, museums, meteorology, agriculture, medical, libraries, processing of food, and storage. Humidity sensors are electronic transducers that turn the total amount of vapor in the surrounding into a quantifiable signal. The humidity content of the atmosphere can be measured using a variety of methods. The most widely utilized parameters to define humidity in all these strategies are absolute humidity (AB) and relative humidity (RH).

3.1.1 Classification of Nanomaterials

Nanostructures including spherical nanoparticles, nanotubes, nanorods, nanowires, dendrimers, quantum dots, and fullerenes are common examples of nanomaterials. Nanomaterials offer a wide range of uses in nanotechnology due to their unique physical and chemical properties than bulk materials. Nanoparticles can be found in two different forms: nanostructures and nanocomposites. Nanostructures can be built from low-dimensional blocks (elementary units) such as one-dimensional (1D), two-dimensional (2D), and three-dimensional (3D). Alagarasi et al. [5] diagrammatically demonstrate the classification of nanomaterials based on dimensions.

3.1.2 *Classification of Sensors*

A sensor is a device that detects changes in physical, chemical, electrical, and magnetic parameters and generates an output as an indication of the quantitative change. The output often be in the form of an electrical or optical signal. Electric current or potential or magnetic or radio sensors, humidity sensors, fluid velocity or flow sensors, pressure sensors, thermal or heat or temperature sensors, proximity sensors, optical sensors, position sensors, chemical sensors, environment sensors, magnetic switch sensors, and so on are some of the most used sensors. Typical examples of the sensors based on their output are displayed in the Table 3.1.

3.2 Types of Humidity Sensors

Humidity sensors are extensively used in various industries including medicine, agriculture, meteorology, museums, libraries, nuclear power plants, food processing, and storage [20]. These sensors are electrical transducers that convert the total amount of vapor in the environment into the signal that can be measured [21]. A variety of methods can be used to determine the humidity content of the atmosphere. "Absolute Humidity" (AB) and "Relative Humidity" (RH) are the most often used parameters to define humidity in all these approaches [22].

The mass of water vapor (m_w) per unit volume of air is measured by absolute humidity (AB). It is commonly measured in g/m^3 and given by the relation

$$\text{Absolute Humidity(AB)} = \frac{m_V}{V} \tag{3.1}$$

Relative humidity (RH) is the percentage ratio of water vapor partial pressure (Pv) to saturation pressure (Ps) at a given temperature [23].

$$\text{Relative Humidity(RH)} = \frac{P_V}{P_S} \tag{3.2}$$

Based on these two properties, humidity sensors are divided into two broad categories such as absolute humidity sensors and relative humidity sensors [24]. All relative humidity sensors are micro-machined, and they work on the concept that any physical feature will change after interacting with water molecules [23–25]. The most popular types of sensors include surface acoustic wave (SAW), impedance, mass sensitive, optical fiber, resistive, and capacitive sensors.

Table 3.1 Types of sensors, measuring parameters, and their output

S. no	Sensor	Measured parameter	Measurement	Devices
1	Temperature	Temperature	Change in temperature	Temperature Sensor ICs (like LM35, DS18B20), Thermistors, Thermocouples, RTD (Resistive Temperature Devices), etc.
2	Proximity	Object	Detects the presence of an object	Mobile Phones, Cars (Parking Sensors), industries (object alignment), Ground Proximity in Aircrafts, etc.
3	Infrared (IR)	Proximity and object detection	Detects the object	Mobile Phones, Robots, Industrial assembly, automobiles etc.
4	Ultrasonic	Distance and velocity of an object	Measure the distance of the object	HC-SR04
5	Light	Intensity of light	Changes in its resistance to measure the intensity of light	Photo Diode and Photo Transistor belong to Analog Digital Light Sensors like BH1750, TSL2561
6	Smoke and gas	Detect any smoke	Which detects any smoke (due to fire) and gives alarm in the form of sound	Smoke sensors
7	Alcohol	Detects alcohol	Identify whether a person is drunk or not	MQ-3
8	Touch	Detect touch of a finger or a stylus	Detect touch of a finger or a stylus	Mobile phones, tablets, laptops, etc.
9	Color	Color identification	Detect the color and produce a square wave whose wavelength matches the color	TCS3200
10	Humidity	Relative humidity estimation	A ratio of water content in air to maximum water holding capacity in air	DHT11 and DHT22
11	Tilt	Detect inclination or orientation	Detect inclination or orientation	Mercury switches

3.2.1 *Humidity Sensing Parameters*

The performance of a humidity sensor could be determined by the following parameters, viz., sensing response, recovery time, actual sensitivity, limit of detection (LOD), linearity, hysteresis, and stability. Resistive humidity sensors are characterized by the following characteristics.

Sensing Response or Point Sensitivity

The humidity sensing response is calculated using the resistance values recorded over the working relative humidity range [26].

The sensing response (S_H) is computed by the following relation:

$$S_H = \frac{|R_{LH} - R_{HH}|}{R_{LH}} \times 100 \tag{3.3}$$

where R_{LH} and R_{HH} are the resistance at lower % RH and higher % RH, respectively.

Real Sensitivity

The slope of the logarithmic resistance *vs.* relative humidity curve in the working relative humidity range was used to determine the sensitivity [20].

Limit of Detection (LOD)

Limit of detection is computed by the following relation [27]:

$$LOD = 3.3 \times \left[\frac{\text{Standard deviation of regression line}}{\text{sensitivity}}\right] \tag{3.4}$$

Linearity (R^2)

Linearity is achieved in the working relative humidity range by plotting logarithmic resistance versus relative humidity. The strong linearity of sensor indicates that it may be utilized in both low and high RH environments, demonstrating that it is a good sensor [28].

Timing Behavior

For the experimental detection of timing parameters, two flasks were utilized, one with a low relative humidity (11% RH) and the other with a high relative humidity (90% RH and 97% RH). The response and recovery periods are then measured when the film shifting from lower to higher RH values rapidly and vice versa (with a switching time ~1 s) [29].

Hysteresis

In the working relative humidity range, the difference between humidification and desiccation processes, Eq. 3.5 can be used to compute the hysteresis loss.

$$\%\text{Hysteresis} = \left[\frac{(Y_{mn} - Y_{mp})}{(Y_{max} - Y_{min})}\right] \times 100 \tag{3.5}$$

where Y_{mn} is the maximum value and Y_{mp} is the minimum value of Y at the mean of the % RH [30].

Stability

The sensing response of nanostructured materials is often tested over time, including evaluation of the sensing response at regular intervals to verify the samples' sensing stability [31].

3.2.2 Resistive-Type Humidity Sensors

These sensors are the most studied devices due to their simple operation, low power consumption, ease of shrinking, and low cost. Resistive-type sensors are mainly detecting the change in resistance due to the variation in humidity [32]. As a result, the sensing materials are deposited on an inert substrate such as glass, aluminum, or ceramic using techniques like spin-coating, drop-casting, spray-coating, dip-coating, printing, and laser transfer deposition. The sensing area and thus the sensitivity is increased by inter-digitizing the electrodes. The sensing material can be made with functional groups that assist water absorption/adsorption by combining metal oxides, ceramics, organic and inorganic compounds, or composites. Otherwise, it can be created in its pure form with no functional groups. Ceramics, metal oxides, polymers, and organic and inorganic hybrids made using various fabrication procedures are often employed to obtain good humidity sensitivity [25]. However, there are several downsides to these sensing materials, including fluctuations of output signals presence of soluble items as well as sensitivity to other impurities and vapors [25].

Humidity Sensors Based on Impedance Method

There are some humidity sensors designed based on the impedance response of a material to variations in humidity when a high-frequency potential is delivered in the kHz to MHz frequency range. Impedance is measured in terms of current fluctuation with changing humidity using impedance spectroscopy [33, 34]. Sensing materials are usually water hygroscopic, have a high porosity, and have a good swelling behavior at ambient temperature, allowing them to give an abundant active surface site [25]. Lv et al. described that humidity sensor is constructed in the same way as resistive sensors: sensing material is placed on an inert substrate with inter-digitized electrodes [25]. They can detect low humidity levels, but they need high-resolution impedance spectroscopy to do so. These sensors can also detect chemicals such as CO and NO_x at a given PPM level [25].

3.2.3 *Capacitive Humidity Sensors*

In capacitive humidity sensors, the capacitance of the detecting material varies when it is subjected to humidity [35, 36]. These materials are generally non-conductive; however they gain proton conductivity after absorbing dampness. As a result, the device is made up of two parts: inter-digitized conducting electrodes and sensing materials as dielectrics, which are designed to expose as much surface area for water adsorption as feasible [37, 38]. To increase water adsorption sites and hence improve sensitivity, sensing materials can be composited with additional proton conducting components such as metals, metal oxides, polymers, and ceramics [39]. The main advantage of this sort of sensor is that it can operate over a wide range of RH and maintains a constant reading over time. The limited distance between the sensor and the signaling circuit, however, is a serious constraint.

3.2.4 *Surface Acoustic Wave (SAW) Humidity Sensors*

These sensors are fabricated based on piezoelectric effect variation of the film [40]. According to Tang et al. [41], a simple SAW sensor consists of an inter-digitized transducer (IDT) on either side of the substrate, with a delay line in the center. The input IDT generates mechanical surface acoustic waves in response to a sinusoidal electrical signal (piezoelectric effect). As the signal goes through the sensor material in the delay line, variations in humidity influence the signal's attributes such as amplitude, frequency, and phase. The second ITD turns these changes into electrical impulses, which are amplified and analyzed later [40]. Wireless control, good stability, high sensitivity, and selectivity are the key advantages of SAW sensors [42]. Maintaining homogeneity on the sensor's surface with an acceptable thickness, on the other hand, is crucial. The majority of these are employed in biomedical analytics [43].

3.2.5 Quartz Crystal Microbalance (QCM) Humidity Sensors

These sensors rely on the mass effect and viscous effect on the film as a result of water adsorption or desorption [38]. According to Zheng et al. the gadget is made up of quartz crystal electrodes coated with water-sensitive materials and the occurrence of mass effect as well as visco-electric change due to adsorption or desorption of water molecules are quantified by the variation of frequency of the resonator [44]. High stability and capacity to detect mass at the sub-nano-gram level are main features of these sensors [45]. However, these sensors could not be used on cells and biomolecular systems that do not respond to elastic changes [46].

3.2.6 Optical Fiber Humidity Sensors

These sensors depend on the changes in optical properties of the sensor film inside the fiber because of moisture exposure [25]. As a result, optical instruments like the Fabry–Perot interferometer and the grating resonator are utilized to track variation of optical properties. A variety of benefits including the capacity to sustain at high temperatures, pressure and at extreme weather conditions are significant benefits of these sensors. Large-scale production, however, is challenging due to the high cost of fabrication and optical measurement instruments [47].

3.3 Nanomaterials Used for Humidity Sensors

The role of sensors is crucial in the prevention of fire explosions in the industries. Sensors detect the harmful gasses produced in the industries. Currently, there are numerous shortcomings in developing sensitive materials in terms of selectivity, stability, and other features; as a result, many researchers and scientists are striving to produce novel materials that will be suitable for sensor applications [8]. The mechanism of sensing materials is largely determined by their microstructures, which necessitates the use of a processing approach to generate materials with the necessary microstructures and morphologies [8].

Humidity measurement is a critical characteristic in a variety of industrial applications, including pharmaceutical, food, medicine, and electronics. Many humidity sensors have been produced based on a reversible interaction between a material and water vapor, and as a result, the developed material also functions as a gas sensor. Optical, gravimetric, capacitive, resistive, piezo-resistive, and magneto-elastic characteristics of certain materials are also used in breathing humidity sensing approaches. Many researchers and experts have claimed that using nanostructures as humidity sensing materials, such as nanoparticles, nanowire, nanotubes, and polymer nanocomposites, might considerably improve the performance of humidity sensors. Because of their huge surface-to-volume ratio, the surface effect becomes dominant in the created nano-structured materials, which is useful in the development of humidity sensors [9]. The representative examples of nanomaterials utilized for humidity sensor applications are displayed in Table 3.2.

Table 3.2 Nanomaterials utilized for fabrication of humidity sensors

S. no.	Nanomaterials	Synthesis method	Application	References
1	Silver nanoparticles	Electrostatic spray deposition	Humidity	[9]
2	Multi-walled CNT	Hot filament chemical vapor deposition (CVD)	Humidity	[10]
3	Metal oxide nanocomposite	Sol–Gel method	Humidity	[11]
4	Silica or SiO_2	Chemical vapor deposition (CVD)	Humidity	[12]
4	Manganese doped nickel oxide	Immersion method	Humidity	[13]
5	Titanium oxide (TiO_2)	Anodization method	Humidity	[14]
6	Carbon dots	Microwave	Humidity and temperature	[15]
7	Silver nanofibers	Chemical method	Humidity, ammonia, and temperature	[16]
8	Graphene flower/zinc oxide	Sol–Gel method	Humidity	[17]
9	Niobium pentoxide-mixed nickel ferrite	Mechano-chemical mixing	Humidity	[18]
10	rGO/ZnO nanorods/Cu	One-pot	Humidity	[19]

3.4 Synthesis, Characterization, and Sensing Response

Nanostructures or nanocomposites with different morphology for sensors have been produced using various methods like sol–gel, solvothermal, hydrothermal, hydrolysis, electrochemical polymerization, co-precipitation, chemical etching, and electrospinning. Structural, microstructural, surface, and chemical properties have been studied using X-ray diffraction (XRD), scanning electron microscopy (SEM), transmission electron microscopy (TEM), X-ray photoelectron spectroscopy (XPS), Fourier transform infrared spectroscopy, Brunauer–Emmett–Teller (BET), etc. The efficiency of the sensor could be influenced by the factors like morphology of the nanostructures, synthesis technique, sensitivity, and stability of the sensor. The variation of performance of the humidity sensors with different factors have given in Table 3.3. Conducting polymer-based nanocomposites surpass other types of nanomaterials in terms of performance and efficiency, according to a review of studies on the humidity sensing capacities of various nanomaterials.

Polyaniline/Yttrium oxide (PYO) composite humidity sensors are produced by physically combining in situ polymerized PANI with Yttrium oxide [20]. Four composites with different compositions are PYO-1 (0.99:0.01), PYO-2 (0.95:0.05), PYO-3 (0.90:0.1), and PYO-4 (0.85:0.15) [20]. FT-IR spectrum investigations revealed a blue shift in the bands of PANI and Y_2O_3 in the PYO composites, which was attributed to the interfacial contact between the N–H group of PANI and the oxygen group of Y_2O_3. Large number of micro-pores on the nanorods noticed in SEM micrographs (Fig. 3.1A). The number of flaws has been increased (Fig. 3.1B). TEM micrographs showed that the nanorods in the PYO-4 composite have a high aspect ratio, indicating that they have a larger surface area for moisture adsorption.

The thin films of the composites used to conduct the sensing measurements. Humidity readings were recorded using a resistive type of sensor and ranged from 11 to 97% relative humidity. The sensing responses of the composites were compared to PANI sample. The PYO-4 composite displayed a maximum sensing response of 99.99%, as well as significant actual sensitivity, least LOD, minimum hysteresis, and maximum stability (Fig. 3.1C). Physical properties including porosity, swelling behavior, and water absorption capacity have been assessed [48, 49].

The cupric oxide/polyaniline nanocomposite was produced via a chemical polymerization method [50]. FTIR investigations emphasized that the blue shift of the quinoid and benzenoid peaks of PANI (Fig. 3.2A) and attributed to the electrostatic interaction of inorganic CuO with organic PANI. The presence of metallic sites in the composite improves the water adsorption capabilities of the composite. SEM analysis revealed that PANI/CuO nanocomposites contain smaller particles than PANI, resulting in more surface area and better vapor adsorption. The humidity sensing measurement and the resistive type of humidity sensing behavior was measured in the RH range of 10–95%. The resistivity of the nanocomposite varied significantly with relative humidity (Fig. 3.2B). The composite had a sensitivity of 4.5 Ω/RH (Fig. 3.2D), with a rapid response time of 40 s and a recovery time of 55 s (Fig. 3.2C).

Table 3.3 Different types of nanostructured humidity sensors, synthesis methods, characterization techniques, and their responses

S. no.	Nanostructured material	Synthesis method	Characterization techniques	Type of sensor	Response, recovery and percentage	References
1	CdS/polyaniline (CdS/PANI) composite	In situ vulcanization process	XRD, FTIR, SEM & XPS	Impedance	95%	[52]
2	PANI/α–Fe_2O_3 composite	In situ polymerization	FTIR, XRD, SEM & UV–VIS	Resistive	70, 90 s	[53]
3	Poly-vanadium–titanium acid/Polyaniline (V-Ti/PANI) composite	Electrochemically polymerizing	SEM, XRD, FT-IR, UV–vis	Impedance	97%	[51]
4	Sulfosuccinic acid-doped polyaniline films	In situ polymerization	TGA, SEM	Resistance	17, 36 s	[54]
5	Graphene oxide/Nafion/indium oxide	In situ polymerization	SEM, TEM, XRD, BET & XPS	Capacitive	97%	[55]
6	Graphene Oxide/Tin Oxide/Polyaniline Ternary Nanocomposite	In situ oxidative polymerization	SEM, TEM, XRD, XPS & FTIR	QCM	7, 2 s	[56]
7	Tin oxide/Reduced Graphene oxide hybrid nanostructure	Hydrothermal	XRD, SEM and TEM	Capacitive	97%	[39]
8	Polyaniline/Poly (vinyl butyral) nanofibers	Electrospinning	SEM and TEM	SAW	90%	[42]
9	Cellulose nanocrystals	Hydrolysis	TEM, FTIR and XRD	QCM	100%	[57]
10	Tin disulfide/Graphene oxide Nanoflower	Solvothermal	XRD, XPS, FT-IR, SEM & TEM	Impedance	0.9, 10 s	[58]
11	Nanodiamond/Graphene oxide nanocomposite film	Modified hummers	SEM and BET	QCM	25, 5 s	[59]
12	Zinc oxide nanorods	Sol–gel	SEM, XRD & FTIR	SAW	90%	[60]

(continued)

Table 3.3 (continued)

S. no.	Nanostructured material	Synthesis method	Characterization techniques	Type of sensor	Response, recovery and percentage	References
13	Tin-substituted Nickel Ferrite	Coprecipitation	XRD, SEM, TEM, UV–vis & FTIR	Optical	101, 53 s	[61]
14	Nickel/Silicon nanocomposites	Wet chemical etching	SEM & TEM	Capacitive	44, 180 s	[62]

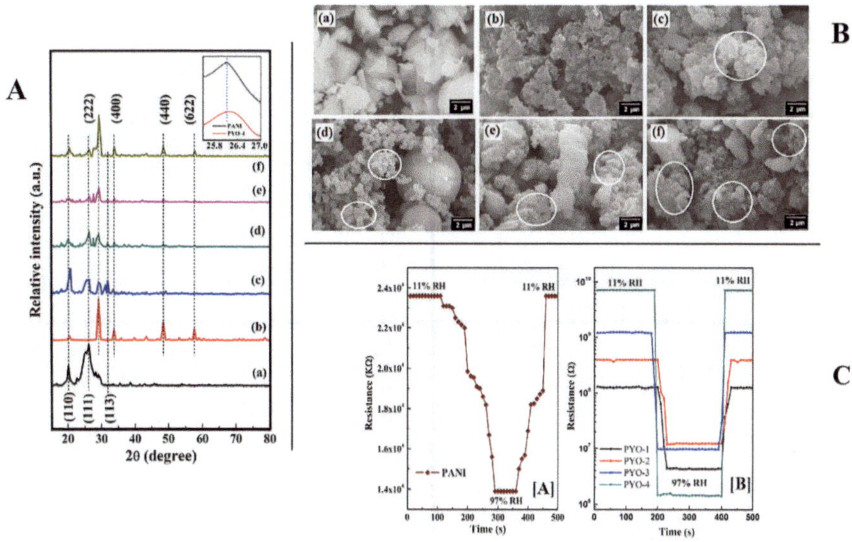

Fig. 3.1 **A** XRD patterns of **a** PANI, **b** Y_2O_3 and **c** PYO-1, **d** PYO-2, **e** PYO-3 **f** PYO-4 composite, **B** SEM images of **a** pristine PANI, **b** Y_2O_3, **c** PYO-1, **d** PYO-2, **e** PYO-3, **f** PYO-4. **C** Response and recovery characteristic curves of (**A**) PANI and (**B**) PTO composites License number-5256310820880 [48, 49]

Thin layer of poly-vanadium–titanium acid/polyaniline (V-Ti/PANI) composite was first produced by dip-coating with poly-vanadium–titanium acid, and then the electrochemical polymerized PANI was coated onto the inter-digitized gold electrode in a sequential manner [51]. FTIR analyses indicated that the PANI band at 1161 cm^{-1} and the poly-vanadium–titanium acid band at 507 cm^{-1} in the composite were relocated to a lower wave number. The interfacial interaction is confirmed by the development of a hydrogen bond between PANI and poly-vanadium–titanium acid (Fig. 3.3A).

The composite's SEM picture reveals little irregular grains that are smaller than PANI. The expanded multi-granular pores and higher surface area confirmed by SEM results aid to improve vapor adsorption (Fig. 3.3B). XRD investigations revealed two broad peaks in the composite, indicating that no more crystalline phases are generated and confirming the formation of composite. A noticeable change of the impedance with the RH (Fig. 3.3C) and a low hysteresis of 5% RH, as well as a fast response time of 8 s and a recovery time of 11 s (Fig. 3.3D) has been reported [51].

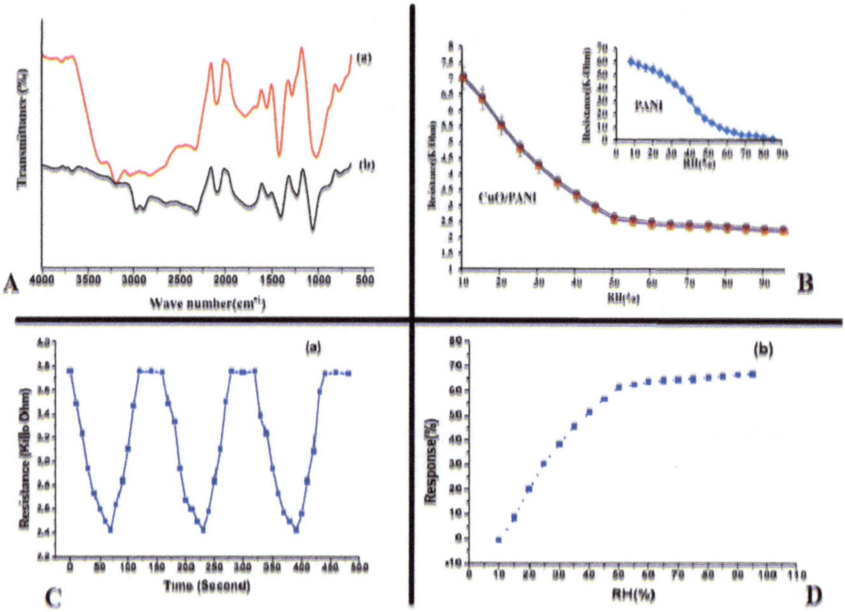

Fig. 3.2 **A** FT-IR spectra of **a** PANI and **b** CuO/PANI nanocomposite, **B** Trends of resistance against % RH of CuO/PANI nanocomposite and PANI (inset). **C** Change in resistance of CuO/PANI nanocomposite with time at 60% RH and **D** Percentage sensing response of CuO/PANI at different relative humidity License number-5256320465111 [50]

3.5 Summary

Nanomaterials are beneficial to the advancement of nanotechnology as well as the production of humidity sensors. The strength of nanoparticles has greatly improved in the field of humidity sensors. Nanomaterials are a fantastic tool for generating effective and efficient environmental monitoring systems due to their unique features. However, much-awaited need still exists to develop fine-tuned nanostructured sensors to further improve their sensitivity and efficiency. The development of extensive applications hinges on the mass manufacture of such sensors. Sensors of this type should, in general, comply with industry standards in terms of design and/or construction.

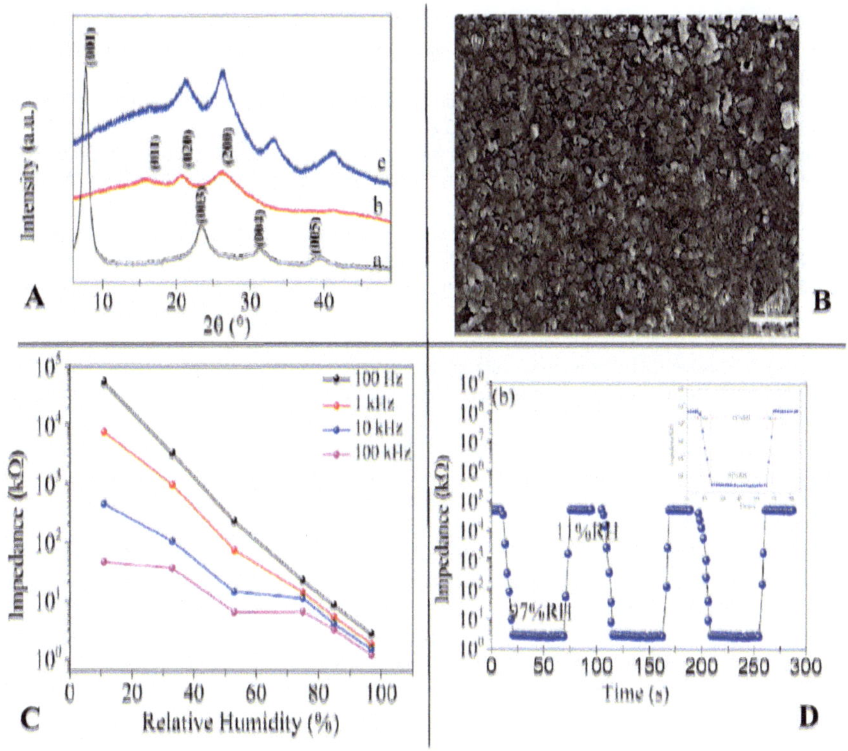

Fig. 3.3 **A** XRD pattern of the **a** poly-vanadium–titanium, **b** pristine PANI, and **c** the multilayer-structured V-Ti /PANI composite with 30 electrodeposition cycles, **B** SEM of multilayer-structured V–Ti/PANI composite with 30 electrodeposition cycles. **C** Variations in impedance of the multilayer-structured V–Ti/PANI composite-based humidity sensor with change in RH (%) measured at various frequencies and **D** Response and recovery properties of the multilayer-structured V–Ti/PANI composite at 100 Hz License number-5256320807991 [51]

References

1. Chani, M.T.S., Khan, S.B.: Applications of Nanocomposites in Humidity sensors and solar cells. Curr. Nanosci.. Nanosci. **16**(4), 504–506 (2020)
2. Sharma, P., Manish, B.: Applications and characteristics of nanomaterials in industrial environment. Int. J. Civ., Struct., Environ. Infrastruct. Eng. Res. Dev. (IJCSEIERD) **3**(4), 63–72 (2013)
3. Baig, N., et al.: Nanomaterials: a review of synthesis methods, properties, recent progress, and challenges. Mater. Adv. **2**, 1821–1871 (2021). https://doi.org/10.1039/d0ma00807a
4. Asha, A.B., Ravin, N.: Nanomaterials properties. Book Chapter-15, Book: II Nanotechnology, Polymer Science and Technology, pp. 343–359 (2020) https://doi.org/10.1016/B978-0-12-816806-6.00015-7
5. Alagarasi, A.: Introduction to Nanomaterials. Chapter, www.nccr.iitm.ac.in
6. Subramanian, A., et al.: Introduction to advanced nanomaterials. Chapter 1, Nanomaterials for Green Chemistry (2018) https://doi.org/10.1016/B978-0-12-813731-4.00001-1

7. Mageswari, A., et al.: Nanomaterials: classification, biological synthesis and characterization. Chapter 2, nanoscience in food and agriculture 3. Sustain. Agri. Rev. **23** (2016). https://doi.org/10.1007/978-3-319-48009-1_2
8. Zeng, W., et al.: Nanomaterials for sensing applications. J. Nanotechnol. (Hindawi) 2083948 (2016). https://doi.org/10.1155/2016/2083948
9. Thiwawong, T., et al.: A Humidity sensor based on silver nanoparticles thin film prepared by electrostatic spray deposition process. Adv. Mater. Sci. Eng. (Hindawi) **2013**, 640428 (2013). https://doi.org/10.1155/2013/640428
10. Cao, C.L., et al.: Humidity sensor based on multi-walled carbon nanotube thin films. J. Nanomater. (Hindawi) **2011**, 707303 (2011). https://doi.org/10.1155/2011/707303
11. Shaheen, K., et al.: Electrical, photocatalytic, and humidity sensing applications of mixed metal oxide nanocomposites. ACS Omega **5**, 7271–7279 (2020). https://doi.org/10.1021/acsomega.9b04074
12. Kano, S., Mekaru, H.: Nonporous inorganic nanoparticles-based humidity sensor: evaluation of humidity hysteresis and response time. Sensors (MDPI) **20**, 3858 (2020). https://doi.org/10.3390/s20143858
13. Parimon, N., et al.: Annealing temperature dependency of structural, optical and electrical characteristics of manganese-doped nickel oxide nanosheet array films for humidity sensing applications. Nanomater. Nanotechnol. (SAGE) **11**, 1–13 (2021). https://doi.org/10.1177/1847980420982788
14. Elham, F., Raheleh, M.: Fabrication of flexible self-powered humidity sensor based on super-hydrophilic titanium oxide nanotube arrays. Sci. Rep. **10**, 13032 (2020). https://doi.org/10.1038/s41598-020-70031-z
15. Rivadeneyra, A., et al.: Carbon dots as sensing layer for printed humidity and temperature sensors. Nanomaterials (MDPI) **10**, 2446 (2020). https://doi.org/10.3390/nano10122446
16. Rashid, H.U., et al.: Synthesis, characterization, and applications of silver nanofibers in humidity, ammonia, and temperature sensing. Micromachines (MDPI) **12**, 682 (2021). https://doi.org/10.3390/mi12060682
17. Saqib, M., et al.: High-performance humidity sensor based on the graphene flower/zinc oxide composite. Nanomaterials (MDPI) **11**, 242 (2021). https://doi.org/10.3390/nano11010242
18. Reddy, L.P.B., et al.: Structural and humidity sensing properties of niobium pentoxide-mixed nickel ferrite prepared by mechano-chemical mixing method. J. Mater. Sci. Mater. Electron. **31**, 21981–21999 (2020). https://doi.org/10.1007/s10854-020-04701-z
19. Kuntal, D., et al.: RGO/ZnO nanorods/Cu based nanocomposite having flower shaped morphology: AC conductivity and humidity sensing response studies at room temperature. J. Mater. Sci. Mater. Electron. **30**, 15544–15552 (2019). https://doi.org/10.1007/s10854-019-01931-8
20. Chethan, B., et al.: Polypyyrole based core-shell structured composite based humidity sensor operable at room temperature. Sens. Actuat. B Chem. **296**, 126639 (2019)
21. Sikarwar, S., Yadav, B.C.: Opto-electronic humidity sensor: a review. Sens. Actuat. A Phys. **233**, 54–70 (2015)
22. Farahani, H., et al.: Humidity sensors principle, mechanism and fabrication technologies: a comprehensive review. Sensors **14**, 7881–7939 (2014)
23. Huang, J., Huang, Q.A.: Micro-machined humidity sensors. In: Huang, Q.A. (ed.) Micro Electro Mechanical Systems, The Journal of Micro and Nano-Manufacturing, pp. 787–816 (2018)
24. Rittersma, Z.M., et al.: Recent achievements in miniaturized humidity sensors—a review of transduction techniques. Sens. Actuat. A Phys. **96**, 196–210 (2002)
25. Lv, C., et al.: Recent advances in graphene-based humidity sensors. J. Nanomater.Nanomater. **9**(3), 422 (2019)
26. Chethan, B., et al.: Nickel substituted cadmium ferrite as room temperature operable humidity sensor. Sens. Actuat. A Phys. **280**, 466–474 (2018)
27. Manjunatha, S., et al.: Room temperature humidity sensing performance of polyaniline–holmium oxide composite. Appl. Phys. A-Mater. **125**, 361 (2019)

28. Manjunatha, S., et al.: Polyaniline based stable humidity sensor operable at room temperature. Physica B Condens. **561**, 170–178 (2019)
29. Delin, W., et al.: Applied novel sensing material graphene/polypyrrole for humidity sensor. Sens. Actuat. B Chem. **181**, 326–331 (2013)
30. Rathi, K., et al.: Impact of doping on GO: fast response—recovery humidity sensor. ACS Omega **2**, 842–851 (2017)
31. Shukla, S.K., et al.: Integrated approach for efficient humidity sensing over zinc oxide and polypyrole composite. Mater. Sci. Eng. C **90**, 325–332 (2018)
32. Lai, J.R., et al.: Micro-sensor for measuring humidity. Rev. Sci. Instrum.Instrum. **39**(8), 1197–1203 (1968)
33. Rheaume, J.M., et al.: A review of recent progress in sensing of gas concentration by impedance change. Ionics **17**, 99–108 (2011)
34. Zhang, D., et al.: Ultrahigh-performance impedance humidity sensor based on layer-by-layer self- assembled tin disulfide/titanium dioxide nano-hybrid film. Sens. Actuat. B Chem. **266**, 52–62 (2018)
35. Gao, W., et al.: Direct laser writing of micro-super-capacitors on hydrated graphite oxide films. Nat. Nanotechnol.Nanotechnol. **6**, 496–500 (2011)
36. Bi, H., et al.: Ultrahigh humidity sensitivity of Graphene oxide. Sci. Rep. **3**, 2714 (2013)
37. Nizhnik, O., et al.: Self-calibrated humidity sensor in CMOS without post processing. Sensors **12**(1), 226–232 (2012)
38. Schubert, P.J., et al.: A polyimide-based capacitive humidity sensor. IEEE Trans. Electron Dev. **32**, 1220–1223 (1985)
39. Zhang, D., et al.: Fabrication and characterization of an ultrasensitive humidity sensor based on metal oxide/graphene hybrid nano-composite. Sens. Actuat. B Chem. **225**, 233–240 (2016)
40. Le, X., et al.: A high performance humidity sensor based on surface acoustic wave and graphene oxide on AlN/Si layered structure. Sens. Actuat. B Chem. **255**, 2454–2461 (2018)
41. Tang, Y., et al.: Highly sensitive surface acoustic wave (SAW) humidity sensors basedon sol–gel SiO_2 films: Investigations on the sensing property and mechanism. Sens. Actuat. B Chem. **215**, 283–291 (2015)
42. Lin, Q., et al.: Highly sensitive Highly sensitive and ultrafast response surface acoustic wave humidity sensor based on electrospun polyaniline/poly(vinyl butyral) nano-fibers. Anal. Chim. Acta **748**, 73–80 (2012)
43. Li, D., et al.: High humidity enhanced surface acoustic wave (SAW) H_2S sensors based on sol–gel CuO films. Sens. Actuat. B Chem. **294**, 55–61 (2019)
44. Zheng, X., et al.: A fast-response and highly linear humidity sensor based on quartz crystal microbalance. Sens. Actuat. B Chem. **283**, 659–666 (2018)
45. Zor, S.D., et al.: QCM humidity sensors based on organic/inorganic nanocomposites of water soluble-conductive poly (diphenylamine sulfonic acid). Int. J. Electrochem. Sci.Electrochem. Sci. **11**, 7976–7989 (2016)
46. Yao, Y., et al.: Facile fabrication of high sensitivity cellulose nanocrystals based QCM humidity sensors with asymmetric electrode structure. Sens. Actuat. B Chem. **302**, 127192 (2020)
47. Shukla, S.K., et al.: Exploring fiber optic approach to sense humid environment over nano-crystalline zinc oxide film. Talanta **80**, 565–571 (2009)
48. Ravikiran et al.: Humidity sensing performance of hybrid nanorods of polyaniline-Yttrium oxide composite prepared by mechanical mixing method. Talanta **215**, 120906. Copyright 2020 Elsevier (2020)
49. Kotresh, S., Ravikiran, Y.T., Raj Prakesh, H.G., Ramana, CH.V.V., Vijaya Kumari, S.C., Thomas, S.: Humidity sensing performance of spin coated polyaniline-carboxymethyl cellulose composite at room temperature. Cellulose **23**, 3177–3186 (2016)
50. Singh, P., et al.: Structurally optimized cupric oxide/polyaniline nanocomposites for efficient humidity sensing. Surf. Interfaces **18**, 100410 (2020)
51. Guo, Y., et al.: Humidity sensing properties of poly-vanadium-titanium acid combined with polyaniline grown in situ by electrochemical polymerization. Sens. Actuat. B Chem. **270**, 80–88 (2018)

52. Chen, Q., et al.: Controlled synthesis and humidity sensing properties of CdS/polyaniline composite based on CdAl layered double hydroxide. Sens. Actuat. B Chem. **254**, 30–35 (2018)
53. Singh, P., et al.: Synthesis and humidity sensing property of α–Fe_2O_3 and polyaniline composite. Mater. Today Proc. **5**, 9118–9125 (2018)
54. Biswas, D., et al.: Investigation on resistive humidity sensing in sulfosuccinic acid doped polyaniline films. Synth. Met. **272**, 116662 (2021)
55. Zhang, D., et al.: Facile fabrication of graphene oxide/Nafion/indium oxide for humidity sensing with highly sensitive capacitance response. Sens. Actuat. B Chem. **292**, 187–195 (2019)
56. Zhang, D., et al.: High-performance QCM humidity sensor based on graphene oxide/tin oxide/polyaniline ternary nano-composite prepared by in-situ oxidative polymerization method. Sens. Actuat. B Chem. **262**, 531–541 (2018)
57. Yao, Y., et al.: Facile fabrication of high sensitivity cellulose nano-crystals based QCM humidity sensors with asymmetric electrode structure. Sens. Actuat. B Chem. **302**, 127192 (2020)
58. Zhang, D., et al.: Fabrication of tin disulfide/graphene oxide nano-flower on flexible substrate for ultrasensitive humidity sensing with ultralow hysteresis and good reversibility. Sens. Actuat. B Chem. **287**, 398–407 (2019)
59. Yao, Y., et al.: Impedance analysis of quartz crystal microbalance humidity sensors based on nano diamond/graphene oxide nano-composite film. Sens. Actuat. B Chem. **211**, 52–58 (2015)
60. Hong, H., et al.: High-sensitivity humidity sensors with ZnO nano-rods based two-port surface acoustic wave delay line. Sens. Actuat. B Chem. **171–172**, 1283–1287 (2012)
61. Tao, B., et al.: Capacitive humidity sensors based on Ni/SiNWs nanocomposites. Sens. Actuat. B Chem. **136**, 144–150 (2009)
62. Manikandan, V., et al.: Fabrication of Tin substituted Nickel Ferrite (Sn–$NiFe_2O_4$) Thin film and its application as Opto-electronic humidity sensor. Sens. Actuat. A Phys. **272**, 267–273 (2010)

Chapter 4
Nanomaterial-Based Sensors for the Detection of Explosives

Nasrin Sultana, Samiran Upadhyaya, and Neelotpal Sen Sarma

Abstract Nanoscience has always been considered as an excellent tool for sensing applications. The detection of explosives using nanomaterials and derived materials has been helpful, owing to the superior characteristics of nanomaterials. Several nanomaterials, such as metal nanoparticles, carbon-based nanomaterials, metal oxide nanoparticles, hybrid nanomaterials and nanoclusters, have been prominently used as explosive detectors. With the immense growth of nanotechnology, portable devices have been widely fabricated, leading to swift detection of explosives, such as nitroaromatic compounds. In this chapter, a detailed discussion of nanomaterials for explosive detection has been presented, along with the available technology and future scopes. Although there are few limitations to the fabrication of explosive detectors, with nanotechnology, improved devices with high selectivity have been developed in recent times.

Keywords Nanomaterial · Metal oxide · Sensing · Explosive

4.1 Introduction

Although explosive-based chemicals are of great concern for human health for the environment, their uses have enormously increased in the last few decades. It has been considered to be a challenge to detect the explosive chemicals to counter terrorist threats, soil contamination and obtain safe groundwater. Based on their properties, explosives can be divided into nitric esters, peroxides, nitro aromatics, derivatives of chloric and perchloric acids, azides and nitramines. In addition, several other

N. Sultana · S. Upadhyaya · N. S. Sarma (✉)
Institute of Advanced Study in Science and Technology, Paschim Boragaon, Guwahati 35, Assam, India
e-mail: neelot@iasst.gov.in

N. Sultana · N. S. Sarma
Academy of Scientific and Innovative Research (AcSIR), Ghaziabad 201002, India

D. Mohanta and P. Chakraborty (eds.), *Nanoscale Matter and Principles for Sensing and Labeling Applications*, Advanced Structured Materials 206,
https://doi.org/10.1007/978-981-99-7848-9_4

compounds are also known that result in explosions, such as acetaldehydes, fulminates, peroxides, tetrazene, ozonides, etc. The most widely used explosive is the nitro aromatics which are organic compounds containing one or more nitro groups ($-NO_2$) in military operations and criminal acts. The one which includes more than one nitro group is considered as the most highly explosive one. Explosives like 2,4,6-trinitrotoluene (TNT), trinitrobenzene (TNB), 2,4,6-trinitrophenol (TNP) or picric acid (PA), tetra nitro-phenyl-methylamine (tetryl), 2,4-dinitrotoluene (DNT), trinitro-triazacyclohexane (RDX), octahydro-1,3,5,7-tetranitro-1,3,5,7-tetrazocine (HMX), triacetonetriperoxide (TATP), hexamethylene triperoxide diamine (HMTD), ethylene glycol dinitrate (EGDN), nitroglycerine (NG), pentaerythritol tetranitrate (PETN), etc., have different properties and can cause damage to the biological systems due to its toxic, mutagenic and carcinogenic nature. Such toxicity, mutagenicity and carcinogenicity result from the penetration of harmful explosives into the human body through skin [1–4]. According to the US Environmental Protection Agency, the maximum allowable limit of 2,4,6-trinitrotoluene (TNT) is about 2 ppb in drinking water [5]. As reported by the New Jersey Department of Health, the permissible limit of picric acid, RDX and HMX are 1 ppb, 30 ppm, and 0.4 ppm, respectively. Most of the other explosives have similar permissible limit in ppm/ppb levels (Hazardous substance fact sheet, New Jersey Department of Health (NJ Health)). Hence, the detection of explosives with high precision is of utmost necessity. The various explosives classified on the basis of different chemical groups are shown in Fig. 4.1.

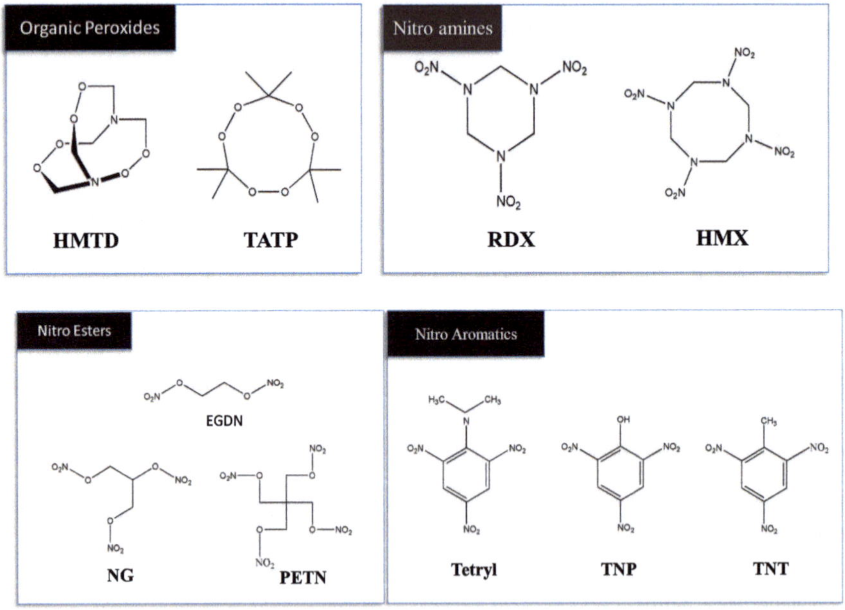

Fig. 4.1 Classifications of explosives based on different chemical groups

Explosives are very reactive and contain a considerable amount of energy which can quickly initiate self-propagation and lead to an explosion [3, 6]. Among all the explosives, the nitro aromatic- and nitro amine-based explosives are the most stable and are composed of benzene rings functionalized with $–CH_3$ or $–NO_2$ groups. These compounds are electron deficient. Thus, electron-rich molecules can easily bind to form stacking complexes via π–π interaction, and detecting such explosives is difficult due to their low volatility. Removing such toxic explosives is very much necessary for ecological cleaning and homeland security. However, detection of such explosives requires innovative ideas and field-deployable tools with high efficiency, fast and reliable techniques and so is a significant demand for the benefit of the society. The most used nitro aromatic-based explosives which are bound with fuels or with themselves to make military explosives are enlisted in Table 4.1. The traditional methods for detecting such explosive traces mostly employed highly developed sophisticated equipment, such as gas chromatography (GC) [8], mass spectrometry (MS) [9], surface-enhanced Raman spectroscopy.

(SERS) [10, 11], ion mobility spectrometry (IMS) [12], Fourier-transform infrared (FTIR) [13], X-ray imaging [14] and thermal neutron analysis [15]. Although these devices are widely used for analysis, they are expensive, bulky and are not convenient for real-time detection and practical use. Researchers have been involved in developing cost-effective, portable, rapid and specific sensors, which can play an essential role in detecting such explosives with socially relevant implications. One such development is towards electrochemical sensors, which have many advantages, including high sensitivity, selectivity, wide linear range, compatibility with minimal space, microfabrication, low-cost instrumentation and microfabrication [16–18]. Such techniques are also suited for field-based applications because of their scan speed rates, low-power equipment and sensitivity [17]. Reducibility of the nitro-groups present in the nitroaromatic explosives facilitated them towards electrochemical detection. In this regard, nanomaterials can play an essential role as they have unique optical and electrical properties and for being potentially selective as simple, sensitive and cost-effective sensors to detect explosives at trace levels. However, many challenges

Table 4.1 Most commonly used nitro aromatics in military [7]

Explosives	Commonly found	Molecular weight (g/mol)	Vapour Pressure at 20 °C (Torr)
Dinitrotoluene (DNT)	Propellants	182.14	1.44×10^{-4}
Trinitrotoluene (TNT)	Amatol, C2, C3, Cyclotol, Octol, Pentolit, PTX-1, PTX-2, Tetrytol	227.1	4.8×10^{-6}
Trinitrophenol (Picric Acid)	Explosive D	229.1	3.1×10^{-8}
Trinitrobenzene (TNB)	–	213.11	2.2×10^{-4}
Tetranitro-N-methylamine (tetryl)	C3,PTX-1, Tetrytol	222.3	8.3×10^{-1}

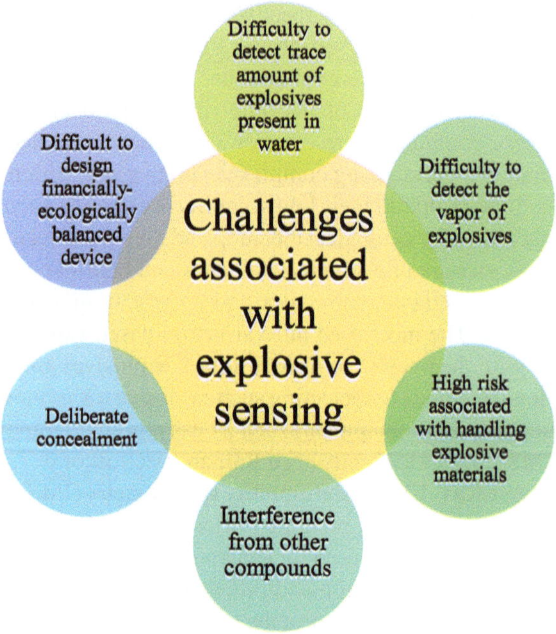

Fig. 4.2 A pictorial representation depicting the common challenges associated with explosive detection

exist, including low vapour pressure, hard detectability and the introduction of novel explosive materials, depicted in Fig. 4.2. The detection of explosive chemicals on the nanoscale is vital, as it can provide extra safety by detecting trace amounts from a distance far away from the source. It is imperative to control the apparatus by an operator. However, there are still limitations regarding trace-based explosive detection due to their selectivity, size, cost and sensitivity [7]. Thus, an innovative prospect was felt necessary for explosive sensing, because of which nanomaterial-derived sensors came into the picture.

Nanoscience and nanotechnology have become crucial bridges connecting the cultural world to the innovative era of device fabrication. Nanomaterials have strong potential to detect explosives with a high surface area-to-volume ratio convergence rather than linear diffusion, catalytic activity, improved selectivity and unique electrical and optical properties, which can be exploited for sensitive molecular adsorption detection as nanoscale dimensions can enhance the physical as well as chemical properties. These materials provide control over the microenvironment. The use of sensors based on nanomaterials gave an effective platform for trace detection of explosives. Thus, the development of advanced nanomaterials provides selective sensitivity towards detecting explosives with single molecule also [3].

The experimental route leading to the formation and utilization of nanomaterials-based explosive sensors is depicted in Fig. 4.3. Although there exist technologies, overcoming the challenges in controlling particle size, growth in fabrication and characterization, stability, sampling and reliable calibration are the fundamental requirements for developing effective sensors. These concerns can be easily addressed

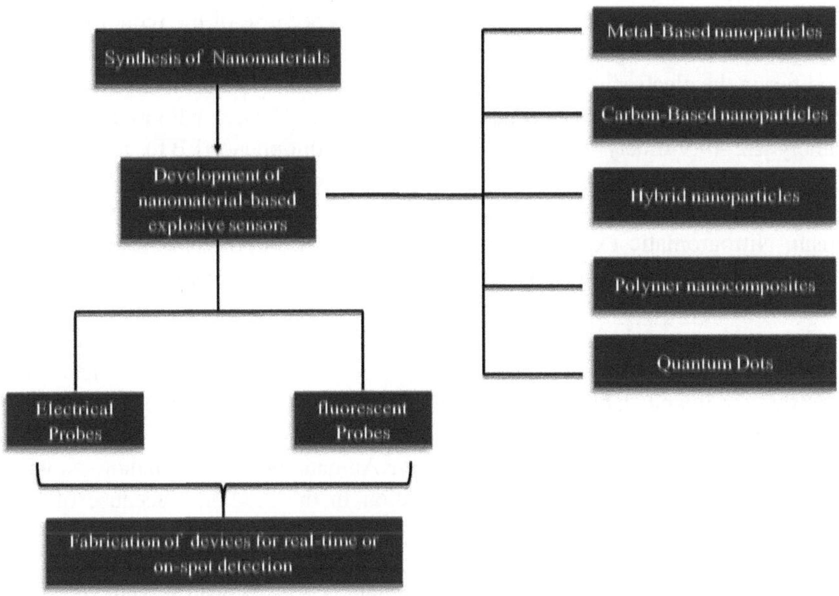

Fig. 4.3 The schematic representation of the experimental routes leading to the formation and utilization of nanomaterials-derived explosive sensors

by electrochemical methods by allowing the growth of nanomaterials with specific potential [19, 20]. So, a cost-effective, easily deployable device with good physical and chemical benefits can be obtained by combining nanomaterials with an electro-analyser as an electrode modifier allowing easy integration of nanosensors with the advancement of conventional explosive detection platform.

4.2 Classifications of Nanomaterials-Based Sensors for the Detection of Explosives

There is an enhancement of physical and chemical properties for the matters in their nanoscale due to the increase in surface area and confinement effects [21]. Nanosensors can be fabricated with the help of nanotechnology for the detection of a variety of analytes, including trace explosives. These nanosensors are designed in such a manner to overcome all the difficulties and limitations associated with the traditional methods. Optimizing such nanomaterials with high surface area and activity will provide high signal-to-noise ratios with extraordinary chemical and physical properties for molecular detection [3]. Nitroaromatic explosive detection can be done in two ways: electrochemically and optically.

In electrochemical reactions, the flow of ions and electrons on the surface of the electrode gave some detectable electronic response. Nitroaromatic explosives can be easily detected by using electrochemical methods because of the reducibility of $-NO_2$ groups. Carbon, porous and metal nanomaterials can also be used as electrode modifier in the electrochemical detection of nitroaromatic explosives [17, 22].

Santos et al. synthesized and characterized a material based on fluorene derivative, namely 2,2′-(9,9-dioctyl-9H-fluorene-2,7-diyl) bisthiophene (FBT), as they show fluorescence in the visible region with photoluminescence quantum yield aiming application in the optoelectronic devices which show turn-off fluorescence mechanism. Nitroaromatic explosives can be detected by fluorescence using the PET (photoinduced electron transfer) process because of their strong electron-accepting nature. Therefore, suitable electron donors like fluorene derivatives can be promising candidates as materials for sensing explosive detection [23].

Thakarda et al. synthesized a new class of luminescent clusters of gold nanostructures that are ligand stabilized. They prepared cetyl trimethyl ammonium bromide-stabilized gold nanorod (Au–CTAB NRs) and polyvinylpyrrolidone-stabilized gold cluster (Au–PVP cluster). They studied these Au nanoclusters and nanorods, which remained unchanged without any aggregation in the medium. Because of these pyrene-grafted nanocomposites, they showed strong emission characteristics. This emission is quenched in the presence of electron-deficient chemical species and can be used for analytical purposes. This luminescent material showed the selective detection and sensing of nitroaromatic explosives such as TNP, DNT and nitrotoluene. It was observed that the LOD (limit of detection) in both solid and solution was of the order of 10 nM [77].

Optically detecting explosives can have advantages over other traditional methods such as high sensitivity, selectivity, portability and cost-effectiveness. Conjugates of polymers and nanomaterials, i.e. carbon dots, metal–organic frameworks (MOFs), quantum dots, etc., polymer–dye composites, can be helpful in forensic science for the detection of explosives based on their fluorescence behaviour. Considerable work has been reported based on doping of fluorophores and electrospinning techniques for the highly sensitive detection of explosives.

4.2.1 Metal-Based Nanomaterials as Explosive Sensors

Metal-based nanoparticles, e.g. Au, Pt, Ti, etc., with high surface area, catalytic effects and improving electrical conductivity can enhance the sensitivity towards detecting explosives electrochemically and optically.

Willner et al., in their work, used Au NPs to improve the sensitivity of detection of explosives such as 2,4,6-trinitrotoluene (TNT) electrochemically. They have used functionalized Au NPs bridged by oligo aniline units to detect the trace of 2,4,6-trinitrotoluene (TNT) in parts per trillion concentrations. This was attained by regulating the interaction of 2,4,6-trinitrotoluene (TNT0 and π-donor cross-linked Au NPs. The selectivity can also has enhanced by imprinting molecular recognition sites into the π-donor cross-linked oligo aniline and Au NPs [24].

In another work, Filanovsky et. al. used a modified carbon-based electrode with mesoporous TiO_2. In this case, mesoporous TiO_2 supports NPs such as Ru, Pt or Au to detect TNT. The main advantage of TiO_2–Au NP and TiO_2–Pt NP is their high selectivity and good linearity between peak currents and concentration of 2,4,6-trinitrotoluene (TNT) [25].

4.2.2 *Carbon-Based Nanomaterials as Explosive Sensors*

Nanomaterials such as graphene, carbon nanotubes, carbon nanoparticles and carbon dots are getting more attention for their application in sensing and energy storage due to their biocompatibility, low cytotoxicity, chemically inertness and unique chemical and physical properties. Carbon-based nanomaterials are known to have specific electronic properties for their use in explosive sensing. Among all the carbon-based nanomaterials, graphene—a two-dimensional (2D) carbon-based structure consisting of sp^2 carbon atoms—is considered as one of the most superior candidates for explosive sensing. It possesses excellent conductivity, high mechanical strength, rapid heterogeneous rate transfer and high elasticity, due to which it is widely explored in both experimental and theoretical fields [20, 26–32]. In 2010, Tang et al. first detected 2,4,6-trinitrotoluene (TNT) explosives electrochemically using graphene substrate as a platform in their studies. As 2,4,6-trinitrotoluene (TNTs) are suitable electron acceptors, they quickly get absorbed on the surface of the electron-rich graphene. They observed signals from the $-NO_2$ group of 2,4,6-trinitrotoluene (TNT), which was deposited on the glassy carbon electrode (GCE) in phosphate-buffered saline (PBS) [33]. Chen et al. determined nitroaromatic explosives such as di-nitrotoluene (DNT), trinitrobenzene (TNB) and dinitrobenzene (DNB) using graphene oxide and their derivatives. They made a graphene oxide platform to detect such explosives electrochemically onto a GCE with the help of square wave stripping voltammetry (SWSV). In that technique, LOD was 8 ppb [34]. Pumera et al. compared the detection of explosives like dinitrobenzene (DNB), 2,4,6-trinitrotoluene (TNT), and di-nitrotoluene (DNT) by graphene oxide-modified glassy-carbon electrode (GCE) with the detection of graphite microparticles and bare glassy-carbon. They found that graphene oxide, in terms of selectivity as well as sensitivity enhances the output results compared to bare glassy-carbon electrode. They also determined the LOD for 2,4,6-trinitrotoluene (TNT) in untreated seawater, which was about 1 $gmol^{-1}$ using graphene-based electrochemical sensors. They also investigated the response of reduced 2,4,6-trinitrotoluene on the surface of mono-, few-, and multilayer graphene nanoribbons and found no significant result compared to micro particle-based electrodes of graphite and graphene nanoribbons [35, 36].

In addition to graphene, carbon nanotubes are also known to impart features for explosive sensing. Zhang et al. theoretically established and used multi-walled carbon nanotubes to detect 2,4,6-trinitrotoluene (TNT) and analysed the interaction between triphenylene (TP) and 2,4,6-trinitrotoluene (TNT) and calculated their electrostatic potential followed by characterization by fluorescence spectra. They also reported that the TP's conjugated structure allows it to easily functionalize onto the carbon nanotube (CNT) surface. TP and TNT are nucleophilic and electrophilic, respectively, which leads to an essential intermolecular charge transfer between them and may help recognize TP to TNT. The LOD is 5 ppb for the TNT at the TP-modified electrode [37].

4.2.3 *Quantum Dots (QDs) for Explosive Sensing*

Quantum dots (QDs) possess fascinating properties such as broad excitation and narrow emission spectra, high stability and high quantum yield. Their quantum confinement makes them one of the most acceptable forms of nanoparticles. They also have unique intrinsic properties, i.e. fluorescence. However, they also show electrochemical properties such as redox and electrocatalytic, leading to interesting bio-sensing labelling capabilities [38, 39]. They can be applied as a luminescent probe to detect a variety of inorganic, biological and organic analytes. Unlike other fluorophores, these are composed of inorganic semiconducting materials like ZnS, CdSe and CdS. The energy involved in electron excitation from the valance band to the conduction band in such semiconductors is constant. The semiconducting materials in the case of quantum dots proceed towards the Bohr radius and thus results in two crucial electronic properties contrasting to that of bulk material; firstly, there is an emission of light when the electron from the conduction band return to the valence band, and conduction band becomes discrete sub-bands after the separation. Secondly, the bandgap no longer remains constant but leads to an increase in it. Therefore, there is a possibility to tune the emission of QDs in the UV and NIR region by switching the morphology, composition or size [40–42]. Detection of explosives electrochemically by QDs related to mass production of devices and cost-effective instruments, i.e. screen-printed electrodes which can be easily infinitesimal forms for in-field and rapid detection. Below, the figure illustrates the comparison of the band gap energy (E_g) between metals, bulk semiconductors and quantum dots based on their sizes (Fig. 4.4).

Tu et al. exemplified the ZnS nanocrystals doped with Mn^{2+} ion to be one of the most susceptible materials for detecting TNT in ultra-trace levels. The photoluminescence of the QDs got quenched because of the interaction of the electron-deficient aromatic rings of TNT and electron-rich amino ligands. The mechanism involved here is the transfer of electrons from the conductive band of ZnS to the LUMO (lowest unoccupied molecular orbital) of TNT anions. To increase the specificity and selectivity of QD-based sensors for detecting explosives, techniques like molecular imprinting can be used [44–46]. In another work, Goldman and co-workers have shown a new QD-based chemo-sensor for the detection of explosives like TNT on nanoscale. The developed sensor comprises a dye-labelled TNT-specific antibody fragment (black hole quencher-10 labelled with trinitrobenzene (TNB-BHQ10)) linked to hybrid ZnS-CdSe QDs by metal-affinity coordination. Such antibody fragments can quench the fluorescence of QDs through the FRET (fluorescence resonance

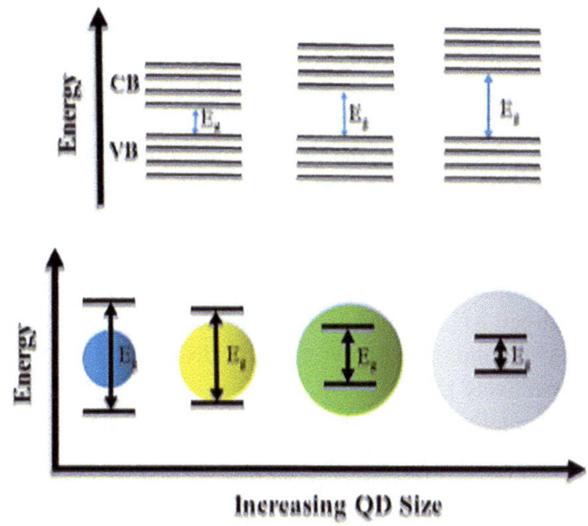

Fig. 4.4 Comparison of the band gap energy (Eg) between metals, bulk semiconductors and semiconductor quantum dots and relationship between the size of semiconductor quantum dots and Eg, resulting in different emission energies [43]

energy transfer) mechanism. However, there are a lot of QDs for the detection of functional groups, the modified one with surface modification or ion-doped shows high sensitivity and specificity [47]. The development of detectors for on-site explosive detection is the main challenge.

Xu et al. developed a tactful sensor for detecting TNT based on quantum dots and molecularly imprinted polymers. To synthesize molecularly imprinted polymer, they have used functional monomer APTES (3-aminopropyltriethoxy silane) and TNP (trinitrophenol). They have used both of these as a dolt impression molecule capped with CdTe quantum dots [48].

A work by Zhang et al. developed a hybrid material from two size variant CdTe QDs for the visual detection of explosives like TNT by ratiometric fluorescence. However, in this process, the green fluorescence of QDs got quenched, and the red fluorescence of QDs in silica NPs remained constant in TNT's presence [49].

Combining QDs with nanostructure or crystalline structure can eventually enhance the physical and chemical properties. Xia et. al. reported a work where composite QDs assimilate Au nanorods (NRs) in the near-infrared region. This compact hybrid assembly is formed by the interaction of carboxyl-terminated QDs and amine-terminated Au NRs. The explosive TNT combines with amino groups of Au NRs and turns on the fluorescence via breaking the compact assembly. In their work, the LOD was found to be ~0.1 nM [50].

4.2.4 Nanoporous Materials as Explosive Sensors

Preparation of inorganic materials which are very sensitive to explosives like nitroaromatic compounds lead to carbon- and silica-based nanoporous materials. The high sensitivity of such materials results from the adsorption of explosives on the large surface area.

Shi et al., in their work, used PDDA (poly (diallyl dimethyl ammonium chloride)) and glassy carbon electrode (GCE) modified with mesoporous SiO_2 as an electrochemical sensor for the detection of nitroaromatic explosives such as DNT, TNT, DNB and TNB in KCl. They applied the DPV technique to detect DNT, DNB, TNB and TNT and found the LOD of about 0.4, 0.3, 0.4 and 0.3 ppb, respectively [51].

Zhang et al., in their work, used SiO_2 of GCE-modified MCM-41 and analysed the volumetric detection of TNB, DNT, DNB and TNT. According to their work, the modified MCM-41 shows better sensitivity towards such nitro aromatics than that of SiO_2, and their LOD for TNB, DNT, DNB and TNT is 0.6, 0.6, 0.4 and 0.4 ppb, respectively [52].

In another work, Trammell et al. used nanoporous organosilica to detect TNT. They used immobilized organosilica, which is both imprinted and nonimprinted benzene-bridged and diethyl benzene-bridged polysilsesquioxane on the surface of GCE for such type of detection in phosphate buffer saline (PBS) solution. There is an enhancement of sensitivity, and less peak tailing of imprinted benzene is higher than that of diethyl benzene due to the speedy release of TNT. However, the imprinted diethyl benzene is best used to detect and absorb aqueous extract of contaminated soil [53].

Although silica materials have advantages because of their uniform pore size and specific surface area, some drawbacks like poor conductivity limit their electrochemical applications. Compared to carbon-based nanoporous materials, they are good conductors and have a very high surface area and uniform pore size. And thus, they can be easily used to detect ultra-trace explosives.

Pumera et al., in their work, used carbon-based nanoporous materials as an electrochemical sensor for the detection of TNT and compared with graphite, carbon nanotubes (CNT) and bare glassy-carbon electrodes. According to their report, carbon-based nanoporous materials exhibit higher heterogeneous electron transfer than CNT and graphite, thus showing higher sensitivity [54]. Zhang et al. reported GCE modified with ordered mesoporous carbon (OMC) for detecting DNB, TNT and DNT with detection limit of 1, 0.2 and 1 ppb, respectively, which is due to the high surface area of OMC of about 745.9 m^2g^{-1}, i.e., higher than that of carbon-based nanomaterials (less than 200 m^2g^{-1}) [55].

4.2.5 Hybrid Nanomaterials-Based Explosive Sensors

Hybrid nanomaterials are the more specially designed class of nanomaterials used for ultra-trace explosive detection and nanomaterials. The most specifically developed properties of the hybrids produced via the Shaojun et al. used platinum nanoparticles assembled on graphene nanosheets to form a Pt–graphene hybrid material for the detection of explosives like TNT as an electrochemical sensor. The main advantage of Pt-decorated graphene nanosheets is their high conductivity and fascinating electrochemical behaviour. These nanoparticles are of small size and uniform in nature. Applying ASV (adsorption stripping voltammetry) to this modified GCE (glassy carbon electrode) showed a wide linear range with a LOD of about 0.3 ppm for TNT [56].

Guo et al. used graphene functionalized with porphyrin to detect DNB, TNB, DNT and TNT. LOD by such hybrid material was found to be 2, 1, 1 and 0.5 ppb for DNB, TNB, DNT and TNT, respectively. The main reason for the high sensitivity of such hybrid materials is the fast charge transfer and large electroactive surface area of porphyrin [57].

Hrapovic et al., in their work reported a Nafion matrix of composites of multi-wall carbon nanotubes and single-wall carbon nanotubes and Cu, Pt and Au nanoparticles. The main advantage of using such material is their stability, low background current and reproducibility with ASV (adsorption stripping voltammetry) and CV (cyclic voltammetry). However, graphene and ionic liquid hybrid materials also show positive results in detecting TNT [58]. This type of hybrid composite is synthesized by using ionic liquids on the surface of 3D graphene nanosheets, leading to the formation of a material with a large specific area and pronounced porosity.

4.3 Polymeric Nanocomposites for Explosive Detection

Polymer is an important material that provides a platform for the efficient application of nanomaterials. Various polymer-based explosives have been known for a long time. However, polymeric nanocomposites have recently attracted the attention of researchers around the globe. Nanoparticles doped within a polymer framework provide added benefits such as enhanced thermal and photo-stability, enhanced electrical properties in solid-state, efficient handling and storage, etc., thus providing better performances than the virgin nanoparticles for explosive sensing applications. Metal nanoparticles, such as Ag- and Au-doped polymer composites, have been widely used in recent years for explosive sensing [59]. Polymer nanocomposites also provide luminescence properties, enabling the development of the materials as fluorescent sensors [60]. In addition, the use of carbon-based nanoparticles, such as graphene, carbon nanotubes and graphene quantum dots, are also frequently used to develop composites with enhanced electrical and optical properties for their application as explosive sensors. Carbon-dots-based polymer composites are known to possess extra stability and enhanced optical properties [61]. Pal et al. reported using carbon dot polymer (polypyrrole) nanocomposite for explosive sensing [62]. Shan et al. also developed porous halide-perovskite polymer composite for efficient detection of explosives with high sensitivity [63]. Quantum dots composites with polymer have also been highly crucial for explosive sensing. Abbasi et al. shown that incorporating CdS quantum dot with N-Methypolypyrrole leads to nanocomposite formation for detecting explosive (picric acid) with high selectivity [64].

The present research group has been actively working on developing polymer nanocomposites for the selective detection of nitroaromatic explosives. Dutta et al. developed several polymer-based nanocomposites with high electron density to detect nitroaromatic explosives such as picric acid, TNT, etc., by adopting the photoluminescence method. They have synthesized polymeric nanocomposite using carbon and silver nanoparticles incorporated in polyvinyl alcohol-g-polythiophene (PVA–g–PTh) as shown in Fig. 4.5. Incorporating conducting nanoparticles in a polymer matrix makes it an electron-rich polymeric material that enhances fluorescent behaviour. However, the fluorescent properties are quenched when PVA–g–PTh meets electron-deficient nitroaromatic compounds [65].

In another work, macromolecular composites were synthesized by incorporating CdTe and CdTe/ZnS core–shell quantum dots within the PVA–polyaniline matrix using MSA (2-mercaptosuccinic acid) as a capping agent to form a sensor for ultra-sensitive and selective sensing of nitro aromatics. The reaction involves a single-step free-radical polymerization technique. The polymer so formed acts as a capable probe

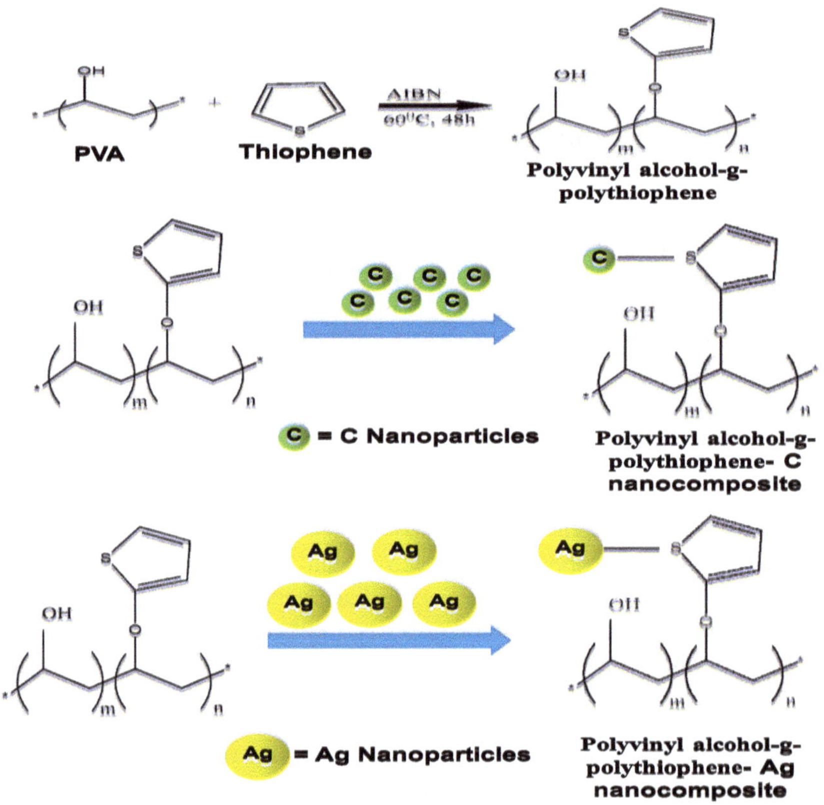

Fig. 4.5 Schematic representation of synthesis of polyvinyl alcohol-g-polythiophene and their nanocomposites

for the detection of picric acid in an aqueous medium. The formation of core–shell quantum dots results in high quantum yield and can decrease toxicity up to a large extent.

The real sample analysis shows a high efficiency with a preferably low limit of detection (LOD), making them essential material for the highly sensitive and selective sensing of picric acid [66]. The interaction between QDs with PA is shown in Fig. 4.6.

In addition, the team also prepared carbon nanoparticle-incorporated silk-fibroin (SF) protein to detect nitroaromatic explosives. The scaffold structure of SF has the excellent property of interconnectivity between the pores, and it does not require any organic solvents during the degumming process. The incorporation of carbon nanoparticles in the scaffold of SF makes it a suitable electrical semiconducting material. Thus, it can be considered as a material of choice to develop vapour sensing devices for the detection of explosives like TNT, TNP, etc. In this work, they developed a cost-effective, highly selective and sensitive chemiresistive sensor to detect vapours of nitroaromatics [67].

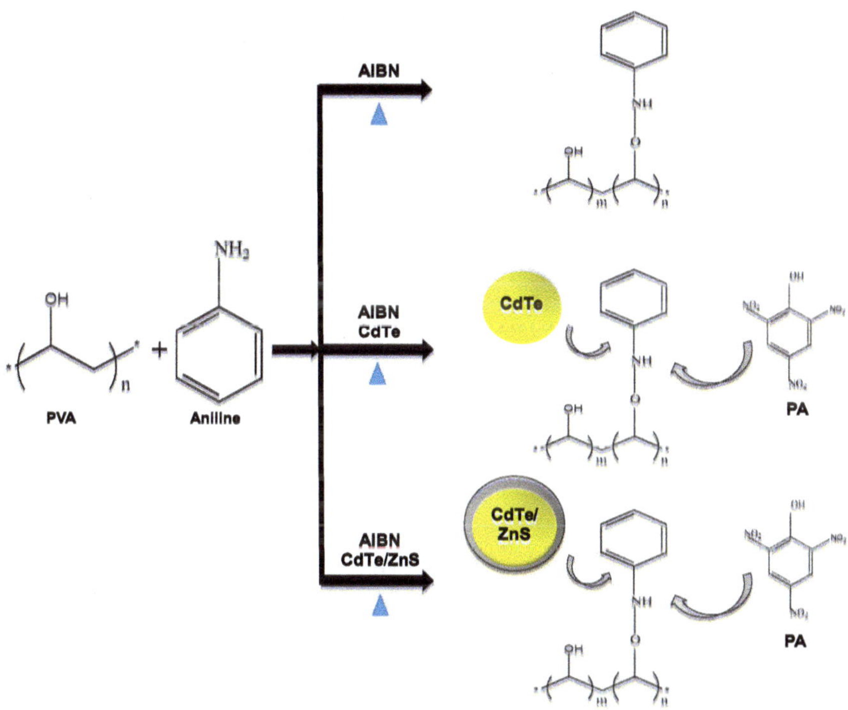

Fig. 4.6 Representation of interaction of CdTe/ZnS core–shell QDs—PVA–polyaniline composite with picric acid

4.4 Recently Developed Nanomaterial-Based Explosive Sensors

With the development of nanotechnology, nanomaterials and their derived composites have become the critical choice for developing explosive sensors that can selectively detect explosives with high accuracy to combat terrorist attacks. Nanomaterials in various forms, such as nanoribbons, nano-arrays, nanorods, etc., have been widely used as explosive sensors. With the growth in nanotechnology applications, there has been steep growth in the new-edge combustible detectors, providing wholesome security for households and the nation overall. Although many notable works have been carried out earlier, few disadvantages were associated with the reported sensors. Dealing with all the disadvantages related to the earlier researches, a few path-breaking reports in recent years have caught the eyes of the researchers. The initially known nanosensors were applied in the form of cantilever beams. The change in various parameters was observed to detect mercury and other vapours, including explosives [68]. After the initial developments of nanomaterials-based explosive sensors, scientists have been actively involved in overcoming the disadvantages associated with the sensors for real-time analysis. In recent times, the path-defining nanomaterial-based techniques, such as fluorescence spectroscopy, surface-enhanced Raman scattering (SERS), switchable ion-mobility spectroscopy (IMS), nanocluster-supported chemiresistive sensing, gas chromatography, artificial olfaction, etc., [69–73] have been developed using nanodispersions and nano-engineered modulated devices, leading to the effective applications of these sensors. With the development in research, nanomaterial-derived flexible explosive sensors [74] and nanomaterial-based redox explosive sensors have also been developed [75].

Kumar et al. developed a 3D Zn-metal–organic framework (MOF) of naphthalene-2,6-dicarboxylic acid (NDA) and 4-amino pyridine (AMP) [Zn(NDA)(AMP)]. This MOF shows a strong emission peak, but it got efficiently quenched in the presence of explosives like trinitrotoluene (TNT), PETN (pentaerythriol tetranitrate), 1,3,5-trinitroperhydro-1,3,5-triazine (RDX), 1,3,5,7-tetranitro-1,3,5,7-tetrazoctane (HMX) and 2,4,6-trinitrophenyl-N-methylnitramine (Tetryl). According to their report, the detection limit of tetryl and PETN is 0.404 and 0.315 μM, respectively [76]. Junaid and co-workers developed a fluorenone-based sensor to rapidly detect 2,4,6-trinitrotoluene (TNT) and 2,4,6-trinitrophenol (TNP). The developed sensor was more selective towards TNP with a limit of detection (LOD) of about 0.5 nM. They have achieved this via electrostatic interaction between TNP and the sensor. They also developed sensors that are selective towards TNT with LOD in the range 5.0–7.0 Nm through fluorescence emission enhancement response [77].

Hong et. al. examined different mono- and di-pyrene derivatives in micelles. The tested pyrene derivatives in micelles displayed a strong excimer band, and the excimer band was quenched in the presence of RDX and TNT. According to their report, the detection limit for RDX was about 334 ppb, and for TNT, it was about 2 ppb [78].

In another work, Nailwal et al. synthesized a porous organic material based on conjugated microporous polymers (CMP) with high surface area and π-conjugated skeleton, which can act as a light-emitting material. THIS CMP is based on truxene core (Tx–CMPs). The synthesized Tx–CMP with aryl linker prevents the aggregation-caused quenching because of the π–π stacking of layers. As the Tx–CMPs are fluorescent and electron-rich, they have been used for the sensitive and selective detection of picric acid in the nanomolar range among all the other nitroaromatic explosives [79].

Rather than TNT, TNP, RDX, etc., picric acid is one of the most threatening pollutants present in the environment due to its high solubility in water. The extensive use of PA in pharmaceutical companies and chemical industries leads to alarming threats to the environment and public health. Thus, detection of PA in trace levels in all the mediums is extremely necessary for civilian and military safety. Dey et al. developed water-soluble 1,8-naphthalimide (ANDI), which have multimodal sensing application of the aggregated nanomaterials for explosives based on nitro compound such as picric acid (PA) in solid, liquid as well as vapour phase. The synthesized material shows ultrasensitive and selective detection of PA with an exceptional quenching constant of $7.23 \times 10^6\ M^{-1}$ is followed up by the limit of detection (LOD) of 149 ppb [80].

Although the development leading to the research carried out in nanomaterials-based explosive sensors is vast and cannot be fully summarized in a page; a few most notable, path-breaking developments in recent years have been presented in Table 4.2 below.

4.5 Conclusions and Future Prospects

Nitroaromatic explosive detection is of utmost importance for homeland safety and security to combat terrorism. Various methods are used for sensing nitroaromatics selectively. Nanoscience has led to the development of scientific innovations for sensing applications. Over the last decade, nanomaterial-based sensors have been of great importance for detecting nitroaromatic explosives because of their easy fabrication, low cost and higher efficiency. Different nanomaterials and their hybrids, conjugates and composites are extensively used for the selective detection of explosives. Although few challenges are associated with designing explosive sensors, nanoparticle-based sensors are considered as useful candidates for the fabrication of devices, which can overcome all the challenges related to explosive sensing. With the development of nanotechnology and the increasing requirement for explosive sensors, more sophisticated and ultra-sensitive devices need unquestionably be designed in the days to come. Nanotechnology will enable the swift and sensitive portable device fabrication for explosive detection and thus will be an essential tool to combat terrorist attacks in more effective ways.

Table 4.2 Some of the recently developed nanoparticle-based explosive sensors

Sl. no	Reported works	Method applied	LOD	Refs.
1	TiO_2 nanosheet based chemiresistive explosive sensor	Electrical technique	–	[81]
2	Tetraphenylethene probe-based silica nanoparticles for detection of explosives	Optical technique	0.01 μM (PA)	[82]
3	Detection of trace amount of nitroaromatics by 1-Pyreneiodide-ligated luminescent gold nanostructures and their forensic applications	Optical technique	10 nM (nitroaromatic-based explosives)	[83]
4	Detection of explosives at B-doped diamond/ graphene nanowall electrodes	Electrical technique	73 ppb (TNT); 270 ppb (TNA)	[84]
5	Chemi-resistive device for detection of explosives based on colloidal PbS quantum dots	Electrical technique	65–16 ppm (nitrobenzene, experimental); 2 ppb (nitrobenzene, theoritical)	[85]
6	AIE-active flexible self-assembled monolayer probe for trace explosive detection	Optical technique	1.2 ppm (TNT), 6 ppm (DNT), 35.7 (nitrobenzene)	[86]
7	Ethylenediamine-bound magnetite nanoparticles as dual function (charge transfer and nanoenzyme activity) colorimetric sensor for explosive detection	Optical technique	Micro range (TNT)	[87]
8	Nucleotide-lanthanide for selective detection of nitroaromatic explosives	Optical technique	5.96 ppb (TNP)	[88]
9	Amine functionalized CedSe/SiO_2 for highly selective and sensitive detection of picric acid	Optical technique	0.5E–07 (PA)	[89]
10	Surface-modified colloidal silica nanoparticles for explosive detection	Optical technique	–	[90]

References

1. Colton, R.J., Russell, J.N.: Making the world a safer place. Science **299**(5611), 1324–1325 (2003)
2. Ma, Y., Huang, S., Wang, L.: Multifunctional inorganic–organic hybrid nanospheres for rapid and selective luminescence detection of TNT in mixed nitroaromatics via magnetic separation. Talanta **116**, 535–540 (2013)
3. Singh, S.: Sensors—an effective approach for the detection of explosives. J. Hazard. Mater. **144**(1–2), 15–28 (2007)
4. Paden, N.E., Smith, E.E., Maul, J.D., Kendall, R.J.: Effects of chronic 2, 4, 6,-trinitrotoluene, 2, 4-dinitrotoluene, and 2, 6-dinitrotoluene exposure on developing bullfrog (Rana catesbeiana) tadpoles. Ecotoxicol. Environ. Saf.. Environ. Saf. **74**(4), 924–928 (2011)
5. Kim, T.H., Lee, B.Y., Jaworski, J., Yokoyama, K., Chung, W.J., Wang, E., Hong, S., Majumdar, A., Lee, S.W.: Selective and sensitive TNT sensors using biomimetic polydiacetylene-coated CNT-FETs. ACS Nano **5**(4), 2824–2830 (2011)
6. Lopez-Lopez, M., García-Ruiz, C.: Infrared and Raman spectroscopy techniques applied to identification of explosives. TrAC, Trends Anal. Chem. **54**, 36–44 (2014)
7. Akhgari, F., Fattahi, H., Oskoei, Y.M.: Recent advances in nanomaterial-based sensors for detection of trace nitroaromatic explosives. Sens. Actuat. B Chem. **221**, 867–878 (2015)
8. Narayanan, A., Varnavski, O.P., Swager, T.M., Goodson, T.: Multiphoton fluorescence quenching of conjugated polymers for TNT detection. J. Phys. Chem. C **112**(4), 881–884 (2008)
9. Mathurin, J.C., Faye, T., Brunot, A., Tabet, J.C., Wells, G., Fuche, C.: High-pressure ion source combined with an in-axis ion trap mass spectrometer. 1. Instrumentation and applications. Anal. Chem. **72**(20), 5055–5062 (2000)
10. Sylvia, J.M., Janni, J.A., Klein, J.D., Spencer, K.M.: Surface-enhanced Raman detection of 2, 4-dinitrotoluene impurity vapor as a marker to locate landmines. Anal. Chem. **72**(23), 5834–5840 (2000)
11. Zachhuber, B., Ramer, G., Hobro, A., Chrysostom, E.T., Lendl, B.: Stand-off Raman spectroscopy: a powerful technique for qualitative and quantitative analysis of inorganic and organic compounds including explosives. Anal. Bioanal. Chem.Bioanal. Chem. **400**(8), 2439–2447 (2011)
12. Eiceman, G.A., Stone, J.A.: Peer Reviewed: Ion Mobility Spectrometers in National Defense (2004)
13. Salinas, Y., Martínez-Máñez, R., Marcos, M.D., Sancenón, F., Costero, A.M., Parra, M., Gil, S.: Optical chemosensors and reagents to detect explosives. Chem. Soc. Rev. **41**(3), 1261–1296 (2012)
14. Hallowell, S.F.: Screening people for illicit substances: a survey of current portal technology. Talanta **54**(3), 447–458 (2001)
15. Vourvopoulos, G., Womble, P.: Pulsed fast/thermal neutron analysis: a technique for explosives detection. Talanta **54**(3), 459–468 (2001)
16. Wang, J.: Electrochemical sensing of explosives. Electroanal.: Int. J. Devoted Fundam. Pract. Asp. Electroanal. **19**(4), 415–423 (2007)
17. Caygill, J.S., Davis, F., Higson, S.P.: Current trends in explosive detection techniques. Talanta **88**, 14–29 (2012)
18. Wang, J.: Electrochemical detection for microscale analytical systems: a review. Talanta **56**(2), 223–231 (2002)
19. Sattayasamitsathit, S., O'Mahony, A.M., Xiao, X., Brozik, S.M., Washburn, C.M., Wheeler, D.R., Gao, W., Minteer, S., Cha, J., Burckel, D.B., Polsky, R.: Highly ordered tailored three-dimensional hierarchical nano/microporous gold–carbon architectures. J. Mater. Chem. **22**(24), 11950–11956 (2012)
20. Pumera, M.: Graphene-based nanomaterials and their electrochemistry. Chem. Soc. Rev. **39**(11), 4146–4157 (2010)

21. Zamborini, F.P., Bao, L., Dasari, R.: Nanoparticles in measurement science. Anal. Chem. **84**(2), 541–576 (2012)
22. Galik, M., O'Mahony, A.M., Wang, J.: Cyclic and square-wave voltammetric signatures of nitro-containing explosives. Electroanalysis **23**(5), 1193–1204 (2011)
23. Santos, A.P.L., Deokaran, G.O., Costa, C.V., Gama, L.I., Júnior, E.G.M., de Assis, A.M., et al.: A "turn-off" fluorescent sensor based on electrospun polycaprolactone nanofibers and fluorene (bisthiophene) derivative for nitroaromatic explosive detection. Forensic Sci. Int. **329**, 111056 (2021)
24. Riskin, M., Tel-Vered, R., Bourenko, T., Granot, E., Willner, I.: Imprinting of molecular recognition sites through electropolymerization of functionalized Au nanoparticles: development of an electrochemical TNT sensor based on π-donor–acceptor interactions. J. Am. Chem. Soc. **130**(30), 9726–9733 (2008)
25. Filanovsky, B., Markovsky, B., Bourenko, T., Perkas, N., Persky, R., Gedanken, A., Aurbach, D.: Carbon electrodes modified with TiO_2/metal nanoparticles and their application for the detection of trinitrotoluene. Adv. Func. Mater. **17**(9), 1487–1492 (2007)
26. Brownson, D.A., Banks, C.E.: Graphene electrochemistry: an overview of potential applications. Analyst **135**(11), 2768–2778 (2010)
27. Pumera, M.: Electrochemistry of graphene: new horizons for sensing and energy storage. Chem. Rec. **9**(4), 211–223 (2009)
28. Pumera, M.: Graphene-based nanomaterials for energy storage. Energy Environ. Sci. **4**(3), 668–674 (2011)
29. Shao, Y., Wang, J., Wu, H., Liu, J., Aksay, I.A., Lin, Y.: Graphene based electrochemical sensors and biosensors: a review. Electroanal.: Int. J. Devoted Fundam. Pract. Asp. Electroanal. **22**(10), 1027–1036 (2010)
30. Geim, A.K., Novoselov, K.S.: The rise of graphene. Nanosci. Technol.: Collect. Rev. Nat. J. 11–19 (2010)
31. Novoselov, K.S., Geim, A.K., Morozov, S.V., Jiang, D., Katsnelson, M.I., Grigorieva, I., Dubonos, S., Firsov, A.A.: Two-dimensional gas of massless Dirac fermions in graphene. Nature **438**(7065), 197–200 (2005)
32. Feng, L., Li, H., Qu, Y., Lü, C.: Detection of TNT based on conjugated polymer encapsulated in mesoporous silica nanoparticles through FRET. Chem. Commun.Commun. **48**(38), 4633–4635 (2012)
33. Tang, L., Feng, H., Cheng, J., Li, J.: Uniform and rich-wrinkled electrophoretic deposited graphene film: a robust electrochemical platform for TNT sensing. Chem. Commun.Commun. **46**(32), 5882–5884 (2010)
34. Chen, T.W., Sheng, Z.H., Wang, K., Wang, F.B., Xia, X.H.: Determination of explosives using electrochemically reduced graphene. Chem. Asian J. **6**(5), 1210–1216 (2011)
35. Chua, C.K., Ambrosi, A., Pumera, M.: Graphene based nanomaterials as electrochemical detectors in Lab-on-a-chip devices. Electrochem. Commun.. Commun. **13**(5), 517–519 (2011)
36. Goh, M.S., Pumera, M.: Graphene-based electrochemical sensor for detection of 2, 4, 6-trinitrotoluene (TNT) in seawater: the comparison of single-, few-, and multilayer graphene nanoribbons and graphite microparticles. Anal. Bioanal. Chem.Bioanal. Chem. **399**(1), 127–131 (2011)
37. Zhang, H.X., Hu, J.S., Yan, C.J., Jiang, L., Wan, L.J.: Functionalized carbon nanotubes as sensitive materials for electrochemical detection of ultra-trace 2, 4, 6-trinitrotoluene. Phys. Chem. Chem. Phys. **8**(30), 3567–3572 (2006)
38. Marin, S., Merkoçi, A.: Direct electrochemical stripping detection of cystic-fibrosis-related DNA linked through cadmium sulfide quantum dots. Nanotechnology **20**(5), 055101 (2009)
39. Marín, S., Pujals, S., Giralt, E., Merkoçi, A.: Electrochemical investigation of cellular uptake of quantum dots decorated with a proline-rich cell penetrating peptide. Bioconjug. Chem.. Chem. **22**(2), 180–185 (2011)
40. Frasco, M.F., Chaniotakis, N.: Semiconductor quantum dots in chemical sensors and biosensors. Sensors **9**(9), 7266–7286 (2009)

41. Costas-Mora, I., Romero, V., Lavilla, I., Bendicho, C.: An overview of recent advances in the application of quantum dots as luminescent probes to inorganic-trace analysis. TrAC, Trends Anal. Chem. **57**, 64–72 (2014)
42. Resch-Genger, U., Grabolle, M., Cavaliere-Jaricot, S., Nitschke, R., Nann, T.: Quantum dots versus organic dyes as fluorescent labels. Nat. Methods **5**(9), 763–775 (2008)
43. To, K.C., Ben-Jaber, S., Parkin, I.P.: Recent developments in the field of explosive trace detection. ACS Nano **14**(9), 10804–10833 (2020)
44. Tu, R., Liu, B., Wang, Z., Gao, D., Wang, F., Fang, Q., Zhang, Z.: Amine-capped ZnS$-Mn^{2+}$ nanocrystals for fluorescence detection of trace TNT explosive. Anal. Chem. **80**(9), 3458–3465 (2008)
45. Pazhanivel, T., Nataraj, D., Devarajan, V.P., Mageshwari, V., Senthil, K., Soundararajan, D.: Improved sensing performance from methionine capped CdTe and CdTe/ZnSquantum dots for the detection of trace amounts of explosive chemicals in liquid media. Anal. Methods **5**(4), 910–916 (2013)
46. Liu, B., Tong, C., Feng, L., Wang, C., He, Y., Lü, C.: Water-soluble polymer functionalized CdTe/ZnS quantum dots: a facile ratiometric fluorescent probe for sensitive and selective detection of nitroaromatic explosives. Chem. Euro. J. **20**(8), 2132–2137 (2014)
47. Goldman, E.R., Medintz, I.L., Whitley, J.L., Hayhurst, A., Clapp, A.R., Uyeda, H.T., Deschamps, J.R., Lassman, M.E., Mattoussi, H.: A hybrid quantum dot−antibody fragment fluorescence resonance energy transfer-based TNT sensor. J. Am. Chem. Soc. **127**(18), 6744–6751 (2005)
48. Xu, S., Lu, H., Li, J., Song, X., Wang, A., Chen, L., Han, S.: Dummy molecularly imprinted polymers-capped CdTe quantum dots for the fluorescent sensing of 2, 4, 6-trinitrotoluene. ACS Appl. Mater. Interfaces **5**(16), 8146–8154 (2013)
49. Zhang, K., Zhou, H., Mei, Q., Wang, S., Guan, G., Liu, R., Zhang, J., Zhang, Z.: Instant visual detection of trinitrotoluene particulates on various surfaces by ratiometric fluorescence of dual-emission quantum dots hybrid. J. Am. Chem. Soc. **133**(22), 8424–8427 (2011)
50. Xia, Y., Song, L., Zhu, C.: Turn-on and near-infrared fluorescent sensing for 2, 4, 6-trinitrotoluene based on hybrid (gold nanorod)−(quantum dots) assembly. Anal. Chem. **83**(4), 1401–1407 (2011)
51. Shi, G., Qu, Y., Zhai, Y., Liu, Y., Sun, Z., Yang, J., Jin, L.: MSU/PDDA n LBL assembled modified sensor for electrochemical detection of ultratrace explosive nitroaromatic compounds. Electrochem. Commun.. Commun. **9**(7), 1719–1724 (2007)
52. Zhang, H.X., Cao, A.M., Hu, J.S., Wan, L.J., Lee, S.T.: Electrochemical sensor for detecting ultratracenitroaromatic compounds using mesoporous SiO_2-modified electrode. Anal. Chem. **78**(6), 1967–1971 (2006)
53. Trammell, S.A., Zeinali, M., Melde, B.J., Charles, P.T., Velez, F.L., Dinderman, M.A., Kusterbeck, A., Markowitz, M.A.: Nanoporousorganosilicas as preconcentration materials for the electrochemical detection of trinitrotoluene. Anal. Chem. **80**(12), 4627–4633 (2008)
54. Poh, H.L., Pumera, M.: Nanoporous carbon materials for electrochemical sensing. Chem. Asian J. **7**(2), 412–416 (2012)
55. Zang, J., Guo, C.X., Hu, F., Yu, L., Li, C.M.: Electrochemical detection of ultratracenitroaromatic explosives using ordered mesoporous carbon. Anal. Chim. Acta **683**(2), 187–191 (2011)
56. Guo, S., Wen, D., Zhai, Y., Dong, S., Wang, E.: Platinum nanoparticle ensemble-on-graphene hybrid nanosheet: one-pot, rapid synthesis, and used as new electrode material for electrochemical sensing. ACS Nano **4**(7), 3959–3968 (2010)
57. Guo, C.X., Lei, Y., Li, C.M.: Porphyrin functionalized graphene for sensitive electrochemical detection of ultratrace explosives. Electroanalysis **23**(4), 885–893 (2011)
58. Guo, C.X., Lu, Z.S., Lei, Y., Li, C.M.: Ionic liquid–graphene composite for ultratrace explosive trinitrotoluene detection. Electrochem. Commun.. Commun. **12**(9), 1237–1240 (2010)
59. Aznar-Gadea, E., Rodríguez-Canto, P.J., Martínez-Pastor, J.P., Lopatynskyi, A., Chegel, V., Abargues, R.: Molecularly imprinted silver nanocomposites for explosive taggant sensing. ACS Appl. Polym. Mater.s **3**(6), 2960–2970 (2021)

60. Ma, Y., Xu, S., Wang, S., Wang, L.: Luminescent molecularly-imprinted polymer nanocomposites for sensitive detection. Trends Anal. Chem. **67**, 209–216 (2015)
61. Liu, J., Li, R., Yang, B.: Carbon dots: a new type of carbon-based nanomaterial with wide applications. ACS Cent. Sci. **6**(12), 2179–22195 (2020)
62. Pal, A., Palassudin, M., Chattopadhyay, A.: Conducting carbon dot-polypyrrole conducting carbon dot-polypyrrole. ACS Appl. Mater. Interfaces **8**(9), 5758–5762 (2016)
63. Shan, X., Zhang, S., Zhou, M., Geske, T., Davis, M., Hao, A., Wang, H., Yu, Z.: Porous Halide Perovskite-polymer nanocomposites for explosive detection with a high sensitivity. Adv. Mater. Interfaces **6**(3), 1801686 (2019)
64. Abbasi, F., Akbarinejad, A., Alizadeh, N.: CdS QDs/N-methylpolypyrrole hybrids as fluorescent probe for ultrasensitive and selective detection of picric acid. Spectrochim. Acta Part A Mol. Biomol. Spectrosc.. Acta Part A Mol. Biomol. Spectrosc. **216**, 230–235 (2019)
65. Dutta, P., Chakravarty, S., Sen Sarma, N.: Detection of nitroaromatic explosives using π-electron rich luminescent polymeric nanocomposites. RSC Adv. **6**(5), 3680–3689 (2016)
66. Dutta, P., Saikia, D., Adhikary, N.C., Sen Sarma, N.: Macromolecular systems with MSA-Capped CdTe and CdTe/ZnS Core/Shell quantum dots as superselective and ultrasensitive optical sensors for picric acid explosive. ACS Appl. Mater. Interfaces **7**(44), 24778–24790 (2016)
67. Chakravarty, S., Bhardwaj, N., Mandal, B.B., Sen Sarma, N.: Silk fibroin–carbon nanoparticle composite scaffolds: a cost effective supramolecular 'turn off'chemiresistor for nitroaromatic explosive vapours. J. Mater. Chem. C **4**(38), 8920–8929 (2016)
68. Thundat, T., Warmack, R.J., Chen, G.Y., Allison, D.P.: Thermal and ambient-induced deflections of scanning force microscope cantilevers. Appl. Phys. Lett. **64**(21), 2894–2896 (1994)
69. Gong, Z., Du, H., Cheng, F., Wang, C., Wang, C., Fan, M.: Fabrication of SERS swab for direct detection of trace explosives in fingerprints. ACS Appl. Mater. Interfaces **6**(24), 21931–21937 (2014)
70. Zhou, Q., Peng, L., Jiang, D., Wang, X., Wang, H., Li, H.: Detection of nitro-based and peroxide-based explosives by fast polarity-switchable ion mobility spectrometer with ion focusing in vicinity of Faraday detector. Sci. Rep. **5**(1), 1–8 (2015)
71. Aluri, G.S., Motayed, A., Davydov, A.V., Oleshko, V.P., Bertness, K.A., Rao, M.V.: Nitro-aromatic explosive sensing using GaN nanowire-titania nanocluster hybrids. IEEE Sens. J. **13**(5), 1883–1888 (2013)
72. Stefanuto, P.H., Perrault, K.A., Focant, J.F., Forbes, S.L.: Fast chromatographic method for explosive profiling. Chromatography **2**(2), 213–224 (2015)
73. Wasilewski, T., Gębicki, J.: Emerging strategies for enhancing detection of explosives by artificial olfaction. Microchem. J.106025–106040 (2021)
74. Liu, W., Song, Z., Zhao, Y., Liu, Y., He, X., Cui, S.: Flexible porous aerogels decorated with Ag nanoparticles as an effective SERS substrate for label-free trace explosives detection. Anal. Methods **12**(33), 4123–4129 (2020)
75. Rani, S., Dilbaghi, N., Kumar, S., Varma, R.S., Malhotra, R.: Rapid redox sensing of p-nitrotoluene in real water samples using silver nanoparticles. Inorg. Chem. Commun.. Chem. Commun. **120**, 108157 (2020)
76. Kumar, A., Sahoo, S.C., Mehta, S.K., Soni, P., Sharma, V., Kataria, R.: Luminescent Zn–MOF for detection of explosives and development of fingerprints. Anal. Methods (2022)
77. Junaid, H.M., Waseem, M.T., Khan, Z.A., Gul, H., Yu, C., Shaikh, A.J., Shahzad, S.A.: Fluorescent and colorimetric sensors for selective detection of TNT and TNP explosives in aqueous medium through fluorescence emission enhancement mechanism. J. Photochem. Photobiol. A: Chem. 113865 (2022)
78. Hong, J.H., Choi, J.H., Cho, D.G.: Simple pyrene derivatives as fluorescence sensors for TNT and RDX in micelles. Bull. Korean Chem. Soc. **35**(11), 3158–3162 (2014)
79. Nailwal, Y., Devi, M., Pal, S.K.: Luminescent conjugated microporous polymers for selective sensing and ultrafast detection of picric acid. ACS Appl. Polym. Mater. (2022)

80. Dey, S., Saha, A., Kumar, P., Kar, C., Chakraborty, S., Sukul, P.K.: Self-assembled nanomaterials of naphthalene monoimide in aqueous medium for multimodal detection of picric acid. J. Photochem. Photobiol. A **423**, 113599 (2022)
81. Li, Y., Zhou, W., Zu, B., Dou, X.: Qualitative detection toward military and improvised explosive vapors by a facile TiO_2 nanosheet-based chemiresistive sensor array. Front. Chem. **8**(29), 1–12 (2020)
82. Nawaz, M.A.H., Meng, L., Zhou, H., Ren, J., Shahzad, S.A., Hayat, A., Yu, C.: Tetraphenylethene probe based fluorescent silica nanoparticles for the selective detection of nitroaromatic explosives. Anal. Methods **13**(6), 825–831 (2021)
83. Thakarda, J., Agrawal, B., Anil, D., Jana, A., Maity, P.: Detection of trace-level nitroaromatic explosives by 1-Pyreneiodide-ligated luminescent gold nanostructures and their forensic applications. Langmuir **36**(50), 15442–15449 (2020)
84. Dettlaff, A., Jakóbczyk, P., Ficek, M., Wilk, B., Szala, M., Wojtas, J., Ossowski, T., Bogdanowicz, R.: Electrochemical determination of nitroaromatic explosives at boron-doped diamond/graphene nanowall electrodes: 2, 4, 6-trinitrotoluene and 2, 4, 6-trinitroanisole in liquid effluents. J. Hazard. Mater. **387**, 121672 (2020)
85. Mitri, F., De Iacovo, A., De Santis, S., Giansante, C., Sotgiu, G., Colace, L.: Chemiresistive device for the detection of nitroaromatic explosives based on colloidal PbS quantum dots. ACS Appl. Electron. Mater. **3**(7), 3234–3239 (2021)
86. Wang, G., Li, M., Wei, Q., Xiong, Y., Li, J., Li, Z., Tang, J., Wei, F., Tu, H.: Design of an AIE-active flexible self-assembled monolayer probe for trace nitroaromatic compound explosive detection. ACS Sens. **6**(5), 1849–1856 (2021)
87. Yardımcı, B., Koç, Ö.K., Üzer, A., Hızal, J., Apak, R.: Ethylenediamine-bound magnetite nanoparticles as dual function colorimetric sensor having charge transfer and nanozyme activity for TNT and tetryl detection. Microchim. Acta. Acta **188**(7), 1–12 (2021)
88. Gao, R., Wang, J., Wang, H., Dong, W., Zhu, J.: Fluorescent nucleotide- lanthanide nanoparticles for highly selective determination of picric acid. Microchim. Acta. Acta **188**(1), 1–8 (2021)
89. Gong, W., Li, H., Gong, X., Zhang, Z., Lu, Z.: Fabrication of amine functionalized cdSe@ SiO_2 nanoparticles as fluorescence nanosensor for highly selective and sensitive detection of picric acid. Spectrochim. Acta Part A Mol. Biomol. Spectrosc.. Acta Part A Mol. Biomol. Spectrosc. **233**, 118221 (2020)
90. Elbasuney, S., El-Sharkawy, Y.H., El-Sayyad, G.S., Gobara, M.: Surface modified colloidal silica nanoparticles: novel aspect for complete identification of explosive materials. Talanta **211**, 120695–120698 (2020)

Chapter 5
Advances in Few-Layered Nanoscale Transition Metal Dichalcogenides in Sensing Application

Ashamoni Neog, Hemanga Jyoti Sarmah, Dambarudhar Mohanta, and Rajib Biswas

Abstract Developing cost-effective and at the same time advanced sensors has always been a challenging task for sensor designers. The choice of the sensing material determines the performance of the final sensor product as it holds paramount importance in the detection and transduction of the external stimuli in question. This fosters the sensor designers to exploit newly discovered or developed materials for different sensing applications. In this regard, layered or two-dimensional (2D) Transition Metal Dichalcogenides (TMDCs) appear to be the next-generation sensing materials. Considering the myriad applications of these materials, this chapter presents a comprehensive overview of various sensors based on these layered materials. Starting from typical sensor classifications, associated sensing mechanisms and parameters, a brief discussion of different types of sensing implementations is procedurally outlined for each selected sensing material.

Keywords TMDC · 2D materials · Sensor

Supplementary Information The online version contains supplementary material available at https://doi.org/10.1007/978-981-99-7848-9_5.

A. Neog · R. Biswas (✉)
Applied Optics and Photonics Laboratory, Department of Physics, Tezpur University, Napaam 784028, Assam, India
e-mail: rajib@tezu.ernet.in

H. J. Sarmah · D. Mohanta (✉)
Nanoscience and Soft Matter Laboratory, Department of Physics, Tezpur University, Napaam 784028, Assam, India
e-mail: best@tezu.ernet.in

H. J. Sarmah
Department of Physics, Morigaon College, Morigaon 782105, Assam, India

D. Mohanta and P. Chakraborty (eds.), *Nanoscale Matter and Principles for Sensing and Labeling Applications*, Advanced Structured Materials 206,
https://doi.org/10.1007/978-981-99-7848-9_5

5.1 Introduction

Sensors are devices which can detect and quantify the presence of any external stimuli. It comprises of a sensing material which can interact with the external stimuli (physical stimuli such as pressure, temperature, chemical stimuli such as the presence of any chemical entity and biological stimuli such as the presence of bacteria, viruses and many more) and a transducer material which manifest the interaction of the sensing material with the external stimuli and convert it to a recognizable signal (mainly electrical or optical) [1–3]. The schematic diagram of a sensor system is shown in Fig. 5.1. Ideally, a sensing material should be chemically stable, also exhibiting large surface area to facilitate a potential platform for the material–analyte interaction [4]. For selective detection of the analytes, it demands target-specific active sites in the sensing material and the material must possess the ability to transduce the binding events into a detectable signal. To be mentioned, match between the effective size of the sensing material with that of the substrate is required to design a sensor with selective detection. These characteristics pertaining to ideal sensing material hold for all types of sensing applications. In this regard, nanomaterials, offering large surface-to-volume ratio, have received tremendous attention due to better sensitivity, selectivity, fast response and miniaturized features [5, 6]. Among several types (e.g. zero-dimensional, one-dimensional and two-dimensional) of nanomaterials, layered or two-dimensional (2D) Transition Metal Dichalcogenides (TMDCs) have received overwhelming response in the last few decades because of their band structure tunability property, large grain size and large surface area. In TMDCs, transition metals, e.g. W, Mo, Ta, Nb and V are sandwiched between chalcogen atoms, e.g. S, Se and Te. The standard chemical formula for these types of materials is MX_2, where M represents the transition metal atom and X represents the chalcogen atom. They possess layered morphology, and each layer has a thickness of ~6–7 Å. The in-plane M–X bonds are predominantly covalent in nature, whereas the outer plane bonds are van der Waals. This allows the crystals to be cleaved along the layer surface. Depending upon combinations of different transition metal elements and chalcogenide elements, they possess different electronic properties, e.g. molybdenum and tungsten disulphides and diselenides are semiconducting, whereas tellurides are semi-metallic [7]. Semiconducting TMDCs are rich in surface chemistry in the sense that the surface properties of the materials can be tuned by controlling the shape and size of the materials. When exfoliated into thin sheets, changes in the orbital interaction and hybridization lead to widening of the band gap and indirect to direct-bandgap conversion in the case of semiconducting TMDCs [8]. Again, reduction in the lateral size of the materials can bring about catalytic properties of the materials which are absent in their bulk counterpart [9]. Being endowed with such attributes, most of the semiconducting TMDCs have found immense application in the field of sensor design and fabrication [5, 10]. Some of the sensors based on layered TMDC are shown in Fig. 5.2.

This chapter specifically encompasses the stratum of such semiconducting TMDCs towards chemical, photo and radiation sensing applications. Based on to

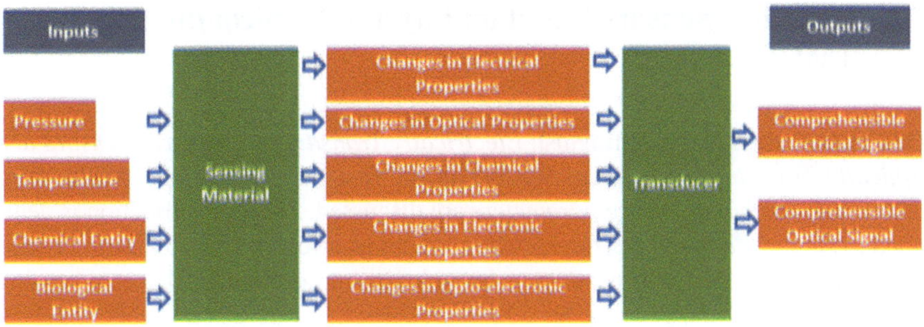

Fig. 5.1 Schematic diagram of a sensor system

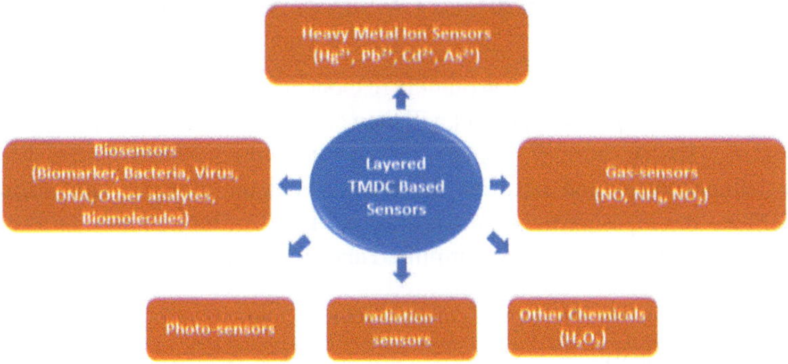

Fig. 5.2 Different types of sensors based on layered transition metal dichalcogenides

what detectable signal the sensing materials convert the external stimuli, sensors may be broadly categorized into electrically transduced sensors and optically transduced sensors. In the case of electrically transduced sensors, external stimuli are converted to electrical signals, whereas, for optically transduced sensors, it is converted into optical signals. In chemical sensing section, heavy metal ion sensors are specifically highlighted where discussions regarding sensing mechanisms, adaptations and parameters have been included. In photosensing section, different photosensing mechanisms and parameters have been discussed. The last section related to dosimetry has illuminated the prospects of the materials regarding radiation sensing.

5.2 Chemical Sensors Based on Layered Semiconducting TMDCs

This section attempts to distinguish *Electrically Transduced Chemical Sensors* and *Optically Transduced Chemical Sensors* in the first two subsections. In the third subsection, *chemical sensing parameters* are discussed. The last subsection is dedicated to *Heavy Metal Ion Sensing/Sensors,* where an effort has been made to present and analyse the reported sensors in a chronological order for each selected sensing material.

5.2.1 Electrically Transduced Chemical Sensor (ETCS)

The sensing material in ETCSs can quickly respond to external stimuli by changing its electrical behaviour. The transducers will then convert the change to some detectable electrical signals like current, voltage, inductance and resistance. There should be a well-defined correlation between the analyte and the signals. This correlation will facilitate the measurements of the concentrations of target analytes. For better realization, the signals are further amplified and filtered [5].

Mechanism of Sensing

There exist several analyte-material interaction mechanisms like van der Waals [11], charge transfer [12], π–π interactions [13], dispersive interaction, chemical reaction, redox reaction, H-bonding [14] etc. Among them, some interactions are prevalent in semiconducting TMDCs. Those interactions depend on how the active sites of the sensing material get modified by incorporation of defects, doping and functionalization. Again, material–analyte interaction changes the physical properties of the sensing material, e.g. conductivity, work function and permittivity etc.; which eventually changes the values of measurable electrical signals like current and voltage. The alteration of the physical parameters of the sensing material can be explained through some fundamental mechanisms like doping level and Schottky barrier inflections, dipole and interfacial layer formation etc. Based on the surface chemistry of the sensing material and relative energies of the material and analyte, the analyte either accepts or donates electrons to the sensing material. This results the alteration of the inherent doping concentration of the working material [15]. As the doping concentration changes, the electronic band structure and hence the optical properties like reflectivity, absorbance etc. and electrical properties like, conductivity, impedance etc. of the sensing material alters accordingly. As this chapter mainly focuses on semiconducting TMDCs, the concept of Schottky barrier is inevitable [16]. When a metal and a semiconductor interact atomically, they form a junction called metal–semiconductor junction and the associated barrier is called Schottky barrier. Schottky barrier depends on the electron affinity of the semiconductor and the work function of the metal. On the other hand, the electron affinity depends on the doping concentration in the semiconductor. So, whenever the doping level of the semiconducting sensing material changes due to the analyte–sensing material interaction, Schottky barrier height varies accordingly. Undoubtedly, this results the manifestation of the physical property of the sensing material, which exhibits the different signals in the output. Proper calibration of the signal output with the concentration of the analyte, helps to quantify the detectable amount.

Adaptations

Multitude device architectures have been adapted to realize electrically transduced sensors. Some of them are briefly described below.

(a) ***Chemiresistor***: A chemiresistor is a two-electrode device. Here, the two electrodes are connected to the sensing material. The electrodes and the sensing materials are supported by an insulator. By analysing the changes in the value of current or resistance of the sensor unit, the presence of different analytes can be confirmed and measurement of the concentration of the same can be done [17, 18].

(b) ***Field Effect Transistor***: It is a three-terminal device. It primarily consists of Source, Gate and Drain electrodes, the channel material and the insulating or oxide layer. The current flows between the Drain and Source terminal through the sensing material. The advantage of using these three terminal devices is that the initial current through the sensing material can be controlled by biasing so that the effect of the analyte can be clearly distinguished. To keep the current through channel in linear range, the Schottky barrier between the metal contacts and the semiconductor sensing material must be optimized. The response of the sensor can be determined by comparing the change in the channel current to the initial channel current [15, 19].

(c) ***Electrochemical Sensors***: In similitude to the electrical devices, there exist two and three electrode-based electrochemical cells. In two electrode-based electrochemical sensors, one electrode acts as reference electrode and the other acts as working electrode. Potential is applied at the working electrode so that it can draw charge particles from the analyte and initiate a flow of current. On the other hand, the reference electrodes should receive the current to maintain the charge flow continuity. Though the method is simple, this is not quite acceptable as the reference electrode cannot be maintained at a constant potential in this case. Three electrode-based systems are however more convenient for this purpose as the reference electrode is fixed at a certain potential, while the current flow through the working electrode and the auxiliary electrode. The current in the cell corresponding to the potential in the working electrode shows the oxidation as well as the reduction peaks. Change in the redox potential eventually determines the concentration of the analytes [20, 21].

(d) ***Photoelectrochemical Sensors***: In photoelectrochemical sensors, electron–hole pairs are generated in the sensing material under the illumination of light. In the case of heavy metal ion sensing, these electrons are consumed by the metal ions and reduced to metal atoms, while the holes are used by other agents in the electrolyte. The response of the sensor becomes better if the sensing material can efficiently generate electrons and holes and the device architectures efficiently transfer the charges to the respective entities before the recombination process takes place [22].

5.2.2 Optically Transduced Chemical Sensor

In such sensors, two types of assays are primarily used—fluorometric assay and colorimetric assay and the sensors are named fluorometric sensors and colorimetric sensors, respectively. A mini discussion regarding fluorometric and colorimetric sensors is provided below.

Fluorometric Sensors

In a particular type of fluorometric assay, when the sensing material is allowed to react with the analyte, it results in a fluorescent material. The resulting material, when excited by a light of a particular wavelength, can give off light having a particular wavelength range. The relative fluorescence units are typically proportional to the amount of analyte being measured. In another type of fluorometric assay, either the sensing material act as a fluorescence quencher and quench the fluorescence activity of the analyte or the analytes quench the intrinsic fluorescence activity of the sensing material [23, 24].

Colorimetric Sensors

Here, the sensing material–analyte interaction results in coloured products. The reaction can be experienced with the naked eye as the product absorbs and then reflects light in the visible range. The optical density of the reaction product is typically proportional to the amount of analyte being measured. The response of the sensor depends on the catalytic property of the sensing material [25].

5.2.3 Chemical Sensing Parameters

Sensing parameters display certain properties of a sensor which help to compare different sensors in question. Some of the most important chemical sensing parameters are discussed below.

Sensitivity, Limit of Detection (LOD), Linear Range and Limit of Quantification

When the result of the transduction is systematically plotted against the concentration of the analytes, then a curve is obtained which is called the calibration curve of the proposed sensor. Usually, for sensing purposes, only the linear portion of the calibration curve is considered where the slope of the curve is constant, which is called the sensitivity of the proposed sensor and is denoted by m. The range with a constant slope has a lower limit and an upper limit. This lower limit is called the LOD and the upper limit as Upper Limit of Detection (ULD). The range between LOD and ULD is called linear range of the proposed sensor. If σ is the standard deviation of the calibration curve, then LOD is given as [12, 26]

$$\mathrm{LOD} = 3\sigma/m,$$

while Limit of Quantification (LOQ) as

$$\mathrm{LOQ} = 10\sigma/m.$$

Selectivity and Specificity

Selectivity is the ability of the sensor to effectively measure the concentration of the target analyte in the presence of interfering substances, whereas specificity is the capacity of the detector to detect only one particular analyte or quality [27].

5.2.4 Heavy Metal Ion Sensing

In this subsection, definition of Heavy Metal Ions (HMIs), importance of the detection of HMI- and TMDC-based sensors and HMI sensors are discussed.

Heavy Metal Ions and Their Effects on Living Body

Literally, the phrase heavy metal represents metal having high atomic weight and density compared to that of water [28]. In practice, the phrase includes transition metals, metalloids, lanthanides, actinides, etc. [29, 30]. Heavy metals are essential for several chemical reactions, biological and physiological functions and are considered as trace elements [31]. However, they are considered to be toxic above some particular limits. Although some of the heavy metal contamination phenomena take place naturally [32], anthropogenic activities have raised the counts exponentially [33]. Human body needs definite supply of these heavy metals, e.g. zinc (Zn), selenium (Se), iron (Fe), copper (Cu), cobalt (Co), chromium (Cr), etc., to maintain proper functioning of the body [31]. Besides, some heavy metals are considered non-essential, e.g. aluminium (Al), cadmium (Cd), germanium (Ge), antinomy (Sb), arsenic (As), gold (Au) and barium (Ba) [34]. Studies carried out in different parts of the world have confirmed carcinogenic and non-carcinogenic effects of As. Toxicity of As depends on individual susceptibilities, age, gender, exposure time, doses and rate [35]. It can cause skin, lung, bladder, liver and prostate cancers, leukaemia and many more. Similarly, non-carcinogenic effects are also there such as neurological memory, intellectual function, diabetes, skin, cardiovascular disorders and reproductive system [36]. Similarly, Cd toxicity may damage kidneys, liver, skeletal system and cardiovascular system, deterioration of sight and hearing [37]. Again, lead (Pb) contamination can adversely affect small children and foetuses. Neurodevelopment problems, reproductive health effects, cardiovascular and renal effects can be seen due to Pb contamination [38]. Mercury (Hg) can affect almost every system of the human body (cardiovascular, pulmonary, immune, nervous, endocrine, digestive and renal systems and reproductive systems) [39]. Hg is endorsed by more than 250 symptoms, making it a challenging task for health workers to diagnose the issue properly. Similarly, Cr [40], Fe [41], Cu [42] and Ni [43] are also on the list.

Layered Semiconducting TMDC-Based Heavy Metal Ion Sensor

It is well known that the semiconducting 2D TMDCs like molybdenum disulphide (MoS_2), molybdenum diselenide ($MoSe_2$), tungsten disulphide (WS_2) and tungsten diselenide (WSe_2) are highly stable in the ambient condition. They have thickness-dependent electronic properties, so their conductivity is tunable. They are mechanically flexible. Again, the surface-to-volume ratio of nanomaterials is thousands of times more than that of their 3D counterpart. Therefore, the sensitivity is incomparable with their 3D counterpart. The *p* orbital of chalcogen atom [44, 45] and the defects [46, 47] are the active sites of the TMDC materials. To increase the selectivity, they can also be functionalized or doped. Below, there is a discussion on HMI sensors highlighting the primary sensing material used in the sensor unit. Subsequently, they are categorized as MoS_2 and WS_2.

(a) **MoS_2**: Being the most highly exploited semiconducting TMDCs, the number of investigations regarding MoS_2-based heavy metal ion sensing is comparatively higher than other materials from this category. Detection of Hg ions by Jiang et al. might be the first attempt to use semiconducting TMDC-based FET sensor for detection of heavy metal ions [15]. It can detect Hg, As and Pb ions and can have a LOD up to 0.1 ppb and a response time of 100–210 s [15, 16]. Improvisation of such sensing schemes was carried out by choosing proper flexible substrates for deposition of MoS_2. This approach helps to obtain flexible multiplex detector. Upto ppb level ultrafast detection (~8 s) of Cd^{2+} was realized with simultaneous monitoring of Hg^{2+} in tap water and Na^{+} in human sweat with the help of this improvisation technique. These investigations lead to the possibilities of MoS_2-based wearable devices featuring multiple heavy metal ion detection [48]. Bare MoS_2 was also used as sensing material for Arsenic (As) detection in a carbon microelectromechanical (C-MEMS) system [18]. Nowadays, different nanostructures [49], heterostructure arrangements [50, 51], functionalized MoS_2 [52], composites [53] and MoS_2 superstructures [54] have been introduced to sensor system to increase the sensitivity and selectivity of the sensors.

(b) ***WS_2***: Very few works have reported heavy metal ion sensing with WS_2 compared to MoS_2. Possibly, the rise of the WS_2-based heavy metal ion sensor started with fluorescence-based sensors [24, 55, 56]. These sensors were based on fluorescence quenching property of WS_2 nanosheets. Subsequently, there has been a progression towards the development of chemiresistive sensors [12, 57]

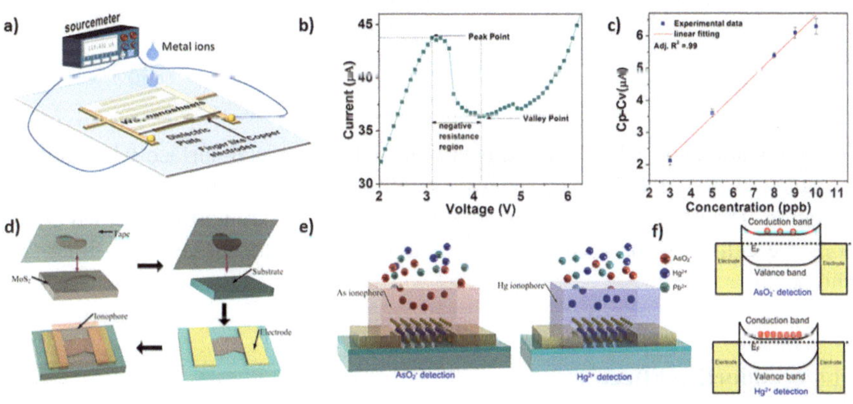

Fig. 5.3 **a** Sensor set-up for detection of metal ions with the help of WS_2 nanosheets, **b** Response of WS_2 nanosheet-based sensor towards metal ions, **c** Calibration curve of WS_2 nanosheets sensor towards the detection of Zn^{2+} ions [12]. (Re-printed with permission from Elsevier©) **d** Fabrication process of sensor based on MoS_2. Mechanically exfoliated MoS_2 flakes were transferred onto the SiO_2/Si substrate. It was followed by fabrication of metal electrodes. Next, the MoS_2 surface was spun coated with ionophore. **e** Detection of arsenite and mercury ions with MoS_2 sensors functionalized with arsenite and mercury ionophore. **f** Band structure of MoS_2 sensors. Decrement of Fermi energy after absorption of arsenite. Increment of Fermi energy after Hg^{2+} absorption (Reprinted under creative common license, Copyright: American Institute of Physics) [16]

and colorimetric sensors based on WS_2 [25]. Recently, grain boundary-rich polycrystalline monolayer WS_2 was used for Hg^{2+} detection [58]. Heavy metal ion sensors based on the other semiconducting 2D TMDCs such as WSe_2 and $MoTe_2$ are rare.

Figure 5.3 shows device architecture, sensing mechanism and sensor output of some of the layered semiconducting TMDC-based HMI sensors.

5.3 Photosensors

Photosensors or photodetectors are the devices which can convert light signals into electrical signals [59]. Some of the widely used materials in photosensors are Si, GaAs, Bi_2Te_3, InGaAs, InGaP, CdS, CdSe, PbS, TlS_2, etc. Some of the well-known light-to-electrical signal conversion mechanisms and the figures of merit of the photodetectors are described below.

5.3.1 Electrical Transduction Mechanisms in Photosensors

There are several means through which light signal is converted into electrical signal. Some of the most conventional mechanisms are briefly stated below.

Photoelectric Effect

It is a phenomenon where electrons are ejected from metal surface when light of suitable wavelength falls on it. The wavelength is normally of the order of ultraviolet radiation. The photodetectors that work on this principle are normally called photo-emissive cells, e.g. photomultipliers, vacuum-type photocells and gas-filled-type photocells.

Photoconductive Effect

Photoconductive effect is the change in conductivity due to the intensity of light. Generation of electron–hole (*e-h*) pairs take place when light of energy greater than the band gap (E_g) is incident on the semiconducting material. When connected to a power source, the effect of change in intensity of the light can be observed in terms of conductivity in the semiconducting material. As the intensity of the light increases, more electron-hole (*e-h*) pairs are generated and hence the current through the material increases, implying increase in conductivity and decrease in resistance and vice versa.

Photothermoelectric Effect

Photothermoelectric effect is the generation of a voltage gradient (ΔV) across a channel made up of a semiconducting material when a light incident on the material leads to temperature gradient (ΔT) across the material. The change in voltage with respect to temperature gradient depends on the Seebeck coefficient of the material.

Photovoltaic Effect

Photovoltaic effect is the generation of a voltage in the system when a light of suitable energy is allowed to incident on the system. It also involves generation of *e-h* pairs as in the case of photoconductive effect; however, the generated electrons in this case are captured by an external circuit. The excited electrons then generate a potential difference or *e.m.f*. [60].

5.3.2 Photosensor Parameters

Like the chemical sensing parameters, there are some sets of parameters exclusively for photosensors. Some of them are as discussed below.

Responsivity

In a photodetector, the ratio of output electrical signal (received either in current or voltage form (I_{out}, V_{out}) to the input optical signal (expressed in terms of incident optical power (P_{in})) is called the responsivity of the photodetector. Though the expressions for responsivity can be different for different systems under study, in general, the current responsivity (R_I) and voltage responsivity (R_v) can be defined, respectively, as

$$R_I = I_{out}/P_{in} \text{ and } R_v = V_{out}/P_{in},$$

Here, the current I_{out} is the photogenerated current and V_{out} is the photogenerated voltage. They are also expressed as I_{ph} and V_{ph}, respectively.

Detectivity

Detectivity of a detector is similar to that of sensitivity of a sensor. It is the minimum power of light that can be detected by a photosensor. Jones [59] suggested that detectivity D of the detector is the inverse of the noise-equivalent power, denoted by P_{NEP}.

$$D = 1/P_{NEP},$$

where P_{NEP} is the power of light detected by the detector when the signal-to-noise ratio is unity. The unit D is W^{-1}. In order to compare the performance of photodetectors more efficiently, Jones put forward another definition, "Specific Detectivity" (D*), where he included the area (A) and electrical bandwidth (Δf); which can be formulated as [59]:

$$D^* = D\sqrt{(A\,\Delta f)}$$

Internal Quantum Efficiency (Quantum Yield)

It represents the number of the *e-h* pairs or charge carriers generated in the active material to the number of photons absorbed or the fraction of incident optical power that contributes to electron–hole pair generation.

External Quantum Efficiency

It is defined as the number of photogenerated charge carriers (coming out as photocurrent from the device) measured across the detector divided by the number of photons incident on the device. If P_{in} is the optical power of the incident light of wavelength λ, then $\frac{P_{in}}{\frac{hc}{\lambda}}$ gives the number of photons incident on the device per unit time. If I is the photocurrent measured in the circuit, then $\frac{I_{out}}{q}$ is the number of charge carriers per unit time. By definition,

$$\eta_{ext} = \frac{I_{out}/q}{P_{in}\lambda/hc}$$

Photoconductive Gain

It is defined as how fast the charge carriers in the photoconductive material transit from one end of the material to another to contribute to the photocurrent before recombination occurs [61].

In the following subsection, photodetectors are discussed on the basis of the primary sensing material used in the sensor. Thus, the subsubsections are named as MoS_2, WS_2, WSe_2, $MoSe_2$ and 2D van der Waals heterostructures.

5.3.3 Layered Semiconducting TMDC-Based Photosensors

MoS_2

Single-layer MoS_2 is a direct band gap semiconductor and has attractive quantum luminescence efficiency; higher than that of bilayer and layers of higher orders [9] and, by the year 2011, researchers were able to achieve the mobility of the order of 200 cm^2/V s [62] for MoS_2-based transistor. As a result, the first-ever phototransistor based on monolayer MoS_2 was built which was able to deliver responsivity of 7.5 mA/W at a back gate voltage of ~50 V and the mobility was ~0.3 cm^2/V s. Though this work created a history, the responsivity as well as the mobility was needed to be improved to make it comparable to that of the Si-based phototransistors. As such monolayer MoS_2 phototransistors with transparent top gate electrode were introduced for the first time to increase the mobility up to 80 cm^2/V s [63]. It was investigated that the photoresponse of monolayer MoS_2 is dominated by photothermoelectric effect rather than separation of electron–hole pair due to incident light [64]. It was realized gradually that to achieve a wider spectral response (from UV (ultraviolet) to NIR (near-infrared)), the layers should be increased from mono to few; it's because the energy band gap broaden up as the number of layer drops in 2D materials, due to quantum confinement effect. This new idea of few layers offered a considerable linear mobility >70 cm^2/V s, wider spectral response (<~900 nm) and higher photoresponsivity (>100 mA/W) [65]. Very high responsivity was observed for monolayer MoS_2-based visible range (400–600 nm) photodetector with maximum photoresponsivity of 880 A/W at a wavelength of 561 nm [66]. Small narrow bandwidth spectral photosensors with relatively high mobility of the order of 10^8 cm^2/V s were developed using graphene-MoS_2 heterostructures [67, 68]. In the following years, fabrication of MoS_2 has evolved from using scotch tape and chemical synthesis to Chemical Vapour Deposition (CVD) technique. Figure 5.4a, b show the schematic of a monolayer MoS_2-based photosensor device [69, 70]. Over time, the photoresponse analysis techniques have also been improved in order to make the fabrication of the sensor more precise and to receive better responses [71]. Now-a-days, research on MoS_2-based photosensing device has increased in unprecedented way. In recent years, notable MoS_2-based photosensors have been observed, including strain-tunable photosensors [72], amorphous MoS_2-based photosensors [73], photosensors based on vertically stacked MoS_2 with different phases [74], and a polarization photodetector based on MoS_2 [75].

WS_2

The first ever FET based on ultrathin WS_2 is claimed to be built by Hwang et al. [76]. They have also analyzed the effect of incident photons on FET performance. It was mathematically calculated by Liu et al. [77] that WS_2 band structure is favorable for FET applications and may outperform Si and other 2D-based FET. It was observed the

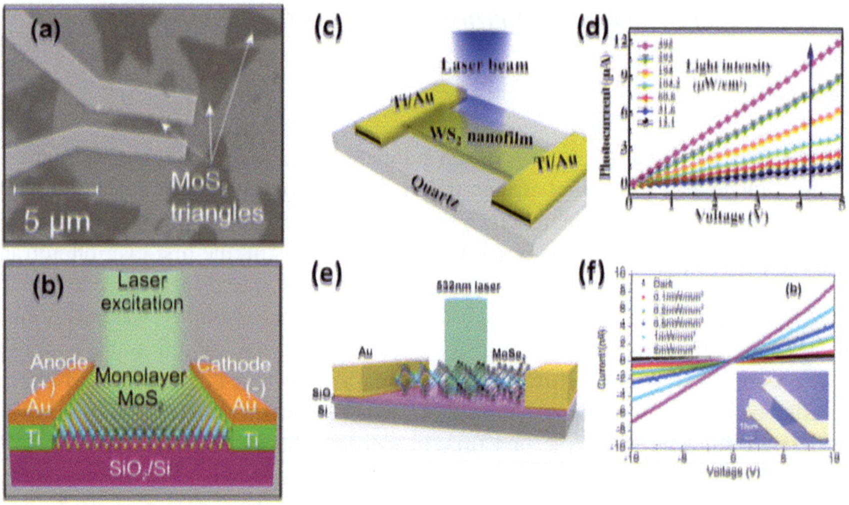

Fig. 5.4 **a** Monolayer MoS_2 triangles and **b** Architecture of monolayer MoS_2-based photosensor (Reprinted under Creative Common License, Copyright: Institute of Physics (IOP) Publications) [70] **c** Schematics of few-layered WS_2 nanofilm-based photodetectors and **d** Photocurrent obtained at different light intensities (Reprinted under Creative Common License, Copyright: Nature) [81] **e** Design architecture of few-layered $MoSe_2$-based photodetectors and **f** Current–voltage characteristics of the same in dark and with 532 nm laser having different fluxes (Reprinted under Creative Common License, Copyright: Royal Society of Chemistry) [87]

photocurrent generated by incident light with shorter wavelengths is higher in few-layered WS_2 photosensors compared to multilayer WS_2 photosensors [78]. WS_2 has been successfully used as window layer in solar cell applications [79]. Narrow band Schottky photodetector (metal–semiconductor–metal photodiode) with responsivity 777 mA/W and high detectivity of 4.94×10^{11} Jones (under 405 nm) was also fabricated [80]. Magnetron-sputtered WS_2 film-based photodetectors have also been investigated. Schematics of few-layered WS_2 nanofilm-based photodetectors and its response towards lights of different intensities are shown in Fig. 5.4c–d [81].

WSe_2

WSe_2 has also been proved to be a favourable photoactive material. Zhang et al. showed that a minimized semiconductor–metal junction contact in WSe_2 phototransistor offers high photoconductivity and specific detectivity [82]. Massicotte et al. then came up with more futuristic approach of using TMDCs in photosensing device by fabricating Graphene- WSe_2 -Graphene van der Waals heterostructure and also explained the photoresponse dynamics [83]. Of late, there has been a surge in van der waals heterostructure-based photosensors that will be discussed in later sections. Recently, in WSe_2-based photodetectors, researchers have tried to add specific features, e.g. flexibility [84], other varieties, e.g. photovoltaic photodetector (self-powered WSe_2–WSe_2 homojunction-based) [85] and configurations, e.g. metal–semiconductor–metal [86].

$MoSe_2$

Xia et al. reported for the first time $MoSe_2$-based single-layered photodetector with photoresponsivity of ~13 mA/W [87]. Later, Abderrahmane et al. improved the photoresponsivity by using few-layered $MoSe_2$ to about 97.1 A/W. In that work, the external quantum efficiency was exceptionally as high as 22,666% for a 532 nm laser excitation [88]. The device architecture and characteristics are shown in Fig. 5.4 (e, f) [87]. Now-a-days, heterostructure-based photosensors are trending in contemporary $MoSe_2$-based photosensor research.

2D van der Waal Heterostructure

In recent years, both homojunction and heterojunction 2D photodetectors have drawn the attention. Different types of heterojunctions that have been investigated are TMDC/TMDC, TMDC/bulk semiconductor, TMDC/other 2D semiconductors, Si–MoS_2-based photodetector possessing responsivity and detectivity of 76.1 A/W and 10^{12} Jones, respectively [89]. MoS_2-organic photodetectors such as MoS_2-pentacene have also been tried out with a fairly good responsivity of 0.27 A/W and a specific detectivity of 4.81×10^{11} Jones for 655 nm wavelength [90]. Self-driven broadband photosensor based on 2D–WS_2/GaAs type-II Zener heterojunction [91] with high responsivity of about 527 mA/W and large specific detectivity of 1.03×10^{14} Jones, avalanche photodiodes based on $MoTe_2$–WS_2–$MoTe_2$ with responsivity 6.02 A/W, external quantum efficiency of 1406% and specific detectivity of 7.24×10^9 Jones [92] are some of the worth mentioning works reported in last few years related to WS_2 heterostructure. Some heterostructured photodiode (WSe_2/MoS_2) have come with added functionalities like image sensor application [93]. Some few-layered $MoSe_2$/WSe_2 heterojunction photodetectors have the ability to extend up to the telecommunication band (1500 nm) with photoresponsivity as good as 127 mA/W [94]. A BP/WSe_2 vertical heterostructure was observed to be polarization sensitive in visible and infrared range with responsivity of ~10^3 A/W and ~5×10^{-1} A/W, respectively. It can offer specific detectivity up-to the order of ~10^{14} and ~10^{10} Jones respectively [95].

5.3.4 Scope in Dosimetry

Another important field of advanced sensing is dosimetry. Apart from application in material modifications and medical sciences, the nuclear plants are getting attention to fulfil the desired energy demand of the world [96]. With this, the importance for detection of nuclear radiation has increased. Besides, the knowledge of effect of radiation (electromagnetic, nuclear, ions) on different materials and devices is necessary for space and other applications [97, 98]. It is an established fact that most of the 2D materials are considerably tolerant to radiation (gamma, protons, electrons) [99]. Therefore, exploitation of 2D materials in this respect is extremely necessary. To validate the approach, graphene field effect transistor (GFET) was treated with different doses of gamma radiation, and it was observed that the Dirac point of graphene had shifted with the dose. The sensitivity was found to be ~1 V/kGy [100]. It is apparent that this work was motivated by the work of Alexandrou et al. [97] on the effect of radiation on GFET. Recently, Tawfik et al. introduced a genius idea of converting the resistivity of graphene in graphene-based gamma and beta radiation sensors to frequency output. The addition of this transduction technique is the unique point of this newly developed sensor [101]. Alteration of magnetic properties has already been witnessed in many TMDCs, because of the irradiation-induced effects. In its pristine state, the TMDCs are non-magnetic in nature, offering mild magnetism in few-layered structures due to edge-atom effect in low-dimensional structures. The evolution of spin density has been witnessed in radiation environment leading to the magnetic effect in the materials at large. Recently, Felix et al. showed the alteration of the B–H curve of monolayer WS_2 after exposure to different gamma dose [102]. The dosimeters based on hysteresis loop signifying different magnetic states would arise due to adequate chalcogen de-atomization, normally observed in radiation environment. Likewise, graphene oxide composite (GOC) based dosimeters have also been fabricated and utilized by researchers for proper monitoring of gamma doses [103]. There, photocurrent of the composite varies with respect to the variation of the gamma radiation dose under the application of constant electrical potential. The effect of gamma radiation is highly irreversible in nature. It has been observed that, like the GOCs the properties of the TMDC compounds can be influenced by gamma and ion radiation [104–107]. Therefore, semiconducting TMDC can be considered potential next-generation nuclear and ion radiation sensing materials.

5.4 Summary and Outlook

Out of several semiconducting TMDCs, MoS_2 and WS_2 have specifically been used in the HMI sensing applications. While several device architectures such as chemi-resistors, FETs, electrochemical cells have been adopted for MoS_2-based HMI sensors; WS_2-based HMI sensors are mostly based on fluorometry and colorimetry. During the last few years, a gradual shift of interest from electrical sensing to electrochemical sensing has been noticed towards 2D MoS_2-based detection of heavy metal ions in aqueous solutions. This may be due to less sophistication associated with electrochemical sensing techniques, as electrical sensing needs an accurate and precisely fabricated sensing unit. The interactions between TMDCs and HMIs are normally governed by van der Waals interactions, or direct chemical bonding. On the other hand, different semiconducting TMDCs such as MoS_2, WS_2, $MoSe_2$ and WSe_2 have been profoundly used in photosensor applications. With the advancement of material processing and fabrication techniques, in a duration of ten years or so, researchers have succeeded in enhancing the carrier mobility of synthesized, standard photosensors by about 10^9 times, responsivity by about 10^6 times and to reach detectivity up to 10^{14} Jones, and excessive quantum efficiency (22,666%). The investigations have already been shifted from a single material to homojunctions and heterojunctions of various kinds. With the development of synthesis methods, integration and measurement techniques, the sensors based on these materials have been setting new milestones every year. Undoubtedly, the world is on the verge of realizing the state-of-the-art engineering of semiconducting TMDCs for wide-ranging sensing applications.

Acknowledgement The authors are thankful to the central Library, Tezpur University for providing access to numerous journals. The financial support received from IUAC, New Delhi (UFR-62312) is highly appreciated. One of the Authors (AN) is thankful to Tezpur University for providing institutional fellowships.

References

1. Banica, F.G.: Chemical Sensors and Biosensors: Fundamentals and Applications. Wiley, New York (2012)
2. Fraden, J.: Handbook of Modern Sensors. Springer, Berlin (2010). ISBN: 978-1-44419-6466-3
3. Joshi, N., Hayasaka, T., Liu, Y., Liu, H., Oliveira, O.N., Lin, L.: A review on chemiresistive room temperature gas sensors based on metal oxide nanostructures, graphene and 2D transition metal dichalcogenides. Microchim. Acta. Acta **185**(4), 1–16 (2018)
4. Gan, X., Zhao, H., Schirhagl, R., Quan, X.: Two-dimensional nanomaterial-based sensors for heavy metal ions. Microchim. Acta. Acta **185**(10), 1–30 (2018)
5. Meng, Z., Stolz, R.M., Mendecki, L., Mirica, K.A.: Electrically transduced chemical sensors based on two-dimensional nanomaterials. Chem. Rev. **119**(1), 478–598 (2019)
6. Choi, S.J., Kim, I.D.: Recent developments in 2D nanomaterials for chemiresistive-type gas sensors. Electron. Mater. Lett. **14**(3), 221–260 (2018)
7. Chhowalla, M., Shin, H.S., Eda, G., Li, L.J., Loh, K.P., Zhang, H.: The chemistry of two-dimensional layered transition metal dichalcogenide nanosheets. Nat. Chem. **5**(4), 263–275 (2013)
8. Mak, K.F., Lee, C., Hone, J., Shan, J., Heinz, T.F.: Atomically thin MoS_2: a new direct-gap semiconductor. Phys. Rev. Lett. **105**(13), 136805 (2010)
9. Splendiani, A., Sun, L., Zhang, Y., Li, T., Kim, J., Chim, C.Y., Galli, G., Wang, F.: Emerging photoluminescence in monolayer MoS_2. Nano Lett. **10**(4), 1271–1275 (2010)
10. Ping, J., Fan, Z., Sindoro, M., Ying, Y., Zhang, H.: Recent advances in sensing applications of two-dimensional transition metal dichalcogenide nanosheets and their composites. Adv. Func. Mater. **27**(19), 1605817 (2017)
11. Lee, J.H., Avsar, A., Jung, J., Tan, J.Y., Watanabe, K., Taniguchi, T., Natarajan, S., Eda, G., Adam, S., Castro Neto, A.H., Özyilmaz, B.: Van der Waals force: a dominant factor for reactivity of graphene. Nano Lett. **15**(1), 319–325 (2015)
12. Neog, A., Biswas, R.: WS_2 nanosheets as a potential candidate towards sensing heavy metal ions: a new dimension of 2D materials. Mater. Res. Bull. **144**, 111471 (2021)

13. Yew, Y.T., Ambrosi, A., Pumera, M.: Nitroaromatic explosives detection using electrochemically exfoliated graphene. Sci. Rep. **6**(1), 1–11 (2016)
14. Ko, M., Aykanat, A., Smith, M.K., Mirica, K.A.: Drawing sensors with ball-milled blends of metal-organic frameworks and graphite. Sensors **17**(10), 2192 (2017)
15. Jiang, S., Cheng, R., Ng, R., Huang, Y., Duan, X.: Highly sensitive detection of mercury (II) ions with few-layer molybdenum disulfide. Nano Res. **8**(1), 257–262 (2015)
16. Li, P., Zhang, D., Sun, Y.E., Chang, H., Liu, J., Yin, N.: Towards intrinsic MoS_2 devices for high performance arsenite sensing. Appl. Phys. Lett. **109**(6), 063110 (2016)
17. Chiu, S.W., Tang, K.T.: Towards a chemiresistive sensor-integrated electronic nose: a review. Sensors **13**(10), 14214–14247 (2013)
18. Majumder, S., Pramanick, B., Mondal, M., Bhattacharyya, T.K.: Carbon electrodes for low power heavy metal sensing using MoS_2 based resistive sensors. In: 2018 IEEE Sensors, pp. 1–4. IEEE (2018)
19. Mao, S., Chang, J., Pu, H., Lu, G., He, Q., Zhang, H., Chen, J.: Two-dimensional nanomaterial-based field-effect transistors for chemical and biological sensing. Chem. Soc. Rev. **46**(22), 6872–6904 (2017)
20. Aswathi, R., Sandhya, K.Y.: Ultrasensitive and selective electrochemical sensing of Hg (II) ions in normal and sea water using solvent exfoliated MoS_2: affinity matters. J. Mater. Chem. A **6**(30), 14602–14613 (2018)
21. Gan, X., Zhao, H., Wong, K.Y., Lei, D.Y., Zhang, Y., Quan, X.: Covalent functionalization of MoS_2 nanosheets synthesized by liquid phase exfoliation to construct electrochemical sensors for Cd (II) detection. Talanta **182**, 38–48 (2018)
22. Shi, J.J., Zhu, J.C., Zhao, M., Wang, Y., Yang, P., He, J.: Ultrasensitive photoelectrochemical aptasensor for lead ion detection based on sensitization effect of CdTe QDs on MoS_2–CdS: Mn nanocomposites by the formation of G-quadruplex structure. Talanta **183**, 237–244 (2018)
23. Wang, Y., Hu, J., Zhuang, Q., Ni, Y.: Label-free fluorescence sensing of lead (II) ions and sulfide ions based on luminescent molybdenum disulfide nanosheets. ACS Sustain. Chem. Eng. **4**(5), 2535–2541 (2016)
24. Zuo, X., Zhang, H., Zhu, Q., Wang, W., Feng, J., Chen, X.: A dual-color fluorescent biosensing platform based on WS_2 nanosheet for detection of Hg^{2+} and Ag^{+}. Biosens. Bioelectron.. Bioelectron. **85**, 464–470 (2016)
25. Tang, Y., Hu, Y., Yang, Y., Liu, B., Wu, Y.: A facile colorimetric sensor for ultrasensitive and selective detection of Lead (II) in environmental and biological samples based on intrinsic peroxidase-mimic activity of WS_2 nanosheets. Anal. Chim. Acta **1106**, 115–125 (2020)
26. Chapter 6, Sensitivity and limit of detection, Editor(s): Massart, D.L., Dijkstra, A., Kaufman, L.: Techniques and Instrumentation in Analytical Chemistry, vol. 1, pp. 143–156. Elsevier. ISSN0167-9244, ISBN9780444417435, https://doi.org/10.1016/S0167-9244(08)700507. https://www.sciencedirect.com/science/article/pii/S0167924408700507
27. Zipser, L.: Selectivity of sensor systems. Sens. Actuat. A **37**, 286–289 (1993)
28. Ferguson, J.E.: The Heavy Elements: Chemistry, Environmental Impact and Health Effects (No. 628.53 F4) (1990)
29. Singh, M.R.: Impurities-heavy metals: IR perspective. Int. J. Phys. Sci. **5**(4), 1045–1058 (2007)
30. Duffus, J.H.: "Heavy Metals"–a meaningless term. Chem. Int.-Newsmag. IUPAC **23**(6), 163–167 (2001)
31. World Health Organization.: Trace Elements in Human Nutrition and Health. World Health Organization (1996)
32. Nriagu, J.O.: A global assessment of natural sources of atmospheric trace metals. Nature **338**, 47–49 (1989)
33. Goyer, R.A., Clarkson, T.W.: Toxic effects of metals. Casarett Doull's Toxicol.: Basic Sci. Poisons **5**, 691–736 (1996)
34. Chang, L.W., Magos, L., Suzuki, T. (eds.): Toxicology of Metals. CRC, Boca Raton, FL (1996)

35. Abernathy, C.O., Liu, Y.P., Longfellow, D., Aposhian, H.V., Beck, B., Fowler, B., Goyer, R., Menzer, R., Rossman, T., Thompson, C., Waalkes, M.: Arsenic: health effects, mechanisms of actions, and research issues. Environ. Health Perspect.Perspect. **107**(7), 593–597 (1999)
36. Hong, Y.S., Song, K.H., Chung, J.Y.: Health effects of chronic arsenic exposure. J. Prev. Med. Public Health **47**(5), 245 (2014)
37. Genchi, G., Sinicropi, M.S., Lauria, G., Carocci, A., Catalano, A.: The effects of cadmium toxicity. Int. J. Environ. Res. Public Health **17**(11), 3782 (2020)
38. Levallois, P., Barn, P., Valcke, M., Gauvin, D., Kosatsky, T.: Public health consequences of lead in drinking water. Curr. Environ. Health Rep. **5**(2), 255–262 (2018)
39. Rice, K.M., Walker, E.M., Jr., Wu, M., Gillette, C., Blough, E.R.: Environmental mercury and its toxic effects. J. Prev. Med. Public Health **47**(2), 74 (2014)
40. Wilbur, S., Abadin, H., Fay, M., Yu, D., Tencza, B., Ingerman, L., Klotzbach, J., James, S.: Toxicological Profile for Chromium (2012)
41. Yuen, H.W., Becker, W.: Iron toxicity. StatPearls [Internet] (2020)
42. Royer, A., Sharman, T.: Copper Toxicity (2020)
43. Das, K.K., Reddy, R.C., Bagoji, I.B., Das, S., Bagali, S., Mullur, L., Khodnapur, J.P., Biradar, M.S.: Primary concept of nickel toxicity–an overview. J. Basic Clin. Physiol. Pharmacol.Pharmacol. **30**(2), 141–152 (2019)
44. Perkins, F.K., Friedman, A.L., Cobas, E., Campbell, P.M., Jernigan, G.G., Jonker, B.T.: Chemical vapor sensing with monolayer MoS_2. Nano Lett. **13**(2), 668–673 (2013)
45. Ovcharenko, R., Dedkov, Y., Voloshina, E.: Adsorption of NO_2 on WSe_2: DFT and photoelectron spectroscopy studies. J. Phys. Condens. MatterCondens. Matter **28**(36), 364003 (2016)
46. Chen, B., Sahin, H., Suslu, A., Ding, L., Bertoni, M.I., Peeters, F.M., Tongay, S.: Environmental changes in $MoTe_2$ excitonic dynamics by defects-activated molecular interaction. ACS Nano **9**(5), 5326–5332 (2015)
47. Feng, Z., Xie, Y., Wu, E., Yu, Y., Zheng, S., Zhang, R., Chen, X., Sun, C., Zhang, H., Pang, W., Liu, J.: Enhanced sensitivity of $MoTe_2$ chemical sensor through light illumination. Micromachines **8**(5), 155 (2017)
48. Li, P., Zhang, D., Wu, Z.: Flexible MoS_2 sensor arrays for high performance label-free ion sensing. Sens. Actuat. A **286**, 51–58 (2019)
49. Hwang, J.H., Islam, M.A., Choi, H., Ko, T.J., Rodriguez, K.L., Chung, H.S., Jung, Y., Lee, W.H.: Improving electrochemical Pb^{2+} detection using a vertically aligned 2D MoS_2 nanofilm. Anal. Chem. **91**(18), 11770–11777 (2019)
50. Mohammadi, A., Heydari-Bafrooei, E., Foroughi, M.M., Mohammadi, M.: Heterostructured Au/MoS_2–MWCNT nanoflowers: a highly efficient support for the electrochemical aptasensing of solvated mercuric ion. Microchem. J.. J. **158**, 105154 (2020)
51. Nigam, A., Goel, N., Bhat, T.N., Rahman, M.T., Dolmanan, S.B., Qiao, Q., Tripathy, S., Kumar, M.: Real time detection of Hg^{2+} ions using MoS_2 functionalized AlGaN/GaN high electron mobility transistor for water quality monitoring. Sens. Actuat. B Chem. **309**, 127832 (2020)
52. Bazylewski, P., Van Middelkoop, S., Divigalpitiya, R., Fanchini, G.: Solid-state chemiresistors from two-dimensional MoS_2 nanosheets functionalized with l-Cysteine for In-line sensing of Part-Per-Billion Cd^{2+} ions in drinking water. ACS Omega **5**(1), 643–649 (2019)
53. Sun, Y.F., Sun, J.H., Wang, J., Pi, Z.X., Wang, L.C., Yang, M., Huang, X.J.: Sensitive and anti-interference stripping voltammetry analysis of Pb (II) in water using flower-like MoS_2/rGO composite with ultra-thin nanosheets. Anal. Chim. Acta **1063**, 64–74 (2019)
54. Xue, T., Qi, K., Hu, C.: Novel SPR sensing platform based on superstructure MoS_2 nanosheets for ultrasensitive detection of mercury ion. Sens. Actuat. B Chem. **284**, 589–594 (2019)
55. Gea, J., Genga, X., Dua, Y., Chena, J., Zhanga, L., Baia, D., Jia, D., Hua, Y., Lia, Z.: Highly sensitive fluorescence detection of mercury(II) ions based on WS_2 nanosheets and T7 exonuclease assisted cyclic enzymatic amplification. Sens. Actuat. B **249**, 189–194 (2017)
56. Feng, B., Liu, X., Zheng, Y., Xiao, Q., Wu, N., Wang, S.: A label-free fluorescent probe for Hg^{2+} based on boron-and nitrogen-doped photoluminescent WS_2. RSC Adv. **6**(55), 49668–49674 (2016)

57. Neog, A., Deb, S., Biswas, R.: Atypical electrical behavior of few layered WS_2 nanosheets based platform subject to heavy metal ion treatment. Mater. Lett. **268**, 127597 (2020)
58. Liu, L., Ye, K., Lin, C., Jia, Z., Xue, T., Nie, A., Cheng, Y., Xiang, J., Mu, C., Wang, B., Wen, F.: Grain-boundary-rich polycrystalline monolayer WS_2 film for attomolar-level Hg^{2+} sensors. Nat. Commun.Commun. **12**(1), 1–8 (2021)
59. Photodetectors. In: Fundamentals of Solid State Engineering. Springer, Boston, MA (2002). https://doi.org/10.1007/0-306-47567-7_16, Print ISBN-978-0-7923-7629-3, online ISBN-978-0-306-47567-2
60. Koppens, F.H.L., Mueller, T., Avouris, P., Ferrari, A.C., Vitiello, M.S., Polini, M.: Photodetectors based on graphene, other two-dimensional materials and hybrid systems. Nat. Nanotechnol.Nanotechnol. **9**(10), 780–793 (2014)
61. Pratiyush, A.S., Krishnamoorthy, S., Muralidharan, R., Rajan, S. and Nath, D.N.: Advances in Ga_2O_3 solar-blind UV photodetectors. In: Gallium Oxide, pp. 369–399. Elsevier (2019)
62. Radisavljevic, B., Radenovic, A., Brivio, J., Giacometti, V., Kis, A.: Single-layer MoS_2 transistors. Nat. Nanotechnol.Nanotechnol. **6**(3), 147–150 (2011)
63. Lee, H.S., Min, S.W., Chang, Y.G., Park, M.K., Nam, T., Kim, H., Kim, J.H., Ryu, S., Im, S.: MoS_2 nanosheet phototransistors with thickness-modulated optical energy gap. Nano Lett. **12**(7), 3695–3700 (2012)
64. Buscema, M., Barkelid, M., Zwiller, V., van der Zant, H.S., Steele, G.A., Castellanos-Gomez, A.: Large and tunable photothermoelectric effect in single-layer MoS_2. Nano Lett. **13**(2), 358–363 (2013)
65. Choi, W., Cho, M.Y., Konar, A., Lee, J.H., Cha, G.B., Hong, S.C., Kim, S., Kim, J., Jena, D., Joo, J., Kim, S.: High-detectivity multilayer MoS_2 phototransistors with spectral response from ultraviolet to infrared. Adv. Mater. **24**(43), 5832–5836 (2012)
66. Lopez-Sanchez, O., Lembke, D., Kayci, M., Radenovic, A., Kis, A.: Ultrasensitive photodetectors based on monolayer MoS_2. Nat. Nanotechnol.Nanotechnol. **8**(7), 497–501 (2013)
67. Roy, K., Padmanabhan, M., Goswami, S., Sai, T.P., Ramalingam, G., Raghavan, S., Ghosh, A.: Graphene–MoS_2 hybrid structures for multifunctional photoresponsive memory devices. Nat. Nanotechnol.Nanotechnol. **8**(11), 826–830 (2013)
68. Zhang, W., Chuu, C.P., Huang, J.K., Chen, C.H., Tsai, M.L., Chang, Y.H., Liang, C.T., Chen, Y.Z., Chueh, Y.L., He, J.H., Chou, M.Y.: Ultrahigh-gain photodetectors based on atomically thin graphene-MoS_2 heterostructures. Sci. Rep. **4**(1), 1–8 (2014)
69. Zhang, W., Huang, J.K., Chen, C.H., Chang, Y.H., Cheng, Y.J., Li, L.J.: High-gain phototransistors based on a CVD MoS_2 monolayer. Adv. Mater. **25**(25), 3456–3461 (2013)
70. Perea-López, N., Lin, Z., Pradhan, N.R., Iñiguez-Rábago, A., Elías, A.L., McCreary, A., Lou, J., Ajayan, P.M., Terrones, H., Balicas, L., Terrones, M.: CVD-grown monolayered MoS_2 as an effective photosensor operating at low-voltage. 2D Mater. **1**(1), 011004 (2014)
71. Wu, C.C., Jariwala, D., Sangwan, V.K., Marks, T.J., Hersam, M.C., Lauhon, L.J.: Elucidating the photoresponse of ultrathin MoS_2 field-effect transistors by scanning photocurrent microscopy. J. Phys. Chem. Lett. **4**(15), 2508–2513 (2013)
72. Gant, P., Huang, P., de Lara, D.P., Guo, D., Frisenda, R., Castellanos-Gomez, A.: A strain tunable single-layer MoS_2 photodetector. Mater. Today **27**, 8–13 (2019)
73. Huang, Z., Zhang, T., Liu, J., Zhang, L., Jin, Y., Wang, J., Jiang, K., Fan, S., Li, Q.: Amorphous MoS_2 photodetector with ultra-broadband response. ACS Appl. Electron. Mater. **1**(7), 1314–1321 (2019)
74. Wang, W., Zeng, X., Warner, J.H., Guo, Z., Hu, Y., Zeng, Y., Lu, J., Jin, W., Wang, S., Lu, J., Zeng, Y.: Photoresponse-bias modulation of a high-performance MoS_2 photodetector with a unique vertically stacked 2H–MoS_2/1T@ 2H–MoS_2 structure. ACS Appl. Mater. Interfaces **12**(29), 33325–33335 (2020)
75. Chen, S., Cao, R., Chen, X., Wu, Q., Zeng, Y., Gao, S., Guo, Z., Zhao, J., Zhang, M., Zhang, H.: Anisotropic plasmonic nanostructure induced polarization photoresponse for MoS_2-based photodetector. Adv. Mater. Interfaces **7**(9), 1902179 (2020)
76. Sik Hwang, W., Remskar, M., Yan, R., Protasenko, V., Tahy, K., Doo Chae, S., Zhao, P., Konar, A., Xing, H., Seabaugh, A., Jena, D.: Transistors with chemically synthesized layered

semiconductor WS_2 exhibiting 105 room temperature modulation and ambipolar behavior. Appl. Phys. Lett. **101**(1), 013107 (2012)
77. Liu, L., Kumar, S.B., Ouyang, Y., Guo, J.: Performance limits of monolayer transition metal dichalcogenide transistors. IEEE Trans. Electron Dev. **58**(9), 3042–3047 (2011)
78. Perea-López, N., Elías, A.L., Berkdemir, A., Castro-Beltran, A., Gutiérrez, H.R., Feng, S., Lv, R., Hayashi, T., López-Urías, F., Ghosh, S., Muchharla, B.: Photosensor device based on few-layered WS_2 films. Adv. Func. Mater. **23**(44), 5511–5517 (2013)
79. Rafiq, M.K.S.B., Amin, N., Alharbi, H.F., Luqman, M., Ayob, A., Alharthi, Y.S., Alharthi, N.H., Bais, B., Akhtaruzzaman, M.: WS_2: a new window layer material for solar cell application. Sci. Rep. **10**(1), 1–11 (2020)
80. Gao, W., Zhang, S., Zhang, F., Wen, P., Zhang, L., Sun, Y., Chen, H., Zheng, Z., Yang, M., Luo, D., Huo, N.: 2D WS_2 based asymmetric schottky photodetector with high performance. Adv. Electron. Mater. **7**(7), 2000964 (2021)
81. Zeng, L., Tao, L., Tang, C., Zhou, B., Long, H., Chai, Y., Lau, S.P., Tsang, Y.H.: High-responsivity UV-vis photodetector based on transferable WS_2 film deposited by magnetron sputtering. Sci. Rep. **6**(1), 1–8 (2016)
82. Zhang, W., Chiu, M.H., Chen, C.H., Chen, W., Li, L.J., Wee, A.T.S.: Role of metal contacts in high-performance phototransistors based on WSe_2 monolayers. ACS Nano **8**(8), 8653–8661 (2014)
83. Massicotte, M., Schmidt, P., Vialla, F., Schädler, K.G., Reserbat-Plantey, A., Watanabe, K., Taniguchi, T., Tielrooij, K.J., Koppens, F.H.: Picosecond photoresponse in van der Waals heterostructures. Nat. Nanotechnol.Nanotechnol. **11**(1), 42–46 (2016)
84. Zheng, Z., Zhang, T., Yao, J., Zhang, Y., Xu, J., Yang, G.: Flexible, transparent and ultra-broadband photodetector based on large-area WSe_2 film for wearable devices. Nanotechnology **27**(22), 225501 (2016)
85. Tan, C., Wang, H., Zhu, X., Gao, W., Li, H., Chen, J., Li, G., Chen, L., Xu, J., Hu, X., Li, L.: A self-powered photovoltaic photodetector based on a lateral WSe_2–WSe_2 homojunction. ACS Appl. Mater. Interfaces **12**(40), 44934–44942 (2020)
86. Zhou, C., Raju, S., Li, B., Chan, M., Chai, Y., Yang, C.Y.: Self-driven metal–semiconductor–metal WSe_2 photodetector with asymmetric contact geometries. Adv. Func. Mater. **28**(45), 1802954 (2018)
87. Xia, J., Huang, X., Liu, L.Z., Wang, M., Wang, L., Huang, B., Zhu, D.D., Li, J.J., Gu, C.Z., Meng, X.M.: CVD synthesis of large-area, highly crystalline $MoSe_2$ atomic layers on diverse substrates and application to photodetectors. Nanoscale **6**(15), 8949–8955 (2014)
88. Abderrahmane, A., Ko, P.J., Thu, T.V., Ishizawa, S., Takamura, T., Sandhu, A.: High photosensitivity few-layered $MoSe_2$ back-gated field-effect phototransistors. Nanotechnology **25**(36), 365202 (2014)
89. Shin, G.H., Park, J., Lee, K.J., Lee, G.B., Jeon, H.B., Choi, Y.K., Yu, K., Choi, S.Y.: Si–MoS_2 vertical heterojunction for a photodetector with high responsivity and low noise equivalent power. ACS Appl. Mater. Interfaces **11**(7), 7626–7634 (2019)
90. Peng, Y., Ding, R., Ren, Q., Xu, S., Sun, L., Wang, Y., Lu, F.: High performance photodiode based on MoS_2/pentacene heterojunction. Appl. Surf. Sci. **459**, 179–184 (2018)
91. Jia, C., Huang, X., Wu, D., Tian, Y., Guo, J., Zhao, Z., Shi, Z., Tian, Y., Jie, J., Li, X.: An ultrasensitive self-driven broadband photodetector based on a 2D-WS_2/GaAs type-II Zener heterojunction. Nanoscale **12**(7), 4435–4444 (2020)
92. Ouyang, T., Wang, X., Liu, S., Chen, H., Deng, S.: A complete two-dimensional avalanche photodiode based on $MoTe_2$- WS_2- $MoTe_2$ heterojunctions with ultralow dark current. Front. Mater. 303 (2021)
93. Ahn, J., Kang, J.H., Park, M.C., Hwang, D.K.: All 2D WSe_2/MoS_2 heterojunction photodiode and its image sensor application. Opt. Lett. **45**(16), 4531–4534 (2020)
94. Xue, H., Wang, Y., Dai, Y., Kim, W., Jussila, H., Qi, M., Susoma, J., Ren, Z., Dai, Q., Zhao, J., Halonen, K.: A $MoSe_2$/WSe_2 heterojunction-based photodetector at telecommunication wavelengths. Adv. Func. Mater. **28**(47), 1804388 (2018)

95. Ye, L., Wang, P., Luo, W., Gong, F., Liao, L., Liu, T., Tong, L., Zang, J., Xu, J., Hu, W.: Highly polarization sensitive infrared photodetector based on black phosphorus-on-WSe_2 photogate vertical heterostructure. Nano Energy **37**, 53–60 (2017)
96. Abu-Khader, M.M.: Recent advances in nuclear power: a review. Prog. Nucl. EnergyNucl. Energy **51**(2), 225–235 (2009)
97. Alexandrou, K., Masurkar, A., Edrees, H., Wishart, J.F., Hao, Y., Petrone, N., Hone, J., Kymissis, I.: Improving the radiation hardness of graphene field effect transistors. Appl. Phys. Lett. **109**(15), 153108 (2016)
98. Neog, A., Biswas, R.: Phase transition and formation of inorganic fullerene type nanostructure in WS_2 nanosheet system under laser excitation. Scripta Mater. **216**, 114729 (2022)
99. Vogl, T., Sripathy, K., Sharma, A., Reddy, P., Sullivan, J., Machacek, J.R., Zhang, L., Karouta, F., Buchler, B.C., Doherty, M.W., Lu, Y.: Radiation tolerance of two-dimensional material-based devices for space applications. Nat. Commun.Commun. **10**(1), 1–10 (2019)
100. Jain, S., Gajarushi, A.S., Gupta, A., Rao, V.R.: A passive gamma radiation dosimeter using graphene field effect transistor. IEEE Sens. J. **20**(6), 2938–2944 (2019)
101. Tawfik, M.W., Sharaf, A., Serry, M.: Monolayer graphene radiation sensor with backend RF ring oscillator transducer. Nanomaterials **12**(3), 305 (2022)
102. Felix, J.F., da Silva, A.F., da Silva, S.W., Qu, F., Qiu, B., Ren, J., de Azevedo, W.M., Henini, M., Huang, C.C.: A comprehensive study on the effects of gamma radiation on the physical properties of a two-dimensional WS_2 monolayer semiconductor. Nanoscale Horizons **5**(2), 259–267 (2020)
103. Feizi, S., Malekie, S., Rahighi, R., Tayyebi, A., Ziaie, F.: Evaluation of dosimetric characteristics of graphene oxide/PVC nanocomposite for gamma radiation applications. Radiochim. Acta. Acta **105**(2), 161–170 (2017)
104. Sarmah, H.J., Mohanta, D., Saha, A.: Perceptible exciton-to-trion conversion and signature of defect mediated vibronic modes and spin relaxation in nanoscale WS_2 exposed to γ-rays. Nanotechnology **31**(28), 285706 (2020)
105. Sarmah, H.J., Mohanta, D., Saha, A.: Highly symmetric and delayed excitonic emission response and space charge-limited current transport in β-irradiated WSe_2 and WS_2 nanoflakes. J. Mater. Res. **36**(4), 870–883 (2021)
106. Isherwood, L.H., Athwal, G., Spencer, B.F., Casiraghi, C., Baidak, A.: Gamma radiation-induced oxidation, doping, and etching of two-dimensional MoS_2 crystals. J. Phys. Chem. C **125**(7), 4211–4222 (2021)
107. Wu, X., Zheng, X., Zhang, G., Chen, X., Ding, J.: Tightly bound trion and bandgap engineering via γ-ray irradiation in the monolayer transition metal dichalcogenide WSe_2. Nanotechnology **32**(30), 3 (2021)

Chapter 6
Light Scattering by One-Dimensional ZnO Nanorods and Their Applications in Optical Sensing

Tanujjal Bora and Waleed S. Mohammed

Abstract Zinc oxide (ZnO) is a versatile metal oxide semiconductor material with remarkable optical and electrical properties. Particularly, the one-dimensional (1D) ZnO nanorods are of great interest because of their superior as well as tunable optical scattering properties and availability of simple synthesis methods using which they can be grown on a variety of substrates. This makes them a top choice material for designing low-cost, sensitive, and robust optical sensors. The chapter discusses the light scattering properties of 1D ZnO nanorods and their utilization in designing optical sensors for the detection of chemical and biological substances. The physical and chemical properties of ZnO and the theory of light scattering by 1D ZnO nanorods are explained. The fundamental sensing mechanism of the optical sensors based on light scattering by ZnO nanorods is described to provide a clear insight into the role of the ZnO nanorods in these optical sensors. A few examples of various optical sensors developed to detect chemical and biological species are given, and the strategies for improving the sensitivity of sensors and reducing the cost are also discussed. A future outlook of such emerging optical sensors is discussed at the end.

Keywords ZnO · Nanoscale · Optical sensing · Light scattering

6.1 Introduction

Rapid industrialization and modernization of our lifestyles have increased our dependencies on various kinds of chemicals. This eventually leads to the environmental pollutions affecting the health of all living organisms including humans [66, 110].

T. Bora (✉)
Center of Excellence in Nanotechnology, School of Engineering and Technology, Asian Institute of Technology, PO Box 4, Klong Luang, Pathum Thani 12120, Thailand
e-mail: tbora@ait.ac.th

W. S. Mohammed
Center of Research in Optoelectronics, Communication and Computational Systems (BU-CROCCS), School of Engineering, Bangkok University, Pathum Thani 12120, Thailand

D. Mohanta and P. Chakraborty (eds.), *Nanoscale Matter and Principles for Sensing and Labeling Applications*, Advanced Structured Materials 206,
https://doi.org/10.1007/978-981-99-7848-9_6

Therefore, monitoring of chemicals or pollutants in the environment, food, and other agricultural products is essential. Additionally, the detection of biomarkers and other biological species for health monitoring and early disease control is becoming more popular these days, since it allows to reduce the morbidity rates ensuring a healthy public life [44]. Hence, the field of sensors is an exponentially growing area where gradually novel concepts are being evolved to detect various target chemicals. A sensor is simply an analytical device which can sense or detect the presence of a chemical or physical species by means of transforming the information partially or completely into a computable signal. For practical applications, a sensor should be able to detect very low concentrations of analytes with high sensitivity, good repeatability, and selectivity, and most importantly, they should be durable and cost-effective.

Numerous kinds of chemical and biological sensors with distinct working principles and sensor materials are available today. In the past few years, the tremendous developments in the field of nanomaterials and nanotechnology have significantly revolutionized the sensor fields, by making the sensors more sensitive, compact, and affordable [4, 83, 93]. Among the different sensor materials, semiconductor metal oxides are widely used to develop sensors. Semiconductor metal oxides exhibit remarkable optical, electronic, mechanical, and thermal properties, and many of them show biocompatibility properties beneficial for the development of biomedical sensors [22]. Zinc oxide (ZnO) is one of these semiconductor metal oxides which have received significant attention of the scientific community. Especially, ZnO nanomaterials are of great interest as they have several benefits, such as bandgap tunability, easy synthesis, biocompatibility, high electron mobility, piezoelectric property, good stability, and large surface-to-volume ratio. As a result, ZnO nanomaterials are used extensively to design sensors based on their electrical, optical, or mechanical properties [38, 63, 87].

Among the various types of ZnO nanomaterial-based sensors, optical sensors are very interesting. Performance of sensors based on electrical conductance is generally affected by the environmental factors, such as moisture content and environmental temperature fluctuations. Optical signals, on the other hand, are less affected by these factors and thus can be used for a more robust sensor design [97]. Light scattering by nanostructured materials is an interesting optical phenomenon, which is affected by the scatterer's geometry and density. Since ZnO nanomaterials can be synthesized to have distinct shapes and sizes using simple wet chemical methods [6], they are ideal candidates to exploit and have controlled light scattering. One-dimensional (1D) ZnO nanorods have shown promising results in terms of light scattering, which can be tuned by varying the length and density of nanorods on a surface [61]. Such systems have recently been utilized to develop optical sensors and used for the detection of various chemical and biological species [14, 28, 95]. This book chapter discusses the physical and chemical properties of ZnO nanomaterials, particularly focusing on the 1D ZnO nanorods structure, and explains the light scattering theory of these nanorods. It then discusses the application of the light scattering property of ZnO nanorods toward the development of highly sensitive and innovative optical sensors for chemical and biological detection. The chapter provides a better insight into the light scattering properties of 1D ZnO nanorods and their potential applications toward designing unique and cost-effective optical sensors. At the end, the chapter also provides a future outlook for this promising concept.

6.2 Zinc Oxide: Physical and Chemical Properties

Zinc oxide (ZnO) is a group II–VI wide and direct bandgap (E_g) semiconductor material with $E_g = 3.37$ eV. It has a hexagonal wurtzite crystal structure having lattice parameters, a = 3.296 Å and c = 5.2065 Å with c/a ratio of about 1.6, as shown in Fig. 6.1. The crystal is composed of alternately stacked and tetrahedrally coordinated Zn^{2+} and O^{2-} ions along the c-axis of the crystal. The crystal shows partial polar features with the (0001) basal plane exhibiting maximum polarity due to the alternating arrangements of Zn^{2+} and O^{2-} lattice points [101]. ZnO also exhibits both piezoelectric and pyroelectric properties due to the lack of inversion symmetry in the tetrahedral coordination of the crystal [73] making it an interesting material for several applications. Due to the polar Zn–O bonds, the crystal planes of ZnO are electrically charged and show different surface energies. Unlike other crystalline materials with polar planes, where surface reconstruction at atomic level occurs to maintain the charge neutrality in the crystal, ZnO polar planes do not show any such surface reconstruction and they are atomically flat and stable [18, 26]. This anomaly of ZnO surface physics is not completely understood till date [72, 75]. ZnO also shows high electron mobility at room temperature and has a high exciton binding energy of 60 meV making it interesting for optoelectronic applications. The various physical and chemical properties of ZnO are summarized in Table 6.1 below.

A range of nanostructures of ZnO have been successfully synthesized, such as nanoparticles [68], nanorods and nanowires [7, 24], nanotubes [55], nanocombs [105, 115], nanobelts [102], nanorings [78], and nanohelices [33, 34]. The various applications where ZnO is widely used mainly include solar cells [15, 51], photonic crystals [18, 85], photodetectors [90], light-emitting diodes (LEDs) [46, 114], photodiodes [62], surface acoustic wave filters [26], photocatalysis [9, 16], gas sensors [87, 99], optical modulator waveguides [106], and varistors [58]. ZnO is also one of the most potential materials for applications in the biomedical field due to its biocompatibility, biodegradability, and non-toxicity [118].

6.2.1 One-Dimensional ZnO Nanorods

One-dimensional (1D) nanostructures of ZnO show remarkable chemical and physical properties and are widely used in many applications [49, 74]. Because of the polar nature of the (0001) crystal plane of ZnO, anisotropic growth of ZnO to obtain 1D nanostructures can be easily achievable using simple solution phase or gas phase synthesis techniques. Wet chemical methods are the most powerful and widely used

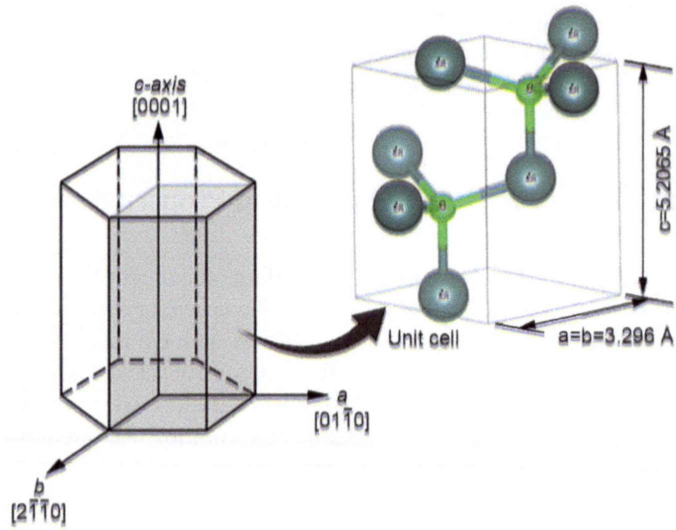

Fig. 6.1 Hexagonal wurtzite crystal structure of ZnO and its unit cell showing the alternating layers of Zn and O atoms along with its lattice parameters

technique due to their low cost and less hazardous conditions, where tuning the various synthesis parameters, the anisotropic growth of ZnO can be well controlled [6]. ZnO nanorods, in particular, have been extensively studied because of their easy formation and wide range of device applications. Controlled growth of ZnO nanorods in aqueous solutions has been initially reported by [98] using thermal decomposition of hexamethylenetetramine (HMT) and zinc nitrate. HMT is a water-soluble organic compound which releases OH^- ions upon thermal decomposition at above 60 °C. These OH^- ions then react with the Zn^{2+} ions in the solution to form ZnO [84]. Due to the polar nature, the growth of ZnO into nanorod form can be well controlled in the hydrothermal process by controlling the reaction parameters, such as pH, reactant concentration, growth temperature, etc. [6, 8]. During the growth of ZnO nanorods in hydrothermal process, HMT plays an important role in defining the final structure. Sugunan et al. [94] proposed that due to the non-polar cheating nature of HMT, it preferentially adsorbs on the non-polar facets of ZnO reducing the supply of Zn^{2+} ions to these surfaces and allows a faster epitaxial growth of the polar (0001) surface. Apart from hydrothermal synthesis, other wet chemical methods for the growth of ZnO nanorods, such as reverse micelle [117], sol–gel technique [81], etc., have also been investigated [108] has reported a non-aqueous route to synthesize extremely thin and relatively monodispersed ZnO nanorods of about 2 nm thickness in organic solvent using oleic acid as capping agent.

Because of the remarkable properties and easy synthesis, ZnO nanorods have become a preferred choice for several optical and optoelectronic devices. ZnO nanorods are among the top choices for short-wavelength LED and laser diode applications [45, 52, 114]. The high exciton binding energy of ZnO (60 meV) along with

Table 6.1 Physical and chemical properties of ZnO

Parameter	Characteristics
Crystal structure	Hexagonal wurtzite
Lattice constant	a = 3.296 Å, c = 5.207 Å
Lattice energy	964 kcal mol^{-1}
Material type	N-type (in common), P-type (rare)
Molar mass	81.408 g·mol^{-1}
Density (ρ)	5.606 g cm^{-3}
Refractive index (n)	2.008
Dielectric constant (ε)	8.75
Bandgap at room temperature (E_g)	3.37 eV (direct)
Electron mobility at room temperature (μ_e)	~200 $cm^2 V^{-1} s^{-1}$
Hole mobility at room temperature (μ_h)	5–30 $cm^2 V^{-1} s^{-1}$
Effective mass of electron (m_e^*)	0.28 m_o ($m_o = 1.602 \times 10^{-19}$ C)
Effective mass of hole (m_h^*)	0.59 m_o
Exciton binding energy	60 meV
Piezoelectric coefficient (D_{xy})	12 pC·N^{-1}
Pyroelectric coefficient (P_i)	6.8×10^{10} A·s^{-1}·cm^{-2}·K^{-1}
Thermal expansion coefficient (α)	$4.3 \times 10^{-6} K^{-1}$ at 20 °C
Cohesive energy	1.89 eV
Heat of fusion	4470 cal·mol^{-1}
Thermal conductivity at room temperature	25 W·m^{-1}·K^{-1}
Melting point	1975 °C
Solubility	1.6 mg·L^{-1} in water at 30 °C
Standard enthalpy of formation	−348.0 kJ·mol^{-1}
Standard molar entropy	43.9 J·K^{-1}·mol^{-1}

its easy synthesis makes it highly suitable for these applications. Single-crystalline ZnO nanorods have been considered an excellent candidate for achieving laser emission at lower cost [19]. UV light sensing is another area where ZnO nanorods have been used widely. UV light detection using both single ZnO nanorods and arrays of ZnO nanorods have been reported [17, 57]. The construction of single ZnO nanorod-based sensors is complex, but the complexity can be reduced in the case of ZnO nanorod arrays. Apart from UV sensors, ZnO nanorods are also considered widely as sensing elements for constructing various chemical and biological sensors [53, 76, 100]. Due to the excellent visible light transparency and high electron mobility, ZnO nanorods have been used extensively for photovoltaic (PV) applications, such as dye-sensitized solar cells [15], perovskite solar cells [91], thin film PV cells [96], organic-based PV cells [109], and Si-based PV cells [79]. Outstanding photocatalytic and antimicrobial properties of ZnO have been reported by several researchers for water purification applications [9, 16, 43]. They have also shown potential in the areas of field emission [116], field effect transistors [2], and power nanogenerators [82]. Because of the wide range of applications, ZnO nanorods are one of the most versatile and important nanomaterials in research as well as in industries.

6.3 Light Scattering by ZnO Nanorods: An Approximate Approach

When light is incident on a sample coated with ZnO nanorods, a part of the light is scattered due to the presence of the nanorods. When light scatters, its angular spectrum expands in a way that corresponds to the scatterer's density and geometry. So, if the scatterers are scarcely placed on the surface of a glass slide, one would expect the minimal scattering signal to be observed. In order to increase this signal, one has to increase the number of scatterers and bring them closer to each other. Exactly as the darker clouds are formed of high density of water droplets in comparison to white ones formed with smaller density.

6.3.1 Side Coupling Through Scattering of Nanorods

In the present case, the scatterers are those tiny rods which are chemically grown on a substrate, typically glass. These ZnO nanorods are grown with enough density on the glass substrate of a finite thickness. As the light shines on the surface, part of the light scatters at larger angles inside the glass substrate (forward scattering). The fraction of this scattered light can sustain angles large enough so that the total internal reflection at the substrate interface is achieved, allowing that fraction of the scattered light to bounce back from the bottom surface of the glass substrate and reflect toward the upper surface with nanorods. Some parts of this light will scatter away from the substrate (loss) and another portion will experience total internal reflection back toward the bottom surface. With every loop, a portion of this light will leak outside, and thus the light power guided inside the glass is expected to reduce exponentially, as illustrated in Fig. 6.2.

In Fig. 6.2a, the incident light experiences backward and forward scattering. If the backward scattering is neglected, the rays that sustain angles larger than the critical angle between the substrate material and the surrounding will reflect back as in Fig. 6.2b. The reflected ray then experiences a loss at the top surface due to the forward scattering by the nanorods as in Fig. 6.2c. One can build a simplified theoretical model in order to help understand the mechanism by assuming that the film of nanorods is a scattering medium composed of a collection of small particles. For one particle shown in Fig. 6.3, the scattering cross section can be expressed as in Eq. 6.1 when considering far-field approximation [13].

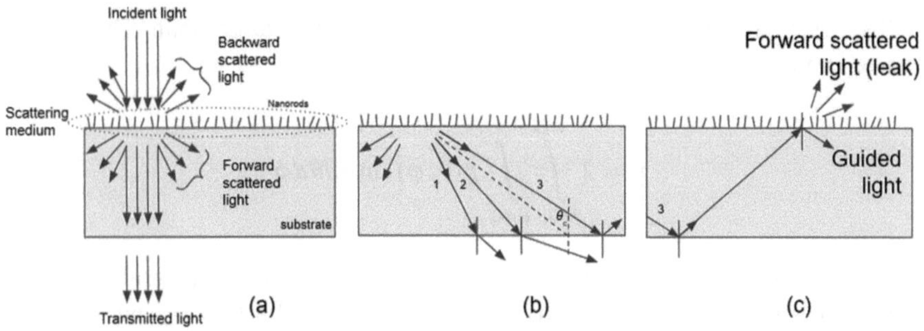

Fig. 6.2 Visualization of side coupling of light through scattering by nanorods grown on the surface of a glass substrate. **a** Demonstration of the forward and backward scattered light as well as the transmitted light by a film of nanorods on the glass surface. **b** Part of the forward scattered light which will be reflected when its angle is larger than the critical angle (θ_c). **c** Part of the reflected light that leaks outside by forward scattering from the nanorods on the glass surface

$$C_{sca} = \int_{\theta=0}^{2\pi} \int_{\phi=0}^{\pi} \frac{|\bar{X}(\theta,\phi)|^2}{k^2} \sin\theta d\theta d\phi \tag{6.1}$$

In Eq. 6.1, the function $\overline{X}(\theta, \phi)$ is the vector scattering amplitude [13] and the constant $k = \frac{2\pi}{\lambda}$ is the wavenumber. In light scattering, it is commonly preferred to use a normalized function $p(\theta, \phi) = \frac{|\overline{X}(\theta,\phi)|^2}{k^2 C_{sca}}$ that is known as the phase function, and in this case Eq. 6.1 takes the following form.

$$1 = \int_{\theta=0}^{2\pi} \int_{\phi=0}^{\pi} p(\theta,\phi) \sin\theta d\theta d\phi \tag{6.2}$$

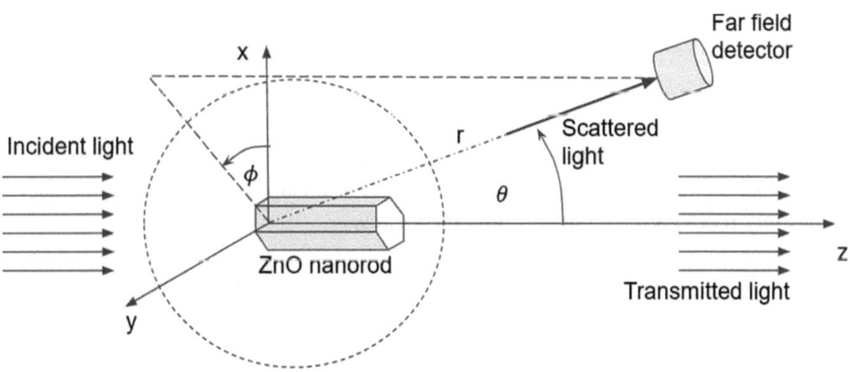

Fig. 6.3 Far-field detection of light scattering by one nanorod element

The fraction of the scattered light that sustains angles larger than critical angle can be expressed as

$$\eta = 2 \int_{\theta=\theta c}^{\pi/2} \int_{\phi=0}^{\pi} p(\theta, \phi) \sin\theta d\theta d\phi \tag{6.3}$$

It can be noted that a factor of 2 is placed in the form of the integral in Eq. 6.3. This is in order to indicate that this relation is sustained for $\theta > \theta_c$ and $\theta < -\theta_c$. Also, the integration over θ has an upper limit of $\pi/2$ as angles larger than $\pi/2$ correspond to backscattering. If the object is assumed to be symmetric around the z-axis (approximating the hexagonal prism in Fig. 6.2 to a cylindrical shape), the fraction of light that is guided is simplified to

$$\xi = 2\pi \int_{\theta=\theta c}^{\pi/2} p(\theta) \sin\theta d\theta \tag{6.4}$$

This brings forward to a reduced cross section that represents the guided light cross section or C_g.

$$C_g = \xi C_{sca} = 2\pi C_{sca} \int_{\theta=\theta c}^{\pi/2} p(\theta) \sin\theta d\theta \tag{6.5}$$

If light is normally incident on the nanorods as in Fig. 6.2, the cross-talk between the rods can be neglected, for simplicity. If this is considered, the ratio of the incident light that can be guided through forward scattering by nanorods having a density of ρ_a nanorods to the incident light per square meter is given by

$$\eta = I_g / I_o = 2\pi \rho_a C_{sca} \int_{\theta=\theta c}^{\pi/2} p(\theta) \sin\theta d\theta \tag{6.6}$$

In Eq. 6.6, I_o is the intensity of the incident light and I_g is the intensity of the guided light. The ratio between the guided light intensity to the incident intensity is commonly referred to as the coupling efficiency.

6.3.2 Sensing Mechanism by Scattering

If one tries to use the sample in Fig. 6.2 to sense the changes in the surrounding, one needs to understand the mechanism by which this can be achieved. One straightforward approach is to examine Eq. 6.5 for which the factors are affecting the change in the environment. One could argue that the change in the surrounding will affect the scattering cross section as well as the phase function. To understand this, one can examine a simpler problem where scattering is a result of surface roughness and the contrast in the refractive index between the substrate and the surrounding, as shown in Fig. 6.4.

When the surrounding index is air (Fig. 6.4a), a large scattering cross section is observed (by the length of the scattering arrows) and a broad phase function is produced (spread of the arrow directions.) When the surrounding refractive index matches that of the substrate (Fig. 6.4c), the two regions, superstrate and substrate, become one region and no scattering is observed. Hence, when the refractive index is in between these two limits, one expects to observe scattering which is less in strength and spreading of the phase function in comparison to air, as shown in Fig. 6.4b. One can build a similar argument for scattering by nanorods. When the surrounding index increases, one expects a reduction in light scattering. Would that mean that the total signal that can be measured by a detector placed at the end of the glass slide would be reduced? To answer this question, one needs to examine the loss mechanism inside the guiding region.

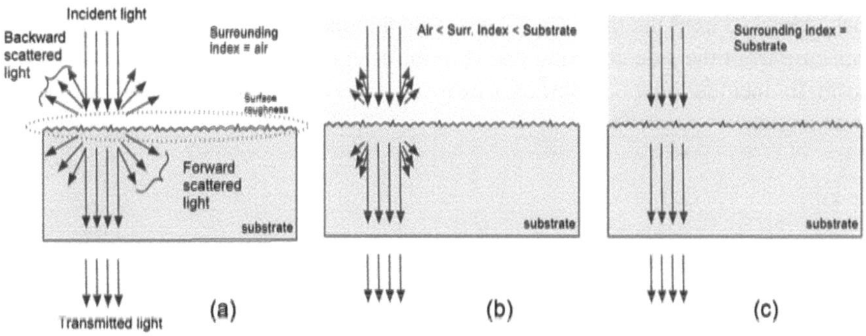

Fig. 6.4 Effect of the change in the surrounding environment on the light scattering by surface roughness. The length of the scattering arrow represents C_{sca} and the directions represent the broadening of the phase function $p(\theta)$. **a** When the surrounding is air, **b** when the surrounding refractive index increases but remains less than the substrate, and **c** when the surrounding index equals the substrate

6.3.3 Loss Mechanism

To examine the loss mechanism, one can hypothetically slice the sample in Fig. 6.2 into thin slices, where the light is assumed to be incident at the first slice only as shown in Fig. 6.5.

At any segment of width Δz along the guiding region, the change of the light power due to scattering by the nanorods on the segment is

$$\Delta P = -\Delta z \alpha P(z) \overset{yields}{\rightarrow} \frac{\Delta P}{P(z)} = -\Delta z \alpha \tag{6.7}$$

In Eq. 6.7, the factor α is the attenuation coefficient due to the nanorods. If only scattering is causing loss from the guiding region surface, then α can be written in terms of the scattering cross section as [scattering by small particles]

$$\alpha = \rho_v C'_{sca} \tag{6.8}$$

In Eq. 6.8, ρ_v is the volume density of nanorods. If one assumes vertically oriented rods, then one can determine that in a unit area, the occupied volume is $V = 1 \times L$, where L is the average height of the nanorods. The number of rods in this unit area is then $N = 1 \times \rho_a$. Hence, the number of nanorods per unit volume is $\rho_v = \frac{N}{V} = \frac{\rho_a}{L}$. It can be noted that in this case where the whole scattering cross section is considered is not the reduced value as in the case of the side coupling in Sect. 6.2 of Fig. 6.5. This is because the entire forward scattering is assumed to contribute to the power loss. Also, we used the term $C\prime_{sca}$ here to distinguish the scattering cross section in this case from the side coupling one (Fig. 6.2). In the side coupling case, light was normally incident on the nanorods; however, in the case of leakage, light is incident at an angle.

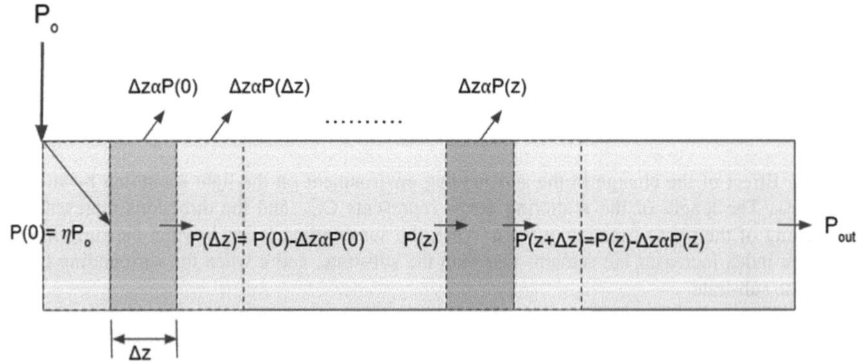

Fig. 6.5 Loss mechanism inside a guiding region coated with scattering medium such as ZnO nanorods. The excitation is assumed to take place at z = 0 location only

If the segment width is infinitesimally small, Eq. 6.8 can be written in differential form as

$$\frac{dP}{P(z)} = -\frac{\rho_a}{L} C'_{sca} dz \tag{6.9}$$

Integration of both sides of the above equation yields the known Beer–Lambert's law for the light transmittance (T)

$$T = P(z)/P_o = \eta \exp\left(-\frac{\rho_a}{L} C'_{sca} z\right) \tag{6.10}$$

where the coupling efficiency $\eta = 2\pi \rho_a C_{sca} \int_{\theta=\theta c}^{\pi/2} p(\theta) sin\theta d\theta$ as it is defined in Eq. 6.6. If the change of the environment caused a change of both scattering coefficients by ΔC_{sca} and $\Delta C'_{sca}$, respectively, then the transmittance is changed by the following equation:

$$\Delta T = \left(\Delta\eta - \frac{\rho_a}{L} z \Delta C'_{sca} \eta\right) \exp\left(-\frac{\rho_a}{L} C'_{sca} z\right) \tag{6.11}$$

If the refractive index of the surrounding increases, from the discussion in Sect. 6.2 of Fig. 6.5, both $\Delta\eta$ and $\Delta C'_{sca}$ will be negative. Hence, if a transmittance is placed at the end-side of the waveguide (glass substrate in this case) in Fig. 6.2, the change of transmittance could be positive or negative according to the following conditions:

$$\frac{|\Delta\eta|}{\eta} > \frac{\rho_a}{L} z |\Delta C'_{sca}| \overset{yields}{\rightarrow} \Delta T < 0 \tag{6.12a}$$

$$\frac{|\Delta\eta|}{\eta} < \frac{\rho_a}{L} z |\Delta C'_{sca}| \overset{yields}{\rightarrow} \Delta T > 0 \tag{6.12b}$$

In many cases, the condition in Eq. 6.12b is the dominant one due to the large length, z, of the glass substrate or waveguide. Hence, typically we expect an increase in the transmittance when the refractive index of the surrounding increases.

6.3.4 Estimation of the Scattering Cross Section

Calculating the scattering cross section of the nanorod geometry in Fig. 6.3 can be a challenging problem. One possible simplified approach is to use the diffraction theory. In this case, one considers the nanorod as an aperture with geometry depending on the way it blocks the incident light. For example, in the case of side coupling, one can consider the geometry in Fig. 6.6a.

The obstacle in Fig. 6.6a can be considered as the absence of an aperture of the same geometry. Hence, the total far field is the subtraction of the field from the

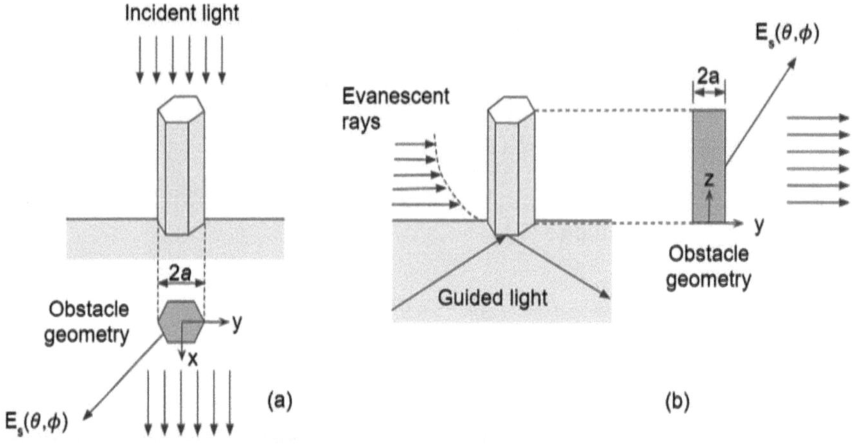

Fig. 6.6 Aperture geometries and scattered far field for the cases of **a** side coupling where light is normally incident on the nanorod and **b** leakage where evanescent rays are incident on the side of the nanorod

aperture and the incident plane wave of amplitude E_o after propagation to a detector placed in an observation plane located at a distance z from the nanorod as in Fig. 6.3.

$$E_{total}(\theta, \phi) = E_o e^{ikz} - E_o \frac{e^{ikr}}{ikr} S(\theta, \phi) \tag{6.13}$$

where the scattering amplitude is defined as

$$S(\theta, \phi) = \frac{k^2}{4\pi} \int_{-\infty}^{\infty} \int_{-\infty}^{\infty} P(x, y) e^{ik \sin\theta (x \cos\phi + y \sin\phi)} (1 + \cos\theta) dx dy \tag{6.14}$$

The expression in Eq. 6.13 is the summation of the incident plane wave after propagation for a distance z along its direction and a scattered field due to the presence of the obstacle. The function $P(x, y)$ is the aperture function (reverse of the obstacle). The cross section is then

$$C_{sca} = \int_{\phi=0}^{\pi} \int_{\vartheta=0}^{2\pi} \frac{|S(\theta, \phi)|^2}{k^2} \sin\theta d\theta d\phi \tag{6.15}$$

The same approach can be applied to the second case; however, the aperture function will be different, as shown in Fig. 6.6b. The main drawback of this approach is that the scatterer element is considered as an obstacle where its refractive index is not considered. The effect of the change in the environment is only considered through the longitudinal propagation constant $k = 2\pi n/\lambda$.

Another possible approach is to calculate the propagation of light through the scatterer using a "beam propagation method" to generate the initial field, $E_i(x, y)$. One can then approximately define a new aperture function

$$P(x, y) = \frac{E_i(x, y) - E_o e^{ikL}}{E_o} \tag{6.16}$$

Here, L is the length of the interaction region between light and the nanorod. Far-field free-space propagation (Eqs. 6.13 and 6.14) can then be used to calculate the scattered wave in the far field. Beam propagation can be implemented using finite difference approach (FDBPM). However, doing so requires sampling the three-dimensional space to solve the scalar Maxwell's equation given by

$$\frac{\partial^2 E}{\partial x^2} + \frac{\partial^2 E}{\partial y^2} + \frac{\partial^2 E}{\partial z^2} + k^2 n^2(x, y, z)E = 0 \tag{6.17}$$

The derivatives in Eq. 6.17 can be written in a discrete form as

$$\frac{\partial^2 E}{\partial x^2} = \frac{1}{\Delta x^2}\left(E_{p-1,l,h} - 2E_{p,l,h} + E_{p+1,l,h}\right) \tag{6.18a}$$

$$\frac{\partial^2 E}{\partial y^2} = \frac{1}{\Delta y^2}\left(E_{p,l-1,h} - 2E_{p,l,h} + E_{p,l+1,h}\right) \tag{6.18b}$$

$$\frac{\partial^2 E}{\partial z^2} = \frac{1}{\Delta z^2}\left(E_{p,l,h-1} - 2E_{p,l,h} + E_{p,l,h+1}\right) \tag{6.18c}$$

In Eq. 6.18a, 6.18b and 6.18c, the constants Δx, Δy, and Δz are the steps in x, y, and z, respectively. The coefficients *p*, *l*, and *h* are integer indices in x, y, and z. Using Eq. 6.17, we can then re-arrange Eq. 6.17 as

$$\begin{aligned} E_{p,l,h+1} = &- E_{p,l,h-1} + \left(\frac{2\Delta z^2}{\Delta x^2} + \frac{2\Delta z^2}{\Delta y^2} + 2 - k^2 n^2_{p,l,h}\Delta z^2\right) E_{p,h,l} \\ &- \frac{\Delta z^2}{\Delta x^2}\left(E_{p+1,l,h} + E_{p-1,l,h}\right) - \frac{\Delta z^2}{\Delta y^2}\left(E_{p,l+1,h} + E_{p,l-1,h}\right) \end{aligned} \tag{6.19}$$

Equation 6.19 is used to evaluate the progress of the electric field along the direction of propagation, assuming an initial condition

$$E_{p,l,-1} = E_{p,l,0} = E_{op,l} \tag{6.20}$$

where $E_{op,l} = E_o(p\Delta x, l\Delta y)$. If a plane wave is considered, then $E_{op,l} = 1$.

6.4 Optical Sensing Using the Light Scattering by ZnO Nanorods

The above section describes the theory of light scattering by ZnO nanorods when grown on a glass substrate. The concept is thought-provoking and practical for the development of cost-effective optical sensors. Besides, ZnO nanorods on glass surface can be grown very easily and the nanorods' length, diameter, and density on the surface can be controlled simply by controlling the synthesis parameters, which allows us to modulate the light scattering properties of the nanorods according to our needs. In the following section, we describe some of the optical sensors developed using 1D ZnO nanorods.

6.4.1 Humidity Sensing

Monitoring humidity is crucial in a wide range of applications, including food processing, agriculture, chemical processes, semiconductor production, and biological analysis. In important applications, if not effectively managed, it could result in catastrophic effects. For instance, humidity monitoring in airplanes is crucial for assessing the risk of leakage due to corrosion, which can result in the deterioration of metallic components and the failure of electrical systems [35]. Typically, measurements in humidity sensing applications are determined by changes in the capacitance, resistance, or thermal conductivity of the sensing materials [23, 89]. In the early 1900s, mechanical methods were used to estimate the relative humidity levels by monitoring moisture adsorption in human hair, silk, or synthetic fibers like nylons. Commercial humidity sensors mostly detect changes in electrical conductivity or capacitance. Humidity sensors based on electrical conductivity utilize non-metallic conductors, such as salt or conductive polymers, whose conductivity changes with water content. Capacitive humidity sensors, on the other hand, utilize dielectric materials whose dielectric constant varies with humidity levels. The first electrolytic humidity sensor based on lithium chloride (LiCl) was introduced in 1938 by Dunmore [25]. Despite having a low working range (40–60% RH), this humidity sensor remained popular until the 1970s. Recent developments in polymers and semiconductors have made humidity sensing more accurate and sophisticated. Ceramic-based humidity sensors were introduced in the 1980s [86, 88]. Around the same time, thin-film capacitive humidity sensors were introduced using metal oxide semiconductors [42]. Capacitive-type humidity sensor using cross-linked poly(methyl methacrylate), PMMA thin films was reported in 1991 [67]. More recently, carbon-based nanomaterials have been used to develop highly sensitive and compact humidity sensors [92]. More specifically, these materials enable development of flexible humidity sensors suitable for wearable technologies.

Nevertheless, most of these electronic humidity sensors are usually insensitive at high humidity levels, where the collection of dew or water droplets could result in charge leakages that compromise the device's performance. In addition, electronic systems are vulnerable to electromagnetic interference (EMI), rendering them unsuitable for application in hazardous and explosive settings. Optical humidity sensors, on the other hand, provide a number of advantages over electrical humidity sensors. In place of electrical current, an optical sensor measures physical quantities using light rays, with detection based on the demodulation of light intensity [70]. As a result, it is immune to EMI and frequently offers increased sensitivity and a larger dynamic range of operation, faster response times, and the capacity to send signals across vast distances for remote operations [21, 60].

Even though optical humidity sensors offer many benefits, their widespread usage is frequently restricted by the need for expensive equipment in the sensing setup. Using a He–Ne laser source, a photodetector, and a lock-in amplifier, recent research [39] demonstrated a tapered plastic optical fiber coated with seeded aluminum doped zinc oxide (ZnO:Al) materials for humidity measurement. In order to store data, [3] described using a laser source, photodetector, and data capture cards connected to PCs. In a different study, [59] demonstrated humidity monitoring utilizing a silica fiber coated with zinc oxide nanorods, a 644 nm laser source, and an optical power meter. A laser source, photodetector, and optical power meter are often required for optical humidity sensors, which are not ideal for the large-scale application of optical sensors for humidity monitoring. The cost of manufacturing, operation, and convenience of deployment of the sensor device must be taken into account in order to increase the commercial deployment of sensors. Due to the adsorption of water molecules on the surface, variations in humidity levels can result in changes in the environment's effective refractive index, which have an impact on the electrical conductivity of the sensing materials (commonly employed in sensing devices) [40]. The complicated refractive index is altered by changes in electrical conductivity, and this alteration influences the optical properties of the sensing materials, making them appropriate for optical sensing applications [61].

ZnO nanorods have been used by a number of researchers to show humidity sensors as extremely sensitive materials for detection. There are many different ways to synthesize nanostructured ZnO, but for sensing applications, 1D nanorod structures have drawn a lot of attention because of their high surface-to-volume ratio [11]. The effectiveness of seeded and unseeded tapered plastic optical fiber coated with ZnO nanostructures in monitoring various uric acid and humidity concentrations was reported by [10]. In terms of sensitivity, response time, and recovery time [41] demonstrated that the humidity-sensing performance of ZnO nanorod structures was nearly two times greater than that of ZnO nanoflakes. The development of ZnO nanorods on glass substrate for electrical resistance measurements versus humidity was published in another work by [50]. As the RH rose from 20 to 90%, the resistance value of the sensor fell.

Recently, an alternative approach to design simple and affordable humidity sensors using ZnO nanorods coated on glass substrates has been demonstrated [111]. The expensive laser light source was replaced by a commercially available green LED light (Fig. 6.7) and ZnO nanorods (the active sensing element) were grown on standard microscopic glass slides using a simple hydrothermal method. The working

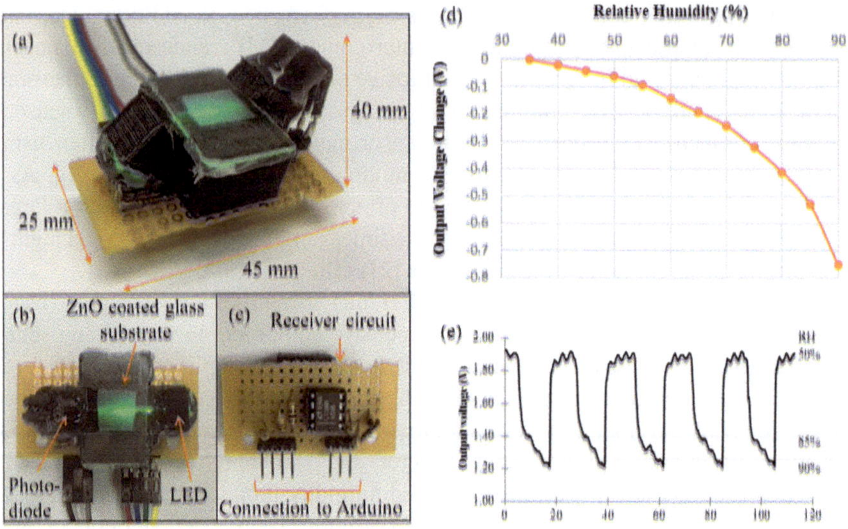

Fig. 6.7 The integrated ZnO nanorod-based humidity sensor chip showing the **a** orthographic view with dimensions, **b** top view, and **c** bottom view. A green LED was used instead of an expensive laser source, while the optical output was detected by using a photodiode. **d** shows the change in the sensor output voltage recorded by an Arduino platform at different RH levels. **e** represents the sensor repeatability performance within the RH range between 50 and 90%

principle of the sensor is simple which is based on the detection of changes in the surrounding refractive index at different relative humidity (RH) levels altering the light scattering behavior of the ZnO nanorods, as described in Sect. 6.3. For the optical signal processing and calibration an Arduino platform was used, circumventing the need for more complex alternatives. Uniform growth of the ZnO nanorods is the key to the robust design of the sensor.

A first-order approximation model was used assuming uniform nanorod growth and the optimum operating conditions for humidity sensing device were estimated and validated experimentally [112]. This approach provides an option to correlate ZnO growth conditions for humidity sensing applications which can set a platform for compact sensors where modulation of light intensity is followed. It was found that the sensor's maximum dynamic range for the humidity sensing occurs for the combination parameters of normalized length (Z) of 0.23 and normalized scattering coefficient (ζ) of 0.3. A reduction in optical transmittance by 38% at humidity levels of 80% with a reference point of 50% humidity was observed. This is the optimum operating condition for the fabricated device to work at the best sensing performance. The experimental results were then matched to the first-order approximation model which correlated reasonably well. The average sensitivity of the sensors was found to be −12 mV/RH% over the measured RH values (35–90%). For RH above 70%, the sensitivity was observed to be higher at −24.6 mV/RH%. In comparison to the typical ambient humidity circumstances (RH 50%), which indicated a speedier recovery time of 2.2 s, an average response time of 3.8 s was found. As opposed to earlier efforts on ZnO-based humidity sensors, the constructed sensor demonstrated improvements such as steady sensing performance with dependable and repeatable signal levels and better response/recovery time.

6.4.2 Chemical Vapor Sensing

Frequent use of chemicals and emission of various kinds of toxins to the environment is a threat to human health. Chemical vapor or gas sensing is therefore important to ensure a healthy working space as well as healthy life. Sensing of chemical vapor is also important for various other fields, such as food industries, agriculture industries, and biomedical applications [5, 47, 113]. In this regard, optical sensors have received considerable attention in recent years; thanks to a number of features such as compactness, high sensitivity, and remote operation. ZnO is one of the most promising materials for the development of such optical sensors [99, 104, 107] because of their remarkable optical, electrical, and mechanical properties, and the possibilities to modulate the morphologies of the ZnO nanostructures (nanorods, nanoparticles, thin films, etc.) using diverse synthesis methods.

Chemical vapors or gas detection using ZnO-based sensors is not a new subject area. Significant developments have happened in this area in the past few decades [48, 119]. Morphology and surface defects are two important factors enabling the high surface area ZnO nanostructures to adsorb more target gas molecules, hence enhancing the sensor response [1]. Some researchers have doped ZnO nanomaterials in order to improve sensor sensitivity by altering the electrical conductance of the materials via bandgap modulation [64, 80]. Germanium (Ge)- and tungsten (W)-doped ZnO nanostructures have been recently reported for CO_2 gas sensing [20], and showed that sensor sensitivity can be affected by doping element type and concentrations. Most of these ZnO-based gas sensors perform best at high temperatures. At high temperature, due to the thermal activation, the carrier concentration near the ZnO surface increases and enhance the gas adsorption sites at the surface. Ammonia (NH_3) vapor sensing response was reported to improve significantly at 150 °C or higher operating temperatures with very low sensor response and recovery time [31, 32], compared to the detection carried out at room temperature conditions [65, 77]. Similarly, the CO_2 gas sensing was enhanced using ZnO thin films operated at 400 °C [36]. However, majority of these ZnO-based chemical vapor or gas sensors are based on the measurements of the electrical conductance of the ZnO nanostructures, and despite having excellent optical properties and other benefits, optical sensing of chemical vapors or gas using ZnO-based optical sensors is limited [37].

Optical gas sensing methods provide several advantages (as described earlier) over electrical sensors, such as high sensitivity, simple design, and higher stability against various surrounding factors. Earlier we have used vertically aligned ZnO nanorod arrays coated on optical fibers and constructed optical sensors to detect chemical

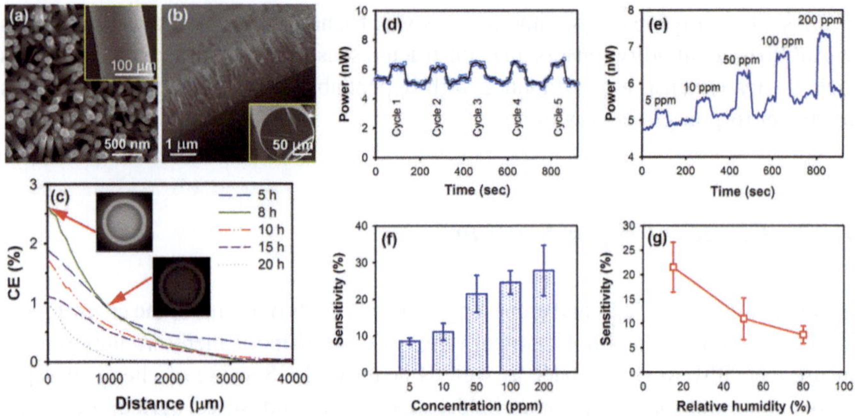

Fig. 6.8 **a, b** SEM micrographs of 1D ZnO nanorods grown on silica optical fibers for chemical vapor sensing, and **c** shows the cladding mode coupling efficiency as a function of ZnO nanorods length varied by varying the growth time of the nanorods from 5 to 20 h. **d** Sensor optical power variations in the presence of 50 ppm methanol vapors, and **e** optical power when methanol concentration was varied between 5 and 200 ppm. **f** indicates the sensor sensitivity calculated for different methanol vapor concentrations and **g** indicates the effect of RH on the sensor sensitivity

vapors based on the light scattering properties of the nanorods [14]. The mean scattering cross section (C_{sca}) increases with the length of the ZnO nanorods coated on the fiber surface and highest light was observed to couple into the cladding layer of the optical fibers (coupling efficiency of 2.65%) at 2.2 μm nanorods length. The sensor demonstrated significant change in the coupled light intensity in the presence of various chemical vapors since scattering of light is dependent on the nature of the surface and highly sensitive to any changes on the surface. Methanol vapor concentration as low as 5 ppm was detected using the nanorod-coated fibers with a sensitivity of 8.6% (Fig. 6.8), which further increased to 27.9% at 200 ppm. The sensor response was investigated for other organic molecules, namely ethanol, toluene, and benzene, and showed reduction in the sensitivity with non-polar organic molecules. A similar approach was used for the detection of CO and CO_2 by coating plastic optical fibers with ZnO nanorods [28]. The presence of nanorods minimizes the coupling of the Fresnel reflection. Exposing the optical sensor to different gas concentrations, an obvious distinguished change in the total reflection can be observed. The sensor demonstrated the detection of CO gas at a concentration as low as 200 ppm at room temperature and exhibited a higher response against CO than CO_2 gas.

6.4.3 Biomarker and Biological Sensing

Detection of biological substances and biomarkers is very important in many areas, such as in food and agriculture industry and in biomedical research. The high increase in the production of processed food made it hard to sustain low or minimal occurrence of bacterial contamination. Conventional techniques such as plating, polymerase chain reaction (PCR), and enzyme-linked immunosorbent assay (ELISA) methods are reliable for bacterial or microorganism detection, but they are time-consuming, laborious, expensive, and need costly lab equipment [27, 54, 103]. Colorimetric detection of biomarkers is another alternative optical strategy which can provide simple, rapid, and on-site detection of biological substances [30, 69]. The technique, however, requires some kind of expensive optical detection systems, such as a spectrophotometer, to achieve sensitive detection at low concentrations. To address these limitations, a low-cost optical sensor solution was recently reported by our group [29]. A thin layer of gold nanoparticles (Au) (around 50 nm) was coated on the tip of a multimode plastic optical fiber and ZnO nanorods were grown on the Au layer through a hydrothermal technique. The sensor showed very fast response within the first 10 s of contacting water sample polluted with *Escherichia coli* (*E. coli*) bacteria. ZnO nanorods used here mainly act as the light scatterer, and the gold layer is added to increase the reflected signal back to the detector beyond the noise level in order to enhance coupling to the fiber and to reduce ambient effect. The sensing mechanism for the sensor counts on the change of the backscattered light intensity due to the presence of E. Coli in the solution surrounding the ZnO nanorods. The increase of the solution effective index as well as the deposition of E. Coli on the nanorods reduces the amount of scattering and hence increases the signal strength. The optical sensor exhibited excellent linear output response for *E. coli* concentrations up to 4000 CFU/ml (CFU: Colony-forming Unit). The proposed design is simple, low cost, and has the potential for rapid detection of microbial and pathogenic contaminations.

In another recent work, we have used the ZnO nanorods and coated Single-mode–Multimode–Single-mode optical fiber sensor for the detection of a volatile organic compound (VOC) biomarker for diabetes (Fig. 6.9), namely, isopropanol (IPA) markers [95]. IPA vapor was considered as one of the key biomarkers for non-invasive diabetes diagnosis [56]. A coreless silica fiber (CSF) coated with ZnO nanorods was connected to a single-mode fiber (SMF) at both ends to achieve an SMF–CSF–SMF structure in order to construct the sensor. The purpose of the CSF is to provide higher light interaction at the interface between the fiber and sensing medium, leading to enhanced sensitivity. The length of the sensor was optimized by numerical modeling to maximize the sensitivity of the device. The single-mode fiber acts as a waveguide for delivering light from the source to the sensing region and then to the detector, which is the spectrometer utilized for recording the intensity spectra. The multimode fiber (MMF) or coreless silica fiber (CSF) coated with ZnO nanorods allows higher interaction of light at the sensor surface, where the cladding layer of the CSF was removed prior to the nanorods coating. The removal of the cladding and the large number of modes allow a higher intensity of the evanescent field at the fiber interface and hence increase the light–matter interaction between the evanescent field and the sensed material [71], leading to the enhancement of sensitivity. When the sensor is exposed to IPA vapors, the refractive index of the medium is affected, resulting change in the propagation constants and modal property inside the fiber. Hence, the multimode interference (MMI) spectrum is altered. The fabricated sensor

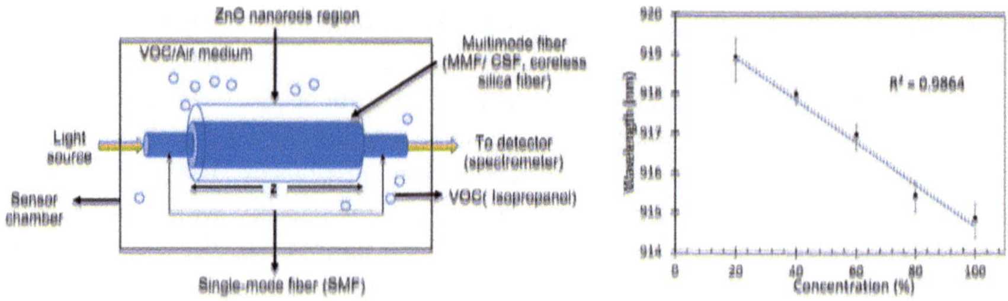

Fig. 6.9 Schematic representation of the ZnO nanorods and coated Single-mode–Multimode–Single-mode optical fiber sensor for the detection of a volatile organic compound (VOC) biomarker (left) and change in the sensor output wavelength change at different IPA concentrations (right)

successfully detected IPA vapors with an excellent sensitivity of 0.053 nm/% IPA vapor. The fabricated fiber sensor showed promise to further detect IPA in exhaled breath, which is one of the VOC biomarkers for diabetes. For the selectivity, surface functionalization can be further utilized and sensitivity can be further enhanced by tapering of multimode fibers [12].

6.5 Conclusions and Outlook

The chapter discusses the light scattering by 1D ZnO nanorods and their applications for optical sensing. Part of the incident light is scattered by the ZnO nanorods, where the amount and direction of the light scattering are strongly influenced by the length, density, and geometry of the nanorods. Because of their 1D structure, it is possible to modulate the light scattering by the ZnO nanorods by arranging them on a surface, such as glass or plastic, where some amount of the scattered light is guided through the substrate by satisfying the total internal reflection conditions. This property of light scattering by ZnO nanorods, which is a function of the surrounding refractive index, is exploited to fabricate optical sensors for the detection of various chemical and biological species. Such optical sensors provide a unique working principle to make highly robust and sensitive sensors. Compared to the traditional electrical conductance-based sensors, optical sensors based on the light scattering by ZnO nanorods show higher signal stability, since the propagating light signal is less susceptible to environmental factors. Researchers have also shown the strategies to develop sensitive, compact, and affordable optical sensors using this concept, where expensive optical components for signal processing are replaced with microcontroller-based processing. However, the concept of using light scattering properties of ZnO nanorods to develop optical sensors is relatively new and most of the developments in this area are currently happening at the proof-of-concept stage validated in laboratory conditions. In order to commercialize these optical sensors, future research should focus on the device level considering other interfacing electronics. Since ZnO nanorods can be grown very easily on flexible substrates, such as paper or textile fibers, future research should also focus on the development of flexible and mobile light scattering-based optical sensors for rapid and onsite detection of chemicals and biological substances. Such technologies are promising for a wide range of applications, e.g., food and water quality monitoring, biomedical devices, health or biomarker monitoring, pollution monitoring, etc.

References

1. Agarwal, S., Rai, P., Gatell, E.N., et al.: Gas sensing properties of ZnO nanostructures (flowers/rods) synthesized by hydrothermal method. Sens. Actuat. B Chem. **292**, 24–31 (2019)
2. Ahmad, R., Tripathy, N., Hahn, Y.-B.: High-performance cholesterol sensor based on the solution-gated field effect transistor fabricated with ZnO nanorods. Biosens. Bioelectron.. Bioelectron. **45**, 281–286 (2013)
3. Aneesh, R., Khijwania, S.K.: Zinc oxide nanoparticle based optical fiber humidity sensor having linear response throughout a large dynamic range. Appl. Opt. **50**, 5310–5314 (2011)
4. Anker, J.N., Hall, W.P., Lyandres, O., et al.: Biosensing with plasmonic nanosensors. In: Nanoscience and Technology, pp. 308–319. Co-Published with Macmillan Publishers Ltd., UK (2009)
5. Arafat, M.M., Dinan, B., Akbar, S.A., Haseeb, A.S.M.A.: Gas sensors based on one dimensional nanostructured metal-oxides: a review. Sensors **12**, 7207–7258 (2012)
6. Baruah, S., Dutta, J.: Hydrothermal growth of ZnO nanostructures. Sci. Technol. Adv. Mater. **10**, 13001 (2009)
7. Baruah, S., Dutta, J.: Effect of seeded substrates on hydrothermally grown ZnO nanorods. J. Sol-Gel Sci. Technol. **50**, 456–464 (2009)
8. Baruah, S., Dutta, J.: PH-dependent growth of zinc oxide nanorods. J. Cryst. GrowthCryst. Growth **311**, 2549–2554 (2009)
9. Baruah, S., Mahmood, M.A., Myint, M.T.Z., et al.: Enhanced visible light photocatalysis through fast crystallization of zinc oxide nanorods. Beilstein J. Nanotechnol.Nanotechnol. **1**, 14–20 (2010)
10. Batumalay, M., Harith, Z., Rafaie, H.A., et al.: Tapered plastic optical fiber coated with ZnO nanostructures for the measurement of uric acid concentrations and changes in relative humidity. Sens. Actuat. A Phys. **210**, 190–196 (2014)
11. Bhattacharya, S., Sasmal, M.: Solution process synthesized ZnO nanoparticle for an intelligent humidity sensor. In: 2015 Second International Conference on Soft Computing and Machine Intelligence (ISCMI), pp. 102–106 (2015)
12. Biazoli, C.R., Silva, S., Franco, M.A.R., et al.: Multimode interference tapered fiber refractive index sensors. Appl. Opt. **51**, 5941–5945 (2012)
13. Bohren, C.F., Huffman, D.R.: Absorption and Scattering of Light by Small Particles. Wiley-VCH, Weinheim (1998)
14. Bora, T., Fallah, H., Chaudhari, M., et al.: Controlled side coupling of light to cladding mode of ZnO nanorod coated optical fibers and its implications for chemical vapor sensing. Sens. Actuat. B Chem. **202**, 543–550 (2014)
15. Bora, T., Kyaw, H.H., Sarkar, S., et al.: Highly efficient ZnO/Au Schottky barrier dye-sensitized solar cells: role of gold nanoparticles on the charge-transfer process. Beilstein J. Nanotechnol.Nanotechnol. **2**, 681–690 (2011)
16. Bora, T., Sathe, P., Laxman, K., et al.: Defect engineered visible light active ZnO nanorods for photocatalytic treatment of water. Catal. Today. Today **284**, 11–18 (2017)
17. Chai, G., Lupan, O., Chow, L., Heinrich, H.: Crossed zinc oxide nanorods for ultraviolet radiation detection. Sens. Actuat. A Phys. **150**, 184–187 (2009)

18. Chen, Y., Bagnall, D., Yao, T.: ZnO as a novel photonic material for the UV region. Mater. Sci. Eng. B **75**, 190–198 (2000)
19. Chu, S., Wang, G., Zhou, W., et al.: Electrically pumped waveguide lasing from ZnO nanowires. Nat. Nanotechnol.Nanotechnol. **6**, 506–510 (2011)
20. Çolak, H., Karaköse, E.: Synthesis and characterization of different dopant (Ge, Nd, W)-doped ZnO nanorods and their CO_2 gas sensing applications. Sens. Actuat. B Chem. **296**, 126629 (2019)
21. Correa-Mena, A.G., González, L.A., Quintero-Rodríguez, L.J., Zaldivar-Huerta, I.E.: Review on integrated optical sensors and its applications. In: 2017 IEEE Mexican Humanitarian Technology Conference (MHTC), pp. 170–173 (2017)
22. Dey, A.: Semiconductor metal oxide gas sensors: a review. Mater. Sci. Eng. B **229**, 206–217 (2018)
23. Dixit, S., Srivastava, A., Shukla, R.K., Srivastava, A.: ZnO thick film based opto-electronic humidity sensor for a wide range of humidity. Opt. Rev. **14**, 186–188 (2007)
24. Djurišić, A.B., Chen, X., Leung, Y.H., Man Ching Ng, A.: ZnO nanostructures: growth, properties and applications. J. Mater. Chem. **22**, 6526–6535 (2012)
25. Dunmore, F.W.: An electric hygrometer and its application to radio meteorography. Bull. Am. Meteorol. Soc.Meteorol. Soc. **19**, 225–243 (1938)
26. Emanetoglu, N.W., Gorla, C., Liu, Y., et al.: Epitaxial ZnO piezoelectric thin films for saw filters. Mater. Sci. Semicond. Process.Semicond. Process. **2**, 247–252 (1999)
27. Eriksson, E., Aspan, A.: Comparison of culture, ELISA and PCR techniques for salmonella detection in faecal samples for cattle, pig and poultry. BMC Vet. Res. **3**, 21 (2007)
28. Fallah, H., Afra, M., Mohajerani, E., Mohammad, W.S.: Utilization of ZnO nanorods growth on a tip of plastic optical fiber toward the realization of low-cost CO and CO_2 gas sensor. J. NanophotonicsNanophotonics **11**, 36023 (2017)
29. Fallah, H., Asadishad, T., Parsanasab, G.M., et al.: Optical fiber biosensor toward E-coli bacterial detection on the pollutant water. Eng J **25**, 1–8 (2021)
30. Fan, Q., Gao, Y., Mazur, F., Chandrawati, R.: Nanoparticle-based colorimetric sensors to detect neurodegenerative disease biomarkers. Biomater. Sci. **9**, 6983–7007 (2021)
31. Ganesh, R.S., Durgadevi, E., Navaneethan, M., et al.: Low temperature ammonia gas sensor based on Mn-doped ZnO nanoparticle decorated microspheres. J. Alloys Compd. **721**, 182–190 (2017)
32. Ganesh, R.S., Navaneethan, M., Patil, V.L., et al.: Sensitivity enhancement of ammonia gas sensor based on Ag/ZnO flower and nanoellipsoids at low temperature. Sens. Actuat. B Chem. **255**, 672–683 (2018)
33. Gao, H., Zhang, X., Zhou, M., et al.: Super-uniform ZnO nanohelices synthesized via thermal evaporation. Solid State Commun.Commun. **140**, 455–458 (2006)
34. Gao, H., Zhang, X.T., Zhou, M.Y., et al.: Growth of novel ZnO nanohelices modified by SiO_2-sheathed ZnO discs. Nanotechnology **18**, 65601 (2007)
35. Ghadiry, M., Gholami, M., Lai, C.K., et al.: Ultra-sensitive humidity sensor based on optical properties of graphene oxide and nano-anatase TiO_2. PLoS ONE **11**, e0153949 (2016)
36. Ghosh, A., Zhang, C., Shi, S., Zhang, H.: High temperature CO2 sensing and its cross-sensitivity towards H_2 and CO gas using calcium doped ZnO thin film coated langasite SAW sensor. Sens. Actuat. B Chem. **301**, 126958 (2019)
37. Gupta, S.P., Pawbake, A.S., Sathe, B.R., et al.: Superior humidity sensor and photodetector of mesoporous ZnO nanosheets at room temperature. Sens. Actuat. B Chem. **293**, 83–92 (2019)
38. Gurlo, A.: Nanosensors: towards morphological control of gas sensing activity. SnO_2, In_2O_3, ZnO and WO_3 case studies. Nanoscale **3**, 154–165 (2011)
39. Harith, Z., Batumalay, M., Irawati, N., et al.: Relative humidity sensor employing tapered plastic optical fiber coated with seeded Al-doped ZnO. Optik (Stuttg) **144**, 257–262 (2017)
40. Harith, Z., Irawati, N., Rafaie, H.A., et al.: Tapered plastic optical fiber coated with Al-doped ZnO nanostructures for detecting relative humidity. IEEE Sens. J. **15**, 845–849 (2015)
41. Ismail, A.S., Mamat, M.H., Malek, M.F., et al.: A study on different morphological structures of zinc oxide nanostructures for humidity sensing application. AIP Conf. Proc. **1733**, 20010 (2016)

42. Jachowicz, R.S., Senturia, S.D.: A thin-film capacitance humidity sensor. Sens. Actuat. **2**, 171–186 (1981)
43. Jaisai, M., Baruah, S., Dutta, J.: Paper modified with ZnO nanorods—antimicrobial studies. Beilstein J. Nanotechnol.Nanotechnol. **3**, 684–691 (2012)
44. Jenkinson, C., Earl, J., Ghaneh, P., et al.: Biomarkers for early diagnosis of pancreatic cancer. Expert Rev. Gastroenterol. Hepatol. **9**, 305–315 (2015)
45. Jha, S.K., Kutsay, O., Bello, I., Lee, S.T.: ZnO nanorod based low turn-on voltage LEDs with wide electroluminescence spectra. J. Lumin. **133**, 222–225 (2013)
46. Jiao, S.J., Zhang, Z.Z., Lu, Y.M., et al.: ZnO p–n junction light-emitting diodes fabricated on sapphire substrates. Appl. Phys. Lett. **88**, 31911 (2006)
47. Jiménez-Cadena, G., Riu, J., Rius, F.X.: Gas sensors based on nanostructured materials. Analyst **132**, 1083–1099 (2007)
48. Kang, Y., Yu, F., Zhang, L., et al.: Review of ZnO-based nanomaterials in gas sensors. Solid State Ionics **360**, 115544 (2021)
49. Kar, S., Pal, B.N., Chaudhuri, S., Chakravorty, D.: One-dimensional ZnO nanostructure arrays: synthesis and characterization. J. Phys. Chem. B **110**, 4605–4611 (2006)
50. Karki, I.B., Nakarmi, J.J., Chatterjee, S.: Synthesis of zincoxide nanorods and its application to humidity sensor. J. Inst. Sci. Technol. **20**, 36–39 (2015)
51. Keis, K., Bauer, C., Boschloo, G., et al.: Nanostructured ZnO electrodes for dye-sensitized solar cell applications. J. Photochem. Photobiol. A Chem. **148**, 57–64 (2002)
52. Könenkamp, R., Word, R.C., Schlegel, C.: Vertical nanowire light-emitting diode. Appl. Phys. Lett. **85**, 6004–6006 (2004)
53. Kumar, M.A., Jung, S., Ji, T.: Protein biosensors based on polymer nanowires, carbon nanotubes and zinc oxide nanorods. Sensors **11**, 5087–5111 (2011)
54. Lantz, P.-G., Hahn-Hägerdal, B., Rådström, P.: Sample preparation methods in PCR-based detection of food pathogens. Trends Food Sci. Technol. **5**, 384–389 (1994)
55. Li, Q., Kumar, V., Li, Y., et al.: Fabrication of ZnO nanorods and nanotubes in aqueous solutions. Chem. Mater. **17**, 1001–1006 (2005)
56. Li, W., Liu, Y., Liu, Y., et al.: Exhaled isopropanol: new potential biomarker in diabetic breathomics and its metabolic correlations with acetone. RSC Adv. **7**, 17480–17488 (2017)
57. Li, Y., Gong, J., Deng, Y.: Hierarchical structured ZnO nanorods on ZnO nanofibers and their photoresponse to UV and visible lights. Sens. Actuat. A Phys. **158**, 176–182 (2010)
58. Lin, Y., Zhang, Z., Tang, Z., et al.: Characterisation of ZnO-based varistors prepared from nanometre Precursor powders. Adv. Mater. Opt. Electron. **9**, 205–209 (1999)
59. Liu, Y., Zhang, Y., Lei, H., et al.: Growth of well-arrayed ZnO nanorods on thinned silica fiber and application for humidity sensing. Opt. Express **20**, 19404–19411 (2012)
60. Lokman, A., Arof, H., Harun, S.W., et al.: Optical fiber relative humidity sensor based on inline Mach-Zehnder interferometer with ZnO nanowires coating. IEEE Sens. J. **16**, 312–316 (2016)
61. Lokman, M.Q., Bin Abdul Rahim, H.R., Harun, S.W., et al.: Light backscattering (e.g. reflectance) by ZnO nanorods on tips of plastic optical fibres with application for humidity and alcohol vapour sensing. Micro. Nano Lett. **11**, 832–836 (2016)
62. Luo, L., Zhang, Y., Mao, S.S., Lin, L.: Fabrication and characterization of ZnO nanowires based UV photodiodes. Sens. Actuat. A Phys. **127**, 201–206 (2006)
63. Lupan, O., Emelchenko, G.A., Ursaki, V.V., et al.: Synthesis and characterization of ZnO nanowires for nanosensor applications. Mater. Res. Bull. **45**, 1026–1032 (2010)
64. Lupan, O., Postica, V., Gröttrup, J., et al.: Hybridization of zinc oxide tetrapods for selective gas sensing applications. ACS Appl. Mater. Interfaces **9**, 4084–4099 (2017)
65. Mani, G.K., Rayappan, J.B.B.: A highly selective and wide range ammonia sensor—nanostructured ZnO: Co thin film. Mater. Sci. Eng. B **191**, 41–50 (2015)
66. Manisalidis, I., Stavropoulou, E., Stavropoulos, A., Bezirtzoglou, E.: Environmental and health impacts of air pollution: a review. Front. Public Heal. **8**, 14 (2020)
67. Matsuguchi, M., Sadaoka, Y., Sakai, Y., et al.: A capacitive-type humidity sensor using cross-linked Poly(methyl methacrylate) thin films. J. Electrochem. Soc.Electrochem. Soc. **138**, 1862 (1991)

68. Meulenkamp, E.A.: Synthesis and growth of ZnO nanoparticles. J. Phys. Chem. B **102**, 5566–5572 (1998)
69. Miao, X., Zhu, Z., Jia, H., et al.: Colorimetric detection of cancer biomarker based on enzyme enrichment and pH sensing. Sens. Actuat. B Chem. **320**, 128435 (2020)
70. Mills, A.: Optical sensors for carbon dioxide and their applications BT—sensors for environment, health and security. In: Baraton, M.-I. (ed.), pp. 347–370. Springer, Netherlands, Dordrecht (2009)
71. Mohammed, W.S., Smith, P.W.E., Gu, X.: All-fiber multimode interference bandpass filter. Opt. Lett. **31**, 2547–2549 (2006)
72. Mora-Fonz, D., Lazauskas, T., Farrow, M.R., et al.: Why are polar surfaces of ZnO stable? Chem. Mater. **29**, 5306–5320 (2017)
73. Morkoç, H., Özgür, Ü.: Zinc Oxide: Fundamentals, Materials and Device Technology, 1st edn. Wiley-VCH, Darmstadt (2008)
74. Özgür, Ü., Alivov, Y.I., Liu, C., et al.: A comprehensive review of ZnO materials and devices. J. Appl. Phys. **98**, 41301 (2005)
75. Pal, S., Jasper-Tönnies, T., Hack, M., Pehlke, E.: Effect of surface reconstruction on the electronic structure of ZnO(0001). Phys. Rev. B **87**, 85445 (2013)
76. Park, J., Choi, S.-W., Kim, S.: Fabrication of a highly sensitive chemical sensor based on ZnO nanorod arrays. Nanoscale Res. Lett. **5**, 353 (2009)
77. Patil, S.L., Chougule, M.A., Sen, S., Patil, V.B.: Measurements on room temperature gas sensing properties of CSA doped polyaniline–ZnO nanocomposites. Measurement **45**, 243–249 (2012)
78. Peng, Y., Bao, L.: Controlled-synthesis of ZnO nanorings. Front. Chem. China **3**, 458–463 (2008)
79. Pietruszka, R., Witkowski, B.S., Gieraltowska, S., et al.: New efficient solar cell structures based on zinc oxide nanorods. Sol. Energy Mater. Sol. Cells **143**, 99–104 (2015)
80. Postica, V., Gröttrup, J., Adelung, R., et al.: Multifunctional materials: a case study of the effects of metal doping on ZnO tetrapods with bismuth and tin oxides. Adv. Funct. Mater. **27** (2017)
81. Qiu, J., Yu, W., Gao, X., Li, X.: Sol–gel assisted ZnO nanorod array template to synthesize TiO_2 nanotube arrays. Nanotechnology **17**, 4695–4698 (2006)
82. Qiu, Y., Zhang, H., Hu, L., et al.: Flexible piezoelectric nanogenerators based on ZnO nanorods grown on common paper substrates. Nanoscale **4**, 6568–6573 (2012)
83. Riu, J., Maroto, A., Rius, F.X.: Nanosensors in environmental analysis. Talanta **69**, 288–301 (2006)
84. Schmidt-Mende, L., MacManus-Driscoll, J.L.: ZnO–nanostructures, defects, and devices. Mater. Today **10**, 40–48 (2007)
85. Seelig, E.W., Tang, B., Yamilov, A., et al.: Self-assembled 3D photonic crystals from ZnO colloidal spheres. Mater. Chem. Phys. **80**, 257–263 (2003)
86. Seiyama, T., Yamazoe, N., Arai, H.: Ceramic humidity sensors. Sens. Actuat. **4**, 85–96 (1983)
87. Sha, R., Basak, A., Maity, P.C., Badhulika, S.: ZnO nano-structured based devices for chemical and optical sensing applications. Sens. Actuat. Rep. **4**, 100098 (2022)
88. Shimizu, Y., Okada, H., Arai, H.: Humidity-sensitive characteristics of porous La–Ti–V–O glass-ceramics. J. Am. Ceram. Soc. **72**, 436–440 (1989)
89. Sin, N.D.M., Mamat, M.H., Musa, M.Z. et al.: Effect of growth duration to the electrical properties of Zn doped SnO_2 thin film toward humidity sensor application. In: 2012 IEEE Business, Engineering and Industrial Applications Colloquium (BEIAC), pp. 86–91 (2012)
90. Soci, C., Zhang, A., Xiang, B., et al.: ZnO nanowire UV photodetectors with high internal gain. Nano Lett. **7**, 1003–1009 (2007)
91. Son, D.-Y., Im, J.-H., Kim, H.-S., Park, N.-G.: 11% Efficient perovskite solar cell based on ZnO nanorods: an effective charge collection system. J. Phys. Chem. C **118**, 16567–16573 (2014)
92. Songkeaw, P., Onlaor, K., Thiwawong, T., Tunhoo, B.: Transparent and flexible humidity sensor based on graphene oxide thin films prepared by electrostatic spray deposition technique. J. Mater. Sci. Mater. Electron. **31**, 12206–12215 (2020)

93. Srivastava, A.K., Dev, A., Karmakar, S.: Nanosensors and nanobiosensors in food and agriculture. Environ. Chem. Lett. **16**, 161–182 (2018)
94. Sugunan, A., Warad, H.C., Boman, M., Dutta, J.: Zinc oxide nanowires in chemical bath on seeded substrates: role of hexamine. J. Sol-Gel Sci. Technol. **39**, 49–56 (2006)
95. Swargiary, K., Metem, P., Kulatumyotin, C., et al.: ZnO Nanorods coated single-mode—multimode—single-mode optical fiber sensor for VOC biomarker detection. Sensors **22** (2022)
96. Tang, Y., Hu, X., Chen, M., et al.: CdSe nanocrystal sensitized ZnO core-shell nanorod array films: preparation and photovoltaic properties. Electrochim. Acta. Acta **54**, 2742–2747 (2009)
97. Tereshchenko, A., Bechelany, M., Viter, R., et al.: Optical biosensors based on ZnO nanostructures: advantages and perspectives: a review. Sens. Actuat. B Chem. **229**, 664–677 (2016)
98. Vayssieres, L., Keis, K., Lindquist, S.-E., Hagfeldt, A.: Purpose-built anisotropic metal oxide material: 3D highly oriented microrod array of ZnO. J. Phys. Chem. B **105**, 3350–3352 (2001)
99. Wan, Q., Li, Q.H., Chen, Y.J., et al.: Fabrication and ethanol sensing characteristics of ZnO nanowire gas sensors. Appl. Phys. Lett. **84**, 3654–3656 (2004)
100. Wang, Y.: ZnO nanorods for gas sensors. In: Ghamsari, M.S., Dhara, S. (eds.) Nanorods and Nanocomposites. IntechOpen, London, Chap. 3 (2020)
101. Wang, Z.L.: ZnO nanowire and nanobelt platform for nanotechnology. Mater. Sci. Eng. R. Rep. **64**, 33–71 (2009)
102. Wen, X., Fang, Y., Pang, Q., et al.: ZnO nanobelt arrays grown directly from and on zinc substrates: synthesis, characterization, and applications. J. Phys. Chem. B **109**, 15303–15308 (2005)
103. Wu, L., Li, G., Xu, X., et al.: Application of nano-ELISA in food analysis: recent advances and challenges. TrAC Trends Anal. Chem. **113**, 140–156 (2019)
104. Wu, W.-Y., Ting, J.-M., Huang, P.-J.: Electrospun ZnO nanowires as gas sensors for ethanol detection. Nanoscale Res. Lett. **4**, 513 (2009)
105. Xu, C.X., Sun, X.W., Dong, Z.L., Yu, M.B.: Self-organized nanocomb of ZnO fabricated by Au-catalyzed vapor-phase transport. J. Cryst. GrowthCryst. Growth **270**, 498–504 (2004)
106. Yao, Y., Wang, W., Zhang, D., Gao, X.: Planar and ridge ZnO optical waveguides produced by 15 MeV C5+ ion irradiation. Opt. Mater. Express **5**, 1498–1504 (2015)
107. Yebo, N.A., Lommens, P., Hens, Z., Baets, R.: An integrated optic ethanol vapor sensor based on a silicon-on-insulator microring resonator coated with a porous ZnO film. Opt. Express **18**, 11859–11866 (2010)
108. Yin, M., Gu, Y., Kuskovsky, I.L., et al.: Zinc oxide quantum rods. J. Am. Chem. Soc. **126**, 6206–6207 (2004)
109. Yin, Z., Wu, S., Zhou, X., et al.: Electrochemical deposition of ZnO nanorods on transparent reduced graphene oxide electrodes for hybrid solar cells. Small **6**, 307–312 (2010)
110. Yuan, Z., Nag, R., Cummins, E.: Human health concerns regarding microplastics in the aquatic environment—from marine to food systems. Sci. Total. Environ. **823**, 153730 (2022)
111. Yusof, H.H.M., Harun, S.W., Dimyati, K., et al.: Low-cost integrated zinc oxide nanorod-based humidity sensors for arduino platform. IEEE Sens. J. **19**, 2442–2449 (2019)
112. Yusof, H.H.M., Harun, S.W., Dimyati, K., et al.: Optical dynamic range maximization for humidity sensing by controlling growth of zinc oxide nanorods. Photonics Nanostruct.–Fundam. Appl. **30**, 57–64 (2018)
113. Zhang, J., Liu, X., Neri, G., Pinna, N.: Nanostructured materials for room-temperature gas sensors. Adv. Mater. **28**, 795–831 (2016)
114. Zhang, X.-M., Lu, M.-Y., Zhang, Y., et al.: Fabrication of a high-brightness blue-light-emitting diode using a ZnO-nanowire array grown on p-GaN thin film. Adv. Mater. **21**, 2767–2770 (2009)
115. Zhang, Y., Song, X., Zheng, J., et al.: Symmetric and asymmetric growth of ZnO hierarchical nanostructures: nanocombs and their optical properties. Nanotechnology **17**, 1916–1921 (2006)
116. Zhao, Q., Xu, X.Y., Song, X.F., et al.: Enhanced field emission from ZnO nanorods via thermal annealing in oxygen. Appl. Phys. Lett. **88**, 33102 (2006)

117. Zhengquan, L., Yi, X., Yujie, X., et al.: Reverse micelle-assisted route to control diameters of ZnO nanorods by selecting different precursors. Chem. Lett. **32**, 760–761 (2003)
118. Zhou, J., Xu, N.S., Wang, Z.L.: Dissolving behavior and stability of ZnO wires in biofluids: a study on biodegradability and biocompatibility of ZnO nanostructures. Adv. Mater. **18**, 2432–2435 (2006)
119. Zhu, L., Zeng, W.: Room-temperature gas sensing of ZnO-based gas sensor: a review. Sens. Actuat. A Phys. **267**, 242–261 (2017)

Chapter 7
Recent Advances in 1D and 2D ZnO Nanostructure-Based Photosensors

Trinayana Deka, Nikhil S. K., Pujita Ningthoukhongjam, Suma Das, and Ranjith G. Nair

Abstract Semiconductor metal oxide nanomaterials have achieved tremendous interest in the last few decades due to their various applications in sensing, photocatalysis, solar cell, fuel cell, etc. Among various semiconductor metal oxides, ZnO is one of the most emerging materials for sensing applications owing to its fascinating physicochemical properties and tunable morphologies. The photo-sensing performance of ZnO mainly depends on the high surface-area-to-volume ratio, porosity, physical and chemical stability, bandgap, high charge carrier generation, and effective separation. Many strategies such as doping, composites, sensitization, and morphological tuning have been reported to overcome the above limitations by improving the photo-response of the sensor. The current review gives an overview of the various factors affecting photo-sensing performance and recent advances in ZnO-based photosensors.

Keywords ZnO · Noble metal · Sensitization · Nanostructures · Photosensors

7.1 Introduction

Photosensors or photodetectors are types of optoelectronic device that convert photon energy to electrical energy and they have gained much attention in the field of sensor applications [1, 2]. Based on the photon detecting capabilities, photodetectors (PDs) can be classified as X-ray (1–10 nm), ultraviolet (10–400 nm), visible (400–700 nm), infrared (700 nm to 1 mm), and terahertz (0.1–1 mm or 0.3–30THz) [1, 2]. Among these, ultraviolet (UV) photosensors play a crucial role in detecting UV radiation in the Earth's atmosphere. It is because UV radiation may either be beneficial or harmful to humanity's survival and progress, depending on the amount of exposure and other fields such as optical communications, missile applications, motion

T. Deka · S. K. Nikhil · P. Ningthoukhongjam · S. Das · R. G. Nair (✉)
Solar Energy Materials Research and Testing Laboratory, Department of Physics, National Institute of Technology Silchar, Silchar, Assam 788010, India
e-mail: rgnair@phy.nits.ac.in

D. Mohanta and P. Chakraborty (eds.), *Nanoscale Matter and Principles for Sensing and Labeling Applications*, Advanced Structured Materials 206,
https://doi.org/10.1007/978-981-99-7848-9_7

monitoring, night vision, video and image processing, biomedicine, and air quality monitoring [1, 3]. In the case of the human body, the production of vitamin D, which promotes bone development, requires moderate UV light exposure to human skin, whether from a natural or artificial source. On the other hand, higher amount of UV radiation consumption may induce cataracts, ageing process, and even lead to life-threatening skin cancer [1, 3–5]. According to ISO standards, the UV band is defined as the spectral range of 10–400 nm [5]. The four primary spectral areas of UV light are UV-A (400–320 nm), UV-B (320–280 nm), UV-C (280–200 nm), and far UV (200–10 nm) [5, 6]. According to studies, the majority of UV light that reaches the atmosphere comes from the sun is UV-A, and thus UV-A radiation detection has risen to prominence [7–9]. However, other than the sunlight, UV radiation may be produced artificially via various sources such as phototherapy, plasma torches, UV light emitting diodes, welding arcs, and xenon mercury arc lamps [5]. As a result, UV radiation must be detected and analyzed, which necessitates the development of efficient and high-performance UV PDs to convert the UV signals to a measurable electrical signal. The working mechanism of UV PDs is briefly explained in the subsections given below:

7.1.1 Working Mechanism and Various Parameters of UV Photodetectors

The working mechanism of UV PDs is based on the photoconductivity effect [4, 7]. The bandgap of photosensitive material should be less than or equal to the energy of the illuminated photon for the detection of a light signal. Absorbing photon energy results in the generation of electron–hole pairs (EHPs) in the photosensitive material. The photogenerated electrons (e^-) and holes (h^+) are then collected as photocurrent in the external circuit. Applying external bias voltage or generating an internal built-in electric field can separate photogenerated e^- and h^+ for photocurrent generation. The response current of a PD is defined as the current produced by the PD. Photoresponse characteristic provides information about the generation of response current per unit incident UV illumination power on a PD. The sensitivity of a PD refers to the information related to the signal-to-noise level when exposed to UV at a certain intensity. Thus, for a high-performance UV PD, both response current and sensitivity must be higher. Another essential characteristic is the photodetection speed of a PD, which determines how quickly a PD detects UV signal or returns to its former condition after switching off the light source. The photodetection speed is primarily defined by two parameters: reaction time and recovery time. So a rapid response, i.e., lower value of response time and recovery time of a UV PD is particularly significant. Additionally, another parameter called specific detectivity of a PD is related to the noise of the signal of a PD. The noise of a signal is caused by the thermal fluctuations and dark current of the PD. Specific detectivity defines the information associated with a weak UV signal detection by PD. Shot noise, Johnson noise, and flicker noise are the three main forms of noises found in photodetectors, with both Johnson and flicker noises originating from temperature fluctuations. The UV PDs used nowadays are based on photosensitive semiconductors as the main working component which are discussed further.

7.1.2 *Current Scenario of Semiconductor-Based UV Photodetectors*

The silicon-based PDs have long been used to detect UV light, but it has a number of drawbacks including high fabrication cost, low photoresponsivity, the necessity for additional filters, and device degradation when exposed to high-energy UV light [10]. In the development of UV PDs, wide-band gap semiconductors such as ZnSe, ZnS, ZnTe, CdSe, CdS, ZnO, TiO_2, SnO_2, GaN, SiC, Ga_2O_3, and diamond are the potential candidates because of their inherent solar-blindness, non-necessity of filters, and many other desirable properties. Among these, Zinc oxide (ZnO), an II–VI semiconductor with a strong exciton binding energy (60 meV) and a wide bandgap (3.36 eV), can be considered a viable photonic material in ultraviolet (UV) spectral applications. Wurtzite structure ZnO is the most conventional ZnO due to its attractive physicochemical properties among three basic crystal structures of ZnO such as zinc blend, rock salt, and wurtzite. Wurtzite ZnO is a hexagonal structure having lattice parameters, a = 0.3296 nm and c = 0.52065 nm and space group P63mc. The Zn and O atoms in the ZnO wurtzite structure are stacked alternately along the c-axis in an hcp pattern with each Zinc atom surrounded by four oxygen atoms and vice versa, forming a tetrahedral non-centrosymmetric structure. As a result, wurtzite ZnO exhibits piezoelectric and pyroelectric capabilities, which are the additional attributes that allow ZnO-based UV PDs to improve photoconductivity. When compared with the bulk ZnO, semiconductor nanostructures have been found to be great candidates for light detection due to their unique physicochemical, optical, electrical, and mechanical properties [11, 12]. Various ZnO nanostructures have been effectively used in a variety of applications till now. The extensive amount of research clearly demonstrates the importance of ZnO morphology or structure in determining the performance. Moreover, the fabrication of ZnO in various dimensions via different techniques supports to obtain feasible photodetectors [4]. ZnO-based nanostructures such as nanowires and nanorods are the most reported nanostructures because of their significant surface-to-volume ratio and quantum size effect [13]. Their outstanding qualities make them ideal material for developing ultrasensitive PDs with rapid photo-response [14]. In addition to this, nanotubes, nanosheets, nanowalls, nanobelts, nanoribbons, and 2D layered and solution-processed nanofilms provide numerous unique beneficial features for photodetection. Owing to the exceptional physical, chemical, optical, and electronic properties, these ZnO-based nanostructures have attracted research attention globally [15–20]. These nanostructures also have a lot of potential for electrical and optoelectronic applications.

The current review focuses on various ZnO nanostructure and their different modification strategies for developing an effective UV PD. In addition to the pristine one, modified ZnO nanostructures via doping, construction of heterojunctions with suitable semiconducting material, and formation of Schottky junction with noble metal/non-metal are briefly discussed in this study for UV PD.

7.2 1D ZnO Nanostructures-Based Photosensor

7.2.1 Pristine 1D ZnO Nanostructures and Their Photo-Sensing Applications

The optical and electrical properties of ZnO are highly dependent on morphology. As such, the sensitivity of the photodetector is also highly influenced by the dimensions and the type of nanostructures of ZnO. Thus, a well-controlled and high-quality ZnO nanostructure is essential for photo-sensing applications.

Soci et al. investigated ZnO nanowires for UV detectors where they have measured photoconductive gain of about 2×10^8 [21]. They reported a significant increase in the photosensitivity and photocarrier lifetime of ZnO nanowires-based PD due to oxygen deficiency. This result was consistent with the photoconductive mechanism driven by oxygen adsorption and desorption at the surface of the nanowire. Meng et al. reported a fiber-based ZnO nanowire for UV Photodetector where they introduced a different technique to achieve a higher on/off ratio through interface and surface gating effects [22]. They reported that when the PD was irradiated by 1.0mW/cm^2 UV light, the on/off ratio reached a value of 1.98×10^8. Cheng et al. observed that the high surface-to-volume ratio of ZnO nanowires has a significant impact on the photo-response characteristics of the PDs. This was attributed to the presence of surface states caused by the breakdown of lattice periodicity on the surface [23]. They showed that the surface states can be modulated by applied bias and illumination intensity. When a negative bias voltage is applied at one end of the detector e^-s are injected into surface states from the electrode. As a result, photogenerated EHPs are separated and transferred efficiently by the surface built-in electric field, leading to a decrease in depletion width and potential barrier height. On the other hand, when both negative bias and light irradiation are applied, electrons can readily pass the electrode/ZnO NW interface, resulting in a strong photoresponsivity. Similarly, Bai et al. developed macroscopic integrated ZnO nanowire UV sensors [24]. They reported in time-resolved photo-response under UV light of 4.5 mWcm^{-2} at 3 V, a significant photo-response current of 14.1 mA while the dark current was at 110 nA, resulting in a 12×10^4 on/off ratio. Conversely, Yanbo Li and his group studied the surface effects of oxygen and water on UV photo-response of ZnO nanowire-based PD [25]. In their study, they found that under high humidity, the current diminishes gradually, whereas under low humidity, the current increases. Changing the humidity during the recovery phase causes a 2–3 order of magnitude shift in the decay time.

H_2O/O_2 absorption on the ZnO nanowires surface trapped an increased number of charges on the surface, and hence raises the dark current, delay in response time, and ultimately reduces the performance of ZnO nanowires-based UV PDs.

Banari et al. grew ZnO nanorod on the ZnO seed layer by two-step synthesis method, where the seed layer was acting as a nucleation center for the growth of ZnO nanorods [26]. The seed layer of various thicknesses of range 50–125 nm was fabricated via sol–gel spin-coating method and then they grew ZnO nanorod via hydrothermal method on the top of it. In their study, the prepared sample having a seed layer of 100 nm thickness showed the highest response current of 38 μA, sensitivity of 5.7, and fastest recovery time of 0.1 s under UV illumination. On the other hand, Cheng et al. reported self-powered ZnO nanorod-based PD grown on indium tin oxide (ITO) substrate by hydrothermal process introducing novel strategy through face-to-face hybrid ZnO nanorod arrays nanostructure [27]. When compared to single-face ZnO nanorod-based UV PDs, they found that the optimum UV PD based on face-to-face hybrid ZnO nanorod arrays nanostructure showed enhanced photoelectric conversion without adding any external DC voltage. They found that ZnO nanorod grown as a steeple top ZnO nanorod (s-ZNRAs) and flat-top nanorod (l-ZNRAs) as hydrothermal growth time was increased (see Fig. 7.1). Under a UV irradiation of 3.06 Wcm^2, the optimum UV PD had a responsivity of 2.45 mAW^{-1} and on/off ratio of 6.6×10^5. Similarly, Su et al. fabricated ZnO nanorod-based UV photosensor through hydrothermal and lithography-based approach where, the nanorod arrays were selectively grown in the space between interdigitated (IDT) electrodes of devices [28]. The reported UV nanorod photosensors had a much higher photo-response in the UV region than a standard ZnO photosensor without nanorod. Moreover, the photoconductive gain of a nanorod-based photosensor improved with increasing UV irradiation time of 18.28 mW/cm^2 UV illuminations. It varied in the range 34.45–5.32×10^2 under UV illumination of different illumination time.

Lao et al. fabricated surface functionalized ZnO nanobelt-based UV sensor by using poly(diallyldimethylammonium chloride) UV-absorbent polymer. They reported a fivefold increase in the UV response [29], which was ascribed to the energy levels introduced by the polymer sitting on top of it. These induced energy levels acted as a "hopping" state and enhanced the excitation probability of e^- to the conduction band (CB).

7.2.2 Doped 1D ZnO Nanostructures and Their Photo-Sensing Applications

ZnO nanostructure's optical and electrical properties can be controlled or enhanced using suitable dopants. Different studies have reported the performance enhancement of the ZnO nanostructures such as nanowires and nanorods through doping [30, 31].

Recently, to attain high optical and electrical properties, Zn^{2+} ions of ZnO nanostructures are mostly replaced with group III elements such as In^{3+}, Al^{3+}, and Ga^{3+}

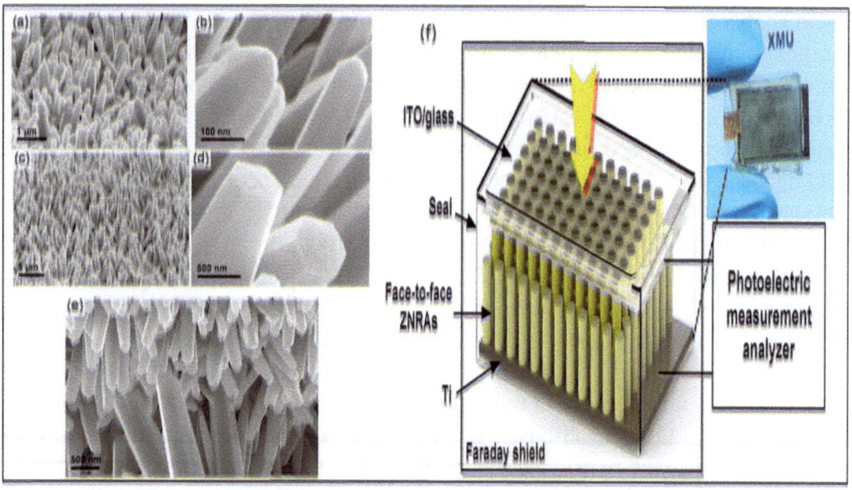

Fig. 7.1 SEM images of **a** *s*-ZNRAs and **c** *l*-ZNRAs and the partially enlarged images of (**b**) *s*-ZNRAs and **d** *l*-ZNRAs; **e** SEM images of face-to-face hybrid *s*-ZNRAs/*l*-ZNRAs nanostructure; **f** Schematic structure of UV PDs based on face-to-face hybrid ZNRAs-based nanostructure and the photograph of the actual photodetector. Chen et al. Nano Energy 65, 104,042 (2019) [27]. *Source* Reprint permission from Elsevier Publisher

[32–40]. Hsu et al. had synthesized Al-doped ZnO (AZO) nanowires using the hydrothermal method [32]. The PDs based on AZO nanowires showed a spectral response twice as that of pristine ZnO in UV and ten times greater in visible light. They also found the quantum efficiency to be maximum for AZO nanowires at 84.9%. The better performance of AZO nanowires was attributed to the occupation of Zn^{2+} by Al^{3+} ions, which generates extra electrons leading to an increase in conductivity. It was also observed that O_2 species acts as a hole sink preventing the recombination process and thus making free electrons available for conduction.

Among the different dopants used for doping in ZnO nanostructures, Gallium (Ga) is the best-preferred element since the atomic radius of Ga^{3+} (0.062 nm) is close to that of Zn^{2+} (0.074 nm). In addition, it exhibits a high conductive nature and low reactivity [41]. Studies have shown that doping ZnO with Ga increases overall electron concentration and provides oxygen adsorption sites thus promoting electron transfer to the surface [35]. Young et al. had reported a Ga-doped ZnO (GZO) nanorod-based flexible PD [36]. They used the chemical bath deposition method for the synthesis. It was observed that the GZO nanorod exhibited a slight thickening and elongation compared to ZnO nanorods. Parameters such as sensitivity, response, and recovery time were improved significantly after Ga doping. As per the defect structure model they proposed, the replacement of Zn^{2+} ions by Ga^{3+} ions results in a low carrier concentration leading GZO to act as a resistive material. This reduced defect density along with improved crystallinity results in better carrier lifetime and longer charge carrier mobility leading to enhanced UV detection. The influence of crystallinity on photodetection is further supported by a study done by

Kumar et al. [37]. They synthesized Ga–Ag co-doped ZnO (GAZO) nanorods-based UV detector using a two-step solution process. The SEM analysis confirmed the formation of hexagonal nanorods which is shown in Fig. 7.2. The average diameter and the crystallite size of the GAZO nanorods were greater than that of pristine ZnO nanorods. The photo-response of GAZO was maximum when compared to other samples. Also, the responsivity and photoconductive gain was found to be the highest in GAZO nanorods. The increase in photo-response was attributed to the highly crystalline nature of the GAZO nanorods. As a result of co doping an increased binding energy shift was observed in the XPS spectra which indicates to extra oxygen vacancies and also better intermolecular charge transfer, which also contributes to an improved photo-response.

Due to its similarity with the ionic radius of Zn and its significant effect on optical and electrical properties, doping ZnO nanorods with Fe has received much attention [40]. Chu et al. reported the influence of Fe doping on improving the photodetection performance of ZnO nanorods [42]. They carried out Fe doping on the ZnO seed layer using a low-temperature solution method. XRD analysis and FESEM confirmed the formation of nanorods. It was found that the Fe-doped ZnO (FZO) nanorods dimension was slightly greater than ZnO nanorods. FZO-based PDs showed a photocurrent with a sensitivity of 471.1. The rise time and decay time were faster when compared to pristine ZnO-based PD. The superior performance of FZO-based PD was explained to be a result of structural and morphological changes induced due to Fe doping. Fe^{3+} having a smaller ionic radius compared to Zn^{2+} easily gets substituted in the Zn^{2+} site.

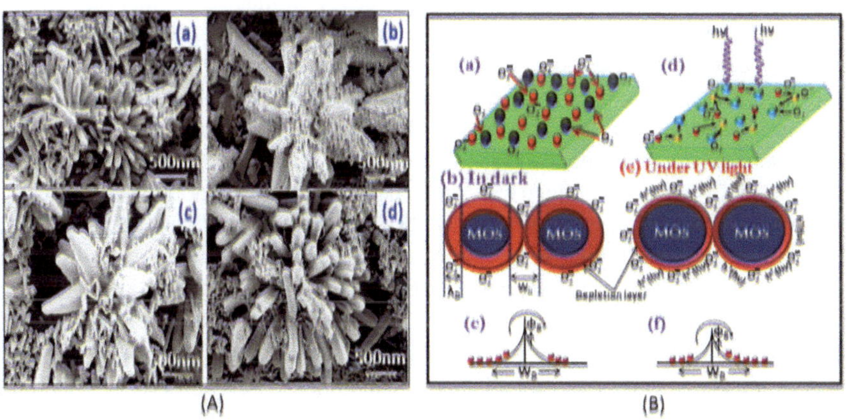

Fig. 7.2 **A** SEM micrographs of **a** ZO, **b** GZO, **c** AZO, and **d** GAZO hetero-structured UV PDs; **B** Schematic representation of **a** the interaction of MOS with oxygen molecules from air, **b** formation of depletion layer due to environmental exposure, **c** variation potential barrier height (ϕ_B) at the interface of two grains in the absence of UV light, **d** conversion of O_2 ions into O molecules in the presence of UV light via UV-photos (h^+), **e** decrease in the thickness of W_D in the presence of UV light, and **f** decrease in the ϕ_B under UV light for the creation of easy path for the electrons from VB to CB. Kumar et al., Applied Surface Science 509, 144,770 (2020) [37]. *Source* Reprint permission from Elsevier Publisher

But in order to keep the charge neutrality, the Zn^{2+} site released electrons. Also doping Fe introduces electrons to the conduction band. These two phenomena enhance the overall electron concentration thus increasing the responsivity. Moreover, the high surface-to-volume ratio of FZO nanorods also improves carrier concentration and introduces new oxygen vacancies thus enhancing the photocurrent response. The influence of Fe concentration on the performance was reported by Nurfani et al. [43]. It was observed that with the increase in Fe concentration, the sensitivity was increased up to 70 times (for 3% Fe) than the pristine ZnO. This can be attributed to the increase in electron concentration and homogenous nature of the film which enhances the transport mechanism in the material.

Another potential dopant for ZnO nanorod is the transition element Mn due to the similarity in ionic radii. It has been reported that Mn-doping increases the surface area and oxygen vacancies of the nanorod [44, 45]. The effect of dopant concentration on the PD parameters was reported by Alshamarti et al. They studied the effect of Mn concentration on ZnO nanorods on PD performance [46]. Mn-doped ZnO nanorod (MZO) was synthesized using microwave-assisted chemical bath deposition. As in the case of transition metal doping, the diameter of the MZO nanorods increases with the increase in doping concentration. Considering *I–V* characteristics, it was observed that the responsivity of the MZO PD increased with the increase in Mn concentration. Similarly, the quantum efficiency was highest for the sample having the highest Mn concentration. These superior performances of MZO samples were due to the introduction of delocalized charges under photoexcitation. Under the irradiance of UV light, charge concentration is increased through free charge carrier generation, thus, enhancing the photo-response. A similar study was reported by Chey et al. in which Mn ions substituted to the Zn sites acted as donor contributors, thereby, increasing carrier concentration and photoresponsivity [47].

Recently, Kim et al. reported a Sb-doped ZnO nanorod for UV sensing application. The samples were prepared using the hydrothermal method. The Sb-doped ZnO UV sensor showed a nine times increase in photocurrent when compared to the pristine ZnO-based UV sensor. The electrical conductivity of the ZnO nanorods was increased due to the Sb dopant-induced vacancies [48]. In other studies, ZnO nanorod-based PDs with Ni and Ti doping were reported by Chu et al. and Ade et al. [49, 50]. Ni-doped ZnO (NZO) nanorods were synthesized via chemical bath deposition on the ZnO seed layer. The sensitivity of the fabricated photodetectors was found to be increasing with the increase in Ni concentration. These NZO-based devices also showed faster rise and decay time. Doping with Ni generates h^+s, due to which, the conductivity of NZO PD is enhanced. Moreover, the high surface-to-diameter ratio provided by Ni doping promotes the adsorption of oxygen molecules by free electron capture leading to better electrical properties [49]. Ade et al. doped ZnO nanorods with Ti at three different doping concentrations (1, 3, 5%) via the SILAR method. They observed that Ti (3%) doped ZnO PD showed better performance compared to other devices. Doping with Ti 3% concentration increases carrier concentration and consequently increases the photocurrent. But at higher doping concentration (at 5%), structural deformities are formed due to the partial replacement of Zn^{2+} by Ti^{4+}. This leads to a decrease in electron mobility and subsequently, the photocurrent is affected [50]. A similar study on the influence of dopant concentration on the charge carriers was reported by Raj et al. [51]. They reported PD based on Nd-doped ZnO nanorods synthesized via the chemical bath deposition technique. They observed that the ZnO doped with 6% Nd shows superior optoelectronic performance compared to the rest of the samples. Substitution of Zn^{2+} with Nd^{3+} results in the formation of more electrons, subsequently allowing more O_2 to be absorbed onto the surface causing improvement in photoconductivity. Various other studies have also been carried out regarding the influence of dopants on the photodetection performance of ZnO nanorods [52–55].

Thus, the photodetection properties of ZnO nanostructures can be enhanced by doping with suitable dopants. This is due to the synergistic effect of the nanostructure morphologies, crystallite size, oxygen vacancies, and the increase in carrier concentration introduced due to doping.

7.2.3 ZnO-Based Heterojunction Nanostructures and Their Photo-Sensing Application

Construction of heterojunction of ZnO with other semiconductors can improve the optical and charge transport properties. As such, hybrid heterostructures with ZnO and other semiconductors are a popular route explored for improved performance for UV detection [56–59]. Many studies have established that structure tuning of ZnO nanostructures can promote the charge transport [60, 61]. However, there are limitations of impeded photocurrent due to the low surface area. Additionally, the preparation of heterojunctions is another way to promote charge separation and charge transfer. As such, many researchers have studied the pairing of 1D ZnO nanostructures like nanorods, nanowires, etc. with other semiconductors like NiO [62, 63], Cu_2O [61, 64, 65], Si [66–68], etc.

NiO, a wide bandgap (3.6 eV) and intrinsic p-type semiconductor, is an attractive candidate to pair with the n-type ZnO to construct highly effective UV sensors. Echresh and co-workers used hydrothermal techniques to prepare heterostructures of p-type NiO and n-type ZnO [62]. The NiO thin film was deposited over well-aligned ZnO nanorods oriented along the c-axis. The detector displayed superior UV detection capabilities with improved performance in terms of photocurrent, threshold voltage, rectification ratio, and light sensitivity. Similarly, Hoa et al. fabricated NiO/ZnO heterostructures with NiO nanosheets and ZnO nanorods via a hydrothermal process [63]. The formation of the heterostructure led to a decrease in trap-state emission and recombination rate. The NiO/ZnO-based photosensor showed 10 times the UV light sensitivity as compared to the pristine ZnO-based photosensor.

ZnO nanorods have also been paired with oxides of copper for enhanced photo-sensing applications [61, 64, 69]. Lin et al. designed Cu_2O/ZnO nanorods photodetectors that were suitable for ambient light detection applications [64]. The heterostructures had a spectral response range from 380 to 570 nm. The formation of the heterojunction with Cu_2O not only expanded the spectral response range but the fabricated photodetectors also showed improved rectifying characteristics with low leakage current and low turn-on voltage. Besides, Liu et al. reported the fabrication of a UV detector based on flower-like Cu_2O and ZnO nanorods heterojunction with good stability and reproducibility [61]. The heterostructure of Cu_2O flowers and ZnO nanorods have an increased effective area and hence provide an increased number of active sites on the surface. Additionally, a study by Ji et al. confirmed the enhanced UV sensing capabilities of ZnO/CuO nanostructure heterojunction [69]. They fabricated PD based on ZnO nanorod arrays decorated with hollow CuO nanospheres which exhibited 32.6 times faster photo-response than the pristine one for UV light detection.

Silicon, particularly *p*-type silicon [67, 68, 70, 71], has been a popular candidate to pair with n-type ZnO nanostructures to enhance photo-sensing properties. Ko et al. used silica-ZnO dome-shaped architecture as the buffer layer for the growth of ZnO nanorod arrays [70]. As-fabricated ZnO nanorod/Si heterostructures were tested for UV photodetection applications and an enhanced UV-induced photocurrent was reported. The intermediate domed architecture reduced light reflection in the UV range and contributed to the improved performance of the heterostructure. Another work reported the fabrication of fast response Si/ZnO nanorods heterostructure-based photosensor [71]. The sensors showed photo-response in both the UV and visible range. Additionally, photo-response in the visible region was found to be enhanced after UV activation. This was attributed to the defect states-trapped e^-s that can be further excited by visible illumination. This increased the carrier concentration and caused the enhanced photo-response.

Liu and co-workers grew ZnO nanorods on GaN thin films via a hydrothermal method and spin-coated graphene quantum dots on the nanorod arrays [72]. They observed that this ternary heterostructure showed high sensitivity and high photoresponsivity for UV detection applications. The graphene quantum dots acted as light absorber and electron donor, increasing the carrier concentration, and leading to enhanced UV detection performance over that of bare ZnO nanorods. SnO_2, another wide bandgap (3.6 eV) semiconductor, is also a candidate to pair with ZnO for UV detection applications. In another study, Fu et al. reported a ZnO/SnO_2 nanorod array with SnO_2 as the shell layer on the ZnO nanorod cores [73]. The core–shell structure showed improved near band-edge photoluminescence emission, photoresponsivity (28.5 A/W), specific detectivity (2.9×10^{14} Jones), response time, and recovery times as compared to bare ZnO nanorods (Fig. 7.3). The SnO_2 shell layer introduces localized states that prevent the photogenerated carriers from being trapped by defect states.

Li et al. inserted a SiO_2 layer in between the ZnO nanowires and NiO. They reported that the insulating layer between the two semiconductor layers served as a barrier that minimized the transmission of low-energy carriers. This resulted in

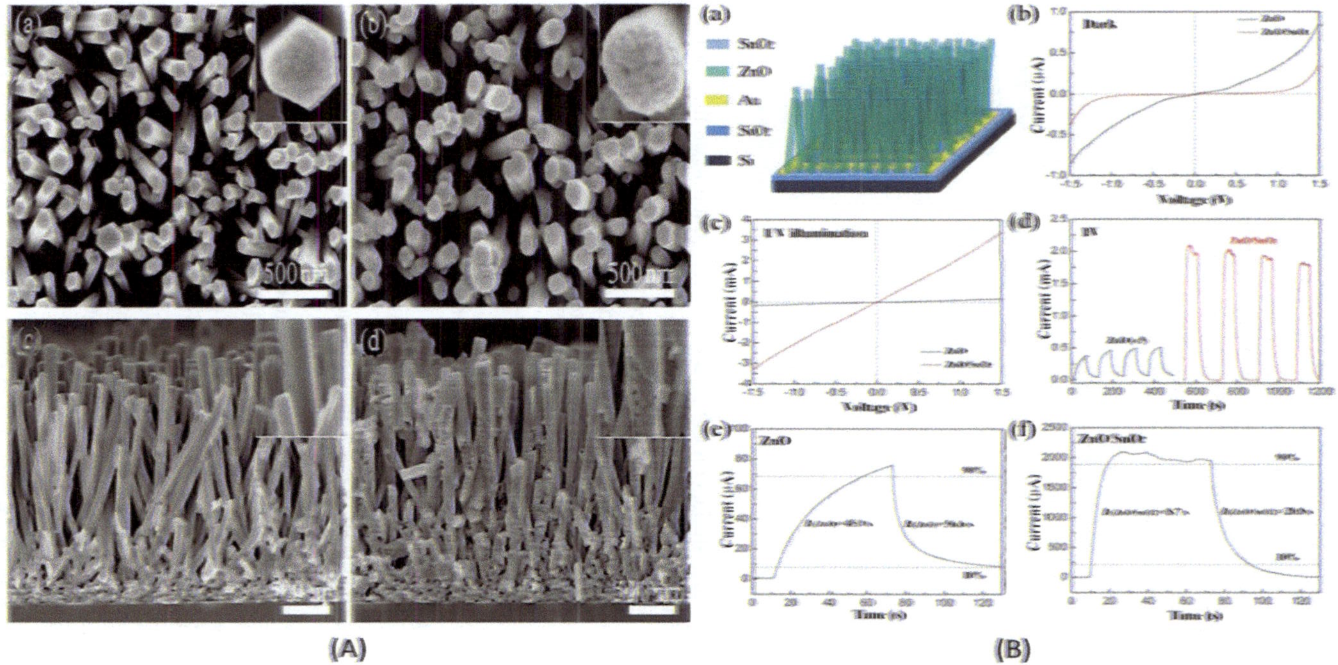

Fig. 7.3 **A** Top view FE-SEM images of **a** bare ZnO NAs and **b** ZnO/SnO_2CSNAs. **c** and **d** are the corresponding cross-sectional view FE-SEM images. The insets are the high-magnification images. Q.-M. Fu, et al. Applied Surface Science 527, 1,469,232 (2020) [69]; **B** Schematic illustration of PD (ZnO/SnO_2). I–V characteristics of PD (ZnO) and PD (ZnO/SnO_2) under **b** dark and **c** UV illumination. **d** Time-dependent photo-response of PD (ZnO) and PD (ZnO/SnO_2) at 1 V. The current of PD (ZnO) is amplified by 5 times. The photo-response time of **e** PD (ZnO) and **f** PD (ZnO/SnO_2). Fu et al., Applied Surface Science 527, 146,923 (2020) [69]. *Source* Reprint permission from Elsevier Publisher

a reduced dark current density (3.5×10^{-7} A/cm^2) and increased UV sensitivity (16.23) as compared to the ZnO/NiO heterostructures without the SiO_2 layer [74]. In a similar vein, Bai and Zhang fabricated ZnO/Cu_2O branched heterojunction arrays, consisting of Cu_2O nanocrystals grown on ZnO nanowire arrays. The PDs based on the branched arrays structure showed superior photo-sensing performance in both UV and visible illumination at zero bias [65]. It was reported that the fabricated heterojunction-based PDs exhibited enhanced photoresponsivity of around 8.2 mA/W in the visible region (400 nm) as compared to the bare ZnO nanowire array. Moreover, Shahkarami et al. grew ZnO nanowires on porous Si and plane Si using hydrothermal methods. The ZnO nanowire/porous Si nanostructure showed lower reflectivity and superior performance as compared to the one with plane silicon [66].

Graphene single layers have been reported to make up for the energy band mismatch between ZnO and Si for smooth transfer of charge carriers [75]. In that context, Liu et al. studied Si/ZnO nanowire heterojunction-based UV photodetectors (PD), with and without an intervening graphene single layer. The detector with graphene exhibited a UV photosensitivity of 1071 which was 7 times better than the detector without graphene. It also showed faster response time (1.02 s) and recovery time (0.34 s). ZnO/SnO_2 nanowires that had been fabricated via the electrospinning technique were tested for UV photodetection application [59]. The nanowire-based PD showed enhanced performance, with high UV sensitivity, good stability, and reversibility.

In addition, various other studies on the effect of heterojunction on the photodetection properties of ZnO nanostructures have been reported [76–79]. Thus, it can be inferred that heterojunction formation can enhance the photodetection properties of ZnO nanostructures through enhanced charge carrier separation, increased charge carrier concentration, and synergistic effect of the two semiconductors.

7.2.4 Schottky Junction-Based ZnO Nanostructures and Its Photo-Sensing Application

Schottky junction-based ZnO UV PDs are another strategy to enhance the PDs performance owing to their relatively low dark current, low noise performance, and efficient photo-response [80]. In the case of Schottky junction, if the work function of the metal is larger than that of ZnO, e^-s in ZnO will flow to the metal, establishing a thermal equilibrium state and aligning the Fermi energy levels. Due to the deficiency of electrons at the surface of ZnO and the accumulation of electrons near the surface of the metal, a space charge area will be formed. The space charge zone has a finite width, a stable internal electric field, and a finite internal potential difference when it is in thermal equilibrium. The band of ZnO bends upward, generating a Schottky barrier that prevents electrons from passing through it. The development of the potential difference is responsible for the separation of photogenerated electrons (e^-) and holes (h^+) in a Schottky barrier-based UV PD. As a result, the height of the

Schottky barrier might impact the UV PDs' properties [14]. As per our knowledge, Fabricius et al. reported the first Schottky junction-based photodiode using a thin sputtered layer between ZnO and Au in 1986 [81]. In this study, the prepared diode showed conventional IV characteristics and sensitivity for wavelengths of less than 400 nm. Cheng et al. described UV PD based on ZnO nanowire Schottky junctions using Au, where 4×10^5, 2.6×10^3 A/W, and 8.5×10^3 were the on/off ratio, responsivity, and photocurrent gain, respectively [82]. A recovery time of 0.28 s was found, during which the photocurrent decreased by three orders of magnitude from its peak. In a similar work, Zheng et al. reported interface engineering of Au/ZnO Schottky junction with TiO_2 inserting layer [83]. The inserting layer modifies the distribution of the interface energy band, decreasing the density of surface states and enhancing the built-in electric field. As a result, the photogenerated EHPs can be efficiently separated, leading to increased photocurrent and decreased dark current.

Zhang et al. reported a sensitive UV photodiode based on graphene-ZnO Schottky junction formation with a barrier height of about 0.623 eV [84]. The photoresponsivity and photoconductive were reported to be 3×10^4 A/W and 10^5, respectively, which were significantly greater than other ZnO nanostructure-based devices. Here, photogenerated electrons transfer from ZnO to graphene after being exposed to UV light, crossing the Schottky barrier to generate photocurrent. Using Schottky-contact and surface functionalization, Zhou et al. improved the response and reset time of a ZnO for UV photo-sensing applications [85]. They reported that under ambient conditions, the UV sensitivity of the nanosensor was increased by four orders of magnitude by using Schottky-contact, and the reset time was substantially decreased from 417 to 0.8 s. Even without rectifying the electronic response of the measuring system, the reset time was lowered to 20 ms by further surface functionalization using function polymers.

Liang et al. reported an n-type ZnO Schottky junction formed with Ag having a Schottky barrier of about 0.84 eV [86]. In this study, they reported a low-frequency photoresponsivity of 1.5A/W, a rise time of 12 ns, and a fall time of 50 ns, at an external bias of 5 V. Similarly, Singh et al. investigated Ag–ZnO nanorods for UV detectors which were grown on ITO substrate by hydrothermal method having the photocurrent and dark current contrast ratio of the prepared nanorods was around 1.67 at 1.0 V biasing [87]. Another efficient Schottky junction of ZnO nanorod with graphene for UV PD was reported by Zhu et al. which favors the separation and transport of the photogenerated electron and hole pairs [88].

7.3 2D ZnO Nanostructures-Based Photosensor

Owing to the exceptional physical, chemical, optical, and electronic properties, 2D frameworks of ZnO typically nanoplates [17], nanosheets [18], nanowalls, and nanodisks have attracted research attention globally. These exceptional properties are due to the outcome of strong quantum confinement of electrons in 2D frameworks and their high specific surface areas [16].

Aside from 1D structures, Young et al. recently reported the photodetection properties of PD based on ZnO nanosheets mediated by a room-temperature solution method, which shows a normalized detectivity of about 2.17×10^9 J at 1 V applied bias [89], whereas Krishnamurthi et al. reported the PD based on ultrathin ZnO nanosheets (of 5 nm thickness) synthesized via a liquid metal-based Van der Waals exfoliation method with enhanced detectivity of about 5.81×10^{15} Jones at an extremely low voltage of 0.05 V [90]. Besides, the PD exhibited external quantum efficiency (EQE) and responsivity, 4.3×10^3%, and 12.64 A/W at a wavelength of 365 nm. In another study, Lee et al. reported the hydrothermal synthesis of thin 2D ZnO nanosheet-based flexible piezo-phototronic ultraviolet (UV) PD [91]. Under bending conditions, the detector exhibited 2, 3, and 4 times higher photosensitivity, responsivity, and EQE, respectively, compared to the normal condition. Moreover, the detectivity of the photodetector under normal and bending conditions had been reported to be 1.98×10^{10} Jones and 8.96×10^{10} Jones, respectively, at a bias of 5 V. Agrawal et al. prepared ZnO nanowalls (honeycomb nanostructures) via a simple and cost-effective hydrothermal approach for photosensitivity [92]. The nanowall structure provides high electron transportation efficacy resulting in large photo-response. The PD demonstrated high photo-response in broadband 200–950 nm. It exhibited maximum photosensitivity (3.2×10^7) and photodetectivity (2.07×10^{15}Jones) with EQE (2.07×10^{15}%) at 300 nm. This result is comparable to some of the existing commercial broadband photodetectors. Alenezi et al. synthesized ZnO hexagonal nanodisks via a one-pot, low-temperature, and controlled synthesis method. They demonstrated a comparison study between MSM ZnO UV detectors based on a single nanodisk and multiple nanodisks. The single nanodisk PD offered a large photoresponsivity (3300 A/W) and high EQE, which is attributed to the absence of grain boundaries, high aspect ratio, and polarity of the exposed surface. Whereas the multiple-nanodisk devices exhibited high photosensitivity of 1500A/W and shorter response and recovery times compared to that of a single nanodisk, due to the presence of nanojunctions in the structure of the multiple-nanodisk devices [93].

Yang et al. reported 2D vertically-aligned Ga-doped ZnO (GZO) nanosheets MSM UV PD by using a low-temperature (90 °C) hydrothermal method [38]. Their results revealed that the fabricated GZO photodetector exhibited a photocurrent to dark current contrast ratio of 15.2 [39]. On the other hand, a similar photodetector has fabricated by Young et al. via a simple aqueous solution method [94]. The fabricated photodetector offers approximately 1.4×10^4 photocurrent and dark-current contrast ratios at 1 V, which is much higher than that of the former one. Alongside, the photodetector shows 2.45 and 4.0 s transient time constants during the rise time and fall time, respectively. In another study by Young et al., high photosensitivity of the order of 5.98×10^5 had been reported using Al-doped ZnO (AZO) nanosheets Metal–Semiconductor-Metal (MSM) UV PD upon UV light illumination [33]. The AZO nanostructures were uniformly grown on the glass substrates through a simple and inexpensive aqueous chemical solution method. The photodetector exhibited fast response and recovery times and good orientation properties. Agrawal et al. also synthesized Al-doped ZnO nanowalls network-based flexible and ultralow voltage UV photodetector via a cost-effective hydrothermal process [34]. It was reported that the charge transport efficiency of nanowalls is far better than other nanostructures due to their larger polar surface area in the conduction path. The device has shown high responsivity and photosensitivity of about 265 and 349, respectively, at just 0.1 V applied bias. Additionally, a prominent UV-sensitive piezoelectric response, having a dark voltage to photovoltage ratio of 24.5, has been demonstrated that has made the device a potential candidate for fast response time, self-powered photodetector [24].

Lee et al. fabricated a flexible PD composed of 2D–ZnO nanosheets/1D-GaN nanorods through two-step growth method [95]. Upon UV light illumination, the fabricated PD exhibited twofold-enhanced photoresponsivity and detectivity than that of GaN nanorods. Along with this, even under the application of tensile strain, the device exhibited enhanced photoresponsivity. This enhancement was attributed to the positive piezo-potential charges induced at the heterojunction interface, which lowers the barrier height and favors the separation and transport of photogenerated carriers at the interface. Shuai et al. synthesized CdS nanowires/ZnO nanowalls heterostructures by in-situ method and investigated their UV–visible photodetection performance [96]. The photodetectors exhibited R_{dark}/R_{light} ratio of 4.3 under the visible light. Compared with pristine CdS nanowires and ZnO nanowalls the photodetectors based on CdS/ZnO heterojunctions demonstrate higher photo response activity. This enhancement in performance was attributed to the charge carrier separation efficiency and fast charge transportation facilitated on account of close contact between CdS NWs and ZnO nanowalls. Dixit et al. demonstrated a high-performance broadband UV-to-NIR (250–850 nm) detection with as-synthesized ZnO-$ZnCr_2O_4$ nanowall-based photodetectors [97]. The dark current was reported to be as low as 0.12 nA. The detector exhibited high sensitivity for the UV region ($\lambda_{ex} = 350$ nm) with the photosensitivity of $\sim 1.28 \times 10^5$, photoresponsivity of 5.49 A/W, photodetectivity of 1.91×10^{13} Jones, and EQE of 1900%.

On the other hand, Li et al. prepared Ag nanorods/nanowires assembled ZnO nanowalls heterostructures by changing synthesis routes via (i) a facile sonication combined with annealing treatment method and (ii) in-situ synthesis and spin-coating method [98, 99]. FESEM and EDS analysis confirmed the growth of vertically-aligned Ag nanorods/tightly anchored Ag nanowires and their uniform distribution over ZnO nanowalls. Upon UV light irradiation, PDs based on Ag nanorods/ZnO nanowalls exhibited higher response and shorter response-recovery time compared to Ag nanowires/ZnO nanowalls at room temperature. Its sensitivity also shows a linear relationship between 10 and 70 °C.

Different ZnO-based nanostructures prepared via different routes and their photosensing characteristics are summarized in Table 7.1.

Table 7.1 ZnO nanostructure-based photo-sensors and their photo-sensing parameters

Configuration	Method	Bias (V)	λ (nm)	Detectivity (Jones)	Rise time/ fall time (s)	Sensitivity	Responsivity (A/ W)	Refs.
Al–ZnO NR	Hydrothermal	5	360	–	–	–	3.61	[32]
Ga–ZnO NR	Chemical bath deposition	1	360	5.84×10^{11}	89/106	–	0.046	[36]
Ga–Ag–ZnO NR	Hydrothermal	5	365	–	13/18	–	1424.5	[37]
Fe–ZnO NR	Chemical bath deposition	3	380	–	–	471.1	0.758	[42]
Fe–ZnO NR	Spray pyrolysis	10	365	–	–	253	–	[43]
Mn–ZnO NR	Microwave-assisted chemical bath deposition	5	365	–	27.07/61.39	–	0.0765	[46]
Mn–ZnO NR	Hydrothermal	5	365	–	2.75/16.8	6.03	0.065	[47]
Sb–ZnO NR	Hydrothermal	0.5	365	–	338/124	–	53.5	[48]
Ni–ZnO NR	Chemical bath deposition	3	380	–	70/43	393.04	2.10	[49]
Ti–ZnO NR	*SILAR	5	365	5.49×10^{10}	–	–	0.30	[50]
Nd–ZnO NR	Chemical bath deposition	5	385	1.18×10^{11}	0.3/1.8	–	0.73	[51]
Cl–ZnO NR	Chemical deposition	0	365	1.12×10^{10}	0.0/0.032	–	0.8	[52]
Co–ZnO NR	Hydrothermal	1	380	7.3×10^{11}	–	–	19.8	[53]
Cu–ZnO NR	Hydrothermal	5	–	–	–	2080	–	[54]
Li–ZnO	Hydrothermal	5	380	–	–	–	34.87	[55]
Cu_2O/ZnO NRs	Chemical Bath Deposition	0	370		0.01/0.02		0.24	[64]

(continued)

Table 7.1 (continued)

Configuration	Method	Bias (V)	λ (nm)	Detectivity (Jones)	Rise time/ fall time (s)	Sensitivity	Responsivity (A/W)	Refs.
ZnO/Cu_2O branched NWs arrays	Hydrothermal	0	400	–	0.14/0.36	1945	0.0082	[65]
Hollow CuO nanospheres/ ZnO NRs	Hydrothermal and Thermal annealing	− 3	254		1.8/ 0.26	4.58	–	[69]
GaN thin film/ZnO NRs	Hydrothermal	10	366	$\sim 10^{12}$	0.1/ 0.12		0.034	[72]
ZnO NRs/Si	Hydrothermal and Plasma Enhanced Chemical Vapor Deposition	-2	365			6.03	–	[70]
ZnO NRs/Si	Hydrothermal	-1	365		2.92/0.24	1.86	–	[71]
ZnO NRs/SnO_2 core shell arrays	Hydrothermal	1	365	2.9×10^{14}	8.7/20.8		28.5	[73]
ZnO/CdS Nanowires	Hydrothermal	0	370	–	0.18/0.32	–	0.354	[77]
ZnO core/ZnSe shell nanowires	Thermal Evaporation	2	365	–	6.2/5.8	233	–	[78]
NiO/ZnO NRs	Chemical Bath Deposition	0	365	5.84×10^{11}	0.002/0.026	110.9	0.074	[79]
ZnO nanosheet	Solution	1	365	2.17×10^{9}	5.37/6.02	–	–	[89]
ZnO ultrathin nanosheet	Liquid metal-based Van der Waals exfoliation	0.05	365	5.81×10^{15}	11.58/73.58	–	12.64	[90]
ZnO nanosheet (Normal)	Hydrothermal	5	–	1.98×10^{10}	–	3	34	[91]
ZnO nanosheet (Bending)	Hydrothermal	5	–	8.96×10^{10}	–	6.57	113	

(continued)

Table 7.1 (continued)

Configuration	Method	Bias (V)	λ (nm)	Detectivity (Jones)	Rise time/ fall time (s)	Sensitivity	Responsivity (A/W)	Refs.
ZnO nanowalls	Hydrothermal	10	300	2.07×10^{15}	–	3.2×10^{7}	115	[92]
ZnO nanowalls	Sonication-annealing	–	365	–	23/81	–	217	[99]
ZnO nanodisk	Hydrothermal	3	365	–	> 7	–	3386	[93]
Ga-doped ZnO nanosheet	Hydrothermal	1	340	–	46/32	> 15.2	2.83×10^{-5}	[38]
Ga-doped ZnO nanosheet	Hydrothermal	1	340	2.24×10^{9}	–	15	2.83×10^{-5}	[39]
Ga (1.35%)-doped ZnO nanosheet	Aqueous solution	1	365	–	2.45/4	1.4×10^{4}	5.75	[94]
Al-doped ZnO nanosheet	Aqueous solution	–	–	–	0.7/2.5	5.98×10^{5}	0.031	[33]
Al-doped ZnO nanowalls	Hydrothermal	0.1		4.5×10^{10}	–	348	0.265	[34]
p–Si/V-doped ZnO nanosheet	–	1	442 & 1064	–	–	8.8	0.120	[18]
CdS nanowires/ZnO nanowalls	In-situ	4.5	365	–	1/1	8.561	–	[96]
ZnO–$ZnCr_2O_4$ nanowalls	Spin coating-annealed		250–850	1.91×10^{13}	–	1.28×10^{5}	5.49	[97]
GaN nanorods/ZnO nanosheet	Hydrothermal	1	–	4.25×10^{10}	–	2.78	0.34	[95]
Ag nanowires decorated ZnO nanowalls	In-situ and Spin coating		365		10/81		285	[98]
Ag nanorods assembled ZnO nanowalls	Sonication-annealing	–	365	–	8/57	–	513	[99]

*Successive ionic layer adsorption and reaction

7.4 Conclusion and Future Scope

Wide band gap semiconductors as UV photo-sensing materials have drawn much attention for various applications in both civil and defence fields, where ZnO seems to be a good candidate due to its excellent physico-chemical properties. In this review, an attempt has been made to give an overview of 1D–2D ZnO nanostructures for photo-sensing applications. In this study, UV photo-response of the various intrinsic ZnO nanostructures as well as modified ZnO with different possible approaches such as doping, heterojunction, and Schottky junction formation is also briefly discussed. The surface-to-volume ratios, interface areas, and contact between surface and environment have a significant impact on the performance of the photodetectors. It was found that the morphology and its modifications significantly affect the performance as a photosensor. It is noted that different kinds of nanorods and nanowires have promising photo-sensing applications. To enhance the photo-sensing performance, some effective approaches have been described, such as surface engineering, interface modification, material and device geometry optimization, semiconductor bandstructure engineering, etc. In conclusion, by combining diverse nanostructures and adopting appropriate device architectures, this review could lead to the development of effective ZnO nanostructure-based photodetectors in the near future.

UV PDs can be further improved in response time by reducing the defects of the ZnO nanostructure. In addition to this optimum surface roughness of the ZnO nanostructure needs to take care of EHP generation and trapping efficiency. The photo-sensing application can be increased by its surface functionalization, optimizing the characteristics of the semiconductor and the piezoelectric behavior of ZnO. Hence, there are scopes of research to improve various parameters in various dynamics to increase the quality of ZnO nanostructures and to use the cost-effective solution path to extend its photo-sensing applications. Another effective scope of improvisation of ZnO quantum dot (QD)-based PDs can be in terms of general, self-powered, and wavelength selective. QDs of various sizes of materials can be blended to create novel self-powered photodetectors and to filter short and long wavelengths of photons for the wavelength selectivity of the photodetectors. The selectivity of self-powered photodetectors may also be investigated using doped QDs, core–shell, and alloyed QDs of various materials.

References

1. Wadhwa, R., Agrawal, A.V., Kumar, M.: J. Phys. D Appl. Phys. **55**, 063002 (2022)
2. Chetia, A., Bera, J., Betal, A., Sahu, S.: Mater. Today Commun. **30**, 103224 (2022)
3. Nabet, B.: Photodetectors. Elsevier (2016)
4. Deka Boruah, B.: Nanoscale Adv. **1**, 2059 (2019)
5. Al Fattah, M.F., Khan, A.A., Anabestani, H., Rana, M.M., Rassel, S., Therrien, J., Ban, D.: Nanoscale **13**, 15526 (2021)
6. Ren, L., Tian, T., Li, Y., Huang, J., Zhao, X.: ACS Appl. Mater. Interfaces **5**, 5861 (2013)
7. Yadav, P.V.K., Ajitha, B., Kumar Reddy, Y.A., Sreedhar, A.: Chemosphere **279**, 130473 (2021)

8. Zou, Y., Zhang, Y., Hu, Y., Gu, H.: Sensors **18**, 2072 (2018)
9. Reddy, Y.A.K., Ajitha, B., Reddeppa, M., Sreedhar, A.: J. Mater. Sci. Mater. Electron. **30**, 20687 (2019)
10. Rasool, A., Santhosh Kumar, M.C., Mamat, M.H., Gopalakrishnan, C., Amiruddin, R.: J. Mater. Sci.: Mater. Electron **31**, 7100 (2020)
11. Guo, Y., Xu, K., Wu, C., Zhao, J., Xie, Y.: Chem. Soc. Rev. **44**, 637 (2015)
12. Thorkelsson, K., Bai, P., Xu, T.: Nano Today **10**, 48 (2015)
13. Pradel, K.C., Wu, W., Zhou, Y., Wen, X., Ding, Y., Wang, Z.L.: Nano Lett. **13**, 2647 (2013)
14. Li, X., Liu, X., Li, Y., Gao, D., Cao, L.: IEEE Access **9**, 11954 (2021)
15. Dai, Z.R., Pan, Z.W., Wang, Z.L.: Adv. Funct. Mater.Funct. Mater. **13**, 9 (2003)
16. Al-Hardan, N.H., Hamid, M.A.A., Shamsudin, R., Othman, N.K.: In: Pyshkin, S.L., Ballato J. (eds.) Optoelectronics—Advanced Device Structures. InTech (2017)
17. Al-Hardan, N.H., Abdullah, M.J., Ahmed, N.M., Yam, F.K., Abdul Aziz, A.: Superlattices Microstruct. **51**, 765 (2012)
18. Dai, Y., Wu, C., Wu, Z., Zhao, Z., Li, L., Lu, Y., Wang, Z.L.: Adv. Sci. **6**, 1900314 (2019)
19. Tiwari, J.N., Tiwari, R.N., Kim, K.S.: Prog. Mater. Sci. **57**, 724 (2012)
20. Li, Z., Xu, K., Wei, F.: Nanotechnol. Rev.. Rev. **7**, 393 (2018)
21. Soci, C., Zhang, A., Xiang, B., Dayeh, S.A., Aplin, D.P.R., Park, J., Bao, X.Y., Lo, Y.H., Wang, D.: Nano Lett. **7**, 1003 (2007)
22. Meng, L., Li, G., Tian, X., Bai, S., Xu, Q., Jia, X., Cui, X., Qin, Y., Wu, W.: ACS Appl. Mater. Interfaces **12**, 1054 (2020)
23. Cheng, B., Xu, J., Ouyang, Z., Xie, C., Su, X., Xiao, Y., Lei, S.: Opt. Express **21**, 29719 (2013)
24. Bai, S., Wu, W., Qin, Y., Cui, N., Bayerl, D.J., Wang, X.: Adv. Funct. Mater.Funct. Mater. **21**, 4464 (2011)
25. Li, Y., Della Valle, F., Simonnet, M., Yamada, I., Delaunay, J.-J.: Appl. Phys. Lett. **94**, 023110 (2009)
26. Banari, M., Memarian, N., Vomiero, A.: Mater. Sci. Eng. B **272**, 115332 (2021)
27. Chen, C., Hu, B., Wang, Z., Lv, X., Zhang, C., Chen, B., San, H., Hofmann, W.: Nano Energy **65**, 104042 (2019)
28. Su, Y.K., Peng, S.M., Ji, L.W., Wu, C.Z., Cheng, W.B., Liu, C.H.: Langmuir **26**, 603 (2010)
29. Lao, C.S., Park, M.-C., Kuang, Q., Deng, Y., Sood, A.K., Polla, D.L., Wang, Z.L.: J. Am. Chem. Soc. **129**, 12096 (2007)
30. Gozeh, B.A., Karabulut, A., Yildiz, A., Yakuphanoglu, F.: J. Alloy. Compd. **732**, 16 (2018)
31. Ben Elkamel, I., Hamdaoui, N., Mezni, A., Ajjel, R., Beji, L.: J. Mater. Sci.: Mater. Electron **30**, 9444 (2019)
32. Hsu, C.-L., Hsu, D.-X., Hsueh, T.-J., Chang, S.-P., Chang, S.-J.: Ceram. Int. **43**, 5434 (2017)
33. Young, S.-J., Liu, Y.-H.: IEEE J. Select. Top. Quantum Electron **23**, 1 (2017)
34. Agrawal, J., Dixit, T., Palani, I.A., Singh, V.: IEEE Sens. Lett. **3**, 1 (2019)
35. Huang, Y., Zhang, H., Dun, G., Li, Y., Liang, R., Yang, Y., Xie, D., Tian, H., Ren, T.-L.: In: 2021 IEEE 16th International Conference on Nano/Micro Engineered and Molecular Systems (NEMS), pp. 680–684 (2021)
36. Young, S.-J., Liu, Y.-H., Shiblee, M.D.N.I., Ahmed, K., Lai, L.-T., Nagahara, L., Thundat, T., Yoshida, T., Arya, S., Furukawa, H., Khosla, A.: ACS Appl. Electron Mater. **2**, 3522 (2020)
37. Guru Sampath Kumar, A., Li, X., Du, Y., Geng, Y., Hong, X.: Appl. Surf. Sci. **509**, 144770 (2020)
38. Yang, C.-C., Su, Y.-K., Hsiao, C.-H., Young, S.-J., Kao, T.-H., Chuang, M.-Y., Huang, Y.-C., Wang, B.-C., Wu, S.-L.: IEEE Photon. Technol. Lett. **26**, 1317 (2014)
39. Yang, C.C., Su, Y.K., Hsiao, C.H., Kao, T.H., Peng, Y.M., Chuang, M.Y., Wang, B.C., Wu, S.L.: J. Electrochem. Soc.Electrochem. Soc. **161**, H399 (2014)
40. Sankar Ganesh, R., Patil, V.L., Durgadevi, E., Navaneethan, M., Ponnusamy, S., Muthamizhchelvan, C., Kawasaki, S., Patil, P.S., Hayakawa, Y.: Chem. Phys. Lett. **734**, 136725 (2019)
41. Young, S.-J., Chiou, C.-L.: Microsyst. Technol.. Technol. **24**, 103 (2018)
42. Chu, Y.-L., Young, S.-J., Ji, L.-W., Tang, I.-T., Chu, T.-T.: Sensors **20**, 3861 (2020)

43. Nurfani, E., Lailani, A., Kesuma, W.A.P., Anrokhi, M.S., Kadja, G.T.M., Rozana, M.: Opt. Mater. **112**, 110768 (2021)
44. Putri, N.A., Fauzia, V., Iwan, S., Roza, L., Umar, A.A., Budi, S.: Appl. Surf. Sci. **439**, 285 (2018)
45. Patil, Y., Pedhekar, R.B., Patil, S., Raghuwanshi, F.C.: Mater. Today: Proc. **28**, 1865 (2020)
46. Alshamarti, H.A., Omran Alkhayatt, A.H.: Mater. Sci. Semicond. Process. **114**, 105068 (2020)
47. Chey, C.O., Liu, X., Alnoor, H., Nur, O., Willander, M.: Physica Status Solidi (RRL)–Rapid Res. Lett. **9**, 87 (2015)
48. Kim, D., Kim, W., Jeon, S., Yong, K.: RSC Adv. **7**, 40539 (2017)
49. Chu, Y.-L., Ji, L.-W., Hsiao, Y.-J., Lu, H.-Y., Young, S.-J., Tang, I.-T., Chu, T.-T., Chen, X.-J.: J. Electrochem. Soc.Electrochem. Soc. **167**, 067506 (2020)
50. Ade, R., Kumar, S.S., Valanarasu, S., Kumar, S.S., Sasikumar, S., Ganesh, V., Bitla, Y., Algarni, H., Yahia, I.S.: Ceram. Int. **47**, 24031 (2021)
51. Poul Raj, I.L., Valanarasu, S., Hariprasad, K., Ponraj, J.S., Chidhambaram, N., Ganesh, V., Ali, H.E., Khairy, Y.: Opt. Mater. **109**, 110396 (2020)
52. Boruah, B.D., Majji, S.N., Misra, A.: Nanoscale **9**, 4536 (2017)
53. Liu, C.-W., Chang, S.-J., Hsiao, C.-H., Lo, K.-Y., Kao, T.-H., Wang, B.-C., Young, S.-J., Tsai, K.-S., Wu, S.-L.: IEEE J. Sel. Top. Quantum Electron. **20**, 89 (2014)
54. Dai, W., Pan, X., Chen, C., Chen, S., Chen, W., Zhang, H., Ye, Z.: RSC Adv. **4**, 31969 (2014)
55. Huang, C.H., Chu, Y.-L., Ji, L.-W., Tang, I-T., Chu, T.-T., Chiou, B.J.: Microsyst. Technol. (2020)
56. Aspoukeh, P.K., Barzinjy, A.A., Hamad, S.M.: Int. Nano Lett. (2021)
57. Witkowski, B.: Acta Phys. Pol. A **134**, 1226 (2018)
58. Spies, M., Monroy, E.: Semicond. Sci. Technol.. Sci. Technol. **34**, 053002 (2019)
59. Luo, L., Zhang, Y., Mao, S.S., Lin, L.: Sens. Actuat. A **127**, 201 (2006)
60. Gu, P., Zhu, X., Yang, D.: J. Alloy. Compd. **815**, 152346 (2020)
61. Liu, J., Yu, N., Qi, Y., Zhao, H., Yuan, Q., Cao, L.: Mater. Res. Express **6**, 015012 (2018)
62. Echresh, A., Chey, C.O., Zargar Shoushtari, M., Khranovskyy, V., Nur, O., Willander, M.: J. Alloys Compd. **632**, 165 (2015)
63. Hoa, L.T., Tien, H.N., Hur, S.H.: Sens. Actuat. A **207**, 20 (2014)
64. Lin, H.-P., Lin, P.-Y., Perng, D.-C.: J. Electrochem. Soc.Electrochem. Soc. **167**, 067507 (2020)
65. Bai, Z., Zhang, Y.: J. Alloy. Compd. **675**, 325 (2016)
66. Shahkarami, M.M.H., Koohsorkhi, J., Fard, H.G.: NANO **12**, 1750044 (2017)
67. Al-Hardan, N.H., Jalar, A., Abdul Hamid, M.A., Keng, L.K., Ahmed, N.M., Shamsudin, R.: Sens. Actuat. A: Phys. **207**, 61 (2014)
68. Huang, H., Fang, G., Mo, X., Yuan, L., Zhou, H., Wang, M., Xiao, H., Zhao, X.: Appl. Phys. Lett. **94**, 063512 (2009)
69. Ji, Y., Jung, U., Xian, Z., Kim, D., Yu, J., Park, J.: Sens. Actuat. A **304**, 111876 (2020)
70. Ko, K.B., Ryu, B.D., Han, M., Hong, C.-H., Doan, T.A., Cuong, T.V.: J. Alloy. Compd. **823**, 153884 (2020)
71. Thongma, S., Tantisantisom, K., Grisdanurak, N., Boonkoom, T.: Sens. Actuat. A **296**, 324 (2019)
72. Liu, D., Li, H.-J., Gao, J., Zhao, S., Zhu, Y., Wang, P., Wang, D., Chen, A., Wang, X., Yang, J.: Nanoscale Res. Lett. **13**, 261 (2018)
73. Fu, Q.-M., Peng, J.-L., Yao, Z.-C., Zhao, H.-Y., Ma, Z.-B., Tao, H., Tu, Y.-F., Tian, Y., Zhou, D., Han, Y.-B.: Appl. Surf. Sci. **527**, 146923 (2020)
74. Li, Y.-R., Wan, C.-Y., Chang, C.-T., Huang, Y.-C., Tsai, W.-L., Lee, I.-C., Cheng, H.-C.: IEEE Electron Dev. Lett. **36**, 850 (2015)
75. Liu, Y., Song, Z., Yuan, S., Xu, L., Xin, Y., Duan, M., Yao, S., Yang, Y., Xia, Z.: Electron. Mater. Lett. **16**, 81 (2020)
76. Lam, K.-T., Hsiao, Y.-J., Ji, L.-W., Fang, T.-H., Hsiao, K.-H., Chu, T.-T.: Nanoscale Res. Lett. **12**, 31 (2017)
77. Bai, Z., Fu, M., Zhang, Y.: J. Mater. Sci. **52**, 1308 (2017)

78. Park, S., Kim, S., Sun, G.-J., Byeon, D.B., Hyun, S.K., Lee, W.I., Lee, C.: J. Alloy. Compd. **658**, 459 (2016)
79. Wei, C., Xu, J., Shi, S., Cao, R., Chen, J., Dong, H., Zhang, X., Yin, S., Li, L.: J. Mater. Chem. C **7**, 9369 (2019)
80. Tian, W., Lu, H., Li, L.: Nano Res. **8**, 382 (2015)
81. Fabricius, H., Skettrup, T., Bisgaard, P.: Appl. Opt. **25**, 2764 (1986)
82. Cheng, G., Wu, X., Liu, B., Li, B., Zhang, X., Du, Z.: Appl. Phys. Lett. **99**, 203105 (2011)
83. Zheng, X., Sun, Y., Qin, H., Ji, Z., Chi, H.: J. Alloy. Compd. **816**, 152537 (2020)
84. Zhang, T.-F., Wu, G.-A., Wang, J.-Z., Yu, Y.-Q., Zhang, D.-Y., Wang, D.-D., Jiang, J.-B., Wang, J.-M., Luo, L.-B.: Nanophotonics **6**, 1073 (2016)
85. Zhou, J., Gu, Y., Hu, Y., Mai, W., Yeh, P.-H., Bao, G., Sood, A.K., Polla, D.L., Wang, Z.L.: Appl. Phys. Lett. **94**, 191103 (2009)
86. Liang, S., Sheng, H., Liu, Y., Huo, Z., Lu, Y., Shen, H.: J. Cryst. GrowthCryst. Growth **225**, 110 (2001)
87. Singh, S., Jit, S., Park, S.-H.: NANO **12**, 1750063 (2017)
88. Zhu, Z., Wang, S., Zhu, Y., Liu, X., Zou, Y., Gu, Y., Ju, D., Zeng, H.: Adv. Mater. Interfaces **5**, 1800136 (2018)
89. Young, S.-J., Liu, Y.-H., Hsiao, C.-H., Chang, S.-J., Wang, B.-C., Kao, T.-H., Tsai, K.-S., Wu, S.-L.: IEEE Trans. Nanotechnol.Nanotechnol. **13**, 238 (2014)
90. Krishnamurthi, V., Ahmed, T., Mohiuddin, M., Zavabeti, A., Pillai, N., McConville, C.F., Mahmood, N., Walia, S.: Adv. Opt. Mater. **9**, 2100449 (2021)
91. Lee, D.J., Ryu, S.R., Kumar, G.M., Ahn, I.-H., Kyhm, J., Kim, D.Y., Ilanchezhiyan, P.: J. Market. Res. **13**, 397 (2021)
92. Agrawal, J., Dixit, T., Palani, I.A., Singh, V.: IEEE Photon. Technol. Lett. **31**, 1639 (2019)
93. Alenezi, M.R., Alshammari, A.S., Alzanki, T.H., Jarowski, P., Henley, S.J., Silva, S.R.P.: Langmuir **30**, 3913 (2014)
94. Young, S.-J., Liu, Y.-H.: Microelectron. Eng. **148**, 14 (2015)
95. Lee, D.J., Ryu, S.R., Kumar, G.M., Cho, H.D., Kim, D.Y., Ilanchezhiyan, P.: Appl. Surf. Sci. **558**, 149896 (2021)
96. Shuai, M., Lingmin, Y., Lei, C., Chun, L., Mingli, Y., Xinhui, F.: J. Alloy. Compd. **827**, 154090 (2020)
97. Dixit, T., Agrawal, J., Ganapathi, K.L., Singh, V., Rao, M.S.R.: IEEE Electron Dev. Lett. **40**, 1143 (2019)
98. Li, C., Fan, X., Yu, L., Cui, L., Yin, M., Li, Y., Nan, N., Liu, N.: Opt. Mater. **103**, 109891 (2020)
99. Li, C., Yu, L., Yuan, X., Li, Y., Ning, N., Cui, L., Ma, S., Kang, W., Fan, X.: J. Alloy. Compd. **830**, 154652 (2020)

Chapter 8
Recent Strategies for Development of ZnO-Based Efficient UV-Photodetectors

Sayan Bayan

Abstract Owing to the quest for achieving efficient and user-friendly electronic components, the scientific community has forwarded tremendous interest in exploring new material systems. In the last few decades, zinc oxide (ZnO) has been investigated comprehensively to achieve futuristic for optoelectronic and sensing gadgets. Among different types of sensors (like gas/humidity/ethanol sensors), this material has demonstrated promising aspects for sensing ultraviolet (UV) light. Since UV detection is an indispensable part in various aspects ranging from health care to defense systems, ZnO has gained immense attention for achieving prototype UV-photodetectors. Beyond the conventional thin films, the nanostructured systems of ZnO open further scope in this direction. The higher surface-to-volume ratio of the ZnO nanostructures not only increases the UV absorption capability but also offers the flexibility for utilization in miniaturized sensors. Consequently, researchers have attempted to modify the material system by altering microstructural/characteristics or by forming heterojunction with other suitable materials to enhance the UV detection features. The present chapter highlights such adopted strategies for improving UV detection characteristics of ZnO nanostructures and emphasizes the outcome achieved through these approaches. The recent pathbreaking results and the future scope of ZnO-based UV detectors have been reviewed extensively on this occasion.

Keywords UV light · ZnO · Nanostructure · Photodetector

8.1 Introduction

With the advent of human civilization, ultraviolet (UV) light detection has been a concern for the scientific community. As far as technology is concerned UV sensing is significant for different ventures like space-to-space communication, combustion

---

S. Bayan (✉)
Department of Physics, Rajiv Gandhi University, Doimukh 791112, Arunachal Pradesh, India
e-mail: sayan.bayan@gmail.com

D. Mohanta and P. Chakraborty (eds.), *Nanoscale Matter and Principles for Sensing and Labeling Applications*, Advanced Structured Materials 206,
https://doi.org/10.1007/978-981-99-7848-9_8

control, missile plume detection, etc. [1, 2]. Further, the sensing or detection of UV radiation is a major issue when its harmful effects on human beings are considered especially in tropical region [2]. For the detection of UV light, the fabrication of metal–insulator-semiconductor (MIS) based photodetectors has been widely explored [1, 2]. However, with the progress in the field of low-dimensional materials, the fabrication of nanostructure-based quality photosensors has been achieved by the scientific community. Compared to the conventional MIS-based photodetectors, the miniaturization and flexibility of the nanostructure-based photodetectors make them efficient for practical application.

With the advancement in synthesis and processing techniques, it has been possible to grow various classes of nanostructured materials (metallic, semiconducting, perovskite, ferroelectric). Consequently, diverse application of nanomaterials right from drug delivery, lasers, photodetectors, and memory devices is now feasible. For example, metallic nanomaterials, such as Ag, Au, and Pd, exhibit size-dependent surface plasmon absorbance bands that are quite advantageous for optoelectronic applications [3, 4]. In addition, plentiful research on the biological application of this category of nanomaterials is gaining worldwide [5, 6]. Based on the success of Si-technology for electronic device application, the semiconductor nanostructured systems (e.g., Ge, GaAs, GaN, ZnO, InGaAsN, etc.) are considered as the founding stones for futuristic electronic and optoelectronic devices [7–10]. Owing to the dimension-dependent optical and electronic properties, the physical properties of semiconductor nanostructures are tunable between their molecular phase and bulk phase. Events like discreteness of energy levels, enhancement in band gap, excitonic binding energy, and oscillator strength are the key features of the semiconductor nanostructures above the bulk counterpart [10–12]. Most of the optoelectronic properties exhibited by the low-dimensional semiconductors are manifestations of these events. To be specific, photodetection by the semiconductor nanostructures is highly affected by the band gap, morphology, and oscillator strength. Low-dimensional metal oxide-based semiconductors with wide band gap has been a choice for nanostructure-based UV-photodetectors. The wide band gap of the material offers good absorption capability in the UV region of the electromagnetic spectrum making the material a potential candidate for UV photodetector application.

8.2 UV-Photodetection with Nanostructures

In general, the UV photoconductivity in n-type wide band gap metal oxide semiconductors can be explained in terms of oxygen adsorption and desorption mechanism upon UV illumination as presented in the scheme of Fig. 8.1A [13]. In normal conditions, the atmospheric oxygen molecules get adsorbed on the surface of such nanomaterial surface and capture the free electrons present in the system (Eq. (8.1)) [13, 14]. This event leads to the formation of a low-conductivity depletion layer near the surface and consequently reduces the channel conduction (often called dark current). The UV illumination (with photon energy above the band gap energy) leads to the formation of electron–hole pairs as per Eq. (8.2). At this stage, the holes tend to migrate to the surface and recombine with the captured electron which results in the release of the absorbed oxygen species (Eq. (8.3)). The unpaired electrons are left in the excited levels and are collected at the electrode which increases the conductivity of the system.

$$O_2(g) + e^- \rightarrow O_2^-(ad) \quad (8.1)$$

$$h\nu \rightarrow e^- + h^+ \quad (8.2)$$

$$h^+ + O_2^-(ad) \rightarrow O_2(g) \quad (8.3)$$

Zinc oxide (ZnO) with a wide band gap (3.37 eV at 300 K) and large exciton binding energy (60 meV), has been a point of attraction for various optoelectronic properties [15–17]. Due to the wide band gap of the materials, it absorbs light in the UV region and thus can be an ideal candidate for UV photodetector with visible blindness. Further, the doping of ZnO with dopants like Mg can lead to the modification in the band gap of the material and make it aspirant for the application of UV sensors or detectors with variable cut-off wavelengths. Further ZnO has additional advantages like a simple fabrication process, cost-effective, and environmentally friendly. In general, UV illumination on ZnO-based PD leads to a rise in the conduction current following the above-mentioned mechanism of increment in conduction electrons as

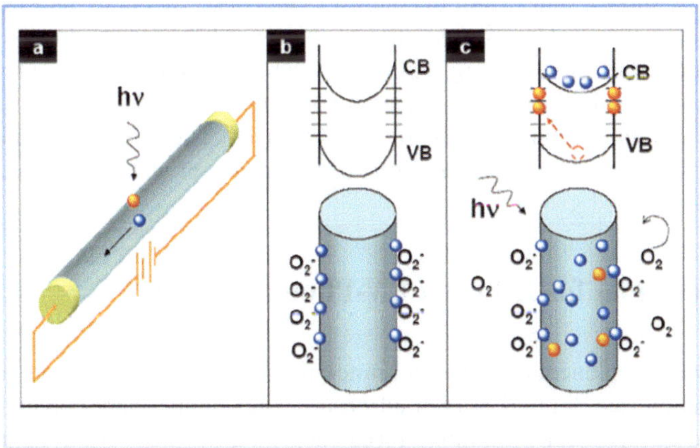

Fig. 8.1 **A** Photoconduction in NW photodetectors: **a** Schematic of an NW photodetector, **b, c** Trapping and photoconduction mechanism in zinc oxide (ZnO) NWs. Reprinted with permission from Soci C., Zhang A., Xiang B., Dayeh S. A., Aplin D. P. R., Park J., Bao X. Y., Lo Y. H., Wang D. (2007) ZnO nanowire UV photodetectors with high internal gain. Nano Lett. 7: 1003–1009. Copyright 2007 American Chemical Society

a result of the desorption of oxygen molecules [18]. Apart from the oxygen adsorption and desorption-related theory, UV-photodetection in ZnO can also be achieved by forming a metal–semiconductor Schottky junction or semiconductor–semiconductor heterojunction [19]. When a metal–semiconductor junction is fabricated with a metal work function higher than the electron affinity of the n-type semiconductor (like ZnO), a Schottky barrier is formed at the interface. The junction can pass current only when a threshold voltage is applied with a particular polarity (i.e., positive on the metal side). In ZnO, an illumination in the UV region leads to the creation of an electron-pair, which not only increases the local conduction but also modifies the Schottky barrier height and the charge distribution. Thus, the current conduction property gets altered during UV illumination and thus the system can work as a UV photodetector. Similarly, in the semiconductor–semiconductor heterojunction, the UV-photodetection can be explained on the basis of UV-induced charge carrier generation and their subsequent separation at the junction/interface.

An interesting feature of photoconductivity studies is the observation of persistent photoconductivity (PPC). PPC is the phenomenon where photoconductivity persists even after the termination of light illumination and originates from the hindrance in the recovery of the initial equilibrium state [18, 20]. It finds several applications like radiation detectors or badges, bistable switching, etc. Researchers have observed PPC in different material systems like ZnO, GaN, Si, WO_3, etc., although the origin of PPC has been a topic of controversy in the absence of any widely accepted theory [20–23]. It is claimed that PPC is sort of a microstructural event and originates from the existence of metastable defect states within the band gap [21, 23]. In contrast to that, PPC has been treated as a surface effect and is assigned to the event of capturing electrons by surface states [14]. In contrast to this observation, in the ZnO nanorod system, the existence of trap states within the band gap has been assigned to the origin of PPC during UV light sensing [18]. It has been proposed that PPC gets enhanced with time due to the increase in electron concentration in the excited state as the bulk electron–hole recombination process is delayed. Therefore, the source of PPC in ZnO is still an issue of consideration due to the lack of widely accepted theory.

8.3 Strategies for Efficient UV-Photodetection

Pure ZnO is often encountered with limited photoresponse due to facts like higher recombination rate, limited oxygen adsorption sites, higher dark conductivity, etc. Hence to overcome such limitations, researchers have adopted various approaches to modify the material configuration. Recently, ZnO has been exploited upon surface/microstructural modification with additional treatments like doping, post-synthesis annealing, etc. Further, the formation of hybrids or heterojunctions with ZnO is found to be another efficient method to achieve photodetectors with improved characteristics. The present chapter will focus on such internal/external features that can be exploited to tune the UV photoresponse from ZnO nanostructures.

8.3.1 Microstructure and Surface Modification

With the advent of different fabrication and processing techniques, it has been possible to alter the microstructure of ZnO nanostructure. The alteration in microstructural characteristics of materials can influence the optical and electrical properties and hence the photoconductivity too. In a particular study, it has been found that the crystallite sizes of the ZnO nanorods fabricated using the solution process (named as S1) and annealing-assisted growth method (S2) can vary as 28 and 9.12 nm, respectively [24]. As can be seen in Fig. 8.2, the growth and decay of photocurrent during UV illumination and termination conditions is different for the nanorods grown using these two different methods. It has been found that the nanorods with 28 nm crystallite size can be characterized by higher growth and decay time (~102 and 70 s, respectively) compared to the nanorods having crystallite size of 9.12 nm (~21 and 9.6 s, respectively). This variation has been explained in terms of variation in oxygen adsorption along the grain boundaries of the crystallites [24]. In general, the oxygen molecules can be adsorbed on the surface as well as the grain boundaries of the nanostructures. Photoconductivity arising from the oxygen molecules situated at the surface is called the surface-related process, while the photoresponse involving oxygen situated at grain boundaries is known as bulk-related process. The oxygen molecules desorbed from the surface get mixed on the ambient easily, while those desorbed from the grain boundaries remain attached to the material as physisorbed species. Thus, the bulk-related process is believed to be faster than the surface-related process [24] and results in the faster photoresponse of the nanorods with lower crystallite size In fact in materials with lower crystallite size, a higher number of oxygen molecules can be adsorbed as the extent of grain boundary gets increased on reducing crystallite size. Thus, bulk-related process dominates over the surface-related process in this case. Nevertheless, the nanorods with slower photoresponse characteristics are suitable for application for UV radiation badges or dosimeters, while nanorods with faster response are attractive for the fabrication of UV sensing application [24]. On different occasions, ZnO nanorods have been fabricated using the annealing-assisted growth at different temperatures of 200, 400, and 600 °C [25]. As predicted from the crystallographic studies, the sample grown at the highest temperature (600 °C) possesses better crystal quality than others. Further photoluminescence (PL) study indicates the improvement in surface quality due to the reduction in surface-related defects. The influence of the microstructure and surface quality is reflected in photoconductivity studies. It has been found that the photosensitivity (photo-to-dark current ratio) is ~20–25 times higher than that of the samples at lower annealing temperatures. In addition, the nanorods grown at higher temperature is found to exhibit faster rise and decay time constants. For these nanorods, the improvement in surface quality due to the minimization of surface defects encourages the bulk-related processes over surface-related processes and results in faster response time [25].

Surface modification technique like vacuum treatment is found to be an efficient method to modify the photoresponse [26]. In particular, nanostructures of ZnO

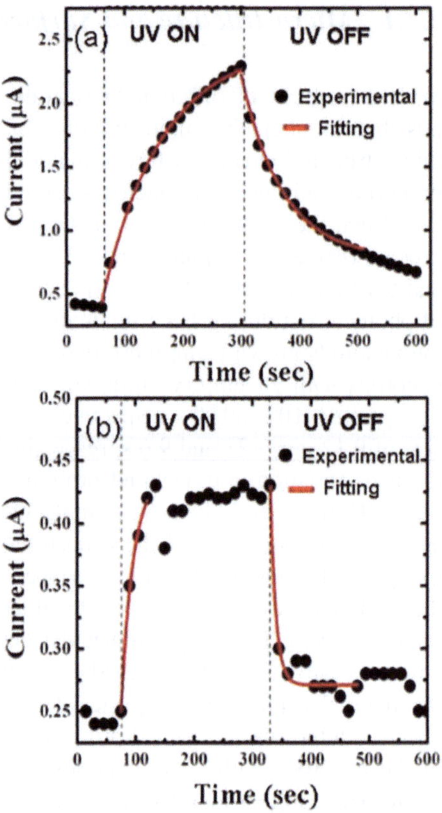

Fig. 8.2 UV photoconductivity of the ZnO nanorods of **a** S1 and **b** S2. Reprinted from Bayan S., Mohanta D., Chakraborty P. (2014) Crystallographic, luminescence and photoconductive characteristics of chemically tailored ZnO nanorods. In: Murli C., Bhattacharyya D., Gadkari S. C. (eds.) DAE Solid State Physics Symposium, India, 2014. AIP Conference Proceedings, 1591, 586, with the permission of AIP Publishing

have been obtained using a simple solid-state grinding process. The photoluminescence (PL) spectrum of such nanostructures reveals a blue emission which has been attributed to the zinc interstitial (Zn_i) related defects present within ZnO. The origin of Zn_i defects has been correlated to the existence of water molecules on ZnO. It has been argued that water molecule upon deposition on the ZnO surface can form a complex of V_{Zn}–H_2O and thus results in the formation of Zn_i defects. The influence of these water molecules on UV detection has been understood upon measuring the photoconductivity in air and vacuum. The observation of a continual rise in photocurrent has been attributed to the desorption of water molecules adsorbed on the ZnO surface under the combined influence of UV light in vacuum conditions. It is expected that the various native defects of ZnO like oxygen vacancies (V_O^+) or zinc interstitial (Zn_i) act as effective sites for the adsorption of water molecules. The water molecules are chemisorbed on the ZnO surface through dissociation on the defect sites. Now through continual UV exposure (~3.3 eV), ZnO can achieve enough energy to detach these chemisorbed water molecules. In such a situation, the intrinsic conductivity will rise due to the increment in conduction band electrons. However, in the normal environment, the desorbed water molecules get re-adsorbed on the surface leading to no alteration in conductivity. In contrast to this in vacuum conditions, the water molecules will be expelled out leading to a higher conductivity [26].

8.3.2 Electronic Structure Modification

Electronic structure modification though elemental doping is a well-recognized route to alter the electrical properties of materials. Doping of semiconductor nanostructures can lead to the improvement of electronic and photoconductive properties [27]. Particularly, ZnO doped with Group III (B, Al, In, Ga) and IV (Sn, Pb) elements can significantly modify the optoelectronic properties. To be specific, the UV-photodetection aspects of Sn-doped ZnO nanostructures have been analyzed in light of the modification in electronic structure. It has been found that in doped samples, several crystal defects become prominent compared to the undoped ZnO. The increment in defects concentration in the doped samples has been assigned to the discrepancy created in the lattice structure due to the incorporation of Sn impurity atoms. As depicted in Fig. 8.3A, a, 2% Sn doping is suitable to improve the photoconductivity in ZnO, while a higher the concentration may quench the photoresponse. Here, the photocurrent is found to vary in a super-linear fashion with the applied voltage. Each I-V curve is found to have two linear regions and can be interpreted in terms of the power-law relationship, $I \propto V^r$, where r represents the slope of the line segments (Fig. 8.3A, b). The super-linear characteristics of the photocurrent (i.e., r greater than 1) at a lower voltage regime indicate the injection of carriers from one of the electrodes as contributed by the hydroxyl or oxygen species present in the surface. Again at higher voltage, the photocurrent is found to vary as $r > 2$, for the undoped as well as 5 at.% Sn-doped ZnO samples. Such effect originates from the presence of trap states and the current is called space-charge-limited current (SCLC) or can be termed as trap-assisted space-charge-limited currents (TASCLC). It has been reported that Sn^{4+} ions in host ZnO would play the role of cationic dopants and will provide electrons to the system [27]. Under UV light exposure, the electrons that have been captured by the defect levels are excited to the conduction band to exhibit better photoconductivity. At higher doping concentration, the photocurrent shows a declining trend. This observation can be assigned to the increment in resistivity due to the aggregation of excess Sn at the grain boundaries of ZnO nanostructures. During the study of stability of photocurrent, it has been found that after an initial rise photocurrent drops down to a moderate value (Fig. 8.3B, a). As can be seen for doped ZnO, the photocurrent increases sharply when compared to the undoped ZnO. Again, during steady illumination condition, the photocurrent shows an exponential drop before getting stabilized. The observation of such anomalous photocurrent has been assigned to the trapping of charge carriers at the ZnO surface. In practice, few desorbed oxygen molecules can get re-adsorbed on the ZnO surface during illumination condition and can lead to the capture of electrons from the conduction band of ZnO. Consequently, the continual decrement in conduction band electron concentration lowers in the photocurrent during steady UV light illumination condition. For further understanding, the photocurrent during illumination has been fitted with a single exponential decay equation ($I = A \exp(-t/\tau) + I_0$, where the τ is the relaxation time constant) as shown in Fig. 8.3B, b. The photocurrent fitted with a single exponential decay equation reveals a decay time constant (~426.98 s), which is lower than that of the one (~944.66 s) and can be assigned to the redistribution of surface state distribution through Sn doping [27].

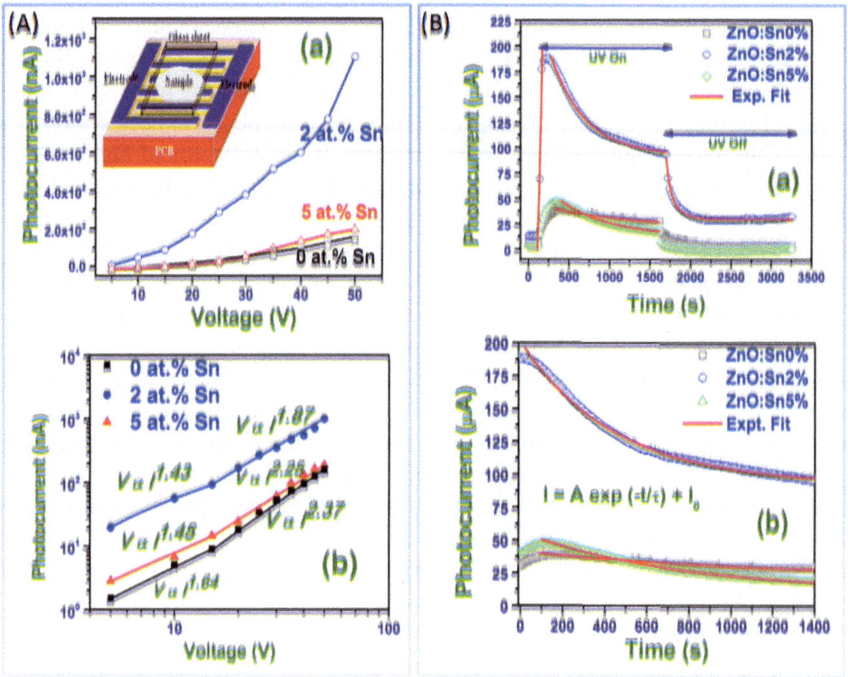

Fig. 8.3 **A** *I–V* characteristics under **a** UV illumination (λ = 365 nm) whereas the inset exhibits an inter-digitated photoconductive cell type device used in the investigation of photoconductivity properties at room temperature, and **b** the same *I-V* curves on logarithmic scale of undoped and Sn-doped ZnO nanostructures. **B, a** Photocurrent rise-decay curves, and **b** decay of photocurrent under steady UV illumination of undoped and Sn-doped ZnO nanostructures at 20 V. The red lines indicate theoretical exponential fitting for all the samples. Reprinted from Efficient UV photosensitive and photoluminescence properties of sol–gel derived Sn doped ZnO nanostructures, 211, Mishra S. K., Bayan S., Shankar R., Chakraborty P., Srivastava R. K., Sensors and Actuators A: Phys. 8–14, Copyright (2014), with permission from Elsevier®

8.3.3 Hybrid or Composite Formation

One of the key ideas of such hybrid materials is to maximize the electron–hole separation by reducing their recombination. Fabrication of heterojunction is regarded as one of the superior techniques to obtain efficient charge carrier separation and hence better photoresponse [19, 28, 29]. For example, a report on the UV-photodetection characteristics of ZnO nanowire /n–Si heterojunction fabricated by a magnetron co-sputtering method can be found in the literature [19]. It has been found upon illuminating ultraviolet light photon-generated electrons in the ZnO side can drift easily toward the Si side exhibiting a good photoresponse property. In fact, heterojunction formation with type II band alignment can efficiently separate the photogenerated charge carriers. The heterojunction of ZnO nanorods with 2D graphitic carbon nitride (g–C_3N_4) nanosheet can be considered as an excellent material in this context. The superior UV photoresponse has been demonstrated by a hybrid heterojunction formed between silver (Ag) nanoparticle decorated g–C_3N_4 nanosheet and ZnO nanorods [29]. As shown in Fig. 8.4A, the hybrid heterojunction of Ag–C_3N_4 and ZnO (ZnAgCN) exhibits improved photoresponse over heterojunction with bare g–C_3N_4 (ZnCN) showing higher photo-to-dark current ratio 10^2. Similarly, the hybrid heterojunction system of g–C_3N_4/CdS/ZnO is demonstrated as an efficient material to demonstrate UV photoconductivity [30]. In this system improved charge carrier separation can be realized due to the formation of step-like band alignment. Again, the loading of Au nanoparticles in g–C_3N_4 nanosheets present in such hybrid heterojunction leads to improved visible light absorption properties yielding UV–visible photodetection characteristics. As shown in Fig. 8.4B, a, photocurrent increases with applied voltage. In this case, the origin of photoconductivity is attributed to carrier separation facilitated by the band alignment of the three semiconductors. The wavelength-dependent photocurrent for Au–C_3N_4/CdS/ZnO (ACCZ) system is found to be higher than that of CdS/ZnO (CZ) indicating the influence of step-like band alignment for carrier separation (Fig. 8.4B, b). Further, the ratio of photocurrent for the two systems shows an elevated value within the 500–580 nm wavelength range. This elevated response is explained by the event of hot electron injection contributed from the surface plasmons of Au nanoparticles [30].

Fabrication of hybrids or composites is a popular technique to achieve improved or modified properties. Metal semiconductor or polymer semiconductor-based composites have been explored for enhancing the UV photoconductivity of ZnO [31–34]. Again, hybrids composed of ZnO nanowire and organic semiconductor is demonstrated to improve the photoconductivity [34]. It has been proposed that the gas-barrier property of polyvinyl alcohol (PVA) can be used to resist oxygen adsorption on the nanorod. Again, the enhancement of PPC upon encapsulating ZnO nanorod with PVA can also be found in the literature [35]. Figure 8.5A shows a comparison of the UV photoresponse of bare ZnO and PVA-capped ZnO nanorod powders. As can be seen in the figure, the current value in bare ZnO doesn't drop quickly on switching off UV light and reaches a value of 58% of the maximum current in 700 s. The non-vanishing nature of the current in the absence of UV light indicates the presence of

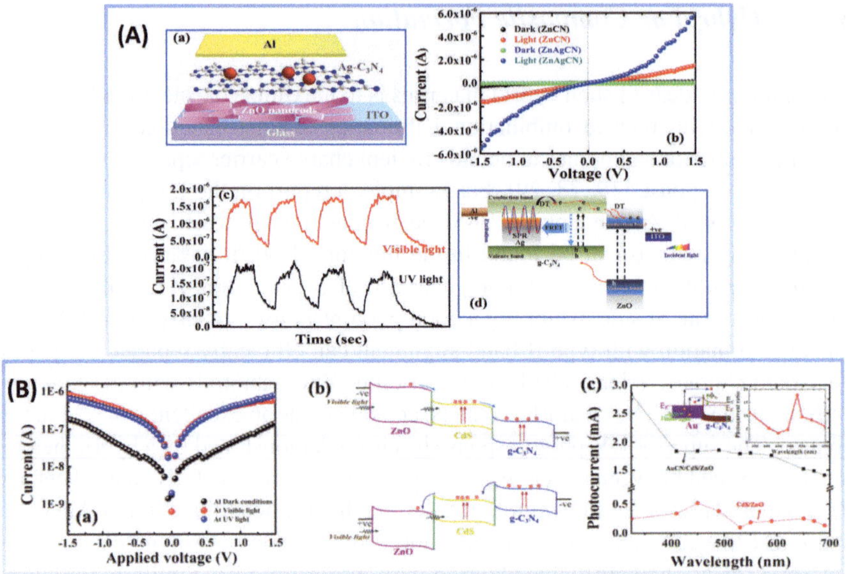

Fig. 8.4 **A, a** Schematic representation of the device used for photoconductivity measurements, **b** *I–V* characteristics in the dark and in the presence of visible light, **c** photocurrent variation under on/off condition of light at a constant bias of −2 V, **d** Schematic view of the mechanism for enhanced photoconduction. Reprinted with permission from Bayan S., Gogurla N., Ghorai A., Ray S. K. (2019) Förster resonance energy transfer mediated charge separation in plasmonic 2D/1D hybrid heterojunctions of Ag–C_3N_4/ZnO for enhanced photodetection. ACS Appl. Nano Mat. 2: 3848–3856. Copyright 2019 American Chemical Society. **B, a** *I–V* characteristics of an AuCN/CdS/ZnO heterojunction device in the dark and under visible/UV light. **b** Schematic mechanism of photoconductivity in the hybrid heterojunction. **c** Photocurrent of ACCZ and CZ devices at a constant bias of 1.5 V with the scheme of hot electron injection from Au nanoparticles to g–C_3N_4. The photocurrent ratio of both devices at 1.5 V bias is included in the inset for comparison. Reproduced from Pal S.., Bayan S, Ray S. K. (2018) Piezo-phototronic mediated enhanced photodetection characteristics of plasmonic Au-g-C_3N_4/CdS/ZnO-based hybrid heterojunctions on a flexible platform. Nanoscale 10: 19,203–19,211, with permission from the Royal Society of Chemistry

PPC in the system. On the other hand, for PVA-coated nanorods the PPC effect is much more prominent as the current value drops to 82% of the maximum current. The enhancement in PPC in this investigation has been treated as a surface event because the coating of the polymer on the nanorods can influence the surface properties rather than the microstructural characteristics. Following this fact, the origin of PPC for bare and PVA-encapsulated ZnO nanorods has been explained using the theory of surface built-in potential. The schematic representation of the mechanism in Fig. 8.5B can be used to describe the phenomenon [35].

It is already established that upon UV exposure, the photo-generated holes recombine with the chemisorbed oxygen species releasing the electron captured by the holes thereby increasing the electrical conductivity. The event of desorption of oxygen species would lower the band bending process leading to the reduction in surface

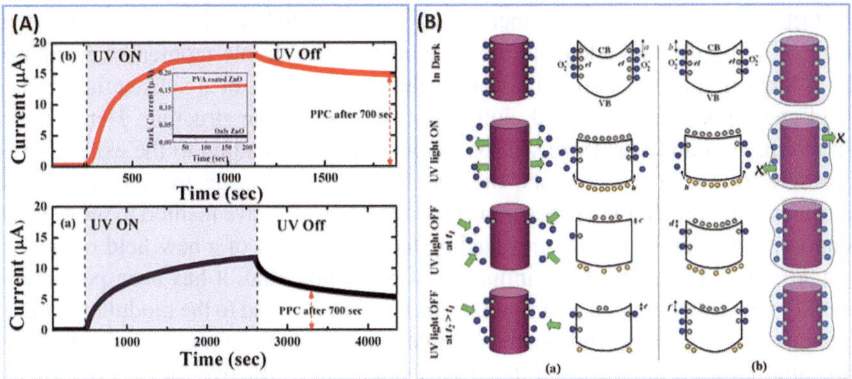

Fig. 8.5 **A** Photoconductivity response of the nanorod powder **a** before and **b** after PVA coating. Inset of (**b**) represents the variation of dark current for the two systems. **B** Proposed scheme of enhancement of PPC. **a** Bare nanorods and **b** PVA-coated nanorods. Reprinted from Enhancement of persistent photoconductivity of ZnO nanorods under polyvinyl alcohol encapsulation, 24, Bayan S., Mishra S. K., Satpati B., Chakraborty P., Mat. Sci. in Semi. Proces. 200–207, Copyright (2014), with permission from Elsevier

built-in potential. On the other hand, the oxygen species can get mixed in the ambient atmosphere upon desorption from the ZnO surface. However, the PVA coating would hinder the mixing of the desorbed oxygen species in the ambient atmosphere. The gas-barrier property of PVA can efficiently limit the diffusion of oxygen species from polymer to ambient. Consequently, the oxygen molecules are situated on adjacent polymer chains around the host nanorod, or even on the ZnO surface through simple physisorption. However, when the UV source is terminated, the system attains the initial condition through the recombination of the holes by the electrons. During this process, the built-in potential is further developed due to the re-adsorption of the oxygen molecules. In general, this process is slightly delayed due to the unavailability of the oxygen molecules that get diffused into the atmosphere. Due to the encapsulation-led confinement of oxygen species within the ZnO–PVA system, the re-adsorption process becomes faster, and hence the development of the surface built-in potential is faster too. Consequently, the carrier recombination process gets hindered largely, and an excess number of uncompensated electrons remains in the sample resulting in the observation of PPC [35].

8.3.4 *Exploiting Piezo-Phototronic Effects*

The demonstration of the piezoelectric effect in ZnO is well known and can be attributed to its non-centrosymmetric crystal structure with tetrahedrally co-ordinated cation and anion atoms (i.e., Zn^{2+} O^{2-} respectively) [36, 37]. On the application of stress on the apex of the tetrahedron, a charge imbalance is created owing to the perturbation of the crystal symmetry. This charge imbalance leads to the development of a dipole with adequate dipole moment. The dipole moments contributed by a large number of unit cells lead to the development of a substantial electric field and a potential is created along the two faces of the structure. Hence under short circuit conditions, there will be a flow of current leading to the exhibition of piezoelectricity. In recent years, the coupling of the piezoelectric effect with the intrinsic semiconducting properties is found to be an effective method to modify and improve the electronic properties leading to the evolution of a new field of "piezo-phototronics" [36]. Under the influence of piezo-potential, it has been possible to modify the semiconductor junction/interface which can lead to the modulation of the charge carriers, e.g., generation, recombination, separation, or transport. Such piezo-potential-induced modification in carrier dynamics is attractive for the photodetection aspect. For example, in the ZnO-based Schottky junction, the application of a compressive strain along the *c*-axis direction of the nanorods would lead to the development of negative piezo-potential at the interface [36]. This effect will increase the Schottky barrier height leading to the enhancement in separation and collection of charge carriers upon illumination. Similarly, in ZnO-based *p–n* junction, the compressive strain along the *c*-axis will lead to the accumulation of negative polarization charges at the interface along the *n*-type side. This effect not only increases the device current but also suppresses the carrier recombination through upward bending during photo-illumination. Thus, the photodetection can be enhanced due to the enhancement in the charge carrier separation. Utilizing the piezo-phototronic property, the improvement in UV-photodetection in a single ZnO micro/nanowire has been demonstrated [38]. In this occasion, the photocurrent can be largely enhanced even at the low-intensity illumination and the responsivity can be enhanced to sub-$mWcm^2$. Furthermore, in ZnO-based hybrid heterojunction, the utilization of the piezo-phototronic effect is found to enhance the UV-photodetection aspects. Particularly, in ZnO/spiro-MeOTAD heterojunction photodetector, with the application of strain, improvement in self-powered photocurrent has been reported. In this case, the responsivity is found to be doubled under strained condition [39].

8.4 Conclusion and Future Prospect

Although the above-mentioned strategies have been found to be effective in tuning the photoresponse of ZnO, exploration of noble approaches may lead to better results. One of the approaches may be the interface modification in the heterojunction. It is known that imperfections at the interface may trap the photo-excited carriers and can lead to a decrement in photoconductivity. Thus, heterojunction formation with proper surface treatment may be a good option to improve the photoresponse of the system. Next hybrid or heterojunction with perovskite materials can be a good option to modify the photoresponse characteristics owing to their excellent optical and electronic properties. Apart from the conventional UV-photodetectors in rigid platforms, special attention should be given to fabricate flexible photodetectors. With the development of modern civilization, smart and lightweight electronic gadgets are highly desirable. In such systems, the flexible UV-photodetectors may be immensely useful. The development of ZnO nanostructures on polymer, or silicon membranes can be an effective way to achieve the photoconductive material for flexible device application. Again, wearable UV-photodetectors have further importance as per as the modern lifestyle is concerned. Finally, self-powered photodetectors can be an excellent choice to meet the technological demands of the modern day [40, 41]. In this context, the piezoelectric or pyroelectric effect can be utilized to separate the photo-excited carriers to achieve photoresponse without any external bias. The piezoelectric effect in ZnO is well known and can be used to tune the electrical property of the material under UV-irradiation. Detailed study regarding the material system and an in-depth understanding of the ongoing mechanism will lead to the fabrication of prototype wearable UV-photodetectors, sensors, and radiation badges that can work in self-powered mode for the next generation.

Acknowledgements The author acknowledges the funding agencies like UGC New Delhi, CSIR New Delhi, DAE India, and IIT Kharagpur for funding the research work in this field. Sincere gratitude is expressed to the publishing agencies ACS, AIP, Elsevier, and RSC for providing the copyright permission.

References

1. Monroy, E., Calle, F., Pau, J.L., et al.: AlGaN-based UV photodetectors. J. Crys. Grow. **230**, 537–543 (2001)
2. Lin, C.-H., Liu, C.W.: Metal-insulator-semiconductor photodetectors. Sensors **10**, 8797–8826 (2010)
3. Zheng, N., Fan, J., Stucky, G.: One-step one-phase synthesis of monodisperse noble-metallic nanoparticles and their colloidal crystals. J. Am. Chem. Soc. **128**, 6550–6551 (2006)
4. Murphy, C.J., Sau, T., Gole, A.M., et al.: Anisotropic metal nanoparticles: synthesis, assembly, and optical applications. J. Phys. Chem. B **109**, 13857–13870 (2005)
5. Liao, H., Nehl, C.L., Hafner, J.H.: Biomedical applications of plasmon resonant metal nanoparticles. Nanomedicine **1**, 201–208 (2006)
6. Gao, J., Gu, H., Xu, B.: Multifunctional magnetic nanoparticles: design, synthesis, and biomedical applications. Acc. Chem. Res. **42**, 1097–1107 (2009)
7. Wang, Y., Herron, N.: Nanometer-sized semiconductor clusters: materials synthesis, quantum size effects, and photophysical properties. J. Phys. Chem. **95**, 525–532 (1991)
8. Brus, L.E.: Electron-electron and electron-hole interactions in small semiconductor crystallites: the size dependence of the lowest excited electronic state. J. Chem. Phys. **80**, 4403–4409 (1984)
9. Nirmal, M., Brus, L.: Luminescence photophysics in semiconductor nanocrystals. Acc. Chem. Res. **32**, 407–414 (1999)
10. Gaponenko, S.V.: Optical Properties of Semiconductor Nanocrystals. United Kingdom (1998)
11. Yoffe, A.D.: Low-dimensional systems: quantum size effects and electronic properties of semiconductor microcrystallites (zero-dimensional systems) and some quasi-two-dimensional systems. Adv. Phys. **42**, 173–262 (1993)
12. Efros Al, L., Efros, A.L.: Interband absorption of light in a semiconductor sphere. Sov. Phys. Semicond. **16**, 772–775 (1982)
13. Zhai, T., Fang, X., Liao, M., et al.: A comprehensive review of one-dimensional metal-oxide nanostructure photodetectors. Sensors **9**, 6504–6529 (2009)

14. Soci, C., Zhang, A., Xiang, B., et al.: ZnO nanowire UV photodetectors with high internal gain. Nano Lett. **7**, 1003–1009 (2007)
15. Schmidt-Mende, L., MacManus-Driscoll, J.L.: ZnO–nanostructures, defects, and devices. Mat. Today. **10**, 40–48 (2007)
16. Su, Y.K., Peng, S.M., Ji, L.W., et al.: Ultraviolet ZnO Nanorod Photosensors. Langmuir **26**, 603–606 (2010)
17. Hasan, K., Alvi, N.H., Lu, J., et al.: Single nanowire-based UV photodetectors for fast switching. Nano. Res. Lett. **6**, 348 (2011)
18. Bayan, S., Mohanta, D.: ZnO nanorod-based UV photodetection and the role of persistent photoconductivity. Phil. Magazine **92**, 3909–3919 (2012)
19. Guo, Z., Dongxu Zhao, D., Liu, Y.: Visible and ultraviolet light alternative photodetector based on ZnO nanowire/n-Si heterojunction. Appl. Phys. Lett. **93**, 163501 (2008)
20. Nayak, J., Kasuya, J., Watanabe, A. et al.: Persistent photoconductivity in ZnO nanorods deposited on electro-deposited seed layers of ZnO. J. Phys. Condens Matter. **20**(1–5), 195222 (2008)
21. Prades, J.D., Hernandez-Ramirez, F., Jimenez-Diaz, R., et al.: The effects of electron–hole separation on the photoconductivity of individual metal oxide nanowires. Nanotechnology **19**, 465501 (2008)
22. Huang, K., Zhang, Q.: Giant persistent photoconductivity of the WO_3 nanowires in vacuum condition. Nano. Res. Lett. **6**, 52 (2011)
23. Lany, S., Zunger, A.: Anion vacancies as a source of persistent photoconductivity in II–VI and chalcopyrite semiconductors. Phys. Rev. B **72**, 035215 (2005)
24. Bayan, S., Mohanta, D., Chakraborty, P.: Crystallographic, luminescence and photoconductive characteristics of chemically tailored ZnO nanorods. In: Murli, C., Bhattacharyya, D., Gadkari, S.C. (eds.) DAE Solid State Physics Symposium, AIP Conference Proceedings, vol. 1591, p. 586. AIP, India (2014)
25. Bayan, S., Mishra, S.K., Chakraborty, P., et al.: Defect-dominated optical emission and enhanced ultraviolet photoconductivity properties of ZnO nanorods synthesized by simple and catalyst-free approach. Appl. Phys. A **115**, 1193–1203 (2014)
26. Bayan, S., Mishra, S.K., Chakraborty, P., et al.: Enhanced vacuum-photoconductivity of chemically synthesized ZnO nanostructures. Phil. Mag. **94**, 914–924 (2014)
27. Mishra, S.K., Bayan, S., Shankar, R., et al.: Efficient UV photosensitive and photoluminescence properties of sol–gel derived Sn doped ZnO nanostructures. Sens. Actuat. A: Phys. **211**, 8–14 (2014)
28. Jeong, S., Choe, M., Kang, J.W., et al.: High-performance photoconductivity and electrical transport of ZnO/ZnS core/shell nanowires for multifunctional nanodevice applications. ACS Appl. Mat. Interfaces **6**, 6170–6176 (2014)
29. Bayan, S., Gogurla, N., Ghorai, A., et al.: Förster resonance energy transfer mediated charge separation in plasmonic 2D/1D hybrid heterojunctions of Ag–C_3N_4/ZnO for enhanced photodetection. ACS Appl. Nano Mat. **2**, 3848–3856 (2019)
30. Pal, S., Bayan, S., Ray, S.K.: Piezo-phototronic mediated enhanced photodetection characteristics of plasmonic Au–g–C_3N_4/CdS/ZnO based hybrid heterojunctions on a flexible platform. Nanoscale **10**, 19203–19211 (2018)
31. Liu, K., Sakurai, M., Liao, M., et al.: Giant improvement of the performance of ZnO nanowire photodetectors by Au nanoparticles. J. Phys. Chem. C **114**, 19835–19839 (2010)
32. Lao, C.S., Park, M.-C., Kuang, Q., et al.: Giant enhancement in UV response of ZnO nanobelts by polymer surface-functionalization. J. Am. Chem. Soc. **114**, 9835–19839 (2010)
33. Gogurla, N., Sinha, A., Santra, S., et al.: Multifunctional Au–ZnO plasmonic nanostructures for enhanced UV photodetector and room temperature NO sensing devices. Sci. Rep. **4**, 6483 (2014)
34. Chen, Q., Ding, H., Wu, Y., et al.: Passivation of surface states in the ZnO nanowire with thermally evaporated copper phthalocyanine for hybrid photodetectors. Nanoscale **5**, 4162–4165 (2013)

35. Bayan, S., Mishra, S.K., Satpati, B., et al.: Enhancement of persistent photoconductivity of ZnO nanorods under polyvinyl alcohol encapsulation. Mat. Sci. in Semi. Proces. **24**, 200–207 (2014)
36. Que, M., Zhou, R., Wang, X. et al.: Progress in piezo-phototronic effect modulated photovoltaics. J. Phys.: Condens. Matter **28**, 433001 (2015)
37. Han, X., Chen, M., Pan, C., et al.: Progress in piezo-phototronic effect enhanced photodetectors. J. Mater. Chem. C **4**, 11341 (2016)
38. Yang, Q., Guo, X., Wang, W.H., et al.: Enhancing sensitivity of a single ZnO Micro-/nanowire photodetector by piezo-phototronic effect. ACS Nano **4**, 6285–6291 (2010)
39. Shen, Y.W., Yan, X.Q., Si, H.N., et al.: Improved photoresponse performance of self-powered ZnO/Spiro-MeOTAD heterojunction ultraviolet photodetector by piezo-phototronic effect. ACS Appl. Mater. Interfaces **8**, 6137–6143 (2016)
40. Xie, Y., Wei, L., Wei, G., et al.: A self-powered UV photodetector based on TiO_2 nanorod arrays. Nano. Res. Lett. **8**, 188 (2013)
41. Bayan, S., Bhattacharya, D., Mitra, R.K., et al.: Self-powered flexible photodetectors based on Ag nanoparticle-loaded g-C_3N_4 nanosheets and PVDF hybrids: role of plasmonic and piezoelectric effects. Nanotechnology **31**, 365401 (2020)

[illegible]

Chapter 9
Sensing Nanomaterials Based on Host–Guest Interactions

Mahesh Pattabiraman, Elamparuthi Ramasamy, and Vijayakumar Ramalingam

Abstract Nanostructures capable of sensing analytes have revolutionized the chemicals-based detection process and enabled chemists to design several cutting-edge technologies applicable in medicine, material science, environmental monitoring, agriculture, energy, and defense sector. The analyte-sensor-transducer-amplifier paradigm employed by chemists in the design of nanosensors have taken chemical detention limits to new limits than that was previously possible with single-molecule detection approach due to the assembling of functional units that serve one of those mentioned roles. In addition, the nanosensing paradigm also allows for the synthesis of low-cost sensing devices. Within the realm of nanosensors, a subset of sensors designed based on the utilization of macrocyclic cavitands that engage in host–guest dynamics is gaining attraction. Cavitands are cyclic oligomeric structures that could encapsulate other smaller molecules within their cavities. which allows for the construction of specific, complex, and functional macroscale devices such as nanosensors. This chapter presents the supramolecular chemistry fundamentals of macrocyclic cavitands and their structural features that render them useful tools in the hands of a nanochemist in designing sensors. Representative examples of nanosensors designed based on the inclusion of complexation dynamics at the heart of its sensing process are discussed. Instances of nanosensors designed with diverse sets of macrocycles such as cyclodextrins, cucurbiturils, calixarenes, and pillarenes are discussed, and their utility in medical diagnostics, thermal sensing, smart materials, and food quality control are given. Toward the end, the chapter presents an envisioned future of the role that macrocyclic cavitands and host–guest chemistry could play in realizing the potential of nanosensors in the aforementioned and newer areas of technology.

M. Pattabiraman (✉)
Department of Chemistry, University of Nebraska Kearney, Kearney, NE 68845, USA
e-mail: pattabiramm2@unk.edu

E. Ramasamy
Department of Chemistry and Biochemistry, The University of Texas at Arlington, Arlington, TX 76019, USA

V. Ramalingam
Department of Biology and Chemistry, SUNY Polytechnic Institute, Utica, NY 13502, USA

D. Mohanta and P. Chakraborty (eds.), *Nanoscale Matter and Principles for Sensing and Labeling Applications*, Advanced Structured Materials 206,
https://doi.org/10.1007/978-981-99-7848-9_9

Keywords Nanosensor · Diagnostic · Cyclodextrin · Calixarene · Macrocyclic

9.1 Introduction

Nanosensors are nanoscale constructs designed to interact with analytes and produce measurable physical signals such as electrical [1, 2], optical [3, 4], or physicochemical output [5–7]. While chemists have historically devised molecular sensors to manifest signal outputs based on the analytical profiles of single molecular entities, the employment of nanostructures in the sensing methodology now offers several unforeseen sensing capabilities. The unique physical, chemical, and electronic properties achieved through organized self-assembly of nanoscale structures have revolutionized the field of chemical sensors [8] and found commercialization [9]. A rational design [10, 11] for sensing molecular system should be composed of two structural components: (a) an interface unit that interacts with the analyte that triggers an electronic change and (b) a transmitter unit that causes discernible physical output signal. In a single-molecule sensor, the theoretical maximum is limited by the output units' electronic properties, which are inherent and non-tunable [12]. Thus the use of molecular sensors for real-time applications typically requires specific equipment that would collect the sensor's signal and amplify it into a human-readable output. Nanosensors, on the other hand, lend themselves to the tunability of sensing and signal output characteristics guided by the physicochemical traits of molecular building blocks [13, 14]. This allows for the designing of devices that can amplify and/or modify a single-molecule sensor's output into brightness and wavelength that is discernible to a human eye in and of itself. In addition, the larger scale allows for the integration of functional components into the sensing construct such as programmability, multi-analyte sensing, and gating [11, 15, 16]. Moreover, such nanomaterials are generally scalable and conveniently produced, which makes such materials cheap and environmentally friendly. It is for this reason that the majority of the sensor design since the turn of this century, especially after 2010, has been predominantly nanomaterials based, as opposed to single-molecule-based systems.

Macrocyclic cavitands [17–21] are cyclic molecules with a hollow space in the molecular center that is wide and deep enough to encapsulate other small molecules through non-bonding interactions such as hydrogen bonding, ion–dipole, $\pi - \pi$ interactions, and hydrophobic forces [22]. The hollow spaces of cavitand hosts and their ability to bind molecular guests have proven to be useful molecular toolkits in the bottom-up [23–26] self-assembly-based nanofabrication approach, as well as their ability to act as a molecular interface to "trap" analytes and initiate a physicochemical process in the sensing mechanism. The use of macrocyclic host–guest interactions for nanofabrication and sensing interface offers several advantages such as reversibility [27], reusability [28], tunability [29], and specificity [30]. Supramolecular chemistry is a relatively new field and knowledge of host–guest dynamics is mostly confined to cyclodextrins (CDs). With the advent of modern organic synthetic, purification methodologies and spectroscopic tools, chemists now have access to many new macrocyclic hosts. The difference in binding strength and directionality offered by chemical functionalities in various hosts and the variation in complex stoichiometry offered by individual oligomers in the same host family leads to a broad range of self-assembly options for nanofabrication and sensing routines. The creative use of macrocyclic cavitands in designing novel sensing materials and devices produced by nanochemists demonstrates the usefulness of host–guest chemistry in enhancing the ability of nanoscience in solving real-life problems. This chapter will present such representative examples of materials and devices that were designed by taking advantage of host–guest chemistry in the construction of nanomaterials and the examples that have utilized macrocyclic hosts in realizing sensing applications.

9.2 Macrocyclic Cavitands

Encapsulation of molecular entities wholly or partially within macrocyclic hosts has been a very efficient and simple strategy for a bottom-up approach to designing materials as well as sensing analytes. The advantage of the host–guest approach over direct interaction approaches derives from the convenience resulting from reversibility, and the qualitative predictability of the complex structure. Formation of host–guest inclusion complex is typically driven by two types of interactions: attractive host–guest and/or solvophobic interactions. A combination of host–guest, guest-guest, host-solvent, and guest-solvent interactions have been skillfully employed by chemists to precisely control molecular architecture. Several hosts are currently available [17, 18, 20, 31] for supramolecular chemists, although a few macrocycles have been employed more frequently than others. Cyclodextrins (CDs) and cucurbiturils (CBs) are arguably the most utilized organic hosts due to their supramolecular versatility, easy synthetic accessibility, and aqueous solubility. Calixarenes (CAs) are the next most utilized host, though to a much lower extent than the former. The latest entrants in this field are pillararenes, which are synthetic hosts made up of hydroquinone units linked by methylene bridges at para positions. Despite their recent entry, pillararenes have found disproportionately higher utility in the design and application of sensing materials.

Each family of the aforementioned macrocycles possesses oligomers that are of comparable dimensions to each other; as a result, they affect similar photochemical outcomes in many instances. At the same time, differences in their chemical functionalities, and therefore different modes of interactions, have led to markedly different photochemistry as well. The chemical structures of common hosts, their dimensions, and comparative 3-D structures rendered through molecular modeling are provided in Fig. 9.1.

Cucurbiturils (CBs) [31, 32] are hosts with repeating glycoluril units that have gained prominence due to their easy synthetic accessibility and strong binding affinity towards cationic guests. CB6, CB7, and CB8 are most commonly used wherein the first two members typically exhibit binary (1:1 host:guest).

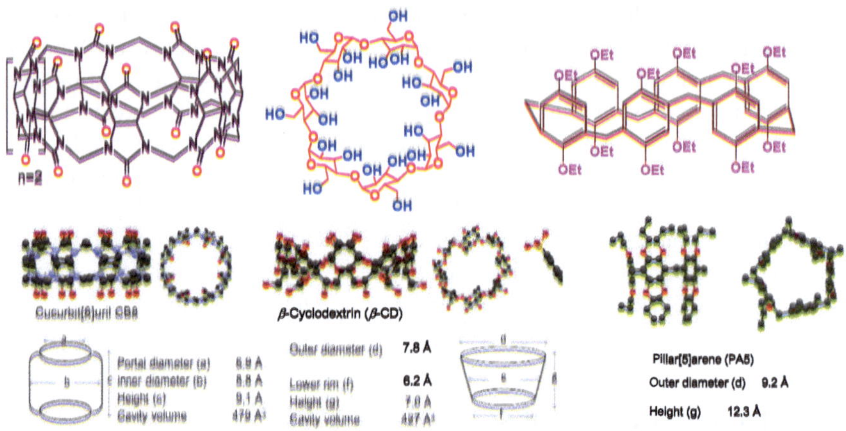

Fig. 9.1 Representation of cavitands used in works reported in this review and their approximate dimensions. Color coding: red (oxygen), blue (nitrogen), green (carbon), yellow (sulfur). Hydrogens are omitted for clarity

binding while the latter can form ternary complexes. The binding affinity of CBs is generally among the highest binding constants noticed for macrocyclic hosts [31]. Binding to cationic guests is predominantly driven by cation-dipole interaction between the oxygen atoms of the carbonyl portals and the guest's positive charge. They also bind to neutral and anionic guests, though at much lower strengths [33]. Cyclodextrins (CDs) [21] are the earliest and most studied family of macrocyclic cavitands. CDs are composed of α-D-glucopyranose monomer units. The hexa-, hepta-, and octameric members of the family are known as α-, β-, and γ-CDs; the host–guest and supramolecular chemistry of these have been extensively studied. The inclusion of guests in the significantly nonpolar cavity of CDs is driven by hydrophobic effects. Composed of hydroxy benzenesulfonic acid units calixarenes are yet another family of cavitands [34]. The hexameric oligomers of all three families have comparable dimensions, and the same is true for the dimensions of other oligomers as well. Pillararenes are a relatively new host and no such comparison could be made based on current knowledge of its host–guest chemistry [18, 35].

9.3 Nanosensor Employing Diverse Host–Guest Interactions

9.3.1 Nanosensors Using Cyclodextrin Hosts

A sensing nanomolecular assembly design involving β-CD host–guest interaction for amplifying analyte binding signal in two stages was reported by Xue and Zink [36] Pores of mesoporous silica nanoparticles containing porcine liver enzyme (PLE, Fig. 9.2a–c) which was entrapped with a β-CD-guest complex acting as a gate. The guest, which is the gating barrier, was an adamantyl-armed triphenylene (Ad-Tp) molecule that would undergo acid-catalyzed cleavage of azide tether anchoring the β-CD to the silica surface. The design of the gating barrier was such that it is permeable for small analytes (blue *, ACA) while larger enzyme substrates were blocked (blue cubes). This nanomolecular assembly was demonstrated to act as a chemical amplifier for sensing the presence of small analytes (ACA), which would pass through the gating barrier to trigger ester hydrolysis, resulting in the formation of acetic acid and hydroxyl cinnamic acid (CA). Formation of acetic acid would activate the acid-catalyzed detaching of the surface-tethered β-CD thereby allowing further enzymatic catalysis of larger substrate molecules (blue cubes), which upon hydrolysis will release more fluorescent products. Thus, the presence of ACA triggers the release of manifold fluorescent products resulting in signal amplification of signal from a small amount of analyte. This nanosensing assembly was demonstrated to fluoresce with a noticeable increase in intensity even if sub-nanomolar concentrations of analyte are present to produce a self-amplified readout.

A self-healing and humidity-sensing conductive nanotube composite centered on the host–guest complexation dynamics was reported by Zhang et al. [37]. The researchers devised it by using single-walled carbon nanotubes (CNT) as the conducting backbone and a pyrene functionalized β-CD (Py-β-CD) as the coupler wherein the pyrene group is physically adsorbed to the CNT surface held by $\pi - \pi$ interaction (Fig. 9.2d). An adamantine polyethylenimine oligomer (PEI-Ad) was used to *crosslink* the CNTs through host–guest interaction between the adamantyl ends of the oligomeric linker guest and β-CD receiver host. This cross-linked self-assembled structure showed self-healing even after physically cutting with the introduction of small amount of water into the medium, which drives the adamantyl linker-β-CD host–guest inclusion process. The self-assembled composite, which was sustained through hydrophobic and hydrogen-bonding interactions (N–H and O–H), showed conductivity through electron transport on its CNT. The same CNT was also responsible for its excellent microwave absorption property because of its ability to generate dielectric and magnetic losses upon microwave excitation. The conductivity of the polymer increased with an increase in moisture due to an increased degree of complexation of adamantyl groups increase driven by hydrophobic forces, serving as the moisture sensor and self-healing. The water-induced cross-linking and conductivity were used to convert this material into a humidity sensor.

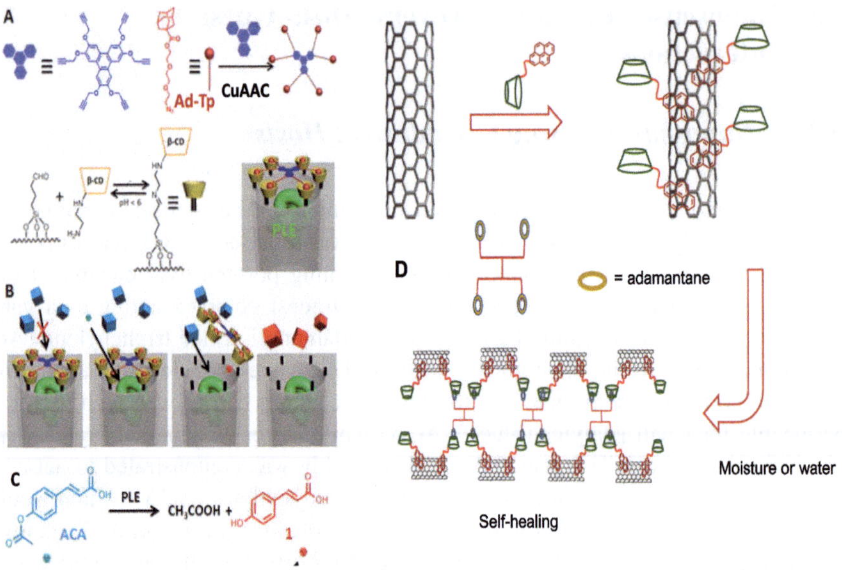

Fig. 9.2 **a** Illustration of the assembly for nanogate-based amplifier. The capping agent is synthesized using a Cu-catalyzed azide—alkyne coupling reaction. **b** The green knot represents the enzyme molecule PLE that is trapped in the pore and the light blue cubes represent an activatable fluorescent dye. The illustration represents mechanism of the self-amplifying chemical sensing process. **c** Chemical reaction affected by analyte ACA. Images used from ref. [36] with permission. **d** Illustration of formation of supramolecular CNT assembly, and schematic illustration of the preparation of the SWNTs-*b*-CD–PEI-Ad complex and demonstration of the healing process for the SWNT-*b*-CD–PEI-Ad. Reproduced from ref. [37]

The use of β-CD-studded gold nanoparticles with functional sensing ability to detect the presence of cholesterol based on guest displacement fluorescent turn-on strategy was reported by Zhao et al. [38]. Gold nanoparticles (AuNPs) were synthesized in the presence of the β-CD cavitand, which acted both as a stabilizer as well as a reducing agent in mild conditions (Fig. 9.3). The Au–NP–β-CD assembly capable of including guests was utilized in harboring photoactive guests such as rhodamine B (RB). The emissive properties of RB and its use in sensory optical application are well-documented in photophysical literature whose fluorescence intensity is dramatically reduced upon encapsulation due to energy transfer to AuNP. Zhao et al. used RB@Au–NP–β-CD system as the sensing assembly wherein the presence of cholesterol in the medium would competitively bind to β-CD cavity and release RB which presents strong fluorescence. Thus, the "turn on" of fluorescence indicated positive identification of cholesterol with a detection limit of 0.15 M despite the presence of interfering analytes such as tryptophan.

An advanced supramolecular sensing system for gene delivery was developed by the group of Davis et al. [39] who was the first to achieve targeted siRNA silencing in human clinical trials. The nanoparticle results from spontaneous self-assembly in one pot when a solution of the siRNA and three other components of the system

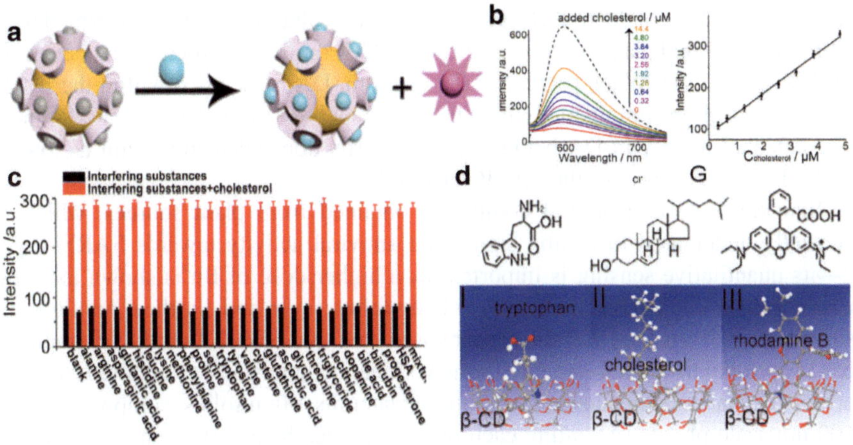

Fig. 9.3 **a** Fluorescent sensing of cholesterol using the composite made of the β-CD@AuNP and RB molecules. **b** Fluorescence spectra of RB in the presence of different concentrations of the AuNPs. Plot of β-CD@AuNP-RB fluorescence intensities versus cholesterol concentrations. **c** Intensity of β-CD@AuNP-RB intensity with cholesterol in presence of interfering analytes. **d** Computed structure of guests bound to β-CD. Images used from ref. [38] and reproduced with permission.

are combined: a CD-containing polymer (CDP), a polyethylene glycol-adamantane steric stabilization agent, (PEG-ad), and a human transferring-polyethylene glycol-adamantane conjugate as a targeting ligand (Fig. 9.4a). The inclusion complex formation between β-CD and PEG-adamantyl and Tf-PEG-adamantyl guests was used to tune the rigidity and stability of the self-assembled system. The transferrin (Tf) on the nanoparticle can bind to transferrin receptors (TfR) on cancer cells, and the nanoparticles can be internalized via receptor-mediated endocytosis. The CDP contains organic groups that are protonated around pH 6, which serves as the "chemical sensing" mechanism that triggers a number of processes that provide escape from endocytic vesicles and nanoparticle release of the nucleic acid into the cytoplasm. The authors optimized the complex formation between cationic poly-CD polymers and siRNA for the formation of homogeneous nanoparticles with a diameter of around 50–70 nm for optimal uptake by the tumors. The modularity and ease of formulation of these complexes demonstrate the potential of self-assembled nanomaterials for the development of functional therapeutic agents. Recently Lee et al. reported the facile synthesis of hexagonal-shaped plate-like polypyrrole (PPY-IC) that was prepared by polymerization of the host–guest pyrrole inclusion complex of β-CD for sensing the neurotransmitter dopamine (DA) [40]. When $FeCl_3$ was added to the PY/β-CD monomer complex, oxidation of pyrrole monomer initiators led to the inclusion of polymerization formation inside the β-CD cavities, followed by PPY-IC growth (Fig. 9.4b). The amount of the monomer complex is speculated to play a crucial role in the morphology of the nanomaterial. The PPY-IC showed good conductivity owing to its polypyrrole backbone. The nanomaterial was conducive for coating on the conductive glass surface. Cyclic voltammetry was performed with a PPY-IC modified glassy carbon electrode (GCE) for the electrochemical detection of dopamine (DA). When present in the medium, DA is included within the β-CD cavity which alters the conductivity of the PPY backbone through redox interference. The catechol moiety in DA being a good electron donor detection limit as low as 0.10 M was registered for the PPy-IC electrode.

Baicalin (Bn) is a natural compound obtained from plant sources, which is currently understood to exhibit significant antiviral activity towards SARS–CoV-2—its quantitative sensing is important as a higher dose of Bn is shown to cause adverse physiological reactions such as body aches, fever, and diarrhea. An ultrasensitive high-performance baicalin sensor based on C_3N_4–SWCNTs/reduced graphene oxide/cyclodextrin metal–organic framework nanocomposite was reported by Zhao et al. [41]. Nanocomposites are multiphase materials, often solids, composed of 10–100 nm scale phases separating each other. The idea behind the nanocomposite is to use building blocks with dimensions in the nanometer range to design and create new materials that combine the advantageous properties of individual phases. The material design involved a novel ultrasensitive Bn electrochemical sensor based on

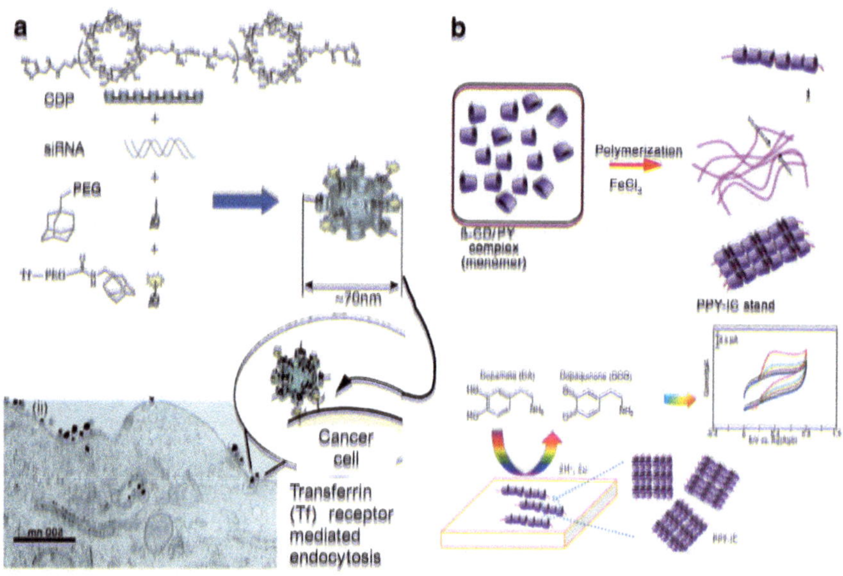

Fig. 9.4 **a** Schematic representation of the one-step self-assembly of four components into a targeted nanoparticle by the electrostatic interaction between siRNA and a cationic cyclodextrin containing polymer (CDP) and their inclusion complexes with adamantane (AD) derivatives on the surface of the nanoparticles; (below) TEM micrograph of nanoparticles entering a cancer cell by receptor-mediated endocytosis. Images used from ref. [39] and with permission. **b** Schematic diagram of the electrochemical reaction of dopamine on the PPY-IC/GCE. Images used from ref. [40] with permission

graphitized carbon-nitride—single-walled carbon nanotube nanocomposites (C_3N_4-SWCNTs), reduced graphene oxide (rGO) and electrodeposited cyclodextrin metal–organic framework (CD-MOF, Fig. 9.5). Under optimal conditions, the sensor exhibited sensitive detection of Bn in a wide linear range of 1×10^{-9} M to 5×10^{-7} M with a LOD of 4.6×10^{-10} M and a sensitivity of 220 A/M, and it showed satisfactory stability and accuracy for detecting Bn in real samples (human serum and bear bile scutellaria eye drops). In addition, the electrochemical reaction site and redox mechanism of Bn were revealed through electrochemical behavior and density functional theory.

9.3.2 Nanosensors Using Cucurbituril Hosts

Cucurbiturils are a versatile family of hosts with excellent supramolecular characteristics for inclusion chemistry. In the context of nanomaterial development, CB[n]s can serve as reducing agents in alkaline aqueous solution and require no additional reagents or heating. This renders CBs to be an eco- and bio-friendly and cost-effective one-pot approach to achieve "green" NPs [42]. Therefore, CB[n] is a powerful tool to easily and precisely regulate particles' shape, size, stability and assembly, interparticle distance, surface functionality, biocompatibility, etc., which are inseparable from their performance for further applications in sensing, SERS, theranostics, and catalysis.

Recently Chen et al.devised a CB-based sensor for imaging of subcutaneous inflammation [43]. Inflammation is one of the most common pathological processes involved in many disease conditions, particularly Alzhimers, cancer, rheumatoid arthritis, heart attack, and stroke. They report a host–guest luminol—AIEgen nanoparticle (SLA NP)-based bioluminescent probe for deep-tissue-imaging of inflamed tissues based on reaction with reactive oxygen species (ROS), which are disproportionately higher in inflammatory conditions (Fig. 9.6a). SLA NPs were synthesized by self-assembly of the binding interaction between three components: adamantane—luminol (AdLum), previously reported aggregation-induced emissive AIEgen – CB7 nanoparticle, and PEG-2000 stabilizer. Luminol emits bioluminescence in the blue region (~425 nm) when oxidized by ROS; the tetraarylethylene unit attached to CB7 in AIEgen-CB7 exhibits red emission (~600 nm) when upon photoexcitation. In ROS-rich environments such as inflamed tissue, the self-assembled nanostructure showed bioluminescence resonance energy transfer (BRET) from luminol (donor) to AIEgen (acceptor), enabling deep-tissue penetration and fluorescent imaging. SLA NPs reported by Chen et al. possessed uniform particle size and excellent stability and biocompatibility for in vitro and in vivo inflammation imaging, superior to the commercially used luminol.

The Enzyme-Linked Immunosorbent Assay (ELISA) serves as an effective mechanism for identifying proteins in medical screenings which functions by pinpointing specific analytes through enzyme-altered antibodies attached to a solid base. Utilizing gold nanoparticles (AuNPs) as light-sensitive markers enables the creation of an

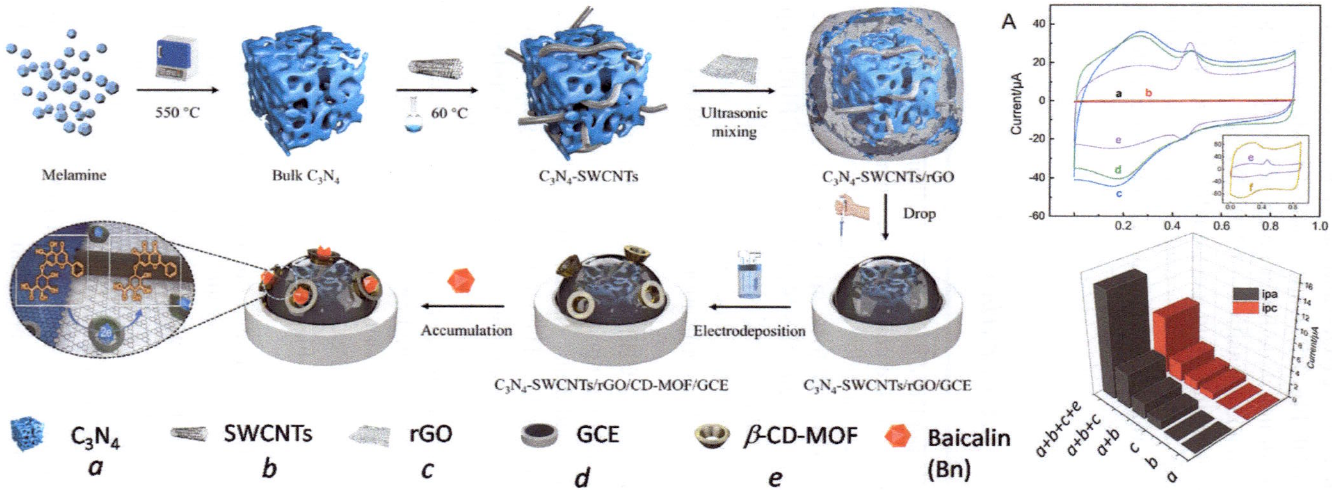

Fig. 9.5 Illustration of the construction process of C3N4-SWCNTs/rGO/CD-MOF/GCE and used for detecting Bn (**A**) CVs and of nanocomposite and quantitative comparison of CV electrode response of composite material versus other individual components for baicalin. Images used from ref. [41] and reproduced with.

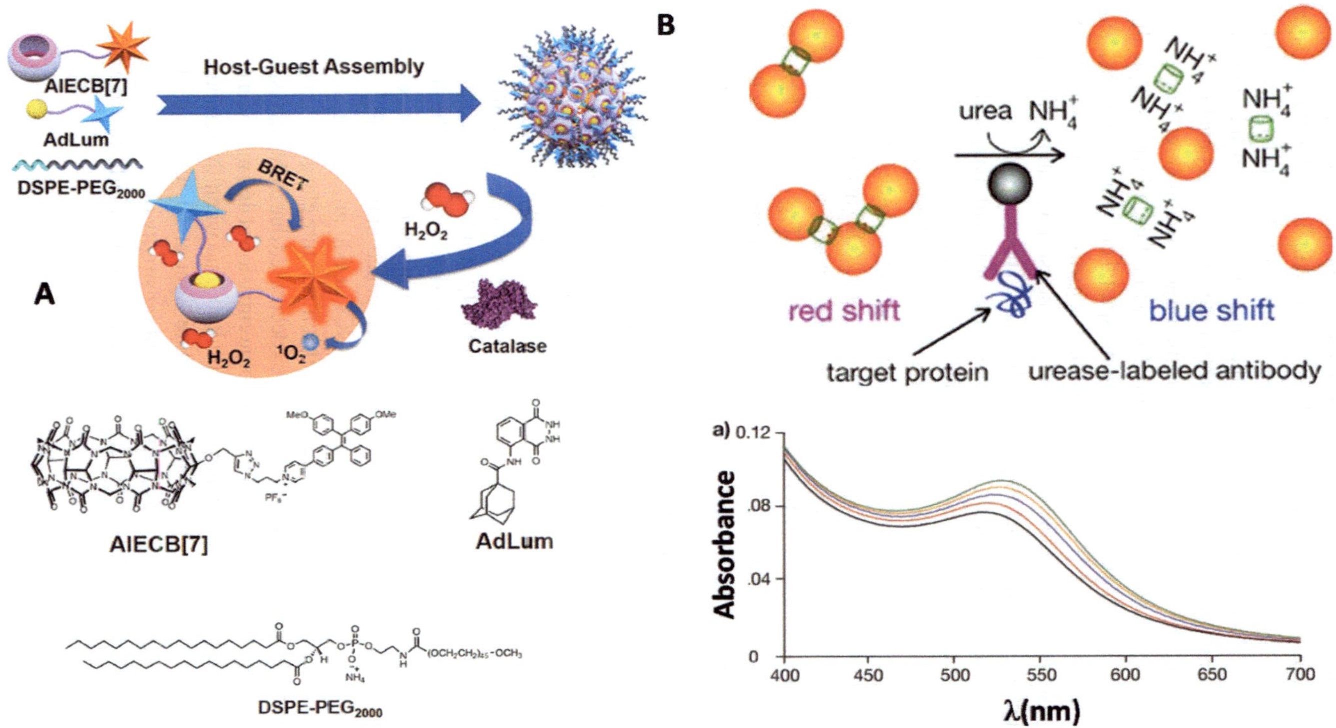

Fig. 9.6 **a** Scheme of the designed SLA NPs and their activation mechanism during in vitro and in vivo inflammation imaging and structures of individual molecular components. Images used from ref. [43] with permission. **b** Schematic representation of the supramolecular assembly of gold nanoparticles by cucurbituril [7] (CB7) reversed by ammonium ions. UV–vis absorption spectra of gold nanoparticle with varying concentrations of CB7. Images used from ref. [44] with permission

extremely responsive method for protein identification by observing changes in the absorbance of plasmon in visible light. Velders and Rica reported the development of light-sensitive markers for magneto-ELISA by coupling AuNPs with CB7 through host–guest chemistry [44]. The presence of ion–dipole interactions between amines and CB7 leads to the formation of ammonium through urease, which subsequently disassembles, causing a blue shift in the absorption spectrum. This enables rapid and precise measurement of enzyme action and the target protein (Fig. 9.6b). The CB7-driven dispersion sensor offers the benefit of negating false positives typically triggered by factors such as ionic strength variations, spatial enzyme limitations, and biodegradation. Leveraging CB7 and AuNPs as light-sensitive markers, and urease-tagged anti-mouse magneto-IgG as the capture agent, mouse-IgG can be effectively identified through competitive immunoassay techniques. This approach has a detection limit (4 pg/mL) that is significantly lower—by three orders of magnitude—than traditional ELISA methods based on colorimetric identification.

Cadaverine (1,5-pentadiammonium) is generated along with the storage of food, which will participate in the formation of carcinogenic nitrosamines, reflecting the metamorphism status of food. The specific recognitions of CB6 and CB7 to cadaverine result in the aggregation of AuNPs, leading to the change of solution color from red to blue, broadening and redshift of the absorption peak (Fig. 9.7a). Therefore, CB[n]s-protected AuNPs can effectively detect cadaverine by visual or spectrophotometric method with a detection limit of 3.9×10^{-6} m, providing a promising tool for applications in food safety. A similar approach to the detection of metformin was also demonstrated by Song et al. [45].

Ma and co-workers developed a thermal and humidity-sensing nanoparticle system by encapsulating two highly blue-emissive thiazolothiazole methyl viologen (TMV) into the cavity of CB [8, 46] The combination resulted in TMV/CB [8] complexes with greenish-yellow emission at 535 nm. The humidity-sensing mechanism derives from the change in complex stoichiometry when dry (ternary) and wet (binary). This change in stoichiometry resulted in spectroscopic changes from greenish-yellow when wet to blue when dry. Similarly, the fluorescence complexes when mixed with thermal-sensitized luminescent Au-nanoclusters yielded a temperature-responsive ratiometric luminescence emission (Fig. 9.7b). The orange emission of glutathione-protected Au–NCs at 600 nm was enhanced at lower temperature and decreased upon heating. With temperature elevation, greenish-yellow emission at 535 nm of TMV/CB [8] complex decreased due to the disassembly of TMV/CB [8], accompanied by the increase of blue emission at 460 nm. As a result, the ratiometric fluorescence colors change from pink to blue when the temperature increases from 0 to 70 °C, and show standard white light at 40 °C. These TMV/CB [8]-AuNCs hybrids exhibit good reversibility and sensitivity of fluorescent color change upon varying the temperature, providing a promising platform for visual temperature detection.

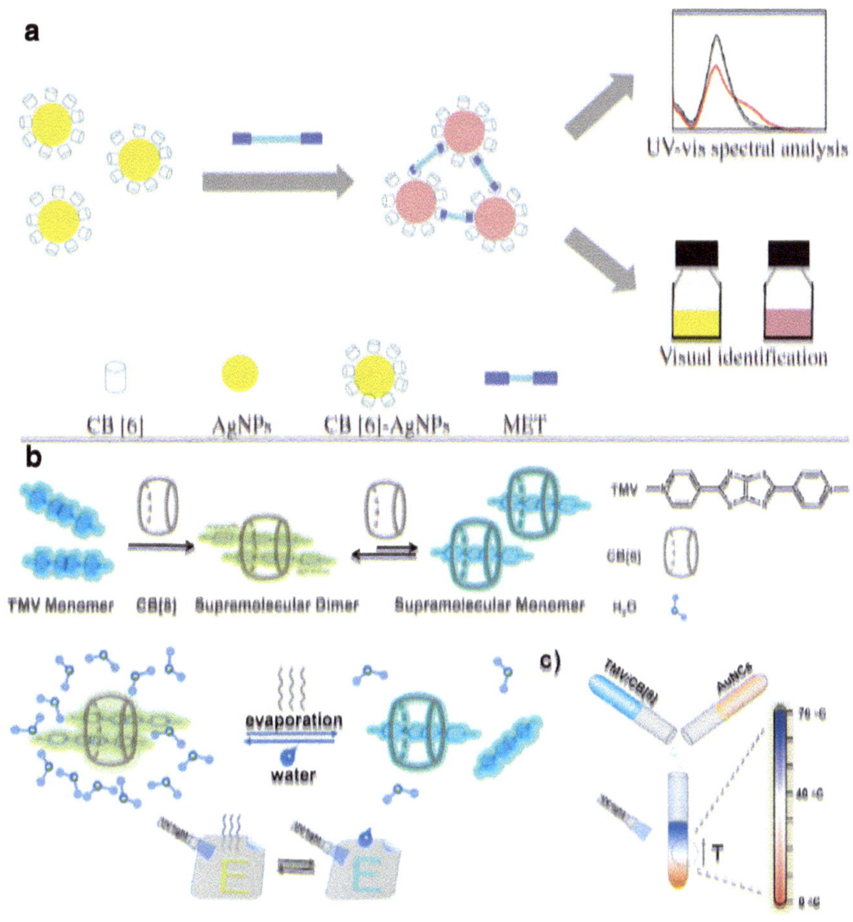

Fig. 9.7 **a** The sensing process for metformin (MET)-mediated aggregation of cucurbituril [6] (CB6) modified silver nanoparticles (AgNPs). Images used from ref. [45] with permission. **b**, **a** Illustration of emissive host—guest dynamic self-assembled motifs of thiazolothiazole methylviologen (TMV) with different equiv. of CB [8] in aqueous solution. Images used from Ref. [46] with permission

9.3.3 Nanosensors Using Macrocyclic Arene Hosts

An original strategy for the self-assembly of synthetic glycoviruses was described by Aoyama et al. [47]. Here, macrocyclic calixarene scaffolds were selectively decorated with sugars on one side and a hydrophobic tail on the other. These amphiphiles self-assemble irreversibly in 5 nm micelles, which can then be agglutinated by hydrogenphosphate ions (Fig. 9.8A). The authors used this affinity for hydrogenphosphate ions to complex plasmid DNA and obtained virus-like particles with a diameter of 50 nm that were able to achieve high transfection efficiency in vitro.

A resorcen-4-arene (RA4)'s inclusion complex-based sensor for selective detection of amphetamines was reported by Masseroni et al. [48]. A nanostructure formed by functionalizing silica nanoparticles with pyrene-derivatized RA4 was shown to be capable of specifically recognizing ecstasy in water was demonstrated (Fig. 9.8B). The sensor presents excimer emission from two pyrene units tethered through a butyl chain from RA4 rims on 1,3-positions, which tend to π-stack upon excitation; the excimer emission from the associated excited state is seen as a distinct broadband at 475 nm, which contrasts with the structured monomer emission between 360 and 420 nm. When an appropriate guest, in the form of amphetamines, is introduced, the excimer emission is quenched. This is due to the physical decoupling of the pyrene units affected by the included guest. When a non-bulky guest (amphetamine) is introduced the emission quenching is nominal, whereas bulkier guests (MDMA and FMA) lead to near-complete quenching. However, the effect is not as pronounced for the single-molecule solution phase sensor due to poor solubility in water and associated impediments of desirable host–guest dynamics. These disadvantages were overcome when the RA4 assembly was immobilized onto a silica gel surface, which was developed into a water-soluble nanostructure. The nanostructure showed enhanced excimer and selectivity toward the analyte due to potential associative interaction between the ammonium group in the analyte and silica surface. The absence of effects from interferents and an efficient electron transfer process occurring after the complexation of ecstasy made this system an efficient fluorescent probe for this popular drug.

9.3.4 Nanosensors Using Pillararene Hosts

Pillararenes are relatively new entrants to the host–guest arena and have a very short history, especially compared to CDs. Though they are far less studied than CBs, CAs, and other hosts in supramolecular chemistry research, they have garnered much higher attention in the nanomaterial context. This is perhaps due to the fact that PAs are poorly soluble in water due to the lack of polar groups while complexation processes in nanoscience often occur in the solid state or slurries where complete solubility of the host is not required.

Montes-Garcia et al. presented a simple procedure for the surface modification of plasmonic nanoparticles (NPs) with a cationic water-soluble ammonium pillar[5]arene (AP[5]A) in order to create selective surface-enhanced Raman scattering (SERS) spectroscopy based sensors [49]. The strategy involved a ligand exchange reaction between the AP[5]A and the stabilizing agent of the as-prepared plasmonic NPs (Fig. 9.9a). The approach could be applied to plasmonic nanoparticles either negatively charged, stabilized by citrate ions (Au spheres), or positively charged, stabilized by cetyltrimethylammonium bromide (Au and Au@Ag nanorods). The SERS performance of all systems was studied as a function of NP size and excitation laser line by using an analyte with no affinity towards the metal surface such as pyrene. The analytical enhancement factor (AEF) for the different systems

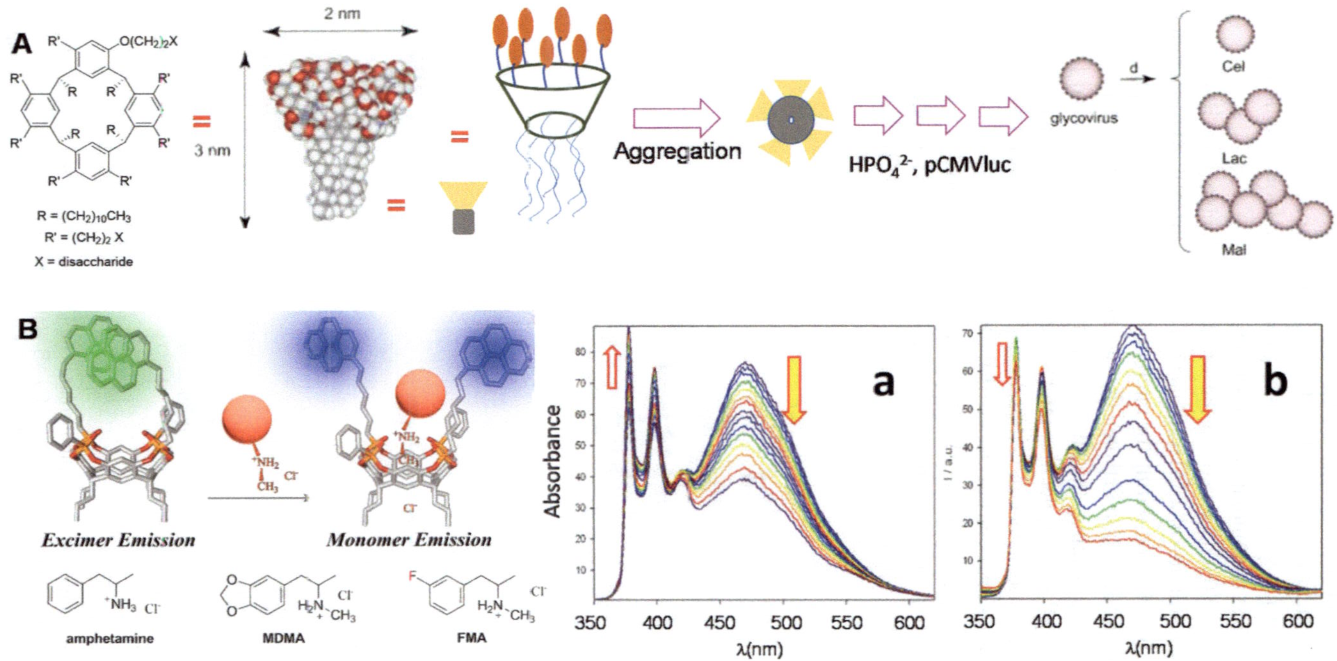

Fig. 9.8 **A** Hierarchical growth of glycocluster amphiphile through nanoparticle (GNP) constructed with macrocyclic arenes to form glycovirus and its aggregates. Images used from ref. [47] and reproduced with permission. **B** Schematic illustration of recognition mode of resorcinarene based chemosensor for detection of dopamine and related analyte. Fluorescence spectra of chemosensor with different additions of FMA (**a**) and MDMA (**b**). Images used from ref. [48] with permission

was estimated between 0.55×10^4 and 1.49×10^5. Finally, the synergistic effect of combining supramolecular chemistry and plasmonic NPs is demonstrated through SERS-based detection, in aqueous media, of molecules with no affinity towards a bare plasmonic substrate such as the contaminant pyrene or the biomolecule pyocyanin with LOD running into nanomolar concentrations.

Chi et al. group developed a sensor for the aminoacid cysteine (Cys) based on thermoresponsive water-soluble pillar[5]arene as well (Fig. 9.9b) [50]. At room temperature, P25 was highly stable, inhibiting the interaction between the cystine probe (CyP) and Cys. The lack of interaction between the two entities showed no responsiveness to Cys. However, upon heating, the encapsulated CyP could be released from P25 and form a Cys…CyP supramolecular interaction resulting in the detection of the analyte. Similarly, a fluorescence indicator displacement (FID)-based chemosensing system for amino acids has been developed on the basis of carboxylato-pillararene [5] PA-based host—guest recognition [51]. In this system, 4-amino-1,8-naphthalimide fluorophores with three different anchors in the 4-position acted as guests, and P4 acted as a host, together constructing the first pillar[n]arene-based FID sensing system for amino acids.

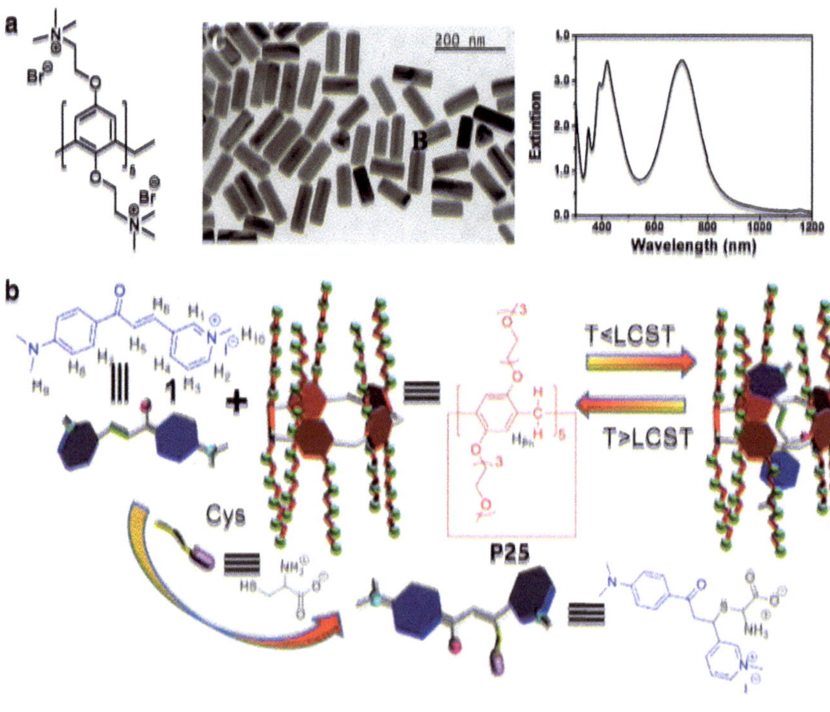

Fig. 9.9 **a** Chemical structure of ammonium pillararene [5] and representative TEM images of CTAB-stabilized Au NRs. Extinction spectrum of CTAB-stabilized Au@Ag NRs in water is shown. Images used from ref. [49] with permission. **b** Working mechanism of the Cys sensor based on pillararene [5] host P25. Images used from ref. [50] and used with permission

A supramolecular tandem strategy, consisting of bis(p-calixarene) (BSC4A) and 6 functionalized AuNP was implemented for the colorimetric sensing of succinylcholine (SuCh) and butyrylcholinesterase (BChE) was reported by Li et al. [52]. BSC4A@AuNP and 6@AuNP displayed good stability and dispersion, with a size of ~5 nm. The presence of SuCh was found to induce aggregation through the complexation of its quaternary ammonium moieties with the cavities of the hosts (Fig. 9.10a), but they did not assemble in the presence of choline (Ch). This resulted in optical changes, visible to the naked eye from red color to blue, indicating the formation of SuCh–BSC4A@AuNP and SuCh–6@AuNP. Furthermore, the hydrolysis of SuCh by BChE promoted the disassembly of the aggregates, providing a dual and reversible sensing platform to assess BChE activity. With the high binding affinity of BSC4A and CP5A towards SuCh, BSC4A–AuNPs, and CP5A–AuNPs could assemble with micromolar level SuCh as SuCh–BSC4A/CP5A–AuNPs. More interestingly, the enzymatic hydrolysis of SuCh by BChE could lead to the disassembly of SuCh–BSC4A/CP5A–AuNPs and provide a sensitive time-dependent color change from blue to red which could be observed by the naked eye and used to monitor BChE activity. As BChE activity is an important biomarker for diseases and poor health conditions, this novel supramolecular tandem colorimetric sensing strategy may have potential use for early diagnosis of diseases.

Yu and co-workers fabricated a graphene oxide (GO) supramolecular hybrid material (GO@P6 ⊃ PyN) utilizing GO and pillar[6]arene (P6) based on host—guest inclusion complexes by hierarchical self-assembly for NIR-light-triggered photoacoustic (PA) and ultrasound (US) signal amplification (Fig. 9.10b) [53]. The host was used as a host to interact with a pyrene-containing amphiphilic molecule PyN through host—guest recognition, obtaining a CO_2-responsive complex (P6 ⊃ PyN). Then, P6 ⊃ PyN can be attached to the surface of GO, forming a hybrid material (GO@P6 ⊃ PyN) via $\pi-\pi$ stacking. Upon NIR light, GO can generate heat, which would trigger the bicarbonate counterions on the surface of Gph@P6 ⊃ PyN to decompose into CO_2. In vitro and in vivo experiments illustrated that the generated CO_2 could act as "molecular boosters" to improve the PA and US signals.

9.4 Conclusion and Future Directions

The possibilities of macrocyclic cavitands in rendering nanomaterials utilitarian and enhancing their applications are limitless. As seen by the examples cited in this chapter, the host–guest chemical dynamics between cavitands and small molecules have been crucial in realizing very interesting applications of sensing nanomaterials. The reversible binding interaction has been shown to be useful in imparting high efficiency, sensitivity, and selectivity in these chemosensory systems. The self-assembly of molecular entities through host–guest interactions leading to complex functional architectures, as demonstrated in some examples here, is a powerful tool for generating modern nanodevices. While host systems such as CAs, CBs, and CDs have been relatively well studied in this context, many other known macrocyclic hosts

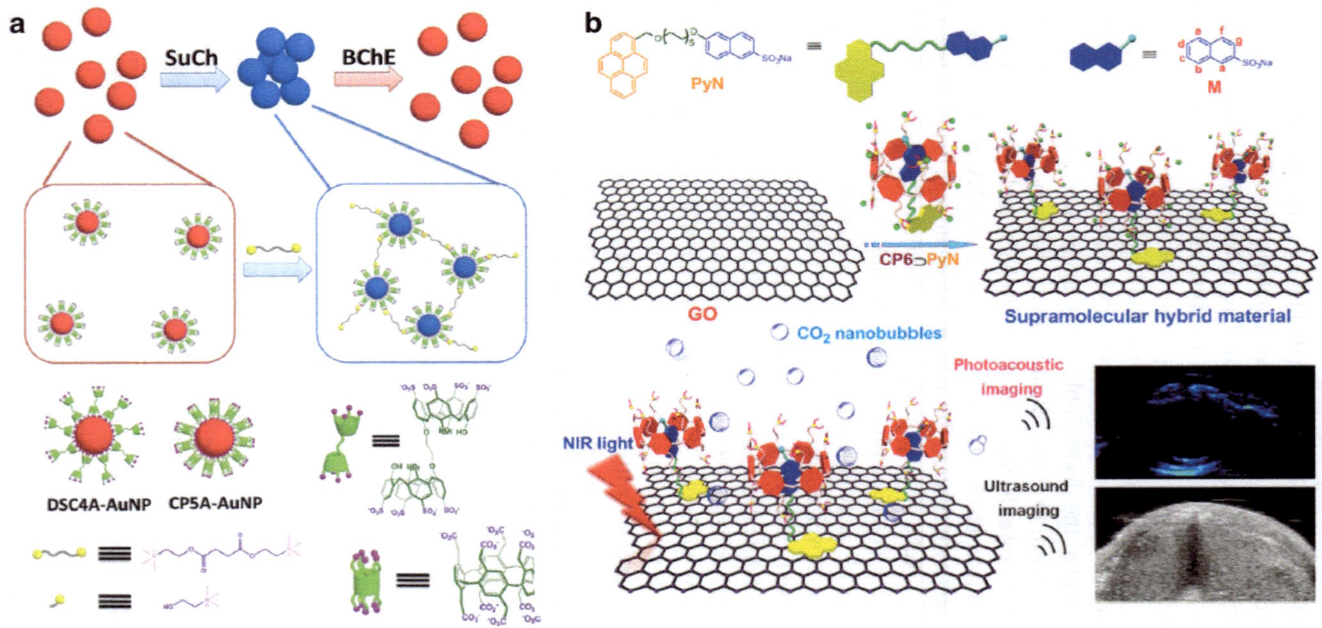

Fig. 9.10 **a** Schematic representation of the supramolecular assembly for colorimetric sensing of succinylcholine (SuCh) and butyrylcholinesterase (BChE). Images from ref. [52] and used with permission. **b** Preparation of the supramolecular hybrid graphene oxide-based material. PA and US imaging can be enhanced through NIR light. Images used from ref. [53] with permission

such as octa acid [54], metallonanocages [55], and prismarenes [56] are yet to be explored by researchers for their nanosensing applications. Despite such examples in literature and many articles that have been referenced, cavitands in nanoscience have remained underutilized in general. As newer macrocyclic hosts emerge, the possibilities of roles in which they could be utilized for meeting real-time applications are sure to expand. It is expected that the availability of more families of macrocycles will provide a powerful arsenal for nanochemists to construct highly sensitive and selective nanosensors for a variety of applications that could be readily commercialized.

Funding The corresponding author's (M. P.) scholarly activity was supported by the National Institute of General Medical Science of the National Institutes of Health (NIH) under award number GM103427. The content is solely the responsibility of the authors and does not necessarily represent the official views of the National Institutes of Health.

References

1. Ravichandran, M., Kanniah, P., Kasi, M.: Silver Nanoparticles as Nanomaterial-Based Nanosensors in Agri-Food Sector. Elsevier B.V. (2021)
2. Satvekar, R.K.: Electrochemical nanobiosensors perspectives for COVID 19 pandemic. J. Electrochem. Sci. Eng. **12**(1), 25–35 (2022)
3. Anh, N.H., et al.: Gold nanoparticle-based optical nanosensors for food and health safety monitoring: recent advances and future perspectives. RSC Adv. **12**(18), 10950–10988 (2022)
4. Shayesteh, O.H., et al.: Nano optical and electrochemical sensors and biosensors for detection of narrow therapeutic index drugs. Microchim. Acta **188**(12), 411 (2021)
5. Cusano, A., et al.: Optochemical Nanosensors, p. 597. CRC Press (2013)
6. Khanna, V.K.: Nanosensors: Physical, Chemical, and Biological, p. 665. CRC Press (2011)
7. Khanna, V.K.: Nanosensors: Physical, Chemical, and Biological, 2nd edn., p. 578. CRC Press (2021)
8. Bogue, R.: Nanosensors: a review of recent progress. Sens. Rev. **28**(1), 12–17 (2008)
9. Deep, A., Kumar, S.: Advances in Nanosensors for Biological and Environmental Analysis. Elsevier (2019)
10. Minamiki, T., Ichikawa, Y., Kurita, R.: The power of assemblies at interfaces: nanosensor platforms based on synthetic receptor membranes. Sensors **20**(8), 2228 (2020)
11. Stelmach, E., et al.: Rational design of nanoptodes architecture—towards multifunctional sensors. Talanta **196**, 226–230 (2019)
12. Gooding, J.J., Gaus, K.: Single-molecule sensors: challenges and opportunities for quantitative analysis. Angew. Chem. Int. Ed. **55**(38), 11354–11366 (2016)
13. Choi, I., Choi, Y.: Plasmonic nanosensors: review and prospect. IEEE J. Sel. Top. Quantum Electron. **18**(3), 1110–1121 (2011)
14. Yang, J., Lin, Y.-S.: Design of tunable terahertz metamaterial sensor with single-and dual-resonance characteristic. Nanomaterials **11**(9), 2212 (2021)
15. Nguyen, T.H.A., et al.: Novel biogenic silver and gold nanoparticles for multifunctional applications: green synthesis, catalytic and antibacterial activity, and colorimetric detection of Fe(III) ions. Chemosphere **287**(Part_3), 132271 (2022)
16. Yan, S., et al.: A high sensitivity nanosensor of ring resonator with inner stubs for multifunctional sensing. Jpn. J. Appl. Phys. **60**(10), 102004 (2021)
17. Guo, D.-S., Liu, Y.: Supramolecular chemistry of p-sulfonatocalix[n]arenes and its biological applications. Acc. Chem. Res. **47**(7), 1925–1934 (2014)

18. Xue, M., et al.: Pillararenes, a new class of macrocycles for supramolecular chemistry. Acc. Chem. Res. **45**(8), 1294–1308 (2012)
19. Ol'ga, A., Samsonenko, D.G., Fedin, V.P.: Supramolecular chemistry of cucurbiturils. Russ. Chem. Rev. **71**(9), 741–760 (2002)
20. Girek, T., Sliwa, W.: Chemistry of Cyclodextrins, p. 162. Wydawnictwo Akademii im. Jana Dlugosza w Czestochowie (2006)
21. Jin, Z.-Y., and Editor.: Cyclodextrin Chemistry: Preparation and Application, p. 267. World Scientific Publishing Co. Pte. Ltd. (2013)
22. Wagner, B.D.: Host-Guest Chemistry: Supramolecular Inclusion in Solution, p. 360. Walter de Gruyter GmbH (2020)
23. Biswas, A., et al.: Advances in top–down and bottom–up surface nanofabrication: techniques, applications & future prospects. Adv. Coll. Interface. Sci. **170**(1–2), 2–27 (2012)
24. Di Palma, G., et al.: Reversible, high-affinity surface capturing of proteins directed by supramolecular assembly. ACS Appl. Mater. Interfaces **11**(9), 8937–8944 (2019)
25. Prabhu, P., et al.: Recent Progress on Nanofabrication of Molecularly Imprinted Polymers. Elsevier B.V. (2018)
26. Maury, P., et al.: Integration of top-down and bottom-up nanofabrication schemes. Mater. Res. Soc. Symp. Proc. **901E**(Assembly at the Nanoscale–Toward Functional Nanostructured Materials), given, Paper #: 0901-Rb12-01 (2006)
27. Yan, D., et al.: Layered host-guest materials with reversible piezochromic luminescence. Angew. Chem. Int. Ed. Engl. **50**(31), 7037–7040 (2011)
28. Park, J., et al.: Extremely rapid self-healable and recyclable supramolecular materials through planetary ball milling and host-guest interactions. Adv. Mater. **32**(39), 2002008 (2020)
29. Wang, Y., et al.: Host–guest materials with room temperature phosphorescence: tunable emission color and thermal printing patterns. SmartMat **1**(1), e1006 (2020)
30. Zhang, S., Boussouar, I., Li, H.: Selective sensing and transport in bionic nanochannel based on macrocyclic host-guest chemistry. Chin. Chem. Lett. **32**(2), 642–648 (2021)
31. Assaf, K.I., Nau, W.M.: Cucurbiturils: from synthesis to high-affinity binding and catalysis. Chem. Soc. Rev. **44**(2), 394–418 (2015)
32. Kim, K., et al.: Functionalized cucurbiturils and their applications. Chem. Soc. Rev. **36**(2), 267–279 (2007)
33. El Haouaj, M., et al.: NMR investigation of the complexation of neutral guests by cucurbituril. J. Chem. Soc., Perkin Trans. **2**(11), 2104–2107 (2001)
34. Rebek, J., Jr.: Host–guest chemistry of calixarene capsules. Chem. Commun. **8**, 637–643 (2000)
35. Ogoshi, T., *Pillararenes*. 2015: Royal Society of Chemistry.
36. Xue, M., Zink, J.I.: An enzymatic chemical amplifier based on mechanized nanoparticles. J. Am. Chem. Soc. **135**(47), 17659–17662 (2013)
37. Zhang, D.L., et al.: An efficient multiple healing conductive composite via host-guest inclusion. Chem. Commun. (Camb.) **51**(29), 6377–6380 (2015)
38. Zhao, Y., et al.: Three-in-one: sensing, self-assembly, and cascade catalysis of cyclodextrin modified gold nanoparticles. J. Am. Chem. Soc. **138**(51), 16645–16654 (2016)
39. Davis, M.E.: The first targeted delivery of siRNA in humans via a self-assembling, cyclodextrin polymer-based nanoparticle: from concept to clinic. Mol. Pharm. **6**(3), 659–668 (2009)
40. Lee, C.-Y., et al.: Facile synthesis of hexagonal-shaped polypyrrole self-assembled particles for the electrochemical detection of dopamine. Appl. Surf. Sci. **363**, 451–458 (2016)
41. Zhao, P., et al.: An ultrasensitive high-performance baicalin sensor based on C_3N_4-SWCNTs/reduced graphene oxide/cyclodextrin metal-organic framework nanocomposite. Sens. Actuat. B Chem. **350**, 130853 (2022)
42. Premkumar, T., Lee, Y., Geckeler, K.E.: Macrocycles as a tool: a facile and one-pot synthesis of silver nanoparticles using cucurbituril designed for cancer therapeutics. Chemistry–A Eur. J. **16**(38), 11563–11566 (2010)
43. Chen, J., et al.: Supramolecular luminol–aiegen nanoparticles for deep-tissue-inflammation imaging. ACS Appl. Nano Mater. **5**(5), 5993–6000 (2022)

44. De La Rica, R., Velders, A.H.: Supramolecular Au nanoparticle assemblies as optical probes for enzyme-linked immunoassays. Small **7**(1), 66–69 (2011)
45. Song, Z., et al.: Visual and spectrophotometric detection of metformin based on the host-guest molecular recognition of cucurbit[6]uril-modified silver nanoparticles. Anal. Bioanal. Chem. **411**(27), 7293–7301 (2019)
46. Jiang, T., et al.: Humidity- and temperature-tunable multicolor luminescence of cucurbit[8]uril-based supramolecular assembly. ACS Appl. Mater. Interfaces **11**(15), 14399–14407 (2019)
47. Aoyama, Y.: Macrocyclic glycoclusters: from amphiphiles through nanoparticles to glycoviruses. Chemistry **10**(3), 588–593 (2004)
48. Masseroni, D., et al.: A fluorescent probe for ecstasy. Chem. Commun. (Camb.) **51**(64), 12799–12802 (2015)
49. Montes-García, V., et al.: Pillar [5] arene-stabilized plasmonic nanoparticles as selective SERS sensors. Isr. J. Chem. **58**(11), 1251–1260 (2018)
50. Chi, X., et al.: Thermo-triggered release of a Cys probe from the cavity of a water-soluble pillar [5] arene. Tetrahedron Lett. **56**(30), 4545–4548 (2015)
51. Bojtár, M., et al.: Amino acid recognition by fine tuning the association constants: tailored naphthalimides in pillar [5] arene-based indicator displacement assays. RSC Adv. **6**(89), 86269–86275 (2016)
52. Li, P.-Y., et al.: Multi-charged bis (p-calixarene)/pillararene functionalized gold nanoparticles for ultra-sensitive sensing of butyrylcholinesterase. Soft Matter **15**(41), 8197–8200 (2019)
53. Yu, G., et al.: A supramolecular hybrid material constructed from graphene oxide and a pillar [6] arene-based host–guest complex as an ultrasound and photoacoustic signal nanoamplifier. Mater. Horiz. **5**(3), 429–435 (2018)
54. Ramamurthy, V.: Photochemistry within a Water-Soluble Organic Capsule. American Chemical Society (2016)
55. Fujita, D., et al.: Protein encapsulation within synthetic molecular hosts. Nat. Commun. **3**, 2093/1–2093/7 (2012)
56. Della Sala, P., et al.: Prismarenes: a new class of macrocyclic hosts obtained by templation in a thermodynamically controlled synthesis. J. Am. Chem. Soc. **142**(4), 1752–1756 (2020)

Chapter 10
Perovskite Nanomaterials as Advanced Optical Sensor

Shahnaz Ahmed, Suman Lahkar, and Swapan K. Dolui

Abstract Perovskite quantum dots (PeQDs) with ultra-small crystal size of a few nanometers are a unique class of luminescent materials having excellent optical properties including narrow and bright emission, color purity, tunable emission, defect tolerance capacity, and very high photoluminescence quantum yield. The physical and chemical nature of perovskite nanocrystals changes with various environmental variables like temperature, moisture, gases, solvents, etc. All these unique features of PeQDs have allowed them to serve as ideal probes for a variety of sensing applications. Among various sensing methods, optical sensing is one of the promising sensing methods due to its simplicity, effective, and low-cost method which occurs via electron transfer, FRET, and ion exchange mechanism. Here, in this chapter, we have discussed various preparation methods of perovskite nanomaterials and their optical sensing mechanism along with their applications which explains the reason why these nanomaterials are been widely used as sensing materials. Although PeQDs are one of the most preferred sensing materials, it has their own limitations which have been discussed in the conclusion section along with their remedies.

Keywords Perovskite nanocrystals · Optical sensing · Photoluminescence · FRET

10.1 Introduction

Perovskite materials are generally represented by the formula ABX_3 (A = inorganic or organic cation; B = Pb, Sn, Ge, etc., and X = halide or oxide ion), where A being larger cation than B coordinated to twelve adjacent "X" anions and each metal cation "B" bonded with six "X" through ionic bonds to form BX_6 octahedra. The structure is best visualized in terms of the BX_6 octahedra, which are corners connected to form a three-dimensional symmetric network. The A cation occupies the hole created by BX_6 octahedral, which is coordinated with 12 X anions [1]. These perovskites

S. Ahmed · S. Lahkar · S. K. Dolui (✉)
Department of Chemical Sciences, Tezpur University, Napaam, Sonitpur, Assam 784028, India
e-mail: dolui@tezu.ernet.in

D. Mohanta and P. Chakraborty (eds.), *Nanoscale Matter and Principles for Sensing and Labeling Applications*, Advanced Structured Materials 206,
https://doi.org/10.1007/978-981-99-7848-9_10

are categorized into different classes based on the A cation and the X anion, e.g., inorganic oxide perovskites, alkaline metal halide perovskites, and organic metal halide perovskites [2]. Inorganic organic hybrid halide perovskite is formed if the A cation is an organic atom such as $CH_3NH_3^+$ or $CH(NH_2)_2^{2+}$, and all inorganic halide perovskite is formed when an inorganic atom such as cesium (Cs) is used as "A" cation. Furthermore, they can be made from zero to three-dimensional nanostructures and used in a variety of applications. Recently, the lead halide perovskite nanocrystals (PeNCs) have drawn massive attention in photovoltaic applications owing to their unique properties such as direct band gap, low exciton diffusion lengths, high absorption coefficient, and high charge carrier mobility [3]. Also, their remarkable optoelectronic properties such as very high photoluminescence quantum yield (PLQY), narrow and intense emission line, tunable emission wavelength, etc., make them a promising fluorescent material among the family of nanomaterials in lighting and display technology and could be an ideal candidate for sensing applications [4, 5].

Systems that transform chemical signals into useful analytical signals are termed chemical sensors. A specific interaction between a sensing material in the sensor and a target analyte produces the signal due to the changes in the physical properties of the sensing probe. Chemical sensors can further be classified into optical and electrochemical sensors based on the detection method used for the analyte. PeNCs are very sensitive in nature as various environmental factors including humidity, temperature, light, polar solvents, gases, ions, etc. can alter their physical and chemical properties. These changes can be reversed under controlled conditions. Recent research has revealed that the sensitive characteristics of perovskite materials can be used to alter this disadvantage into an advantage by using them as promising sensory materials [6]. When exposed to an external stimulus, the physical parameters of metal halide perovskites such as absorbance, photoluminescence (PL) emission intensity and wavelength, as well as electrical conductivity are modified. Therefore, the perovskite nanomaterials can be employed as both optical and electrochemical sensing material based on the technique used for the detection of target molecules. Different techniques such as photoluminescence, photoelectrochemistry (PEC), electrochemiluminescence (ECL), and chemoresistance (CR) have been established for the sensing application [7–9]. In this chapter, we mainly focus on the use of PeNCs as a probe for optical sensory material using optical emission or absorption techniques for the sensitive and selective detection of various target analytes including inorganic metal ions, volatile organic compounds, organic explosives, and gas molecules. Furthermore, the significant applications of perovskite nanomaterials in humidity and temperature sensing are also discussed.

10.2 Synthetic Strategies of Perovskite Nanocrystals

In recent years, simple approaches for synthesizing PeNCs of various dimensions, sizes, and shapes have arrested a lot of scientific interest. The synthesis method can be broadly classified into top-down and bottom-up approaches. The top-down

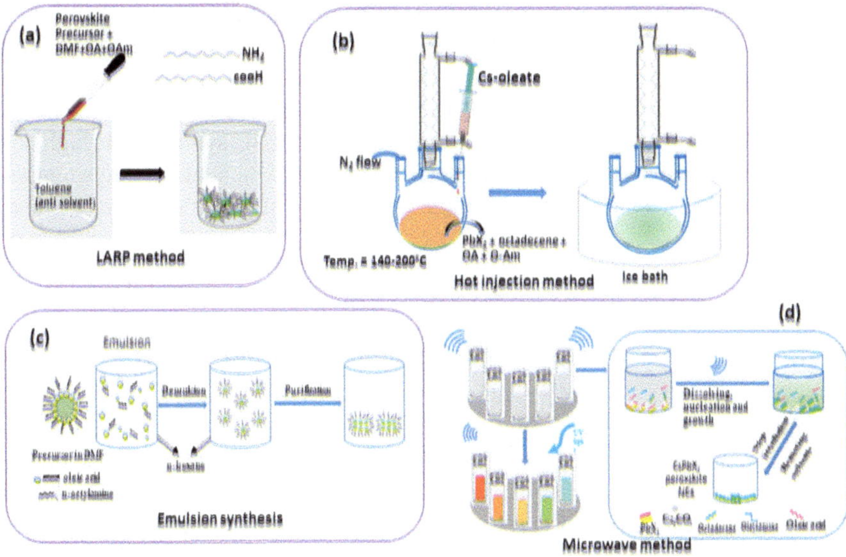

Fig. 10.1 Schematic diagram of different synthetic approaches of perovskite nanocrystals, **a** LARP method, **b** Hot injection method, **c** Emulsion synthesis, **d** Microwave synthesis

method involves the mechanical (ball-milling) or chemical (e.g., chemical exfoliation) fragmentation of macromolecules, whereas the bottom-up approach begins with molecules and ions and proceeds through gas- or liquid-phase chemical reactions [10]. The liquid-phase technique or wet chemical process has been found to be the most effective for the construction of colloidal PeNCs (Fig. 10.1) which are discussed below.

10.2.1 Hot Injection Method (HI)

HI or high-temperature method is conventionally used for the synthesis of metal chalcogenide nanocrystals (NCs) [11]. In this method, a cation precursor, e.g., Cs^+ or MA^+ in oleate form is rapidly injected into a hot solution of metal halide precursor and organic ligands under an inert atmosphere. Upon immediate injection, fast nucleation occurs with the simultaneous formation of PeNCs. By achieving a separation between the nucleation and growth stages, this technique allows for the synthesis of PeNCs with a restricted size distribution. Controlling several parameters such as reaction temperature, the ratio between the surfactants and precursors, precursor concentration, and time of reaction, the size and shape of colloidal PeNCs can be tuned. This HI method was modified by Protesescu et al. to synthesize all inorganic cesium lead halide PeNCs by injecting Cs-oleate precursor into a PbX_2 ($X = Br, Cl, I$) solution containing oleylamine (OAm), Oleic acid (OA), trioctylphosphine (TOPO), and octadecene (ODE) at high temperature (140–200 °C) under an inert atmosphere [12]. The resulting $CsPbX_3$ PeNCs had size distribution in the range of 4–15 nm with a tunable band gap covering the visible spectrum of 410–700 nm. A very high PLQY of up to 90% was achieved for $CsPbX_3$ PeNCs by this method [12]. Although the HI approach provides the finest morphological control in PeNC synthesis, its high cost makes it unsuitable for mass production due to the requirement of very high temperature and inert atmosphere. Consequently, the need for alternate synthesis methods has emerged.

10.2.2 Ligand-Assisted Re-Precipitation Method (LARP)

The low-cost LARP technique is the most often utilized approach to synthesize PeNCs at room temperature. Hybrid organic–inorganic perovskite nanocrystals were first synthesized by Schmidt et al. in which a 6 nm sized stable colloidal solution of $MAPbBr_3$ PeNCs with PLQY value of 20% was formed by using a long chain alkyl ammonium bromide (OABr) with OA and ODE as ligand [13]. Following this literature, LARP and emulsion synthesis have been developed. Zhang et al. introduced the LARP technique to produce $MAPbX_3$ (X = Cl, Br, I) PeNCs with PLQY 50–70% [14]. In the LARP method, perovskite precursors and ligands were dissolved in DMF (polar solvent). Then a definite volume of the precursor solution was added into toluene (non-polar anti-solvent) with vigorous stirring. The anti-solvent induces the crystallization of perovskite nanoparticles with bright fluorescent color due to the lowering of solubility. In order to manage the sizes and shapes of PeNCs, the choice of ligands and solvents used in precursors is crucial.

10.2.3 Emulsion Synthesis

This method is divided into two basic steps: emulsion formation and demulsion. A pair of immiscible solvents, e.g., DMF and n-hexane along with surfactant (OA) and the precursors of perovskites ($PbBr_2$ and CH_3NH_3Br) were used to create an emulsion system. To break the emulsion, demulsifier (e.g., tert-butanol, acetone) was added. Then, the mixing of precursors could cause a change in solubility which further accelerates the nucleation and crystallization of PeNCs. This approach improves the crystallization control process of PeNCs, yielding monodisperse $CH_3NH_3PbBr_3$ quantum dots with adjustable sizes ranging from 2 to 8 nm with absolute PLQYs of 80–92% [15].

10.2.4 Microwave and Ultrasonic Method

Song et al. were the first to demonstrate ultrasound-assisted synthesis of PeNCs [16]. These methods are mainly used to produce large-scale synthesis of PeNCs. A polar solvent-free synthesis approach is provided by this method. Long et al. described a microwave-assisted technique to produce high-quality $CsPbX_3$ PeNCs with a controlled morphology [17]. Microwave heating provides various advantages such as reduced energy costs and high heating rates that produce uniform nanoparticles with excellent photophysical characteristics. Here, a heterogeneous solid–liquid reaction system of ODE, OA, OAm, PbX_2, Cs_2CO_3 was poured into a glass vessel and then put in a microwave oven. The $CsPbX_3$ PeNCs were produced using microwave irradiation for approximately 4 min. The size and shape of the NCs formed by this method were microwave irradiation time dependent. When the irradiation time was 2 min, irregular nanocubes of $CsPbI_3$ PeNCs with 4.9–5.3 nm size were formed. The size of the NCs increased to 11 nm when the irradiation time was increased to 7 min.

10.3 Optical Sensor and Its Mechanism

In optical sensor, the change in optical parameter associated with the concentration of analyte is measured. An optical sensor typically composed of a chemical recognition system (termed as receptor) is connected to a transducer which converts the receptor's signal into a quantifiable signal that can be processed via amplification, filtering, recording, and so on. The receptor interacts with target analytes and recognizes parameters such as the concentration of the analyte, pH, and so on. It generates an optical signal proportional to the magnitude of these parameters. The principle of optical sensing is mainly based on light absorption and emission. Light emission or luminescence-based sensing methods that involve fluorescence quenching have an advantage over light absorption because of their excellent selectivity and sensitivity (a thousand times greater than the absorption-based method) [18]. In luminescence-based quenching method, the decrease in intensity can be discussed using the Stern–Volmer equation:

$$F_o/F = 1 + K_{SV}[C] \tag{10.1}$$

where F_o and F represent the PL emission intensities of fluorophore before and after the addition of the analyte, respectively, [C] is the concentration of quencher and K_{SV} is the Stern–Volmer quenching constant which indicates the sensitivity of receptor molecule to the quencher.

The optical sensing mechanism is mainly categorized into static and dynamic quenching. Dynamic quenching is further divided into various other effects, for example, (i) FRET (Forster resonance energy transfer), (ii) Electron transfer, (iii) IFE (Inner filter effect), and so forth [19]. As mentioned below, we have highlighted the mechanism responsible for the sensing of various analytes by considering the perovskite crystal structure.

10.3.1 *Static Quenching*

Fluorophores can form non-fluorescent ground-state compounds with quenchers. This is known as static quenching because it does not depend on diffusion or molecular collisions. The absorption or emission spectra of the fluorophore are changed after the introduction of the target analyte in static quenching [19]. In the case of PeNCs, this can be best understood through an ion exchange mechanism.

Ion Exchange

Perovskites possess very fast anion exchange properties in the presence of different halide salts. This halide ion exchange property of ABX_3-type perovskite can cause emission wavelength shift which covers the whole visible spectrum. This is an ideal property for colorimetric sensing and utilizing this property many researchers applied PeNCs for visual detection applications which will be discussed in the next section. Similarly, B site cation exchange with the analyte molecule also takes place which tunes the composition of PeNCs resulting in a change in fluorescence intensity [20].

10.3.2 *Dynamic Quenching*

When the molecular interaction between the fluorophore and the quencher occurs through diffusive encounters, then it is called dynamic quenching. As it affects the excited states of the donor, no alteration in the absorption spectra is detected.

FRET (Forster Resonance Energy Transfer)

The emission spectrum of the donor overlaps with the absorption spectrum of the acceptor when the two moieties are in the range of 15–60 Å distance [19]. The energy transfer taking place due to this process is termed as FRET. A dipole–dipole interaction connects the donor and acceptor. The distance between the donor and acceptor as well as the degree of spectral overlap define the magnitude of energy transfer. $CsPbBr_3$/PMMA fiber membrane designed using the electro-spinning method was used as a FRET-based fluorescence detector for metal ions, biological proteins, and pH sensing. After attaching cyclam with Cu^{2+}, the FRET process occurs between $CsPbBr_3$ and cyclam-Cu^{2+} leading to the quenching of fluorescence and detection of Cu^{2+} with LOD (Limit of detection) 10^{-15} M [21].

Electron Transfer

Electron transfer is possible when the donor and acceptor moieties' valence band and conduction band energy levels are aligned. The gap between the energy levels of the two plays the most important role in the electron transfer process. Highly efficient charge separation and exciton diffusion properties of perovskites accelerate the electron transfer from the electron donor perovskite to the electron-deficient analyte. The PL quenching can occur by electron transfer from the excited state of perovskite to the quencher which has been used for sensing metal ions, explosives, etc. [9, 22].

Both cations and anions can be detected by PeNCs, using the static (ion exchange) and dynamic quenching mechanism. In ABX_3 type perovskites, the composition can be tuned by adjusting the B site cation and the X anions, which can change the luminescence property such as wavelength shift or quenching of emission peak of PeNCs. Based on these cation and anion exchange principles of perovskite material, several works of metal sensing and gas sensing have been reported which are discussed in the next section. On exposure to gases, e.g., amine vapors and NH_3 gas molecules reversible H-bonding interaction between perovskites and the gas molecules takes place, leading to a change in optical properties of PeNCs. Also, the gas molecules can intercalate into the perovskite crystal lattices to form an intermediate complex which can revert to perovskite before the irreversible decomposition of PeNCs. This reversible nature greatly enhances the reproducibility of perovskite-based sensing systems. Analytical methods such as XPS, EDX, and XRD further give the evidence of ion exchange mechanism of PeNCs in the presence of sensing elements.

Aside from the compositional changes, on exposure to external stimulus electron transfer and emission energy, transfer takes place from perovskite to the electron-deficient analyte which is facilitated by the effective charge separation and exciton diffusion in perovskite nanomaterials. Various toxic gases, explosives, and metal ions have been detected by utilizing the electron transfer-induced PL quenching of PeNCs. Perovskite nanomaterials can adsorb small molecules or ions on their surface, which induces new surface defect states for non-radiative recombination pathways resulting in the decrease of PL intensity via electron transfer from PeNC to target analyte, leading to a dynamic quenching process. Various reports of metal sensing have been published using this property of PeNC as discussed in the next section. Also, gas molecules are easily absorbed on the surfaces of PeNC to adjust the surface trap states and defects of perovskites [20].

Furthermore, the instability of PeNCs on exposure to humidity can alter the photoluminescence property of perovskites. This property is reversible under controlled conditions. A reversible phase transformation of perovskites generally occurs with a change in temperature, which can be reflected in their PL spectrum. Therefore, PeNC can be used as a suitable fluorescent temperature and humidity sensor [22].

10.4 Optical Sensing Applications of Perovskite Nanocrystals

Various materials such as organic dyes, polymer nanocomposite, metal oxide nanoparticles, and metal chalcogenide quantum dots are used for optical sensing applications. However, these materials possess limitations including weak PL intensity, low PLQY value, low color purity, self-quenching phenomenon at higher concentration, less sensitivity, wide emission line, and pH dependency [20, 23, 24]. PeNCs as an alternative are extremely useful due to their remarkable optical properties and display a number of advantages over their contemporaries. Highly luminescent characteristics of ABX_3 type perovskite nanomaterials (also termed as perovskite quantum dots) with very high PLQY, tunable optical properties, ion exchange nature, and strong fluorescence emission enable them to function as optical sensors. The detection technique is mainly based on the change in the emission peak of the fluorophores (either turn on or turn off) after absorbing the target molecules. The interaction of target analytes with the surface of PeNCs results in a significant change in the emission intensity. Table 10.1 summarizes the various sensing applications of perovskite nanomaterials and nanocomposites based on the work performed in the last few years.

10.4.1 Metal Ion Sensing

Pollution caused by heavy-metal ions has now become a worldwide concern due to their numerous negative effects on human health. It is critical to detect biological and environmental metal ions quickly and selectively. Various toxic heavy-metal ions, e.g., mercury, chromium, lead, arsenic, etc. are detrimental to human health and the environment. Therefore, it is important to measure such harmful metal ions using simple and sensitive methods. Benefitting from the advantageous features of fluorescent perovskite quantum dots, it can act as a photoluminescence probe for sensitive and selective detection of metal ions under various conditions. Advanced analytical techniques such as atomic absorption spectroscopy (AAS), inductively coupled plasma-mass spectrometry (ICP-MS), mass spectroscopy (MS), X-ray fluorescence spectroscopy (R-FS), and potentiometric approaches are used in standard procedures for trace heavy-metal detection [25]. Optical sensing provides a simple and low-cost method in comparison to the above-mentioned standard method.

The ability of Metal halide perovskite (MHP) to quickly transfer anion can be the most distinguishing property of MHP NCs. By adjusting the halide composition of MHP emission wavelength range can be tuned from blue to red color which allows them to be used in the process of visual colorimetric detection. $CsPbBr_3$/cellulose-based nanocomposite synthesized via hot injection method was used for colorimetric detection of Cl^- and I^- in tap water by Park and his co-workers [26]. The colorimetric sensing the ability of this nanocomposite was mainly based on the anion

Table 10.1 Different analytical applications of perovskite nanomaterials toward optical sensing

Material	Target analyte	Performance/LOD	Performance range	References
$CsPbBr_3$/cellulose	I-Cl-	2.56 mM 4.11 mM	0.0001 to 1 M	[26]
$CH_3NH_3PbBr_3$	Hg^{2+}	0.124×10^{-9} M	$0–100 \times 10^{-9}$ M	[27]
$CsPbBr_3$	Cu^{2+}, Yb^{3+}	2×10^{-9} M	$2 \times 10^{-9}–2 \times 10^{-6}$ M	[9]
$CsPbBr_3$	Cu^{2+}	0.1 nM	0 to 100 nM	[28]
$Cs_3Bi2Br9:Eu^{3+}$	Cu^{2+}	10 nM	5 nM to 3 μM	[29]
$CsCuCl_3$	Pb^{2+}	10^{-7} M	1×10^{-7} M–1.5×10^{-6} M	[47]
$CH_3NH_3PbBr_3$@MOF-5	Cd^{2+}, Cu^{2+}	$2 \times 10^{-9}–2 \times 10^{-6}$ M	$2 \times 10^{-3}–100 \times 10^{-3}$ M (Cd^{2+}) 20×10^{-5} to 200×10^{-5} M (Cu^{2+})	[48]
$CH_3NH_3PbBr_3$	Amine vapors (primary, secondary, tertiary, and aromatic amine)	62 ppb (DEA)	–	[30]
$HPbBr_3/PbBr_2@SiO_2$ nanospheres	Methyl amine (MA) gas	70 ppb	–	[32]
$CsPbBr_3$	H_2S	0.18 μM	0–100 μM	[49]
$CsPbBr_3$	HCl vapor	5 ppm		
$CH_3NH_3PbBr_3$	SO_2 gas	155 ppb	2–10 ppm	[35]
$MAPbBr_3$	Humidity	0.68% RH	7–98% humidity	[36]
Eu^{3+} doped $CsPbBr_3$	Temperature	$S_a = 0.0224\ K^{-1}$ $S_r = 2.25\%\ K^{-1}$	93 K to 383 K	[37]
Tb^{3+} doped $CsPbI_3$	Temperature	$S_a = 0.0398\ K^{-1}$ $S_r = 7.12\%\ K^{-1}$	293–393 K	[38]

(continued)

Table 10.1 (continued)

Material	Target analyte	Performance/LOD	Performance range	References
$CsPbBr_3$/Eu-BTC nanohybrid	Temperature	$S_r = 3.9\%\ °C^{-1}$ temperature resolution of 0.004 °C	20–100 °C	[39]
$CsPbBr_2I$	TNP	5 µL of 0.1 mM solution of TNP quenches the PL up to 54%	–	[42]
$CH_3NH_3PbBr_3$	TNP	10^{-15} M	10^{-3} M–10^{-15} M	[44]
$CsPbBr_3$ and $CsPbI_3$	Picric acid (PA)	0.8 nM ($CsPbBr_3$) 1.9 nM ($CsPbI_3$)		[41]
$CsPbBr_3$, layered $(PEA)_2Cs_2Pb_3Br_{10}$, and $CsPbBr_3$ nanocrystals	DNT vapor	50% recovery of initial PL intensity in 1 min	–	[45]
$MAPbBr_3$ nanocrystals embedded in a poly (vinylidene fluoride)	2,4,6-trinitrotoluene (TNT), 1,3,5-trinitroperhydro-1,3,5-triazine (RDX), and trinitroglycerin (TNG)		–	[46]

exchange property of lead halide perovskites. The composite displayed good stability under various humid conditions and enhanced photoluminescence due to the strong adhesion of the $CsPbBr_3$ PQDs to the porous cellulose fibers. The nanocomposite probe showed a fast response with the introduction of Cl^- and I^- with the shift in PL peak. The LOD of I^- and Cl^- was found to be 2.56 mM and 4.11 mM, respectively, for this sensor [26]. A $CH_3NH_3PbBr_3$ PeQD-based nanosensor was designed by Qiang and his group for the determination of ultra-trace Hg^{2+} [27]. In this study, Hg^{2+} was shown to dramatically reduce the PL of PeQDs via a surface ion exchange process (Fig. 10.2c). XPS and EDX analysis showed the partial replacement of Pb^{2+} with Hg^{2+} on the surface of $CH_3NH_3PbBr_3$ PeQD. PL intensity continuously decreased with the addition of Hg^{2+} within the concentration range of 0–1000 nM along with a notable blue shift from 520 to 507 nm. The author explained this on the basis of cation exchange on the surface of PeQDs. There is a good linear relationship between the plot of I_0/I and Hg^{2+} concentration in the range of 0–100 nM (Fig. 10.2b).

Sheng and his co-workers reported all inorganic colloidal $CsPbBr_3$ PeQD for the detection of both Cu^{2+} and Yb^{3+} in organic mediums like cyclohexane, various bio-oils, and industrial oils [9]. The LOD was found to be as low as 2×10^{-9} M with

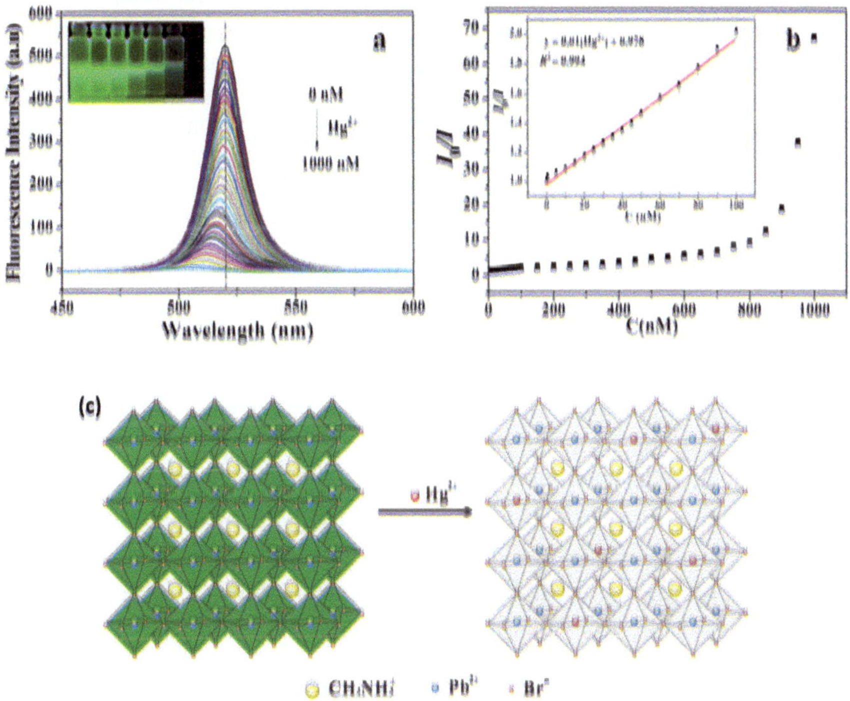

Fig. 10.2 **a** Fluorescence spectra of $CH_3NH_3PbBr_3$ QDs upon the addition of different amounts of Hg^{2+}. **b** Linear fitting curve of I_0/I with respect to Hg^{2+} concentration over the range of 0–100 nM, **c** The illustration of the ion exchange process on the surface of $CH_3NH_3PbBr_3$ QDs (Reprinted from ref. [27] Copyright (2017), with permission from Elsevier)

high sensitivity, selectivity, and rapid response. On the basis of both experimental and density functional theory (DFT) calculations, the processes of PL quenching or enhancement were examined. Adsorption of Cu^{2+} on the surface of PeQDs, followed by charge transfer caused the PL quenching of the fluorophore. They confirmed from EPR (electron paramagnetic resonance) analysis that Cu^{2+} rather than Cu^{+} coordinated to the surface of PeQDs was responsible for PL quenching which is different from the case of cadmium chalcogenide (CdS) probing system for Cu^{2+} sensing. The reduction of Cu^{2+} to Cu^{+} was the main reason for PL quenching in a CdS-based system. From the DFT calculations operated on the electronic properties of the system, "Br" contributed to the valence band maximum of $CsPbBr_3$, and the conduction band minimum was contributed by Pb. At the valence band maximal edge of $CsPbBr_3$ QD, the Cu^{2+} addition resulted in the emergence of various new states or non-radiative recombination pathways due to the formation of Br–Cu–OOCH complex on the surface of PeQDs. These non-radiative recombination pathways were mainly responsible for the fluorescence quenching of the probe whereas no such surface defect states were induced in the case of Yb^{3+}. In addition to the crystal size and structure, the surface plane smoothness also plays a pivotal role in the selective detection of metal ions. Their work offers a potentially strong tool for identifying the specific metal ions in biological and industrial organic solution systems. Another research work also employed $CsPbBr_3$ PeQDs synthesized by hot injection method for Cu^{2+} sensing in the organic phase with an LOD value of 0.1 nM within the range of 0–100 nM. They explained the PL quenching phenomenon based on electron transfer from PeQDs to Cu^{2+} and the process was confirmed by the photoluminescence decay lifetime and absorption spectrum [28].

Based on the same electron transfer mechanism (Fig. 10.3), Ding et al. established an optical sensing medium for Cu^{2+} by employing a nontoxic europium-doped lead-free $Cs_3Bi_2Br_9$ Perovskite Quantum Dots via a modified ligand-assisted re-precipitation method. The resulting $Cs_3Bi_2Br_9$:Eu^{3+} nanocomposite showed blue emissions of exciton emission and red emission of Eu^{3+} ions which were mostly generated from efficient electron transfer from the PeQDs to Eu^{3+}. Eu^{3+} doping significantly increased the water stability of $Cs_3Bi_2Br_9$ PeQDs. $Cs_3Bi_2Br_9$:Eu^{3+} PeQDs exhibited a detection limit of 10 nM for Cu^{2+} ion having a linear range from 5 nM to 3 μM. Overall, doping of lanthanide ions into $Cs_3Bi_2Br_9$ PeQDs has the potential to improve their optical performance and broaden their applications [29].

10.4.2 Detection of Volatile Organic Compounds

Volatile organic compounds (VOC) are toxic to human health and environment and tracing them has become one of the most discussed topics. Metal halide perovskite nanocrystals exhibit excellent gas-sensing properties. Wu et al. designed a low-cost $CH_3NH_3PbBr_3$ PeNC-based fluorescence sensor to detect several amine vapors (primary, secondary, tertiary, and aromatic amine). They exposed the amine vapors to the $CH_3NH_3PbBr_3$ perovskite film deposited using two-step method. A color change

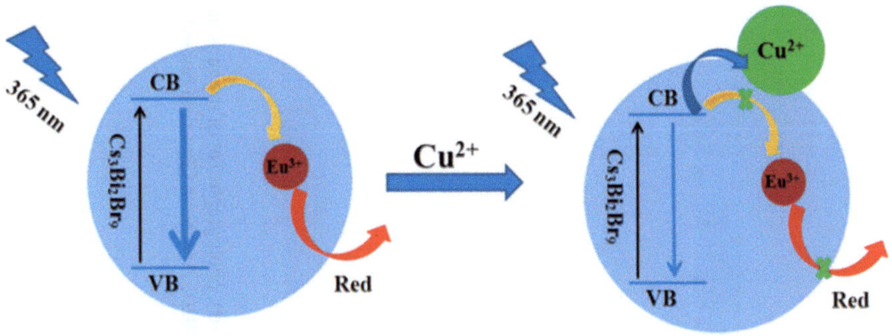

Fig. 10.3 Schematic diagram of PL quenching of Cu^{2+} due to electron transfer (Reprinted with permission from ref. [29]. Copyright {2019} American Chemical Society)

of the film from yellow to white was visible on exposure to aliphatic amines. In the case of aniline (aromatic amines) no color change was observed due to the less donor ability of amine which is in conjugation with the aromatic ring. A very quick response (<4 s) of PL quenching was observed after the introduction of amines [30]. Kim and his co-workers also utilized $CH_3NH_3PbBr_3$ perovskite for amine vapor ($EtNH_2$, Et_2NH, Et_3N) sensing (response time <1 s) and found a reversible recovery of quenched PL intensity in case of DEA and TEA. When $CH_3NH_3PbBr_3$ films were exposed to Et_3N vapor, they recovered completely in 1 s, whereas Et_2NH took longer (15 s) and only partially recovered. X-ray diffraction (XRD) measurements and quantum chemical calculations showed that fluorescence quenching is caused by reversible or irreversible hydrogen bonding interactions between perovskite and aliphatic amines [31].

Recently, methyl amine (MA) gas sensing was performed by fluorescent turn on and wavelength shift-based colorimetric method [32]. Because of the formation, $MAPbBr_3$ nanocrystals (NCs) in hollow silica nanospheres by the interaction between MA gas and $HPbBr_3/PbBr_2@SiO_2$ nanospheres showed the fluorescence turn-on and wavelength-shift phenomena with a LOD of 70 ppb. The non-fluorescent $HPbBr_3/PbBr_2@SiO_2$ converted into fluorescent $MAPbBr_3$ on interaction with MA, leading to a PL turn on sensing system for MA gas. Furthermore, when the MA concentration increases from 1.0 to 95 ppm, the emission wavelength red-shifted from 478.7 to 510.6 nm, implying the possibility of colorimetric sensing (Fig. 10.4c, e). The increase of the crystal size during the evolution of NCs is the main reason for redshift of PL maxima. The $MAPbBr_3$ film could be reused by giving a heat treatment for 5 min at 150 °C (Fig. 10.4d).

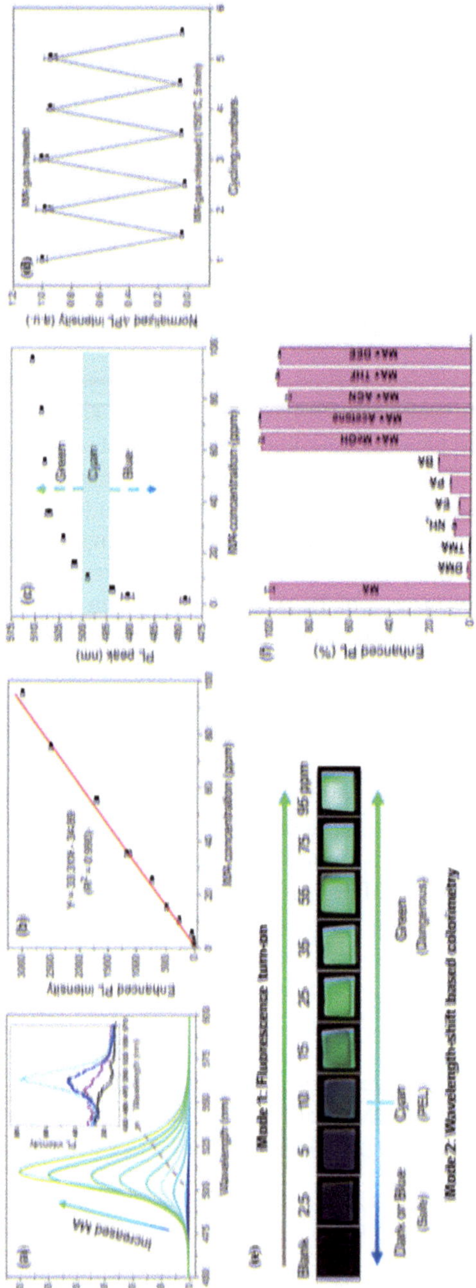

Fig. 10.4 Emission spectrum **a** enhanced photoluminescence intensity **(b)**, PL peak **(c)**, The reusability of the sensing film **(d)**, photographs **(e)** of the sensing film with MA concentration, and selectivity **f** tests of the sensing film. **(Adapted with permission from ref. [32]. Copyright {2020} American Chemical Society)**

10.4.3 Gas Sensing

In many different sectors, gas sensors are essential for monitoring and detecting harmful inert gases, as well as assuring safety and air quality. Metal halide Perovskite NCs exhibit excellent gas-sensing properties. Zhao et al. reported optical bleaching of $MAPbI_3$ films after they were exposed to ammonia gas molecules and observed a rapid reversible color change upon removing NH_3 gas molecules. XRD measurements demonstrated a phase transformation of $MAPbI_3$ films either because of the formation of intercalation compound or coordination compound with NH_3 molecules. The UV–vis absorption intensity of the perovskite film became zero after the interaction with the ammonia gas. Long-time exposure to the gas molecules did not revert back to the original brown color of the $MAPbI_3$ films [33].

The discharge of gaseous HCl has been a source of tremendous worry in terms of environmental protection due to the very poisonous and corrosive nature of HCl. Furthermore, because of its extremely volatile nature, the use of an HCl solution results in the production of HCl vapor. $CsPbBr_3$ PeNCs synthesized using a hot injection method were utilized as optical probes for the sensing of HCl vapor with a LOD as low as 5 ppm by Xiao and his groups [34]. They described the sensing process using the anion exchange between Br^- of the PeNC and Cl^- of the HCl. On interaction with HCl gas (5 ppm) for various durations, there was a gradual blue shift to 495 nm from the peak centered at 513 nm after 20 min of exposure. Then the gas ion exchange reaction reached the dynamic equilibrium suggesting the formation of $CsPb(Br/Cl)_3$ without the change in morphology and crystal structure of $CsPbBr_3$ PeNCs confirmed by the XRD and TEM image. An increase in the HCl content reduced the reaction time and increased the observed shift in PL peak and absorption wavelengths. Because reverse anion exchange of $CsPb(Br/Cl)_3$ perovskite nanocrystals with HBr vapor was possible, the interactions with HCl were reversible. The selectivity of this probe was determined by interacting with various volatile compounds such as HNO_3, CH_3COOH, NH_4OH, and water vapor. No PL peak shift is observed indicating its high selectivity.

You et al. designed a highly luminescent water-stable $CH_3NH_3PbBr_3$ PeNCs by encapsulating them with super hydrophobic silica aerogel. The porous structure of as-synthesized nanocomposite makes them ideal for gas sensing and finally constructed a 3D printed gas cell, a unique, sensitive, selective, and reversible fluorescence sensor for SO_2 gas with LOD 155 ppb. With the addition of 1000 ppm SO_2 gas, the quantum yield of SiO_2-aerogel-$CH_3NH_3PbBr_3$ NCs (66.4%) was significantly reduced to 26.6%, indicating their capacity to recognize SO_2 gas [35].

10.4.4 Humidity and Thermal Sensing

Certain circumstances, such as a water-sensitive organic reaction system or a storage environment for reagents, often impose a particularly strict restriction on water content. Therefore, humidity measurement is necessary for a variety of applications including structural health monitoring, food processing and storage, medicinal, ecological, and agricultural uses, and even aerospace exploration. Many materials, e.g., metal oxide, polymer nanocomposite, and metal–organic frameworks are applied for humidity sensing. However, they have significant drawbacks including difficult synthesis, low sensitivity, and a limited operating range. Drawing inspiration from the advantageous properties of PeNCs, the $MAPbBr_3$ was utilized as a humidity sensor which was synthesized by a simple grinding of $PbBr_2$ and MABr in 20% relative humidity. With the increase of humidity within the range of 7–98%, the PL intensity at 530 nm was quenched gradually due to the instability of PeNCs in the presence of moisture. The process was reversible when the interaction of $MAPbBr_3$ with water molecules was for a short time [36].

Among various temperature detection techniques fluorescence intensity ratio (FIR)-based method has gained a lot of interest recently because of the benefits such as non-contact, high measurement accuracy, and applicability for a variety of situations. In comparison to single-wavelength temperature-dependent emitters the FIR approach, involving the fluorescence intensity ratio of two temperature-dependent emitters, typically provides better self-referencing sensing performance. Highly stable dual emitting Eu^{3+} doped $CsPbBr_3$ PeNCs embedded into tellurite glasses were prepared by an in-situ glass crystallization method for optical thermometer use [37]. The two emission centers were originated from the exciton recombination of $CsPbBr_3$ (green emission) and Eu^{3+} (red emission) under the excitation wavelength of 394 nm. FIR of $CsPbBr_3$ PeNC and Eu^{3+} emission in the nanocomposite displayed good temperature sensing properties in the temperature range of 93–383 K, owing to their fundamentally distinct thermal-quenching characteristics. The absolute (S_a) and relative temperature (S_r) sensitivity were found to be as high as 0.0224 K^{-1} and 2.25% K^{-1}, respectively. Another report stated that Tb^{3+} doped $CsPbI_3$ PeNCs in borosilicate glass as an optical probe for temperature sensing [38]. The absolute and relative temperature sensitivities were determined to be 0.0398 K^{-1} and 7.12% K^{-1}, respectively. The nanocomposite displayed a noticeable color change from orange-red to yellow-green in the temperature range of 293–393 K. Also, Liu et al. utilized FIR-based technique to fabricate a self-calibrated dual emissive optical thermometer composed of $CsPbBr_3$ perovskite quantum dot encapsulated into a lanthanide metal–organic framework (Eu-BTC) [39]. Different temperature-dependent luminescence responses originating from the PL of $CsPbBr_3$ PeNCs at 528 nm and the $^5D_0$–$^7F_2$ emission of Eu^{3+} at 618 nm were examined within the temperature limit of 20–100 °C. The resulting $CsPbBr_3$/Eu–BTC nanohybrid showed a high S_r of 3.9% $°C^{-1}$ and an outstanding temperature resolution of 0.004 °C in the measured temperature limit indicating that it is preferable as a ratiometric luminous thermometer for temperature sensing. As shown in Fig. 10.5, the PL intensity at 618 nm was increased with temperature which might be attributed to the high luminescence efficiency of the embedded guest $CsPbBr_3$ PeQD and the host skeleton limiting effect of the Eu-BTC matrix.

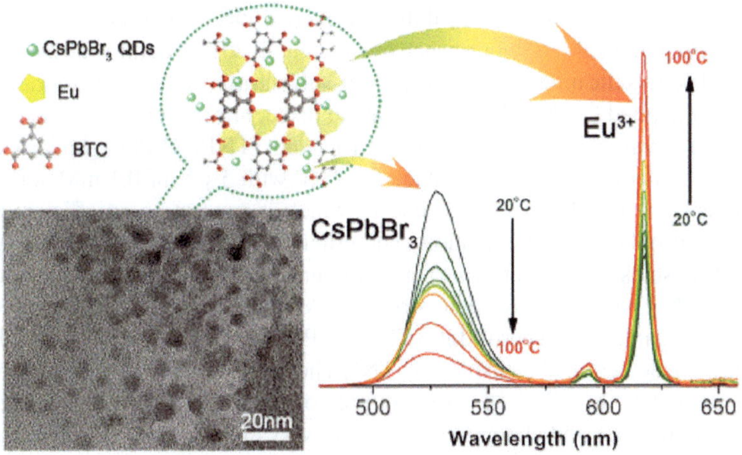

Fig. 10.5 Temperature-dependent PL response of $CsPbBr_3$@Eu–BTC composite at 528 and 618 nm (Reprinted with permission from ref. [39]. Copyright {2020} American Chemical Society)

10.4.5 Explosive Detection

Different methods like surface-enhanced Raman spectroscopy (SERS), gas chromatography–mass spectrometry (GC–MS), liquid chromatography–tandem mass spectrometry (LC–MS), X-ray diffraction, ion mobility spectroscopy, and electrochemical methods have been used for decades to detect trace amount of explosives. However their applications for speedy on-site detection are greatly limited due to the expensive cost and complex operation [40]. Thus, a lot of research work centers on the finding of cost-effective detection methods for explosive materials.

The ease of synthesis procedures and more intense fluorescence due to their ultrahigh PLQY of halide perovskites can potentially lead to detect explosive nitroaromatics at a low cost compared to the other fluorophores. Fluorescence quenching of the PeNCs after absorption of particles or vapors of the explosives via electron transfer is the primary explosive mechanism which is illustrated in Fig. 10.6. Chen et al. reported PeNCs, $CsPbBr_3$/$CsPbI_3$ synthesized by typical hot injection method. As synthesized PeNCs act as a fluorescent sensor for selective discrimination of picric acid (PA) against other nitroaromatic explosives [41]. They showed an interesting result where $CsPbBr_3$ shows an LOD value of 0.8 nM for PA and $CsPbI_3$ shows an LOD value of 1.9 nM for PA. According to them the variety of detection range between the PeNCs may be the luminous efficiency. Higher quantum yields might bring better PL properties and a higher photon-to-electron conversion efficiency and hence lower detection limit. Along with UV–visible and XPS analysis, also from density functional theory (DFT) calculations electrostatic interaction assisted electron transferring between PA and PeQD has been confirmed. DFT calculation provides that the LUMO energy level of strong electrophilic PA ($E_{LUMO} = -3.897$ eV) is lower than that of $CsPbBr_3$ ($E_{LUMO} = -3.6$ eV) and $CsPbI_3$ (E_{LUMO}

= −3.75 eV), signifying the feasibility of excited electrons transferred from the PQDs to PA and resulting fluorescence quenching. In 2017, Aamir et al. synthesized $CsPbBr_2I$ perovskite microcrystals using anti-solvent-assisted crystallization process [42]. They applied $CsPbBr_2I$ microcrystals to detect nitrophenols-2-nitrophenol (2-NP), 3-nitrophenol (3-NP), and 2,4,6-trinitrophenol (TNP). $CsPbBr_2I$ microcrystals showed a higher quenching percentage for TNP with 5 μL of 0.1 mM solution of TNP quenches the PL up to 54% due to which it can act as an efficient sensor to detect nitrophenol contamination. The efficient PL quenching was assigned to either Π–Π stacking (benzene ring) or hydrogen bonding arising from the hydroxyl group of TNP with the perovskite structure. Muthu et al. reported a luminescent organic–inorganic hybrid perovskite ($CH_3NH_3PbBr_3$) nanoparticles for the selective and sensitive detection of an explosive, TNP. They can detect TNP noticeably within the range of 10^{-3} M–10^{-15} M. In many previous works it has been assumed that as exciton diffusion and charge separation are very efficient in perovskite, electron transfer could be possible from perovskites to the analytes. So, this might be acting as the major pathway for fluorescence quenching. Muthu and his groups come with the conclusion that the presence of hydrogen bonding between the hydroxyl group and the three nitro groups plays a key role in the detection mechanism making TNP a strong electron acceptor and quenches the PL intensity of donor efficiently [43, 44]. Harwell et al. examined the PL quenching response of hybrid perovskite films $CsPbBr_3$, layered $(PEA)_2Cs_2Pb_3Br_{10}$, and $CsPbBr_3$ nanocrystals with 3D-0D dimension region toward the vapor of dinitrotoluene (DNT) and achieved strong PL responses to DNT molecules at concentrations in the parts per billion range by 0D perovskite nanocrystals [45]. Upon decreasing dimension surface area increases for the reaction to occur more efficiently and hence response is also larger and faster. Even after desorption of vapors, PeNC film is again able to reach a new equilibrium concentration with the clean carrier gas. Thus, they presented 0 D PeNC nanofilms with long-term repeatability and stability toward the DNT vapor sensing.

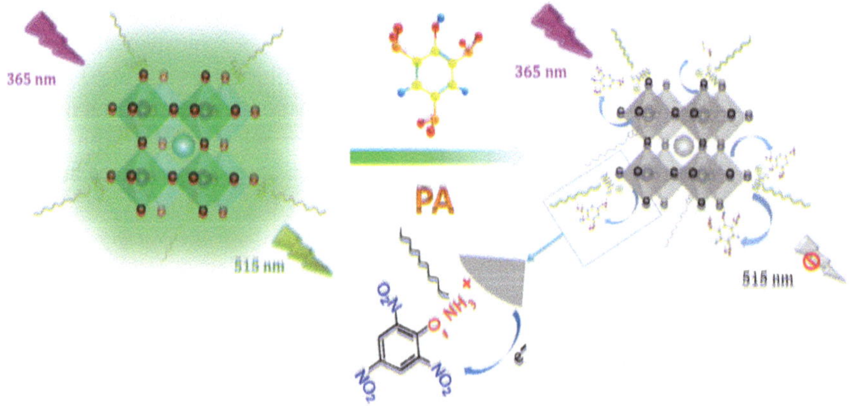

Fig. 10.6 Schematic diagram to depict the sensitive fluorescence detection of PA (Reprinted from ref. [41] Copyright (2020), with permission from Elsevier)

Shan et al. reported very stable porous nanocomposites consisting of $MAPbBr_3$ nanocrystals embedded in a poly (vinylidene fluoride) (PVDF) polymer matrix which shows no fluorescence degradation upon soaking in water/alcohol [46]. They were able to indicate strong fluorescence quenching upon exposure to a few nanograms of various explosives such as 2,4,6-trinitrotoluene (TNT), 1,3,5-trinitroperhydro-1,3,5-triazine (RDX), and trinitroglycerin (TNG) and with a low cost. From their research works they came up with a new perception that an increase in trap density of $MAPbBr_3$ PeNCs after exposure to the organo nitro compounds mainly leads to fluorescence quenching in perovskite. Transient absorption spectroscopy (TAS) analysis manifests that TNT has introduced more trap states into the perovskite and shortened its charge carrier lifetime, which majorly contributes to the observed fluorescence quenching in perovskite-TNT.

10.5 Conclusion and Future Challenges

In this chapter, various features and current research progresses of perovskite nanocrystals in metal ion detection, explosive, gas sensing, VOC detection, humidity, and temperature sensor have been summarized. This optical sensor takes advantage of both the unique and excellent optical properties of PeQDs and their sensitive nature, enabling them to be efficient sensing systems and thereby exhibiting great selectivity and sensitivity. The sensing mechanism of different perovskite nanomaterials as the optical sensor has also been explained by different processes such as ion exchange, resonance energy transfer, electron transfer, and compositional change. External factors like temperature and humidity result in a reversible phase transformation and degradation of perovskite with a change in the optical properties, making them suitable humidity and temperature sensors. Also, various interactions of PeQDs such as adsorption and intercalation with analytes (toxic gases, explosives, metal ions) can either modify the surface trap states or form an intermediate complex, leading to a change in their optical parameters. All these changes can be reversed under controlled conditions which further enhances the reusability of perovskite-based sensing probes. Despite their great potential as sensor material, the long-term stability issue in aqueous media stands as an obstacle in the race for promising sensing materials. So, the key focus is to further improve their long-term stability. One such step to reduce this limitation is by passivation of metal halide perovskite with polymer material, graphene, and encapsulation of PeNCs into a porous matrix such as metal–organic framework that can create a protective shield against water molecule attachments. Sensors with certain functional groups or molecularly imprinted polymers (MIPs) incorporating PeNCs are being expected to be designed to boost the sensors' affinity for target analytes. Another drawback lies in the toxic property of lead which can be removed by substituting Pb with Sn, Bi, and Cu. However, lead-free perovskite systems have their own limitations as they have low PLQY, thereby hindering their sensing ability. Thus various metal ion doping methods are being used to enhance the PLQY. So, we can conclude that further thrust in the research of PeNC will help it to be the best sensing probe among its contemporary competitors.

References

1. Baikie, T., Fang, Y., Kadro, J.M., Schreyer, M., Wei, F., Mhaisalkar, S. G., et al.: Synthesis and crystal chemistry of the hybrid perovskite (CH 3 NH 3) PbI 3 for solid-state sensitised solar cell applications. J. Mater. Chem. A **1**(18), 5628–5641 (2013)
2. Assirey, E.A.R.: Perovskite synthesis, properties and their related biochemical and industrial application. Saudi Pharm. J. **27**(6), 817–829 (2019)
3. Li, B., Zhang, Y., Fu, L., Yu, T., Zhou, S., Zhang, L., Yin, L.: Surface passivation engineering strategy to fully inorganic cubic CsPbI3 perovskites for high-performance solar cells. Nat. Commun. **9**, 1–8 (2018)
4. Shi, Z., Li, Y., Zhang, Y., Chen, Y., Li, X., Wu, D., Xu, T., Shan, C., Du, G.: High-efficiency, and air-stable perovskite quantum dots light-emitting diodes with an all-inorganic heterostructure. Nano Lett. **17**, 313–321 (2017)
5. Zhang, H., Wang, X., Liao, Q., Xu, Z., Li, H., Zheng, Fu, H.: Embedding perovskite nanocrystals into a polymer matrix for tunable luminescence probes in cell imaging. Adv. Funct. Mater. **27**, 1604382 (2017)
6. Huangfu, C., Feng, L.: High-performance fluorescent sensor based on $CsPbBr_3$ quantum dots for rapid analysis of total polar materials in edible oils. Sens. Actuat. B: Chem. 130193 (2021)
7. Ren, K., Huang, L., Yue, S., Lu, S., Liu, K., Azam, M., Wang, Z.: Turning a disadvantage into an advantage: synthesizing high-quality organometallic halide perovskite nanosheet arrays for humidity sensors. J. Mater. Chem. C **5**(10), 2504–2508 (2017)
8. Weng, Z., Qin, J., Umar, A.A., Wang, J., Zhang, X., Wang, H., et al.: Lead-free $Cs_2BiAgBr_6$ double perovskite-based humidity sensor with superfast recovery time. Adv. Funct. Mater. **29**(24), 1902234 (2019)
9. Sheng, X., Liu, Y., Wang, Y., Li, Y., Wang, X., Wang, X., et al.: Cesium lead halide perovskite quantum dots as a photoluminescence probe for metal ions. Adv. Mater. **29**(37), 1700150 (2017)
10. Wang, S., Yousefi Amin, A.A., Wu, L., Cao, M., Zhang, Q., Ameri, T.: Perovskite nanocrystals: synthesis, stability, and optoelectronic applications. Small Struct. **2**(3), 2000124 (2021)
11. Murray, C., Norris, D.J., Bawendi, M.G.: Synthesis and characterization of nearly monodisperse CdE (E = sulfur, selenium, tellurium) semiconductor nanocrystallites. J. Am. Chem. Soc. **115**(19), 8706–8715 (1993)
12. Protesescu, L., Yakunin, S., Bodnarchuk, M.I., Krieg, F., Caputo, R., Hendon, C.H., Yang, R.X., Walsh, A., Kovalenko, M.V.: Nanocrystals of cesium lead halide perovskites (CsPbX3, X = Cl, Br, and I): n ovel optoelectronic materials showing bright emission with wide color gamut. Nano Lett. **15**, 3692–3696 (2015)
13. Schmidt, L.C., Pertegás, A., González-Carrero, S., Malinkiewicz, O., Agouram, S., Minguez Espallargas, G., et al.: Nontemplate synthesis of CH3NH3PbBr3 perovskite nanoparticles. J. Am. Chem. Soc. **136**(3), 850–853 (2014)
14. Zhang, F., Zhong, H., Chen, C., Wu, X. G., Hu, X., Huang, H., et al.: Brightly luminescent and color tunable colloidal CH3NH3PbX3 (X = Br, I, Cl) quantum dots: potential alternatives for display technology. ACS Nano **9**(4), 4533–4542 (2015)
15. Huang, H., Zhao, F., Liu, L., Zhang, F., Wu, X.G., Shi, L., et al.: Emulsion synthesis of size-tunable CH3NH3PbBr3 quantum dots: an alternative route toward efficient light-emitting diodes. ACS Appl. Mater. Interfaces **7**(51), 28128–28133 (2015)
16. Jang, D.M., Kim, D.H., Park, K., Park, J., Lee, J.W., Song, J.K.: Ultrasound synthesis of lead halide perovskite nanocrystals. J. Mater. Chem. C **4**(45), 10625–10629 (2016)
17. Long, Z., Ren, H., Sun, J., Ouyang, J., Na, N.: High-throughput and tunable synthesis of colloidal $CsPbX_3$ perovskite nanocrystals in a heterogeneous system by microwave irradiation. Chem. Commun. **53**(71), 9914–9917 (2017)

18. Lobnik, A., Turel, M., Urek, S.K.: Optical chemical sensors: design and applications. Adv. Chem. Sens. 3–28 (2012)
19. Lakowicz, J.R. (ed.).: Principles of Fluorescence Spectroscopy. Springer Science & Business Media (2013)
20. Halali, V.V., Sanjayan, C.G., Suvina, V., Sakar, M., Balakrishna, R.G.: Perovskite nanomaterials as optical and electrochemical sensors. Inorg. Chem. Front. **7**(14), 2702–2725 (2020)
21. Wang, Y., Zhu, Y., Huang, J., Cai, J., Zhu, J., Yang, X., et al.: Perovskite quantum dots encapsulated in electrospun fiber membranes as multifunctional supersensitive sensors for biomolecules, metal ions and pH. Nanoscale Horiz. **2**(4), 225–232 (2017)
22. Zhu, Z., Sun, Q., Zhang, Z., Dai, J., Xing, G., Li, S., et al.: Metal halide perovskites: stability and sensing-ability. J. Mater. Chem. C **6**(38), 10121–10137 (2018)
23. Resch-Genger, U., Grabolle, M., Cavaliere-Jaricot, S., Nitschke, R., Nann, T.: Quantum dots versus organic dyes as fluorescent labels. Nat. Methods **5**(9), 763–775 (2008)
24. Moseley, P.T.: Progress in the development of semiconducting metal oxide gas sensors: a review. Meas. Sci. Technol. **28**(8), 082001 (2017)
25. Malik, L.A., Bashir, A., Qureashi, A., Pandith, A.H.: Detection and removal of heavy metal ions: a review. Environ. Chem. Lett. **17**(4), 1495–1521 (2019)
26. Park, B., Kang, S.M., Lee, G.W., Kwak, C.H., Rethinasabapathy, M., Huh, Y.S.: Fabrication of $CsPbBr_3$ Perovskite quantum dots/cellulose-based colorimetric sensor: dual-responsive on-site detection of chloride and iodide ions. Ind. Eng. Chem. Res. **59**(2), 793–801 (2019)
27. Lu, L.Q., Tan, T., Tian, X.K., Li, Y., Deng, P.: Visual and sensitive fluorescent sensing for ultratrace mercury ions by perovskite quantum dots. Anal. Chim. Acta **986**, 109–114 (2017)
28. Liu, Y., Tang, X., Zhu, T., Deng, M., Ikechukwu, I. P., Huang, W., et al.: All-inorganic $CsPbBr_3$ perovskite quantum dots as a photoluminescent probe for ultrasensitive Cu 2+ detection. J. Mater. Chem. C, **6**(17), 4793–4799 (2018)
29. Ding, N., Zhou, D., Pan, G., Xu, W., Chen, X., Li, D., et al.: Europium-doped lead-free $Cs_3Bi_2Br_9$ perovskite quantum dots and ultrasensitive Cu2+ detection. ACS Sustain. Chem. Eng. **7**(9), 8397–8404 (2019)
30. Wu, P., He, Q., Zhu, D., Jiang, H., Jiao, Z., Zhang, Y., et al.: Highly efficient fluorescent and colorimetric sensing of organic amine vapors based on organometal halide perovskite nanostructures. Anal. Methods **9**(25), 3804–3809 (2017)
31. Kim, S.H., Kirakosyan, A., Choi, J., Kim, J.H.: Detection of volatile organic compounds (VOCs), aliphatic amines, using highly fluorescent organic-inorganic hybrid perovskite nanoparticles. Dyes Pigm. **147**, 1–5 (2017)
32. Huang, Y., Wang, S., Zhu, Y., Li, F., Jin, J., Dong, J., et al.: Dual-mode of fluorescence turn-on and wavelength-shift for methylamine gas sensing based on space-confined growth of methylammonium lead tribromide perovskite nanocrystals. Anal. Chem. **92**(8), 5661 5665 (2020)
33. Zhao, Y., Zhu, K.: Optical bleaching of perovskite (CH_3 NH_3) PbI_3 through room-temperature phase transformation induced by ammonia. Chem. Commun. **50**(13), 1605–1607 (2014)
34. Chen, X., Hu, H., Xia, Z., Gao, W., Gou, W., Qu, Y., Ma, Y.: CsPbBr 3 perovskite nanocrystals as highly selective and sensitive spectrochemical probes for gaseous HCl detection. J. Mater. Chem. C **5**(2), 309–313 (2017)
35. You, X., Wu, J., Chi, Y.: Superhydrophobic silica aerogels encapsulated fluorescent perovskite quantum Dots for reversible sensing of SO2 in a 3D-printed gas cell. Anal. Chem. **91**, 5058–5066 (2019)
36. Xu, W., Li, F., Cai, Z., Wang, Y., Luo, F., Chen, X.: An ultrasensitive and reversible fluorescence sensor of humidity using perovskite $CH_3NH_3PbBr_3$. J. Mater. Chem. C **4**(41), 9651–9655 (2016)
37. Li, X., Yu, Y., Hong, J., Feng, Z., Guan, X., Chen, D., Zheng, Z.: Optical temperature sensing of Eu^{3+}-doped oxyhalide glasses containing CsPbBr 3 perovskite quantum dots. J. Lumin. **219**, 116897 (2020)
38. Yang, Q., Zhao, L., Fang, Z., Yang, Z., Cao, J., Cai, Y., et al.: Transparent perovskite glass-ceramics for visual optical thermometry. J. Rare Earths **39**(6), 712–717 (2021)

39. Liu, J., Zhao, Y., Li, X., Wu, J., Han, Y., Zhang, X., Xu, Y.: Dual-emissive $CsPbBr_3$@ Eu-BTC composite for self-calibrating temperature sensing application. Cryst. Growth Des. **20**(1), 454–459 (2019)
40. Aznar, E., Sanchez-Alarcon, I., Cantó, P.J.R., Perez-Pla, F., Martínez-Pastor, J.P., Abargues, R.: Molecularly imprinted nanocomposites of $CsPbBr_3$ nanocrystals: an approach towards fast and selective gas sensor of explosive taggants. J. Mater. Chem. C (2022)
41. Chen, X., Sun, C., Liu, Y., Yu, L., Zhang, K., Asiri, A.M., et al.: All-inorganic perovskite quantum dots $CsPbX_3$ (Br/I) for highly sensitive and selective detection of explosive picric acid. Chem. Eng. J. **379**, 122360 (2020)
42. Aamir, M., Khan, M.D., Sher, M., Bhosale, S.V., Malik, M.A., Akhtar, J., Revaprasadu, N.: A facile route to cesium lead bromoiodide perovskite microcrystals and their potential application as sensors for nitrophenol explosives. Eur. J. Inorg. Chem. **2017**(31), 3755–3760 (2017)
43. Xing, G., Mathews, N., Sun, S., Lim, S.S., Lam, Y.M., Grätzel, M., et al.: Long-range balanced electron-and hole-transport lengths in organic-inorganic CH3NH3PbI3. Science **342**(6156), 344–347 (2013)
44. Muthu, C., Nagamma, S.R., Nair, V.C.: Luminescent hybrid perovskite nanoparticles as a new platform for selective detection of 2, 4, 6-trinitrophenol. RSC Adv. **4**(99), 55908–55911 (2014)
45. Harwell, J.R., Glackin, J.M.E., Davis, N.J.L.K., Gillanders, R.N., Credgington, D., Turnbull, G.A., Samuel, I.D.W.: Sensing of explosive vapor by hybrid perovskites: effect of dimensionality. APL Mater. **8**(7), 071106 (2020)
46. Shan, X., Zhang, S., Zhou, M., Geske, T., Davis, M., Hao, A., et al.: Porous halide perovskite–polymer nanocomposites for explosive detection with a high sensitivity. Adv. Mater. Interfaces **6**(3), 1801686 (2019)
47. Aamir, M., Sher, M., Malik, M.A., Akhtar, J., Revaprasadu, N.: A chemodosimetric approach for the selective detection of Pb^{2+} ions using a cesium-based perovskite. New J. Chem. **40**, 9719 (2016)
48. Zhang, D., Xu, Y., Liu, Q., Xia, Z.: Encapsulation of $CH_3NH_3PbBr_3$ perovskite quantum dots in MOF-5 microcrystals as a stable platform for temperature and aqueous heavy metal ion detection. Inorg. Chem.. Chem. **57**, 4613–4619 (2018)
49. Chen, C., Cai, Q., Luo, F., Dong, N., Guo, L., Qiu, B., Lin, Z.: Sensitive fluorescent sensor for hydrogen sulfide in rat brain microdialysis via $CsPbBr_3$ quantum dots. Anal. Chem. **91**, 15915–15921 (2019)

Chapter 11
Metal–Organic Frameworks and Their Composites for Sensing Applications

Rituraj Dutta and Kakoli Doloi

Abstract Metal–organic frameworks (MOFs) are microporous novel coordination polymers with extraordinary structural geometry of metal centers or nodes associated with ligands. Different organic molecules such as benzene 1,4-dicarboxylic acid, biphenyl-4,4'-dicarboxylic acid, benzene 1,3,5-tricarboxylic acid, and azoles are commonly used as bridging ligands in MOFs. MOFs possess significant pore parameters with the ability to functionalize different chemical compounds to enrich their widespread research attention in the field of electrochromic, catalytic, storage, sequestration, and sensing applications. Functionalized MOFs are promising nanomaterials to probe sensing devices for the detection of toxic gases such as H_2S, NH_3, and CO_2 and biomolecules like protein, enzymes, and antibodies with high selectivity as well as sensitivity. Luminescent MOFs are promising candidates for optical sensing applications. In this chapter, the sensing performance of MOFs and their composite nanostructures have been discussed including electrochemical sensing, gas sensing, optical sensing, and biosensing applications for the detection of different hazardous gases, solvent molecules as well as biomolecules.

Keywords Metal–organic frameworks · Electrochemical sensing · Gas sensing · Luminescent MOFs · Biosensing

11.1 Introduction

Metal–organic frameworks (MOFs) are highly porous crystalline hybrid compounds with designable periodic topology consisting of unsaturated metal cations or nodes (secondary building units) coordinated with organic linker moieties [1, 2]. MOFs have versatile structures based on the functionalization and association of different linker

---

R. Dutta (✉) · K. Doloi
Department of Physics, Tezpur University, PO: Napaam, Tezpur 784028, Assam, India
e-mail: riturajdutta2009@gmail.com

R. Dutta
Department of Physics, Namrup College, PO: Parbatpur, Namrup 786623, Assam, India

D. Mohanta and P. Chakraborty (eds.), *Nanoscale Matter and Principles for Sensing and Labeling Applications*, Advanced Structured Materials 206,
https://doi.org/10.1007/978-981-99-7848-9_11

molecules with the inorganic metal nodes [3, 4]. Due to tunable pore parameters and excellent coordination geometry, MOFs possess significant research interest in electrochemistry, catalysis, gas storage, drug delivery, sensing, and clean energy applications [5, 6]. In the synthesis of MOFs sulfate, acetate, nitrate, chloride, oxide, etc. metals are treated as precursor materials to form a variety of metal nodes for possible metal-linker combinations in MOFs. Different MOF structures can be formed by substituting variant linker molecules in metal clusters for commercial purpose. One of the well-known MOFs MOF-5 (ZnBDC-MOF) is a combination of tetrahedral Zn_4O nodes coordinated with terephthalic acid (benzene 1,4-dicarboxylic acid). Isoraticular MOFs (IRMOFs) are the category of MOFs having similar topological symmetry with asymmetric pore parameters that can be formed by bond elongation of carbon network in the organic ligands. The pore features of MOFs are controlled by the functionality of the benzene rings in the organic ligands and the optimization of the length of the carbon network. MOFs are typical host materials to incorporate tiny guest molecules or ions within their pores and their phase dynamics as well as the pore textures can be maintained by selective guest–host interaction [7, 8]. The topological features of MOFs can be controlled by selective synthesis procedures. The experimental temperature and pressure, polarity of the solvents, concentration, and pH of the precursor materials optimize the characteristics of the metal nodes in MOFs. Hydrothermal and solvothermal procedures are common and easy-to-handle methods for MOF synthesis. In both methods, different carboxylic acid groups are conjugated with metal precursors in solvent media like DMF, and vacuum treatment under ambient temperature is needed to complete the reaction mechanism for the formation of MOFs. Some limitations of these processes involve the lesser amount of yield formation, use of toxic solvents like DMF, time-consuming reaction, and need for excess temperature [9]. The mechanochemical synthesis procedure is an effective approach for solventless MOF synthesis by keeping ball bearings inside a container with stoichiometric amounts of precursors of MOFs at lesser reaction time. Microwave-assisted synthesis is an efficient way for the synthesis of MOF that deals with the molecular interaction of solvent with impregnating electromagnetic waves at a quicker reaction rate [10]. Ren et al. [11] investigated the synthesis of octahedral UiO-66 framework following microwave-assisted approach that exhibited 1.26 wt% of H_2 storage capacity. Several researchers adopted an electrochemical synthesis process for synthesizing MOFs where the linker molecules were mixed in an electrolyte solution and metal ions were generated by the dissolution mechanism of the electrode materials associated with the reaction process. Electrochemical synthesis of MIL-100 (Fe) was reported by Campagnol et al. [12] by incorporation of 1,3,5-benzenetricarboxylic acid in ethanol-contained Milli-Q water using an solid-state electrochemical cell. Deposition of UiO-66 framework on Zr foil by electrochemical approach was reported by Stassen et al. [13] by applying 80 mA current at 383 K. Other well-known synthesis approaches for the synthesis of MOFs include sonochemical procedure and vapor diffusion procedure, etc. MOFs can be classified into four different groups as per their structural compositions as well as surface properties.

11.1.1 Rigid Frameworks

Rigid MOFs typically possess highly porous robust topology and can display type-I isotherms upon gas adsorption. As reported by Kaye et al. [14], ZnBDC-MOF (MOF-5) reveals BET surface area of 3800 m^2/g, and other Zn-based MOFs MOF-177 and MOF-200 exhibit surface area of 4746 m^2g^{-1} and 6260 m^2g^{-1}, respectively [15, 16]. Different ligand-pairs can be linked with the Zn_4O metal clusters to generate a UMCM-2 framework that reveals a 5200 m^2g^{-1} BET surface area with 1.4–1.8 nm micropore diameter and 2.4–3.0 nm mesopore diameter [17].

11.1.2 Dynamic or Flexible Frameworks

Flexible MOFs change their structural topology upon incorporation of external materials. Reorientation or rearrangement of bond angles due to alteration in molecular interaction upon incorporation of guest molecules can be attributed to the softness of the framework materials. Some flexible MOFs such as MIL-8 and MIL-5 reveal drastic changes in their unit cell parameters upon adsorption and desorption of gas molecules. After adsorption of guest molecules Al-based MOF (MIL-53) exhibits diamond-like 1D structures upon incorporation of guest molecules caused by the alteration of Al–OH-Al bond angle [18]. Al $(bdc\text{-}OH)_x$ dynamic framework can selectively adsorb P-xylene isomers for efficient drug delivery applications [19].

11.1.3 Lewis Acid Frameworks

In these MOFs, the unsaturated metal clusters or nodes contain specific binding sites for interaction with guest molecules for significant gas storage applications. CuBTC-MOF or HKUST-1 contains unsaturated paddle wheel Cu metal sites where guest molecules can be embedded in the Cu^{2+} axial sites for heterogeneous catalysis applications [20, 21]. Cr-based framework MIL-101 contains unsaturated Cr^{+3} nodes that exhibit large cavities that can intensify the diffusion rate of substrates for effective catalytic applications [22].

11.1.4 Surface Functionalized MOFs

In these MOFs, various functional groups can be substituted on the framework surfaces to enhance their adsorption capability. Functional groups such as hydroxyl and arylamine are typically used for grafting or functionalization on framework materials by rearranging the linker molecules or coordinating with the metal clusters to enrich the CO_2 adsorption capability [23]. Different amine-functionalized MOFs USO-2-Ni and USO-3-In-A exhibit enhanced CO_2 adsorption characteristics after surface functionalization as compared to the pristine ones [24].

11.1.5 Electrically Conducting MOFs

Though conventional MOFs are proven to be versatile due to their captivating properties most of them are found to be electrical insulators, which in turn hinder their use in electrochemical or electrical sensing applications. In the last decade strategies have been applied to introduce conductivity to the framework by different means which includes decorating or incorporating the MOF framework with conducting materials like conducting polymers, metal nanoparticles, and carbon dots, etc. [25]. Apart from inserting foreign material into its porous network studies on intrinsically conducting MOF have also come into the scenario. This led to the development of π-conjugated organic linkers with highly delocalized charge carriers which provided long-range charge transfer pathways [26]. Such intrinsically conducting MOFs were classified according to the conduction mechanism as through space, through bond, and hopping transport mechanism. Such conduction pathways were obtained by considering some design strategies like selecting a compatible pair of redox-active metal centers and π-conjugated organic linkers (benzene, triphenylene, naphthalene based), by incrementing redox activity through mixed valence metals at the nodes, using mixed metals in the framework, etc. [27–29].

Due to their excellent designable topology and high surface area as well as pore volume MOFs are fascinating hybrid nanomaterials for widespread applications in view of industrial and environmental approaches. By choosing selective and specific combinations of linker moieties with metal clusters variety of MOF structures can be developed for energy storage, gas storage, catalysis, sensing, gas sequestration, dye removal as well as biomedical applications. In this chapter, the sensing applications of different MOFs and their composites will be discussed in view of electrochemical sensors, gas sensors, optical sensors as well as biosensors.

11.2 MOF-Based Electrochemical Sensors

MOF-based materials are widely used for electrochemical sensing applications due to their remarkable ultrahigh porosity, adjustable surface topology, and tunable metal-linker interaction. For the detection of specific target molecules in electrochemical sensing, particular linker molecules as well as functional groups can be adjusted with metal nodes during the synthesis of MOFs. Li et al. [30] reported the hydrothermal synthesis of Co-based MOF nanosheet on Ni foam using terephthalic acid as an organic ligand to be utilized as an electrochemical glucose sensor. The fabricated electrochemical sensor revealed a sensitivity of 10,886 $\mu A\ mM^{-1}\ cm^{-2}$ with a detection limit of 1.3 nM at an alkaline state. Wu et al. [26] reported water-stable Cu-based MOF, fabricated on glassy-carbon electrodes for simultaneous detection of L-tryptophan and ascorbic acid at bio-mimic conditions. Zhang et al. [31] reported the synthesis of 2D conductive Co-MOF and the formation of MOF nanosheets by following the multi-layer Langmuir–Blodgett technique. The fabricated nanosheets significantly intensified the electroreduction performance of H_2O_2 with compatible accuracy and sensitivity. However, different studies on MOFs reveal that pristine MOFs typically do not exhibit good performance in electrochemical sensing owing to their poor conductivity and weak mechanical stability as well as less stability in solvent media. Therefore, it is necessary to combine pristine MOFs with carbon-based materials, conducting polymers, and metal nanoparticles to overcome these issues. Various research groups reported the use of carbon nanomaterials such as graphene nanosheets, carbon nanofibers, carbon nanotubes, and graphene aerogels to prepare efficient MOF composites for electrochemical sensing applications [31]. Guo and his group investigated the catalytic behavior of the ZIF-67 framework functionalized with 1D carbon fiber that exhibited remarkably sensitive response to detect L-cysteine and nitrobenzene with excellent selectivity and stability due to the presence of more active sites in the composite nanostructure [32]. Studies on MOF-based graphene nanostructures have been carried out by several research groups to investigate their electrocatalytic activities and sensing performance. Feng and his group reported the surface modification of Cu-based MOF by impregnating n-ERGO for enhanced electron transfer through hydrogen bonding and π–π interaction that showed superior electrocatalytic performance toward reduction of H_2O_2 with greater sensitivity [33]. Yang and his coworkers reported the synthesis of Cr-based MOF (MIL-101) functionalized with reduced graphene oxide for the detection of 4-nonylphenol [34]. Cao et al. [35] carried out the experimental work on immobilization and dispersion of HKUST-1 nanoparticles on the surface of exfoliated GO nanosheets at 300 K that exhibited superior electrochemical response with enhanced adsorption capability and sensitivity to detect 8-hydroxy-20-deoxyguanosine.

MOF-derived carbon materials can be synthesized by carbonization of MOFs at high temperatures for their potential applications in electrochemical sensing for the determination of toxic materials and heavy metal ions. Following pyrolysis of ZIF polyhedra, Gai et al. [36] synthesized porous carbon materials under nitrogen treatment for the detection of uric acid, dopamine, and ascorbic acid solutions. Xu et al. [37] studied the effect of temperature on the carbonization of ZIF-8 to derive porous carbon structures. At the calcination temperature of 100 °C the MOF-derived carbon materials exhibited excellent electrochemical performance with sensing capability to determine catechol, uric acid, and hydroquinone.

Composites of MOF with reduced graphene oxide rGo, graphene oxide (GO), and carbon-based materials were seen to be used in detecting organic analytes such as isomers of hydroxybenzene. Isomers like catechol, hydroquinone, resorcinol, etc. are difficult to detect simultaneously due to their similar chemical formula and structure by the commercial electrodes. However, when the electrodes were modified with MOF-based composites simultaneous detection of isomers became possible with excellent selectivity. Such work was done by Hailong Wang et al. where they modified the carbon paste electrode with Mil-101(Cr) and rGO composites [38]. The modified electrode identified the two isomers while used in a mixture solution of catechol (CC) and hydroquinone (HQ) with LOD lower as 4.14 μM for CC and 0.66 μM for HQ. A similar attempt was made by Qi Chen et al. MOF-199 (Cu) and GO composite through solvothermal synthesis route. They electrodeposited the MOF/ GO composite into a glassy-carbon electrode where GO was reduced. By performing differential pulse voltammetry, they obtained LOD lower than the work by Wang and the electrode could detect in the linear range of 0.1 μM–566 μM [39].

11.3 MOF-Based Gas Sensors

MOFs are significant nanomaterials for gas sensing applications due to their extraordinary porous structure and designable surface topology. Liu et al. [40] reported the synthesis of Cu-based MOF for oxygen sensing application that exhibited 99.7% quenching efficiency at 1 bar of oxygen with compatible permeability and excellent photochemical stability. Gong et al. [41] investigated the photoresponse properties of methyl viologen-contained MOF as an effective oxygen sensor and observed that the color of the MOF changed in an oxygen atmosphere indicating fast detection of O_2 in a selective manner. As reported, the nanorod film of ZnO@ZIF-8 was fabricated on $KMnO_4$ glass substrate for facile detection of H_2 over CO_2 with high sensitivity of the core–shell structure [42]. MOF-derived materials are widely used for the selective detection of hazardous gases like H_2S, NO, CO_2, CO, etc. Xu et al. [43] reported Pb-based amphoteric MOF as a colorimetric sensor for the detection of H_2S as the color of the MOF changes in the presence of H_2S due to the interaction of H_2S species with the Pb metal centers. Nagarkar et al. [44] investigated the fluorescence properties of NO_2-functionalized MOF for the detection of H_2S gas through the selective interaction of H_2S with the MOF. Campbell et al. [45] investigated the charge delocalization in 2D Cu-based MOF with a conductivity of 0.2 S cm^{-1} and fabricated a chemiresistive sensing device for easy detection of ammonia. Meng et al. [46] reported a bimetallic conductive 2D Ni-based MOF with excellent surface topology and sufficient active sites for the fabrication of a chemiresistive device for nitric oxide sensing. Reed et al. [47] reported a Fe-based MOF that exhibited significant adsorption capability for CO gas with reversible changes in spin states for CO binding. Gassensmith, et al. [48] synthesized a cyclodextrin-based framework for efficient electrochemical detection of CO_2, where the interaction of CO_2 molecules with the MOF was monitored by the hydroxyl groups present in the MOF, resulting in a sharp decrease in the electrochemical impedance spectra of the MOF. Wu et al. [49] synthesized a Mg-based MOF with 1,2,4,5-benzenetetracarboxylic acid as an organic linker that revealed blue-purple emission with complete fluorescence quenching at 0.8 vol% of CS_2 exposure.

11.4 MOF-Based Optical Sensors

MOFs are prominent materials for optical sensing applications as the commonly used aromatic linkers are capable of luminescent emission and adsorption. MOF-based optical sensors can be demonstrated by the luminescence quenching of aromatic fluorophores as well as lanthanide ions. The coordination of metal clusters or nodes with linker molecules and intramolecular energy exchange from linkers to metal nodes can influence the sensing performance of luminescent MOFs for metal ion sensing. Yang et al. [50] reported the synthesis of Al-based fluorescent MOF MIL-53 for selective detection of Fe^{3+} ions in aqueous medium.

The cation transfer between the Fe^{3+} ion and the Al^{3+} ion of the framework resulted in fluorescence quenching of the MIL-53 framework as shown in Fig. 11.1. In this process, the strong fluorescent framework MIL-53(Al) can be transformed into the weak fluorescent material MIL-53(Fe) through the exchange of cations between Fe^{3+} ions and Al^{3+} ions present in MIL-53(Fe) and MIL-53(Al) frameworks, respectively. The quenching mechanism of Fe^{3+} ions on the fluorescent MIL-53(Al) framework resulted in the smooth and sensitive detection of Fe^{3+} ions in aqueous media at 0.9 μM detection limit in the selective range of 3–200 μM. The cation exchange phenomenon in the chemically stable MIL-53(Al) framework can be adopted for selective and quantitative detection of Fe^{3+} ions in human urine with remarkable spike recovery in the range of 98.2%–106.2%. Zheng et al. [51] reported a Cd-based MOF with selective fluorescence quenching response for adsorption as well as detection of Cu^{2+} ions. The quenching effect was dominated by the energy transfer mechanism from the organic linkers to the Cu^{2+} ions due to the coordination of Cu^{2+} ions with the ligands. Lu and Hupp reported the development of a Fabry-Pérot sensing device by fabricating ZIF-8 homogeneous films on Si substrate that exhibited effective sensing response to the volumetric determination of ethanol in water medium [52]. Water-stable Eu-based MOF was synthesized by Wang et al. [53] with extensive channels of hydrogen bonds consisting of H_2O and dimethyl ammonium for the fabrication of optical sensors with a fluorescence quenching effect for the sensitive detection of Cu^{2+} ions. Shahat et al. [55] demonstrated a UiO-66 framework-based colorimetric sensing device for simultaneous colorimetric determination and quantification of toxic metal ions such as Pb (II), Hg (II), Bi (III), and Cd (II) up to the detection range of 10 ^{10} M in aquatic media. Kreno et al. [56] reported a strategy for signal amplification in optical sensors based on CuBTC-MOF coated with Ag nanoparticles. They had observed 14-fold signal amplification for the detection of CO_2 by LSPR spectroscopy.

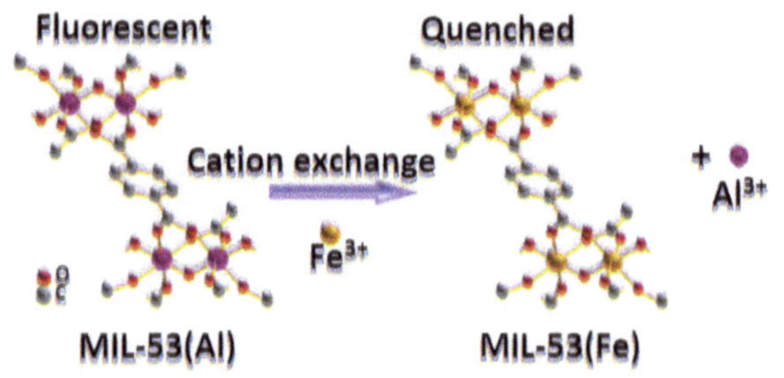

Fig. 11.1 Cation exchange phenomenon in MIL-53 as a fluorescent sensor for detection of Fe^{3+}. Reprinted with permission from Reference [44] Copyright 2013, American Chemical Society

11.5 MOF-Based Biosensors

In recent years, functional MOFs have provided widespread advancements in the field of biosensing applications for the detection of biomolecules, antibodies, oligonucleotides, and enzymes. Due to the tunable adsorption capabilities of MOFs, various functional groups can be combined for the sensitive determination of electroactive nanomaterials for their possible applications as biosensors. Different research groups have developed MOF-based biosensors for selective enzyme detection. Wang et al. [57] developed NiPd and glucose oxidase (GOx) functionalized ZIF-8 framework-based effective biosensors for colorimetric glucose detection. Koo et al. [58] reported the synthesis of Pd nanoparticle functionalized MOF as an effective synergistic catalyst to enhance sensing performance. Figure 11.2a illustrates the diffusion of Pd^{2+} ions inside the cavity of the ZIF-8 framework and their reduction to Pd nanoparticles that can be homogeneously distributed as selective catalysts over the sensing material. The ZIF-8 framework consists of Zn metal clusters with 2-methylimidazole as organic linkers. The synthesized pristine ZIF-8 framework exhibited cubic topology with a 50–200 nm pore diameter. As depicted in Fig. 11.2a, well-dispersed Pd nanoparticles can be incorporated in the pores of ZIF-8 by infiltration in the solution process followed by the reduction of Pd^{2+} ions to generate metallic Pd nanoparticles. Pd-incorporated ZnO-WO_3 nanostructures were developed by electrospinning the solution containing Pd@ZIF-8, AMH, and PVP materials followed by calcination in the air as depicted in Fig. 11.2b. During the calcination process, ZnO was formed by the oxidization of the ZIF-8 framework, and the Pd@ZIF-8 particles were confined with the WO_3 caused by the decomposition of PVP as well as the crystallization of porous WO_3 structures. Also, mesopores are formed in the vicinity of the WO_3 network due to the thermal degradation of the organic ligands comprising the framework. The synthesized Pd-incorporated ZnO-WO_3 catalyst revealed high accessibility for biosensing and enhanced performance in the detection of toluene.

Shu et al. [59] reported the synthesis of hemin-containing Ni-based MOF for selective detection of hydrogen peroxide. The efficiency of the organophosphate-based enzyme was enhanced by encapsulating it in terbium-based MOF as reported by Mehta et al. [60] for sensitive determination of pesticides as well as to investigate their catalytic degradation. Zhao et al. [61] reported the synthesis of a water-stable mesoporous Al-based MOF (PCN-333) that can be utilized for the encapsulation

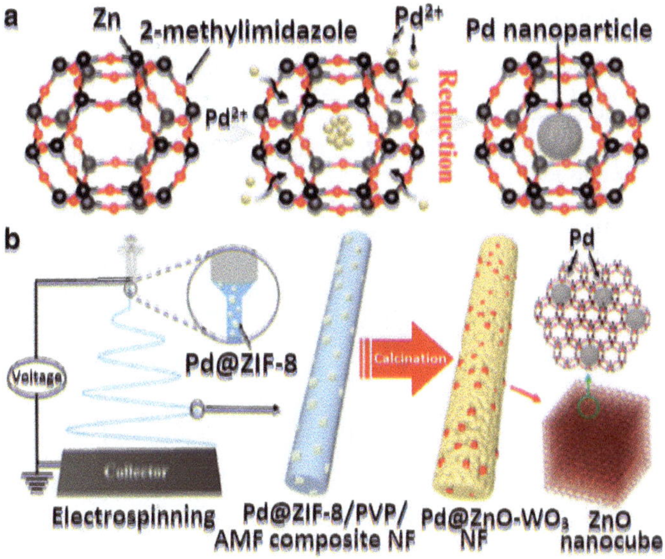

Fig. 11.2 **a** Encapsulation of Pd nanoparticles in ZIF-8 cavity **b** Synthesis procedure of Pd-incorporated ZnO-WO_3 nanostructures. Reprinted with permission from Reference [51] Copyright 2016, American Chemical Society

of cholesterol oxidase and HRP enzymes for cholesterol detection. Hou et al. [62] developed a fluorescence biosensor for the systematic visual sensing of alkaline phosphatase using an amino-functionalized Cu-based MOF. Different electroactive conducting materials, such as carbon nanotubes, conducting polymer, ionic liquid, graphene, and porous carbon can be incorporated in MOF pores for enhancement in sensing performance as electrochemical biosensors. Uzak et al. [63] developed an electrochemical glucose sensor using reduced graphene oxide and Pt-embedded Zn-MOF nanostructures. Zhang et al. [64] synthesized a Zr-based MOF (PCN-222) for the fabrication of a photoelectrochemical sensor for the detection of phosphoprotein as depicted in Fig. 11.3a. The light-harvesting TCPP linkers in PCN-222 can increase the photocurrent response to dopamine in oxygen-dissolved media. The dopamine material can inhibit the electron–hole charge recombination process leading to the increase in the efficiency of photoelectric response. As illustrated in Fig. 11.3b, the fabricated signal probes revealed a compatible response towards selective detection of phosphoprotein assays due to the steric hindrance phenomenon caused by the interaction of phosphate species with the Zr-O metal nodes in the binding sites of the highly porous PCN-222 framework. This type of MOF-based electrochemical sensors can be used in therapeutics as well as to detect and quantify biomolecules in clinical diagnostics. A colorimetric biosensor was developed by Yin et al. [65] to entrap bovine hemoglobin within the ZIF-8 framework for the detection of H_2O_2 using 4-aminoantipyrine and observed the catalytic performance of the system to be enhanced up to 423% as compared to that in case of free bovine hemoglobin. Wu

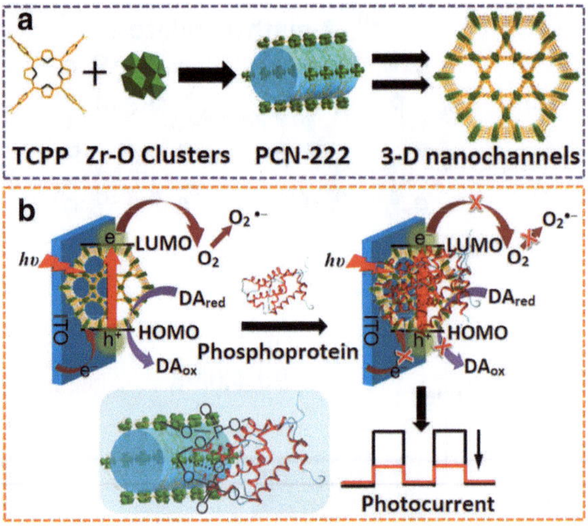

Fig. 11.3 **a** Synthesis of PCN-222-based nanochannels, **b** Fabrication of sensing electrodes for detection of phosphoprotein with photocurrent response. Reprinted with permission from Reference [57] Copyright 2016, American Chemical Society

et al. [66] reported the colorimetric probing of a glucose biosensor by encapsulating HRP and GOx enzymes within the ZIF-8 framework.

11.6 MOF-Based SERS Sensors

Surface-enhanced Raman scattering (SERS) is an amplified version of Raman spectroscopy used as a highly sensitive analytical tool. Here, the increment in Raman signals is governed by two phenomena—electromagnetic mechanism and the chemical enhancement mechanism. SERS is a consequence of electromagnetic waves encountering surface plasmons which gives rise to an enlarged local field [67]. An intense SERS signal is reliant on the nanoscale roughened surface of the substrate. MOFs as well as their composites have been explored as SERS substrates for the analysis of some particular analytes that are found to be arduous to detect using conventional SERS substrates [68]. The scope for post-synthetic modification in MOFs allows it to decorate or functionalize with desired groups which in turn makes it selective towards specific target. It has been found to be designed for analytes like polycyclic aromatic hydrocarbons (PAHs) [69], volatile organic compounds (VOCs) [70], different gases [71], etc. In 2013, Tsung-Han Yu et al. reported a series of MIL MOFs with metal clusters of Cr, Fe, and Al which showed intense SERS signals in interaction with methyl orange dye. It was the first time MOF without any modification could detect an analyte with an enhanced SERS signal. On further investigation, the mechanism of that enhancement was found to be due to the interaction of methyl orange with the metal oxide nodes of the MOF [72]. Later, MIL-100 (Fe) was used to detect toluene with a detection limit of 2.7×10^{-2} μM [73]. Since MOFs are made up of aromatic molecules and metal nodes, they can be used for various sizes of analytes from VOCs to polar groups through metal-heteroatomic bonds, $\pi - \pi$ interaction, etc.

Composites of MOFs with metal nanoparticles (MNP) are proved to show synergistic behavior to obtain an improved performance of the SERS substrate. The role of MNPs and MOF in SERS analysis depends on the way in which MNPs are loaded into MOF. Either MNPs can be loaded inside the pores of the MOF or MNPs can be anchored to the surface of the MOF. In both ways, MOF provides a good stabilizing platform for MNP as it has a large surface area which prevents the overaccumulation of particles. While MOFs are decorated on the outer surface it facilitates the Raman hot spots as well as the adsorption active sites for targets in the close proximity. Such composites can be prepared through many synthesis methods like chemical vapor deposition (CVD), solution impregnation, solid grinding, in-situ growth of MOF over MNP, etc. [74–76]. The group of He Xu has functionalized MOF UiO-66 on Silver (Ag) NPs through the hydrothermal method followed by Ag reduction on its surface. They used the substrate for detecting di-(2-ethylhexyl) phthalate (DEHP) in plastics as the composite provides good adsorption sites for DEHP and Ag increased the hot spots for better sensitivity. They found an enrichment factor of 1.1×10^7 with a lower limit of detection 3×10^{-12} mol/L (R2 = 0.9950) [77]. Yanshu Zhang et al. prepared a gold (Au) NP core and MOF-74 shell through one-pot synthesis. They used the composite for 4-nitrothiophenol detection with a very low LOD of 69 nmol·L^{-1}. The mechanism behind the detection of the analyte involved two reactions; the first was nitration of the analyte due to its interaction with surface plasmons of gold NPs, and the second reaction was nitrothiophenol reduced into 4-thioaminophenol upon laser irradiation due to the presence of MOF-74 [78].

Some of the chemical substances don't show any affinity towards noble metals which is an essential criterion for SERS analysis. For such chemicals, MNP-incorporated MOF proved to be a suitable substrates for getting adsorbed closely to the SERS active sites. For example, acetamiprid is difficult to detect with bare metal substrate because of its low affinity towards metals. The group of Xiaolin Cao reported a series of MOF-based SERS substrates with Au NP embedded into MOF-199, Uio-66, and Uio-67 for the detection of acetamiprid. All three composites showed excellent SERS- behavior as the MOF adsorbed the analytes through $\pi - \pi$ interaction. Composites have lower LODs of 20 nM, 9 nM, and 20 nM out of which AuNP and MOF-199 composite was superior in SERS activity [79]. MOF-based SERS substrates were also used in biomedical and clinical applications like the detection of alpafetoprotein (AFP), glucose, lactose, etc. For instance, MIL-101(Cr) was used as a host for the in-situ growth of Au NP for the detection of glucose and lactate by immobilizing glucose and lactate oxidase to form an enzymatic nanocomposite. Further, those nano enzymes were used for monitoring the alteration in the physiological and pathological conditions of living brains due to change in glucose and lactate levels [80].

11.7 Conclusion and Future Prospects

MOFs are attaining promising research interest in recent years due to their extraordinary porous structure, tunable phase behavior, and selective functionality with a wide range of chemical compounds. In this chapter, MOFs and their composites have been discussed as potential sensing materials for different sensing applications. MOFs fabricated with carbon-based materials such as carbon nanotubes, porous carbon, and graphene oxide can be designed as effective electrochemical sensors for sensing chemical compounds and solvent molecules. MOF-derived composites can be used for the fabrication of gas sensors for the quantification of toxic gases such as H_2S, CO_2, NH_3, and CO. Luminescent MOFs can be used to develop optical sensors for the colorimetric and volumetric determination of metal ions, gas molecules, and biomolecules. MOF-based biosensors can be fabricated for the systematic detection of enzymes, antibodies, proteins, and hydrocarbons for biomedical applications.

Synthesis of MOFs with selective metal clusters with physical and chemical stability and low toxicity associated with different organic ligands may be of fascinating research interest in sensing applications. Swift Heavy Ion (SHI) irradiation on MOFs can enlarge their surface areas by inducing certain defects in a controlled way leading to an increase in their functionality to entrap guest molecules for the fabrication of compatible sensing devices. Low energy ion irradiation can also induce defects and disorders in MOF-based composites to influence their pore parameters to enrich their sensing performance. Inelastic Neutron Scattering experiments in MOFs can reveal their phonon-assisted vibrations related to the linker moieties that can help in demonstrating their functionality with chemical compounds for different sensing applications.

Acknowledgements Here we would like to offer sincere gratitude to our former supervisor Late Prof. Ashok Kumar for his guidance, suggestions, and motivations during the whole work. We will be ever grateful for his assistance.

References

1. Prestipino, C., Regli, L., Vitillo, J.G., Bonino, F., Damin, A., Lamberti, C., Zecchina, A., Solari, P.L., Kongshaug, K.O., Bordiga, S.: Local structure of framework Cu (II) in HKUST-1 metallorganic framework: spectroscopic characterization upon activation and interaction with adsorbates. Chem. Mater. **18**, 1337–1346 (2006)
2. Tranchemontagne, D.J., Hunt, J.R., Yaghi, O.M.: Room temperature synthesis of metal-organic frameworks: MOF-5, MOF-74, MOF-177, MOF-199, and IRMOF-0. Tetrahedron **64**, 8553–8557 (2008)
3. Bohrmanand, J.A., Carreon, M.A.: Synthesis and CO_2/CH_4 separation performance of Bio-MOF-1 membranes. Chem. Commun. **48**(42), 5130–5132 (2012)
4. Valenzano, L., Civalleri, B., Chavan, S., Bordiga, S., Nilsen, M.H., Jakobsen, S., Lillerud, K.P., Lamberti, C.: Disclosing the complex structure of UiO-66 metal organic framework: a synergic combination of experiment and theory. Chem. Mater. **23**, 1700–1718 (2011)

5. Eddaoudi, M., Kim, J., Rosi, N., Vodak, D., Wachter, J., O'Keeffe, M., Yaghi, O.M.: Systematic design of pore size and functionality in isoreticular MOFs and their application in methane storage. Science **295**, 469–472 (2002)
6. Hong, D.Y., Hwang, Y.K., Serre, C., Ferey, G., Chang, J.S.: Porous chromium terephthalate MIL-101 with coordinatively unsaturated sites: surface functionalization, encapsulation, sorption and catalysis. Adv. Func. Mater. **19**, 1537–1552 (2009)
7. Chen, Y., Hu, Z., Gupta, K.M., Jiang, J.: Ionic liquid/metal–organic framework composite for CO_2 Capture: a computational investigation. J. Phys. Chem. C **115**(44), 21736–21742 (2011)
8. Fujie, K., Otsubo, K., Ikeda, R., Yamada, T., Kitagawa, H.: Low temperature ionic conductor: ionic liquid incorporated within a metal–organic framework. Chem. Sci. **6**, 4306–4310 (2015)
9. Bauer, S., Serre, C., Devic, T., Horcajada, P., Marrot, J., Férey, G., Stock, N.: High-throughput assisted rationalization of the formation of metal organic frameworks in the iron (III) aminoterephthalate solvothermal system. Inorg. Chem. **47**, 7568–7576 (2008)
10. Choi, J.Y., Kim, J., Jhung, S.H., Kim, H.K., Chang, J.-S., Chae, H.K.: Microwave synthesis of a porous metal-organic framework, zinc terephthalate MOF-5. Bull. Korean Chem. Soc. **27**, 1523–1524 (2006)
11. Ren, J., Segakweng, T., Langmi, H., Musyoka, N., North, B., Mathe, M., Bessarabov, D.: Microwave-assisted modulated synthesis of zirconium-based metal–organic framework (Zr-MOF) for hydrogen storage applications. Int. J. Mater. Res. **105**, 516–519 (2014)
12. Campagnol, N., Van Assche, T., Boudewijns, T., Denayer, J., Binnemans, K., De Vos, D., Fransaer, J.: High pressure, high temperature electrochemical synthesis of metal–organic frameworks: films of MIL-100 (Fe) and HKUST-1 in different morphologies. J. Mater. Chem. A **1**, 5827–5830 (2013)
13. Stassen, I., Styles, M., Van Assche, T., Campagnol, N., Fransaer, J., Denayer, J., Tan, J.-C., Falcaro, P., De Vos, D., Ameloot, R.: Electrochemical film deposition of the zirconium metal–organic framework UiO-66 and application in a miniaturized sorbent trap. Chem. Mater. **27**, 1801–1807 (2015)
14. Kaye, S.S., Dailly, A., Yaghi, O.M., Long, J.R.: Impact of preparation and handling on the hydrogen storage properties of Zn_4O (1,4-benzenedicarboxylate)$_3$ (MOF-5). J. Am. Chem. Soc. **129**, 14176–14177 (2007)
15. Furukawa, H., Ko, N., Go, Y.B., Aratani, N., Choi, S.B., Choi, E., Yazaydin, A.O., Snurr, R.Q., O'Keeffe, M., Kim, J., Yaghi, O.M.: Ultrahigh porosity in metal-organic frameworks. Science **329**, 424–428 (2010)
16. Chae, H.K., Siberio-Perez, D.Y., Kim, J., Go, Y., Eddaoudi, M., Matzger, A.J., O'Keeffe, M., Yaghi, O.M.: A route to high surface area, porosity and inclusion of large molecules in crystals. Nature **427**, 523–527 (2004)
17. Ferey, G., Mellot-Draznieks, C., Serre, C., Millange, F., Dutour, J., Surble, S., Margiolaki, I.A.: Chromium terephthalate-based solid with unusually large pore volumes and surface area. Science **309**, 2040–2042 (2005)
18. Horcajada, P., Serre, C., Maurin, G., Ramsahye, N.A., Balas, F., Vallet-Regi, M., Sebban, M., Taulelle, F., Ferey, G.: Flexible porous metal-organic frameworks for a controlled drug delivery. J. Am. Chem. Soc. **130**, 6774–6780 (2008)
19. Finsy, V., Kirschhock, C.E.A., Vedts, G., Maes, M., Alaerts, L., De Vos, D.E., Baron, G.V., Denayer, J.F.M.: Framework breathing in the vapour-phase adsorption and separation of xylene isomers with the metal–organic framework MIL-53. Chem.-A Euro. J. **15**, 7724–7731 (2009)
20. Alaerts, L., Seguin, E., Poelman, H., Thibault-Starzyk, F., Jacobs, P.A., De Vos, D.E.: Probing the Lewis acidity and catalytic activity of the metal-organic framework [Cu-3(btc)(2)](BTC = benzene-1,3,5-tricarboxylate). Chem.-A Euro. J. **12**, 7353–7363 (2006)
21. Kramer, M., Ulrich, S.B., Kaskel, S.: Synthesis and properties of the metal-organic framework $Mo_3(BTC)_2$(TUDMOF-1). J. Mater. Chem. **16**, 2245–2248 (2006)
22. Kitaura, R., Onoyama, G., Sakamoto, H., Matsuda, R., Noro, S., Kitagawa, S.: Immobilization of a metallo schiff base into a microporous coordination polymer. Angew. Chem. Int. Ed. **43**, 2684–2687 (2004)

23. Millward, A., Yaghi, O.: Metal-organic frameworks with exceptionally high capacity for storage of carbon dioxide at room temperature. J. Am. Chem. Soc. **127**, 17998–17999 (2005)
24. Arstad, B., Fjellvåg, H., Kongshaug, K., Swang, O., Blom, R.: Amine functionalised metal organic frameworks (MOFs) as adsorbents for carbon dioxide. Adsorption **14**, 755–762 (2008)
25. Meng, H., Han, Y., Zhou, C., Jiang, Q., Shi, X., Zhan, C., Zhang, R.: Conductive metal–organic frameworks: design, synthesis, and applications. Small Methods **4**(10), 2000396–2000405 (2020)
26. Zhao, Q., Li, S.-H., Chai, R.-L., Ren, X., Zhang, C.: Two-dimensional conductive metal-organic frameworks based on truxene. ACS Appl. Mater. Interfaces **12**(6), 7504–7509 (2020)
27. Park, J.G., Aubrey, M.L., Oktawiec, J., Chakarawet, K., Darago, L.E., Grandjean, F., Long, G.J., Long, J.R.: Charge delocalization and bulk electronic conductivity in the mixed-valence metal-organic framework Fe(1,2,3-Triazolate)2(BF4)x. J. Am. Chem. Soc. **140**(27), 8526–8534 (2018)
28. Dou, J.-H., Sun, L., Ge, Y., Li, W., Hendon, C.H., Li, J., Gul, S., Yano, J., Stach, E.A., Dincă, M.: Signature of metallic behavior in the metal-organic frameworks M3(Hexaiminobenzene)2 (M=Ni, Cu). J. Am. Chem. Soc. **139**(39), 13608–13611 (2017)
29. Chen, D., Xing, H., Su, Z., Wang, C.: Electrical conductivity and electroluminescence of a new anthracene-based metal-organic framework with π-conjugated zigzag chains. Chem. Commun. **52**(10), 2019–20222 (2016)
30. Li, Y., Xie, M., Zhang, X., Liu, Q., Lin, D., Xu, C., Xie, F., Sun, X.: Co-MOF nanosheet array: a high-performance electrochemical sensor for non-enzymatic glucose detection. Sens. Actuat. B: Chem. **278**, 126–132 (2019)
31. Wu, Q., Ma, J.G., Li, H., Chen, D.M., Gu, W., Yang, G.M., Cheng, P.: Metal–organic framework biosensor with high stability and selectivity in a bio-mimic environment. Chem. Commun. **51**, 9161–9164 (2015)
32. Zhang, M., Wang, G., Zheng, B., Li, L., Lv, B., Cao, H., Chen, M.: 3-Layer conductive metal–organic nanosheets as electrocatalysts to enable an ultralow detection limit of H_2O_2. Nanoscale **11**, 5058–5063 (2019)
33. Jahan, M., Liu, Z., Loh, K.P.: A graphene oxide and copper-centered metal organic framework composite as a tri-functional catalyst for HER, OER, and ORR. Adv. Funct. Mater. **23**, 5363–5372 (2013)
34. Zhang, Y., Bo, X., Nsabimana, A., Han, C., Li, M., Guo, L.: Electrocatalytically active cobalt-based metal–organic framework with incorporated macroporous carbon composite for electrochemical applications. J. Mater. Chem. A **3**, 732–738 (2015)
35. Li, C., Zhang, T., Zhao, J., Liu, H., Zheng, B., Gu, Y., Yan, X., Li, Y., Lu, N., Zhang, Z., Feng, G.: Boosted sensor performance by surface modification of bifunctional rht-Type metal-organic framework with nanosized electrochemically reduced graphene oxide. ACS Appl. Mater. Interfaces **9**, 2984–2994 (2017)
36. Zhang, Y., Yan, P., Wan, Q., Yang, N.: Integration of chromium terephthalate metal-organic frameworks with reduced graphene oxide for voltammetry of 4-nonylphenol. Carbon **134**, 540–547 (2018)
37. Cao, G., Wu, C., Tang, Y., Wan, C.D.: Ultrasmall HKUST-1 nanoparticles decorated graphite nanosheets for highly sensitive electrochemical sensing of DNA damage biomarker 8-hydroxy-2'-deoxyguanosine. Anal. Chim. Acta **1058**, 80–88 (2019)
38. Gai, P., Zhang, H., Zhang, Y., Liu, W., Zhu, G., Zhang, X.: Simultaneous electrochemical detection of ascorbic acid, dopamine and uric acid based on nitrogen doped porous carbon nanopolyhedra. J. Chen. J. Mater. Chem. B **1**, 2742–2749 (2013)
39. Xu, J., Xia, J., Zhang, F., Wang, Z.: An electrochemical sensor based on metal-organic framework-derived porous carbon with high degree of graphitization for electroanalysis of various substances. Electrochim. Acta **251**, 71–80 (2017)
40. Wang, H., Hu, Q., Meng, Y., Jin, Z., Fang, Z., Fu, Q., Gao, W., Xu, L., Song, Y., Lu, F.: Efficient detection of hazardous catechol and hydroquinone with MOF-rGO modified carbon paste electrode. J. Hazard. Mater. **355**(5), 151–157 (2018)

41. Chen, Q., Li, X., Min, X., Cheng, D., Zhoua, J., Li, Y., Xie, Z., Liua, P., Cai, W., Zhang, C.: Determination of catechol and hydroquinone with high sensitivity using MOF-graphene composites modified electrode. J. Electroanal. Chem. **789**, 114–122 (2017)
42. Liu, S.Y., Qi, X.L., Lin, R.B., Cheng, X.N., Liao, P.Q., Zhang, J.P., Chen, X.M.: Porous Cu(I) triazolate framework and derived hybrid membrane with exceptionally high sensing efficiency for gaseous oxygen. Adv. Funct. Mater. **24**, 5866–5872 (2014)
43. Gong, Y.N., Lu, T.B.: A highly stable dynamic fluorescent metal–organic framework for selective sensing of nitroaromatic explosives. Chem. Commun. **49**, 7711–7713 (2013)
44. Wu, X., Xiong, S., Mao, Z., Hu, S., Long, X.: A designed ZnO@ZIF-8 core-shell nanorod film as a gas sensor with excellent selectivity for H_2 over CO. Chem. Eur. J. **23**(33), 7969–7975 (2017)
45. Cui, J., Wong, Y.L., Zeller, M., Hunter, A.D., Xu, Z.: Pd uptake and H_2S sensing by an amphoteric metal-organic framework with a soft core and rigid side arms. Angew. Chem., Int. Ed. **53**, 14438–14442 (2014)
46. Nagarkar, S.S., Desai, A.V., Ghosh, S.K.: Nitro ($-NO_2$) Functionalized metal–organic framework as a reaction based fluorescence turn-on probe for rapid and selective H_2S detection. Chem. Eur. J. **21**, 9994–9997 (2015)
47. Campbell, M.G., Sheberla, D., Liu, S.F., Swager, T.M., Dincǎ, M.: Cu_3(hexaiminotriphenylene)$_2$: an electrically conductive 2D metal-organic framework for chemiresistive sensing. Angew. Chem. Int. Ed. **54**, 4349–4352 (2015)
48. Meng, Z., Aykanat, A., Mirica, K.A.: Welding metallophthalocyanines into bimetallic molecular meshes for ultrasensitive, low-power chemiresistive detection of gases. J. Am. Chem. Soc. **141**, 2046–2053 (2019)
49. Reed, D.A., Xiao, D.J., Gonzalez, M.I., Darago, L.E., Herm, Z.R., Grandjean, F., Long, J.R.: Reversible CO scavenging via adsorbate-dependent spin state transitions in an iron (II)–triazolate metal–organic framework. J. Am. Chem. Soc. **138**, 5594–5602 (2016)
50. Gassensmith, J.J., Kim, J.Y., Holcroft, J.M., Farha, O.K., Stoddart, J.F., Hupp, J.T., Jeong, N.C.: A metal-organic framework-based material for electrochemical sensing of carbon dioxide. J. Am. Chem. Soc. **136**, 8277–8282 (2014)
51. Wu, Z.F., Tan, B., Feng, M.L., Lan, A.J., Huang, X.Y.: A magnesium MOF as a sensitive fluorescence sensor for CS_2 and nitroaromatic compounds. J. Mater. Chem. A **2**, 6426–6431 (2014)
52. Yang, C.X., Ren, H.B., Yan, X.P.: Fluorescent metal–organic framework MIL-53(Al) for highly selective and sensitive detection of Fe^{3+} in aqueous solution. Anal. Chem. **85**, 7441–7446 (2013)
53. Zheng, T.T., Zhao, J., Fang, Z.W., Li, M.T., Sun, C.Y., Li, X., Wang, X.L., Su, Z.M.: A luminescent metal organic framework with high sensitivity for detecting and removing copper ions from simulated biological fluids. Dalton Trans. **46**, 2456–2461 (2017)
54. Lu, G., Hupp, J.T.: Metal–organic frameworks as sensors: a ZIF-8 based Fabry-Pérot device as a selective sensor for chemical vapors and gases. J. Am. Chem. Soc. **132**, 7832–7833 (2010)
55. Wang, X., Qin, T., Bao, S.S., Zhang, Y.C., Shen, X., Zheng, L.M., Zhu, D.R.: Facile synthesis of a water stable 3D Eu–MOF showing high proton conductivity and its application as a sensitive luminescent sensor for Cu^{2+} ions. J. Mater. Chem. A **4**, 16484–16489 (2016)
56. Shahat, A., Hassan, H.M.A., Azzazy, H.M.E.: Optical metal–organic framework sensor for selective discrimination of some toxic metal ions in water. Anal. Chim. Acta **793**, 90–98 (2013)
57. Kreno, L.E., Hupp, J.T., Van Duyne, R.P.: Metal–organic framework thin film for enhanced localized surface plasmon resonance gas sensing. Anal. Chem. **82**, 8042–8046 (2010)
58. Wang, Q., Zhang, X., Huang, L., Zhang, Z., Dong, S.: GOx@ZIF-8(NiPd) nanoflower: an artificial enzyme system for tandem catalysis. Angew. Chem. Int. Ed.. Chem. Int. Ed. **56**, 16082–16085 (2017)
59. Koo, W.T., Choi, S.J., Kim, S.J., Jang, J.S., Tuller, H.L., Kim, I.D.: Heterogeneous sensitization of metal-organic framework driven Metal@Metal oxide complex catalysts on an oxide nanofiber scaffold toward superior gas sensors. J. Am. Chem. Soc. **138**, 13431–13437 (2016)

60. Shu, Y., Chen, J.Y., Xu, Z., Jin, D.Q., Xu, Q., Hu, X.Y.: Nickel metal-organic framework nanosheet/hemin composite as biomimetic peroxidase for electrocatalytic reduction of H_2O_2. J. Electroanal. Chem. **845**, 137–143 (2019)
61. Mehta, J., Dhaka, S., Bhardwaj, N., Paul, A.K., Dayananda, S., Lee, S.E., Kim, K.H., Deep, A.: Application of an enzyme encapsulated metal-organic framework composite for convenient sensing and degradation of methyl parathion. Sens. Actuat. B **290**, 267–274 (2019)
62. Zhao, M.Y., Li, Y., Ma, X.J., Xia, M.F., Zhang, Y.D.: Adsorption of cholesterol oxidase and entrapment of horseradish peroxidase in metal-organic frameworks for the colorimetric biosensing of cholesterol. Talanta **200**, 293–299 (2019)
63. Hou, L., Qin, Y.X., Li, J.Y., Qin, S.Y., Huang, Y.L., Lin, T.R., Guo, L.Q., Ye, F.G., Zhao, S.L.: A ratiometric multicolor fluorescence biosensor for visual detection of alkaline phosphatase activity via a smartphone. Biosens. Bioelectron. **143**, 111605 (2019)
64. Uzak, D., Atiroglu, A., Atiroglu, V., Cakiroglu, B., Ozacar, M.: Reduced graphene oxide/pt nanoparticles/Zn-MOF-74 nanomaterial for a glucose biosensor construction. Electroanalysis **32**, 510–519 (2020)
65. Zhang, G.Y., Zhuang, Y.H., Shan, D., Su, G.F., Cosnier, S., Zhang, X.J.: Zirconium-based porphyrinic metal-organic framework (PCN-222): enhanced photoelectrochemical response and its application for label-free phosphoprotein detection. Anal. Chem. **88**(22), 11207–11212 (2016)
66. Yin, Y., Gao, C., Xiao, Q., Lin, G., Lin, Z., Cai, Z., Yang, H.: Protein-metal organic framework hybrid composites with intrinsic peroxidase-like activity as a colorimetric biosensing platform. ACS Appl. Mater. Interfaces **8**(42), 29052–29061 (2016)
67. Wu, X., Ge, J., Yang, C., Hou, M., Liu, Z.: Facile synthesis of multiple enzymecontaining metal–organic frameworks in a biomolecule-friendly environment. Chem. Commun. **51**(69), 13408–13411 (2015)
68. Graham, D., van Duyne, R., Ren, B.: Surface-enhanced raman scattering. Analyst **141**, 4995 (2016)
69. Huang, C., Li, A., Chen, X., Wang, T.: Understanding the role of metal-organic frameworks in surface-enhanced raman scattering application. Small **16**, 2004802 (2020)
70. Li, D., Cao, X., Zhang, Q., Ren, X., Jiang, L., Li, D., Deng, W., Liu, H.: Facile: In situ synthesis of core-shell MOF@Ag nanoparticle composites on screen-printed electrodes for ultrasensitive SERS detection of polycyclic aromatic hydrocarbons. J. Mater. Chem. A **7**, 14108–14117 (2019)
71. Kreno, L.E., Greeneltch, N.G., Farha, O.K., Hupp, J.T., Van Duyne, R.P.: SERS of molecules that do not adsorb on Ag surfaces: a metal-organic framework-based functionalization strategy. Analyst **139**, 4073–4080 (2014)
72. Yang, K., Zong, S., Zhang, Y., Qian, Z., Liu, Y., Zhu, K., Li, L., Li, N., Wang, Z., Cui, Y.: Array-assisted sers microfluidic chips for highly sensitive and multiplex gas sensing. ACS Appl. Mater. Interfaces **12**, 1395–1403 (2020)
73. Yu, T.H., Ho, C.H., Wu, C.Y., Chien, C.H., Lin, C.H., Lee, S.: Metal-organic frameworks: a novel SERS substrate. J. Raman Spectrosc. **44**, 1506–1511 (2013)
74. Sun, H., Cong, S., Zheng, Z., Wang, Z., Chen, Z., Zhao, Z.: Metal-organic frameworks as surface enhanced raman scattering substrates with high tailorability. J. Am. Chem. Soc. **141**, 870–878 (2019)
75. Fu, J., Lai, H., Zhang, Z., Li, G.: UiO-66 metal-organic frameworks/gold nanoparticles-based substrates for SERS analysis of food samples. Anal. Chim. Acta **1161**, 338464 (2021)
76. Hu, Y., Cheng, H., Zhao, X., Wu, J., Muhammad, F., Lin, S., He, J., Zhou, L., Zhang, C., Deng, Y., et al.: Surface-enhanced raman scattering active gold nanoparticles with enzyme-mimicking activities for measuring glucose and lactate in living tissues. ACS Nano **11**, 5558–5566 (2017)
77. Jiang, Z., Gao, P., Yang, L., Huang, C., Li, Y.: Facile in situ synthesis of silver nanoparticles on the surface of metal-organic framework for ultrasensitive surface-enhanced raman scattering detection of dopamine. Anal. Chem. **87**, 12177–12182 (2015)
78. Zhang, Y., Hu, Y., Li, G., Zhang, R.: Composite prepared from gold nanoparticles and a metal organic framework (type MOF-74) for determination of 4-nitrothiophenol by surface-enhanced Raman spectroscopy. Microchim. Acta **186**, 477–487 (2019)

79. Xu, H., Zhu, J., Cheng, Y., Cai, D.: Functionalized UIO-66@Ag nanoparticles substrate for rapid and ultrasensitive SERS detection of di-(2-ethylhexyl) phthalate in plastics. Sens. Actuat. B Chem. **349**, 130793 (2021)
80. Cao, X., Hong, S., Jiang, Z., She, Y., Wang, S., Zhang, C., Li, H., Jin, F., Jin, M., Wang, J.: SERS-active metal–organic frameworks with embedded gold nanoparticles. Analyst **142**, 2640–2647

Chapter 12
Electroactive Polymer-Based Nanostructures and Nanocomposites for Sensing Applications

Bitupon Boruah, Sandeepan Borah, and Madhuryya Deka

Abstract Electroactive polymers (EAPs) are an intriguing group of materials suitable for next-generation sensing applications. Sensor technologies have revolutionized and simplified the living and working environment of human beings through their applications in almost all fields by detecting/collecting signals in various forms. Sensors improve our ability to comprehend and report on the environment around us. Herein we have reviewed the major types of EAP based sensors made by using polypyrrole (PPy), polyaniline (PANI), and poly (3,4-ethylene dioxythiophene) (PEDOT). Conducting polymers can be utilized for various gas, humidity, temperature, and pressure sensors due to their key characteristics including low weight, high stain capability, good flexibility, low voltage applicability, room temperature operability, rapid response, and high energy efficiency. In recent decades, polymer electrolyte-based nanostructures and nanocomposites have received significant attention for sensing applications. Due to their superior mechanical and detecting response qualities, polymer electrolyte membranes can possess highly responsive sensors for practical applications.

Keywords Conducting polymer · Nanostructure · Electrolyte · Sensor

12.1 Introduction

In the presence of applied electric fields, the polymers that exhibit shape and dimensional variations are categorized as electroactive polymers (EAPs). EAP materials have several appealing properties, such as low weight, high stain capability, flexibility, etc. [1]. The development of nanocomposites based on EAPs can be considered as the most significant topic of concern in the field of sensors. For reliable sensing applications, some EAP materials possess the reverse effect of converting mechanical

B. Boruah · S. Borah · M. Deka (✉)
Department of Physics, The Assam Kaziranga University, Jorhat 785006, Assam, India
e-mail: madhuryya@kazirangauniversity.in

D. Mohanta and P. Chakraborty (eds.), *Nanoscale Matter and Principles for Sensing and Labeling Applications*, Advanced Structured Materials 206,
https://doi.org/10.1007/978-981-99-7848-9_12

strain to an electrical signal [2]. According to their mode of activation, EAP materials are divided into two main categories: ionic EAPs which are activated by Coulomb force, and electronic EAPs which are activated when the mobility or diffusion of ions takes place. Electronic EAP materials have the ability to maintain induced displacement when subjected to D.C. voltage, enabling them to be declared as potential candidates for different uses. In addition, these materials require a strong activation field (>10 V/μm) near the breakdown limit and show rapid response, and greater mechanical energy density [3]. On the other hand, ionic EAP materials such as gels, carbon nanotubes, conductive polymers, etc., require an electrolyte for operations in the air that have low response time and electrochemical coupling [2]. The major advantages of ionic EAPs, such as conducting polymers, are their distinctive potentiality of bio-stability and low voltage requirements [4]. Voltages as low as 1–2 V are sufficient to activate ionic EAPs [5]. The need of electrolytes and the occurrence of electrolysis (at >1.23 V) is a serious issue in ionic EAPs. In order to work effectively in open-air environments, ionic EAPs require encapsulated or protective layers. Ionic EAPs, excluding carbon nanotubes and conducting polymers, do not retain strain when applied to the dc activation field. For sensing applications, nanostructures and nanocomposites based on different polymers have gained academic and industrial interest in the last few decades. Among those sensors, EAP based-sensors is mostly studied by researchers due to their low voltage applicability, room (air) condition operability, rapid response, good flexibility, and high energy convention efficiency [6, 7]. Research on various sensors like gas sensors, humidity sensors, and pressure sensors based on EAPs has been extensively studied, but reviews on such EAP-based sensor applications are very rare. This chapter summarises the recent advancements made by a variety of EAP-based sensors and their working mechanisms, along with applications that might be helpful for further research in this field.

12.2 Sensor as the Key

A sensor is an input device or a module that detects a change in physical, electrical, chemical, or other quantities and generates optical and electrical signals as an output. Basically, sensors are split into two groups: active and passive [8]. Active sensors require an external excitation signal in order to function properly. On the other hand, a passive sensor directly produces an electric signal without any additional source of power. The method for detecting electrical, biological, chemical, radioactive, and other reactions is also used to classify the sensors in another way [9]. The phenomena of conversion, such as photoelectric, thermoelectric, electrochemical, electromagnetic, and thermotic that follow the input/output procedure provide the basis for the next categorization of sensing devices [10]. Moreover, sensors can be divided into analog and digital types. A continuous output signal, usually voltage, is generated by analog sensors. While the digital sensors typically capture binary data, which is utilized for conversion and transmission. Sensors that have good sensitivity, high flexibility, rapid response, and long lifetime are required for a variety of disciplines, including touch interface, health monitoring, electronic skin, and many more [11].

12.2.1 Various Types of Sensors

Different kinds of sensors are utilized for constant monitoring of physical properties like pressure, temperature, gas, humidity, heartbeat & pulse, color, speed, etc. For a variety of applications, temperature & thermocouple sensors, proximity sensors, ultrasonic sensors, image sensors, pressure sensors, light sensors, color sensors, touch sensors, metal detectors, flow and level sensors, smoke & fog sensors, gas sensors, alcohol & ethanol sensors, humidity sensors, soil moisture and rain sensors, etc. are used. In the current chapter, we specifically covered pressure, humidity, and gas sensors that are based on EAP materials.

Gas sensor: It is an electronic tool that measures the existence or concentration of gases by detecting the inimitable breakdown voltage and identifies different types of gases like LPG, methane, propane, butane, carbon dioxide, carbon monoxide, etc. in the atmosphere [12].

Humidity sensor: The sensors that sense and determine the air moisture content in the atmosphere are specified as humidity sensors. The relative humidity which depends on temperature significantly helps the humidity sensor to monitor the air temperature.

Pressure sensor: A pressure sensor, equipped with a pressure-sensitive element, is a device that detects and regulates the pressure exerted on a surface and converts it to an electrical signal. Pressure sensors are vaguely utilized in medical, industrial, consumer, automotive, and building systems that need accurate and stable measurements of pressure for reliable operation.

12.3 Electroactive Polymer for Sensing

Ionic EAP materials are able to function as sensors due to the electromechanical characteristics that they possess. An ion reservoir is highly imperative for the ion transportation mechanism in such EAPs.

Materials employed in the production of ionic EAP, such as polypyrrole (PPy), polyaniline (PANI), poly (3,4-ethylene dioxythiophene) (PEDOT) that contain ions, have some similar properties as follows: (i) The migration of ions will be caused by the stress or strain applied to this material, which disrupts the charge distribution. (ii) For sensing applications, the kinetics of ion transportation through a separator or into the electrodes will influence the final results [13].

12.3.1 Conductive Polymer-Based Sensors

Organic polymers that are inherently conducting because of their conjugate π electrons with a comparatively high and reversible ion storage capacity are known as conducting polymers or intrinsically conducting polymers (ICPs). They have the structural features of the combination of monomeric units and can undergo reduction and oxidation reactions [14]. A huge amount of study has been done in the domain of conducting polymers [15–17]. The mechanism of mechanical sensing and actuation is influenced by the incorporation and separation of charge into and from the polymer matrix. Conducting polymers like polypyrrole (PPy), and polyaniline (PANI) are excellent choices for integration into the design of sensors and actuators as they can modulate their resistivity and generate electrical signals in response to external stimuli. Because of their thermal and chemical stability in a variety of environmental circumstances, they have the greatest possible applications. Ionic EAPs are synthesized by using a variety of materials. Among these materials, conducting polymers are a promising choice for sensing applications due to their exceptional qualities, including flexibility, ease in the synthesis process, and cheap, low working temperature with remarkable sensitivity when subjected to different gases. The development of sensors heavily relies on characteristics like selectivity, durability, sensitivity, detection range, reaction, and recovery time, etc. Among the classes of conjugated polymers, polyaniline is a suitable material for sensor applications, since its conductivity is strongly sensitive to chemical vapor. The greater surface area of polyaniline nanofibers exhibits superior interaction relative to conventional materials. The charge conduction phenomena of polyaniline are determined by its potential to carry charge carriers along the polymer backbone and by enabling the charge carrier to hop throughout the polymer chain. The conductivity of polyaniline is sensitive to both of these processes [18]. Additionally PANI is an appealing material for sensor applications due to its ambient temperature sensitivity, structural modification, flexibility of deposition, etc. A study on the development and operation of ammonia gas sensors was conducted by K. Crowley [19] using an inkjet printer to deposit a film of polyaniline nanoparticles. Layer width and thermal characteristics were used to describe the assembled conductive films on the alternating electrode array. It was observed that the nanoscale PANI sensors were extremely sensitive to gaseous ammonia, which is also resistant to moisture in the relative humidity range of 35–98%. Using PANI nanofibers with highly distributed gold nanoparticles (AuNPs), C. Liu and colleagues [20] created a chemiresistive sensor that allows researchers to examine volatile sulfur compounds (VSC) in the human breath. Small AuNPs are produced on PANI nanowires as a consequence of the redox reaction between chloroauric acid ($HAuCl_4$) and the emeraldine form of PANI. The PANI/AuNPs sensors offer excellent H_2S and CH_3SH gas sensing capabilities. PANI-based composites are also a promising candidate for the detection of ecologically toxic gasses, including NO_2, H_2S, CO, LPG, etc. One of the most harmful pollutants to the environment, NO_2, has a number of negative impacts on the health of living organisms since it can cause airway burns at high exposure levels. As a result, nanocomposite gas sensors based on CP toward NO_2 have drawn a significant amount of interest and have been thoroughly investigated by numerous researchers. Polyaniline (PANI) thin film functioning as a nitrogen dioxide gas sensor at ambient temperature has been explained by Sonker and co-workers [21]. Through a chemical oxidative polymerization process using ammonium persulfate (APS) as the oxidizing agent and aniline as the monomer, the synthesized film displayed the greatest sensitivity of ~12.10% compared to 100 ppm NO_2 with quick reaction and recovery times of ~11 s and 7 min respectively. Polyaniline composites can also be used as H_2S gas sensors due to their ease of preparation, excellent performance, room temperature operability, and H_2S detection capability. Suhail et al. [22] also created a thin film based on PANI that contained functionalized single-wall carbon nanotubes (f-SWCNTs) in a range of concentrations to create a hydrogen sulfide (H_2S) gas sensor. The f-SWCNT-doped PANI films have a quick response and recovery time. This happens as a result of the formation of conduction pathways and the hopping of electrons via conducting routes of the carbon nanotubes. The maximum ionic conductivity is found to be about 6.22 Scm^{-1}. The centrally hollow core structure of f-SWCNT present in PANI could assist the absorption of H_2S gas. In the PANI film, the large surface area of f-SWCNT offers additional interaction sites. In contrast, it was found that both the response and recovery time of the sensor decreased as the operating temperature increased. The maximum sensitivity of 60 ppm was found at 50 °C whose value decreases as the operating temperature rises more. Humidity sensors made up of conducting polymers can be used to approximate the value of relative humidity in a variety of circumstances. A polyaniline nano fibers-based humidity sensor was developed by Zeng et al. [23] and tested its responsiveness to humidity. When the relative humidity is low, this sensor responds by reducing the electrical resistance, however, at higher RH, it reverses its response by increasing the electrical resistance. The 'reversed behaviour' is thought to be caused by the nanofiber's distinctive structure, which is susceptible to deformation during polymer swelling. In the existence of water vapour, conductivity of PANI changes due to the proton effect [23]. The conductivity of PANI nanofibers is a result of the combination of the "proton effect" and the "swelling effect". At low relative humidity, the proton effect prevailed and a decrease in electrical resistance was observed. The polymer swelling reversed the response of the sensor and increased the sensor resistance at high relative humidity. This unique behaviour is the result of the nanostructure being deformed by the molecules of water that have been adsorbed to the surface. The micro aggregate PANI sensor displayed superior performance. B. Chethan and colleagues [24] investigate the effectiveness of a core–shell structured Polypyrrole/Tantalum pentoxide (PTO) composite in terms of its ability to sense humidity. The PTO-4 composite based humidity sensor showed sensitivity of 99.99% (0.044/% RH) with response and recovery times of 6 and 7 s, respectively. Guo et al. [25] introduce a unique stretchable humidity sensor made from polyaniline composite fibers. After 5000 cycles of stretching, the "twining spring" arrangement of composite of fibers keeps its electrical conductivity steady up to a strain of 200%.

The pressure sensor is one of the most widely used components in modern sensing systems. These systems have significant factors in the operation and monitoring of a wide range of applications, from the biomedical to the environmental. A wide range of sensing techniques, including piezoresistivity, piezoelectricity, and triboelectricity have been developed for flexible pressure sensors [26]. As active material, Liu et al. utilized polyaniline nanofiber-wrapped nonwoven fabric to fabricate flexible, high-performance pressure sensors with a bottom interdigitated textile electrode [27]. The resulting pressure sensor exhibited exceptional performance due to the presence of ordered structures, irregular surface configuration of the PANI-coated fabric, and high conductivity of the electrodes. Moreover, it showed an extremely high sensitivity of 46.48 kPa^{-1} over a broad linear range (<4.5 kPa), a quick response time (7/16 ms), and a low detection limit (0.46 Pa). Considering these benefits, the pressure sensor is used to demonstrate wearable electronics to monitor human physiological data and spatial distribution of subtle pressure, which is vital for health monitoring. Figure 12.1 depicts the modelling procedure of a pressure sensor using PANI. The PANI customized nonwoven fabric serves as the piezoresistive layer for the fiber-based flexible pressure sensor in which the bottom is layered with screen-printed interdigital electrodes.

Polypyrrole (PPy) is a comparatively air-stable intrinsically conductive polymer (ICP) with high electrical conductivity, and excellent environmental, electrical, and chemical stability. Due to their distinct physicochemical characteristics, PPy nanostructures and their hybrid composites have been considered as promising gas sensing materials. The ammonia sensing capabilities of one-dimensional nanostructured PPy and its nanocomposites with metal oxides, as well as the impact of the nanofiller on

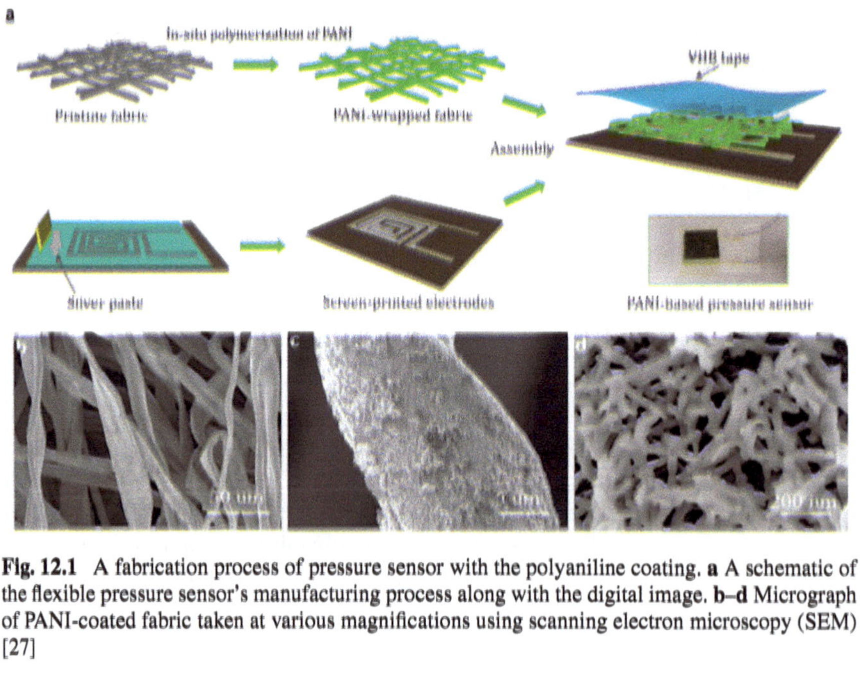

Fig. 12.1 A fabrication process of pressure sensor with the polyaniline coating. **a** A schematic of the flexible pressure sensor's manufacturing process along with the digital image. **b–d** Micrograph of PANI-coated fabric taken at various magnifications using scanning electron microscopy (SEM) [27]

the functionality of the sensor were investigated by Joulazadeh et al. [28]. With regard to methanol and ethanol vapors, the interaction of metal oxides with PPy and its nanocomposites displayed a strong selectivity for ammonia. The excellent selectivity value of 1.04 (ppm^{-1}) and linearity response for the PPy/ZnO sensor in the presence of ammonia vapor (10–80 ppm) show tremendous potential in ammonia detection. When exposed to ammonia, it was noticed that the PPy/ZnO nanocomposite sensor's response was significantly more prominent and retainable than those of PPy (0.33 ppm^{-1}) and PPy/SnO_2 (0.79 ppm^{-1}), as shown in Fig. 12.2. The cyclability and linearity of this sensor point to its tremendous potential for ammonia detection.

On a silicon substrate, Chartuprayoon et al. [29] fabricated an ammonia sensor based on a single nanowire of polypyrrole. Anodic alumina (AAO) was used as a template to create high-density PPy nanowires with small diameters, and the results showed great responsiveness to ammonia concentrations as low as 1.5–77 ppm [30]. The rapid and effective diffusion of target gases through sensing wires is induced by the inflated surface-to-volume ratio of high-density PPy. However, a moderate response time (~60 s at 77 ppm) was observed for this sensor. The PPy nanowire sensor has a sensitivity of 0.06% per ppm ammonia. According to the literature [31], the typical redox reaction that takes place between PPy and ammonia is considered a major cause of ammonia sensing. The mathematical expression for the redox reaction that happens between PPy and ammonia is as follows.

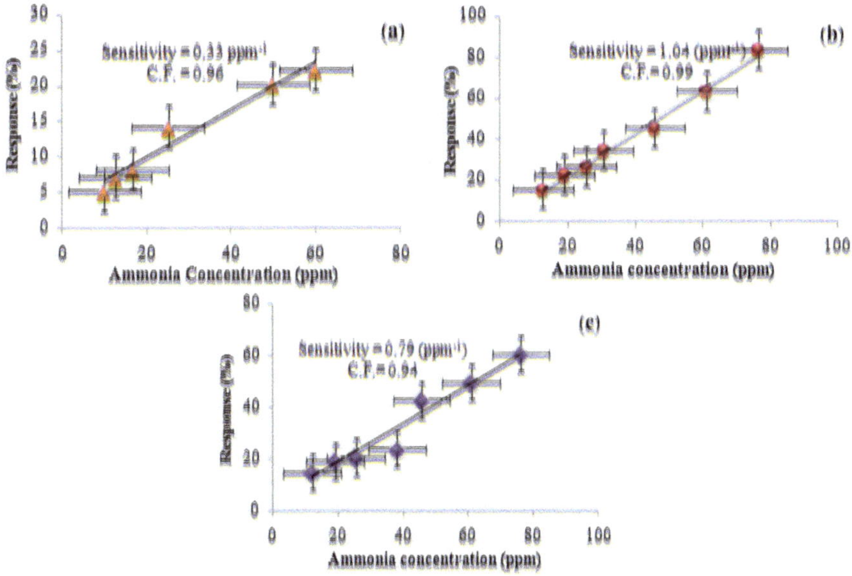

Fig. 12.2 The response of the **a** PPy **b** PPy/ZnO, and **c** PPy/SnO_2-based sensors versus the ammonia concentration [28]

$$PPyH^+ + NH_3 \rightarrow PPy^0 + NH_4^+(\text{Adsorption}) \tag{12.1}$$

$$PPy^0 + NH_4^+ \rightarrow PPyH^+ + NH_3(\text{Desorption}) \tag{12.2}$$

When p-type PPy is deprotonated in the vicinity of reducing gas, the hole density in the PPy backbone decreases, increasing the base resistance of the sensor [32]. Garg et al. [33] suggested the model of the redox reaction to explain the possible H_2S sensing process of PPy. According to the author's vision, the proton transfer from H_2S to PPy via the redox reaction is the fundamental mechanism behind the sensation. The detailed redox reaction is listed as follows.

$$PPyH^+ + H_2S \leftrightarrow PPyH^+.\,H_2S \tag{12.3}$$

$$PPyH^+.\,H_2S \rightarrow PPyH^+ + H_2S + H_3O^+ \tag{12.4}$$

The base resistance decreases as a result of the redox reaction between PPy and H_2S, which enhances the doping level of PPy. The schematic illustration of PPy based fundamental gas sensing mechanism is shown in Fig. 12.2. The ferrocene (Fc) modified polypyrrole (PPy) by direct incorporation during polymerization increases the sensitivity of carbon monoxide (CO) gas sensors [34]. A sulfonate anion-doped nanostructures of conducting polypyrrole was used to create a selective dimethyl sulfoxide (DMSO) gas sensor by S. Pirsa and his co workers [35]. A sulfonated doped polypyrrole specimen has been synthesized in the presence of an oxidizing agent by polymerizing pyrrole on the substrate of polyester fibers. A fast response time (<1 s) was observed for PPy-S based gas sensors. The $PPyHSO_3$ exhibited a lower detection range compared to DMSO and was utilized as a highly selective DMSO detector. The observation revealed a linearity of calibration curve in the range of 0.1–500 g. Sulfonate anions were added to the PPy to create a conductive material appropriate for use in chemiresistor sensors (Fig. 12.3).

The incorporation of PVDF/PPy nanofibers into electrospun 2D–Bi_2S_3 was reported by Veeralingam et al. [36] as a flexible basis for supersensitive stress, strain, and temperature monitoring. PPy nanoparticles make it easier for electrons to go through the membrane, which causes the current to vary rapidly with pressure. The manufactured PVDF/PPy membrane's tensile strength is enhanced using Bi_2S_3-based fillers to support a larger range of applied pressure. A normalized current versus time plot was utilized in order to study the response time. With a response time of 0.04 s and sensitivity of 1.51 kPa^{-1} throughout a large linear range of 1–50 kPa, a pressure sensor was successfully constructed. The gauge factor (GF) and reaction time of the BS-PVDF/PPy composite, which was used as a strain sensor in the range of 3.1–61.5%, were 45.45 and 0.1 s respectively. By using the following formula, the manufactured sensor's sensitivity (S) was calculated.

$$S = (\Delta I / I_0)/\Delta P \tag{12.5}$$

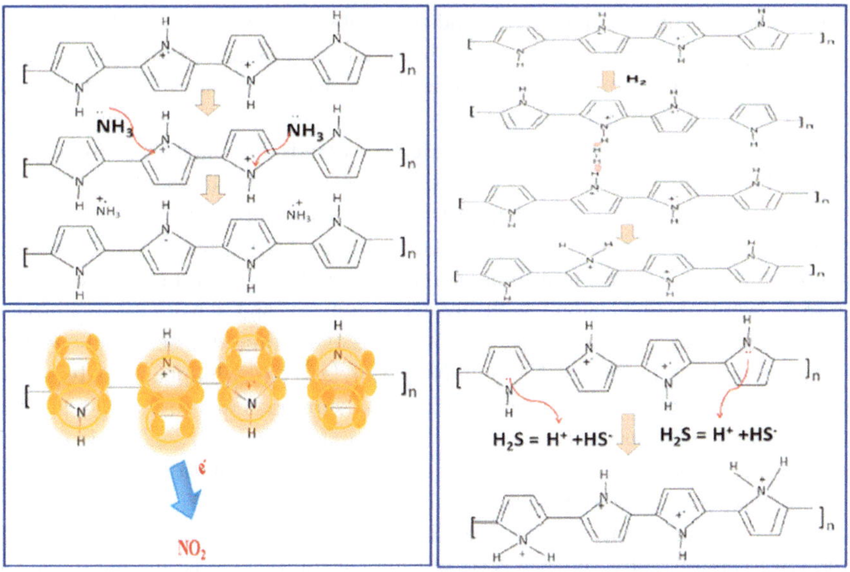

Fig. 12.3 Gas sensing mechanism of PPy [30]

where I_0, ΔI, and ΔP represent the initial current, change in current, and relative change in pressure respectively. For the precise identification of a variety of medical disorders, accurate body temperature monitoring is essential. Due to the high Seebeck coefficient and low thermal conductivity, Bi_2S_3 has attracted much as a promising thermoelectric material. The sensor was placed on the temperature-controlled plate in order to carry out the electrical characterization of the device by varying the temperature from 24 to 48 °C.

In addition, the temperature-dependent current versus voltage characteristics was investigated by changing the voltage from −1 V to +1 V throughout a temperature range of 24–48 °C in order to study the charge transfer process in the BS-integrated PVDF/PPy sensor. The linear regression curve was achieved with 57.76 $°C^{-1}$ sensitivity [36]. Multi-walled carbon nanotubes (MWNTs)/PPy based AG piezo-resistive pressure sensor was created by Gunasekara et al. [37] by layering a small block of MWNTs laminated AG with PPy through in situ polymerization while using SDBS (sodium dodecyl benzenesulfonate) as a dopant and surface-active agent. There has been a significant amount of interest in the advancement of extremely sensitive, relatively inexpensive pressure sensors with superior detection accuracy due to the need for physical motion detection and monitoring devices in many industries. Two different coating methods were utilized in the manufacturing of the piezo-resistive pressure sensor. Multi-walled carbon nanotubes (MWNTs) were used for the first coating, and aerogel (AG) substrate, was used for the subsequent polypyrrole (PPy) coating with in-situ polymerization method. Within the pressure range of 0.45–4 kPa, the obtained MWNTs/PPy AG pressure sensor exhibits great sensitivity (34.64 kPa^{-1}

with 1 kPa), adequate conductivity, and reasonable stability over 5000 continuous pressure cycles. When there are no extra influences of pressure, piezo-resistive functionality of pressure-sensing devices indicates inadequate connectivity between the conductive fillers, which results in a high level of resistance. The resistance steadily reduces as a result of pressing, which gradually expands the contact regions between the conductive fillers. Nearly 0.2 kPa was the lowest detection threshold of MWNTs/PPy AG. Karmakar et al. [38] demonstrated the detecting properties of the silver-polypyrrole nanocomposite (Ag–PPy) based nitrogen dioxide gas sensor. At ambient temperature, the Ag–PPy thin films incorporated with 15% p-toluenesulfonic acid (PTSA) were obtained along with greater specificity and sensitivity to nitrogen dioxide gas. With 148 s response and 500 s recovery time, the sensor showed its response value up to 68% at 100 ppm NO_2. Ethylene dioxythiophene) polystyrene sulfonate (PEDOT:PSS) is a significant sensing substance due to its high conductivity, superior processability, inexpensive, poor redox potential, and ability to work at ambient temperature. Lv et al. found that adding ferric chloride ($FeCl_3$) to poly (3, 4-ethylene-dioxythiophene)-poly(styrene sulfonate) (PEDOT: PSS), thin films improved their sensitivity in detecting NH_3 gas [39]. $FeCl_3$ plays two vital roles that allow it to significantly enhance NH_3 sensing capability. One is that the hydrogen ion (H^+), which is created by the hydrolysis of $FeCl_3$, could be provided by $FeCl_3$ additive ink. Water and H^+ ions diffuse into the PEDOT: PSS film as the $FeCl_3$ ink is printed onto its surface which allows the film to swell. Because of the swelling of the film, the distance between neighbouring PEDOT-rich domains has increased. Consequently, the charge carrier's mobility has decreased. When compared to pure PEDOT: PSS film for sensing 50 ppm NH_3 at 20 °C, raised the response value up to tenfold, or around 44%. The NH_3 concentration as low as 0.5 ppm was detected by a thin film sensor based on $FeCl_3$–PEDOT: PSS at ambient temperature, with a response value of 7.6%. As a promising sensor material suited for NO assessment, Zampettia et al. [40] investigated a composite sensor made of a nanofibrous covering of TiO_2 coated with a fine layer of PEDOT-PSS. As a result, it was proved that a PEDOT-based sensor can detect NO. According to some research, the presence of such a polymer in nanoshaped forms might allow the detection of NO_2 levels. The faster removal of electrons from the sensing material is made possible by NO_2, a more powerful electron-withdrawing molecule than NO, leading to changes in the conductance of the material. As a result of their high sensitivity and ability to detect NO_2 at concentrations as low as 100 parts per billion, titanium dioxide (TiO_2) nanofibers coated with vapour-phase polymerized PEDOT have been developed. The electronic structure of CP was altered by the establishment of p–n junctions between the TiO_2 nanofibers and PEDOT-PSS layer, which led to an improvement in gas sensitivity. When the nanocomposite is exposed to NO_2, there is an increase in the conductivity of the material, which can lead to a reduction in the size of the depletion region. Demonstrated TiO_2–PEDOT:PSS nanofibers displayed a strong response to NO_2 that was concentration-dependent and reversible. Lee et al. [41] used the pressure lamination approach to build and investigate the comparison of non-sandwiched sensors (AgNW and AgNW/polymer sensors) and sandwich-structured sensors (sandwiched AgNW and AgNW/polymer sensors). The PEDOT:PSS layer enhanced the stability and sensitivity of the sensor by suppressing the protrusion of AgNWs from PDMS. In the low-pressure region, the sandwich-structured AgNW/polymer sensor seemed to have the maximum sensitivity (2.59 kPa^{-1}) and gauge factor (37.8). It also has the highest initial resistance.

Because of their inexpensive and low-power requirement, humidity sensing materials that are based on polymers have emerged as a leading contender for use in the sensor technology to detect humidity. The fabrication capability and consistency are massive concerns for the implementation of polymer sensing materials. The manufacturing capability of an inkjet printable humidity sensor was enhanced when poly (3, 4-ethylene dioxythiophene): poly (styrene sulfonate) (PEDOT: PSS) was developed by Lee et al. [42].

Water molecules have a strong influence on the proton transport mechanism of PEDOT: PSS, hence, it was chosen as a baseline material for humidity sensing. Due to their large interfacial area, silicon dioxide (SiO_2) and aluminium zinc oxide (AZO) nanoparticles are naturally good at adsorbing water molecules and utilize to achieve more stable performance. Sensitivity was improved by doping PEDOT: PSS sensing films with nanoparticles of aluminium-doped zinc oxide (AZO) and silicon dioxide with a diameter of 30 nm and 20 nm, respectively. After introducing SiO_2 nanoparticles, PEDOT: PSS/SiO_2 becomes more resistant. Humidity sensors made of PEDOT: PSS/SiO_2 typically have a resistance 20 times greater than those made of pure PEDOT: PSS. Figure 12.4 depicts the sensor responses from 20% RH to 80% RH. The change of resistance (ΔR) was calculated as $[(R-R_0)/R_0] \times 100\%$, where R_0 was chosen as the initial environment value of resistance. At 80% relative humidity (RH), the ΔR of a PEDOT:PSS/SiO_2 sensor was over 1200%, that of a PEDOT:PSS/AZO sensor was 1000%, and a pure PEDOT:PSS sensor was approximately 600% (Fig. 12.5).

All the above-mentioned sensors using copolymers with their sensitivity values are presented in Table 12.1.

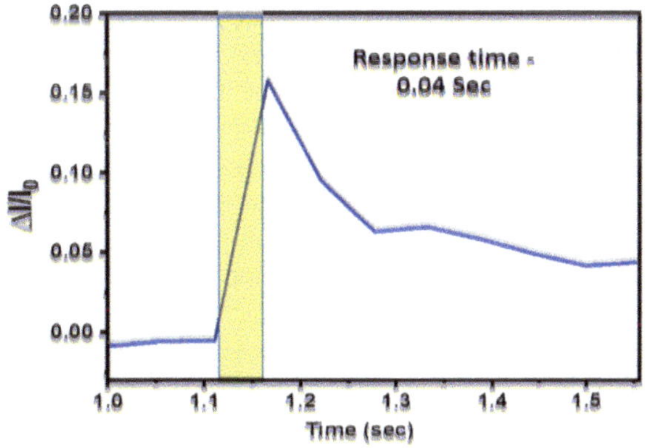

Fig. 12.4 A normalized plot of current versus time that shows a response time of 0.04 s [35]

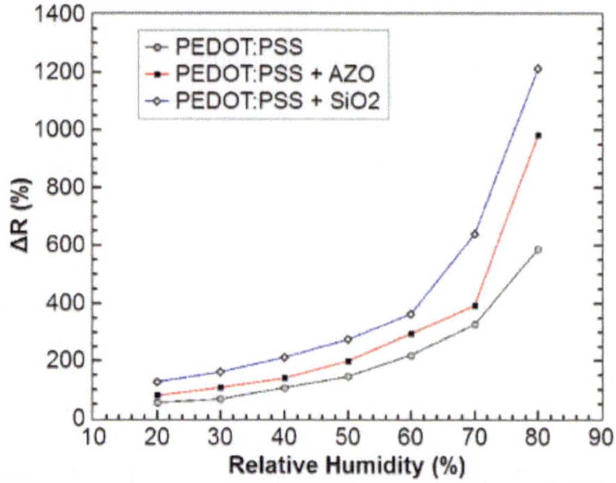

Fig. 12.5 Color online DR versus RH for PEDOT sensors, PEDOT:PSS blended with AZO and SiO_2 nanoparticles [42]

Polymer Electrolyte-Based Sensors

Researchers have been looking into a novel class of sensors based on polymer electrolyte (PE) which is a kind of ionic EAPs. Resembling conducting polymers, polymer electrolytes have received much interest in recent research on various sensor technologies. Many researchers are focused on creating a sensor based on a polymer electrolyte that can detect volatiles quickly and accurately at the start of their decay process. The development of polymer electrolyte based sensing devices such as gas sensor, humidity sensors, flexible, lightweight pressure sensors grab the attention of researchers particularly because of their excellent sensitivity and durability [43–46]. The results obtained from various characterizations methods demonstrate that the interconnected porous structure in PEs has a significant impact on the sensitivity of various sensing devices. The ratio of synthesis, morphology, and heat treatment mechanism of PEs are all correlated with these pore geometries. Additionally, there is a connection between PE conductivity and sensor sensitivity. The sensitivity, durability, stability, response time, detection time, and recovery time of every sensor are some of the factors that determine its performance characteristics. Sensors based on polymer electrolyte exhibit good sensing responses to the most ecologically harmful gasses, including NO_2, CO, and H_2S etc. Nitrogen oxide is one of the most hazardous substances to human health. Our environment is severely impacted by acid rain, which is specifically caused by nitrogen dioxide. Therefore, in order to protect the environment, it is crucial to develop nitrogen oxide sensors. The chemiluminescence method [47], biosensors, optical sensors [48], surface acoustic wave gas sensors [49], electrochemical sensors [50], etc. can be used to measure the amount of nitrogen dioxide in a gaseous phase. Luo and colleagues [51] synthesized a poly-vinylidene fluoride (PVDF) matrix in a 1-methyl-2-pyrrolidone (NMP) solvent to immobilize 1-3-methylimidazolium tetrafluoroborate ([EMIM] [BF4]), which is the basis for a new electrochemical amperometric nitrogen dioxide (NO_2) sensor. The solid polymer electrolyte (SPE) based amperometric NO_2 sensor measures the current generated

Table 12.1 Different conductive polymer-based sensors and their sensitivity

Sensors based on conductive polymer								
Sensors based on PANI			Sensors based on PPy			Sensors based on PEDOT		
Gas sensor sensitivity	Humidity sensor sensitivity	Pressure sensor sensitivity	Gas sensor sensitivity	Humidity sensor sensitivity	Pressure sensor sensitivity	Gas sensor sensitivity	Humidity sensor sensitivity	Pressure sensor sensitivity
12.10–100 ppm of NO_2 [21]	<50% RH [23]	46.48 kPa^{-1} [26]	1.04 (ppm^{-1}) for NH_3 [27]	99.99% RH [24]	1.51 kPa^{-1} [35]	50 ppm NH_3 [38]	80% RH [41]	2.59 kPa^{-1} [40]

by the electrochemical reduction of NO_2 at the electrode. The following equations explain the commonly used NO_2 reduction process at the working electrode.

$$NO_2 + e^- \rightarrow NO_2^- \tag{12.6}$$

$$NO_2^- + 2H^+ + e^- \rightarrow NO + H_2O \tag{12.7}$$

An in-depth investigation comprising the identification and measurement of the reaction products is necessary to provide a more precise representation of the reduction mechanism.

Water is most likely oxidized at the auxiliary electrode:

$$H_2O^- \rightarrow 2H^+ + 2e^- + O_2 \tag{12.8}$$

Consequently, the entire reaction in the sensor can be expressed as

$$NO_2^- \rightarrow NO + O_2 \tag{12.9}$$

The reduction process of NO_2 is responsible for regulating the flow of current through the sensor. Protons are produced when water undergoes an oxidation reaction, and the rate at which they are transferred from the auxilary electrode to the working electrode is high enough. During NO_2 detection, the compositions of the sensor are unaltered. The data demonstrates that the interdigital electrode and solid polymer electrolyte (SPE) form a two-layer sensor configuration that reliably and repeatable response to NO_2. This SPE has the highest gas sensitivity, which is 0.00275 V/ppm. Kubersky et al. [52] developed a novel SPE with glassy carbon to create an amperometric NO_2 sensor. The ionic liquid 1-ethyl-3-methylimidazolium bis(trifluoromethylsulfonyl)imide [C_2mim][NTf_2] immobilized in a poly(vinylidene fluoride) matrix [PVDF] serves as the basis for the electrolyte. The sensor exhibited a linear behavior over the entire investigation range from 0 to10 ppm and its sensitivity was found to be 0.66×10^{-6} A/ppm. The surface area of the working electrode affected the sensor's noise and response stability. On the basis of variations in the electrical conductivity with water vapour polymeric electrolyte solutions have also been utilized in humidity sensor devices. Dai et al. [53] synthesized a superfast response humidity sensor that relies on mercaptopropyl polyhedral oligomeric silsesquioxane (MPOSS), pentaerythritol tetra(3-mercaptopropionate) (PETMP) gel polymer electrolytes. The resulting moisture sensor exhibits ultra-fast response/recovery times (0.19/0.30 s) making it a good candidate for monitoring respiration. The quick response mechanism of this humidity sensor was studied using complex impedance spectroscopy in conjunction with a quartz crystal microbalance. For detecting pressure, piezoelectric effect provides the foundation for the development of pressure sensors. Due to its attractive characteristics, such as high response time and low energy requirement, they have received an ever-increasing amount of interest. Zhao et al. created a piezoelectric MXene/poly(vinylidene fluoride) (PVDF) hybrid film-based extremely sensitive flexible pressure sensor [54]. The piezoelectric coefficient of MXene/PVDF hybrid film hits the peak of 43 pC/N with a small load of 0.4 wt%. Excellent voltage sensitivity up to 0.0480 V/N was provided by the sensor based on the hybrid MXene/PVDF film. At the same time, the unique hybrid film sensor showed steady state operation capabilities under cyclic stress and a rapid recovery time of 3.1 ms. The structural characteristics of piezoelectric sensors are simple and diverse. They have great advantages in integration, implantation, and miniaturization due to the high level of flexibility and structural stability, confirming the applicability of reliable pressure sensors. The effects of carbon monoxide (CO) on human health are exceedingly dangerous [55]. Normally, the odourless and colourless CO gas cannot be detected by humans. In order to detect CO quickly and precisely, especially at low concentrations, it is crucial to have sensors that are highly selective and sensitive. Electrochemical gas sensors are appealing as a high performance CO sensor among others since they can work at ambient temperature and use cation-conducting polymers (like Nafion) as the electrolyte material [55]. The CO sensing capabilities of EC(nPt/SnO_2(Tm) sensors were studied by Hyodo et al. [56], particularly at 30 °C in humid synthetic air (57%RH). The electrochemical CO sensor utilized SnO_2 loaded with Pt as an electrode and an anionic conductive polymer electrolyte. The sensing-electrode potential was estimated by the net potential generated from electrochemical CO (or H_2) oxidation and O_2 reduction. Electrochemically, the rate of oxidation of the CO molecule on the mono-dispersive and oxidized Pt species was significantly faster than the rate of oxidation of the H_2 molecule. After being heated in H_2 at 250 °C, the surface of Pt-loaded SnO_2 shrank. This significantly improved both the CO and H_2 responses and lowered the CO selectivity against H_2. This result was most likely caused by reduction of Pt species with metal surfaces that were very effective against the anodic reaction of CO and H_2. By integrating mechanically doped graphene with SPEs, Liu et al. [57] established piezoionic-powered strain and touch sensors. The piezoionic effect in SPE has an effect on the Dirac point voltage of the SPE-coated graphene field-effect transistor (S-GFET), which moves to the left or right depending on whether the material is being compressed or stretched. Dirac point voltage of S-GFET is influenced by piezoionic effect, which moves to the left or right depending on whether the material is being compressed or stretched. It is possible to generate various S-GFET Dirac point voltages for a piezoionic-powered strain sensor by mechanically tuning the piezoionic doping of graphene. The strain sensor exhibited a high GF of ~−30 with remarkable stability. In addition, the S-GFET sensor is also suitable for touch sensing application. Touch sensing is a crucial part of many human–machine interface systems [58]. Hydrogen is used especially in the petrochemical and energy sector as a valuable industrial raw material. Its leaking can cause explosions and hydrogen embrittlement frequently result from its penetration into metal and alloy systems [58]. Quantifying the amount of hydrogen generated during a reaction process allows one to investigate the kinetics of numerous reactions. The procedure and hazard control systems for hydrogen are essential for ensuring the effectiveness of hydrogen sensors due to its industrial importance and operational safety [58]. Using a polymer electrolyte membrane fuel cell, Lu et al.

[57] created an amperometric hydrogen gas sensor. In addition to being more precise, the linear response of amperometric sensors makes them a potential candidate for practical application.

To prevent leaks, corrosion, and volatilization, solid polymer electrolytes are superior to liquid electrolytes. SPE-based amperometric hydrogen sensors allow miniaturization because of their lighter weight and smaller dimension. Due to its high proton conducting properties, Nafion has found widespread application in electrochemical gas sensors. The sensor is able to detect and measure hydrogen levels in air or the inert environment at room temperature. The steady-state currents in the sensor exhibit a linear response to hydrogen concentrations from 560 to 11,500 ppm (v/v), with a fixed potential of 0.15 V. The simple miniature hydrogen sensor possesses high sensitivity, fast response time, broad linear range, and long-term durability. Hydrogen concentrations in gas mixtures can be monitored online using it. Ramesh et al. [59] synthesized an amperometric hydrogen sensor using polymer electrolyte membrane fuel cell. It contains a proton-conducting solid polymer electrolyte which is synthesized by blending PVA with H_3PO_4. The sensor's sensitivity depends strongly on the conductivity of the polymer electrolytes. The electrochemical processes occurring at the electrodes are:

$$H_2 \rightarrow 2H^+ + 2e^-(\text{anode}) \quad (12.10)$$

$$2H^+ \rightarrow \frac{1}{2}O_2 + 2e^- \rightarrow H_2O(\text{cathode}) \quad (12.11)$$

12.3.2 *NO_2 Gas Sensor*

Fabrication of different gas sensors operable at room temperature using nanostructured electroactive polymers is a very essential topic of concern for sensing applications nowadays. Understanding the change in physical parameters of conducting polymers is most important for sensor applications. A PANI-based NO_2 sensor was introduced by Sonker et al. [21]. When subjected to NO_2, the resistivity of PANI falls off steadily and exponentially. The gas thus appears to serve as the dopant for the PANI. When the NO_2 gas was switched off and injected fresh air into the test chamber, the initial resistance was attained. The resistance of NO_2 gas sensors varies as a result of gas adsorption on the surface of PANI film. Furthermore, gas or vapor response is mostly determined by gas kinetics on the active sensor layer. Holes are the majority charge carriers of polyaniline. However, when an oxidizing gas NO_2 interacts with p-type PANI, it attracts electrons and reduces the value of resistance. The major charge carrier density (holes) in these materials rises because of the electron-accepting properties of NO_2, which causes a rise in conductivity. The possible explanation of the reaction mechanism can be stated as follows.

$$NO_{2(gas)} + e^- \rightarrow NO^-_{2(ads)} \tag{12.12}$$

The above reaction increases the resistance of the material by decreasing the electron concentration on the PANI surface. Moreover, the following reaction takes place between $NO^-{}_{2(ads)}$ and $O^-{}_{(ads)}$ as

$$NO^-_{2(ads)} + O^-_{(ads)} \rightarrow 2O^-_{(ads)} + NO_2 \tag{12.13}$$

Thus, the cycle of NO_2 formation and breakdown continues as:

$$NO_{2(g)} + e^- \rightarrow NO^-_{(ads)} \tag{12.14}$$

The sensing response to NO_2 gas was shown to be enhanced with the enlarged PANI surface.

12.3.3 Humidity Sensor

Over the past forty years, researchers and scientists have studied and implemented polymeric humidity sensors in industrial and commercial applications. The majority of these sensors are made using thin sheets of porous electroactive polymers. Based on changes in electrical conductivity with moisture, ion-conducting polymeric materials have been utilized in humidity sensors. Polymer electrolytes with polymer cations or anions as well as inorganic salts with polymers are used as raw ingredients in the fabrication process. The functionality of the sensor is determined by the ability of the films to absorb water both physically and chemically. In addition to this, polymer electrolytes exposed to high concentrations of humidity have a severe resistance change inadequacy. There are many applications where measuring and controlling humidity are important, such as in paper, food, and electronic industries, household environments, medical equipment, etc.

The investigations of the morphological, as well as structural characteristics of the sensing process of the composite surface, are displayed in Fig. 12.6 [24]. As a result of the self-ionization of water vapor, protons (H^+) and hydroxide ions (OH^-) are produced and dispersed rapidly in the preliminary phase.

$$H_2O \Leftrightarrow H^+ + OH^- \tag{12.15}$$

At low relative humidity (RH), H_2O molecules are chemically adsorbed on the active sites of the sensing materials. A chemisorbed layer is formed when OH^- ions are adsorbed onto metal cations that have a high density of charge carrier and a large electric field. The small increase in conductivity is.

caused by the leftover H^+ ions as they move from one location to another. When relative humidity (RH) is raised, each water molecule forms an immobile physisorbed layer by hydrogen bonding with two hydroxyl groups of the chemisorbed layer.

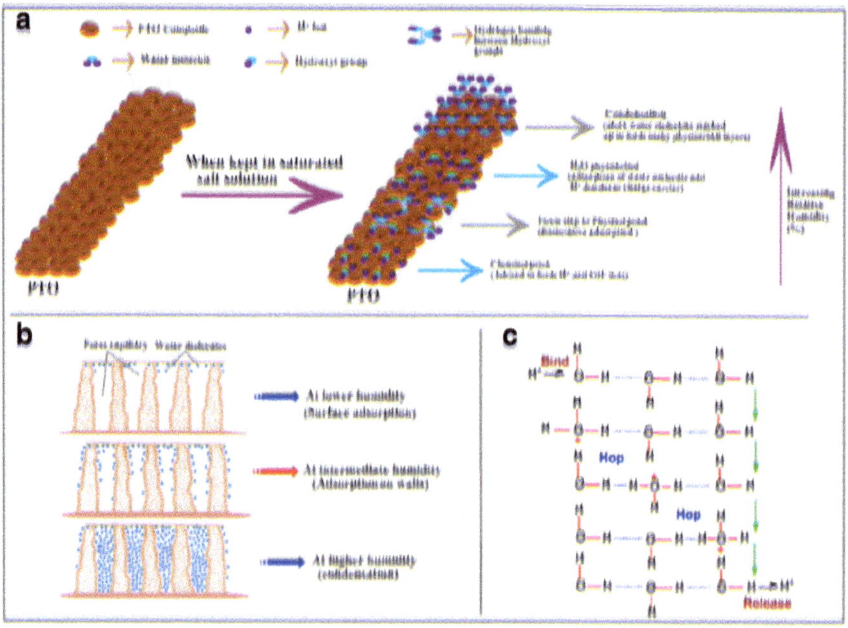

Fig. 12.6 **a** Schematic diagram of the humidity detection process, **b** Water molecule adsorption mechanism onto a sensor film, **c** Grotthus mechanism [24]

$$2H_2O \rightarrow H_3O^+ + OH^- \tag{12.16}$$

Additional physisorbed layers will grow on top of the initial layers as the relative humidity continues to rise. These layers are free to flow and will act in the same manner as a bulk liquid. Under those circumstances, H_3O^+ is formed when H_2O creates a single bond with the $-OH$ group. As a result, the H_3O^+ ionizes into H_2O and H^+ via the Grotthus mechanism. This strategy is illustrated in Fig. 12.6c.

$$H_3O^+ \rightarrow H_2O + H^+ \tag{12.17}$$

The hydrogen bonds of water molecules allow the extra protons to diffuse across it by forming links with nearby molecules. According to X-ray diffraction analysis, it was observed that the composite material has significant imperfections that promote higher specific charge density and a greater electrostatic field, causing protons to be generated as charge carriers. Figure 12.6b illustrates the condensation as a result of further increases in relative humidity.

12.3.4 *Pressure Sensor*

Understanding the atomic structure of piezoelectric polymer materials is essential to know the detection mechanism of pressure sensors. Employing PVDF film, Tanaka et al. [60] created a haptic finger sensor for assessing the state of the skin. An indicator of skin roughness and stiffness was created by converting the signal variance and power spectral density dispersion in the frequency domain. Therefore, there was a reliable agreement between the examined data and clinical evaluations of different skin conditions. The sensor is built on a thin stainless steel plate, with layers of sponge rubber, PVDF foil and acetate film. The force exerted by the sensor on the skin is measured by a strain gauge installed within the device. This composite sensor fits over the tip of the fingers and can directly be applied on the skin surface of the target. Through a bandpass filter (70 Hz to 25 kHz) and a strain gauge amplifier, the output signals from the PVDF foil and the strain gauge are, respectively, captured by a digital oscilloscope. A digital oscilloscope with a sampling rate of 0.02 ms receives the output signal from the PVDF secondary film after measuring it for a duration of 100.0 ms.

12.4 Conclusion and Outlook

Electroactive polymers are considered as intriguing materials that play a crucial role in the development of next-generation sensing systems. Among EAPs, conducting polymers are mostly used in sensor applications. The unique property of the conducting polymer can be utilized for various gas, humidity, pressure sensors, etc. Moreover, chemical, electrical, optical as well as physical aspects of the conducting polymers make them an ideal material for the purpose of highly sensitive and stable sensing. With excellent detection response and mechanical properties, polymer electrolyte membranes are capable of possessing highly responsive sensors. Because of their nanoscale morphology, the PEs offers better sensitivity and response time. Fast recovery time and consistent performance are significant features of PE-based sensors which demonstrate their viability in several sensor-related technologies.

References

1. Bae, J.-H., Chang, S.-H.: PVDF based ferroelectric polymers and dielectric elastomers for sensors and actuator applications: a review. Funct. Compos. Struct. **1**(1), 012003 (2019). https://doi.org/10.1088/2631-6331/ab0f48
2. Bar-Cohen, Y., Zhang, Q.: Electroactive polymer actuators and sensors. MRS Bull. **33**(3), 173–181 (2008). https://doi.org/10.1557/mrs2008.42

3. Thetpraphi, K., Le, M.Q., Houachtia, A., Cottinet, P.-J., Petit, L., Audigier, D., Kuhn, J., Moretto, G., Capsal, J.-F.: Surface correction control based on plasticized multilayer P(VDF-TrFE-CFE) actuator—live mirror. Adv. Opt. Mater. **7**, 1900210 (2019). https://doi.org/10.1002/adom.201900210
4. Gaur, A., Maiti, P.: Reinforced polymers for electroactive devices. In: Gutiérrez, T.J. (eds) Reactive and Functional Polymers, vol. 2, pp. 325–347. Springer, Berlin (2020). https://doi.org/10.1007/978-3-030-45135-6_11
5. Bar-Cohen, Y., Kim, K.J., Choi, H.R., Madden, J.D.W.: Electroactive polymer materials. Smart Mater. Struct. **16**(2) (2007). https://doi.org/10.1088/0964-1726/16/2/E01
6. Kamamichi, N., Maeba, T., Yamakita, M., Mukai, T.: Fabrication of bucky gel actuator/sensor devices based on printing method. In: IEEE/RSJ International Conference on Intelligent Robots and Systems, pp. 582–587 (2008). https://doi.org/10.1109/IROS.2008.4651100
7. York, A., Dunn, J., Seelecke, S.: Systematic approach to development of pressure sensors using dielectric electro-active polymer membranes. Smart Mater. Struct. **22**, 094015 (2013). https://doi.org/10.1088/0964-1726/22/9/094015
8. MohdIsa, W., Hunt, A., HosseinNia, S.H.: Sensing and self-sensing actuation methods for ionic polymer-metal composite (IPMC): a review. Sensors **19**(18), 3967 (2019). https://doi.org/10.3390/s19183967
9. Upadhyayula, V.K.K.: Functionalized gold nanoparticle supported sensory mechanisms applied in detection of chemical and biological threat agents: a review. Anal. Chim. Acta **715**, 1–18 (2012). https://doi.org/10.1016/j.aca.2011.12.008
10. Pearton, S.J., Kang, B.S., Kim, S., Ren, F., Gila, B.P., Abernathy, C.R., Lin, J., Chu, S.N.G.: GaN-based diodes and transistors for chemical, gas, biological and pressure sensing. J. Phys.: Condens. Matter **16**, R961–R994 (2004). https://doi.org/10.1088/0953-8984/16/29/R02
11. Zhang, S., Wang, C., Ding, L., Zhang, L., Chen, J., Huang, H., Jiang, D., Chen, Z., Pan, G.: Flexible bifunctional sensor based on porous copper nanowires ion gels composite films towards high-resolution stress/deformation detection. J. Mater. Chem. C **8**, 4081–4092 (2020). https://doi.org/10.1039/C9TC06091J
12. Abdullah, M.M., Singh, P., Ikram, S.: Chapter 5—Recent developments in nanostructured metal oxide-based electrochemical sensors. In: Pal, K., Gomes, F. (eds.) Micro and Nano Technologies, Nanofabrication for Smart Nanosensor Applications, pp. 123–134. Elsevier (2020). https://doi.org/10.1016/B978-0-12-820702-4.00005-2
13. Wang, T., Farajollahi, M., Choi, Y.S., Lin, I.-T., Marshall, J.E., Thompson, N.M., Narayan, S.K., Madden, J.D.W., Smoukov, K.S.: Electroactive polymers for sensing. Interface Focus **6**(4), 20160026 (2016)
14. Tsakova, V., Seeber, R.: Conducting polymers in electrochemical sensing: factors influencing the electroanalytical signal. Anal. Bioanal. Chem.Bioanal. Chem. **408**, 7231–7241 (2016). https://doi.org/10.1007/s00216-016-9774-7
15. Liu, C., Tai, H., Zhang, P., et al.: A high-performance flexible gas sensor based on self-assembled PANI-CeO_2 nanocomposite thin film for trace-level NH_3 detection at room temperature. Sens. Actuat. B-Chem. 261, 587–597 (2018). https://doi.org/10.1016/j.snb.2017.12.022
16. Zhang, R., Peng, B., Yuan, Y.: Flexible printed humidity sensor based on poly(3,4-ethylenedioxythiophene)/reduced graphene oxide/Au nanoparticles with high performance. Compos. Sci. Technol. **168**, 118–125 (2018). https://doi.org/10.1016/j.compscitech.2018.09.013
17. Luo, C., Liu, N., Zhang, H., Liu, W., Yue, Y., Wang, S., Rao, J., Yang, C., Su, J., Jiang, X., Gao, Y.: A new approach for ultrahigh-performance piezoresistive sensor based on wrinkled PPy film with electrospun PVA nanowires as spacer. Nano Energy **41**, 527–534 (2017). https://doi.org/10.1016/j.nanoen.2017.10.007
18. Virji, S., Huang, J., Kaner, R.B., Weiller, B.H.: Polyaniline nanofiber gas sensors: examination of response mechanisms. Nano Lett. **4**(3), 491–496 (2004). https://doi.org/10.1021/nl035122e
19. Crowley, K., Morrin, A., Hernandez, A., O'Malley, E., Whitten, P.G., Wallace, G.G., Smyth, M.R., Killard, A.J.: Fabrication of an ammonia gas sensor using inkjet-printed polyaniline nanoparticles. Talanta **77**, 710–717 (2008). https://doi.org/10.1016/j.talanta.2008.07.022

20. Liu, C., Hayashi, K., Toko, K.: Au nanoparticles decorated polyaniline nanofiber sensor for detecting volatile sulfur compounds in expired breath. Sens. Actuat. B **161**, 504–509 (2012). https://doi.org/10.1016/j.snb.2011.10.068
21. Sonker, R.K., Yadav, B.C., Dzhardimalieva, G.I.: Preparation and properties of nanostructured PANI thin film and its application as low temperature NO_2 Sensor. J. Inorg. Organomet. Polym. Mater.Inorg. Organomet. Polym. Mater. **26**, 1428–1433 (2016). https://doi.org/10.1007/s10904-016-0439-y
22. Suhail, M.H., Abdullah, O.G., Kadhim, G.A.: Hydrogen sulfide sensors based on PANI/f-SWCNT polymer nanocomposite thin films prepared by electrochemical polymerization. J. Sci.: Adv. Mater. Dev. **4**(1), 143–149 (2019). https://doi.org/10.1016/j.jsamd.2018.11.006
23. Zen, F.W., Liu, X.X., Diamond, D., Lau, K.T.: Humidity sensors based on polyaniline nanofibers. Sens. Actuat. B **143**, 530–534 (2010). https://doi.org/10.1016/j.snb.2009.09.050
24. Chethana, B., Raj Prakasha, H.G., Ravikiranb, Y.T., Vijayakumaric, S.C., Thomas, S.: Polypyyrole based core-shell structured composite based humidity Sensor operable at room temperature. Sens. Actuat. B Chem. **296**, 126639 (2019). https://doi.org/10.1016/j.snb.2019.126639
25. Guo, Y.N., Gao, Z.Y., Wang, X.X., Sun, L., Yan, X., Yan, S.Y., Long, Y.Z., Han, W.P.: A highly stretchable humidity sensor based on spandex covered yarns and nanostructured polyaniline. RSC Adv. **8**, 1078–1082 (2018). https://doi.org/10.1039/C7RA10474J
26. Chen, W., Yan, X.: Progress in achieving high-performance piezoresistive and capacitive flexible pressure sensors: a review. J. Mater. Sci. Technol. **43**, 175–218 (2020). https://doi.org/10.1016/j.jmst.2019.11.010
27. Liu, K., Zhou, Z., Yan, X., Meng, X., Tang, H., Qu, K., Gao, Y., Li, Y., Yu, J., Li, L.: Polyaniline nanofiber wrapped fabric for high performance flexible pressure sensors. Polymers **11**, 1120 (2019). https://doi.org/10.3390/polym11071120
28. Joulazadeh, M., Navarchian, A.H.: Ammonia detection of one-dimensional nano-structured polypyrrole/ metal oxide nanocomposites sensors. Synth. Met. **210**, 404–411 (2015). https://doi.org/10.1016/j.synthmet.2015.10.026
29. Chartuprayoon, N., Hangarter, C.M., Rheem, Y., Jung, H., Myung, N.V.: Wafer-scale fabrication of single polypyrrole nanoribbon-based ammonia sensor. J. Phys. Chem. C **114**, 11103–11108 (2010). https://doi.org/10.1021/jp102858w
30. Das, M., Roy, S.: Polypyrrole and associated hybrid nanocomposites as chemiresistive gas sensors: a comprehensive review. Mater. Sci. Semicond. Process.Semicond. Process. **121**, 105332 (2021). https://doi.org/10.1016/j.mssp.2020.105332
31. Wang, Y., Jia, W., Strout, T., Schempf, A., Zhang, H., Li, B., Cui, J., Lei, Y.: Ammonia gas sensor using polypyrrole-coated TiO_2/ZnO nanofibers. **21**(12), 1432–1438 (2009). https://doi.org/10.1002/elan.200904584
32. Hamouma, O., Kaur, N., Oukila, D., Mahajanc, A., Chehimi, M.M.: Paper strips coated with polypyrrole-wrapped carbon nanotube composites for chemi-resistive gas sensing. Synth. Met. **258**, 116223 (2019). https://doi.org/10.1016/j.synthmet.2019.116223
33. Garg, R., Kumar, V., Kumar, D., Chakarvarti, S.K.: Polypyrrole microwires as toxic gas sensors for ammonia and hydrogen sulphide. J. Sens. Instrum. **3**, 1–13 (2015). https://doi.org/10.7726/jsi.2015.1001
34. Radhakrishnan, S., Paul, S.: Conducting polypyrrole modified with ferrocene for applications in carbon monoxide sensors. Sens. Actuat. B Chem. **125**, 60–65 (2007). https://doi.org/10.1016/j.snb.2007.01.038
35. Pirsa, S., Alizadeh, N.: A selective DMSO gas sensor based on nanostructured conducting polypyrrole doped with sulfonate anion. Sens. Actuat. B Chem. **168**, 303–309 (2012). https://doi.org/10.1016/j.snb.2012.04.027
36. Veeralingam, S., Badhulika, S.: Bi_2S_3/PVDF/Ppy-based freestanding, wearable, transient nanomembrane for ultrasensitive pressure, strain, and temperature sensing. ACS Appl. Bio Mater. **4**, 14–23 (2021). https://doi.org/10.1021/acsabm.0c01399
37. Gunasekara, D.S.W., He, Y., Liu, H., Liu, L.: Smart wearable, highly sensitive pressure sensor with MWNTs/PPy aerogel composite. Fibers Polym. **22**, 2102–2111 (2021). https://doi.org/10.1007/s12221-021-0787-2

38. Karmakar, N., Fernandes, R., Jainb, S., Patil, U.V., Shimpi, N.G., Bhat, N.V., Kothari, D.C.: Room temperature NO_2 gas sensing properties of p-toluenesulfonic acid doped silver-polypyrrole nanocomposite. Sens. Actuat. B **242**, 118–126 (2017). https://doi.org/10.1016/j.snb.2016.11.039
39. Lv, D., Chen, W., Shenb, W., Peng, M., Zhang, X., Wang, R., Xu, L., Xu, W., Song, W., Tan, R.: Enhanced flexible room temperature ammonia sensor based on PEDOT: PSS thin film with $FeCl_3$ additives prepared by inkjet printing. Sens. Actuat. B Chem. **298**, 126890 (2019). https://doi.org/10.1016/j.snb.2019.126890
40. Zampetti, E., Pantalei, S., Muzyczuk, A., Bearzotti, A., Cesarea, F.D., Spinella, C., Macagnano, A.: A high sensitive NO_2 gas sensor based on PEDOT–PSS/TiO_2 nanofibres. Sens. Actuat. B **176**, 390–398 (2013). https://doi.org/10.1016/j.snb.2012.10.005
41. Lee, J., Kim, G., Shin, D.-K., Park, J.: Solution-processed resistive pressure sensors based on sandwich structures using silver nanowires and conductive polymer. IEEE Sens. J. **18**(24), 9919–9924 (2018). https://doi.org/10.1109/JSEN.2018.2876142
42. Lee, C.-H., Chuang, W.-Y., Lin, S.-H., Wu, W.-J., Lin, C.-T.: A printable humidity sensing material based on conductive polymer and nanoparticles composites. Jpn. J. Appl. Phys. **52**, 05DA08 (2013). https://doi.org/10.7567/JJAP.52.05DA08
43. Sakthivel, M., Weppner, W.: Development of a hydrogen sensor based on solid polymer electrolyte membranes. Sens. Actuat. B Chem. **113**, 998–1004 (2006). https://doi.org/10.1016/j.snb.2005.04.036
44. Kim, E., Kim, S.Y., Jo, G., Kim, S., Park, M.J.: Calorimetric and resistive polymer electrolyte thin films for realtime humidity sensors. ACS Appl. Mater. Interfaces **4**, 5179–5187 (2012). https://doi.org/10.1021/am3011115
45. Yu, C., Wang, Y., Hua, K., Xing, W., Lu, T.: Electrochemical H_2S sensor with H_2SO_4 pretreated nafion membrane as solid polymer electrolyte. Sens. Actuat. B Chem. **86**(2–3), 259–265 (2002). https://doi.org/10.1016/S0925-4005(02)00200-9
46. Hadhy, R.A., Kadem, B.Y.: Gel polymer electrolyte based on SiC:PVA composites for body motion sensor and paper based application. In: 12th International Conference on Developments in eSystems Engineering (DeSE), pp. 730–735 (2019). https://doi.org/10.1109/DeSE.2019.00136.
47. Breuninger, C., Meixner, F.X., Kesselmeier, J.: Field investigations of nitrogen dioxide (NO_2) exchange between plants and the atmosphere. Atmos. Chem. Phys. **13**, 773–790 (2013). https://doi.org/10.5194/acp-13-773-2013
48. Do, J.-S., Shieh, R.-Y.: Electrochemical nitrogen dioxide gas sensor based on solid polymeric electrolyte. Sens. Actuat. B Chem. **37**(1–2), 19–26 (1996). https://doi.org/10.1016/S0925-4005(97)80068-8
49. Luo, J., Feng, X., Kan, H., Li, H., Fu, C.: One-dimensional Bi_2S_3 nanobelts-based surface acoustic wave sensor for NO_2 detection at room temperature. IEEE Sens. J. **21**(2), 1404–1408 (2021). https://doi.org/10.1109/JSEN.2020.3020301
50. Mawrence, R., Munniks, S., Valente, J.: Calibration of electrochemical sensors for nitrogen dioxide gas detection using unmanned aerial vehicles. Sens. **20**(24), 7332 (2020). https://doi.org/10.3390/s20247332
51. Luo, R.-B., Li, H.-B., Du, B., Zhou, S.-S., Chen, Y.-H.: A Printed and flexible NO_2 sensor based on a solid polymer electrolyte. Front. Chem. **7**, 286 (2019). https://doi.org/10.3389/fchem.2019.00286
52. Kubersky, P., Hamácek, A., Nespurek, S., Soukupa, R., Vik, R.: Effect of the geometry of a working electrode on the behavior of a planar amperometric NO_2 sensor based on solid polymer electrolyte. Sens. Actuat. B Chem. **187**, 546–552 (2013). https://doi.org/10.1016/j.snb.2013.03.081
53. Dai, J., Zhao, H., Lin, X., Liu, S., Fei, T., Zhang, T.: Design strategy for ultrafast-response humidity sensors based on gel polymer electrolytes and application for detecting respiration. Sens. Actuat. B Chem. **304**, 127270 (2020). https://doi.org/10.1016/j.snb.2019.127270
54. Zhao, Q., Yang, L., Ma, Y., Huang, H., He, H., Ji, H., Wang, Z., Qiu, J.: Highly sensitive, reliable and flexible pressure sensor based on piezoelectric PVDF hybrid film using MXene

nanosheet reinforcement. J. Alloy. Compd. **886**, 161069 (2021). https://doi.org/10.1016/j.jallcom.2021.161069

55. Kleinman, M.T.: Environmental Toxicants: Human Exposures and Their Health Effects, Carbon Monoxide, 4th edn., pp. 445–486 (2020). https://doi.org/10.1002/9781119438922.ch12
56. Hyodo, T., Takamori, M., Goto, T., Ueda, T., Shimizu, Y.: Potentiometric CO sensors using anion-conducting polymer electrolyte: Effects of the kinds of noble metal-loaded metal oxides as sensing-electrode materials on CO-sensing properties. Sens. Actuat. B Chem. **287**, 42–52 (2019). https://doi.org/10.1016/j.snb.2019.02.036
57. Liu, D.-S., Ryu, H., Khan, U., Wu, C., Jung, J.-H., Wu, J., Wang, Z., Kim, S.-W.: Piezoionic-powered graphene strain sensor based on solid polymer electrolyte. Nano Energy **81**, 105610 (2021). https://doi.org/10.1016/j.nanoen.2020.105610
58. Lu, X., Wu, S., Wang, L., Su, Z.: Solid-state amperometric hydrogen sensor based on polymer electrolyte membrane fuel cell. Sens. Actuat. B Chem. **107**(2), 812–817 (2005). https://doi.org/10.1016/j.snb.2004.12.022
59. Ramesh, C., Velayutham, G., Murugesan, N., Ganesan, V., Dhathathreyan, K.S., Periaswami, G.: An improved polymer electrolyte-based amperometric hydrogen sensor. J. Solid State Electrochem.Electrochem. **7**, 511–516 (2003). https://doi.org/10.1007/s10008-002-0293-4
60. Tanaka, M., Leveque, J.L., Tagami, H., Kikuchi, K., Chonan, S.: The "Haptic Finger" anew device for monitoring skin condition. Skin Res. Technol. **9**, 131–136 (2003). https://doi.org/10.1034/j.1600-0846.2003.00031.x

Chapter 13
Nano-Reinforced Polymers and Polymer Nanocomposites

Mehdi Sahihi and Fahmi Bedoui

Abstract The increasing trend in the mission towards the development of reinforced polymers lies in exploring their optimal properties, such as stiffness and load. Nano-reinforced materials are the answer to these requirements. The chapter has introduced the basic aspects of these nano-reinforced polymeric compounds, methods of synthesis, properties and potential applications. Furthermore, given a wide variety of options in terms of nanoparticle compositions, sizes, surface treatment, etc. in the ocean of available polymers, it is almost impossible to perform a judicious experimental screening of all probable combinations in a reasonable timeframe. Thus, computational modeling methods, discussed at the end of this chapter, are very important to predict and investigate the properties of such polymer nanocomposites.

Keywords Polymers · Nanocomposites · Sensing · MD Simulation

13.1 Introduction

Pure polymers usually have low mechanical, thermal, electrical and barrier properties and need to be reinforced with fillers to extend their applications. On the other hand, the emergence of practical synthesis and assembly of a diversity of nano-structured systems provides the opportunity to develop optimal designs for specific applications. These opportunities are widely recognized, for exploring and exploiting the unusual physical aspects of nano-structured materials in such technological fields, as energy, electronics, biology, pharma, water and environment. New combinations of nanoscale

M. Sahihi (✉)
Université Clermont Auvergne, CNRS, Clermont Auvergne INP, Institut de Chimie de Clermont-Ferrand, 63000 Clermont-Ferrand, France
e-mail: falsemehdi.sahihi@gmail.com; mehdi.sahihi@uca.fr

F. Bedoui
Roberval Laboratory, Université de Technologie de Compiègne, Alliance Sorbonne Université, 60200 Compiègne, France

D. Mohanta and P. Chakraborty (eds.), *Nanoscale Matter and Principles for Sensing and Labeling Applications*, Advanced Structured Materials 206,
https://doi.org/10.1007/978-981-99-7848-9_13

fillers and polymers can lead to produce polymer nanocomposites with enhanced properties. This enhancement is due to the "nano-effect" which defines as high degree of contact and large interfacial area between nano-fillers and polymers [1, 2]. The interfacial area is related to the geometry, surface chemistry, aspect ratio, and size of nanomaterials. Requirement of less volume-fraction of nano-fillers as compared to that of conventional composites helps the nano reinforced polymers preserve their transparency and low density. For the first time, the term "nanocomposites" was proposed by Theng in 1970 [3].

In general, there are three different types of polymer nanocomposites that differ based on the geometry of the used nano-filler (Fig. 13.1 and Table 13.1):

1. Polymer nanocomposites with one-dimensional nano-fillers: These nano-fillers are plate-like materials with one dimension less than 100 nm.
2. Polymer nanocomposites with two-dimensional nano-fillers: These nano-fillers are rod-like materials with two dimensions less than 100 nm.

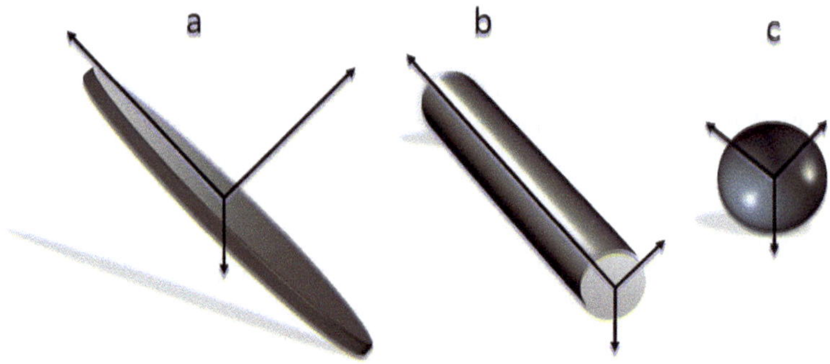

Fig. 13.1 Schematic representation of **a** 1D, **b** 2D and **c** 3D nanofillers

Table 13.1 Examples of different types of nanofillers and their corresponding applications

Nanofiller	Type	Application	Refs.
Nanographene platelets (NGPs)	1D	Nanoelectronics, Nanocomposites, Hydrogen supercapacitors	[4, 5]
ZnO nanosheets	1D	Photonic crystals, Photodetectors, Varistors, Solar cells, Transparent conductive films, Bulk acoustic wave devices	[6, 7]
Nanotubes	2D	Creep resistant polymer nanocomposites	[8]
Nanowires	2D	High conductive nanocomposites	[9]
Graphene	2D	Gas barrier	[10, 11]
Nanometal oxides	3D	Scratch-resistant coatings	[12]
Quantum dots (QDs)	3D	Semiconductors	[13]

3. Polymer nanocomposites with three-dimensional nano-fillers: These nano-fillers are spherical or cube-shaped materials with all three dimensions less than 100 nm.

These isotropic and anisotropic nanofillers can be dispersed uniformly and lead to producing ultra large interfacial area per unit volume between nanofillers and the polymer. Moreover, molecular dimensions of the distance between the nanofillers are approachable at their very low contents. Hence, this is a morphology controlled by nanoscale topographies.

The mechanism of reinforcing not only depends on the type of the filler, but also is related to the amount, distribution and the chemical natures of the polymer and nano-filler. The choice of the nano-filler depends on the purpose of application and the requirement of changing some original properties of the polymer.

13.2 Synthesis of Polymer Nanocomposites

Uniform dispersion of nanofillers (without their aggregation) in polymer matrix is the critical step in the synthesis of nano-reinforced polymers. The extremely small separation between the nanofillers causes very high van der Waals or electrostatic interactions between the nanoparticles and it is very difficult to insert the nanofillers within a polymer matrix keeping them at close distance as well as distinctly separated. The most common methods to provide these conditions are described briefly as follows:

13.2.1 Ultrasonication-Assisted Solution Mixing

In the ultrasonication-assisted method, nanofillers are solved in the solvent and then dispersed in the polymer matrix using an ultrasonic apparatus. Then, the solvent is evaporated to prepare the polymer nanocomposite. Ultrasonic energy is higher than the energy of interaction between the nanoparticles in the aggregates. This method was used to prepare reduced graphene oxide (rGO)-polystyrene nanocomposite [14]. Although this method is very fast, although its high sonication power can damage the one dimension or two dimensions of the nanofillers [15]. Furthermore, the solvent should be completely removed in order to avoid the degradation in the quality of the desired nanocomposites.

13.2.2 Shear Mixing

Unlike the ultrasonication method, shear mixing is a less intensive method to prepare nanocomposites and could be done by a magnetic stir. In this method, rotating the stirrer generates shear force that is used to separate the aggregates of nanofillers. The required sheared energy is completely dependent on the solvent and low viscosity solvents are usually appropriate to be used in this method. The shear mixing is too weak and only could be used to disperse some layered materials [16].

13.2.3 Three Roll Milling

In the three-roller milling method, nanofillers are dispersed by the shearing force produced by a machine composed of three cylindrical rollers with different rotating speeds. High-shear force generated in this method allows to use high content of nanofiller with little or no solvent. This method has been used for different nanofillers like nanoclays [17, 18] and CNTs [19, 20]. Because the distance between the adjacent rollers is more than 1 μm, and this method may not be suitable to prepare polymer nanocomposites with three dimensional nanofillers.

13.2.4 Ball Milling

Different types of nanofillers, like silica nanoparticles [21], graphene nanosheets [22] and CNTs [23] have been dispersed in thermoplastic and thermosetting media using the ball milling method. In this method, hard ball collision is used to produce huge shear force and for grinding and mixing powders in a closed space. Ball milling method is an appropriate way to disperse nanofillers in liquid monomers of thermosets, soluble thermoplastics, solid thermoplastic matrices [24, 25].

13.2.5 Double-Screw Extrusion

The double-screw extrusion is another solvent free method to fabricate thermoplastic based nanocomposites [26]. In this method, the twin screws with high-speed rotations prepare the required forces to disperse different types of nanofillers like nanosilica [27], CNTs [28], and GNP [29]. Table 13.2 summarizes the characteristics of the above methods for the synthesis of polymer nanocomposites.

Table 13.2 Characteristics of the methods for synthesis of polymer nanocomposites

Method	Type of filler	Type of matrix	Solvent
Ultrasonication-assisted solution mixing	1D, 2D, 3D	Liquid or viscous monomers or oligomers	Required
Shear mixing	1D	Liquid or viscous monomers or oligomers	Required
Three roll milling	1D, 2D	Liquid or viscous monomers or oligomers	Not required
Ball milling	1D, 2D, 3D	Liquid or solid thermoplastics	Not required
Double-screw extrusion	1D, 2D, 3D	Solid thermoplastics	Not required

13.3 Characterization of Polymer Nanocomposites

Various techniques can be used to characterize the polymer nanocomposites e.g., wide-angle x-ray diffraction (WAXD), small angle X-ray scattering (SAXS), scanning electron microscopy (SEM), transmission electron microscopy (TEM), thermal gravimetric analysis (TGA), atomic force microscopy (AFM), solid-state nuclear magnetic resonance (NMR) and dynamic mechanical thermal analyzing (DMTA). In nano-reinforced polymers, it is highly desirable to have a clear view on how the nano-reinforcement is dispersed in the matrix, how it interacts with the surrounding materials and how it affects the macroscopic properties of the studied composite. This information is spread-out through the different hierarchical materials length-scales. Therefore, not a single technique could bring answers to all of them. It is rather a combination of multiple complimentary techniques that are used, sometimes separately or in combination. For nonreinforcement morphological characterizations, x-ray diffraction, electron microscopy (scanning and transmission modes), etc. are most widely used techniques. Wide-Angle X-Ray Scattering (WAXS) is an accessible and a very simple method that is frequently used to investigate the structure of polymer nanocomposites [30, 31] and sometimes to study the kinetics of polymer melt [32]. In the case of nano-reinforced polymers, this technique is highly used to investigate simultaneously the structural organization of both the matrix and nanofiller. It allows through the nanofiller structural investigation to have access to their morphological distribution (exfoliation and intercalation of nano-platelets [30]). Small Angle X-ray Scattering (SAXS) technique is used to investigate the materials' microstructure at higher length-scale compared to WAXS. It helps develop a statistical investigation of nanoparticle distributions in nanocomposite materials and investigate such distribution in terms dispersity and/or aggregation nature within the matrix. More rich information will be obtained if SAXS and WAXD are used simultaneously [33], as such direct correlation between each component's structure and the obtained microstructure could be unambiguously derived. In some cases, temperature resolved SAXS and WAXS is used and allows to have a clear view on the interaction between the nanofiller and the surrounding matrix by tracking the structural parameters change as function of temperature. In this case, the effect of the nanofiller on the different polymer transition, crystallization kinetics, and melt rheology could be investigated. All these information requires expert handling of the obtained data in terms of signal integration and phase peaks identification.

For morphological investigation, electron microscopy methods may be more explicit. Depending on nano-filler size, SEM or TEM could be chosen. The interest of nano-reinforcemnt is mostly due to the high surface-to-volume ratio. Such ratio leads to the use of smaller size fillers. Therefore, TEM techniques are the best tools for these materials. TEM allows the understanding of internal structure of the nanocomposites and spatial distribution of different material components. When facing lack of electron contrast, staining materials could be used to enhance TEM imaging. Furthermore, a combination of XRD and TEM characterizes the structures of polymer nanocomposites. XRD is a suitable method to recognize intercalated structures such as layered silicate/polymer nanocomposites. When it comes to the characterization of surface properties of such species, NMR could also be suited as it can track through the different chemical bonds, the nature of the surface chemistry of such species. NMR technique is often used to obtain better understanding about surface chemistry, phase structure and morphology of the polymer nanocomposites [33, 34].

At larger scales and when investigating materials macroscopic behavior is required, besides other experimental techniques. Mechanical and thermal properties are of high interest. For mechanical behavior local and global information may be required. Standard mechanical testing (tensile, torsion, etc.) could be used for static investigation. Dynamic behavior analysis could be investigated through dynamic mechanical and thermal analysis techniques. In this case along the mechanical response, the correlation of the mechanical behavior with the thermal material's properties could be investigated. At a local scale Atomic Force Microscopy (AFM) could be also used. AFM could be used also for microstructural investigation. Thermal behavior characterization and investigation of the thermal stability of polymer nanocomposites can be done using the TGA method [35]. Distribution of nanoparticles in polymers is also investigated using the AFM method. The resonant frequencies of the AFM-tip for polymer crystal and nanofillers are different and the tapping frequency is increased during the measurement to distinguish nanofillers from the polymer matrix [36]. X-ray Photoelectron Spectroscopy (XPS) is also another efficient method to characterize the nanocomposites. In this method, the material is irradiated with a beam of energetic monochromatic x-rays and the kinetic energy and the intensity of photo-electrons that essentially escape from the top 1 to 10 nm of the material are measured simultaneously. This quantitative spectroscopic procedure investigates the elemental composition, chemical state, and electronic state of the elements. Moreover, the structural changes of the nanofillers due to their chemical interactions can be also considered to quantify nano-reinforcement distribution within the polymeric matrices. From all of the above-mentioned methods we cannot recognize the retaining of nanofiller cluster core in the composite. Therefore, extended X-ray absorption fine structure (EXAFS) analysis as a powerful technique is used to investigate the local atomic environment of atoms independent of the state of the sample.

13.4 Properties of Polymer Nanocomposites

Various properties of the polymers (like mechanical, thermal, gas barrier, ionic conductivity and magnetic properties, etc.) can be enhanced by incorporating nanoscale fillers. Some of the most important properties of the polymers that are improved by nano-reinforcement have been briefly discussed.

13.4.1 Mechanical Properties

It turns from the abundant data in the literature that nano-reinforcement offers interesting mechanical properties making polymeric materials a good candidate for many load-bearing application where virgin polymers were automatically excluded. Unlike the microscale particulate composites, the mechanical properties of nanocomposites depend strongly on the dispersion state of the nanofillers, apart from the shape, and size of the filler particles and matrix material [37, 38]. Young's modulus (or stiffness), yield stress, ultimate stress and strain, fracture toughness, and shock resistance can be mentioned as mechanical properties [39, 40]. The nanosized dimensions of the fillers and the strong polymer–filler interactions are the main reasons for improvement in the mechanical performances of nanocomposites. Many reports of this subject have been provided [41–44], for example by the incorporation of constant silica volume fraction into poly(methyl methacrylate) (PMMA) [41]. It is concluded that by reducing the particle size and keeping constant silica volume fraction, particle-polymer interaction is enhanced. Such interactions induce mobility restriction of the polymeric layers around the particles. The reduced mobility could be interpreted as stiff layers around the polymer playing the role of optimal transmitter of stress between the soft and stiff domains. As a consequence of this mechanism, for a volume fraction of 4% of silica, the gain in modulus of the nanocomposite is almost doubled by decreasing the diameter from 500 (18%) to 15 nm (34%) (Fig. 13.2). Furthermore, the surface modification changes the particle dispersion in the matrix and affects the mechanical performances of the final nanocomposite [41].

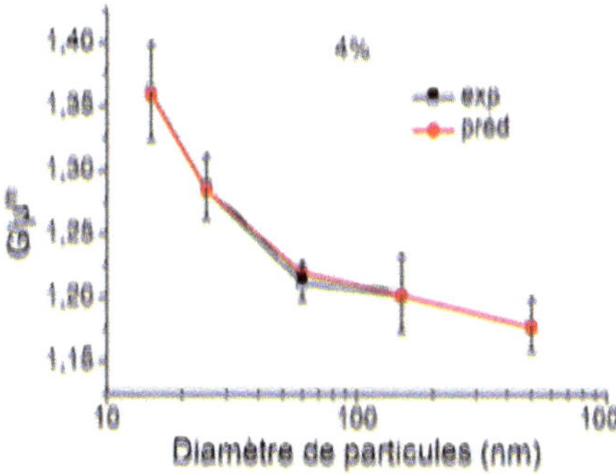

Fig. 13.2 Size effect in silica nanoparticle reinforced PMMA at constant volume fraction of 4%

13.4.2 Rheological Properties

Addition of nanofillers can improve the viscosimetric, viscoelastic and rheological properties (such as high specific strength, stiffness as well as plasticity) of the polymer and its nanocomposites [45, 46]. It has been shown that storage modulus (G′) and loss modulus (G″) raise with varying formulations of the carbon nanotube or TiO_2 incorporated in the polymer matrix [47, 48]. Moreover, the rheological performance of the nanocomposites depends on the content of the filler, filler-to-matrix bonding and filler configuration. In general, increased storage modulus and loss modulus for different nanocomposites can be explained by the interaction between the polymer and nanofillers that block the movement of the polymer chains.

13.4.3 Thermal Stability

During the thermal stabilization of a polymer, it is very important to understand its thermal decomposition process appropriately. This understanding helps us develop a rational technology for polymer processing, produce fire-safe polymeric supplies, certify polymers usage, improve new technologies for effective energy managing, and conduct the thermal reprocessing of discarded plastics. It was shown that the introduction of a nanofiller into a polymer matrix will result in the useful improvement of thermal stability [49]. The thermal stabilization effect of the nanofillers depends on its type and unique characteristics. For example, for the CNTs it could be attributed to its more interfacial interactions with the polymer which in turn would lead to an increase in the activation energy of decomposition [50]. This observation can be attributed to a barrier effect of nanofillers implying a hindered transport of polymer chains.

13.4.4 Magnetic and Electric Properties

The polymer nanocomposites display new magnetic and electric properties due to the combined action of quantum-sized effects, strain and surface effects, interface between nanofillers and the polymer matrix, polymer matrix properties and morphologies of nanofillers. The combination of magnetic inorganic components and polymers will produce interesting properties and can be used for a number of applications. Some of the polymers such as poly(vinylidene fluoride) (PVDF) have inherent piezoelectric properties that can be improved by magnetic nanofillers like magnetite [51, 52]. PVDF has three main crystalline phases, i.e., α, β, and γ. The α-phase is nonpolar with trans-gauche conformation ($TG^{+}TG^{-}$) and the β-phase is polar with an all-trans molecular conformation (TTTT). The β phase of PVDF shows inherent piezo-, pyro-, and ferroelectric properties. The γ-phase (TTTGTTTG′

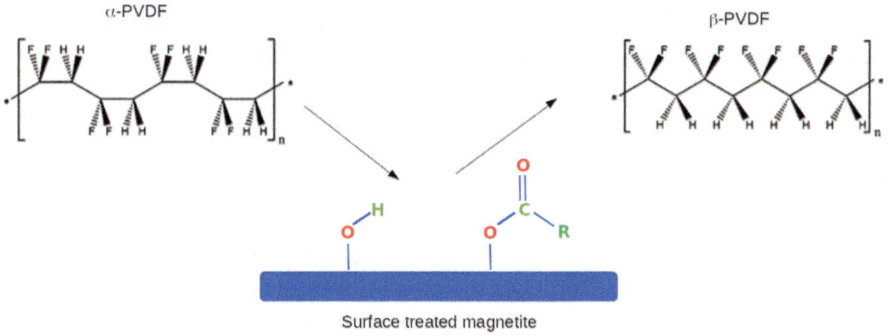

Fig. 13.3 $\alpha \rightarrow \beta$ phase transition of the magnetite reinforced PVDF system

conformation) is a second polar version of PVDF with weaker piezoelectric effect than β-phase. Normally, PVDF exists as a nonpolar α-phase that could be changed to the β-phase by thermal, mechanical, or electric treatments [52]. Previously investigations have shown that the β-phase will be formed by the addition of organically modified silicates [53], Lucentite STN nanoclay [54], Ni–Zn ferrite nanoparticles [55], Ni nanoparticles [56] or magnetic oxides such as $La_{0.7}Sr_{0.3}MnO_3$ and Fe_3O_4 [51, 52, 57]. These effects are usually attributed to interaction between the electron density of the nanofiller and the positive charged density of CH_2 groups of PVDF (Fig. 13.3).

13.5 Simulation of Polymer Nanocomposites

Recent experimental works on nanocomposites, made at constant volume-fraction of combination of the nanofillers and polymers, demonstrate the dependency of a material's macroscopic response on the size and surface chemistry of the nanofillers. Given an exceedingly large variety of options in terms of nanoparticle compositions, sizes, surface-treatment, etc. for all available polymers, it is almost impossible to perform experimental screening of reasonable combinations in a realistic timeframe. Thus, it is an urgent need to control these developments using rational and optimal design. Therefore, it would be a reasonable choice to perform an a priori screening using computer simulation of nanofillers within a variety of polymer matrices to establish interfacial properties, property-dependencies on nanofiller size and system composition, thereby leading to improved mechanical and electrical responses. Indeed, fast-screening and cost-effective development methodology suited for larger systems covering larger combinations is desired.

Computer simulation can be classified into five major categories by considering the used fundamental theories for calculation of the properties and structure of polymer nanocomposites: Quantum mechanical simulation, Monte Carlo

(MC) simulation, Molecular dynamics (MD) simulation, Mesoscopic simulation, and Continuum simulation (or macroscopic scale simulation).

13.5.1 Quantum Mechanical Simulation

The quantum mechanical simulation is based on the first-principle 'density functional theory' (DFT). The DFT determines the properties of many-electron system using functionals that are dependent on the density of electrons and electron correlation. The density functional theory uses the electron density which is a function of the position coordinate (r) as an input and produced energy as the output.

Generally, the DFT method is computationally expensive and only used to simulate the systems with around 100 atoms.

13.5.2 MC Simulation

The fundamentals of the Monte Carlo (MC) simulation are based on relying on repeated random sampling and statistical analysis to compute the results. In MC calculations, it is not needed to calculate the force between the particles. A typical Monte Carlo simulation includes one or more input variables, one or more output variables and a mathematical model between inputs and outputs. Different types of MC simulations have been used to model polymer nanocomposites e.g., classical MC based on random walking [58], metropolis MC [59], coarse-grained MC [60], kinetic MC [61] and off-lattice MC simulation [62].

13.5.3 MD Simulation

Molecular dynamics (MD) simulations offer the capability to forecast properties by commencing with an atomic-scale portrayal of the system under investigation. In essence, this approach relies on the premise that, within typical conditions (excluding extremes in temperature or pressure), the movements of atoms within a molecular system or material can be replicated through numerical solutions of Newton's laws of motion within a computer environment [63]. Executing this method necessitates the development of an atomistic model detailing the system's atomic arrangement, along with a technique to compute all the interactions occurring among the constituent atoms. These interactions can be theoretically approached at various levels, spanning from quantum–mechanical electronic structure methods to analytical force fields, which may either derive from semi-empirical origins or be calibrated based on electronic structure computations [63].

Within MD simulations, two categories of interactions are distinguished:

(i) Bonded interactions encompass the interplay between atoms within a nanofiller or polymer and are described through diverse potential terms that account for changes in bond length, bond angles, and dihedrals.
(ii) Non-bonded interactions, on the other hand, are characterized by van der Waals and Coulomb potentials.

Lately, the MD simulation technique has found utility in exploring the physical characteristics and structural alterations of α-PVDF (alpha phase polyvinylidene fluoride) when coupled with magnetite nanoparticles possessing varying surface chemistries [52]. Experimental evidence has validated that the incorporation of magnetite nanoparticles enhances the contribution of the β phase, which boasts the highest piezoelectric coefficient among all PVDF crystalline forms. Nevertheless, a comprehensive report detailing the precise mechanisms and the profound impact of Fe_3O_4 nanoparticles on the ferroelectric properties and structural transitions (phase changes) of PVDF has been conspicuously absent.

Examination of MD simulation trajectories has illuminated that hydrogen bond interactions occurring between the hydroxyl (OH) groups on the surface of magnetite and PVDF polymer chains serve to stabilize the system and facilitate the transformation from the α phase to the β phase (as depicted in Fig. 13.3). Furthermore, the replacement of 20% of the hydroxyl groups with carboxylic acids has been observed to curtail the extent of this phase transition.

In a separate investigation, MD simulations were employed to scrutinize the impact of silica nanoparticle dimensions on the "macroscopic" mechanical behavior and interfacial interactions within model nano-reinforced polymers [64]. The findings from this study indicated that smaller nanoparticles prompt the formation of organized polymeric layers at the interface.

13.5.4 Mesoscopic Simulation

The mesoscopic simulation combines microscopic and macroscopic models. This method is fast but has a lower accuracy than experimental or quantum mechanical simulations. Its timescale is in the range of microseconds to seconds and its length scale is in the range of micrometer to millimeter.

13.5.5 Continuum Simulation

This method uses continuity equations to investigate the macroscopic solid mechanics. The continuity equations are usually too complicated and are implied in expensive software. However, this simulation method can be used for very large systems in very long-time scales. Unlike composite materials, nano-reinforced polymers present a size dependent behavior. Therefore, standard micromechanics fails to predict their behavior. To handle the size effects into the micromechanical modeling, interfacial interaction is taken on account through different methodology. The first consists on adding an interphase between the nanoparticles and the matrix. This solution is costly as it requires full identification of the interphase properties (thickness and mechanical properties). The other rout that is still being investigated is to consider the interaction as a coherent interfacial interaction. This later case is technically easier as it requires the he introduction of only two surfaces linear elastic parameters as introduced through the equilibrium equation of Young–Laplace [65, 66]. Without full understanding of the physics behind the identified surface elastic parameters, this approach could not help predict nano-reinforced behavior.

13.6 Some of Applications of Polymer Nanocomposites

13.6.1 Application of Biodegradable Polymer Nanocomposites in Electronics Industry

Traditional petroleum-derived polymers like polypropylene (PP), polystyrene (PS), poly(ethylene terephthalate) (PET), polyimide (PI), and epoxy that have been used in electronic devices cause serious environmental issues due to their long time of degradation in the natural environment. Therefore, using the biodegradable polymers is one of the effective solutions to overcome this issue [67, 68]. Furthermore, most of the biodegradable polymers are appropriate for the implantable chips, electronic skin, biomedical diagnosis, and wearable devices. Thus, biodegradable polymer nanocomposites will have all the advantages of nanocomposites and biodegradable polymers to develop the next generation of green electronics.

Biodegradable polymers are generally classified into natural and synthetic polymers. Protein, cellulose, chitosan and starch are some of the most common natural biodegradable polymers. Also, poly(ε-caprolactone) (PCL), poly(butylene succinate) (PBS), polylactide (PLA) and polyhydroxyalkanoates are some of synthetic biodegradable polymers. Figure 13.4 represents the chemical structure of some of the most famous biodegradable polymers of both classes of biodegradable polymers.

PLA is a tough transparent polymer with a tensile strength of 50MPa, Young's modulus of 3GPa, and elongation at break of 4–5% [69, 70]. This polymer has been used as substrate and/or dielectric layer for the assembly of memristors and transistors [71–74]. On the other hand, carbon-based materials, such as CNTs, graphene and CQDs, have large aspect ratio, exceptional conductivity and high thermal stability that make them an appropriate choice for electronic devices. Therefore, PLA nanocomposites with carbon based nanofillers have been used in sensors, biosensors, harvesters and electrodes [75–88].

PCL is hydrophobic with a high degree of crystallinity and has been proven by the United States Food and Drug Administration (USFDA) to be exploitable for the tissue engineering applications. Moreover, it shows good chemical resistance,

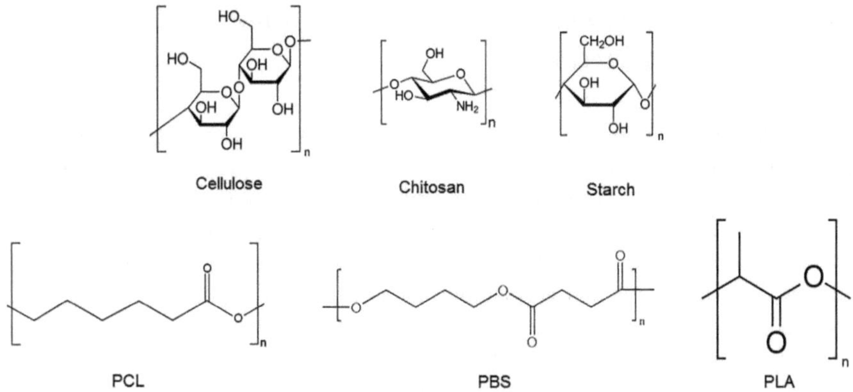

Fig. 13.4 The chemical structure of some of the most famous natural and synthetic biodegradable polymers

high ductility, low melting point and low viscosity. However, PCL's low Young's modulus and strength limit its application in electronic devices. Therefore, PCL reinforcement with nanofillers is an option to enhance its properties. PCL/CNTs nanocomposites was developed to quantitative identify different organic vapors in different concentrations, prepare conductive multifilament fibers and to use as sensing phase to monitor the temperature [89–91]. Also, reinforced PCL with graphene and silver nanowires have been used to prepare the self-healing strain sensor [92].

Cellulose is the most abundant of natural biodegradable polymer and can be used for different applications such as transistors, actuators, sensors and memories by different types of nanofillers (CNTs, Fe_2O_3, ionic liquids, graphene oxide, etc.) [93–105].

Chitosan is a unique cationic polysaccharide with amine and hydroxyl groups in its molecular chains that cause strong inter- and intra-molecular hydrogen bonding and a rigid crystalline structure [106].

Despite various advantages and applications of biodegradable polymer nanocomposites, there are still many challenges to overcome to incorporate them into electronic devices. For example, the biggest issue is the duration of these biodegradable electronic devices and change of electric properties of these nanocomposites in the presence of moisture. The second issue is their low thermal stability that makes these biodegradable polymer nanocomposites vulnerable during the preparation process of electronic devices.

However, despite these abovementioned concerns, biodegradable polymer nanocomposites are receiving more and more attention from researchers due to the urgent need of sustainable electric and electronic industry.

13.6.2 Application of Polymer Nanocomposites in Photodetectors

A photodetector (PD) mostly converts a light signal to an electrical signal for different applications such as medical imaging, light controlled switcher, sensors for physical health, information and communication, and display technology [107]. The core part of a PD is an active layer of materials which can fulfill the functions efficiently and swiftly. Currently, the commercial PDs mainly use crystalline inorganic semiconductors, such as silicon and germanium. Due to the new application scenarios for optoelectronic devices, like electronic skins, virtual reality and wearable devices, new device characteristics including low cost, flexibility, integrated circuit (IC) integrability and transparency should be considered in the development of PDs and novel materials with proper properties should be presented in the PDs. In this regard, the polymer nanocomposites are of particular interest for their unique properties. Indeed, these compounds found vast application potential in optoelectronic devices like solar cells and PDs [108, 109]. For example, the polymer nanocomposite of poly(3-hexylthiophene-2,5-diyl) (P3HT) and [6]-phenyl C_{61} butyric acid methyl ester (PCBM) is used in organic solar cells [109], and PDs [110]. The polymer nanocomposite of P3HT and lead sulfide (PbS) inorganic nanocrystals produce high-performing photoconductors [111].

13.6.3 Application of Polymer Nanocomposites in Pressure Sensors

Pressure sensor can convert mechanical deformation into electrical signals and have extensive application in touch screen, memories, wearable electronics, and robots [112–115]. Usual microelectromechanical system-based pressure sensors are normally based on a rigid silicon substrate and have limited applications for flexible electronic skin. On the contrary, pressure sensors based on polymers or polymer nanocomposites are highly flexible and stretchable and have a wonderful potential for future wearable and robotic applications [116, 117]. For example, pressure sensor based on the silver nanowires (AgNWs)-polydimethylsiloxane (PDMS) electrode shows high pressure sensitivity [118]. A highly sensitive pressure sensor based on polyurethane (PU) and sea-urchin shaped metal nanoparticles (SSNPs) has been resented and shows excellent transparency [119]. Recently, a bio-inspired graphite/PDMS polymer nanocomposite has been fabricated for piezoresistive pressure sensors [120]. Finally, polymer nanocomposite-based pressure sensors have a high potential to be expanded to different types of wearable applications like wearable pressure pulse-sensing devices, e-skin, and phonation rehabilitation.

13.6.4 *Application of Polymer Nanocomposites in Energy Storage Devices*

Nowadays, the development of renewable energy storage resources to store and output energy in a stable environment is in urgent demand because of increasing energy crisis of traditional fossil fuel consumption. Recently, a lot of efforts have been made to the research of new materials for batteries and supercapacitors as the main energy storage systems. Polymer nanocomposites are one category of the new materials that are being used in electrodes, electrolytes and separators, which can improve the electrochemical performance and physical property of energy storage devices. For example, Wu et al. successfully fabricated an electrode with single-wall CNT film as current collector and polyimide as active materials [121]. Also, Li et al. designed a type of solid electrolyte consisting 1-vinyl-3-methylimidazolium bis(trifluoromethylsulfonyl)imide [[VMIM]TFSI] monomer, polyethylene oxide, lithium bis(trifluoromethanesulfonyl)imide, azodiisobutyronitrile, and poly(ethyleneglycol) diacrylate [122]. Moreover, PVA-ceramic composites are promising materials that have been introduced for separators of supercapacitors [123].

13.7 Summary and Future Outlook

The chapter has discussed the basic aspects, properties, synthesis and characterization methods, applications and computational modelling methods of polymer nanocomposites. As mentioned, properties of nano-reinforced polymers not only depend on the interfacial interaction between the nanofiller and the polymeric matrix, but also are related to the size, alignment, profile and dispersion of nanofillers. Although polymer nanocomposites have a unique place in modern technology, these are still in the early phase of development. With more focused research in this field, this new generation of nanocomposite materials will be useful in various fields. For example, nanofillers with biodegradable nature and less toxicity (like nanocellulose) can be used as an alternative for carbon-based nanofillers in implant or tissue engineering scaffolds.

References

1. Crosby, A.J., Lee, J.: Polymer nanocomposites: the “Nano” effect on mechanical properties. Polym. Rev.. Rev. **47**, 217–229 (2007)
2. Fu, S., Sun, Z., Huang, P., et al.: Some basic aspects of polymer nanocomposites: a critical review. Nano Mater. Sci. **1**, 2–30 (2019)
3. Theng, B.K.G.: Interactions of clay minerals with organic polymers. Some practical applications. Clays Clay Miner **18**, 357–362 (1970)

4. Shen, J., Hu, Y., Li, C., et al.: Synthesis of amphiphilic graphene nanoplatelets. Small **5**, 82–85 (2009)
5. Li, B., Zhong, W.H.: Review on polymer/graphite nanoplatelet nanocomposites. J. Mater. Sci. **46**, 5595–5614 (2011)
6. Bai, W., Zhu, X., Zhu, Z., et al.: Synthesis of zinc oxide nanosheet thin films and their improved field emission and photoluminescence properties by annealing processing. Appl. Surf. Sci. **254**, 6483–6488 (2008)
7. Mani, G.K., Rayappan, J.B.B.: A simple and template free synthesis of branched ZnO nanoarchitectures for sensor applications. RSC Adv. **4**, 64075–64084 (2014)
8. Yang, J., Zhang, Z., Friedrich, K., et al.: Creep resistant polymer nanocomposites reinforced with multiwalled carbon nanotubes. Macromol. Rapid Commun.. Rapid Commun. **28**, 955–961 (2007)
9. Lonjon, A., Laffont, L., Demont, P., et al.: Structural and electrical properties of gold nanowires/P(VDF-TrFE) nanocomposites. J. Phys. D **43**, 345401 (2010)
10. Wang, X., Song, M.: Toughening of polymers by graphene. Nanomater. Energy **2**, 265–278 (2013)
11. Paszkiewicz, S., Pawelec, I., Szymczyk, A., et al.: Thermoplastic elastomers containing 2D nanofillers: montmorillonite, graphene nanoplatelets and oxidized graphene platelets. Polish J. Chem. Technol. **17**, 74–81 (2015)
12. Vengatesan, M.R., Mittal, V. (eds.): Nanoparticle- and Nanofiber-Based Polymer Nanocomposites: An Overview. Wiley-VCH, Weinheim (2016)
13. Kumar, A.P., Depan, D., Singh Tomer, N., et al.: Nanoscale particles for polymer degradation and stabilization-tends and future perspectives. Prog. Polym. Sci.Polym. Sci. **34**, 479–515 (2009)
14. Stankovich, S., Dikin, D.A., Dommett, G.H.B., et al.: Graphene-based composite materials. Nature **442**, 282–286 (2006)
15. Lu, K.L., Lago, R.M., Chen, Y.K., et al.: Mechanical damage of carbon nanotubes by ultrasound. Carbon **34**, 814–816 (1996)
16. Backes, C., Higgins, T.M., Kelly, A., et al.: Guidelines for exfoliation, characterization and processing of layered materials produced by liquid exfoliation. Chem. Mater. **29**, 243–255 (2017)
17. Dalir, H., Farahani, R.D., Nhim, V., et al.: Preparation of highly exfoliated polyester–clay nanocomposites: process–property correlations. Langmuir **28**, 791–803 (2012)
18. Park, J.J., Lee, J.Y.: Effect of nano-sized layered silicate on AC electrical treeing behavior of epoxy/layered silicate nanocomposite in needle-plate electrodes. Mater. Chem. Phys. **141**, 776–780 (2013)
19. Gojny, F.H., Wichmann, M.H.G., Köpke, U., et al.: Carbon nanotube-reinforced epoxy-composites: enhanced stiffness and fracture toughness at low nanotube content. Compos. Sci. Technol. **64**, 2363–2371 (2004)
20. Thostenson, E.T., Chou, T.W.: Processing-structure-multi-functional property relationship in carbon nanotube/epoxy composites. Carbon **44**, 3022–3029 (2006)
21. Castrillo, P.D., Olmos, D., Amador, D.R., et al.: Real dispersion of isolated fumed silica nanoparticles in highly filled PMMA prepared by high energy ball milling. J. Colloid Interface Sci. **308**, 318–324 (2007)
22. Gu, J., Li, N., Tian, L., et al.: High thermal conductivity graphite nanoplatelet/UHMWPE nanocomposites. RSC Adv. **5**, 36334–36339 (2015)
23. Gupta, T.K., Singh, B.P., Mathur, R.B., et al.: Multi-walled carbon nanotube–graphene–polyaniline multiphase nanocomposite with superior electromagnetic shielding effectiveness. Nanoscale **6**, 842–851 (2014)
24. Gu, J., Guo, Y., Lv, Z., et al.: Highly thermally conductive POSS-g-SiCp/UHMWPE composites with excellent dielectric properties and thermal stabilities. Compos. Part A Appl. Sci. Manuf. **78**, 95–101 (2015)
25. Gu, J., Xie, C., Li, H., et al.: Thermal percolation behavior of graphene nanoplatelets/polyphenylene sulfide thermal conductivity composites. Polym. Compos.. Compos. **35**, 1087–1092 (2014)

26. Fawaz, J., Mittal, V.: Synthesis Techniques for Polymer Nanocomposites, p. 130. Wiley-VCH, Weinheim (2014)
27. Zou, H., Wu, S., Shen, J.: Polymer/silica nanocomposites: preparation, characterization, properties, and applications. Chem. Rev. **108**, 3893–3957 (2008)
28. Villmow, T., Pötschke, P., Pegel, S., et al.: Influence of twin-screw extrusion conditions on the dispersion of multi-walled carbon nanotubes in a poly(lactic acid) matrix. Polymer **49**, 3500–3509 (2008)
29. Kim, I.H., Jeong, Y.G.: Polylactide/exfoliated graphite nanocomposites with enhanced thermal stability, mechanical modulus, and electrical conductivity. J. Polym. Sci. Part B Polym. Phys. **48**, 850–858 (2010)
30. Giannelis, E.P., Krishnamoorti, R., Manias, E., et al.: Polymer-silicate nanocomposites: model systems for confined polymers and polymer brushes. In: Granick, S., Binder, K., de Gennes, P.G., et al. (eds.) Polymers in Confined Environments, pp. 107–147. Springer, Berlin, Heidelberg (1999)
31. Giannelis, E.P.: Polymer layered silicate nanocomposites. Adv. Mater. **8**, 1–9 (1996)
32. Vaia, R.A., Jant, K.D., Kramer, E.J.: Microstructural evaluation of melt-intercalated polymer-organically modified layered silicate nanocomposites. Chem. Mater. **8**, 2628–2635 (1996)
33. Mathias, L.J., Davis, R.D., Jarrett, W.L.: Observation of alpha and gamma crystal forms and amorphous regions of nylon 6 clay nanocomposites using solid-state15N nuclear magnetic resonance. Macromolecules **32**, 7958–7960 (2012)
34. Vanderhart, D.L., Asano, A., Gilman, J.W.: NMR measurements related to clay dispersion quality and organic-modifier stability in nylon-6/clay nanocomposites. Macromolecules **34**, 3819–3822 (2001)
35. Higgins, B.A., Brittain, W.J.: Polycarbonate carbon nanofiber composites. Eur. Polym. J.Polym. J. **41**, 889–893 (2005)
36. Yu, J., Liu, R.Y.F., Poon, B., et al.: Polymers with palladium nanoparticles as active membrane materials. J. Appl. Polym. Sci.Polym. Sci. **92**, 749–756 (2004)
37. Long, Y., Shanks, R.A.: PP–elastomer–filler hybrids. I. Processing, microstructure, and mechanical properties. J. Appl. Polym. Sci. **61**(11), 1877–1885 (1996)
38. Misra, R.D.K., Nerikar, P., Bertrand, K., et al.: Some aspects of surface deformation and fracture of 5–20% calcium carbonate-reinforced polyethylene composites. Mater. Sci. Eng. A **384**(1–2), 284–298 (2004)
39. Tjong, S.C.: Structural and mechanical properties of polymer nanocomposites. Mater. Sci. Eng. **53**, 73–197 (2006)
40. Alexandre, M., Dubois, P.: Polymer-layered silicate nanocomposites: preparation, properties and uses of a new class of materials. Mater. Sci. Eng. **28**(1–2), 1–63 (2000)
41. Blivi, A.S., Bedoui, F., Weigand, S., et al.: Multiscale analysis of nanoparticles size effects on thermal, elastic, and viscoelastic properties of nano-reinforced polymers. Polym. Eng. Sci.. Eng. Sci. **60**, 1773–1784 (2020)
42. Gelineau, P., Weigand, S., Cauvin, L., et al.: Compatibility effects of modified montmorillonite on elastic and visco-elastic properties of nano-reinforced Poly(lactic acid): experimental and modeling study. Polym. Test.. Test. **70**, 441–448 (2018)
43. Blivi, A.S., Benhui, F., Bai, J., et al.: Experimental evidence of size effect in nano-reinforced polymers: case of silica reinforced PMMA. Polym. Test.. Test. **56**, 337–343 (2016)
44. Gelineau, P., Stepień, M., Weigand, S., et al.: Elastic properties prediction of nano-clay reinforced polymer using multi-scale modeling based on a multi-scale characterization. Mech. Mater. **89**, 12–22 (2015)
45. Hsissou, R., El Bouchti, M., El Harfi, A.: Elaboration and viscosimetric, viscoelastic and rheological studies of a new hexafunctional polyepoxide polymer: hexaglycidyl ethylene of methylene dianiline. J. Mater. Environ. Sci. **8**(12), 4349–4361 (2017)
46. Louisy, E., Bellayer, S., Fontaine, G., et al.: Novel hybrid poly(l-lactic acid) from titanium oxo-cluster via reactive extrusion polymerization. Eur. Polym. J.Polym. J. **122**, 109238 (2020)
47. Grich, M., El Gouri, M., Ziraoui, R., et al.: Thermal and rheological study of blended carbon nanotube/epoxy resin nanocomposites. J. Mater. Environ. Sci. **5**(2), 374–379 (2014)

48. Hsissou, R., Dagdag, O., Berradi, M., et al.: Development rheological and anti-corrosion property of epoxy polymer and its composite. Heliyon **5**, e02789 (2019)
49. Pavlidou, S., Papaspyrides, C.D.: A review on polymer-layered silicate nanocomposites. Prog. Polym. Sci.Polym. Sci. **33**, 1119–1198 (2008)
50. Marosfői, B.B., Szabó, A., Marosi, G., et al.: Thermal and spectroscopic characterization of polypropylene-carbon nanotube composites. J. Therm. Anal. Calorim.Calorim. **86**, 669–673 (2006)
51. Jayakumar, O.D., Mandal, B.P., Majeed, J., et al.: Inorganic–organic multiferroic hybrid films of Fe_3O_4 and PVDF with significant magneto-dielectric coupling. J. Mater. Chem. C **1**, 3710–3715 (2013)
52. Sahihi, M., Jaramillo-Botero, A., Goddard, W.A., III., et al.: Interfacial interactions in a model composite material: insights into $\alpha \rightarrow \beta$ phase transition of the magnetite reinforced Poly(Vinylidene Fluoride) systems by all-atom molecular dynamics simulation. J. Phys. Chem. C **125**(39), 21635–21644 (2021)
53. Priya, L., Jog, J.P.: Poly (vinylidene fluoride)/clay nanocomposites prepared by melt intercalation: crystallization and dynamic mechanical behavior studies. J. Polym. Sci. Part B Polym. Phys. **40**, 1682–1689 (2002)
54. Buckley, J., Cebe, P., Cherdack, D., et al.: Nanocomposites of poly (vinylidene fluoride) with organically modified silicate. Polymer **47**, 2411–2422 (2006)
55. Andrew, J.S., Clarke, D.R.: Enhanced ferroelectric phase content of polyvinylidene difluoride fibers with the addition of magnetic nanoparticles. Langmuir **24**, 8435–8438 (2008)
56. Mandal, B.P., Vasundhara, K., Abdelhamid, E., et al.: Improvement of magnetodielectric coupling by surface functionalization of nickel nanoparticles in Ni and polyvinylidene fluoride nanohybrids. J. Phys. Chem. C **118**, 20819–20825 (2014)
57. Thirmal, C., Nayek, C., Murugavel, P., et al.: Magnetic, dielectric and magnetodielectric properties of PVDF-La0. 7Sr0. $3MnO_3$ polymer nanocomposite film. AIP Adv. **3**, 112109 (2013)
58. Termonia, Y.: Polymer chain properties and thermodynamic stability in oriented-platelet nanocomposites. Phys. Rev. E Stat. Nonlin. Soft Matter Phys.Nonlin. Soft Matter Phys. **88**, 012603 (2013)
59. Nam, S., Cho, H.W., Lim, S., et al.: Enhancement of electrical and thermomechanical properties of silver nanowire composites by the introduction of nonconductive nanoparticles: experiment and simulation. ACS Nano **7**, 851–856 (2013)
60. Goikoetxea, M., Reyes, Y., de las Heras Alarcón, C.M., et al.: Transformation of waterborne hybrid polymer particles into films: morphology development and modeling. Polymer **53**, 1098–1108 (2012)
61. Rosenthal, L., Greve, H., Zaporojtchenko, V., et al.: Formation of magnetic nanocolumns during vapor phase deposition of a metal-polymer nanocomposite: experiments and kinetic Monte Carlo simulations. J. Appl. Phys. **114**, 044305 (2013)
62. Duong, H.M., Yamamoto, N., Bui, K., et al.: Morphology effects on nonisotropic thermal conduction of aligned single-walled and multi-walled carbon nanotubes in polymer nanocomposites. J. Phys. Chem. C **114**, 8851–8860 (2010)
63. Frenkel, D., Berend, S.: Understanding Molecular Simulation: From Algorithms to Applications, vol. 1, pp. 1–638. Elsevier (formerly published by Academic Press) (2002)
64. Bedoui, F., Jaramillo-Botero, A., Pascal, T.A., et al.: Focus on the deformation mechanism at the interfacial layer in nano-reinforced polymers: a molecular dynamics study of silica—poly(methyl methacrylate) nano-composite. Mech. Mater. **159**, 103903 (2021)
65. Gurtin, M.E., Murdoch, A.I.: A continuum theory of elastic material surfaces. Arch. Rat. Mech. Anal. **57**, 291–323 (1975)
66. Povstenko, Y.Z.: Theoretical investigation of phenomena caused by heterogeneous surface tension in solids. J. Mech. Phys. Sol. **41**, 1499–1514 (1993)
67. Irimia-Vladu, M.: Green electronics: biodegradable and biocompatible materials and devices for sustainable future. Chem. Soc. Rev. **43**, 588–610 (2014)

68. Irimia-Vladu, M., Glowacki, E.D., Voss, G., et al.: Green and biodegradable electronics. Mater. Today **15**, 340–346 (2012)
69. Raquez, J.M., Habibi, Y., Murariu, M., et al.: Polylactide (PLA)-based nanocomposites. Prog. Polym. Sci.Polym. Sci. **38**, 1504–1542 (2013)
70. Hamad, K., Kaseem, M., Ayyoob, M., et al.: Polylactic acid blends: the future of green, light and tough. Prog. Polym. Sci.Polym. Sci. **85**, 83–127 (2018)
71. Bettinger, C.J., Bao, Z.N.: Organic thin-film transistors fabricated on resorbable biomaterial substrates. Adv. Mater. **22**, 651–655 (2010)
72. Wu, X.H., Ma, Y., Zhang, G.Q., et al.: Thermally stable, biocompatible, and flexible organic field-effect transistors and their application in temperature sensing arrays for artificial skin. Adv. Funct. Mater.Funct. Mater. **25**, 2138–2146 (2015)
73. Wu, W., Han, S.T., Venkatesh, S., et al.: Biodegradable skin-inspired nonvolatile resistive switching memory based on gold nanoparticles embedded alkali lignin. Org. Electron. **59**, 382–388 (2018)
74. Fang, S.L., Liu, W.H., Li, X., et al.: Biodegradable transient resistive random-access memory based on MoO_3/MgO/MoO_3 stack. Appl. Phys. Lett. **115**, 244102 (2019)
75. Devaux, E., Aubry, C., Campagne, C., et al.: PLA/carbon nanotubes multifilament yarns for relative humidity textile sensor. J. Eng. Fibers Fabr. **6**, 13–24 (2011)
76. Skaria, E., Patel, B.A., Flint, M.S., et al.: Poly(lactic acid)/carbon nanotube composite microneedle arrays for dermal biosensing. Anal. Chem. **91**, 4436–4443 (2019)
77. Miller, C., Stiglich, M., Livingstone, M., et al.: Impedance-based biosensing of *Pseudomonas putida* via solution blow spun PLA: MWCNT composite nanofibers. Micromachines **10**, 876 (2019)
78. Miller, C.L., Stafford, G., Sigmon, N., et al.: Conductive nonwoven carbon nanotube–PLA composite nanofibers towards wound sensors via solution blow spinning. IEEE Trans. Nanobiosci.Nanobiosci. **18**, 244–247 (2019)
79. Xu, Y.L., Jin, L. He, X.B., et al.: Glowing stereocomplex biopolymers are generating power: polylactide/carbon quantum dot hybrid nanofibers with high piezoresponse and multicolor luminescence. J. Mater. Chem. A **7**, 1810–1823 (2019)
80. Cardoso, R.M., Castro, S.V.F., Silva, M.N.T., et al.: 3D-printed flexible device combining sampling and detection of explosives. Sens. Actua. B Chem. **292**, 308–313 (2019)
81. Sathies, T., Senthil, P., Prakash, C.: Application of 3D printed PLA-carbon black conductive polymer composite in solvent sensing. Mater. Res. Express **6**, 115349 (2019)
82. Silva, A.L., Correa, M.M., de Oliveira, G.C., et al.: Development and application of a routine robust graphite/poly(lactic acid) composite electrode for the fast simultaneous determination of Pb^{2+} and Cd^{2+} in jewelry by square wave anodic stripping voltammetry. New J. Chem. **42**, 19537–19547 (2018)
83. Wu, X.J., Jiang, H., Zheng, J.S., et al.: Highly sensitive recognition of cancer cells by electrochemical biosensor based on the interface of gold nanoparticles/polylactide nanocomposites. J. Electroanal. Chem.Electroanal. Chem. **656**, 174–178 (2011)
84. Wang, J.J., Yu, J.S., Bai, D.Y., et al.: Biodegradable, flexible, and transparent conducting silver nanowires/polylactide film with high performance for optoelectronic devices. Polymers **12**, 604 (2020)
85. Mallick, S., Ahmad, Z., Touati, F., et al.: PLA–TiO2 nanocomposites: thermal, morphological, structural, and humidity sensing properties. Ceram. Int. **44**, 16507–16513 (2018)
86. Malakooti, M.H., Julc, F., Sodano, H.A.: Printed nanocomposite energy harvesters with controlled alignment of barium titanate nanowires. ACS Appl. Mater. Interfaces **10**, 38359–38367 (2018)
87. Nascimento, M., Inacio, P., Paixao, T., et al.: Embedded fiber sensors to monitor temperature and strain of polymeric parts fabricated by additive manufacturing and reinforced with NiTi wires. Sensors **20**, 1122 (2020)
88. Bowers, D.T., Tanes, M.L., Das, A., et al.: Spatiotemporal oxygen sensing using dual emissive boron dye polylactide nanofibers. ACS Nano **8**, 12080–12091 (2014)

89. Castro, M., Lu, J.B., Bruzaud, S., et al.: Carbon nanotubes/poly(ε-caprolactone) composite vapour sensors. Carbon **47**, 1930–1942 (2009)
90. Krucinska, I., Surma, B., Chrzanowski, M., et al.: Application of melt-blown technology for the manufacture of temperature-sensitive nonwoven fabrics composed of polymer blends PP/PCL loaded with multiwall carbon nanotubes. J. Appl. Polym. Sci.Polym. Sci. **127**, 869–878 (2013)
91. Rana, V.K., Akhtar, S., Chatterjee, S., et al.: Chitosan and chitosan-*co*-poly(ε-caprolactone) grafted multiwalled carbon nanotube transducers for vapor sensing. J. Nanosci. Nanotechnol.Nanosci. Nanotechnol. **14**, 2425–2435 (2014)
92. Liu, S.Q., Lin, Y., Wei, Y., et al.: A high performance self-healing strain sensor with synergetic networks of poly(ε-caprolactone) microspheres, graphene and silver nanowires. Compos. Sci. Technol. **146**, 110–118 (2017)
93. Faraji, S., Danesh, E., Tate, D.J., et al.: Cyanoethyl cellulose-based nanocomposite dielectric for low-voltage, solution-processed organic field-effect transistors (OFETs). J. Phys. D Appl. Phys. **49**, 185102 (2016)
94. Wang, X., Yu, C.J.: Flexible low-voltage paper transistors harnessing ion gel/cellulose fiber composites. J. Mater. Res. **35**, 940–948 (2020)
95. Cai, G.F., Ciou, J.H., Liu, Y.Z., et al.: Leaf-inspired multiresponsive MXene-based actuator for programmable smart devices. Sci. Adv. **5**, 7956 (2019)
96. Sun, Z.Z., Du, S.Q., Li, F., et al.: High-performance cellulose based nanocomposite soft actuators with porous high-conductivity electrode doped by graphene-coated carbon nanosheet. Cellulose **25**, 5807–5819 (2018)
97. Wang, F., Jin, Z., Zheng, S.H., et al.: High-fidelity bioelectronic muscular actuator based on porous carboxylate bacterial cellulose membrane. Sens. Actuat. B Chem. **250**, 402–411 (2017)
98. Kim, S.S., Kee, C.D.: Electro-active polymer actuator based on PVDF with bacterial cellulose nano-whiskers (BCNW) via electrospinning method. Int. J. Precis. Eng. Manuf. **15**, 315–321 (2014)
99. Li, J.Z., Zhang, J.Q., Sun, H., et al.: An optical fiber sensor based on carboxymethyl cellulose/carbon nanotubes composite film for simultaneous measurement of relative humidity and temperature. Opt. Commun.Commun. **467**, 125740 (2020)
100. Sadasivuni, K.K., Ponnamma, D., Ko, H.U., et al.: Flexible NO_2 sensors from renewable cellulose nanocrystals/iron oxide composites. Sens. Actuat. B Chem. **233**, 633–638 (2016)
101. Mehta, P., Vedachalam, S., Sathyaraj, G., et al.: Fast sensing ammonia at room temperature with proline ionic liquid incorporated cellulose acetate membranes. J. Mol. Liq. **305**, 112820 (2020)
102. Zhao, W., Qu, X.Y., Xu, Q., et al.: Ultrastretchable, self-healable, and wearable epidermal sensors based on ultralong Ag nanowires composited binary-networked hydrogels. Adv. Electron. Mater. **6**, 2000267 (2020)
103. Souza, F.G., Oliveira, G.E., Anzai, T., et al.: A sensor for acid concentration based on cellulose paper sheets modified with polyaniline nanoparticles. Macromol. Mater. Eng.. Mater. Eng. **294**, 739–748 (2009)
104. Meng, J.J., Pan, P., Yang, Z.C., et al.: Degradable and highly sensitive CB-based pressure sensor with applications for speech recognition and human motion monitoring. J. Mater. Sci. **55**, 10084–10094 (2020)
105. Liu, T., Wu, W., Liao, K.N., et al.: Fabrication of carboxymethyl cellulose and graphene oxide bio-nanocomposites for flexible nonvolatile resistive switching memory devices. Carbohydr. Polym.. Polym. **214**, 213–220 (2019)
106. Wei, D.W., Sun, W.Y., Qian, W.P., et al.: The synthesis of chitosan-based silver nanoparticles and their antibacterial activity. Carbohydr. Res.. Res. **344**, 2375–2382 (2009)
107. de Arquer, F.P.G., Armin, A., Meredith, P., et al.: Solution-processed semiconductors for next-generation photodetectors. Nat. Rev. Mater. **2**, 16100 (2017)
108. Li, C., Huang, W.C., Gao, L.F., et al.: Recent advances in solution-processed photodetectors based on inorganic and hybrid photo-active materials. Nanoscale **12**, 2201–2227 (2020)

109. Heeger, A.J.: 25th anniversary article: bulk heterojunction solar cells: understanding the mechanism of operation. Adv. Mater. **26**, 10–28 (2014)
110. Chen, H.Y., Lo, M.K.F., Yang, G.W., et al.: Nanoparticle-assisted high photoconductive gain in composites of polymer and fullerene. Nat. Nanotechnol.Nanotechnol. **3**, 543–547 (2008)
111. Sun, Z.H., Li, J.H., Yan, F.: Highly sensitive organic near-infrared phototransistors based on poly(3-hexylthiophene) and PbS quantum dots. J. Mater. Chem. **22**, 21673–21678 (2012)
112. Ho, D.H., Sun, Q., Kim, S.Y., et al.: Stretchable and multimodal all graphene electronic skin. Adv. Mater. **28**, 2601 (2016)
113. Lou, Z., Chen, S., Wang, L., et al.: An ultra-sensitive and rapid response speed graphene pressure sensor for electronic skin and health monitoring. Nano Energy **23**, 7 (2016)
114. Mu, C., Song, Y., Huang, W., et al.: Flexible normal-tangential force sensor with opposite resistance responding for highly sensitive artificial skin. Adv. Funct. Mater.Funct. Mater. **28**, 1707503 (2018)
115. Nela, L., Tang, J., Cao, Q., et al.: Large-area high-performance flexible pressure sensor with carbon nanotube active matrix for electronic skin. Nano Lett. **18**, 2054 (2018)
116. Luo, N., Huang, Y., Liu, J., et al.: Hollow-structured graphene–silicone-composite-based piezoresistive sensors: decoupled property tuning and bending reliability. Adv. Mater. **29**, 1702675 (2017)
117. Qiu, A., Li, P., Yang, Z., et al.: A path beyond metal and silicon: polymer/nanomaterial composites for stretchable strain sensors. Adv. Funct. Mater.Funct. Mater. **29**, 1806306 (2019)
118. Shuai, X., Zhu, P., Zeng, W., et al.: Highly sensitive flexible pressure sensor based on silver nanowires-embedded polydimethylsiloxane electrode with microarray structure. ACS Appl. Mater. Interfaces **9**, 26314 (2017)
119. Lee, D., Lee, H., Jeong, Y., et al.: Highly sensitive, transparent, and durable pressure sensors based on sea-urchin shaped metal nanoparticles. Adv. Mater. **28**, 9364 (2016)
120. Sun, Q.J., Zhao, X.H., Zhou, Y., et al.: Fingertip-skin-inspired highly sensitive and multifunctional sensor with hierarchically structured conductive graphite/polydimethylsiloxane foams. Adv. Funct. Mater.Funct. Mater. **29**, 1808829 (2019)
121. Wu, H.P., Shevlin, S.A., Meng, Q.H., et al.: Flexible and binder-free organic cathode for high-performance lithium-ion batteries. Adv. Mater. **26**(20), 3338–3343 (2014)
122. Li, Y.H., Sun, Z.J., Shi, L., et al.: Poly(ionic liquid)-polyethylene oxide semi-interpenetrating polymer network solid electrolyte for safe lithium metal batteries. Chem. Eng. J. **375**, 121925 (2019)
123. Bon, C.Y., Mohammed, L., Kim, S., et al.: Flexible poly(vinyl alcohol)-ceramic composite separators for supercapacitor applications. J. Ind. Eng. Chem. **68**, 173–179 (2018)

Chapter 14
Carbon Dots and Their Sensing Behavior in Organic Medium

Kiranjyoti Mohan, Anindita Bora, and Swapan Kumar Dolui

Abstract A carbon-based fluorescent nanomaterial, called carbon dot (CD), has emerged as a hot research area in the recent years due to its unique optical properties. This article tries to shed light on the photoluminescent (PL) behavior of CDs and their application as sensors. Different synthetic routes of CDs and some of the most probable causes of PL are mentioned along with various ways of surface modifications to tune these PL properties. Two types of sensing mechanisms—static and dynamic quenching—are discussed in detail with special emphasis on inner filter effect (IFE), fluorescence resonance energy transfer (FRET), photo-induced electron transfer (PET), etc. Application of CDs in sensing of various kinds of cations, anions and small molecules are shown. Finally, as an example, the proper way to design a sensor based on quenching of CDs' blue PL emission in the vicinity of 2,4,6-trinitrophenol (TNP) molecules in an organic medium is elaborated. The sensor is reasonably stable and selective towards TNP in contrast to other nitrophenol analogues.

Keywords Carbon dots · Hydrophobic · Photoluminescence · Sensor · Trinitrophenol detection

List of Abbreviations

CD	carbon dot
PL	photoluminescence
IFE	inner filter effect
FRET	fluorescence resonance energy transfer
PET	photo-induced electron transfer

K. Mohan
Department of Chemistry, Rangapara College, Rangapara, Sonitpur, Assam 784505, India

A. Bora · S. K. Dolui (✉)
Department of Chemical Sciences, Tezpur University, Napaam, Sonitpur, Assam 784028, India
e-mail: dolui@tezu.ernet.in

D. Mohanta and P. Chakraborty (eds.), *Nanoscale Matter and Principles for Sensing and Labeling Applications*, Advanced Structured Materials 206,
https://doi.org/10.1007/978-981-99-7848-9_14

DMF	N,N-dimethylformamide
THF	tetrahydrofuran
HOMO	highest occupied molecular orbital
LUMO	lowest unoccupied molecular orbital
NAC	nitroaromatic compound
TNP	2,4,6-trinitrophenol
2,4-DNP	2,4-dinitrophenol
p-NP	para-nitrophenol

14.1 Introduction

Carbon dots (CDs) are the newest member of the carbon nanoparticle family with size less than 10 nm (Fig. 14.1) [1]. They were serendipitously discovered in 2004 by Xu and co-workers during the purification of single walled carbon nanotubes [2]. CDs are popularly described as quasi zero-dimensional nanoparticles with a diverse combination of aromatic (sp^2 hybridized graphene-type islands rich in π-electrons) and aliphatic (sp^3 hybridized diamond-type inclusions) regions with different surface functional groups such as amino, carbonyl, carboxyl, hydroxyl, epoxy and ether [3]. This structure gives rise to several interesting optical properties in CDs like fluorescence which can either be excitation wavelength dependent or independent, up-conversion and down-conversion, etc. It has been observed that by varying the π-conjugation and the presence of different emissive traps on the surface of CDs, their optical properties can be manipulated to accommodate different applications like sensing and bioimaging. The strong photoluminescence (PL) of CDs is, in fact, one of their most appealing properties, both from fundamental and application point of view. As PL based sensors, both organic fluorescent dyes and inorganic semi-conductor quantum dots are widely used due to their high quantum yields but both categories have different disadvantages. Organic dyes require elaborate preparation measures and have poor photo-stability and water-solubility. On the other hand, inorganic quantum dots are not biocompatible. A middle way is the use of CDs, which show good quantum yield with the additional advantages of being photo-stable, negligibly toxic, and biocompatible.

14.2 Synthesis Method

The CDs can be tailor made to suit the application during the synthesis process itself. Hence, their synthesis is a major field of research. The conditions used during the synthesis like precursors, temperature and time affect the PL behavior and hydrophilicity of CDs. CDs are very easy to prepare, and they can also be

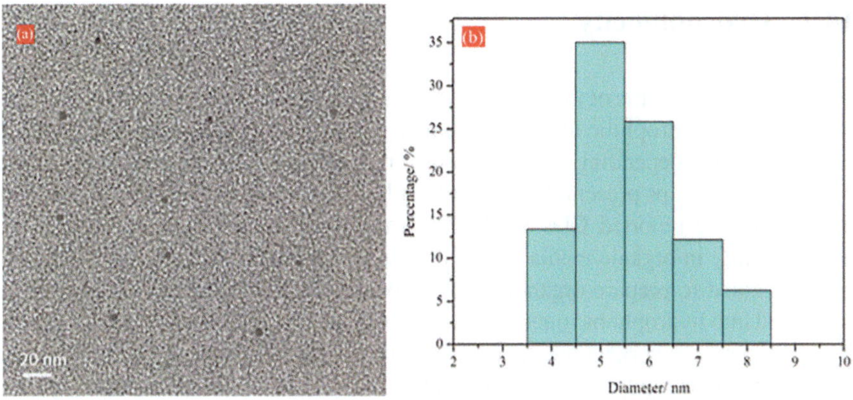

Fig. 14.1 **a** TEM micrograph of CDs and **b** the corresponding DLS number distribution [4]

synthesized from a variety of precursors in an inexpensive manner. The methods for synthesis are generally divided into two groups based on their approaches: top-down and bottom-up.

14.2.1 Top-Down

In the top-down approach, materials like graphite, carbon nanotubes, graphene oxide, carbon fibers, etc. are used as carbon sources. These are large molecules that are broken down via ultrasonication, laser ablation and acid treatment, etc. However, this approach uses comparatively expensive precursors at high temperatures for longer durations of time, which are disadvantageous [5].

14.2.2 Bottom-Up

In the bottom-up approach, small organic molecules are used as carbon sources. Saccharides, citric acid, and amino acids are some of the popular carbon sources. For example, we have reported the hydrothermal synthesis of CDs from citric acid, with maximum PL emission at in the blue region around 440 nm when excited at a wavelength of 360 nm [6]. Eggs [7], pomelo peel [8] and potato starch [9] have also been carbonized to produce CDs through synthetic methods such as hydrothermal reaction [10], ultrasonic [11] and microwave irradiation [12].

14.3 Hydrophilicity

An interesting outcome of the presence of O containing functional groups is that they make the CDs hydrophilic in nature, and this, in turn is a big reason for their biocompatibility. Thus, depending on their surface morphology, i.e., the type and number of functional groups present, CDs can either be hydrophilic or hydrophobic, though majority of the reported CDs are highly water soluble. But for their widespread use, solubility in organic media is also of equal importance. Certain strategies have been designed to prepare organophilic/hydrophobic CDs. Hydrophilic CDs can be converted into hydrophobic ones by covalently attaching nonpolar surface modifiers. Bourlinos and co-workers prepared hydrophobic CDs by using organic ammonium and used them as catalysts in organic synthesis [13]. Varisco and co-workers synthesized hydrophobic CDs by functionalization them with hydrophobic dodecylamine [14]. Use of organic solvent during the synthesis offers a one-step reaction route for producing hydrophobic CDs. Prikhozhdenko and co-workers solvothermally prepared hydrophobic CDs in non-aqueous solution [15]. We have also reported the successful synthesis of hydrophobic CDs by a solvothermal method using gallic acid and N,N-dimethylformamide (DMF) as the precursors. These CDs were used as sensors for 2,4,6-trinitrophenol (TNP) in organic medium [4].

14.4 Origin of Fluorescence

Since their fluorescent properties form the basis for the application of CDs in optical sensors, it is imperative to know about the origin of fluorescence phenomenon. CDs show two electronic transitions/absorption bands at 235 nm and 338 nm due to $\pi-\pi^*$ transition of aromatic C=C bonds and $n-\pi^*$ transitions of C=O [16, 17]. After absorbance, fluorescence is the most obvious way for CDs to return to the ground state. This occurs, in maximum cases, with the emission of blue light (Fig. 14.2). Fluorescence based sensing has many benefits like excellent sensitivity, simple sensor design, low cost and quick response. CDs prepared by different routes that exhibit different optical behavior, thus a single effect can't describe the PL property of all CDs. Many mechanisms have been reported that successfully explain the PL behavior of a particular type of CD. For instance, the most widely accepted mechanism is the presence of surface defect states. These defects give rise to excitation wavelength dependent PL emission. On the other hand, excitation wavelength independent PL emission is shown by the intrinsic state emission. Intrinsic state emission is controlled by the carbon core, whereas surface state emission is controlled by the surface functional groups [18]. In addition to the composition, quantum size effect, i.e., optical selection of differently sized nanoparticles, also affects the PL properties of CDs. PL properties vary sensitively with the size of CDs. The emission is mainly caused by fragments of quantum-sized graphite structure [19]. As the size of the fragment increases, the highest occupied molecular orbital

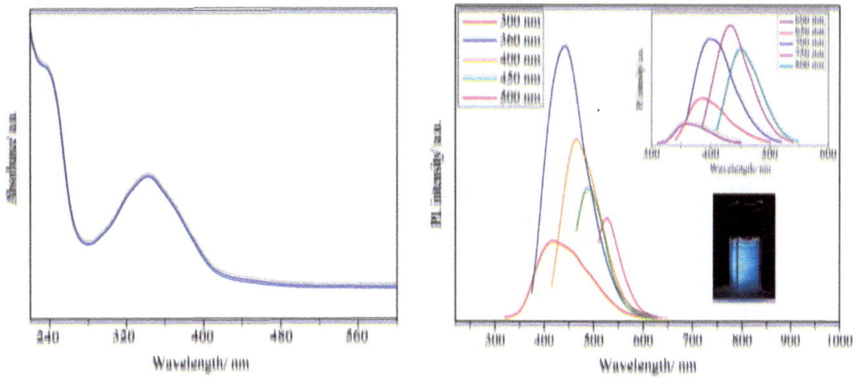

Fig. 14.2 **a** UV–visible absorption spectra of CD and **b** down-converted PL spectra of CDs at different excitation wavelengths (inset, top: up-converted PL spectra of CDs and inset, bottom: a digital photograph of blue coloured emission for CDs excited at 360 nm) [6]

(HOMO)—lowest unoccupied molecular orbital (LUMO) gap gradually decreases and the emission wavelength redshifts. To tune the optical properties and obtain the desired emission, surface modification is often used. Surface modification can be of two types: covalent and non-covalent [20]. Covalent methods include amide coupling, silylation, esterification, sulfonylation and copolymerization reactions. For example, Gao and co-workers applied esterification reaction to prepare quaternary ammonium-functionalized CDs. The resulting hydrogen bonding between the modified CDs and hydroxyl groups of TNP was used for sensing TNP via fluorescence quenching method [21]. Non-covalent methods include modifications via π interactions, complexation reactions, etc. For example, Jiang and co-workers developed Si-CD@dopamine through π−π interactions between Si-CDs and dopamine and applied them for detection of Ag^+. In the presence of dopamine, Ag^+ was reduced to Ag nanoparticles and this led to the emergence of a new peak at 438 nm [22]. Another strategy for emission tuning is heteroatom doping by substituting C with other metals and/or non-metals of different valencies. Doping changes the electronic and surface properties of CDs. Common dopants include N, S, P, B, etc. For example, Hu and co-workers doped CDs with N and Cl (N,C-CDs) and used them for sensing curcumin as the presence of curcumin quenches the fluorescence of CDs [23].

14.5 Sensing Mechanism

The CD sensors work on the "turn on"-"turn off" system. In the presence of an analyte, the fluorescence of the CDs is quenched, and this forms the basis for their sensing platform. Ideally, any phenomenon that can cause a change in the fluorescence intensity or wavelength or lifetime related to an analyte can be used as a foundation for sensors. Practically, fluorescence quenching can be static or dynamic quenching.

14.5.1 Static Quenching

In static quenching mechanism, CDs interact with a quencher/analyte to form a non-fluorescent complex in the ground state. Due to the formation of this complex, the absorption spectrum of the CDs is altered. For example, Yue and co-workers prepared EGTA-modified CDs by a two-step hydrothermal process and used them for detection of Ca^{2+} ions [24]. The CDs showed blue fluorescence which was quenched in the presence of Ca^{2+} ions due to static quenching. Rise in temperature has a deteriorating effect on static quenching. High temperature can destabilize the complex to diminish the quenching.

14.5.2 Dynamic Quenching

In dynamic quenching mechanism, once the CDs are excited, they come back to their ground state by colliding with the quenchers either by charge transfer or energy transfer. This causes a decrease in the PL emission intensity, which can be calculated by using the Stern–Volmer equation given below:

$$\frac{\mathrm{i_o}}{i} - 1 = \mathrm{K_{SV}[C]} \tag{14.1}$$

where $\mathrm{i_o}$ and i are the PL intensity before and after addition of the analyte, K_{SV} is the Stern–Volmer quenching constant, and [C] is the molar concentration of the analyte. K_{SV} represents the quenching efficiency. Dynamic quenching gives rise to a number of effects like the inner filter effect (IFE), fluorescence resonance energy transfer (FRET), photo-induced electron transfer (PET), etc. In contrast to static quenching, the absorption spectrum of CDs is not altered by the quenchers and temperature, rather, has a positive influence on dynamic quenching [25].

IFE: IFE is a re-absorption phenomenon where the excitation light and/or the fluorescent light emitted by the CDs is absorbed by the analyte, causing a decrease in its excitation or emission intensity. In such cases, the analyte/absorber act as a filter. Thus by tuning either the CD or the quencher concentration, a sensing strategy can be developed. It is, however, a pre-requisite that there is an overlap between the absorber absorption spectrum and the CD excitation/emission spectrum. For example, Tang and co-workers obtained CDs hydrothermally from houttuynia cordata, whose excitation wavelength at 320 nm overlapped with the absorption wavelength of metronidazole at 319 nm [26]. They used this overlapping to develop an approach for detection of metronidazole based on IFE, with the metronidazole acting as an IFE absorber. In another instance, Wang and co-workers prepared CDs from wool by microwave irradiation. Then they developed an IFE based strategy between CDs and Ag nanoparticles to determine glyphosate. Due to overlapping of CD emission spectrum with

Ag excitation spectrum, the PL emission of CDs is quenched, but the introduction of glyphosate aggregates the Ag nanoparticles. As a result, IFE diminishes. The re-emergence of blue light of CDs confirms the presence of glyphosate [27].

FRET: In FRET, interaction takes place between two fluorescent moieties when they are less than 10 nm apart. Energy transfer takes from a donor which has higher energy to an acceptor which has lower energy. As a result, the fluorescence intensity of the donor decreases or quenches and that of the acceptor increases. For example, Dong and co-workers synthesized N and S doped CDs hydrothermally from 3-aminothiophenol and used them for detection of glutathione. Initially, in a mixture of N,S-CDs and Au nanoparticles, fluorescence quenching took place by energy transfer from N,S-CDs (donor) to Au (acceptor) via FRET process. But by adding glutathione into the mixture, the fluorescence of N,S-CDs was recovered [28].

PET: In PET, a complex is formed when a redox reaction occurs between the excited CDs and the analyte which can donate or accept an electron. Fluorescence is quenched due to the formation of this complex. For example, Lan and co-workers prepared blue emitting CDs hydrothermally from citric acid. When modified with 2-(diphenylphosphino) ethylamine, the blue light of CDs-P was quenched via electron transfer from lone pair of P to CDs. However, on addition of hydrogen peroxide, it reacted with diphenylphosphine to produce diphenylphosphine oxide. The new complex cancelled the PET effect and blue emission of CDs was recovered [29].

14.6 Sensing Applications

The CDs act as sensors for a wide range of ions and small molecules. CD based sensing of some analytes is presented below.

14.6.1 Cations

The CDs have been successfully used for determination of various metal ions like Fe^{3+}, Cu^{2+}, Ag^{+}, Hg^{2+} and Cr^{6+}. Most of them employ a very simple sensor design of fluorescence quenching by direct interaction of the metal cations with CDs (turn off). Various functional groups present on the surface of CDs like carboxylate and amino groups can coordinate with the metal ions, and their fluorescence is quenched via IFE, FRET or any other process mentioned above. For instance, N-doped CDs prepared hydrothermally from fermented rice provide a highly sensitive and selective sensor for the detection of Fe^{3+} with a limit of detection (LOD) of 0.1 μM [30]. CDs synthesized using mesoporous silica spheres as nanoreactors were used to detect Cu^{2+} based on a charge transfer quenching mechanism. The addition of l-cysteine, however, could reverse the quenching process [31]. Hg^{2+} ions could quench the fluorescence of CDs prepared by pyrolysis of ethylenediamine tetracetic acid salts via electron transfer [32]. In another instance, Hg^{2+} ions turned off the PL emission of thioctic acid-CDs by an electron transfer effect, while the introduction of thiophanate methyl turned on the PL by forming thiophanate methyl-Hg complexes [33].

14.6.2 Anions

The CDs have also been successful in determination of several anions like I^-, PO_4^{3-} and S^{2-}. In contrast to cations, which directly quench the CD fluorescence, anions sensing mostly occurs by a slightly different strategy of fluorescence quenching and recovery. CDs are initially integrated with other quenchers like metal cations. The fluorescence of these quenched CD complexes can be recovered by the introduction of anions (turn off-turn on), which have more affinity for the cations than CDs. For instance, the PL emission of CDs prepared from l-asparagine by pyrolysis reduced drastically when Ag nanoparticles were formed on its surface. However, the emission was recovered after the addition of S^{2-} [34]. In another case, fluorescence of CDs was quenched by adjustment with Eu^{3+}. In the presence of PO_4^{3-}, the fluorescence was turned on [35].

14.6.3 Small Molecules

The CDs serve as efficient sensing platforms for small molecules like monosaccharides, thiols, drugs, vitamins and explosive nitroaromatics. For detection of some molecules, an elaborate strategy may be required. For instance, for detection of glucose, post functionalization of CDs, prepared by microwave heating of citric acid and urea solution, was done by addition of $AgNO_3$ to produce CD-Ag nanoparticle complex. This complex did not show any emission because the emission was quenched by Ag nanoparticles via an energy transfer process from CDs to the Ag nanoparticles. However, the presence of glucose and glucose oxidase enzyme in its vicinity led to the production of H_2O_2, by enzymatic oxidation of glucose.

The H_2O_2 etched Ag nanoparticles to Ag^+ ions, thus restoring the fluorescence of CDs [36]. For detection of cysteine, CDs were prepared hydrothermally from p-phenylenediamine. These CDs didn't show emission due to PET-caused fluorescence quenching because of the presence of lone pair of electrons of the pyridinic nitrogen. However, the presence of cysteine caused oxidation of CDs leading to a decrease in the number of lone pair of electrons. This led to recovery of red coloured PL emission of CDs [37]. For TNP detection in an organic medium, hydrophobic CDs synthesized from ascorbic acid and hexadecyl amine in ethanol were dispersed in tetrahydrofuran (THF). The emission of CDs at 495 nm was quenched by TNP via IFE [38]. Glutathione was detected using nitrogen and phosphorus co-doped CD-gold nanoparticle (NPCD–AuNP) composites. The composite formation led to PL

Table 14.1 Comparison of sensing responses of CDs towards different analytes

Sl. No	Analyte	Limit of detection/μM	References
1	Fe^{3+}	0.1	[30]
2	Cu^{2+}	0.023	[31]
3	Hg^{2+}	0.0042	[32]
4	Hg^{2+} and thiophanate methyl	0.033 and 0.0076	[33]
5	S^{2-}	0.01	[34]
6	PO_4^{3-}	0.051	[35]
7	$O_2^{\cdot-}$	0.0044	[41]
8	Glucose	1.39	[36]
9	Cystine	0.0004	[37]
10	TNP	1.8	[38]
11	Glutathione	0.1	[39]
12	Tetracyclines antibiotics	0.16	[40]
13	TNP and metanil yellow	0.06 and 0.45	[42]

quenching by IFE which was recovered in the presence of glutathione [39]. For tetracyclines antibiotics detection, the bright blue PL emission of hydrothermally prepared copper doped CDs (Cu-CDs) was quenched by these antibiotics by IFE [40].

Comparison of sensing responses of CDs towards various cations, anions and small molecules are shown in Table 14.1.

14.7 Detection of TNP in Organic Medium

Till this point, a general idea of CDs as sensing element is discussed. From this point onwards, the focus will be on designing a specific sensing platform for TNP detection. TNP is also popularly called as picric acid. It is a strong organic acid and used in various chemical laboratories and industries like dye, leather, and pharmaceuticals [43]. But TNP is explosive in nature. Because of this, it poses a serious security concern. Additionally, it has mutagenic properties and overexposure to this compound can cause strong irritation to eyes, skin, and respiratory system [44]. Water is the most common medium for TNP contamination, however, organic media like THF and DMF also act as carriers for TNP. In the following sections, we will look more closely into different aspects behind the application of hydrophobic CDs in TNP sensing in an organic medium [4].

14.7.1 Synthesis of Hydrophobic CDs

0.6 g gallic acid is dissolved in 20 mL DMF and 0.17 mL ethylenediamine is added to it. The reaction mixture is then transferred to a Teflon lined autoclave and heated at 180 °C for 5 h. After completion of the solvothermal reaction, hydrophobic CDs are obtained.

14.7.2 Sensing Protocol

The protocol for studying the quenching effect of TNP on the fluorescence intensity of CDs is very simple and straight-forward. Since we are considering an organic medium, a 10 mL solution of 0.01 mg mL^{-1} CDs is prepared in DMF. The PL emission spectrum of this solution excited at 365 nm wavelength is considered as the reference, and all PL emission intensities measured subsequently are matched with it. The PL emission of CDs in DMF displays blue colored emission peaked at 426 nm. Meanwhile, TNP is added to a batch of CD solutions in calculated quantities to obtain mixtures of TNP and CD in DMF having TNP in the concentration range of 1–250 μM. The next step is incubation for which the solution mixture is kept undisturbed at room temperature for 10 min. Post that, the emission intensities of the solutions are measured.

14.7.3 Sensing Behavior

The presence of TNP causes the quenching of the PL intensity of CDs (Fig. 14.3a). Gradual increase in concentration of TNP leads to further quenching, i.e., higher the concentration more the quenching. Stern–Volmer equation is frequently applied to analyze the quenching behavior of fluorophores. It is plotted as a graph of the relative PL intensities against different concentrations of TNP, known as the Stern–Volmer plot. From the linear part of the graph (0–100 μM), the quenching efficiency is calculated to be $2.6 \times 10^4\ M^{-1}$ with LOD value of 0.75 μM (Fig. 14.3b). The Stern–Volmer plot, at higher concentrations of TNP (>100 μM), shows an exponential increase.

14.7.4 Stability of the Sensor

The stability of the sensing material is very important for it to be applied effectively. Also, it is important to have information about the time required for quenching. For this purpose, a simple experiment can be designed based on the variation in

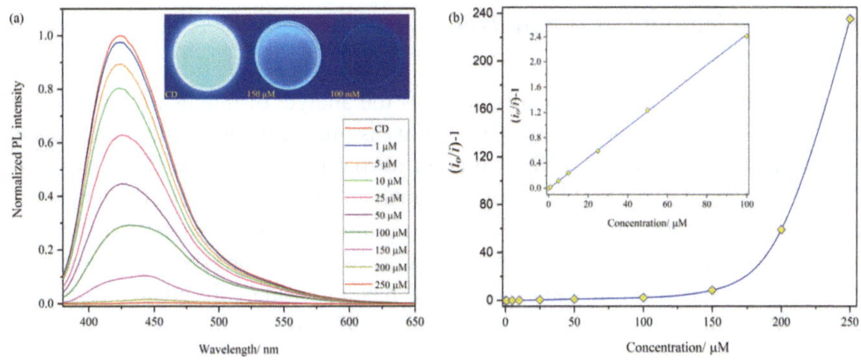

Fig. 14.3 **a** Variation of PL emission of CDs with addition of TNP of different concentrations (inset: digital photographs of quenching of blue emission of CDs with addition of TNP at λ_{ex} = 365 nm) and **b** Stern–Volmer plot for TNP sensing at different concentrations (inset: linear range from 0 to 100 μM) [4]

incubation time. After adding TNP to CD solution (final concentration = 100 μM), the incubation time is varied from 1 to 120 min and PL intensities are measured at regular intervals. As seen from Fig. 14.4, the addition of TNP causes the quenching process of the CDs fluorescence within 1 min and remains as such for the remaining 120 min. This demonstrates the rapidity and excellent stability of the sensing design.

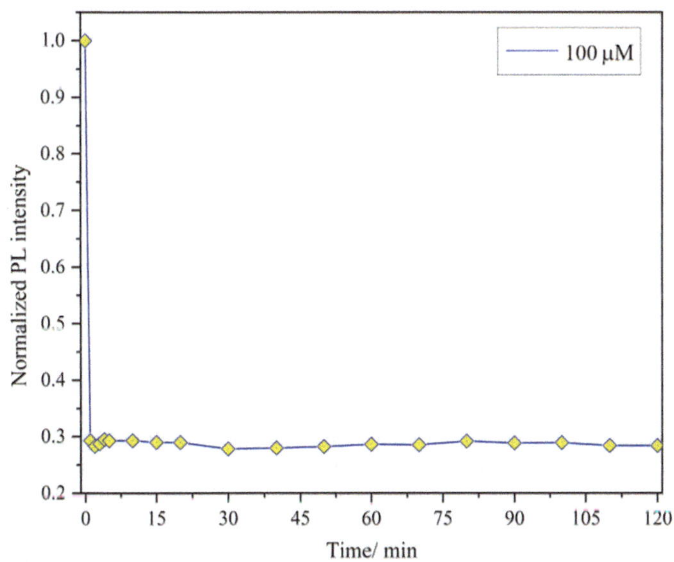

Fig. 14.4 Stability of CDs sensing platform [4]

14.7.5 Selectivity of the Sensor

A sensor should be highly selective towards the analyte of interest, otherwise it is of no use. For this reason, their sensing ability is often compared with other related compounds. In the case of TNP, two other nitroaromatic compounds (NACs) viz. 2,4-dinitrophenol (2,4-DNP) and para-nitrophenol (p-NP) are taken for comparison purpose. The PL quenching plots of these two compounds and the comparative Stern–Volmer plots of all the NACs are shown in Fig. 14.5. The corresponding quenching constants are tabulated in Table 14.2. TNP has a considerably greater K_{SV} value that suggests that the selectivity of CDs towards TNP is very high in comparison to others.

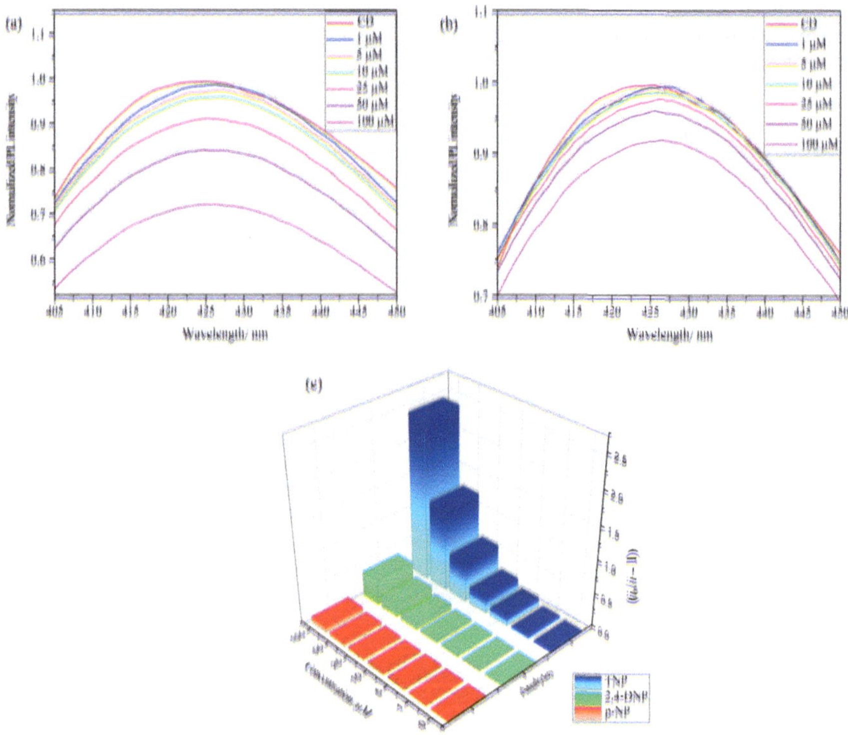

Fig. 14.5 Variation of PL emission of CDs with addition of **a** 2,4-DNP and **b** p-NP, and **c** variation of $((i_o/i)-1)$ for TNP, 2,4-DNP and p-NP in the linear range from 0 to 100 μM [4]

Table 14.2 Fluorescence quenching constant (KSV) and correlation values of different analytes [4]

Analyte	K_{SV}/M^{-1}	Correlation coefficient
TNP	2.42×10^4	0.9998
2,4-DNP	3.7×10^3	0.9992
p-NP	8.0×10^2	0.9946

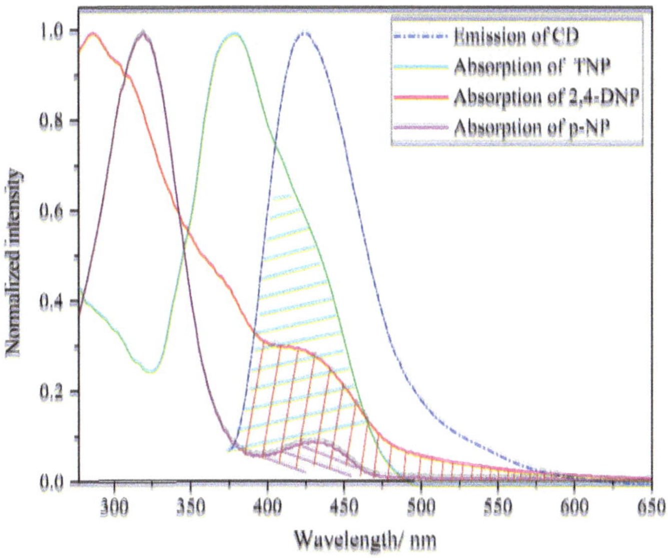

Fig. 14.6 UV–visible spectra of TNP, 2,4-DNP and p-NP, and their overlap with the CD emission spectral feature [4]

14.7.6 PL Quenching Mechanism

As already discussed above, the most common mechanisms that explain the PL quenching phenomenon include IFE, FRET, PET, etc. If we look at the emission spectrum of the CDs and the absorption spectrum of TNP, substantial spectral overlap can be seen around 400–450 nm (Fig. 14.6), so there is high possibility for the TNP molecules to re-absorb the emission of CDs. This overlap satisfies the condition for IFE. IFE also helps to explain the selectivity of CDs towards TNP. In the same figure, we observe gradually reducing spectral overlap between the CD emission spectrum and the 2,4-DNP and p-NP absorption spectra.

14.8 Conclusion and Outlook

This article describes various aspects of a CD sensor. The properties that are required for a specific analyte detection system can be tailor made right during the synthesis process. Fluorescence is the primary property on which these sensing platforms are designed. Quenching of CD fluorescence in the vicinity of different analytes occurs due to several phenomena like IFE, FRET, PET, etc. We have also explained, citing the case of TNP detection in DMF, all the stages that are required to be performed to demonstrate the practical application of CD based fluorescence sensor.

Although CDs have seen tremendous success as an eco-friendly alternative to quantum dots, there are still some areas that need to be focused on, including: (1) improving the synthesis process to obtain better quantum yield, (2) large scale production of CDs with reproducibility, (3) understanding the exact mechanism of PL for effective sensor design and (4) improvement in selectivity and accuracy of CDs.

References

1. Tuerhong, M., Xu, Y., Yin, X.B.: Review on carbon dots and their applications. Chin. J. Anal. Chem. **45**(1), 139–150 (2017)
2. Xu, X., Ray, R., Gu, Y., Ploehn, H.J., Gearheart, L., Raker, K., Scrivens, W.A.: Electrophoretic analysis and purification of fluorescent single-walled carbon nanotube fragments. J. Am. Chem. Soc. **126**(40), 12736–12737 (2004)
3. Demchenko, A.P., Dekaliuk, M.O.: Novel fluorescent carbonic nanomaterials for sensing and imaging. Methods Appl. Fluoresc. **1**(4), 042001 (2013)
4. Bora, A., Mohan, K., Dolui, S.K.: Carbon dots as cosensitizers in dye-sensitized solar cells and fluorescence chemosensors for 2,4,6-trinitrophenol detection. Ind. Eng. Chem. Res. **58**(51), 22771–22778 (2019)
5. Nazri, N.A.A., Azeman, N.H., Luo, Y., Bakar, A.A.A.: Carbon quantum dots for optical sensor applications: a review. Opt. Laser Technol. **139**, 106928 (2021)
6. Bora, A., Mohan, K., Doley, S., Goswami, P., Dolui, S.K.: Broadening the sunlight response region with carbon dot sensitized TiO_2 as a support for a Pt catalyst in the methanol oxidation reaction. Catal. Sci. Technol. **8**(16), 4180–4192 (2018)
7. Wang, J., Wang, C.F., Chen, S.: Amphiphilic egg-derived carbon dots: rapid plasma fabrication, pyrolysis process, and multicolor printing patterns. Angew. Chem. Int. Ed. **51**(37), 9297–9301 (2012)
8. Lu, W., Luo, Y., Zhang, Y., Asiri, A.M., Liu, S., Qin, X., Chang, G., Al-Youbi, A.O., Sun, X.: Economical, green synthesis of fluorescent carbon nanoparticles and their use as probes for sensitive and selective detection of mercury(II) ions. Anal. Chem. **84**(12), 5351–5357 (2012)
9. Zheng, J.X., Liu, X.H., Yang, Y.Z., Liu, X.G., Xu, B.S.: Rapid and green synthesis of fluorescent carbon dots from starch for white light-emitting diodes. New Carbon Mater. **33**(3), 276–288 (2018)
10. Zhu, S., Meng, Q., Wang, L., Zhang, J., Song, Y., Jin, H., Zhang, K., Sun, H., Wang, H., Yang, B.: Highly photoluminescent carbon dots for multicolor patterning, sensors, and bioimaging. Angew. Chem.—Int. Ed. **52**(14), 3953–3957 (2013)
11. Li, H., He, X., Liu, Y., Huang, H., Lian, S., Lee, S.-T., Kang, Z.: One-step ultrasonic synthesis of water-soluble carbon nanoparticles with excellent photoluminescent properties. Carbon **49**(2), 605–609 (2010)
12. Zhu, H., Wang, X., Li, Y., Wang, Z., Yang, F., Yang, X.: Microwave synthesis of fluorescent carbon nanoparticles with electrochemiluminescence properties. Chem. Commun. **0**(34), 5118–5120 (2009)
13. Bourlinos, A.B., Stassinopoulos, A., Anglos, D., Zboril, R., Georgakilas, V., Giannelis, E.P.: Photoluminescent carbogenic dots. Chem. Mater. **20**(14), 4539–4541 (2008)
14. Varisco, M., Zufferey, D., Ruggi, A., Zhang, Y., Erni, R., Mamula, O.: Synthesis of hydrophilic and hydrophobic carbon quantum dots from waste of wine fermentation. R. Soc. Open Sci. **4**(12), 170900 (2017)
15. Prikhozhdenko, E.S., Bratashov, D.N., Mitrofanova, A.N., Sapelkin, A.V., Yashchenok, A.M., Sukhorukov, G.B., Goryacheva, I.Y.: Solvothermal synthesis of hydrophobic carbon dots in reversed micelles. J. Nanopart. Res. **20**(9), 1–11 (2018)

16. Wang, Y., Kalytchuk, S., Zhang, Y., Shi, H., Kershaw, S.V., Rogach, A.L.: Thickness-dependent full-color emission tunability in a flexible carbon dot ionogel. J. Phys. Chem. Lett. **5**(8), 1412–1420 (2014)
17. Suzuki, K., Malfatti, L., Takahashi, M., Carboni, D., Messina, F., Tokudome, Y., Takemoto, M., Innocenzi, P.: Design of carbon dots photoluminescence through organo-functional silane grafting for solid-state emitting devices. Sci. Rep. **7**(1), 5469 (2017)
18. Xu, D., Lin, Q., Chang, H.T.: Recent advances and sensing applications of carbon dots. Small Methods **4**(4), 1900387 (2020)
19. Li, H., He, X., Kang, Z., Huang, H., Liu, Y., Liu, J., Lian, S., Tsang, C.H.A., Yang, X., Lee, S.T.: Water-soluble fluorescent carbon quantum dots and photocatalyst design. Angew. Chem.—Int. Ed. **49**(26), 4430–4434 (2010)
20. Yan, F., Jiang, Y., Sun, X., Bai, Z., Zhang, Y., Zhou, X.: Surface modification and chemical functionalization of carbon dots: a review. Microchim. Acta **185**(9), 1–34 (2018)
21. Gao, K., Guo, Y., Niu, Q., Han, L., Zhou, L., Wang, L.: Quaternary ammonium-functionalized carbon dots for sensitive and selective detection of 2,4,6-trinitrophenol in aqueous medium. Sens. Actuators, B Chem. **262**, 298–305 (2018)
22. Jiang, Y., Wang, Z., Dai, Z.: Preparation of silicon–carbon-based dots@ dopamine and its application in intracellular Ag^+ detection and cell imaging. ACS Appl. Mater. Interfaces. **8**(6), 3644–3650 (2016)
23. Spangler, C., Schaeferling, M., Wolfbeis, O.S.: Fluorescent probes for microdetermination of inorganic phosphates and biophosphates. Microchim. Acta **161**(1), 1–39 (2008)
24. Yue, J., Li, L., Cao, L., Zan, M., Yang, D., Wang, Z., Chang, Z., Mei, Q., Miao, P., Dong, W.F.: Two-step hydrothermal preparation of carbon dots for calcium ion detection. ACS Appl. Mater. Interfaces. **11**(47), 44566–44572 (2019)
25. Molaei, M.J.: Principles, mechanisms, and application of carbon quantum dots in sensors: a review. Anal. Methods **12**(10), 1266–1287 (2020)
26. Tang, J., Zhang, Y., Liu, Y., Liu, D., Qin, H., Lian, N.: Carbon quantum dots as a fluorophore for "inner filter effect" detection of metronidazole in pharmaceutical preparations. RSC Adv. **9**(65), 38174–38182 (2019)
27. Wang, L., Bi, Y., Hou, J., Li, H., Xu, Y., Wang, B., Ding, H., Ding, L.: Facile, green and clean one-step synthesis of carbon dots from wool: application as a sensor for glyphosate detection based on the inner filter effect. Talanta **160**, 268–275 (2016)
28. Dong, W., Wang, R., Gong, X., Dong, C.: An efficient turn-on fluorescence biosensor for the detection of glutathione based on FRET between N,S dual-doped carbon dots and gold nanoparticles. Anal. Bioanal. Chem. **411**(25), 6687–6695 (2019)
29. Lan, M., Di, Y., Zhu, X., Ng, T.W., Xia, J., Liu, W., Meng, X., Wang, P., Lee, C.S., Zhang, W.: A carbon dot-based fluorescence turn-on sensor for hydrogen peroxide with a photo-induced electron transfer mechanism. Chem. Commun. **51**(85), 15574–15577 (2015)
30. Chen, Y., Shan, X., Jiang, D., Li, X.: One-step hydrothermal synthesis of N-doped fluorescent carbon dots from fermented rice with highly selective characteristics for label-free detection of Fe^{3+} ions and as fluorescent ink. Anal. Sci. 20P433 (2021)
31. Zong, J., Yang, X., Trinchi, A., Hardin, S., Cole, I., Zhu, Y., Li, C., Muster, T., Wei, G.: Carbon dots as fluorescent probes for "off–on" detection of Cu^{2+} and l-cysteine in aqueous solution. Biosens. Bioelectron. **51**, 330–335 (2014)
32. Zhou, L., Lin, Y., Huang, Z., Ren, J., Qu, X.: Carbon nanodots as fluorescence probes for rapid, sensitive, and label-free detection of Hg^{2+} and biothiols in complex matrices. Chem. Commun. **48**(8), 1147–1149 (2012)
33. Wang, S., Chen, H., Xie, H., Wei, L., Xu, L., Zhang, L., Lan, W., Zhou, C., She, Y., Fu, H.: A novel thioctic acid-carbon dots fluorescence sensor for the detection of Hg^{2+} and thiophanate methyl via S-Hg affinity. Food Chem. **346**, 128923 (2021)
34. Sinduja, B., John, S.A.: Silver nanoparticles capped with carbon dots as a fluorescent probe for the highly sensitive "off–on" sensing of sulfide ions in water. Anal. Bioanal. Chem. **411**(12), 2597–2605 (2019)

35. Zhao, H.X., Liu, L.Q., De Liu, Z., Wang, Y., Zhao, X.J., Huang, C.Z.: Highly selective detection of phosphate in very complicated matrixes with an off–on fluorescent probe of europium-adjusted carbon dots. Chem. Commun. **47**(9), 2604–2606 (2011)
36. Ma, J.L., Yin, B.C., Wu, X., Ye, B.C.: Simple and cost-effective glucose detection based on carbon nanodots supported on silver nanoparticles. Anal. Chem. **89**(2), 1323–1328 (2017)
37. Kalaiyarasan, G., Hemlata, C., Joseph, J.: Fluorescence turn-on, specific detection of cystine in human blood plasma and urine samples by nitrogen-doped carbon quantum dots. ACS Omega **4**(1), 1007–1014 (2019)
38. Cheng, F., An, X., Zheng, C., Cao, S.: Green synthesis of fluorescent hydrophobic carbon quantum dots and their use for 2,4,6-trinitrophenol detection. RSC Adv. **5**(113), 93360–93363 (2015)
39. Le, T.H., Kim, J.H., Park, S.J.: "Turn on" fluorescence sensor of glutathione based on inner filter effect of Co-doped carbon dot/gold nanoparticle composites. Int. J. Mol. Sci. **23**(1), 190 (2021)
40. Guo, J., Lu, W., Zhang, H., Meng, Y., Du, F., Shuang, S., Dong, C.: Copper doped carbon dots as the multi-functional fluorescent sensing platform for tetracyclines and pH. Sens. Actuators, B Chem. **330**, 129360 (2021)
41. Yue, J., Peng, J., Yu, L., Sun, M., Sun, Z., Tan, H., Wang, S.: Superoxide anion turns on the fluorescence of carbon dots-ferric complex for sensing. Microchem. J. **168**, 106412 (2021)
42. Abbasi, A., Shakir, M.: Simple one-step solid-state synthesis of highly crystalline N doped carbon dots as selective turn off-sensor for picric acid and metanil yellow. J. Fluoresc. 1–8 (2022)
43. Volwiler, E.H.: Medicinals and dyes. Ind. Eng. Chem. **18**(12), 1336–1337 (1926)
44. Yin, X., Meng, S., Xie, J.: Heterobimetallic Mg–Ag coordination polymer with luminescence and 2,4,6-trinitrophenol sensing properties. Polyhedron **139**, 262–266 (2018)

Chapter 15
An Overview of Carbon-Based Nanomaterials and Their Derivatives for Different Sensing Applications

Kunal Biswas and Yugal Kishore Mohanta

Abstract Carbon based nanomaterials (CNM), henceforth mentioned as CNMs, are the classes of materials which consist of basic carbon entities as the prime architecture in the composition. Usually, CNMs comprise 2D graphene sheets, 1D carbon nanotubes, 0D carbon quantum dots, and 3D carbon nanoflowers. The alterations in the electronic configurations in the different nano-scaled systems invoke the different sensing applications and their responses, ranging from electrochemical, biosensing, amperometry, piezoelectric sensing, etc., elicited differentially from CNMs like Carbon nanotubes, Graphene sheets, Carbon dots, Carbon nanoflowers, etc., which would be highlighted with recent trend of work in this domain. Besides that, the real-life applications of CNMs in few agricultural applications would also be explained with past and recent trends with their discoveries. The underlying specific sensing properties at the atomic and electronic dimensions of different classes of CNMs would be unraveled in this chapter to showcase the significant sensing attributes existing in the inherent carbon-based nanomaterials and its modified parts for specific type of sensing applications.

Keywords CNMs · C-dot · Graphene · Electronic confinement · Physico-chemical · Sensing

---

K. Biswas (✉)
Centre for Nanoscience and Nanotechnology, Sathyabama Institute of Science and Technology, Jeppiaar Nagar, Rajiv Gandhi Salai, Chennai 600119, India
e-mail: kunal.sapiens@gmail.com

Y. K. Mohanta
Department of Applied Biology, University of Science and Technology Meghalaya, Ri-Bhoi 793101, India

D. Mohanta and P. Chakraborty (eds.), *Nanoscale Matter and Principles for Sensing and Labeling Applications*, Advanced Structured Materials 206,
https://doi.org/10.1007/978-981-99-7848-9_15

15.1 Introduction

Owing to the different bio-compatible and natural components of carbonaceous entities, the different allotropes of carbon-based nanomaterials have shown to exhibit tremendous extraordinary physico-chemical phenomena [1]. Among the different classes of carbon-based nanomaterials, different shapes dictate different inherent properties among them [2]. Some of them are 0D carbon dots [3], which are 0D systems. Others are 2D sheets, the electrons in them are confined in one dimension, and are free to move in other two dimensions [4]. The examples of the sheets are graphene and planar 2D carbon sheets. In the case of 1D nano-scale systems, carbon nanotubes find tremendous applications, because of their pointed edges and extremely designed chiralities [5]. The nature of rolling in their structure governs the different electronic and allied physico-chemical properties associated with them [6, 7]. Similarly, 3D carbon nanoflowers also are coming in nowadays, which exhibit extraordinary physical and chemical nature in their native structures [8]. The rate of functionalization either by means of chemical origin or biological origin plays a critical role in determining the augmented sensing and other applications in the carbon-based nano-scale systems [9]. All the mentioned nano-scaled carbonaceous elements have been showing different sensing applications ranging from electrochemical, electronic, humidity, biosensing, impedimetric, piezoelectric, etc. [10–12]. The relative interactions between the flanking groups residing over the native and functionalized architecture of the carbonaceous nano-scaled systems and the sensing analyte determine the mode of operation of different sensing applications and their underlying working principles. In the area of agricultural to biomedical sciences, several studies have been performed using different classes of CNMs for varied applications [13, 14]. The use of CNMs in agricultural and environmental sectors comprises preliminarily water and soil de-contamination strategies using CNMs like SWCNTs, graphene Oxides, etc. based anti-microbial agents [15, 16], gradual release of pesticides and fertilizers from nano-encapsulations including SWCNTS [17], energy harvesting means like enhanced sensing devices like solar, wind harvesting at an augmented rate [18], sensing of herbicides, pesticides, heavy metals in water bodies, food products, soil moisture conditions has been studied using different CNM-based sensors [19, 20], which are required collectively for improved agricultural productivity. Also, the applications of CNMs in different biomedical sciences like Targeted Drug Delivery (TDD), molecular and tissue-based theranostics, and cellular imaging have been extensively studied in the last few years [21–23]. Owing to the extraordinary surface properties, bio-compatibility, lesser cost of production, and environmental benign conditions, the use of CNM-based drug delivery applications to associated biomedical applications has been extensively used by researchers during several years [24, 25].

15.2 Carbonaceous Nanomaterials for Sensing

15.2.1 Graphene-Based Sensing Applications: A 2D Sensing Platform

This is due to the planar sheets' architecture of graphene; the same has been exploited to sense different analytes from bio-analytes, antigens, metal analytes, etc. Owing to the extraordinary surface area of ~2630 m^2/gram, the surface electrons play a significant role in sensing applications of target analytes [26]. It has been reported previously by several groups that graphene-based nanocomposites exhibited extraordinary humidity sensing applications [27]. Similarly, piezoelectric sensing applications for Graphene-based systems have also been exhibited by several research groups [28]. The high surface-to-volume ratio and huge amount of surface electrons play a critical role in determining the myriad sensing applications [29, 30]. In one of the previous studies, graphene-based piezoelectric sensors have been designed for tactile sensing [31]. These types of sensors have limitations of requirements of high-voltage operations and are energy inefficient in nature, when employed in applications in tactile skin type sensing [32]. The sensing device works on the principle of extended gate configurations, wherein the GFET is related to a piezoelectric–metal–insulator–metal (MIM). In this study, the pressure over the MIM device generates a piezo-potential value which mediates and helps in modulation of channel current in GFET systems. The sensitivity response has been recorded to be ~4.55×10^{-3} kPa^{-1}. This sort of low-voltage application at ~100 mV has excited myriad applications in wearable electronics [32].

In another study, graphene-based electronic based sensing system has been shown by previous studies, which have been predominantly synthesized by conventional chemical routes like Hummer's method and allied for a purely 'bottom-up' mediated synthesis and fabrication of the graphene sheets, which has been employed extensively in different electrochemical and electronic sensing systems [33]. PEDOT nanocomposites and graphene systems have shown electronic-based sensing mechanism, which has exhibited a robust rate of selectivity and sensitivity responses in different applications [34].

It has been reported that graphene and its different doped forms by chemical dopants like metallic or polymeric compounds in the interstellar space of the graphene systems have exhibited an exponential rate of humidity sensing [35]. The doping of polydopamine over the GO sheets on a quartz–crystal microbalance has shown excellent humidity sensing applications. It has shown high stability, repeatability and fast response and recovery time. It has shown ultrasensitive rate of humidity detection in human sweat and breath, which has been shown by bimodal exponential kinetical adsorption model of sensors [36].

The 2D graphene systems plays also a crucial role in the applications of catalysis. Owing to the tunable properties of the interactions between graphene and the associated catalysts, it has been finding numerous applications from the substrate interactions and the different ranges of catalytic applications. Several ranges of applications of 2D graphene systems in the context of effective catalysis fall in oxygen reduction reactions, water splitting reactions, etc. [37]. Also, the implications of 2D graphene systems have been extensively been used in recent times in the area of molecular imaging and diagnosis remedy. Owing to the ultra-high surface area and unique physico-chemical properties, along with the augmented charge-carrying phenomenon, the 2D graphene systems have potentially been explored to tag and label at the nano-scale into bio-macromolecules of interest for probing the fate of the tagged nano-organ-specific entities using relevant microscopy from exterior. In this regard, Graphene Quantum Dots (GQDs) plays a significant role in tracing the fate of the tagged biomolecules when studying in the in vivo model organism for some medical and drug-related analyses [38]. It has also been extensively studied in last few years that due to the presence of higher surface-to-volume ratio and augmented sensing capabilities, the 2D graphene systems and their different doped and chemically functionalized forms were extensively used in the applications of sensors and micro-fabrications for the detection of heavy metals, toxic contaminants, etc. from the water bodies, etc., thereby replacing the conventional mode of sensing mechanism using bulk counterparts of materials bearing compromised selectivity and sensitivity [39, 40].

The underlying electrochemical properties of the myriad ranges of carbon-based nanomaterials (CNMs) exhibit significant attributes in terms of different applications. The widespread applications of CNMs like graphene materials owing to their high surface-to-volume ratio, higher pore structures, augmented sensitivity, higher surface electrons, etc. make the material under CNMs an ideal candidate for the applications in electrode manufacturing and in supercapacitors [41, 42]. Moreover, properties like high potential windows for water splitting applications, higher conductance, higher stability, lower or constant background current with the higher electron-transfer kinetics along with the higher charge storage capacities are the preferable properties of the carbon nanomaterials making them a suitable material for electrochemical applications, in supercapacitors, batteries, etc. In comparison to the conventional counterparts of the carbon materials, the charge storage properties and associated electron transfer kinetics in the bulk counterparts are negligible and minimized due to the properties like lower surface-to-volume ratio and compromised sensing phenomenon. Similarly, carbon nanotubes and their conjugated forms have been thoroughly employed in the recent years for the diverse electrochemical applications like in supercapacitors, batteries, and in the storage systems. In one of the studies, electrospun carbon nanofibre–carbon nanotube-coated polyaniline composites exhibiting augmented and improved electrochemical properties have been used in supercapacitors [43]. In that study, the composite electrode material of carbon nanotubes exhibited a capacitance value of ~1119 F g^{-1} at 1 A g^{-1} compared to the bare CNF films, with a specific capacitance of 278 F g^{-1}. The composite material used as the electrode exhibits the excellent cyclic stability attributes with the retention percentage ~98% even after the 2000 cycles at a current density of ~10 A g^{-1}.

15.2.2 Carbon Nanotube-Based Sensing Applications: A 1D Sensing Platform

It is because of the 1D carbon-based nano-scaled systems, and the rate and nature of chirality in the tubular architecture, the sensing applications of the tubes find numerous applications from electrochemical to biosensing applications. Single-walled to multiwalled carbon nanotubes due to their native rolling aspects determine the differential sensing applications against the same and different target analytes.

In one of the previous studies, phosphomolybdic acid-functionalized carbon nanotubes have been designed for low power detection of NH_3 and NO_2 gases at room temperature [44]. The nature of chirality and surface functionality in the carbon nanotube architecture determines the rate of gas sensing selectivity and sensitivity at the room temperature conditions. In one of the studies, the incorporation of brucine in the carbon nanotube (brucine/SWNTs) architecture has been shown to sense electrochemically the presence of hydroxylamine in the surrounding environment. The detection limit for the detection of hydroxylamine (HA) has come out to be ~0.021 μM based on the 3-sigma criterion [45]. The designed carbon nanotube-functionalized sensor has been used in the determination of Hydroxylamine (HA) in the hydroxyurea tablets and pralidoxime iodide injections with optimal results.

The sensing mechanisms involved with the structure of carbon nanotubes depend upon the architectural differences of the tubes known as chirality. The orientations (rolling angle) determine the electrical conductivity of the carbon nanotube structures. Based on their rolling factors, it behaves as semi-conducting or metallic. However, most nanotubes exhibit semi-conductive bandgap ranging 0.1–0.2 eV [46]. The controlling of the bandgap determines the type of electronic devices. The doping and the incorporation of the extrinsic atoms and elements into the pristine structures of the CNTs make the system behave in a different pattern. Such tuning of the bandgap plays a crucial role in finding different modes of applications from biosensing to electrochemical devices.

Different types of CNT structures, which are dependent upon the chirality features (n,m), dictate the different mode of sensing applications [47]. For example, there are several types of sensors manufactured on the basis of rolling dimensions like the ionization sensors, sorption sensors, capacitance sensors, resonance frequency shift gas sensors, etc. The mechanism involved with all the possible interacting CNT-based sensors with the adsorbing molecules lies in the fact of charge transfer interactions taking place between the adsorbed molecules and the surface area of the CNTs (both SWCNTs and MWCNTs). Upon functionalization of the pristine CNTs, the adsorbing molecules like the –COOH, –OH, $-NH_2$, etc. play a significant role in sensing the surrounding environmental milieu in an augmented fashion.

The different sensing potentialities and applications of carbon-based nanomaterials have been depicted in a schematic fashion in Fig. 15.1.

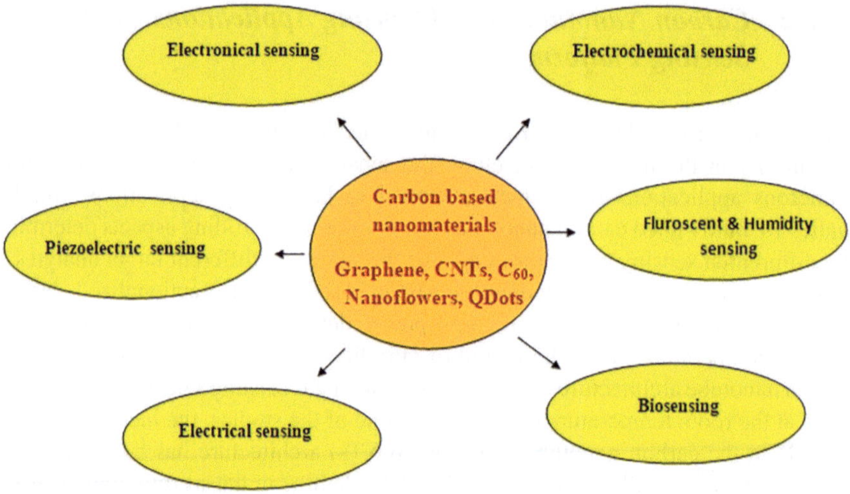

Fig. 15.1 Schematic of the myriad of sensing applications of carbon-based nanomaterials

15.2.3 Fullerene Based Sensing Platform: A 0D Sensing Platform

Owing to the 0D carbon-based nano-scale systems, the electronic configurations within the system play a significant role in acting as sensing platforms. The native structure of the fullerene (C_{60}) molecules, along with their different doped and functionalized versions, plays a significant factor in sensing different analytes. The electrochemical, physico-chemical, and biophysical aspects of the fullerene molecules determine the major breakthrough in finding sensing applications. One of the fullerene molecules (C_{20}) has been studied using ab initio and experimental means for the detection and removal of picric acid and allied toxic chemical contaminants [48, 49]. It has been studied that textile sensors doped with Fullerene/reduced graphene sensors for the detection of sweat characterization [50]. Humidity sensing platform has been designed based on the metal nanoparticle loaded fullerene molecules [51]. It has been reported that unfolded fullerene carbon nanoparticles have played a significant role determining chromium (VI) ions in the water systems using colorimetric means [52]. There have been studies which involve the interface fabrication of MWCNT-Fullerene nanocomposites for the detection of food additive vanillin [53].

15.2.4 Nanoflower for Sensing Applications: A 3D Sensing Platform

It is very much interesting to explore that 3D nanoflowers play a significant role in determining the different sensing applications. It is because of the electronic freedom in all the three dimensions, the interface of the underlying electronic phenomenon, and the target analyte electronic properties that the sensing phenomenon gets optimized and robust. It has been reported that tumor-targeting cholesterol decorated DNA nanoflowers have been extensively studied for the intracellular radiometric sensing [54]. Also, a study has been performed whereby the electrochemical detection between the interaction between daunorubicin and DNA molecule has been extensively studied by hybrid nanoflower-based electrodes [55]. It has been studied that horseradish peroxidase enzyme-encapsulated nanoflower structure has been used in acting as potential signal generation colorimetric biosensors [56]. A study focusses on the applications of Metal Organic Framework (MOF-Au@Pt) nanoflowers for the electrochemical detection of H_2O_2 in living cells [57]. Also, in one of the studies, the doping of N-doped porous molybdenum carbide nanoflowers has been studied for organophosphorus pesticide detection [58].

15.2.5 Carbon Dots as Sensing Platforms: A 0D Sensing Platform

Owing to the restrictions in the electronic movement in all the three axes, the 0D dots made up of carbon-based materials play a crucial role in multimodal imaging and tagging agents. The bottom-up synthesized carbon dots play a significant role in eliciting different sensing applications from electrochemical to piezoelectric, electronic sensing, etc. In one of the recent studies, carbon dots doped with copper atoms has been used for the multifunctional fluorescence detection of tetracycline and pH [59]. In one of the studies, a mechanism of On–Off–On-based fluorescent chemo sensor based on N/P co-doped carbon dots for the detection of microcystine LR has been reported [60]. In one of the further studies, Glutathione-modified N-doped carbon dots has been used in the sensitive and selective dopamine detection mechanism [61]. In one of the interesting studies, polyethyleneimine-functionalized carbon dots has been employed as a fluorescent probe for the detection and assessment of doxorubicin hydrochloride using inner filter effect [62]. A suitable comparison table enumerating different carbon-based sensing materials with different response rates and limits of detections employing different transducing sensing materials has been shown in Table 15.1.

Table 15.1 A comprehensive comparison table for different types of carbon-based nanomaterial (CNMs) sensing categories

Sl. No.	Sensing material	Sensor type	LOD	References
1	Graphene oxide	Electrochemical	1.5 nM	[9]
2	Multi-walled carbon nanotubes	Electrochemical	8.4 ± 0.3 nM	[10]
3	Reduced graphene oxide nanosheets	Electrical	0.001 nM (1 pM)	[11]
4	Electrolyte gated graphene FET	Electrical	0.3 nM	[33]
5	SWCNT/$C60Br_{24}$ composites	Chemiresistive	~50 ppb	[46]
6	MWCNTs/MnO_2/PANi Nanowires	Electrochemical	4.42×10^{-13} M	[65]
7	C-dots (CDs)	Conductometric	50 ppb	[92]

15.3 Working Principle of Carbon-Based Sensing Interface

The mechanism of action of the working principle of different carbon-based nanomaterials in the sensing attributes has been enumerated in this section. The analyte, transducer, and the amplifications of the received signals upon target-sensing interface have been played by the underlying physical phenomenon of electrons to sense the bio-analyte. The selectivity and sensitivity of the inherent carbon-based nanomaterials toward the different sensing applications mediated by transducing entity act as a mode of principle for the sensing applications for different sensing applications [63].

The sensing system exhibited by CNMs has been manifested for the detection of several biomolecules. The underlying electronic, physico-chemical, and chemical properties along with the extraordinary surface-to-volume ratio of the carbon nanomaterials make them an ideal candidate for choosing CNMs as preferred sensing materials. The schematic of the CNM sensing over different applications has been shown in Fig. 15.2. The higher surface area like 2630 m^2/gram, higher conductivity, and higher mechanical stability along with underlying physico-chemical properties make the CNMs an ideal platform for sensing and labeling different analytes from biomolecular to organ-specific imaging applications. The electronic transfer rate along with the conductance phenomenon taking place between the CNMs and the target organs [64] makes them a potential candidate for the charge transfer activities between the interacting species.

Such molecular level interactions taking place between the chosen CNMs and the interacting molecules make the interplay of the CNMs and the organ-specific interactions feasible and result in the production of calculatable sensing responses.

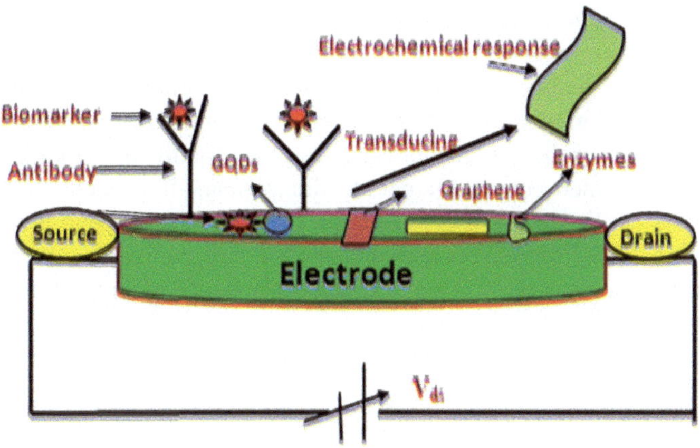

Fig. 15.2 Schematic of the carbon nanomaterial sensing mechanism

15.4 Carbon-Based Nanomaterials for Biomedical Applications

Since carbon-based nanomaterials have unique properties, such as electrical, thermal, mechanical, optical, and structural diversity, they have attracted huge attention as sensing materials in biomedical applications. With the aid of these potential and unique peculiar properties, carbon-based nanomaterials including graphene oxide (GO), carbon nanotubes (CNT), fullerene (C_{60}), and graphene quantum dots (GQDs) have been broadly investigated in biomedical applications. Some most recent development of carbon-based nanomaterials in biosensing has been studied by several authors, whereby the intrinsic properties of the different CNMs were explored for their unique optical, photoluminescence, electrical, and electronic activities [65]. The biosensing attributes of the different classes of CNMs depends upon their available surface functional groups, underlying electronic mobility, surface-to-volume ratio, etc. [66]. The transducer element, which is primarily the different tailor-made carbon elements, aids in transducing the received signals (bio-analyte) in the receptor for conversion to the optimal and desired output signals which are measurable in nature. In this case, different varieties of CNMs play a pivotal role in the ultrafast and sensitive response rates for different bio-analytes, making the biosensor adaptable and acceptable for myriad biosensing applications.

15.4.1 Carbon Nanotubes (CNTs) as Biosensors

Historically, CNTs have been explored as brand new nanoprobes because of their exceptional electronic, mechanical, optical, and structural properties [67]. They are exceptionally suitable for biosensing applications in the healthcare industry due to their high conductivity, excellent chemical stability, and sensitivities [68], and rapid electron-transfer rates [69] make them extraordinarily fit for biosensing applications in the healthcare industry. The most important and primary element of carbon nanotubes-based biosensors is the surface immobilization of biomolecules which therefore insist the identification and most importantly the signal transduction process. Based on their potential target identification and transduction mechanisms, these currently utilized biosensors are immensely enumerated into electronic and electrochemical CNT biosensors and optical CNT biosensors. Moreover, the CNTs have been illustrious as most promising materials for developing electron transfer potentials, which accomplish them earmark for connecting electrochemical and electronic biosensors [70–75]. The graphical illustration of the CNT sensing mechanism has been shown in Fig. 15.3.

Numerous CNT-glucose biosensors have been designed and implemented in the real-time field with the glucose oxidase conjugation. Carbon Nanotube non-woven Textures (CNTFs) are utilized to sense glucose from a glucose oxidase-impregnated polyvinyl liquor solution [76]. In addition, the impact of surface chemistry and the structure of glucose oxidase-coated MWCNT have been complemented in electrochemical glucose detection [77]. For more versatile applications, CNTs electrochemical biosensors have been anticipated for detecting nitric oxide and sensing epinephrine [78, 79].

As of late, the label-free discovery of person proteins' efflux from Escherichia coli (microbes) and Pichia pastoris (yeast) in genuine time was achievable by utilizing SWCNT [80], whereas [81] fortuitously developed a MWCNT-based device

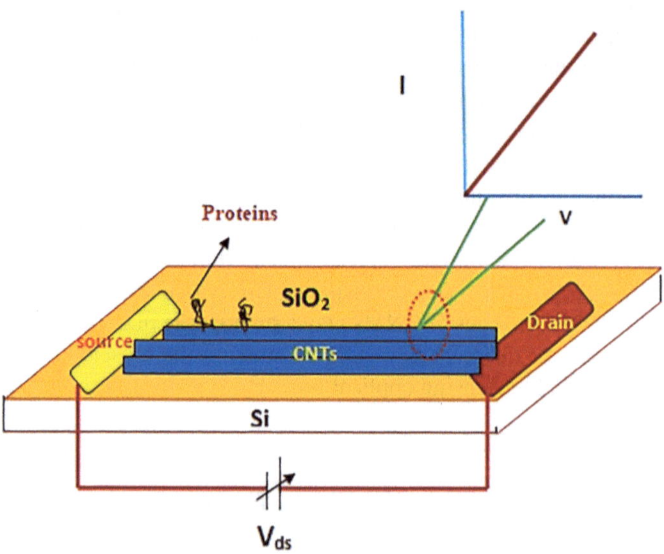

Fig. 15.3 Schematic illustration of the CNT-based sensing system

detecting arginase-1. The Zhou group has detailed the DNA-mediated surface-enhanced Raman scrambling (SERS) property of SWNTs, which allowed the ultra-sensitive location of a wide run of *ct*DNA in human blood [82]. Moreover, their photophysical properties, such as fluorescence outflow within the NIR locale and great photo solidness, make SWCNTs successful optical tests in biomedicine. With growing use of CNTs, the single-stranded DNA-functionalized SWCNTs were planned to counter to the lipid substance within the endosomal lumen of live cells and from the NIR photoluminescence of the SWCNTs, the lipid substance was measured through solvatochromic shift [83].

15.4.2 Graphene Oxide as Biosensor

Graphene oxide from the carbon materials group is accomplished by connecting with the probe and/or for the transduction of a peculiar response dynamically toward the target molecules which establish GO as potential candidates to use in biosensor technology in the biomedical field. Most importantly, this transduction process is completed by the techniques such as fluorescence, Raman scattering, and electrochemical reaction that recommended strongly for using GO as biosensors [84, 85], and some recent works on the progress of GO-based biosensing applications has been discussed. Graphene nanomaterials have been largely utilized for the fussy single- and double-stranded DNA electrochemical sensing [86, 87], which could be associated with the significant and strong electrochemical properties of graphene. In addition, with the possible ionic interaction between the adversely charged –COOH group and the emphatically charged nucleobases, and the vigorous π–π stacking between the nucleobases and honeycomb carbon system, the graphene nanomaterials works more efficiently during the real-time application in the biomedical field. Zhang and co-workers created carboxyl (–COOH)-functionalized GO and polyaniline (PANI)-altered GO and effectively recognized DNA by means of DPV [88]. Moreover, Johnson and co-workers built a label-free DNA biosensor associated with graphene field impact transistors (GEFTs) functionalized with single-stranded test DNA, and this gadget advertised a wide expository run [89]. The schematic illustration of the graphene-based sensing system has been shown in Fig. 15.4.

By the same group, a sensor device based on gold nanoparticle-decorated GEFTs (Au NP-Gr-FETs) was manufactured by the physical vapor testimony strategy and the thiol-functionalized Au NP-Gr-FETs able to identify DNA and displayed tall specificity against no complementary DNA [90].

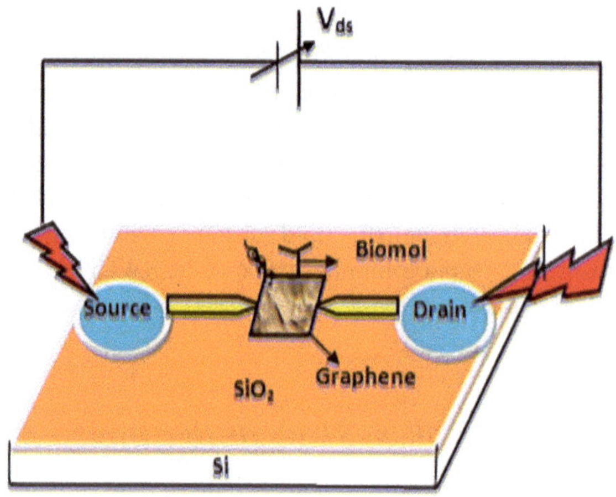

Fig. 15.4 Schematic illustration of the graphene-based sensing system

15.4.3 Graphene Quantum Dots (GQDs) as Biosensor

In the development and techno vent attitude of scientists, the carbon nanomaterials are very deeply investigated to explore its deep-rooted potentials like GQD-based biosensors which have broadly been advanced for various practical applications in clinical analysis as well as disease diagnosis. On the basis of excellent holding potentials such as electro chemiluminescence (ECL), photoluminescence (PL), and electrochemical behaviors of GQD, these have been extensively utilized for discerning biomacromolecules including DNA, RNA, proteins, and glucose molecules with improved distinction and sensitivity [91, 92]. Owing to the potential properties of GQDs, there have developed of DNA probe-functionalized reduced GQDs to detect DNA, based on the Forster Resonance Energy Transfer (FRET) fluorescence sensing method [93]. Zhang et al. (2017) developed pyrene-1-boronic acid (PBA) functionalized GQD for glucose sensing. The Wei group arranged an electrochemifluorescent polyvinyl liquor (PVA)/GQD nanofiber for exceedingly touchy and specific location of both H_2O_2 and glucose [94]. The location of cancer cells in early organize of illness has gotten to be a perquisite worldview, and Wang et al. (2016) planned Pd nanoparticle-enhanced N-doped GQD (NGQD) for cancer detection. The NGQD@NC@Pd HNS hybrid material performed critical electrochemical reduction of H_2O_2 which inspires the plausibility of discovery of different living cancer cells [95].

15.5 Carbon-Based Nanomaterials for Agricultural Applications

In the current scenario of the world's agricultural development and need to establish the sustainable systems for food security, nanotechnology has a great role in the present era and the use of carbon nanomaterials (CNMs) in agriculture has proven to be a boon for the extensive development in the crop improvements. Upon harnessing the different unique sensing attributes of the different classes of CNMs, the crop improvement, crop protection from the invading pathogens, from natural calamity, etc., CNM-based biosensing systems and associated computational systems aid in the augmentation of crop production and protection from widespread spoilage from pathogens, thereby playing a significant role in the early sensing of crop spoilage and deterioration for betterment of productivity and overall crop management. A schematic illustration of the sensing applications in the agricultural domain has been shown in Fig. 15.5.

The objective of the current chapter is to have a perception on the potential applications of CNM-based sensors in the agricultural developments. Subsequently, several distinctive carbon-based NMs (e.g., carbon nanotubes, fullerenes, and graphene oxides) have been applied to the earthbound plants and related soil organisms, due to their conceivable applications in controlling plant developments [96, 97].

15.5.1 Carbon Nanotubes (CNTs) Sensing in Agriculture

Due to potential physiochemical properties and sustainable applications, CNT-based biosensors have excellent momentous recognitions, such as high sensitivity, strong luminescence properties, quick response, and immense stability [98]. Various and unique types of sensors are marked for monitoring the existing pollutant/species in

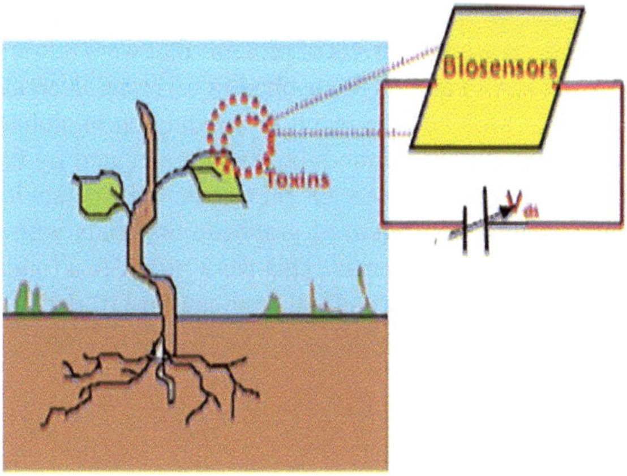

Fig. 15.5 Schematic illustration of biosensors in agricultural applications

the medium. Moreover, the peculiar applications of CNT biosensors are employed to detect aromatic and organic compounds and halogenated pesticides. The CNT biosensors are used in soil diagnosis to determine the corresponding activity of existing supportive and adverse microbes based on discreet oxygen consumption. CNT-based optical nano-biosensors are being investigated within the agricultural as well as food processing sector and nourishment preparing to screen the real-time location of pathogenic microbes [99], pesticides and organophosphate agents [100], proteins, toxic materials, etc. [101]. The 1D properties of CNTs speed up the ultrasensitive discovery of analyte since of surface particles, and minor trouble within the chemical environment that can furiously modify the electrical or optical properties [102]. Sensors are mainly needed to monitor the activities and the effects so that the immediate precaution can be taken. Besides the sensing applications of CNTs in the agriculture sectors, there different other potentials applications have been observed in plants such as germination and growth, seedings, root enhancement, and shooting [103]. It is noteworthy that CNMs have promising capacity to improve seed germination, plant development, supplement take-up, and natural product quality. Among the other CNMs, CNTs are broadly explored and have appeared promising positive impacts, with low to decently high dosages of CNTs progressing by and large plant development [104].

15.5.2 Graphene Oxide (GO) in Agricultural Sensing

The potential applications of graphene-based nanomaterials in agriculture are receiving much attention, but their potential risks to food safety and crop quality remain unknown. A recent study has revealed that graphene oxide (GO), graphene quantum dots (GQDs), and reduced graphene oxide (rGO) translocated from wheat stems to grains and formed large nanomaterial aggregates. In addition, the nanomaterials decreased wheat grain globulin, prolamin, amylose, and amylopectin contents by 8–28%, 11–25%, 5–34%, and 23–37%, respectively, and decreased mineral elements, while upregulating soluble sugar content by 19–36%; rGO, on the other hand, decreased the levels of proteins that were nutrient reservoirs to a greater extent than GO. rGO and GQD resulted in a greater decrease in mineral content in grains than GO, probably the result of a reduction in alpha-amylase inhibitor expression, and consequently, calmodulin was upregulated by the ABC transporter. GQD and rGO had greater effects on proteomic and metabolomic profiles than GO, which suggests that non-oxidized graphene materials with a low oxidation content have greater adverse outcomes. This work provides a foundation for future nanomaterial development and agricultural applications [105].

15.6 Limitations and Challenges of Carbonaceous Nanomaterials in Sensing

Owing to the extraordinary role and nature of basic organic molecule viz. carbon molecule, the same has been used in the allotropic formation of different structures ranging from sheets to rods to smaller dots. It has been due to the different bottom-up techniques and principles the electronic properties get guided with their native structures. The different shapes and forms of the carbonaceous structure dictate its myriad sensing applications ranging from electronical to humidity sensing.

The major challenges lying with the use of such carbonaceous architectures is predominantly the lack of production of the materials in laboratory conditions as well as the inherent purity of the synthesized materials. It has been found while testing with the biological systems and in vivo conditions that the carbon-based materials elicit severe deleterious toxicity to the biological systems, rendering their applications limited in different sensing platforms while employing in in vivo conditions.

15.7 Conclusion and Outlook

The breakthrough has been achieved in the domain of sensing using the carbon-based nanomaterials in a robust manner during the last decades. The use of different architectures of carbon nanomaterials like graphene sheets has been extensive in the sensing applications of globulin proteins, fibrillar proteins, etc., which has been broadly categorized under biosensing. The use of rod structured carbon nanotubes has been studied extensively for the detection of biomarkers in cancers, DNA sensing, etc. Similarly, the different architectures like nanoflowers, buckyball structures, etc. make the sensing exponentially faster and at a much affordable and sensitive detection limit regime. This approach of using the nano-scaled carbon materials paves the way in achieving a robust, augmented, sensitive, and selective analyte detection for all sorts of sensor fabrication in upcoming days, thereby outcompeting the conventionally available bulk-scaled sensors which bear compromised sensitivity and selectivity with delayed response time.

Acknowledgements The authors are thankful to Sathyabama Institute of Science & Technology, Chennai, and University of Science and Technology, Meghalaya, India, for providing the research facility.

Authors Contributions K. B. drafted and revised the manuscript, and Y. K. M. performed the revisions and necessary modifications for the submission of the book chapter.

Conflict of Interest The authors have no known conflict of interest.

References

1. Niknam, F.: Recent advances carbon nanomaterials as a new agent for breast cancer treatment. Adv. Appl. NanoBio Technol. 72–81 (2022)
2. Kościk, I., Jankowski, D., Jagusiak, A.: Carbon nanomaterials for theranostic use. C **8**(1), 3 (2022)
3. Li, K., Jiang, Y., Rao, W., Li, Y., Liu, X., Zhang, J., Xu, X., Lin, K.: Cooperative coupling strategy for constructing 0D/2D carbon nitride composites with strengthened chemical interaction for enhanced photocatalytic applications. Chem. Eng. J. **431**, 134075 (2022)
4. Sajna, M.S., Simon, S.M., Unnikrishnan, N.V., Sadasivuni, K.K.: An overview of graphene-based 2D/3D nanostructures for photocatalytic applications. Top. Catal. 1–25 (2022)
5. Chatzichristos, A., Hassan, J.: Current understanding of water properties inside carbon nanotubes. Nanomaterials **12**(1), 174 (2022)
6. Gaviria Rojas, W.A., Hersam, M.C.: Chirality-enriched carbon nanotubes for next-generation computing. Adv. Mater. **32**(41), 1905654 (2020)
7. Amir, D., Hendler-Neumark, A., Wulf, V., Ehrlich, R., Bisker, G.: Oncometabolite fingerprinting using fluorescent single-walled carbon nanotubes. Adv. Mater. Interfaces 2101591 (2021)
8. Zheng, Y., Chen, S., Zhang, K.A., Zhu, J., Xu, J., Zhang, C., Liu, T.: Ultrasound-triggered assembly of covalent triazine framework for synthesizing heteroatom-doped carbon nanoflowers boosting metal-free bifunctional electrocatalysis. ACS Appl. Mater. Interfaces. **13**(11), 13328–13337 (2021)
9. Cheraghi, S., Taher, M.A., Karimi-Maleh, H., Karimi, F., Shabani-Nooshabadi, M., Alizadeh, M., Al-Othman, A., Erk, N., Raman, P.K.Y., Karaman, C.: Novel enzymatic graphene oxide based biosensor for the detection of glutathione in biological body fluids. Chemosphere **287**, 132187 (2022)
10. Bravo, I., Prata, M., Torrinha, Á., Delerue-Matos, C., Lorenzo, E., Morais, S.: Laccase bioconjugate and multi-walled carbon nanotubes-based biosensor for bisphenol A analysis. Bioelectrochemistry **144**, 108033 (2022)
11. Biswas, K., Sett, A., Mondal, M., Tripathy, S., Bandyopadhyay, J., De, D., Bhattacharya, T.K.: Reduced graphene oxide nanosheets for selective picomolar detection of Bovine serum albumin. IEEE Trans. NanoBioscience (2021)
12. Li, J.W., Huang, C.Y., Chen, K.Y., Chen, J.X., Hsu, X.Y., Chen, Y.F., Kuo, C.F.J., Cheng, C.C., Suen, M.C., Chiu, C.W.: Enhanced piezoelectric properties of poly (vinylidenefluoride-co-trifluoroethylene)/carbon-based nanomaterial composite films for pressure sensing applications. Polymers **12**(12), 2999 (2020)
13. Majeed, N., Panigrahi, K.C., Sukla, L.B., John, R., Panigrahy, M.: Application of carbon nanomaterials in plant biotechnology. Mater. Today: Proc. **30**, 340–345 (2020)
14. Cui, X., Xu, S., Wang, X., Chen, C.: The nano-bio interaction and biomedical applications of carbon nanomaterials. Carbon **138**, 436–450 (2018)
15. Biswas, K., De, D., Bandyopadhyay, J., Sen, P.: Differential antibacterial response exhibited by graphene nanosheets toward gram-positive bacterium *Staphylococcus aureus*. IET Nanobiotechnol. **12**(6), 733–740 (2018)
16. Sah, U., Sharma, K., Chaudhri, N., Sankar, M., Gopinath, P.: Antimicrobial photodynamic therapy: single-walled carbon nanotube (SWCNT)-porphyrin conjugate for visible light mediated inactivation of *Staphylococcus aureus*. Colloids Surf., B **162**, 108–117 (2018)
17. Ali, S., Mehmood, A., Khan, N.: Uptake, translocation, and consequences of nanomaterials on plant growth and stress adaptation. J. Nanomater. **2021** (2021).
18. Wang, P., Jian, M., Zhang, C., Wu, M., Ling, X., Zhang, J., Wei, B., Yang, L.: Highly stable graphene-based flexible hybrid transparent conductive electrodes for organic solar cells. Adv. Mater. Interfaces **9**(3), 2101442 (2022)
19. Baig, N., Waheed, A., Sajid, M., Khan, I., Kawde, A.N., Sohail, M.: Porous graphene-based electrodes: advances in electrochemical sensing of environmental contaminants. Trends Environ. Anal. Chem. **30**, e00120 (2021)

20. Rabai, S., Teniou, A., Catanante, G., Benounis, M., Marty, J.L., Rhouati, A.: Fabrication of AuNPs/MWCNTS/chitosan nanocomposite for the electrochemical aptasensing of cadmium in water. Sensors **22**(1), 105 (2022)
21. Debnath, S.K., Srivastava, R.: Drug delivery with carbon-based nanomaterials as versatile nanocarriers: progress and prospects. Front. Nanotechnol. **3**, 15 (2021)
22. Haghighi, M., Khoshfetrat, A.: Au-O-MWCNTs and TiO_2-O-MWCNTs as efficient nanocarriers for dexamethasone: adsorption isotherms and kinetic studies. Int. J. Chem. Eng. **2021** (2021)
23. Ma, K., Li, W., Zhu, G., Chi, H., Yin, Y., Li, Y., Zong, Y., Guo, Z., Wang, L., Xu, W., Cui, C.: PEGylated DOX-coated nano graphene oxide as pH-responsive multifunctional nanocarrier for targeted drug delivery. J. Drug Target. **29**(8), 884–891 (2021)
24. Lage, T., Rodrigues, R.O., Catarino, S., Gallo, J., Bañobre-López, M., Minas, G.: Graphene-based magnetic nanoparticles for theranostics: an overview for their potential in clinical application. Nanomaterials **11**(5), 1073 (2021)
25. Riley, P.R., Narayan, R.J.: Recent advances in carbon nanomaterials for biomedical applications: a review. Curr. Opin. Biomed. Eng. **17**, 100262 (2021)
26. Karaphun, A., Phrompet, C., Tuichai, W., Chanlek, N., Sriwong, C., Ruttanapun, C.: The influence of annealing on a large specific surface area and enhancing electrochemical properties of reduced graphene oxide to improve the performance of the active electrode of supercapacitor devices. Mater. Sci. Eng., B **264**, 114941 (2021)
27. Duan, Z.H., Zhao, Q.N., Li, C.Z., Wang, S., Jiang, Y.D., Zhang, Y.J., Liu, B.H., Tai, H.L.: Enhanced positive humidity sensitive behavior of p-reduced graphene oxide decorated with n-WS2 nanoparticles. Rare Met. **40**(7), 1762–1767 (2021)
28. Song, S., Li, Y., Wang, Q., Zhang, C.: Boosting piezoelectric performance with a new selective laser sintering 3D printable PVDF/graphene nanocomposite. Compos. A Appl. Sci. Manuf. **147**, 106452 (2021)
29. Noorinezhad, E., Merati, A.A., Moazeni, N.: Enhancement of chromic-piezoelectric sensitivity responses of polyvinylidene fluoride/polydiacetylene nanofibers using graphene oxide. J. Polym. Res. **28**(12), 1–11 (2021)
30. Xia, S.Y., Long, Y., Huang, Z., Zi, Y., Tao, L.Q., Li, C.H., Sun, H., Li, J.: Laser-induced graphene (LIG)-based pressure sensor and triboelectric nanogenerator towards high-performance self-powered measurement-control combined system. Nano Energy 107099 (2022)
31. Mohammed, M.K., Al-Nafiey, A., Al-Dahash, G.: Manufacturing graphene and graphene-based nanocomposite for piezoelectric pressure sensor application: a review. Nano Biomed Eng. **13**, 27–35 (2021)
32. Yogeswaran, N., Navaraj, W.T., Gupta, S., Liu, F., Vinciguerra, V., Lorenzelli, L., et al.: Piezoelectric graphene field effect transistor pressure sensors for tactile sensing. Appl Phys Lett. **113** (2018). https://doi.org/10.1063/1.5030545
33. Ohno, Y., Maehashi, K., Yamashiro, Y., Matsumoto, K.: Electrolyte-gated graphene field-effect transistors for detecting pH and protein adsorption. Nano Lett. **9**(9), 3318–3322 (2009)
34. Sun, B., Pang, J., Cheng, Q., Zhang, S., Li, Y., Zhang, C., et al.: Synthesis of wafer-scale graphene with chemical vapor deposition for electronic device applications. Adv. Mater. Technol. **6**, 2000744 (2021). https://doi.org/10.1002/admt.202000744
35. Sett, A., Biswas, K., Majumder, S., Datta, A., Bhattacharyya, T.K.: Graphene and Its Nanocomposites Based Humidity Sensors: Recent Trends and Challenges. In: Intech (2021)
36. Zhang, D., Song, X., Wang, Z., Chen, H.: Ultra-highly sensitive humidity sensing by polydopamine/graphene oxide nanostructure on quartz crystal microbalance. Appl. Surf. Sci. **538**, 147816 (2021)
37. Zhu, S., Wang, J., Fan, W.: Graphene-based catalysis for biomass conversion. Catal. Sci. Technol. **5**(8), 3845–3858 (2015)
38. Rodwihok, C., Tam, T.V., Choi, W.M., Suwannakaew, M., Woo, S.W., Wongratanaphisan, D., Kim, H.S.: Preparation and characterization of photoluminescent graphene quantum dots from watermelon rind waste for the detection of ferric ions and cellular bio-imaging applications. Nanomaterials **12**(4), 702 (2022)

39. Velusamy, S., Roy, A., Sundaram, S., Kumar Mallick, T.: A review on heavy metal ions and containing dyes removal through graphene oxide-based adsorption strategies for textile wastewater treatment. Chem. Rec. **21**(7), 1570–1610 (2021)
40. Khan, Z.U., Khan, W.U., Ullah, B., Ali, W., Ahmad, B., Yap, P.S.: Graphene oxide/PVC composite papers functionalized with p-phenylenediamine as high-performance sorbent for the removal of heavy metal ions. J. Environ. Chem. Eng. **9**(5), 105916 (2021)
41. Zhang, H., Yang, D., Lau, A., Ma, T., Lin, H., Jia, B.: Hybridized graphene for supercapacitors: beyond the limitation of pure graphene. Small **17**(12), 2007311 (2021)
42. Lim, T., Suk, J.W.: Effect of the particle size of graphene oxide powders on the electrochemical performance of graphene-based supercapacitors. Funct. Compos. Struct. **3**(1), 015005 (2021)
43. Agyemang, F.O., Tomboc, G.M., Kwofie, S., Kim, H.: Electrospun carbon nanofiber-carbon nanotubes coated polyaniline composites with improved electrochemical properties for supercapacitors. Electrochim. Acta **259**, 1110–1119 (2018)
44. Zhu, Y., Chen, C., Wu, S., Cheng, R., Gao, J., Yu, Y., et al.: Phosphomolybdic acid-decorated carbon nanotubes for low-power sensing of NH_3 and NO_2 at room temperature. ACS Appl. Nano Mater. **4**, 1976–1984 (2021). https://doi.org/10.1021/acsanm.0c03333
45. Xi, W., Zhai, J., Zhang, Y., Tian, L., Zhang, Z.: Integrating brucine with carbon nanotubes toward electrochemical sensing of hydroxylamine. Microchim. Acta **187**, 1–9 (2020)
46. Zhou, J., Bagheri, M., Järvinen, T., Pravda Bartus, C., Kukovecz, A., Komsa, H.P., Kordas, K.: C60Br 24/SWCNT: a highly sensitive medium to detect H_2S via inhomogeneous carrier doping. ACS Appl. Mater. Interfaces. **13**(49), 59067–59075 (2021)
47. Nißler, R., Kurth, L., Li, H., Spreinat, A., Kuhlemann, I., Flavel, B.S., Kruss, S.: Sensing with chirality-pure near-infrared fluorescent carbon nanotubes. Anal. Chem. **93**(16), 6446–6455 (2021)
48. Sarvestani, M.J., Charehjou, P.: Fullerene (C20) as a potential adsorbent and sensor for the removal and detection of picric acid contaminant: DFT studies. Cent. Asian J. Environ. Sci. Technol. Innov. **2**, 12–19 (2021)
49. Khan, S., Naushad, M., Al-Gheethi, A., Iqbal, J.: Engineered nanoparticles for removal of pollutants from wastewater: status and prospects of nanotechnology for remediation strategies. J. Environ. Chem. Eng. **9**(5), 106160 (2021)
50. Zhang, J., Zhou, Q., Cao, J., Wu, W., Zhang, H., Shi, Y., et al.: Flexible textile ion sensors based on reduced graphene oxide/fullerene and their potential applications of sweat characterization. Cellulose. **28AD**(5), 3123–3133 (2021)
51. Tang, K., Chen, X., Ding, X., Zhao, X., Yu, X., Yu, X., et al.: Humidity sensitivity enhancement effects of metal nanoparticles loaded fullerene. Sens. Actuators B Chem. **329**, 129086 (2021)
52. Babazadeh, S., Bisauriya, R., Carbone, M., Roselli, L., Cecchetti, D., Bauer, E.M., et al.: Colorimetric detection of chromium(VI) ions in water using unfolded-fullerene carbon nanoparticles Vi (2021)
53. Taouri, L., Bourouina, M., Bourouina-Bacha, S., Hauchard, D.: Fullerene-MWCNT nanostructured-based electrochemical sensor for the detection of Vanillin as food additive. J. Food Compos. Anal. **100**, 103811 (2021)
54. Kim, N., Kim, E., Kim, H., Thomas, M.R., Najer, A., Stevens, M.M.: Tumor-targeting cholesterol-decorated DNA nanoflowers for intracellular ratiometric aptasensing. Adv. Mater. **33**, 2007738 (2021). https://doi.org/10.1002/adma.202007738
55. Findik, M., Bingol, H., Erdem, A.: Electrochemical detection of interaction between daunorubicin and DNA by hybrid nanoflowers modified graphite electrodes. Sens. Actuators B **329**, 129120 (2021)
56. Zeng, R., Wang, J., Wang, Q., Tang, D., Lin, Y.: Horseradish peroxidase-encapsulated DNA nanoflowers: an innovative signal-generation tag for colorimetric biosensor. Talanta **221**, 121600 (2021)
57. Wang, H., Chen, W., Chen, Q., Liu, N., Cheng, H., Li, T.: Metal-organic framework (MOF)-Au@Pt nanoflowers composite material for electrochemical sensing of H_2O_2 in living cells. J. Electroanal. Chem. **897**, 115603 (2021)

58. Zhou, L., Zhou, X., Zhao, C., Liu, Y., Li, Y., Ma, L., et al.: N-doped porous molybdenum carbide nanoflowers: a novel sensing platform for organophosphorus pesticides detecting. Microchem. J. **165**, 106169 (2021)
59. Guo, J., Lu, W., Zhang, H., Meng, Y., Du, F., Shuang, S., et al.: Copper doped carbon dots as the multi-functional fluorescent sensing platform for tetracyclines and pH. Sens. Actuators B Chem. **330**, 129360 (2021)
60. Wang, X., Yu, J., Ji, W., Arabi, M., Fu, L., Li, B., et al.: On–off–on fluorescent chemosensors based on N/P-codoped carbon dots for detection of microcystin-LR. ACS Appl Nano Mater. **4**, 6852–6860 (2021). https://doi.org/10.1021/acsanm.1c00921
61. Wang, L., Jana, J., Chung, J.S., Hur, S.H.: Glutathione modified N-doped carbon dots for sensitive and selective dopamine detection. Dye Pigm. 109028 (2020)
62. Yang, M., Yan, Y., Liu, E., Hu, X., Hao, H., Fan, J.: Polyethyleneimine-functionalized carbon dots as a fluorescent probe for doxorubicin hydrochloride by an inner filter effect. Opt Mater (Amst). **112**, 110743 (2021)
63. Speranza, G.: Carbon nanomaterials: synthesis, functionalization and sensing applications. Nanomaterials **11**(4), 967 (2021)
64. Gaur, M., Misra, C., Yadav, A.B., Swaroop, S., Maolmhuaidh, F.Ó., Bechelany, M., Barhoum, A.: Biomedical applications of carbon nanomaterials: fullerenes, quantum dots, nanotubes, nanofibers, and graphene. Materials **14**(20), 5978 (2021)
65. Tran, L.T., Tran, H.V., Tran, T., Nguyen, N.T., Van Bui, D., Tran, P.Q., Van Chu, T.: A highly sensitive electrochemical DNA sensor based on nanostructured electrode of multi-walled carbon nanotubes/manganese dioxide nano-flowers-like/polyaniline nanowires nanocomposite. J. Electrochem. Soc. **168**(5), 057518 (2021)
66. Zan, X., Bai, H.: Novel carbon nanomaterials based flexible electrochemical biosensors. J. Electrochem. Soc. **168**(2), 027504 (2021)
67. Tîlmaciu, C.-M., Morris, M.C.: Carbon nanotube biosensors. Front Chem. **3**, 59 (2015). https://doi.org/10.3389/fchem.2015.00059
68. Zhao, Q., Gan, Z., Zhuang, Q.: Electrochemical sensors based on carbon nanotubes. Electroanal. An. Int. J. Devoted Fundam. Pract. Asp. Electroanal. **14**, 1609–1613 (2002)
69. Lin, Y., Lu, F., Tu, Y., Ren, Z.: Glucose biosensors based on carbon nanotube nanoelectrode ensembles. Nano Lett. **4**, 191–195 (2004)
70. Jacobs, C.B., Peairs, M.J., Venton, B.J.: Review: carbon nanotube based electrochemical sensors for biomolecules. Anal. Chim. Acta **662**, 105–127 (2010). https://doi.org/10.1016/j.aca.2010.01.009
71. Holzinger, M., Le Goff, A., Cosnier, S.: Nanomaterials for biosensing applications: a review. Front Chem. **2** (2014)
72. Kumar, S., Ahlawat, W., Kumar, R., Dilbaghi, N.: Graphene, carbon nanotubes, zinc oxide and gold as elite nanomaterials for fabrication of biosensors for healthcare. Biosens. Bioelectron. **70**, 498–503 (2015)
73. 29. Wang, Y., Wei, H., Lu, Y., Wei, S., Wujcik, E.K., Guo, Z.: Multifunctional carbon nanostructures for advanced energy storage applications. Nanomaterials **5** (2015)
74. Yang, N., Chen, X., Ren, T., Zhang, P., Yang, D.: Carbon nanotube based biosensors. Sens. Actuators B Chem. **207**, 690–715 (2015). https://doi.org/10.1016/j.snb.2014.10.040
75. Zribi, B., Roy, E., Pallandre, A., Chebil, S., Koubaa, M., Mejri, N., et al.: A microfluidic electrochemical biosensor based on multiwall carbon nanotube/ferrocene for genomic DNA detection of Mycobacterium tuberculosis in clinical isolates. Biomicrofluidics **10**, 14115 (2016)
76. Zhu, L., Deng, C., Chen, P., Dong, X., Su, Y.H., Yuan, Y. et al.: Glucose oxidase biosensors based on carbon nanotube non-woven fabrics. Carbon N Y. **67** (2014)
77. González-Gaitán, C., Ruiz-Rosas, R., Morallón, E., Cazorla-Amorós, D.: Effects of the surface chemistry and structure of carbon nanotubes on the coating of glucose oxidase and electrochemical biosensors performance. RSC Adv. **7**, 26867–26878 (2017). https://doi.org/10.1039/C7RA02380D

78. Ulissi, Z.W., Sen, F., Gong, X., Sen, S., Iverson, N., Boghossian, A.A., et al.: Spatiotemporal intracellular nitric oxide signaling captured using internalized, near-infrared fluorescent carbon nanotube nanosensors. Nano Lett. **14**, 4887–4894 (2014)
79. Mphuthi, N.G., Adekunle, A.S., Ebenso, E.E.: Electrocatalytic oxidation of epinephrine and norepinephrine at metal oxide doped phthalocyanine/MWCNT composite sensor. Sci. Rep. **6**, 26938 (2016). https://doi.org/10.1038/srep26938
80. Landry, M.P., Ando, H., Chen, A.Y., Cao, J., Kottadiel, V.I., Chio, L., et al.: Single-molecule detection of protein efflux from microorganisms using fluorescent single-walled carbon nanotube sensor arrays. Nat. Nanotechnol. **12**, 368–377 (2017)
81. Baldo, S., Buccheri, S., Ballo, A., Camarda, M., La Magna, A., Castagna, M.E., et al.: Carbon nanotube-based sensing devices for human arginase-1 detection. Sens. Bio-Sens. Res. **7**, 168–173 (2016). https://doi.org/10.1016/j.sbsr.2015.11.011
82. Zhou, Q., Zheng, J., Qing, Z., Zheng, M., Yang, J., Yang, S., et al.: Detection of circulating tumor DNA in human blood via DNA-mediated surface-enhanced raman spectroscopy of single-walled carbon nanotubes. Anal. Chem. **88**, 4759–4765 (2016)
83. Jena, P.V., Roxbury, D., Galassi, T.V., Akkari, L., Horoszko, C.P., Iaea, D.B., et al.: A carbon nanotube optical reporter maps endolysosomal lipid flux. ACS Nano **11**, 10689–10703 (2017)
84. Kim, S.-W., Kyung Lee, Y., Yeon Lee, J., Hee Hong, J., Khang, D.: PEGylated anticancer-carbon nanotubes complex targeting mitochondria of lung cancer cells. Nanotechnology **28**, 465102 (2017)
85. Suvarnaphaet, P., Pechprasarn, S.: Graphene-based materials for biosensors: a review. Sensors **17** (2017)
86. Liu, Y., Dong, X., Chen, P.: Biological and chemical sensors based on graphene materials. Chem. Soc. Rev. **41**, 2283–2307 (2012). https://doi.org/10.1039/C1CS15270J
87. Tang, L., Wang, Y., Li, J.: The graphene/nucleic acid nanobiointerface. Chem. Soc. Rev. **44**, 6954–6980 (2015). https://doi.org/10.1039/C4CS00519H
88. Cheng, D., Yang, L., Li, X., Zhou, J., Chen, Q., Yan, S., et al.: An electrochemical DNA sensing platform using carboxyl functionalized graphene as the electrode modified material. J. Electrochem. Soc. **164**, H345–H351 (2017)
89. Ping, J., Vishnubhotla, R., Vrudhula, A., Johnson, A.T.C.: Scalable production of high-sensitivity, label-free DNA biosensors based on back-gated graphene field effect transistors. ACS Nano **10**, 8700–8704 (2016). https://doi.org/10.1021/acsnano.6b04110
90. Gao, Z., Kang, H., Naylor, C.H., Streller, F., Ducos, P., Serrano, M.D., et al.: Scalable production of sensor arrays based on high-mobility hybrid graphene field effect transistors. ACS Appl. Mater. Interfaces **8**, 27546–27552 (2016). https://doi.org/10.1021/acsami.6b09238
91. Xie, R., Wang, Z., Zhou, W., Liu, Y., Fan, L., Li, Y., et al.: Graphene quantum dots as smart probes for biosensing. Anal. Methods **8**, 4001–4016 (2016). https://doi.org/10.1039/C6AY00289G
92. Sawalha, S., Moulaee, K., Nocito, G., Silvestri, A., Petralia, S., Prato, M., Bettini, S., Valli, L., Conoci, S., Neri, G.: Carbon-dots conductometric sensor for high performance gas sensing. Carbon Trends **5**, 100105 (2021)
93. Qian, Z.S., Shan, X.Y., Chai, L.J., Ma, J.J., Chen, J.R., Feng, H.: A universal fluorescence sensing strategy based on biocompatible graphene quantum dots and graphene oxide for the detection of DNA. Nanoscale **6**, 5671–5674 (2014). https://doi.org/10.1039/C3NR06583A
94. Zhang, D.-Y., Zheng, Y., Tan, C.-P., Sun, J.-H., Zhang, W., Ji, L.-N., et al.: Graphene oxide decorated with Ru(II)-polyethylene glycol complex for lysosome-targeted imaging and photodynamic/photothermal therapy. ACS Appl. Mater. Interfaces **9**, 6761–6771 (2017)
95. Xi, J., Xie, C., Zhang, Y., Wang, L., Xiao, J., Duan, X., et al.: Pd nanoparticles decorated N-doped graphene quantum dots@N-doped carbon hollow nanospheres with high electrochemical sensing performance in cancer detection. ACS Appl. Mater. Interfaces **8**, 22563–22573 (2016)
96. Khot, L.R., Sankaran, S., Maja, J.M., Ehsani, R., Schuster, E.W.: Applications of nanomaterials in agricultural production and crop protection: a review. Crop Prot. **35**, 64–70 (2012)

97. Mukherjee, A., Majumdar, S., Servin, A.D., Pagano, L., Dhankher, O.P., White, J.C.: Carbon nanomaterials in agriculture: a critical review. Front. Plant Sci. **7**, 1–16 (2016)
98. Patel, D.K., Kim, H., Dutta, S.D., Ganguly, K., Lim, K.: Carbon nanotubes-based nanomaterials and their agricultural and biotechnological applications. Materials (Basel). **13**, 1679 (2020)
99. Sanvicens, N., Pastells, C., Pascual, N., Marco, M.-P.: Nanoparticle-based biosensors for detection of pathogenic bacteria. Trends Anal. Chem. **28**, 1243–1252 (2009)
100. Simonian, A.L., Good, T.A., Wang, S.-S., Wild, J.R.: Nanoparticle-based optical biosensors for the direct detection of organophosphate chemical warfare agents and pesticides. Anal. Chim. Acta **534**, 69–77 (2005)
101. Nam, J.-M., Thaxton, C.S., Mirkin, C.A.: Nanoparticle-based bio-bar codes for the ultrasensitive detection of proteins. Science **301**, 1884–1886 (2003)
102. Kruss, S., Hilmer, A.J., Zhang, J., Reuel, N.F., Mu, B., Strano, M.S.: Carbon nanotubes as optical biomedical sensors. Adv. Drug Deliv. Rev. **65**, 1933–1950 (2013)
103. Chen, J., Zeng, X., Yang, W., Xie, H., Ashraf, U., Mo, Z., Liu, J., Li, G., Li, W.: Seed priming with multiwall carbon nanotubes (MWCNTs) modulates seed germination and early growth of maize under cadmium (Cd) toxicity. J. Soil Sci. Plant Nutr. **21**(3), 1793–1805 (2021)
104. Mathew, S., Tiwari, D.K., Tripathi, D.: Interaction of carbon nanotubes with plant system: a review. Carbon Letters **31**(2), 167–176 (2021)
105. Li, X., Mu, L., Li, D., Ouyang, S., He, C., Hu, X.: Effects of the size and oxidation of graphene oxide on crop quality and specific molecular pathways. Carbon N Y. **140**, 352–361 (2018). https://doi.org/10.1016/j.carbon.2018.08.063

Chapter 16
Development of Carbon Dots and Nanohybrids for Biosensing and Bioimaging Relevance

Vijay Bhooshan Kumar and Dambarudhar Mohanta

Abstract In the last few decades, fluorescent studies have become one of the most effective methods for biosensing and bioimaging. Novel fluorescent nanomaterials with tunable properties are currently in great demand for the rapid and practical analysis of biological samples. Until now, the impact of the chemical modification of the passivating molecule on the fabrication of a carbon dots (CDs) nanomaterial-based sensing probe has not been fully explored. Therefore, this book chapter examined several different precursors (such as protein, polyethylene glycol, urea, thiourea, cysteine, and glycine) to develop eco-friendly CD and doped CD formulations by bottom-up chemical synthesis approach for smart biosensors. CDs and doped CDs have received special attention in terms of low toxicity, high solubility in many solvents, chemical modification, and mechanical properties, large active surface area, abundant functional groups, and edge sites for many different organic molecules/drugs. In addition, we have investigated the properties of CDs in medicine with regards to bioimaging, switchable luminescence, and biosensors.

Keywords CDs · Synthesis and characterization · Application · Biosensor · Live cell imaging · Bioimaging · Nanomedicine

16.1 Introduction

Nanomaterials are broadly classified as organic and inorganic. Since the discovery of carbon allotropes such as fullerenes [1], carbon nanotubes [2], and graphene [3], that have attracted significant attention due to their exclusive properties and impact on the range of applications, CDs are emerging as a novel nanomaterial and exhibit

V. B. Kumar (✉)
Faculty of Life Science, Tel Aviv University, 6997801 Tel Aviv, Israel
e-mail: vijaybhushan86@gmail.com

D. Mohanta
Department of Physics, Tezpur University, Tezpur, Assam 784028, India

D. Mohanta and P. Chakraborty (eds.), *Nanoscale Matter and Principles for Sensing and Labeling Applications*, Advanced Structured Materials 206,
https://doi.org/10.1007/978-981-99-7848-9_16

unique properties originating from zero-dimensional morphological features and a high surface area and specific physicochemical properties, making them promising building blocks and platforms for scientific studies and technological development [4, 5]. As typical examples, CDs and doped CDs demonstrate various applications in photovoltaic devices [6], photocatalysis [7], batteries [8], supercapacitors [9], sensors [10], fuel cells [11], nanomedicine [12], bioimaging [13], catalysis [14], and optoelectronics [15], and exhibit superior performances compared to other conventional nanomaterials, such as metal/metal oxide and organic semiconductors [16]. Therefore, highly fluorescent CDs have acquired increasing research attention in the last few years [13, 17–19]. CDs are quasi-zero-dimensional nanomaterials, which have less than 10 nm size in all three dimensions [13]. Recently, doped CD nanomaterials (N, P, B, Ga, In, Ag, Au, etc.) have also gained significant attention due to their interesting optical and electrical properties [5, 20, 21].

Various starting materials have been used to synthesize CDs and doped CDs via deferent routes over the past decade years [22, 23]. Popular approaches include microwave-mediated synthesis, hydrothermal synthesis, laser ablation of graphite, electrooxidation of graphite, oxidation of candle soot, sonochemistry, and the thermal cracking of organic compounds [22]. Among them, hydrothermal synthesis is the most used approach for CDs and doped CD synthesis because it is simple and efficient. Furthermore, the surface and carbo-genic core of CDs can be easily modified by versatile engineering methods to endow the CDs with desired physicochemical and optical properties. Despite the growing attention to the biological applications of CDs, a review for this book chapter is focusing on their applications in bioimaging and sensor. In this book chapter, we provide an overview of the recent research on the synthesis and design of CDs/doped CDs including their preparation, unique optical properties, and chemical modification to enhanced bioimaging and sensor applications.

16.2 Preparation of CDs and Doped CDs

Recently, researchers have developed two main routes for the synthesis of CDs: (i) top-down method and (ii) bottom-up approach (Fig. 16.1). The top-down method refers to breaking down larger carbon structures via chemical oxidation, discharge, electrochemical oxidation, and ultrasonic methods [24]. However, this top-down method has a few drawbacks, including harsh reaction conditions, expensive materials, and long reaction time. Alternatively, the bottom-up approach refers to converting carbon-based small molecule/structures into CDs of the desired size. These bottom-up methods include the chemical synthesis routes: sonochemical, microwave, hydrothermal thermal decomposition, pyrolysis, carbonization synthesis, and solvothermal. Here, we will discuss a primarily bottom-up approach for the synthesis of CDs obtained from small organic molecules [9].

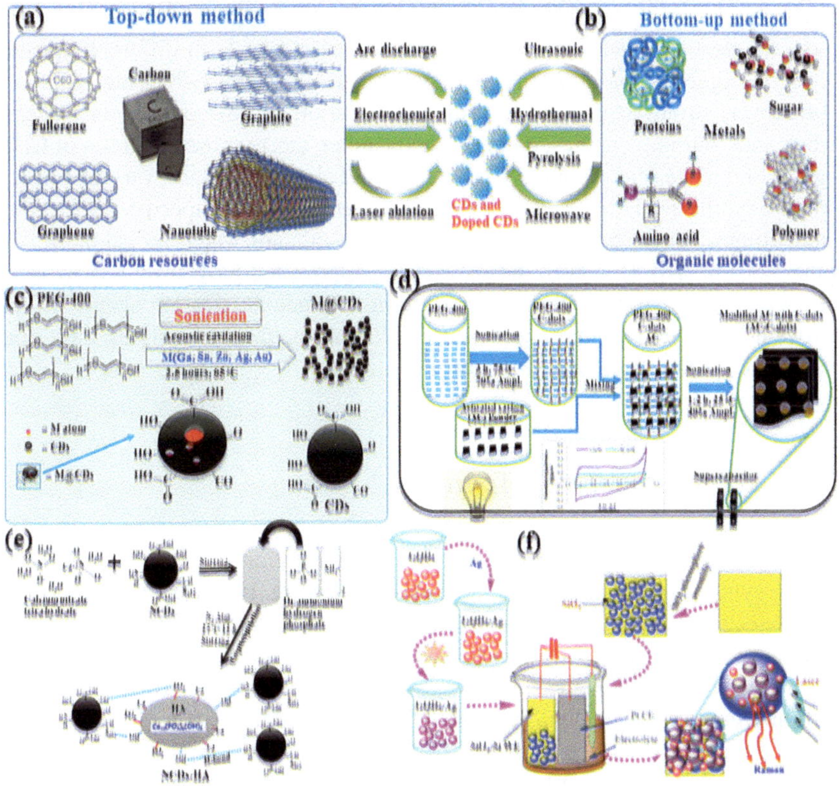

Fig. 16.1 Schematic presentation of the **a** top-down and **b** bottom-up approaches for the synthesis of CDs [29]. **c** Synthesis procedure for metal-doped CDs from molten metal [30]. **d** CD decoration on modified activated carbon substrate for supercapacitor [31]. **e** Schematic illustration of the synthetic route of NCDs-HA nanoparticles and possible binding mechanism of NCDs with HA nanoparticles [32]. **f** Silver-doped CDs [33]. Reproduced with copyright permission

16.2.1 Synthesis of CDs by Top-Down Approach

This CD preparation method mainly includes the following methods: arc-discharge, laser ablation, electrochemical or strong acid oxidation, etc. [9, 12, 24, 25]. The first synthetic procedure used for the formulation of CDs was the arc-discharge method [26]. In the arc-discharge procedure, two new classes of nanomaterials are produced from the crude soot. One of them is a short, tubular single/multiwall carbon nanotube, while the other is a mixture of CDs or fluorescent carbon nanomaterials [26]. CDs are also synthesized by laser ablation of a carbon target in water vapor with argon as carrier gas followed by surface passivation by attaching simple organic species (Fig. 16.1a). However, researchers have also used different carbon-based nanomaterials such as graphite, nanocarbons, and coal to prepare CD by laser ablation method [19]. The electrochemical oxidation method is also a popular method to synthesize CDs using graphite as the working electrode along with platinum (Pt) as counter electrode and Ag/AgCl as the reference electrode in different aqueous electrolytes aqueous solution of H_2SO4, phosphate buffer, NaH_2PO_4, etc. [27].

Moreover, strong acid was used for the oxidation of different carbon sources (carbon soot, burning candles, combustion chamber, etc.) for the synthesis of CDs. Finally, solid nitric acid produces CDs. In all these above processes, various sizes of CDs were synthesized [22]. Then, they were separated by a high-speed centrifugation system using water–organic solvent mixture (water–ethanol, water–propanol, etc.) or another solvent [28]. However, these top-down approaches have several drawbacks such as series of experiments, time-consuming, surface passivating agents for the improvement of fluorescent properties, need of strong acid or electrolytes, long-time synthesis procedure, and finally not cost-effective.

16.2.2 CD by Bottom-Up Approach

To address the above drawbacks, researchers have developed by the bottom-up or chemical synthesis procedure for the synthesis of CDs (Fig. 16.1b). This methodology is mainly established on the polymerization reaction from small organic molecules followed by carbonization of organic molecules for the formation of CDs or doped CDs. These methods are microwave-assisted pyrolysis, hydrothermal, ultrasonic, dehydration/direct heating, pyrolysis, etc. The carbon-containing molecules such as glucose, fructose, sucrose, ascorbic acid, chitosan, glycerol, and ethylene diamine tetra acetic acid are used as carbon-based precursors for the preparation of CDs by the above methods (Fig. 16.1b) [22]. Some inexpensive bio-precursors are also reported to be used for the preparation of biocompatible water-soluble fluorescent CD by this approach. Hydrothermal is the most facile and self-passivated one-pot method for the preparation of water-soluble fluorescent CD containing large numbers of oxygenase groups. In this method, carbonization of small organic molecules takes place at an elevated temperature (150–200 °C) and pressure in the presence of water or liquid organic molecules [34]. In this vein, the sonochemical method is also developed to prepare CD using PEG-400 as a carbon source and as both solvent and coating agent. Briefly, 12 mL of polyethylene glycol (PEG 400) is placed in a test tube which is dipped in water/silicon oil bath, then an ultrasonic transducer is immersed up to 1 cm above the bottom of the tube and subjected to ultrasound irradiation [20]. The simple experimental setup of sonochemical synthesis of CDs is depicted in Fig. 16.1c. It also demonstrates the preparation of different metal-doped CDs (M = Ga, Sn, Zn, Au, Ag, etc.) and CD decorated activated carbon substrates for storing the energy in the form of supercapacitor (Fig. 16.1d) [24]. It is worth mentioning that the metal doping of the CDs was performed from the metal in their liquid state. Kumar et al. thoroughly investigated the effect of different experimental parameters such as temperature (4–150 °C), sonication time (3 min, 30 min, 1 h, 2 h, 2.5 h, and 3 h), and ultrasound amplitude (40–70%) [20].

16.2.3 Doped CD-Based Nanohybrids

The CD-metal-based nanohybrids are mainly prepared using three basic approaches: (1) physical mixing, (2) chemical reduction, and (3) hydrothermal/solvothermal methods. The physical technique is mainly prepared by mixing CD with the dispersion of metal nanoparticles by mechanical shearing followed by sonication. Kumar et al. prepared TiO_2/Sn@CDs nanohybrid by adding TiO_2 nanoparticles distribution into the aqueous solution of Sn@CD under ultrasonic conditions for 20 min and followed by vacuum drying at 80 °C for 12 h [35]. Lai et al. prepared Cu_2O/CDs nanohybrid using the ultrasonic method [36]. In contrast, Kumar et al. prepared hybrid nanostructured CD decorated Fe_3O_4 by ultrasonic treatment [37]. The prepared CDs and Fe_3O_4 nanospheres were dispersed in an aqueous solution by ultrasonication for 30 min. The mixture is then dried under vacuum at 25 °C for 12 h. Finally, the product is washed with ethanol followed by distilled water several times and finally dried at 25 °C for 12 h to obtain the desired nanohybrid [37].

On the other hand, Tang et al. have prepared CDs-BiVO4 nanosphere by refluxing the dispersion of $BiVO_4$ powder in an aqueous solution of CD at 90 °C for three hours. Gold-CD nanomaterials were synthesized by Mandal et al. by using anisotropic gold nano nanomaterials. Kumar et al. fabricated N-doped carbon dots-hydroxyapatite nanohybrid only by stirring at room temperature in water for bone tissue engineering. Kumar et al. have synthesized the N-doped CDs (NCDs) via hydrothermal reaction at 175 °C, followed by decomposition, polymerization, condensation, and carbonization of the BSA polymer [34]. This method finally yielded an NCD material with a complex surface structure and very high quantum yields (~44%) [38]. Using the same hydrothermal method, boron and phosphorus-doped CDs were also synthesized [5]. In order to synthesize N-doped CDs, a temperature-dependent mechanism has been proposed [38]. Using these NCDs, it was possible to functionalize hydroxyapatite nanomaterials (NCDs-HA) in water by stirring the hydroxyapatite (HA) aqueous solution with the NCDs at room temperature [32]. A schematic illustration of the functionalization HA nanoparticles with synthesized NCD materials in an aqueous solution is shown in Fig. 16.1e. NCDs were acting as stabilizing agents in three different compositions of HA particles containing 1–5% NCDs [32]. Furthermore, using an electrochemical method, Ge et al. were able to prepare aqueous GQD (graphene quantum dots, which are the same as CDs) (0.1 mg/mL) [39]. After this step, measured volumes of 0.1 M $AgNO_3$ solution (5–60 ml) were added to CDs (GQD) solutions (2 ml), respectively, and the mixtures were stirred for one hour to form homogeneous solutions of Ag-GQDs [39]. Following the preparation of the mixture (Fig. 16.1f), UV light at a wavelength of 365 nm (18 W) was used to illuminate it for three hours from a distance of 20 mm [29, 39]. Upon completion of the photochemical reaction, the color of the solutions changed from yellow to brownish red (depending on the volume ratio between GQDs and Ag ions), indicating the formation of silver-doped CDs (Ag-CDs). Ag-CD nanoparticles deposited on silica were engineered to form highly active surfaces that were enhanced by Raman scattering (Fig. 16.1f) for sensor applications. Similarly, microwave method is also developed for the preparation of CDs. In this method, 500 W microwave power is irradiated in the liquid reaction medium for 2–10 min and the reaction medium gradually transformed to dark brown from colorless solution [24]. It is a simple approach for the large-scale synthesis of water-soluble green carbon nanodots from different kinds of food waste-derived carbon resources. Again, Lai et al. synthesized highly emissive CD by pyrolysis of glycerol under reflux (~230 °C, 1 atmosphere air pressure) conditions for 30 min. In this work, the researchers also prepared PEG-modified CD using the same procedure [27]. In the final step, CD is formed from the aromatic clusters at a critical concentration of super-saturation point. Recently, Kumar et al. have reported the formation of ultrafine CDs [20], Sn-doped CDs, and Ga-doped C-dots [40] by ultrasonic cavitation of PEG-400 as a precursor. This synthesis procedure extended for the formation of both Sn@CDs@Sn nanoparticles and Sn@C-dots [41]. The structural and textural properties of the produced Sn@C-dots, as a primary carbon particle and secondary coating for metals, are closely linked to the reaction parameters of the sonolytic synthesis [41].

16.3 Physical and Chemical Properties of CDs

CDs are 2–10 nm in size, with sp^2 hybridized graphitic core functionalized with polar oxygenator groups. In UV–Vis's absorption spectrum, two types of absorbance peaks are observed at around 235–240 nm and 280–360 nm due to the $\pi-\pi^*$ transition of sp^2 core carbon and $n-\pi^*$ transition carbonyl functional groups of CDs. Kumar et al. synthesized NCDs from protein (bovine serum albumin) and analyzed their fluorescence properties at different excitation wavelengths ($\lambda ex = 330$, 350, 370, 390, 410, 430, 450, and 470 nm). The fluorescence emission spectra of the N-doped CDs showed narrow bands, with the maximum emission centered between 450 and 490 nm (Fig. 16.2a). The maximum fluorescence emission intensity was detected at the excitation wavelength of 390 nm, revealing the maximum intensity emission peak at 462 nm, implying a relatively narrow size distribution of the particles. The fluorescence was affected by metal doping, whereas the lattice parameter did not change parameters [30]. In addition, Kumar et al. also analyzed the graphitic nature of carbon by Raman spectroscopy of the N-doped CDs. This analysis revealed the presence of functional groups on the surface of N-doped CDs. Figure 16.2b illustrates the Raman spectra of N-doped CDs, with two major bands, one at the D-band at about 1346 cm^{-1}, attributed to the presence of sp^3 carbon defects, and another at the G-band at about 1566 cm^{-1}, attributed to the carbon G-band as well as the stretching vibration of C–C bonds (sp^2). These results indicate the presence of graphitic carbons in the N-doped CDs [38].

The XPS images shown in Fig. 16.2c confirm the presence of carbon, nitrogen, and oxygen on the surface of the N-doped CD particles. XPS spectral analysis confirmed the presence of essential elements like carbon and oxygen in the structure of CD, as

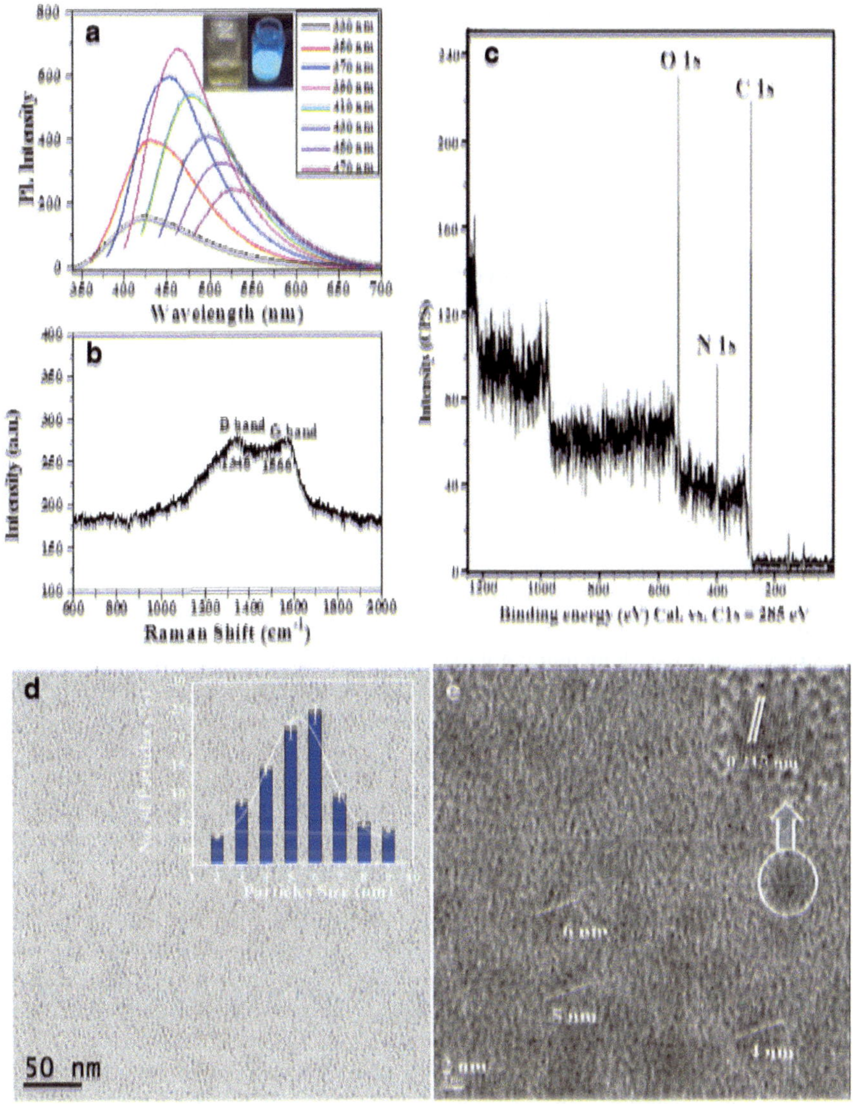

Fig. 16.2 **a** Fluorescence spectra of N-doped CD aqueous solution. **b** Raman spectra. **c** XPS full-scan range of N-doped CDs. **d** Transmission electron microscopy (TEM) images of a cluster of N-doped CDs. **e** Higher resolution image of the N-doped CDs. Inset: the crystal lattice of a single particle. Copyright with permission from Ref. [38] (2017), from American Chemical Society

the peaks at 384 and 532 eV indicate the binding energy of these elements (Fig. 16.2c). The high-resolution XPS spectra of C1s and O1s revealed C–C, C = C, C = O, C–O, and C–OH in the structure of CD. However, researchers have also reported nitrogen-containing groups in CD, which may appear due to nitrogen-containing precursors or doping or a functionalizing agent in CD. The HR-TEM images showed spherical N-doped CD particles with a relatively narrow size distribution (Fig. 16.2d, e). The average size of the N-doped CDs was about $\sim$5 nm, which agrees with the previous study's findings. Furthermore, the particle size of N-doped CDs was also confirmed by DLS, which is in reasonable agreement with the TEM analysis (3−8 nm). Also, CDs and N-doped CDs were chemically and physically stable at pHs 3.5, 5.0, 7.0, and 10.0. The XPS spectra of N@CDs (Fig. 16.2c) have three influential bands at around 288.66, 398.99, and 532.61 eV, which correspond to C1s, N1s, and O1s, respectively. The XPS analysis revealed that the nitrogen presents in the N@CD samples originated from the amino acid present in BSA. The hydrothermal reaction (>170 °C) causes fragmentation of the BSA polymer, which leads to the formation of a small-chain oligomer of the amino acid [38].

Kumar et al. have studied the morphology of the various metal-doped CDs (Ga, Sn, Ag, Zn, Au, etc.) by high-resolution TEM. The average particle size of synthesized metal-doped CDs is reported in the range of 4–7 nm. The shapes of the new CD nanoparticles were spherical with disk shape and particle size of 3–7 nm. EDS and XPS determined the elemental composition of the CDs; only C and O with traces of doped metal were found, as expected. The fluorescence was affected by metal doping, whereas the lattice parameter did not change parameters [5, 30].

16.4 Applications

CDs and CD-based nanomaterials have a unique blend of properties. CDs are very promising for many applications in the physical, chemical, and biological worlds. They offer the perspective of a C-based equivalent of optically active quantum dots (QDs) with the additional benefits of C-based materials, as the absence of toxic elements (typically heavy metal atoms), which limit, in particular, the use of QDs in bio-oriented applications. The bright emission of CDs, combined with the marked electron-donor capability, can be exploited in optoelectronic devices. The sensitivity and emission to metal ions in the solution can be measured with CD-based nanosensors and can recognize different metal/nonmetal ions. Their non-toxicity is a key advantage to perform in vivo and in vitro bioimaging experiments and drug delivery experiments. Indeed, a wide range of applications of CDs was reported, for example, antiviral, antimicrobial, antiparasitic, and antibacterial properties, and CDs were even used as nano weapons against mosquitos or as a component of fluorescent inks. So far, most CD-based devices have been developed and tested just in a laboratory environment. However, their commercial and industrial exploitation should not be too far away. The following paragraphs illustrate the technological versatility of CDs through a selection of different possible applications.

16.4.1 Nano Sensing Applications

In the literature, nanosensors based on CDs were developed by two different strategies: 1. Nanosensors can consist of "pure" CDs, as synthesized, or passivated through specific target groups. 2. The other oneis the functional CD nanocomposite, which can bind the target chemical or other nano/micro-materials. The use of CDs as "raw" sensors is quite common in the literature. Differently prepared CDs can be applied to detect various metal cations as Hg^{2+}, Cu^{2+}, Fe^{3+}, or small molecules or macromolecules in solution, even in a biological context [42, 43]. Sensing occurs due to the emission of quenching/enhancing due to the detachment, for instance, of the nanoparticle from the dot's surface induced by the analyte's presence. Cui et al. have developed CD-labeled oligodeoxyribonucleotide (ODN) and graphene oxide to improve the sensing performance of mercury ion detection (Fig. 16.3a) [44]. In the system, the fluorescence of ODN-CDs was quenched by graphene oxide via fluorescence resonance energy transfer (Fig. 16.3b). Upon addition of Hg^{2+}, the fluorescence was recovered by releasing CDs from GO due to the formation of T-Hg^{2+}-T duplex. The CD-based materials can sense a dynamic range of 5–200 nM of Hg^{2+} detection (Fig. 16.3b) [45]. Wang et al. utilized a similar fluorescence approach to detect Hg^{2+} ions by novel polyaniline-carbon dots (PANI/CDs) nanocomposites [45]. The addition of an Hg^{2+} ion, the solid binding affinity between the Hg^{2+} ion, and the amino groups of polyanilines made the CDs break away and release from the PANI, which resulted in the restoration of CDs' fluorescence [45]. CDs worked as energy donors, and polyaniline as a FRET acceptor. Therefore, deformation happened more quickly when it interacted with the nitrogen atoms of PANI. The sensor possessed great sensitivity and selectivity to Hg^{2+} detection, with a linear range of 0.05–1.0 mM and a limit of detection (LOD) of 0.8 nM[45]. Typically, when CDs are mixed with a specific analyte, their luminescence is quenched (decreases), allowing to detect the analyte with high sensitivity and, often, high selectivity, as obtained by a proper design of the surface functional groups, interacting only with a given type of metal cation. As a result, the detection sensitivity can be very high, and the minimum concentration of detectable ions can be as low as femtomolar [44, 45].

A fluorescent NCD-based sensing system was also recently developed by Guo et al. for the detection of environmental pollutants such as enrofloxacin in water solutions [46]. Based on this principle, a method for detecting enrofloxacin (ENR) utilizing the N-CDs-Cu^{2+} system was developed and applied to the analysis of ENR in tap water and river water. Figure 16.3c illustrates the mechanism for the detection of ENR through the sensing system [46]. Based on Fig. 16.3d, the fluorescence intensity of the N-CDs was dramatically reduced after the addition of Cu^{2+}. The PL intensity of the N-CDs-Cu^{2+} system was then restored after the addition of ENR. In Fig. 16.3d, it can be observed that the fluorescence intensity of the N-CDs gradually decreased with the gradual addition of Cu^{2+}, and the maximum wavelength was shifted to blue. This can be seen in the inset of Fig. 16.3d, where F_0 and F are the fluorescence intensities of N-CDs before and after adding Cu^{2+} [46]. The results will demonstrate the effectiveness of the proposed sensing system in detecting ENR in real water samples.

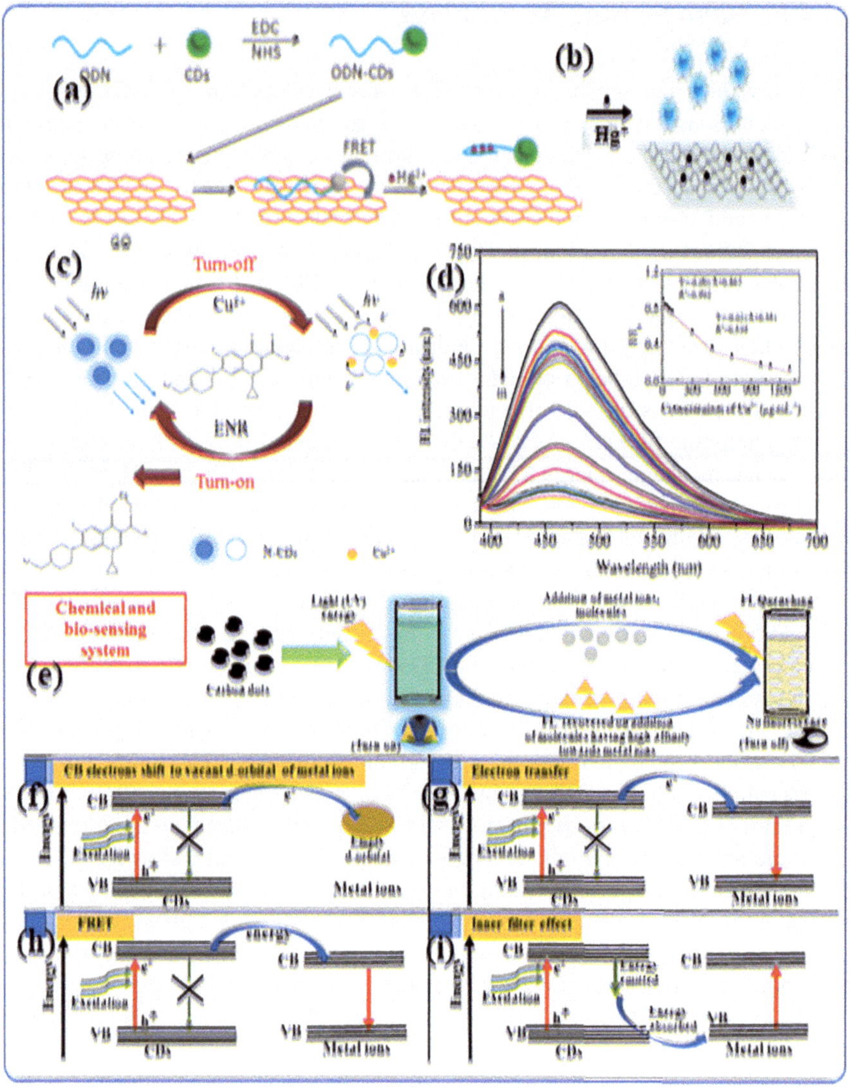

Fig. 16.3 **a** Schematic design of GO quenched ODN labeled carbon dots and the GO-CDs-ODN system for Hg detection due to the fluorescence recovery. **b** Representation of the fluorescence off–on mechanism of PANI-CDs probe for the detection of Hg^{2+}. Copyright with permission (2015) from Ref. [44] Elsevier. **c** Figure illustrating the process of ENR sensing using a fluorescent probe. **d** The fluorescence spectra of N-CDs after adding various amounts of Cu^{2+} ions (ranging from 0 to 1300 $\mu g{\cdot}mL^{-1}$). See inset for the relationship between F/F0 and Cu^{2+} concentration. Copyright with permission (2019) from Ref. [46] Elsevier. **e** Role of CDs in chemical and biosensing through fluorescence quenching: CDs change their fluorescence intensity because of many different mechanisms, including **f** Electrons in the conduction band of CD shift to the low-lying vacant d orbital of metal ions, **g** Transfer of electrons from the CB of CDs to the CB of metal ions, **h** FRET (fluorescence resonance energy transfer), and **i** An inner filter effect. Copyright with permission (2019) from Ref. [47] Springer Nature

Furthermore, Cui et al. highlight the advantages of nanoscale materials for sensing, biosensing, and chemical sensing due to their unique properties, including enhanced photostability, low cytotoxicity, and better aqueous solubility [48]. Their fluorescence properties are altered by a variety of mechanisms, including resonance energy transfer, inner filter effect, photo-induced electron, and charge transfer [49]. CD-based nanomaterials are capable of sensing several biomolecules and intracellular metal ions, including H_2O_2, Fe^{3+}, $C_6H_{12}O_6$, Vitamin B12, L-cysteine, galactose, etc. through changes in their fluorescence intensity (Fig. 16.3e–i) [47]. Upon preparation of the CDs, they exhibited peroxidase-like activity which could be used for the detection of H_2O_2 in the presence of 2, 2^1-azino-bis(3-ethylbenzothiazoline-6-sulphonic acid (ABTS) with a detection limit of 20 μM. Upon interaction between metal ions and surface functional groups of CDs, a new electron–hole recombination is initiated as a result of an energy transfer route, which changes the nature of the CD's fluorescence (Fig. 16.3e–i) [47]. This property of CDs can be utilized for the detection of different types of metal ions (Hg^{2+}, Ag^{+}, Cu^{2+}, Fe^{3+}).

Moreover, the main reason for quenching is much debated. It may involve different fundamental interactions, such as a charge transfer from the dot to the ions, which hinders radiative recombination, or an energy transfer from the CDs to the acceptor, quenching the emission of CDs and enhancing the emission of the latter. Because of such an imperfect understanding, the engineering of these applications still proceeds mainly through trial-and-error routes.

In the human body, improper glucose transport will correlate with diabetes and some cancers. Thus, monitoring glucose is critical to the prevention, diagnosis, and treatment of diseases. A methodology exploiting boronic acid has flourished in recognizing and detecting glucose. Chemical compounds possessing boronic acid motifs showed a strong affinity and a noticeable specificity toward glucose levels. Precisely, boronic acid will bind to 1,2- and 1,3-diols to form cyclic boronate ester. Researchers developed a series of CD-based nanomaterial with boronic acid functional groups on the surface, applying this principle [42, 50, 51]. For example, Qu et al. prepared fluorescent carbon dots with carboxylic groups on the surface, which attached with phenylboronic amino acid (APBA) as presented in Fig. 16.4a [51]. The CD functionalized boronic acid could react with glucose under covalent interactions, and the fluorescence was quenched. Qu et al. archived the linear range of 0.1–10 mM and a LOD of 5.0 mM for glucose detection [51]. An APBA-GQD exhibits a rapid and stable response to glucose in an aqueous solution. Glucose molecules have an effective quenching effect on the APBA-GQD system, where the attached APBA functional groups function as glucose receptors. The fluorescent spectra of the APBA-GQDs are shown in Fig. 16.4b as a function of glucose concentration [51]. Interestingly, Fig. 16.4c illustrates that disaccharides have a very different effect on APBA-GQD fluorescence compared to monosaccharides. The sugars sucrose, lactose, and maltose tend to enhance rather than quench the fluorescence of APBA-GQDs [51]. It is well discussed how APBA-GQDs exhibit good glucose specificity and have different responses to monosaccharides and disaccharides. As reported by Li et al., electrostatic attraction between CDs and N, N0-4,40-bis(benzyl-2-boronic acid)bi-pyridinium dibromide (BBV) resulted in the formation of a ground-state complex, leading to a decrease in the fluorescence (FL) intensity of CDs [50].

The boronic acids can be converted to tetrahedral anionic glucoboronate esters upon the addition of glucose, which effectively neutralizes the cationic bipyridinium's net charge, thus greatly diminishing its quenching efficiency and recovering the fluorescence intensity of CD-based materials [50]. These two analyses were applied synthesis and boronic acid modification of CDs separately. The synthesis and

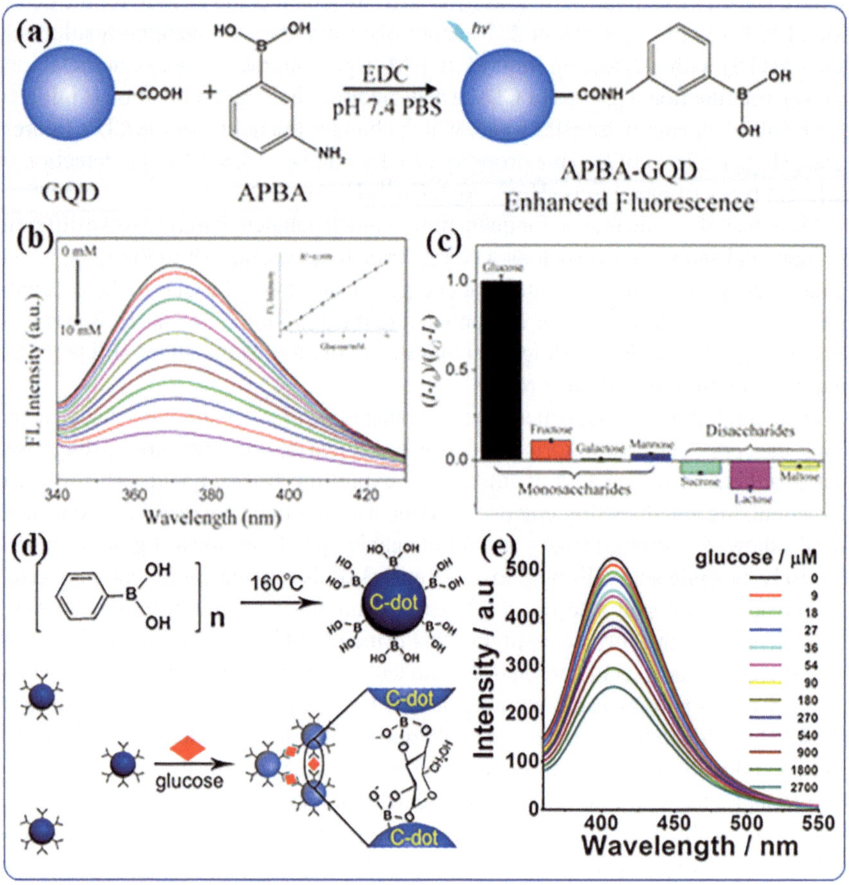

Fig. 16.4 **a** Schematic presentation of fluorescent CD functionalization with APBA. **b** Fluorescence characteristics of the APBA-GQDs as a function of glucose concentration. An inset shows a linear plot of the intensity of fluorescence quenching as a function of glucose concentration. **c** The relative intensity of fluorescence between different saccharides $(I-I_0)/(I_G-I_0)$. Copyright with permission from Ref. [51] (2013), from Royal Society of Chemistry. **d** Synthesized modification integration strategy for fabrication of boronic acid functionalized CDs, the working principle (fluorescence-off) for glucose detection, and the **e** corresponding fluorescence spectra of CDs at various glucose concentrations. Copyright with permission from Ref. [52] (2014), from American Chemical Society

modification of CDs could also be integrated into one step using specific synthetic precursors. Shen et al. also reported the facile fabrication of boronic acid functionalized CDs as a single precursor for sensing the glucose (Fig. 16.4d) [52]. Shen et al. found that increase in glucose concentrations, then the intensity of fluorescence gradually decreased [52]. The CD probe demonstrated a compassionate response to glucose detection in serum samples with a linear range of 9–900 mM and a LOD of 1.5 mM as shown in Fig. 16.4e. Moreover, the CD-based sensor showed a selective response to glucose detection compared with metal ions. It did not show any other fluorescence response against other saccharides, coexisting interferents in serum. Later, Misra et al. prepared CDs using citric acid as a precursor and aminophenyl boronic acid as a surface passivation agent and then applied them for glucose detection [53].

16.4.2 Biosensing and Biospecificity

In addition to the typical sensor, they can also be used CDs in biological sensing applications due to their excellent electronic and electrochemical properties. CDs are utilized for photoluminescence, electrochemiluminescence, sensors, and electrochemical sensors [54]. By utilizing the turn off/turn on photoluminescence properties of CDs, biological ions, biomolecules (sugar, antibodies, other sugars, amino acid), and proteins can be detected [52, 55, 56]. Cu^{2+} and Fe^{3+} ions are play a crucial role in the detection of diseases [57, 58]. Hence, CDs can be used for detecting such ions which are co-factors for specific disease.

Recently, Liu et al. have reported the facile synthesis of green fluorescence CDs with the help of new passivating agent poly (acrylate sodium) [59]. Furthermore, they did intracellular experiments and showed that folic acid (FA)-based CD probe could accurately distinguish folate receptor (FR)-positive cancer cells in cell mixture with normal cells through turn-on mode as depicted in Fig. 16.5a [59]. Furthermore, Jia et al. have reported forming magneto fluorescent Mn-CDs from manganese (II) phthalocyanine for cancer cell biosensors. Significantly, the Mn-CDs can not only effectively produce ROS but also highly catalyze the generation of oxygen. Thus, Mn-CDs will be a good biosensing material for tumor hypoxia and improved PDT [60]. In addition to the above work, we have also noticed that specific binding of the mannose units to *Escherichia coli* allowed quantitative detection of the bacteria at levels down to 450 colony forming units (CFU)/mL in the lab [61]. This experiment facilitated the application of the mannose-CDs for bacterial analyses of actual examples. Later, Chen and the group reported on the sensing of *E. coli* O157: H7 using water-soluble CDs as a probe [61]. It is observed that the *E. coli* caused an enhancement effect on the fluorescence emission of CDs with a linear range of 105–108 CFU/ml [61]. Modifying carbon dots with receptors was necessary and widely exploited by other researchers to detect bacterial sensitively and selectively. Nandi et al. presented a simple, analytical platform for bacterial detection and imaging based on the attachment of amphiphilic carbon dots to bacterial cells, as shown in

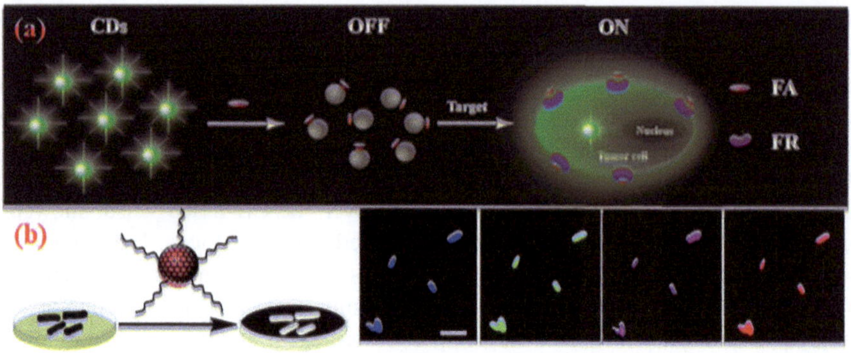

Fig. 16.5 **a** Schematic representation of the designed fluorescence-based biosensor for turn-on probe for imaging overexpressed cancer cells. Copyright with permission from Ref. [59], from Elsevier. **b** Detection methodology is based upon labeling bacteria with fluorescent CDs and the sensing of multicolor fluorescence microscopy images of *Escherichia coli*. Copyright with permission from Ref. [62], from Royal Society of Chemistry

Fig. 16.5b [62]. CDs functionalized with hydrocarbon chains readily bind to bacterial cells following short incubation and enabled bacteria detection through fluorescence spectroscopy and microscopy [62]. The dependence of intensity and spectral position of carbon dots' fluorescence upon bacterial species could offer a tool for distinguishing among bacteria even in mixed bacterial populations.

16.4.3 Bioimaging

Organic and inorganic fluorophore use in bioimaging is limited due to cytotoxicity and limited fluorescence, and conjugation with antibodies is needed for cellular uptake. Due to excellent optical properties, biocompatibility, and efficiency of cellular uptake, CDs are growing as an alternative for organic and inorganic fluorophores for bioimaging [63]. Kumar et al. illustrate a strategy for low-cost, one-pot synthesis of N-doped CDs. The low toxicity and suitability of the N-doped CDs for fluorescence imaging are validated using zebrafish as a model [38]. Strong fluorescence emission from zebrafish embryos and larvae confirms the distribution of N-doped CDs in zebrafish. Furthermore, the retention of N-doped CDs is very stable, long-lasting, and with no detectable toxicity [38]. It validated the concept by soaking zebrafish embryos in N-doped CDs. In Fig. 16.6a, one can notice the zebrafish embryos incubated with 6 mg mL^{-1} N-doped CD aqueous solution at 3- and 6-h post fertilization (hpf). The fluorescence indicates that N-doped CD nanoparticles quickly and spontaneously enter zebrafish embryos across the chorion and the germ ring around the yolk sac during soaking because of their smaller size. As reported previously, [64] the pore size of the ZF embryo chorion is 0.17 μm, which is much larger than N-doped CDs (~5 nm) [34, 64]. Therefore, N-doped CDs can easily penetrate ZF embryos through the pores of the chorion. It is noteworthy that zebrafish embryos exposed to 6 mg mL^{-1} N-doped CD solution at 24 hpf showed strong fluorescence, selectively in the yolk sac, trunk, and tail region, indicating the tissue-dependent affinity of N-doped CDs. The N-doped CD fluorescence emission in other parts of the zebrafish embryo at 24 hpf is very weak (Fig. 16.6b) [38].

Similarly, at 48 hpf, when the embryos hatched into larvae, the fluorescence emission of N@CDs in ZF larvae was primarily in the yolk sac. Furthermore, fluorescence was observed in the head (mainly eyeball) and on the longitudinal melanophore strip at the rear part of the body (Fig. 16.6c) [38]. These observations confirm that N-doped CDs have no toxic effect on embryo development because of their stable fluorescence emission and biocompatibility. Thus, bioimaging of zebrafish indicates that N-doped CDs may have potential imaging applications in elucidating different aspects of lipoprotein and nutritional biology in a zebrafish yolk lipid transport and metabolism model [65, 66].

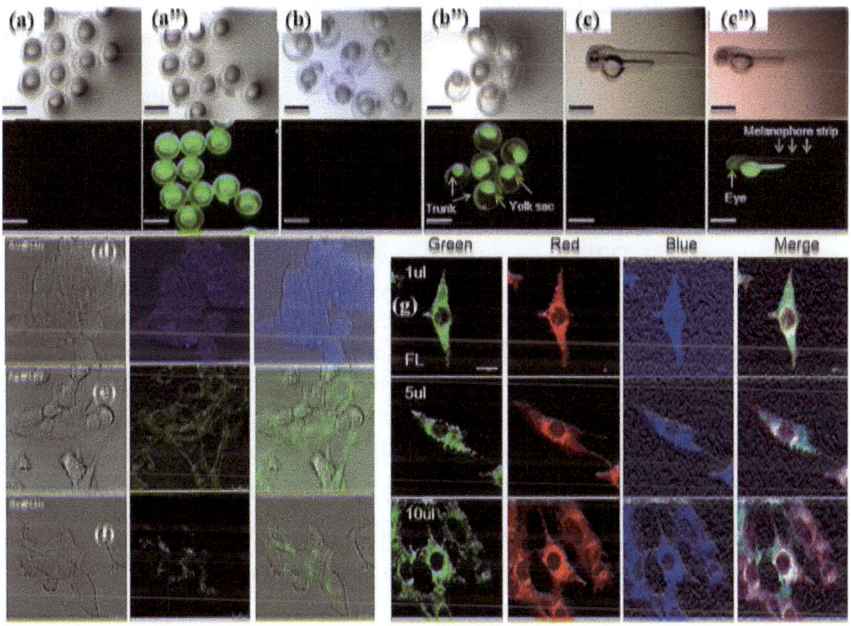

Fig. 16.6 **a** Bright-field and fluorescence images of zebrafish embryos and larvae after soaking in 6 mg mL^{-1} N-doped CD solution for 2.5 h at different time intervals: (**a, a″**) 6; (**b, b″**) 24; and (**c, c″**) 48 hpf. **a**, **b**, and **c** are the control experiment. The scale bar is 2.0 mm. Reproduced with Copyright [38] permission from the American Chemical Society. **d–f** Confocal microscope image of neuronal cells (SH-SY5Y) after incubation with formulated samples. Top: bright-field imaging; middle: fluorescent imaging; and bottom: overlap image. Scale bar = 10 μm. Reproduced from Ref. [21], Copyright (2019), with permission from Elsevier. **j** Confocal fluorescence images of C6 glioblastoma cells treated with CDs are merged into the three colors of green (488 nm EX), red (594 nm EX), and blue (358 nm EX). The scale bar is 10 μm. Reproduced with Copyright [73] permission from the American Chemical Society

The biological application of CDs and promising results were also reported when formulated through sonochemical method compared to other methods [67]. This is due to excellent biocompatibility of CDs generated through sonication, whereas cytotoxicity is induced in chemical methods due to use of toxic chemical and experimental conditions. Sun et al. employed the CDs as a fluorescent tag for cell imaging applications [68]. Then CDs are widely used for intracellular imaging of various types of cells including HeLa cells, HepG2 cells, and neuronal cells such as PC12 and SH-SY5Y cells (Fig. 16.6d–f) [5, 67, 69]. Due to superior photostability, biocompatibility, and bright fluorescence CDs also used for real time molecular tracking in live cells utilizing its optical properties [70]. Zheng et al. utilized the CDs to conjugate with protein for real-time monitoring [71]. Compared to cell imaging and cellular uptake, in vivo imaging fluorescent probes should be able to absorb and emit long wavelengths, and transparent to biological tissues. CDs are a good choice of material for in vivo imaging due to multicolor fluorescence properties. Yang et al. reported the in vivo imaging using CDs through injection into mice [72].

A group of researchers has designed CDs (C-dots) with diameters ranging 3–4 nm based on a variety of carbon nanotube starting materials, including a single-walled carbon nanotube, a multiwalled carbon nanotube, and a graphite starting material [74]. Additionally, the in vivo biodistribution and toxicology of the CD dots are studied in mice over an extended period of time; no obvious signs of toxicity for the CDs are seen in the treated animals. Three CDs were subcutaneously administered to a naked mouse in three different locations. After an intraperitoneal injection of 1% pentobarbital, the mouse was anesthetized and imaged using a Maestro in vivo optical imaging system [74]. Different excitation wavelengths were applied during in vivo imaging of the mouse, including blue, green, yellow, orange, red, deep red, and NIR light with center wavelengths of 455, 523, 595, 605, 635, 661, and 704 nm, respectively. Under all different excitation light settings, the subcutaneously injected spots (red) on the mouse were evident under spectral unmixing to differentiate them from background autofluorescence (green). This work synthesized fluorescent nanoprobes that are promising for biomedical imaging applications [74]. It has been reported by other researchers that functionalized CDs can also be used for bioimaging and biosensor applications [75–77].

Additionally, Kim et al. recently reported luminescent carbon dots passivated with nitrogen-rich molecules, such as N-isopropylacrylamide (NIPAAM), with a QY of 94%; this is the highest QY to date [73]. The NIPAMM@CDs were synthesized directly from CA, EDA, and NIPAAM by hydrothermal reaction. Furthermore, in order to test its usefulness as a fluorescent marker, C6 glioblastoma cells and NIPAAM@CDs were cultured together and both were subsequently analyzed by confocal bioimaging and flow cytometry [73]. It was observed that multicolor imaging and labeling of NIPAMM@C-Dots for C6 glioblastoma cells was effective. Figure 16.6g illustrates C6 cells incubated with NIPAAM@C-Dots illuminated with blue, green, and red colors under laser EX at 358, 488, and 594 nm. The exact subcellular localization site of the NIPAAM@CDs must be investigated in more detail, but it is apparent that the NIPAAM@CDs form a granular inclusion in abundant cytoplasmic vacuole-like structures in C6 cells sparing the nucleus, and the pattern of photoluminescence under the whole EX WL is totally combined (Fig. 16.6g) [73]. Based on these findings, functionalized CDs could be a potentially useful multicolor label for long-term cellular imaging as they possess a wide dosing window as well as relatively good biocompatibility. The robustness of these CDs under photoirradiation and chemical inertness of these CDs is consistent with the strong photostability of functionalized CDs. Due to these characteristics, N-doped CDs are suitable for imaging and sensing applications.

16.5 Conclusions and Future Prospective

In the coming years, CDs and CD-based materials will be widely used in biosensing electronic devices due to their biocompatibility, non-photo blinking properties, and excellent fluorescence properties. It is possible that they may be used in a variety of applications such as biosensors, organic light-emitting devices, energy storage devices, and organic photovoltaics. Although there has been rapid development of intriguing applications, rational design of biosensors and synthesis of desired functional CDs are still challenging. CD-based materials synthesized by top-down or bottom-up methods are usually self-assembled according to the self-assembly process. In the one-pot synthesis method, which is still in its infancy of synthesizing CDs and functionalized CDs, the introduction of different functional groups into CDs and the chemical doping of CDs are more controllable. There are many challenges that must be addressed before CD-based materials can be adopted widely, the most significant of then being developing manufacturing processes that effectively control color and quantum yield. As a result of large-scale CD production, CDs used in commercial applications will grow rapidly. It should be possible to extend the spectrum of functionalized CD-derived materials to all spectral regions, particularly in the near-IR [77–79]. It is expected that the extension of the spectrum will allow organic CDs to find applications in biosensors, sensors, and bioelectronics.

Here, we have reported on the developed CD-based materials that have gained significant attention in a variety of fields due to their inherent physical and chemical properties. This book chapter discusses the synthesis of CDs as well as biocompatibility, bioimaging, sensors, and biosensors of small molecules. CD doping has improved the properties of small molecules in terms of photoluminescence, bioimaging, and biosensing. In determining the sensitivity and selectivity of CDs and doped CDs for their future applications, precursors played an important role. Additionally, it is important to control and tune the size, shape, and surface modifications of CDs to meet the above requirements. Future research should focus on developing CDs with a broad visible emission. Also, researchers have recently developed low-cost, highly sensitive electronic biosensors based on CDs. In the future, CDs will need to be functionalized with appropriate dopants, which are essential for CDs to play an active role in biosensors and electronic devices. It is anticipated that further research will enhance the utilization and function of CDs, thereby improving the selectivity and sensitivity of electronic biosensors. In biosensing applications, achieving maximum sensitivity while maintaining selectivity can be challenging and needs more and focused investigation.

References

1. Taylor, R., Walton, D.R.M.: The chemistry of the fullerenes. Nature **363**, 685–693 (1993). https://doi.org/10.1002/9783527619214
2. Wang, Q.H., Strano, M.S.: Carbon nanotubes: a bright future for defects. Nat. Publ. Gr. **5**(10), 812–813 (2013). https://doi.org/10.1038/nchem.1768
3. Singh, V., Joung, D., Zhai, L., Das, S., Khondaker, S.I., Seal, S.: Graphene based materials: past, present and future. Prog. Mater. Sci. **56**(8), 1178–1271 (2011). https://doi.org/10.1016/j.pmatsci.2011.03.003
4. Li, H., Kang, Z., Liu, Y., Lee, S.-T.: Carbon nanodots: synthesis, properties and applications. J. Mater. Chem. **22**(46), 24230 (2012). https://doi.org/10.1039/c2jm34690g
5. Kumar, V.B., Kumar, R., Friedman, O., Golan, Y., Gedanken, A., Shefi, O.: One-pot hydrothermal synthesis of elements (B, N, P)-doped fluorescent carbon dots for cell labelling, differentiation and outgrowth of neuronal cells. ChemistrySelect **4**(14), 4222–4232 (2019). https://doi.org/10.1002/slct.201900581
6. Wang, R., Lu, K.-Q., Tang, Z.-R., Xu, Y.-J.: Recent progress in carbon quantum dots: synthesis, properties and applications in photocatalysis. J. Mater. Chem. A **5**(8), 3717–3734 (2017). https://doi.org/10.1039/C6TA08660H
7. Yu, X., Liu, J., Yu, Y., Zuo, S., Li, B.: Preparation and visible light photocatalytic activity of carbon quantum dots/TiO_2 nanosheet composites. Carbon N. Y. **68**, 718–724 (2014). https://doi.org/10.1016/j.carbon.2013.11.053
8. Kumar, V.B., Tang, J., Lee, K.J., Pol, V.G., Gedanken, A.: In situ sonochemical synthesis of luminescent Sn@C-dots and a hybrid Sn@C-dots@Sn anode for lithium-ion batteries. RSC Adv. **6**(70), (2016). https://doi.org/10.1039/c6ra09926b
9. Khadijeh Nekoueian, S., Amiri, M., Sillanpa, M., Marken, F., Boukherroub, R., Sabine: Chem Soc Rev Carbon-based quantum particles: an electroanalytical and biomedical perspective. Chem. Soc. Rev. **48**, 4281–4316, (2019). https://doi.org/10.1039/c8cs00445e
10. Yan, F., Zou, Y., Wang, M., Mu, X., Yang, N., Chen, L.: Highly photoluminescent carbon dots-based fluorescent chemosensors for sensitive and selective detection of mercury ions and application of imaging in living cells. Sens. Actuators, B Chem. **192**, 488–495 (2014). https://doi.org/10.1016/j.snb.2013.11.041
11. Lawrence, K., et al.: Functionalized carbon nanoparticles, blacks and soots as electron-transfer building blocks and conduits. Chem. Asian J. 1226–1241 (2014). https://doi.org/10.1002/asia.201301657
12. Gayen, B., Palchoudhury, S., Chowdhury, J.: Review article carbon dots: a mystic star in the world of nanoscience. J. Nanomater. **2019**, 19 (2019)
13. Kumar, V.B., Sahu, A.K., Mohsin, A.S.M., Li, X., Gedanken, A.: Refractive-index tuning of highly fluorescent carbon dots. ACS Appl. Mater. Interfaces **9**(34), 28930–28938 (2017). https://doi.org/10.1021/acsami.7b08985
14. Tangy, A., Kumar, V.B., Pulidindi, I.N., Kinel-Tahan, Y., Yehoshua, Y., Gedanken, A.: In-situ transesterification of *Chlorella vulgaris* using carbon-dot functionalized strontium oxide as a heterogeneous catalyst under microwave irradiation. Energy Fuels **30**(12), 10602–10610 (2016). https://doi.org/10.1021/acs.energyfuels.6b02519
15. Kumar, V.B., Sahu, A.K., Mohsin, A.S.M., Li, X., Gedanken, A.: Refractive-index tuning of highly fluorescent carbon dots. ACS Appl. Mater. Interfaces **9**(34), (2017). https://doi.org/10.1021/acsami.7b08985

16. Zhang, J., Yu, S.H.: Carbon dots: large-scale synthesis, sensing and bioimaging. Mater. Today **19**(7), 382–393 (2016). https://doi.org/10.1016/j.mattod.2015.11.008
17. Zheng, X.T., Ananthanarayanan, A., Luo, K.Q., Chen, P.: Glowing graphene quantum dots and carbon dots: properties, syntheses, and biological applications. Small **11**(14), 1620–1636 (2015). https://doi.org/10.1002/smll.201402648
18. Luo, P.G., et al.: Carbon-based quantum dots for fluorescence imaging of cells and tissues. RSC Adv. **4**(21), 10791–10807 (2014). https://doi.org/10.1039/C3RA47683A
19. Baker, S.N., Baker, G.: Luminescent carbon nanodots: emergent nanolights. Angew. Chemie Int. Ed. **49**(38), 6726–6744 (2010). https://doi.org/10.1002/anie.200906623
20. Kumar, V.B., Porat, Z., Gedanken, A.: Facile one-step sonochemical synthesis of ultrafine and stable fluorescent C-dots. Ultrason. Sonochem. **28**, 367–375 (2016). https://doi.org/10.1016/j.ultsonch.2015.08.005
21. Kumar, V.B., Kumar, R., Gedanken, A., Shefi, O.: Fluorescent metal-doped carbon dots for neuronal manipulations. Ultrason. Sonochem. **52**, 205–213 (2019). https://doi.org/10.1016/j.ultsonch.2018.11.017
22. El-Shafey, A.M.: Carbon dots: discovery, structure, fluorescent properties, and applications. Green Process. Synth. **10**, 134–156 (2021)
23. Kainth, S., Maity, B., Soumen, B.: RSC Advances passivated fluorescent carbon dots. RSC Adv. **10**, 36253–36264 (2020). https://doi.org/10.1039/d0ra06512a
24. Kumar, R., Kumar, V.B., Gedanken, A.: Sonochemical synthesis of carbon dots, mechanism, effect of parameters, and catalytic, energy, biomedical and tissue engineering applications. Ultrason. Sonochem. **64**(September 2019), 105009 (2020). https://doi.org/10.1016/j.ultsonch.2020.105009
25. De Medeiros, T.V., Manioudakis, J., Noun, F., Macairan, J.R., Victoria, F., Naccache, R.: Microwave-assisted synthesis of carbon dots and their applications. J. Mater. Chem. C **7**(24), 7175–7195 (2019). https://doi.org/10.1039/c9tc01640f
26. Xu, X., et al.: Electrophoretic analysis and purification of fluorescent single-walled carbon nanotube fragments. J. Am. Chem. Soc. **126**(40), 12736–12737 (2004). https://doi.org/10.1021/ja040082h
27. Liu, M., Xu, Y., Niu, F., Gooding, J.J., Liu, J.: Carbon quantum dots directly generated from electrochemical oxidation of graphite electrodes in alkaline alcohols and the applications for specific ferric ion detection and cell imaging. Analyst **141**(9), 2657–2664 (2016). https://doi.org/10.1039/c5an02231b
28. De, B., Karak, N.: Recent progress in carbon dot-metal based nanohybrids for photochemical and electrochemical applications. J. Mater. Chem. A **5**(5), 1826–1859 (2017). https://doi.org/10.1039/C6TA10220D
29. Kumar, V.B., Porat, Z.: Synthesis of doped/hybrid carbon dots and their biomedical application. Nanomaterials **12**, 898 (2022)
30. Kumar, V.B., Kumar, R., Gedanken, A., Shefi, O.: Fluorescent metal-doped carbon dots for neuronal manipulations. Ultrason. Sonochem. **52**(November 2018), 205–213 (2019). https://doi.org/10.1016/j.ultsonch.2018.11.017
31. Kumar, V.B., et al.: Activated carbon modified with carbon nanodots as novel electrode material for supercapacitors. J. Phys. Chem. C **120**(25), 13406–13413 (2016). https://doi.org/10.1021/acs.jpcc.6b04045
32. Khajuria, D.K., Kumar, V.B., Gigi, D., Gedanken, A., Karasik, D.: Accelerated bone regeneration by nitrogen-doped carbon dots functionalized with hydroxyapatite nanoparticles. ACS Appl. Mater. Interfaces **10**(23), 19373–19385 (2018). https://doi.org/10.1021/acsami.8b02792
33. El-Shabasy, R.M., Elsadek, M.F., Ahmed, B.M., Farahat, M.F., Mosleh, K.M., Taher, M.M.: Recent developments in carbon quantum dots: properties, fabrication techniques, and bio-applications. Processes **9**(2), 1–24 (2021). https://doi.org/10.3390/pr9020388
34. Kumar, V.B., Sheinberger, J., Porat, Z., Shav-Tal, Y., Gedanken, A.: A hydrothermal reaction of an aqueous solution of BSA yields highly fluorescent N doped C-dots used for imaging of live mammalian cells. J. Mater. Chem. B **4**(17), 2913–2920 (2016). https://doi.org/10.1039/c6tb00519e

35. Kumar, V.B., Perkas, N., Porat, Z., Gedanken, A.: Solar-light-driven photocatalytic activity of novel Sn@C-Dots-Modified TiO_2 catalyst. ChemistrySelect **2**(23), 6683–6688 (2017). https://doi.org/10.1002/slct.201701375
36. Lai, C.W., Hsiao, Y.H., Peng, Y.K., Chou, P.T.: Facile synthesis of highly emissive carbon dots from pyrolysis of glycerol; gram scale production of carbon dots/$mSiO_2$ for cell imaging and drug release. J. Mater. Chem. **22**(29), 14403–14409 (2012). https://doi.org/10.1039/c2jm32206d
37. Kumar, V.B., et al.: Ultrafine highly magnetic fluorescent γ-Fe_2O_3/NCD nanocomposites for neuronal manipulations. ACS Omega **3**(2), 1897–1903 (2018). https://doi.org/10.1021/acsomega.7b01666
38. Khajuria, D.K., Kumar, V.B., Karasik, D., Gedanken, A.: Fluorescent nanoparticles with tissue-dependent affinity for live zebrafish imaging. ACS Appl. Mater. Interfaces **9**(22), 18557–18565 (2017). https://doi.org/10.1021/acsami.7b04668
39. Ge, J., Li, Y., Wang, J., Pu, Y., Xue, W., Liu, X.: Green synthesis of graphene quantum dots and silver nanoparticles compounds with excellent surface enhanced Raman scattering performance. J. Alloys Compd. **663**, 166–171 (2016). https://doi.org/10.1016/j.jallcom.2015.12.055
40. Kumar, V.B., Perelshtein, I., Lipovsky, A., Porat, Z., Gedanken, A.: The sonochemical synthesis of Ga@C-dots particles. RSC Adv. **5**(32), 25533–25540 (2015). https://doi.org/10.1039/c5ra01101a
41. Kumar, V.B., Pol, V., Tang, J., Lee, K.J., Gedanken, A.: In situ sonochemical synthesis of luminescent Sn@C-dots and hybrid Sn@C-dots@Sn anode for lithium-ion batteries. RSC Adv. **6**, 66256–66265 (2016). https://doi.org/10.1039/C6RA09926B
42. Ngo, Y.L.T., Nguyen, P.L., Jana, J., Choi, W.M., Chung, J.S., Hur, S.H.: Simple paper-based colorimetric and fluorescent glucose sensor using N-doped carbon dots and metal oxide hybrid structures. Anal. Chim. Acta **1147**(xxxx), 187–198 (2021). https://doi.org/10.1016/j.aca.2020.11.023
43. Li, P., Li, S.F.Y.: Recent advances in fluorescence probes based on carbon dots for sensing and speciation of heavy metals. Nanophotonics **10**(2), 877–908 (2021). https://doi.org/10.1515/nanoph-2020-0507
44. Cui, X., et al.: A fluorescent biosensor based on carbon dots-labeled oligodeoxyribonucleotide and graphene oxide for mercury(II) detection. Biosens. Bioelectron. **63**, 506–512 (2015). https://doi.org/10.1016/j.bios.2014.07.085
45. Wang, X., Zhang, J., Zou, W., Wang, R.: Facile synthesis of polyaniline/carbon dot nanocomposites and their application as a fluorescent probe to detect mercury. RSC Adv. **5**, 41914–41919 (2015). https://doi.org/10.1039/b000000x
46. Guo, X., et al.: Fluorescent carbon dots based sensing system for detection of enrofloxacin in water solutions. Spectrochim. Acta Part A Mol. Biomol. Spectrosc. **219**, 15–22 (2019). https://doi.org/10.1016/j.saa.2019.02.017
47. Sharma, A., Das, J.: Small molecules derived carbon dots : synthesis and applications in sensing, catalysis, imaging, and biomedicine. J. Nanobiotechnol. 1–24 (2019). https://doi.org/10.1186/s12951-019-0525-8
48. Cui, L., Ren, X., Sun, M., Liu, H., Xia, L.: Carbon dots: synthesis, properties and applications. Nanomaterials **11**, 3419 (2021). https://doi.org/10.3390/nano11123419
49. Han, M., et al.: Recent progress on the photocatalysis of carbon dots: Classification, mechanism and applications. Nano Today **19**, 201–218 (2018). https://doi.org/10.1016/j.nantod.2018.02.008
50. Li, Y.H., Zhang, L., Huang, J., Liang, R.P., Qiu, J.D.: Fluorescent graphene quantum dots with a boronic acid appended bipyridinium salt to sense monosaccharides in aqueous solution. Chem. Commun. **49**(45), 5180–5182 (2013). https://doi.org/10.1039/c3cc40652k
51. Zhi-bei, Q., et al.: Boronic acid functionalized graphene quantum dots as fluorescent probe for selective and sensitive glucose determination in microdialysate. Chem. Commun. **49**, 9830–9832 (2013). https://doi.org/10.1039/b000000x

52. Shen, P., Xia, Y.: Synthesis-modification integration: one-step fabrication of boronic acid functionalized carbon dots for fluorescent blood sugar sensing. Anal. Chem. **86**(11), 5323–5329 (2014). https://doi.org/10.1021/ac5001338
53. Kiran, S., Misra, R.D.K.: Mechanism of intracellular detection of glucose through nonenzymatic and boronic acid functionalized carbon dots. J. Biomed. Mater. Res. Part A **103**(9), 2888–2897 (2015). https://doi.org/10.1002/jbm.a.35421
54. Sun, H., Wu, L., Wei, W., Qu, X.: Recent advances in graphene quantum dots for sensing. Mater. Today **16**(11), 433–442 (2013). https://doi.org/10.1016/J.MATTOD.2013.10.020
55. Ma, N., Jiang, W., Li, T., Zhang, Z., Qi, H., Yang, M.: Fluorescence aggregation assay for the protein biomarker mucin 1 using carbon dot-labeled antibodies and aptamers. Microchim. Acta **182**(1–2), 443–447 (2015). https://doi.org/10.1007/s00604-014-1386-3
56. Wang, Y., Qi, W., Song, Y.: Antibody-free detection of protein phosphorylation using intrinsic peroxidase-like activity of platinum/carbon dot hybrid nanoparticles. Chem. Commun. **52**(51), 7994–7997 (2016). https://doi.org/10.1039/C6CC02771G
57. Zhu, A., Qu, Q., Shao, X., Kong, B., Tian, Y.: Carbon-dot-based dual-emission nanohybrid produces a ratiometric fluorescent sensor for in vivo imaging of cellular copper ions. Angew. Chemie Int. Ed. **51**(29), 7185–7189 (2012). https://doi.org/10.1002/anie.201109089
58. Liu, Y., et al.: Red emission B, N, S-*co*-doped carbon dots for colorimetric and fluorescent dual mode detection of Fe^{3+} ions in complex biological fluids and living cells. ACS Appl. Mater. Interfaces **9**(14), 12663–12672 (2017). https://doi.org/10.1021/acsami.6b15746
59. Liu, Q., et al.: Distinguish cancer cells based on targeting turn-on fluorescence imaging by folate functionalized green emitting carbon dots. Biosens. Bioelectron. **64**, 119–125 (2015). https://doi.org/10.1016/j.bios.2014.08.052
60. Jia, Q., et al.: A magnetofluorescent carbon dot assembly as an acidic H_2O_2-driven oxygenerator to regulate tumor hypoxia for simultaneous bimodal imaging and enhanced photodynamic therapy. Adv. Mater. **30**(13), 1706090 (2018). https://doi.org/10.1002/adma.201706090
61. Wang, N., Wang, Y., Guo, T., Yang, T., Chen, M., Wang, J.: Green preparation of carbon dots with papaya as carbon source for effective fluorescent sensing of Iron (III) and *Escherichia coli*. Biosens. Bioelectron. **85**, 68–75 (2016). https://doi.org/10.1016/j.bios.2016.04.089
62. Nandi, S., Ritenberg, M., Jelinek, R.: Bacterial detection with amphiphilic carbon dots. Analyst **140**(12), 4232–4237 (2015). https://doi.org/10.1039/c5an00471c
63. Zhu, S., et al.: Highly photoluminescent carbon dots for multicolor patterning, sensors, and bioimaging. Angew. Chemie Int. Ed. **52**(14), 3953–3957 (2013). https://doi.org/10.1002/anie.201300519
64. Fernando, K.A.S., et al.: Carbon quantum dots and applications in photocatalytic energy conversion. ACS Appl. Mater. Interfaces **7**(16), 8363–8376 (2015). https://doi.org/10.1021/acsami.5b00448
65. Park, S.Y., et al.: Advanced carbon dots via plasma-induced surface functionalization for fluorescent and bio-medical applications. Nanoscale **9**(26), 9210–9217 (2017). https://doi.org/10.1039/c7nr03026f
66. Li, Z., Ptak, D., Zhang, L., Walls, E.K., Zhong, W., Leung, Y.F.: Phenylthiourea specifically reduces zebrafish eye size. PLoS One **7**(6), 1–14 (2012). https://doi.org/10.1371/journal.pone.0040132
67. Kumar, V.B., Natan, M., Jacobi, G., Porat, Z., Banin, E., Gedanken, A.: Ga@C-dots as an antibacterial agent for the eradication of *Pseudomonas aeruginosa*. Int. J. Nanomed. **12**, 725–730 (2017). https://doi.org/10.2147/IJN.S116150
68. Sun, Y.-P., et al.: Quantum-sized carbon dots for bright and colorful photoluminescence. J. Am. Chem. Soc. **128**(24), 7756–7757 (2006). https://doi.org/10.1021/ja062677d
69. Cao, L., et al.: Carbon dots for multiphoton bioimaging. (2007). https://doi.org/10.1021/JA073527L
70. Kumar, V.B., Dolitzky, A., Michaeli, S., Gedanken, A.: Antiparasitic ointment based on a biocompatibile carbon dot nanocomposite. ACS Appl. Nano Mater. **1**(4), 1784–1791 (2018). https://doi.org/10.1021/acsanm.8b00213

71. Zheng, F.-F., Zhang, P.-H., Xi, Y., Chen, J.-J., Li, L.-L., Zhu, J.-J.: Aptamer/graphene quantum dots nanocomposite capped fluorescent mesoporous silica nanoparticles for intracellular drug delivery and real-time monitoring of drug release. Anal. Chem. **87**(23), 11739–11745 (2015). https://doi.org/10.1021/acs.analchem.5b03131
72. Yang, S.-T., et al.: Carbon dots for optical imaging in vivo. J. Am. Chem. Soc. **131**(32), 11308–11309 (2009). https://doi.org/10.1021/ja904843x
73. Kim, M.C., et al.: Highly photoluminescent N-isopropylacrylamide (NIPAAM) passivated carbon dots for multicolor bioimaging applications. Eur. Polym. J. **98**(November 2017), 191–198 (2018). https://doi.org/10.1016/j.eurpolymj.2017.11.018
74. Tao, H., et al.: In vivo NIR fluorescence imaging, biodistribution, and toxicology of photoluminescent carbon dots produced from carbon nanotubes and graphite. Small **8**(2), 281–290 (2012). https://doi.org/10.1002/smll.201101706
75. Huang, G., Su, C., Wang, L., Fei, Y., Yang, J.: The application of nucleic acid probe-based fluorescent sensing and imaging in cancer diagnosis and therapy. Front. Chem. **9**(June), 1–10 (2021). https://doi.org/10.3389/fchem.2021.705458
76. Zhang, J., et al.: Structure of the zein protein as treated with subcritical water. Int. J. Food Prop. **21**(1), 128–138 (2018). https://doi.org/10.1080/10942912.2017.1414839
77. Gao, T., et al.: A peptide nucleic acid–regulated fluorescence resonance energy transfer DNA assay based on the use of carbon dots and gold nanoparticles. Microchim. Acta **187**(7), (2020). https://doi.org/10.1007/s00604-020-04357-w
78. Peng, C., et al.: Afterglow carbon dots: from fundamentals to applications. Research **2021**, 1–27 (2021). https://doi.org/10.34133/2021/6098925
79. Jiang, X., Zong, S., Chen, C., Zhang, Y., Wang, Z., Cui, Y.: Gold-carbon dots for the intracellular imaging of cancer-derived exosomes. Nanotechnology **29**(17), 175701 (2018). https://doi.org/10.1088/1361-6528/aaaf14

Chapter 17
Flexible and Wearable Chemical Sensor Based on Graphene Derivatives

Hemen Kalita, Anurag Kashyap, Rajesh Ghosh, and Biswajit Dehingia

Abstract Flexible and wearable sensors have attained tremendous attention in the last few years due to their rising demand for real-time monitoring. Starting from monitoring human health to the surrounding environment, wearable sensors have drawn the attention of researchers worldwide. Nanomaterials with superior electrical, mechanical and chemical properties with high environmental stability have shown high demand for the fabrication of wearable chemical sensors. Traditional materials used for chemical sensors have limitations in terms of robustness and flexibility that hinder their applications as flexible sensors. Various advanced materials have already been investigated to overcome the shortcomings of the traditional chemical sensors with desired performance, commercial viability and productivity at the industrial level. Graphene and its derivatives have evolved to be one of the superior and reliable nanomaterials for the fabrication of wearable sensors owing to their excellent properties as well as cost-effective synthesis with high yield. Chemically derived graphene has tunable properties which have been widely utilized to enhance the performance of the sensors that could be operated in a normal environment. Moreover, the abundant functional groups make graphene derivatives more suitable for surface functionalization. This chapter discussed different forms of graphene and their composite-based flexible chemical sensors and their sensing performances. The potentialities of graphene derivatives in developing smart wearable sensors are also briefly summarized. The challenges and future scope of flexible chemical wearable sensors have also been highlighted.

Keywords Graphene · Graphene derivatives · Graphene composites · Gas sensing · Flexible sensor

---

H. Kalita (✉) · A. Kashyap · R. Ghosh · B. Dehingia
Department of Physics, Gauhati University, Guwahati, Assam 781014, India
e-mail: hemenkalita@gauhati.ac.in

D. Mohanta and P. Chakraborty (eds.), *Nanoscale Matter and Principles for Sensing and Labeling Applications*, Advanced Structured Materials 206,
https://doi.org/10.1007/978-981-99-7848-9_17

17.1 Introduction

The detection of environmental pollutant chemicals emitted from various industries, vehicles, agriculture, medical sector, etc. is a major concern for all living beings. The emission of hazardous chemical gases as by–products such as oxides of carbon (CO_x), nitrogen (NO_x), sulfur (SO_x) and methane (CH_4) and vapor by–products such as common indoor pollutants like ammonia (NH_3), formaldehyde, acetone, etc. have increased in a significant manner and are becoming a threat to human life as well as to the environment [1]. Excess amounts of carbon dioxide (CO_2) and carbon monoxide (CO) can cause dizziness, unconsciousness or even death, whereas gases such as ammonia are corrosive and hazardous to human health. It causes injuries to human eyes, skin, respiratory tract, liver and kidneys if the concentration exceeds a certain limit [2]. Similarly, formaldehyde is a carcinogen as labeled by the International Agency for Research on Cancer (IARC) [3]. Continuous exposure of formaldehyde to a human being may cause cancer [3]. On the other hand, highly volatile compounds such as acetone have been termed as biomarkers for the detection of diabetes that can be detected through breath analysis [4]. Therefore, continuous monitoring of these chemicals needs highly sensitive and selective sensing devices. Though solid–state sensors are commonly used to detect these chemicals, a few disadvantages hinder their applications in a few critical fields. The development and improvement of traditional chemical sensors require materials that are cost–effective, flexible and robust in nature. Flexible sensors have the advantage of stretchability and bendability which makes them suitable for use on surfaces of random shapes. Silicon–based materials and organic semiconductors have been investigated for flexible sensors. However due to their poor stability and flexibility, there is a need for flexible and stretchable sensors with advanced materials which can eliminate the above–mentioned limitations.

The rise of thin electronic devices has widened the sphere of flexible and wearable electronics from foldable circuits and gadgets to advanced biomedical devices [5]. Wearable electronics can be fixed on any surface, human bodies or can be embedded in clothes. These smart electronics have drawn massive attention in recent years due to their versatile potentialities [6, 7]. Wearable sensors are predicted to have a major performance in monitoring human health, the surrounding environment and all areas, which are in connection with the Internet of Things (IoT) [7]. The significant advancement of wearable electronic devices demands materials that are compatible with the flexibility and stretchability of the devices. The channel materials and the substrates play critical roles in sensors. Flexible sensors demand substrates with better mechanical properties like bendability and stretchability. Typical rigid materials such as silicon (Si), indium tin oxide (ITO), and germanium (Ge), and alloys such as gallium arsenide (GaAs) and cadmium sulfide (CdS) used in traditional electronics are unable to be utilized in flexible devices. Their poor mechanical properties and brittleness hinder their applications in these bendable electronic devices [6]. Intensive efforts have been made in recent years to search for advanced materials, which shows good compatibility with flexible electronic devices. With a high tensile strength of 130 GPa and a high Young's modulus of elasticity of 1 TPa [8], graphene has come out as a versatile candidate for utilization in flexible and wearable devices. Since its discovery in 2004, several graphene derivatives including graphene oxide (GO), reduced graphene oxide (rGO), graphene quantum dots (GQDs), graphene nanoribbons (GNR), hybrid graphene structures, etc. have been tested extensively in different areas [9–12]. Along with unique mechanical properties, graphene also possesses high carrier mobility (350,000 $cm^2/(V\ s)$) [13], large specific surface area [14], high thermal conductivity [15] and very good optical transparency (~97.7%) [16]. All these unique properties of graphene make it suitable for a range of applications such as energy storage devices, chemical sensors, biomedical applications, aerospace applications, photodetectors, light emitting diodes (LEDs), transparent conducting electrodes (TCEs), and flexible devices [7, 17]. Though pristine graphene has numerous advantages, it has a few limitations as well. The sensing performance of a material can be increased through functionalization and doping. As intrinsic graphene has fewer active sites for doping and functionalization, there is a need for the hybrid structure of graphene (Fig. 17.1).

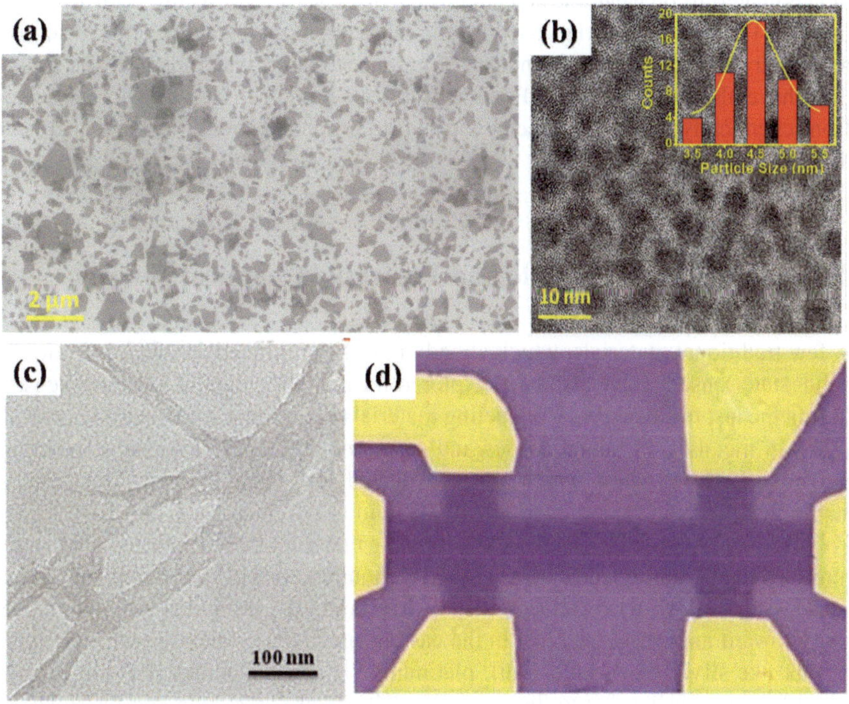

Fig. 17.1 Different derivatives of graphene: **a** FESEM image of graphene oxide (GO) sheets (*Reprinted with permission from* [9]). **b** TEM image of graphene quantum dots (GQDs) (*Reprinted with permission from* [10]) and **c** graphene nanoribbons (GNRs) (*Reprinted with permission from* [11]). **d** First graphene–based chemical sensor (*Reprinted with permission from* [17])

The derivatives of graphene have gained massive attention due to ease of fabrication and functionalization. Graphene can be oxidized with oxygen functional groups which enhance its applicability in different chemicals as well as gas sensors. The surface functional groups of graphene help in detecting different types of chemicals, toxic vapors and gas molecules that come in the proximity of the sensing layers. Apart from that, the electronic property, that has a vital role in sensing applications, can be tuned by reducing the oxygen functional groups present in the GO surfaces. The presence of oxygen functional groups also provides immense capabilities for surface modifications of graphene with other materials like metal nanoparticles, metal oxides, polymers and other materials which enhances the sensitivity and selectivity as well as induces flexibility of the graphene derivatives [7]. Owing to their transformed morphologies, the hybrid structures of graphene derivatives with other materials have been utilized for highly selective and sensitive sensors for different hazardous chemicals.

In this chapter, we are discussing flexible and stretchable chemical sensors based on graphene derivatives for the detection of harmful chemicals and gases. The roles of nanomaterials on wearable and flexible chemical sensors have also been discussed in brief. Traditional materials used for the fabrication of wearable chemical sensors with their limitations and advantages of graphene–based materials over the limitations have also been included. Pristine graphene–based sensors, sensors fabricated from chemically derived graphene and hybrid graphene structures have also been discussed.

17.2 Traditional Flexible and Wearable Chemical Sensors

A few traditional materials, which already have been utilized for the fabrication of solid–state sensors, have also been explored for the fabrication of wearable sensors during the last few decades. Conducting materials including carbon–based materials [18–20], metals [21], metal oxides and semiconductors [22, 23], and conducting polymers [24] are a few which are investigated for wearable applications due to stable flexibility and good conductivity. Common polymers such as PEDOT:PSS, Polypyrrole, PANI and PVP are used as sensing materials due to their room temperature sensing, flexibility and ease of chemical modifications [25]. But polymer–based sensors have a few disadvantages such as irreversibility, instability and poor selectivity toward analyte gases [26]. In the case of metal–based flexible sensors, noble metals like silver (Ag), gold (Au), platinum (Pt) and palladium (Pd) are investigated for gas sensing due to their excellent electrical conduction and resistance to oxidation [7]. All the noble metals are used as composites with metal oxides, polymers and carbon–based materials to enhance the sensitivity [25]. But the fabrication process of noble metal nanoparticle–based sensors is costly. Besides, metal oxides are extensively investigated for gas sensors. Metal oxides such as tin oxide (SnO_x), zinc oxide (ZnO), copper oxide (CuO), tungsten oxide (WO_3) and nickel oxide (NiO) are commonly used for sensitive and selective gas sensors [25]. But the major disadvantage of metal oxide–based sensors is they require high operating temperatures for sensing which hinders their applications in wearable sensors. By fabricating hybrid or composite structures with other materials, the operating temperature of metal oxide–based sensors can be reduced. There are several reports on metal oxide–based hybrid/composite gas sensors for selective and sensitive detection of analyte gases at varying temperatures. Similarly, conducting polymers and noble metals are also used to design composite structures for performance enhancement. Carbon nanotube (CNT)–based sensors are also investigated for selective gas sensing applications. But CNT–based sensors have less selectivity which is a major drawback of CNT for gas sensing applications. Therefore, to overcome the shortcomings of the conventional sensing materials, there is a need to explore advanced nanomaterials used for the development of wearable gas sensors.

17.3 Role of Nanomaterials in Flexible and Wearable Chemical Sensors

Nanomaterials due to their excellent electrical, superficial, mechanical, thermal and biological properties make them a suitable candidate for flexible chemical sensors. With decreasing size, the specific surface area increases, which enhances the number of active sites for interaction with other molecules [27]. So, as compared to bulk materials, nano counterparts are suitable for channel materials in chemical sensors. Cost–effective synthesis methods, and suitable mechanical and chemical properties of nanomaterials have attracted researchers to utilize them in stretchable chemical sensors. Due to size tuneability, the properties can be varied which enhances the selectivity and sensitivity of the nanomaterials. By varying the size, the same nanomaterial can be used for selective sensing of different test gases and vapors. The fabrication methods and transconductance principle are other two important factors that need special attention during the selection of nanomaterials as sensing materials. With the rise in printed electronics, print compatible materials are more favorable for the synthesis of flexible and wearable chemical sensors. Printing of material requires solution processable materials where nanomaterials have advantages over bulk conjugate [28]. Nanomaterials have better solution processibility which is necessary for printing applications [29]. Nano scaled metal oxides such as ZnO, SnO_2, In_2O_3, TiO_2 and NiO, emerging advanced nanomaterials like 0D quantum dots, 1D nanowires and nanorods, 2D nano structures of various materials such as transition metal dichalcogenides (TMDs), carbon based materials, semiconductors and metal nanoparticles have been extensively explored for flexible chemical sensors in last few years [30–33]. The better adhesive nature of these materials on flexible substrates is another important factor that makes them eligible for flexible chemical sensors [28]. Among them, graphene and its derivatives have drawn massive attention for its extraordinary electromechanical and chemical properties which have been discussed in the following section.

17.3.1 Advantage of Graphene—The Miracle Material

Graphene has significant advantages over traditional sensing materials used for gas sensors. Graphene-based sensors provide immense flexibility, mechanical durability and chemical stability than traditional inorganic sensing materials used for fabricating gas sensors [7]. Graphene has good chemical and mechanical stability which is essential for sensing applications. Cost–effective synthesis methods, high sensitivity, better selectivity and a wide working range have made graphene and its derivatives better candidates for the fabrication of wearable smart sensors [7, 25, 34]. Graphene and its derivatives can be synthesized using various chemical as well as physical routes with high yield. Graphene oxide has abundant oxygen functional groups through which its properties can be tuned easily. Moreover, due to the presence of oxygen, graphene oxide can be functionalized easily with various functional groups as compared to pristine graphene. Graphene-based composites show better sensitivity toward different toxic gases and VOCs at room temperature and very low operating temperatures as well [7]. Cho et al. have fabricated a composite structure of graphene with MoS_2 on a PI substrate which has shown good mechanical stability after 5,000 bending cycles [35]. The electrical properties of graphene remain unaltered upon application of mechanical strain which is a major advantage of graphene for manufacturing stretchable sensors [36]. The high signal–to–noise ratio of graphene is another parameter which makes graphene more prominent as compared to other conventional materials [37]. All the above–mentioned advantages of graphene and its derivatives make them potential candidates over traditional materials used for gas sensing.

17.4 Different Derivatives of Graphene for Flexible and Wearable Chemical Sensors

The pristine or unmodified graphene embedded into a flexible device structure promises to deliver an ultrasensitive gas sensing response. The ability to dope and functionalize the surface of graphene leads to the development of different derivatives of graphene. Thus, researchers have developed different forms or derivatives of graphene by suitably modifying the graphene surface with the aid of several other compatible nanomaterials. Graphene and its different derivatives have found promising applications in the fabrication of efficient chemical sensors. The performance of different derivatives of graphene toward developing flexible chemical sensors investigated so far for gas sensing purposes has been summarized as follows.

17.4.1 Pristine Graphene

Flexible sensors using graphene materials as primary sensing elements have been fabricated onto different polymer substrates including polyethylene terephthalate (PET), polyimide (PI) and many others, which show excellent mechanical robustness and high optical transparency. Lee et al. fabricated a flexible chemical sensor using CVD–synthesized graphene fabricated onto a pre–patterned flexible PET substrate [38]. The fabricated device exhibited an excellent sensing response toward NO_2 under bending conditions. The flexible sensor demonstrated a significant increase in resistance by 7.7% when strained at 1.39% in NO_2 ambiance. The overall response of the flexible sensor toward NO_2 was 23% (relaxed) and 25% (strained), respectively. This study demonstrated great promise for the fabricated graphene/PET sensor toward developing effective wearable chemical sensors. Similarly, Choi and others demonstrated a flexible and transparent NO_2 sensor using CVD–synthesized graphene onto a polyethersulfone (PES) substrate [39]. The flexible sensor was fabricated by transfer of single layer graphene (SLG) as the sensing channel in between two bilayer graphene (BLG) used as a heater over the PES substrate. The transfer process was accomplished by using a sacrificial supporting layer of poly(methyl methacryate) (PMMA). The sensor was then patterned with Cr/Au electrodes, and the resultant structure showed a high transmittance value of more than 90%. The device made with SLG channel and BLG heater junction displayed a suggestively faster response and complete recovery toward a range of NO_2 concentrations in comparison to the conventional solid–state sensors. Upon a bending strain of 1.4%, the embedded sensor exhibited excellent gas sensitivity of 12% on exposure to 1 ppm of NO_2. The consistent performance exhibited by the sensor–heater junction devices under different bending conditions presents good promise for the design and development of smart wearable sensors. Hong et al. also reported a self–activated flexible NO_2 sensor using three–layer graphene (3LG) synthesized by the CVD method [40]. In this work, the 3LG was patterned onto a flexible, transparent and thermally stable PI substrate using a PMMA layer to fabricate the sensor. The 3LG was used as both the sensing layer and the electrode to fabricate the sensing device. The all–graphene sensor displayed an ohmic behavior that enabled a reliable and reproducible sensing performance of the fabricated sensor when tested in consecutive pulses of NO_2. Upon exposure to 5 ppm of NO_2, the fabricated device showed a low resistive response of nearly 13%, but under high bending conditions, only ~3% degradation was observed. Moreover, the 3LG sensor demonstrated good selectivity and outstanding reliability in sensing response in humid conditions (~50% RH), which showed its potential for developing smart and wearable electronics (Fig. 17.2).

These findings placed graphene in the category of most potential material for the development of effective and highly efficient nanomaterial for flexible and wearable chemical sensors. However, in order to provide the devices with maximum efficiency, the graphene sensors must be modified as the pristine graphene suffers from certain limitations. For instance, the graphene surface is adapted to several gas molecules at a particular instant of time, which often limits its selectivity for a specific target gas.

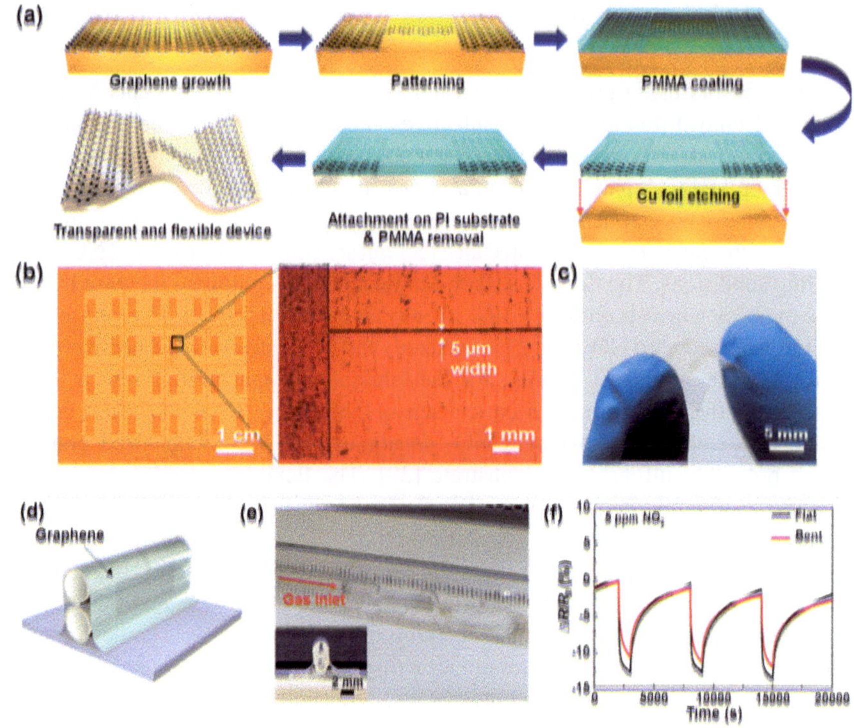

Fig. 17.2 **a** Fabrication steps involved for developing the all–graphene sensor. **b** Microscopic images of graphene patterned on a Cu foil. **c** Digital image of the fabricated device on PI substrate. **d** Schematic of the sensor attached on ball pen leads and **e** Optical image showing the sensing setup for the device under bending strain. Inset: side view of the device with a bending radius of 1 mm and **f** Sensing performance under bending strain and relaxed condition. (*Reprinted with permission from* [40])

Moreover, the lack of dangling bonds and chemically inert structure of graphene also restrict the surface toward interaction with gas molecules limiting its sensitivity.

17.4.2 Chemically Modified Graphene

As discussed in the earlier section, the limitations often arising in the pristine form of graphene could be overcome by forming different derivatives of graphene. The chemically synthesized or modified form of graphene aroused wide attention among researchers for designing functionalized graphene surfaces to provide strong interaction with the targeted gas molecules with high sensitivity detection and selectivity. The oxygen functionalities in chemically synthesized graphene including GO and rGO present good flexibility in selectively tuning the sensing performance toward different gas molecules. Dua et al. demonstrated a flexible chemiresistive sensor by inkjet printing of rGO platelets synthesized by ascorbic acid reduction of exfoliated GO onto a PET substrate [41]. The sensing devices were fabricated onto a flexible PET substrate for the detection of different vapors of oxidizing and reducing gases. The rGO/PET device showed very sensitive responses toward different vapors, while a particularly selective response was observed on exposure to the electron–accepting vapors. The sensor also exhibited ppb level detection of different vapors, which shows its potential for highly sensitive detection of aggressive chemical vapors. In a different report, Ghosh et al. fabricated a robust and cost–effective rGO sensor based on a paper substrate for high–sensitive NH_3 detection [42]. The device was fabricated by using a Whatman filter paper to which the rGO dispersion was coated to develop the low–cost flexible sensing device. The sensing response of the rGO–coated device was observed to be highly sensitive toward NH_3 with a detection limit estimated to be around 430 ppb. The rGO–coated device showed a reliable and highly selective sensing response toward NH_3 and in addition, good stability under high humidity (RH ~ 100%) conditions was also observed (Fig. 17.3). The sensor in a convex bending condition (40°) was reported to observe negligible variation in response as obtained in flat conditions. Hassinen et al. used recyclable paper as the substrate for the facile fabrication of a low–cost and disposable sensor for the NO_2 molecule detection using rGO as the sensing material [43]. The sensing material (rGO) was prepared by chemically or electrochemically reducing GO that is synthesized following the modified Hummers' method. The thin GO flakes were reduced chemically using both hydrazine (H–rGO) and ascorbic acid (A–rGO), respectively, while the thicker GO flakes were electrochemically reduced. For the device fabrication, silver nanoparticle–based electrodes were patterned onto the paper substrate following the inkjet–printing technique. The rGO films were spray–coated onto the disposable paper substrate, and the copper wires were attached for carrying out the sensing measurements. The H–rGO and A–rGO sensors exhibited excellent sensing performance with similar response behavior although the device recovery was very poor. Furthermore, the H–rGO device demonstrated the best sensing performance with faster UV–assisted recovery among all the sensors tested at a 10 ppm concentration of NO_2. On varying the NO_2 concentration from 1 to 100 ppm, the H–rGO sensor showed an exponential variation in both resistive and capacitive responses. The work suggested facile fabrication for low–cost flexible rGO sensors for sensitive NO_2 detection without the prior requirement of sophisticated techniques.

Thus, chemically synthesized graphene has been considered to have much potential for tuning the gas sensing performance of graphene devices. Moreover, researchers have also been looking into the performance enhancement of graphene in terms of precise control and tailoring over the sensing parameters. Therefore, different other derivatives of graphene have also been investigated for this purpose.

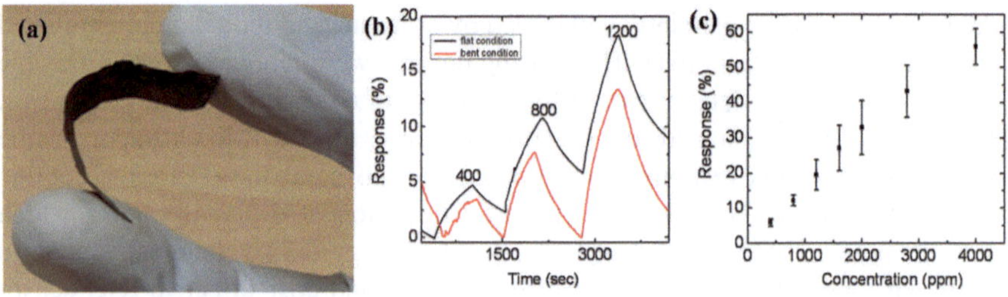

Fig. 17.3 **a** Photograph of a flexible rGO paper in bending condition. **b** Comparative sensing performance of the flexible rGO sensor toward three different NH_3 concentrations (400–1200 ppm) under flat and bending conditions. **c** Sensing responses of three different samples toward seven different concentrations of NH_3 (400–4000 ppm). (*Reprinted with permission from* [42])

17.4.3 Graphene Composites

Graphene hybrid structures with other nanomaterials or nanoparticles significantly improve the sensing performances for designing highly sensitive and highly selective monitoring devices. Chung et al. reported a flexible H_2 sensor using Pd nanoparticle–modified graphene sheets [44]. The CVD method was used to synthesize the single–layer graphene on a Cu foil and was transferred onto a PET substrate by using PMMA as a sacrificial layer. The Pd nanoparticles were deposited onto graphene by thermal evaporation and then the gold (Au) wires were connected with silver paste to measure the gas sensing response. The graphene sensor exhibited an increasing response with varying Pd thickness with a maximum response of about ~32.9% at 3 nm thickness for a concentration of 1000 ppm. The sensor also showed very negligible variation in sensing response for a curved geometry with a bending radius of 3 mm (Fig. 17.4). The formation of palladium hydride during adsorption explains the sensing mechanism of the modified graphene sensor. Moreover, the strong bonding between Pd and H_2 molecules results in incomplete recovery of the sensors, which could be improved by optimizing the deposition thickness and density of Pd NPs on graphene. Huang et al. reported a fully printed chemically modified graphene sensor fabricated by gravure printing of sulfonated rGO (S–rGO) modified with Ag nanoparticles [45]. The Ag–S–rGO sheets were prepared by uniform deposition of Ag nanoparticles of size 10–20 nm to the sulfonated rGO solution. The sensing layer of Ag–S–rGO was directly deposited onto a flexible PI substrate pre–patterned with Ag IDEs. The Ag–S–rGO hybrids exhibited superior sensitivity toward 500 ppb to 50 ppm of NO_2 concentrations with a response value increasing from 3.4 to 74.6% (Fig. 17.5). The response and recovery times were also observed within 12 s and 20 s respectively for the sensor. In addition, the response of the Ag–S–rGO hybrids was found to be very stable at different RH, and thus has a negligible influence of RH on its sensing. Further, the tests conducted at different bending cycles of the sensor also showed excellent electromechanical stability of the printed Ag–S–rGO hybrids. Cho and others reported the effect of Pd and Al nanoparticle decoration on the sensing performance of graphene by exposing it to different analytes (NO_2 and NH_3) [46].

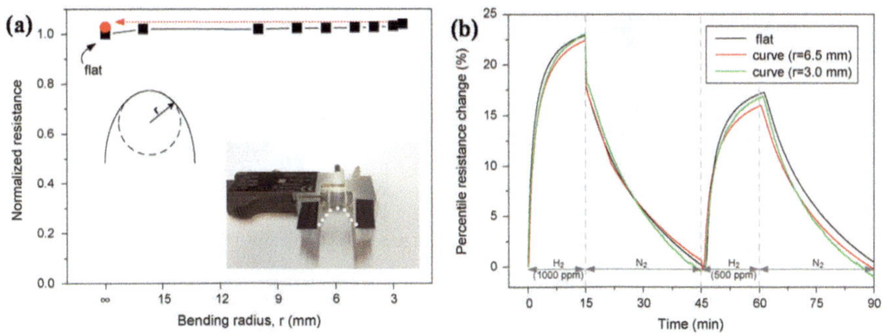

Fig. 17.4 **a** The performance of the Pd nanoparticle–modified graphene device under bending conditions as shown in inset with respect to bending radius. **b** The device sensing performance toward 500 and 1000 ppm H_2 under different bending radii. (*Reprinted with permission from* [44])

The multilayer graphene films obtained by the CVD method were transferred onto a PI substrate and the Au/Ti electrodes were patterned into an array structure onto the deposited graphene films. It was observed that the metal decoration enhanced the sensing performance and the selectivity of the patterned graphene toward both NO_2 and NH_3. The selectivity of the sensor toward NH_3 was significantly improved on decorating with Pd nanoparticles, while the Al nanoparticles improved the NO_2 sensitivity of the graphene sensors. The bending test shows a slight variation in the sensing response of the metal–decorated graphene after 5000 bending cycles, though no significant variation was observed with increasing bending radius. This study suggests a simple route for fabricating flexible graphene–based sensors with precise control over the selectivity for practical implementation.

Graphene hybrid structures with vertical ZnO nanorods were also reported to demonstrate a flexible gas sensor for ppm–level detection of ethanol vapors with high sensitivity [47]. The hybrid structure is composed of a bottom conducting layer of ZnO, vertical ZnO nanorods as the channel material and conductive graphene layer as the top electrode (as shown in Fig. 17.6a–d). The optical transmittance of the hybrid architecture fabricated onto a glass substrate was observed to be larger than ~70% indicating the potential use in transparent devices. On exposure to ethanol vapors, the hybrid architecture exhibited very high response (R_a/R_g) of ~9 and 90 for 10 ppm and 50 ppm, respectively. Moreover, the excellent stability of the device response up to 100 cycles of tensile strain ~1.3% indicates its great potential for flexible gas sensors. Vertical carbon nanotubes (CNTs)/reduced graphene hybrids fabricated onto a PI substrate have been used to demonstrate room temperature detection of NO_2 [48]. The vertical CNT arrays were patterned by the plasma–enhanced CVD method over the spin–coated GO sheets deposited onto a SiO_2/Si substrate. Later, the film was detached from the substrate by etching in a dilute HF solution and transferred to an Au/PI flexible substrate. On exposure to NO_2 at concentrations from 0.5 to 10 ppm, the hybrid film showed an enhanced response as compared to a reduced film. Under extreme bending conditions also, the hybrid film showed very little deterioration in the sensing response indicating the good promise of the hybrid sensor for wearable

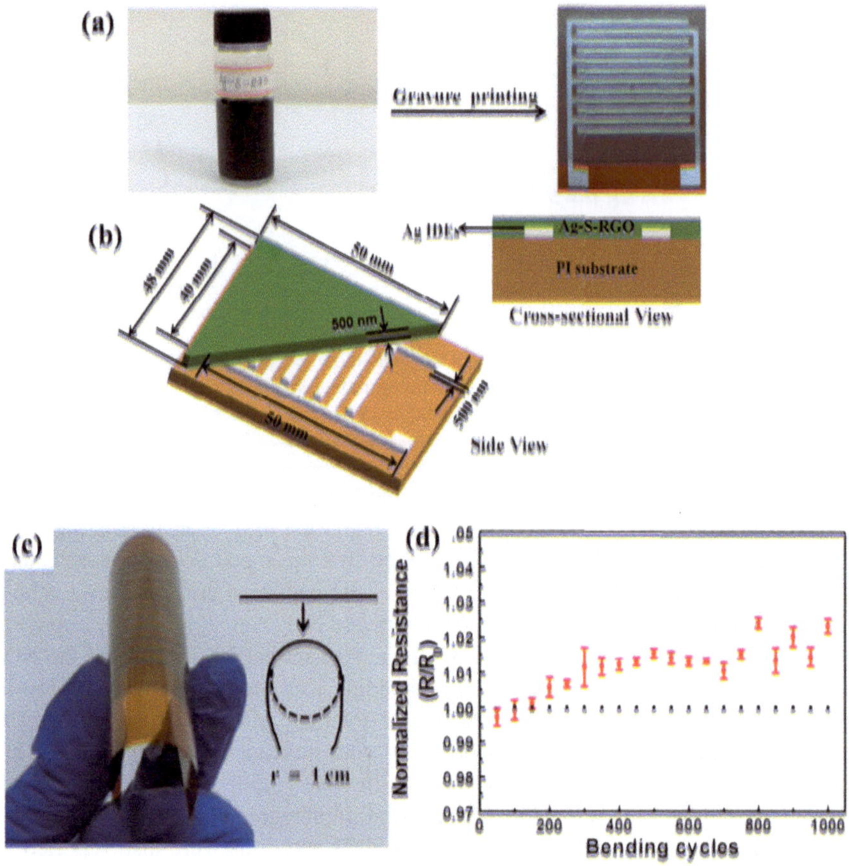

Fig. 17.5 **a–b** Digital images of Ag–S–rGO ink and Ag–S–rGO sensing layer printed on Ag–IDEs fabricated PI substrate, respectively. **c** The printed Ag–S–rGO sensor in bending conditions and **d** its response under number of bending cycles. (*Reprinted with permission from* [45])

sensors. Poly(3,4–ethylenedioxythiophene): poly(styrenesulfonate) (PEDOT:PSS) conducting polymer and graphene–based composites were fabricated onto a flexible substrate to demonstrate highly sensitive and selective sensing performance toward NH_3 detection [49]. The composite films were dispersed uniformly following the inkjet printing technique onto a flexible, transparent substrate. The sensor showed good selectivity and repeatable response toward NH_3 over a variety of VOCs investigated at room temperature (Fig. 17.6e–g). Interestingly, the flexible composite sensor exhibited increasing responses while varying bending angles from 10° to 70°, which is attributed to the swelling process within the polymer/graphene matrix during the diffusion of NH_3 molecules. Moreover, the low–temperature processing and high sensing performance of the sensing device would be useful for developing flexible gas sensors using graphene/polymer composites.

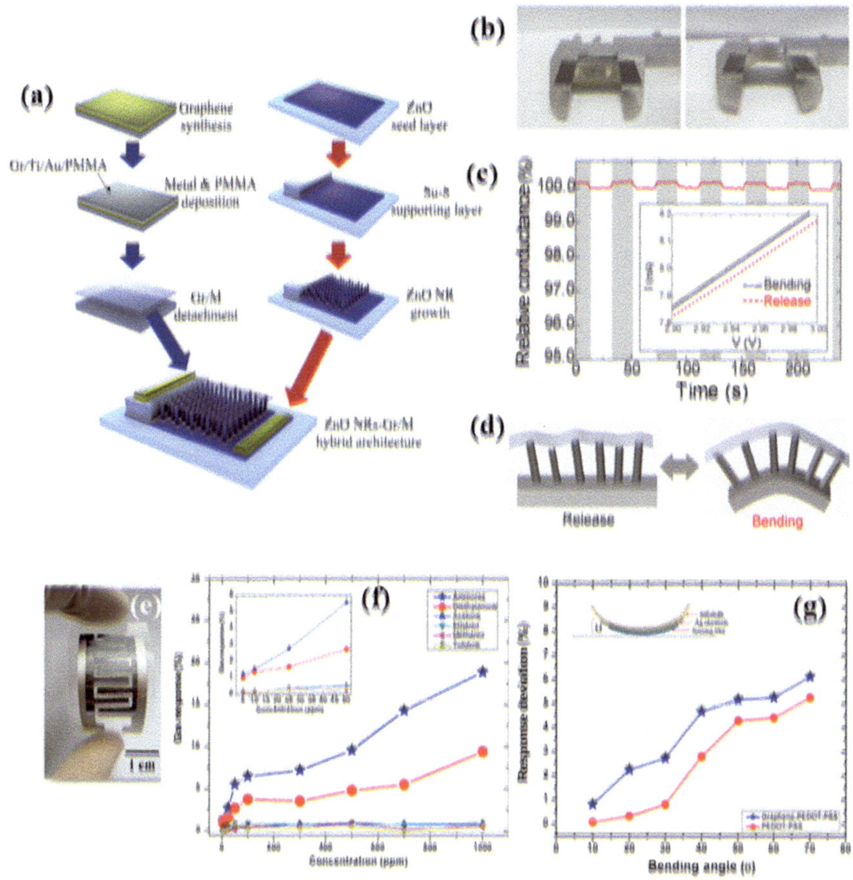

Fig. 17.6 **a** Schematic illustration of steps involved in synthesis ZnO NRs–Gr/M hybrid architectures. **b** The flexible ZnO NRs–Gr/M hybrid device on a SUS foil in relaxed and bending condition. **c** Resistive changes of the device under repeated bending–unbending cycles. (Inset) I–V curves under release and bending conditions. **d** Schematic illustration of the device in release and bending conditions *(Reprinted with permission from* [47]). **e** Photograph of the flexible graphene–PEDOT:PSS sensing device. **f** Sensing response of flexible graphene–PEDOT:PSS gas sensor toward different VOCs at different concentrations at RT. **g** Device response toward bending condition at concentration 500 ppm of NH_3. (*Reprinted with permission from* [49])

Therefore, it could be stated that graphene–based derivatives exhibit gas sensing with desired performance for use as wearable sensors. Although there has been much development in the gas sensing field using graphene and its derivatives as potential nanomaterialsas shown in Table 17.1, a lot of scopes still remain in this arena of research. The flexibility of graphene–based chemical sensors is yet to meet the desired performance and demand for industrial applications.

Table 17.1 Comparison of gas sensing performances of graphene–based flexible gas sensors

Sensing material	Substrate	Gas analytes	Detection range	Response	Response/ recovery time	LOD	References
Pristine graphene	PET	NO_2	200 ppm	$\Delta G/G_a$=23%	–	–	[38]
Pristine graphene	PES	NO_2	1 ppm	$\Delta R/R_a$=12%	14 s/95 s	–	[39]
Pristine graphene	PI	NO_2	5 ppm	$\Delta R/R_a$=13%	89 s/579 s	1 ppm	[40]
rGO	PET	NO_2	100 ppm to 500 ppb	–	–	500 ppb	[41]
rGO	Whatman paper	NH_3	400–4000 ppm	$\Delta R/R_{N_2}$=5–55%	–	430 ppb	[42]
rGO	Porous paper	NO_2	1–100 ppm	–	–	Sub-ppm	[43]
Pd-graphene	PET	H_2	20–1000 ppm	$\Delta R/R_a$~33% (1000 ppm)	–	–	[44]
AgNPs–sulfonated rGO	PI	NO_2	0.5–50 ppm	$\Delta R/R_a$=3.4–74.6%	12 s/20 s (50 ppm)	–	[45]
Pd, Al decorated graphene	PI	NO_2, NH_3	5–100 ppm (NH_3), 1.2–5 ppm (NO_2)	–	–	–	[46]
Vertically aligned ZnO NR-graphene	Stainless steel (SUS) foil	Ethanol	10–50 ppm	R_a/R_g~9–90	–	–	[47]
CNT–rGO	PI	NO_2	0.5–10 ppm	–	–	–	[48]
Graphene–PEDOT:PSS composite		NH_3	5–1000 ppm	$\Delta R/R_a$=18.9% (1000 ppm)	–	<10 ppm	[49]

17.5 Gas Sensing Mechanism of Graphene–Based Sensors

Graphene–based gas sensors work on the principle of change in electrical conductivity upon molecular interaction of gas molecules to be detected. The change in electrical conductivity is accompanied by the change in carrier concentration upon interaction with the gas molecules [17]. The gas molecules interact with the graphene surface, whether pristine or functionalized, through different physical or chemical interaction depending upon the nature of the gas species. Due to ballistic transport of charge carriers and large specific surface area exhibited by graphene materials, even a small change in the carrier concentration by the gas adsorbates can induce significant change in the electrical conductivity of graphene [50, 51]. Being a p–type material, graphene–based sensing materials show either an increase or decrease in conductivity upon adsorption of different oxidizing or reducing gases [12]. Graphene–based materials possess different electrical properties based upon its different structural forms, e.g., doped graphene, graphene oxide (GO), reduced graphene oxide (rGO) and composites of graphene. Generally, in pristine graphene, the interaction with gas adsorbates is mostly dominated by physical adsorption. On the other hand, doping with heteroatoms into graphene materials facilitates both physical adsorption and chemical interaction at the active defect sites. Moreover, in other graphene materials, GO and rGO, the chemical interaction dominates the sensing mechanism due to the presence of numerous oxygen functional groups on the graphene surface. Therefore, based on the structural form of graphene, different sensing mechanisms are being reported by researchers to explain the detection of different gas molecules.

17.6 Future Scope and Conclusion

There has been great progress in graphene–based wearable sensors in the last decade due to their superior chemical, electrical, optical and mechanical properties. Along with pristine graphene–based toxic chemical sensors, chemically modified graphene like graphene oxide and reduced graphene oxide–based sensors have also been discussed. Sensors fabricated from graphene derivatives and composite with other materials are also investigated for wearable and flexible sensors. It is observed that a sensor made of chemically derived graphene and a hybrid composite of graphene shows better sensitivity and selectivity as compared to its pristine counterparts. The two important parameters of a sensor, i.e., sensitivity and selectivity depend upon various factors starting from synthesis condition to the concentration of the dopants. Hence, it can be concluded that the proper combination of graphene with suitable dopants plays a crucial role in the sensitivity of the sensors mainly due to their enhanced morphological effects. Doping in graphene creates defects in the planner structure of graphene which enhances the number of active sites for adsorbing the molecules of the test gases. However, along with the improvement of the performance of graphene–based wearable sensors, research should also focus on fully integrated wearable devices that can replace the traditional equipment. Flexible graphene sheets can fulfill most requirements for the fabrication of flexible electronic devices. Flexible graphene–based sensors have tremendous potential for the development of wearable, transparent and stretchable sensors that can be integrated on any kind of surface. For example, graphene–based RFID sensor technologies are significantly growing in several flexible devices including cell phones. The RFID–based sensors have tremendous importance in connection with the Internet of Things (IoT) sector. As the IoT is included in every field of health care to space exploration, artificial intelligence, etc., the research on wearable sensors based on graphene as well as other nanomaterials will have very high demands in the coming years. Stability and durability are two major challenges for advanced materials including 2D graphene under fluctuating surrounding environments. The integration of the sensor with wireless technology for IoT application is another challenge for future research as well as commercialization of the sensors. There are other challenges such as finding lightweight and low–cost materials for the fabrication of flexible and wearable sensors on a large scale to compensate the traditional solid–state–based sensors used in chemical and biological applications.

References

1. Nurazzi, N.M., Abdullah, N., Demon, S.Z.N., Halim, N.A., Azmi, A.F.M., Knight, V.F., Mohamad, I.S.: The frontiers of functionalized graphene-based nanocomposites as chemical sensors. Nanotechnol. Rev. **10**, 330–369 (2021). https://doi.org/10.1515/NTREV-2021-0030
2. Manisalidis, I., Stavropoulou, E., Stavropoulos, A., Bezirtzoglou, E.: Environmental and health impacts of air pollution: a review. Front. Public Health **8** (2020)
3. Kim, K.H., Jahan, S.A., Lee, J.T.: Exposure to formaldehyde and its potential human health hazards. **29**, 277–299 (2011). https://doi.org/10.1080/10590501.2011.629972
4. Wang, Z., Wang, C.: Is breath acetone a biomarker of diabetes? A historical review on breath acetone measurements. J. Breath Res. **7** (2013)
5. Das, T., Sharma, B.K., Katiyar, A.K., Ahn, J.H.: Graphene-based flexible and wearable electronics. J. Semiconductors **39** (2018). https://doi.org/10.1088/1674-4926/39/1/011007
6. Kim, H., Ahn, J.H.: Graphene for flexible and wearable device applications. Carbon N. Y. **120**, 244–257 (2017). https://doi.org/10.1016/j.carbon.2017.05.041
7. Singh, E., Meyyappan, M., Nalwa, H.S.: Flexible graphene-based wearable gas and chemical sensors. ACS Appl. Mater. Interfaces **9**, 34544–34586 (2017). https://doi.org/10.1021/acsami.7b07063
8. Lee, C., Wei, X., Kysar, J.W., Hone, J: Measurement of the elastic properties and intrinsic strength of monolayer graphene. Science (1979) **321**, 385–388 (2008). https://doi.org/10.1126/SCIENCE.1157996
9. Palaparthy, V.S., Kalita, H., Surya, S.G., Baghini, M.S., Aslam, M.: Graphene oxide based soil moisture microsensor for in situ agriculture applications. Sens. Actuators, B Chem. **273**, 1660–1669 (2018). https://doi.org/10.1016/j.snb.2018.07.077
10. Kalita, H., Harikrishnan, V., Shinde, D.B., Pillai, V.K., Aslam, M.: Hysteresis and charge trapping in graphene quantum dots. Appl. Phys. Lett. **102**, 1–5 (2013). https://doi.org/10.1063/1.4800236
11. Shinde, D.B., Debgupta, J., Kushwaha, A., Aslam, M., Pillai, V.K.: Electrochemical unzipping of multi-walled carbon nanotubes for facile synthesis of high-quality graphene nanoribbons. J. Am. Chem. Soc. **133**, 4168–4171 (2011). https://doi.org/10.1021/JA1101739

12. Ghosh, R., Aslam, M., Kalita, H.: Graphene derivatives for chemiresistive gas sensors: a review. Mater. Today Commun. **30**, 103182 (2022)
13. Novoselov, K.S., Geim, A.K., Morozov, S.V., Jiang, D., Zhang, Y., Dubonos, S.V., Grigorieva, I.V., Firsov, A.A.: Electric field in atomically thin carbon films. Science (1979) **306**, 666–669 (2004). https://doi.org/10.1126/science.1102896
14. Rashid, A., Yusoff, M., Dai, L., Cheng, H.-M., Liu, J.: Graphene based energy devices. Nanoscale (2015). https://doi.org/10.1039/c5nr90062j
15. Bolotin, K.I., Sikes, K.J., Jiang, Z., Klima, M., Fudenberg, G., Hone, J., Kim, P., Stormer, H.L.: Ultrahigh electron mobility in suspended graphene. Solid State Commun. **146**, 351–355 (2008). https://doi.org/10.1016/J.SSC.2008.02.024
16. Nair, R.R., Blake, P., Grigorenko, A.N., Novoselov, K.S., Booth, T.J., Stauber, T., Peres, N.M.R., Geim, A.K.: Fine structure constant defines visual transparency of graphene. https://doi.org/10.1126/science.1156965
17. Schedin, F., Geim, A.K., Morozov, S.V., Hill, E.W., Blake, P., Katsnelson, M.I., Novoselov, K.S.: Detection of individual gas molecules adsorbed on graphene. Nat. Mater. **6**, 652–655 (2007). https://doi.org/10.1038/NMAT1967
18. Wu, R., Ma, L., Hou, C., Meng, Z., Guo, W., Yu, W., Yu, R., Hu, F., Liu, X.Y.: Silk composite electronic textile sensor for high space precision 2D combo temperature-pressure sensing. Small **15**, 1901558 (2019). https://doi.org/10.1002/SMLL.201901558
19. Gu, X., Chen, X., Tang, X., Zhou, Z., Huang, T., Yang, Y., Ling, J.: Pure-silk fibroin hydrogel with stable aligned micropattern toward peripheral nerve regeneration. Nanotechnol. Rev. **10**, 10–19 (2021). https://doi.org/10.1515/NTREV-2021-0002/HTML
20. Heo, J.S., Eom, J., Kim, Y.H., Park, S.K.: Recent progress of textile-based wearable electronics: a comprehensive review of materials, devices, and applications. Small **14** (2018). https://doi.org/10.1002/SMLL.201703034
21. Kim, K.K., Hong, S., Cho, H.M., Lee, J., Suh, Y.D., Ham, J., Ko, S.H.: Highly sensitive and stretchable multidimensional strain sensor with prestrained anisotropic metal nanowire percolation networks. Nano Lett. **15**, 5240–5247 (2015). https://doi.org/10.1021/ACS.NANOLETT.5B01505
22. Wang, B., Thukral, A., Xie, Z., Liu, L., Zhang, X., Huang, W., Yu, X., Yu, C., Marks, T.J., Facchetti, A.: Flexible and stretchable metal oxide nanofiber networks for multimodal and monolithically integrated wearable electronics. Nat. Commun. **11**(1) (2020). https://doi.org/10.1038/s41467-020-16268-8
23. Rim, Y.S., Bae, S.-H., Chen, H., Yang, J.L., Kim, J., Andrews, A.M., Weiss, P.S., Yang, Y., Tseng, H.-R.: Printable ultrathin metal oxide semiconductor-based conformal biosensors. ACS Nano **9**, 12174–12181 (2015). https://doi.org/10.1021/ACSNANO.5B05325
24. Chen, J., Peng, Q., Thundat, T., Zeng, H.: Stretchable, injectable, and self-healing conductive hydrogel enabled by multiple hydrogen bonding toward wearable electronics. Chem. Mater. **31**, 4553–4563 (2019). https://doi.org/10.1021/ACS.CHEMMATER.9B01239
25. Bag, A., Lee, N.E.: Recent advancements in development of wearable gas sensors. Adv. Mater. Technol. **6**, 1–37 (2021). https://doi.org/10.1002/admt.202000883
26. Liu, X., Cheng, S., Liu, H., Hu, S., Zhang, D., Ning, H.: A survey on gas sensing technology. Sensors (Switzerland) **12**, 9635–9665 (2012)
27. Mamun, M.A. Al., Yuce, M.R.: Recent progress in nanomaterial enabled chemical sensors for wearable environmental monitoring applications. Adv. Funct. Mater. (2020)
28. Han, S.T., Peng, H., Sun, Q., Venkatesh, S., Chung, K.S., Lau, S.C., Zhou, Y., Roy, V.A.L.: An overview of the development of flexible sensors. Adv. Mater. **29** (2017)
29. Park, S., Kim, H., Kim, J.H., Yeo, W.H.: Advanced nanomaterials, printing processes, and applications for flexible hybrid electronics. Materials **13**, 3587 (2020). https://doi.org/10.3390/MA13163587
30. Mirzaei, A., Leonardi, S.G., Neri, G.: Detection of hazardous volatile organic compounds (VOCs) by metal oxide nanostructures-based gas sensors: a review. Ceram. Int. **42**, 15119–15141 (2016). https://doi.org/10.1016/J.CERAMINT.2016.06.145

31. Galstyan, V.: "Quantum dots: perspectives in next-generation chemical gas sensors" - a review. Anal. Chim. Acta **1152**, 238192 (2021). https://doi.org/10.1016/J.ACA.2020.12.067
32. Donarelli, M., Ottaviano, L.: 2D materials for gas sensing applications: a review on graphene oxide, MoS2, WS2 and phosphorene. Sensors (Basel) **18** (2018). https://doi.org/10.3390/S18113638
33. Yang, S., Jiang, C., Wei, S.h.: Gas sensing in 2D materials. Appl. Phys. Rev. **4**, 021304 (2017). https://doi.org/10.1063/1.4983310
34. Qiao, Y., Li, X., Hirtz, T., Deng, G., Wei, Y., Li, M., Ji, S., Wu, Q., Jian, J., Wu, F., Shen, Y., Tian, H., Yang, Y., Ren, T.L.: Graphene-based wearable sensors. Nanoscale **11**, 18923–18945 (2019). https://doi.org/10.1039/C9NR05532K
35. Cho, B., Yoon, J., Lim, S.K., Kim, A.R., Kim, D.H., Park, S.G., Kwon, J.D., Lee, Y.J., Lee, K.H., Lee, B.H., Ko, H.C., Hahm, M.G.: Chemical sensing of 2D graphene/MoS_2 heterostructure device. ACS Appl. Mater. Interfaces **7**, 16775–16780 (2015). https://doi.org/10.1021/ACSAMI.5B04541
36. Graphene Sensors and their Properties and Benefits. https://www.azonano.com/article.aspx?ArticleID=5465. Accessed 9 Nov 2021
37. Huang, X.M., Liu, L.Z., Zhou, S., Zhao, J.J.: Physical properties and device applications of graphene oxide. Front. Phys. **15**, 1–70 (2020). https://doi.org/10.1007/s11467-019-0937-9
38. Lee, C., Ahn, J., Lee, K.B., Kim, D., Kim, J.: Graphene-based flexible NO_2 chemical sensors. Thin Solid Films **520**, 5459–5462 (2012). https://doi.org/10.1016/j.tsf.2012.03.095
39. Choi, H., Choi, J.S., Kim, J.-S., Choe, J.-H., Chung, K.H., Shin, J.-W., Kim, J.T., Youn, D.-H., Kim, K.-C., Lee, J.-I., Choi, S.-Y., Kim, P., Choi, C.-G., Yu, Y.-J.: Flexible and transparent gas molecule sensor integrated with sensing and heating graphene layers. Small **10**, 3685–3691 (2014). https://doi.org/10.1002/smll.201400434
40. Kim, Y.H., Kim, S.J., Kim, Y.-J., Shim, Y.-S., Kim, S.Y., Hong, B.H., Jang, H.W.: Self-activated transparent all-graphene gas sensor with endurance to humidity and mechanical bending. ACS Nano **9**, 10453–10460 (2015)
41. Dua, V., Surwade, S.P., Ammu, S., Agnihotra, S.R., Jain, S., Roberts, K.E., Park, S., Ruoff, R.S., Manohar, S.K.: All-organic vapor sensor using inkjet-printed reduced graphene oxide. Angew. Chem. Int. Ed. **49**, 2154–2157 (2010)
42. Ghosh, R., Singh, A., Santra, S., Ray, S.K., Chandra, A., Guha, P.K.: Highly sensitive large-area multi-layered graphene-based flexible ammonia sensor. Sens. Actuators, B Chem. **205**, 67–73 (2014). https://doi.org/10.1016/j.snb.2014.08.044
43. Hassinen, J., Kauppila, J., Leiro, J., Määttänen, A., Ihalainen, P., Peltonen, J., Lukkari, J.: Low-cost reduced graphene oxide-based conductometric nitrogen dioxide-sensitive sensor on paper. Anal. Bioanal. Chem. **405**, 3611–3617 (2013)
44. Chung, M.G., Kim, D.H., Seo, D.K., Kim, T., Im, H.U., Lee, H.M., Yoo, J.B., Hong, S.H., Kang, T.J., Kim, Y.H.: Flexible hydrogen sensors using graphene with palladium nanoparticle decoration. Sens. Actuators, B Chem. **169**, 387–392 (2012). https://doi.org/10.1016/j.snb.2012.05.031
45. Huang, L., Wang, Z., Zhang, J., Pu, J., Lin, Y., Xu, S., Shen, L., Chen, Q., Shi, W.: Fully printed, rapid-response sensors based on chemically modified graphene for detecting NO_2 at room temperature. ACS Appl. Mater. Interfaces **6**, 7426–7433 (2014). https://doi.org/10.1021/am500843p
46. Cho, B., Yoon, J., Hahm, M.G., Kim, D.H., Kim, A.R., Kahng, Y.H., Park, S.W., Lee, Y.J., Park, S.G., Kwon, J.D., Kim, C.S., Song, M., Jeong, Y., Nam, K.S., Ko, H.C.: Graphene-based gas sensor: metal decoration effect and application to a flexible device. J. Mater. Chem. C **2**, 5280–5285 (2014). https://doi.org/10.1039/c4tc00510d
47. Yi, J., Lee, J.M., Park, W., il,: Vertically aligned ZnO nanorods and graphene hybrid architectures for high-sensitive flexible gas sensors. Sens. Actuators, B Chem. **155**, 264–269 (2011). https://doi.org/10.1016/J.SNB.2010.12.033
48. Jeong, H.Y., Lee, D.S., Choi, H.K., Lee, D.H., Kim, J.E., Lee, J.Y., Lee, W.J., Kim, S.O., Choi, S.Y.: Flexible room-temperature NO_2 gas sensors based on carbon nanotubes/reduced graphene hybrid films. Appl. Phys. Lett. **96**, 2012–2015 (2010). https://doi.org/10.1063/1.3432446

49. Seekaew, Y., Lokavee, S., Phokharatkul, D., Wisitsoraat, A., Kerdcharoen, T., Wongchoosuk, C.: Low-cost and flexible printed graphene–PEDOT:PSS gas sensor for ammonia detection. Org. Electron. **15**, 2971–2981 (2014). https://doi.org/10.1016/j.orgel.2014.08.044
50. Lu, G., Ocola, L.E., Chen, J.: Reduced graphene oxide for room-temperature gas sensors. Nanotechnology **20**, 445502 (2009). https://doi.org/10.1088/0957-4484/20/44/445502
51. Yuan, W., Shi, G.: Graphene-based gas sensors. J. Mater. Chem. A **1**, 10078–10091 (2013)

Chapter 18
Pulsed Laser-Mediated Phototherapeutic Mechanisms for Biomedical Applications

L. Sophia Jacquline, Pooja Naik, and Junaid Masud Laskar

Abstract Plasmonics has recently facilitated promising developments in the fields of energy, optics, biology, and medicine. Plasmonic nanoparticles (NPs) have attracted a lot of attention in the past few decades due to their highly tunable optical sensitivity and ability to change properties with an external light source. The non-radiative conversion of electromagnetic (EM) radiation into heat energy and acoustic vibrations by plasmonic nanostructures has triggered strong interest due to the wide range of potential applications, particularly in the field of biotechnology and biomedicine. Fundamental properties of plasmonic NPs must be better understood and regulated to realise their full potential for their clinical applications such as photo-based diagnosis and treatment.

The theory and physics of plasmonic photothermal and photoacoustic conversions are introduced initially, followed by a discussion of methods to enhance efficiency by modifying optical parameters—the focus of this chapter. Furthermore, recent applications are used to determine how plasmonic NPs can be tuned to give maximum output based on the requirements of the application. Overall, this chapter summarizes recent advances in understanding the principles of plasmonic NPs, their applications in biomedicine and biotechnology, and potential future avenues.

Keywords Plasmons · Localised surface plasmon resonance (LSPR) · Metal nanoparticles · Photothermal effects · Photoacoustic effects · Mie scattering · Optical tuning

L. S. Jacquline · P. Naik · J. M. Laskar (✉)
Nanophotonics Laboratory, Department of Physics and Nanotechnology, SRM Institute of Science and Technology, Kattankulathur 603203, Chennai, Tamil Nadu, India
e-mail: jmlaskar@gmail.com; junaidmb@srmist.edu.in

D. Mohanta and P. Chakraborty (eds.), *Nanoscale Matter and Principles for Sensing and Labeling Applications*, Advanced Structured Materials 206,
https://doi.org/10.1007/978-981-99-7848-9_18

18.1 Introduction

Pathogens such as bacteria, viruses, parasites, and fungi are continuing to evolve genetically and acquiring treatment resistance, which poses a global threat to civilisation and the environment. Despite the myriad antimicrobial methods that have been developed to combat infectious diseases, there is a surge in multidrug-resistant bacterial infections, which is a major source of concern for human health globally [1]. Therefore, a new drug is usually required to combat new or mutated strains of microorganisms. However, as evidenced by COVID-19, a novel corona virus disease that resulted in the loss of lives and livelihoods, relying solely on drug development process is not always feasible. Since the release of the cowpox vaccine in 1976, we have relied on vaccines to treat diseases. But drug development necessitates the use of chemicals which may have several potential side effects, and hence, the drug needs time to be developed, tested, and obtain regulatory approval for usage. As a result, rather than relying on a single source, it is essential that we identify and employ chemical-free, non-contact approaches to inactivate such pathogen-causing diseases. One approach for addressing these concerns is implementing electromagnetic (EM) wave-based therapy.

The use of EM radiation in the treatment of pathogen-induced disorders has been practiced since the period of ancient Egyptian, Indian, and Chinese civilisations. Initially, sunlight was used to treat skin conditions including psoriasis and skin cancer [2, 3]. Following this, in the eighteenth century, sunlight was utilized to treat ailments such as tuberculosis, paralysis, muscle weakness, rickets, and scurvy. The significance of using EM radiation in medicine was recognised at the beginning of the twentieth century with the discovery of using sunlight to cure lupus vulgaris, a type of tuberculosis [3, 4], which marked the beginning of modern phototherapy. "Phototherapy", as the name suggests, refers to the use of light for therapeutic purposes. Among the developed phototherapeutic treatments, photothermal therapy (PTT) and photoacoustic therapy (PAT) has been widely studied. Several applications for plasmonic nanomaterials in phototherapy have been proposed [5–8].

PTT is a type of phototherapy that deactivates pathogens by converting the EM radiation into thermal energy. PAT, on the other hand, utilizes thermoelastic expansion produced due to EM radiation to generate shock waves that disintegrate the pathogen's exterior layer known as the protein capsid [9]. PTT and PAT have several advantages over other existing techniques including (a) being minimally invasive, (b) having a wide antibacterial spectrum, (c) not causing mutation, and (d) being remotely controllable. While the two phototherapeutic methods have significant advantages over conventional pathogen-combating strategies, they have a few drawbacks. One of the disadvantages is the restricted efficacy due to the limited penetration depth of light. Despite this, recent advancements in nanotechnology have enabled the synthesis and fabrication of systems that can substantially improve the impact of EM radiation in biotechnology and biomedicine.

The materials that have been researched to enhance the effect of phototherapy, particularly in PTT and PAT, are organic nanomaterials [9], polymers [7], and metallic nanostructures [10]. Metallic nanoparticles (MNPs) have gained huge interest over the past few decades owing to their novel ability to confine and enhance EM field on a sub-wavelength scale. MNPs possess five times the absorption compared to that of the organic ones and hence are favoured over other materials [11]. Plasmonic MNPs such as gold (Au) and silver (Ag) NPs can produce strong localised surface plasmon resonance (LSPR) when the incident light energy matches the resonance frequency of the free charge carriers, i.e., the electrons [12]. Such novel light-matter interaction in MNPs and the many phenomena that resulted from them, such as conversion of photonic energy into thermal energy, drew a lot of attention.

This book chapter presents a systematic review of the underlying physics of plasmonics and their biomedical applications. It also presents a detailed explanation on how the efficiency of plasmonic photo-induced therapy can be increased by tuning the geometry of plasmonic systems.

18.2 Plasmonics

18.2.1 Bulk Plasmons and Surface Plasmons (SPs)

Plasmons are collective oscillations of free electrons in a conductor or metal generated by EM radiation with reference to fixed positive atom cores. Due to charge displacement, incident light causes polarisation in a bulk material (Fig. 18.1a), resulting in the generation of forward and backward EM waves known as "bulk plasmons". The energy of these bulk plasmons is given by [13]

$$E_{bulk\ plasmons} = \hbar\omega_p \tag{18.1}$$

where $\hbar$ is Planck's constant and ω_p is the plasmon frequency, the frequency of collective oscillation of free electrons which is unique to every material. Plasmon frequency is a function of electron density (n),

$$\omega_p = \sqrt{\frac{ne^2}{m_e\varepsilon_0}} \tag{18.2}$$

and m_e, e, ε_0 are electron effective mass, electron charge, and permittivity of free space, respectively. Plasmon frequency physically indicates how easily electrons can migrate in response to the incident EM field.

When the bulk metal dimension is reduced to a thin film, travelling surface charge waves are produced at the metal–dielectric interface, known as surface plasmon polaritons (SPPs) [13] (Fig. 18.1b). The metal–dielectric interface imposes further limitations on the frequency at which the metal electrons can oscillate. The frequency (ω_{SPR}) at which the surface plasmon resonance (SPR) occurs is given by

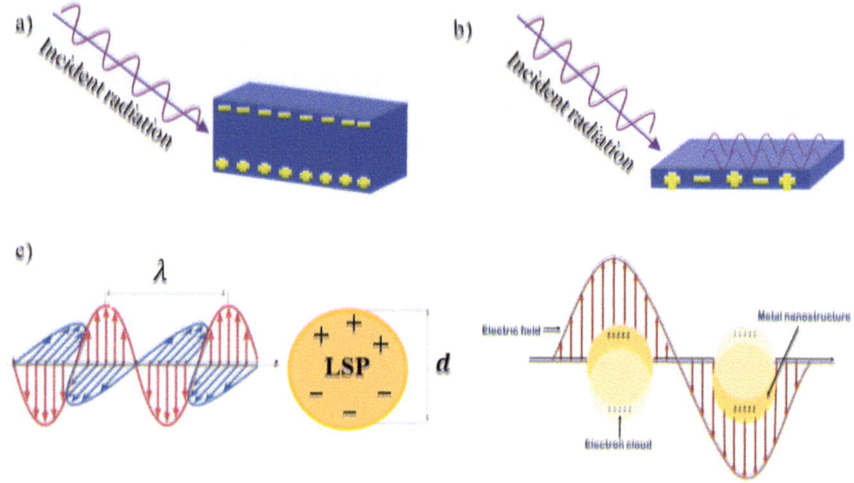

Fig. 18.1 Illustration of **a** bulk plasmons inside a solid material, **b** surface plasmons in a thin film, and **c** Localised Surface Plasmon Resonance (LSPR) in MNPs. Figure modified and reprinted with permission from [14]

$$\omega_{SPR} = \frac{\omega_p}{\sqrt{1 + \varepsilon_m}} \tag{18.3}$$

The equation indicates that the SPR frequency depends on the dielectric constant of the media surrounding the plasmonic MNPs (ε_m) and plasmon frequency. The energy of surface plasmons is less compared to that of the bulk plasmons, which quantise the oscillation of electrons inside the material volume.

18.2.2 Localised Surface Plasmon Resonance (LSPR)

When the size of a metallic particle is reduced to nanoscale, the electron mobility is spatially constrained, which has a considerable impact on the physical, chemical, and optical characteristics of metals. When the dimensions of metallic materials are much smaller than the incident wavelength of light, the new properties that come from spatial restrictions are significantly different from those of bulk materials (confinement effect) [15]. As a result, when the size of the NP (d) is smaller than the wavelength of light (λ), the incident electric field will be constant across the NPs, causing a uniform displacement of the electron density (Fig. 18.1c). The free electrons on the particle surface are stimulated when MNP is irradiated by light, and the local electron cloud is asymmetrically dispersed over the MNP [16]. The Coulombic restoring force between the negatively charged electrons and the positive nuclei are produced because of charge separation, resulting in a series of back-and-forth oscillations of the electron cloud on the particle surface. This collective coherent oscillation of conduction band electrons within the MNPs at the metal–dielectric interface is known as LSPR.

When the incident light frequency matches with the LSPR frequency of plasmonic MNPs, the NPs absorb significant amount of light, resulting in highly amplified and localised electric fields around the particle surface. The frequency (ω_{LSPR}) at which this strong absorption takes place is also a function of the dielectric constant of the media surrounding the MNPs. The basis for the plasmon resonance of the electrons and their resultant effects can also be visualised by understanding polarisability. For a spherical MNP, the quasi-static polarisability (α) is mathematically defined as [17]

$$\alpha(\lambda) = 4\pi\varepsilon_0 r^3 \frac{\varepsilon_p(\lambda) - \varepsilon_m}{\varepsilon_p(\lambda) + 2\varepsilon_m} \tag{18.4}$$

where ε_p and r are the dielectric constant and radius of the metal nanosphere respectively, and ε is the wavelength-dependent complex dielectric function:

$$\varepsilon(\lambda) = Re[\varepsilon(\lambda)] + iIm[\varepsilon(\lambda)] \tag{18.5}$$

The resonance frequency is defined by the real part of the dielectric constant ($Re[\varepsilon(\lambda)]$), while the imaginary dielectric component ($Im[\varepsilon(\lambda)]$) characterises losses within the plasmonic system. Radiative damping, thermal loss from metal heating, and structural disorders and defects can all contribute to losses in plasmonic particles. Figure 18.2 shows the dependence of real and imaginary dielectric counterparts on the incident light wavelength. The graph (Fig. 18.2b) illustrates that Au nanospheres yield high losses in the visible spectral range, but Ag nanospheres induce minor losses. This occurs because Ag nanospheres conduct interband transitions above SPR frequency, limiting LSP decay. In the case of Au nanospheres, however, there is a substantial drop of plasmon intensity [14].

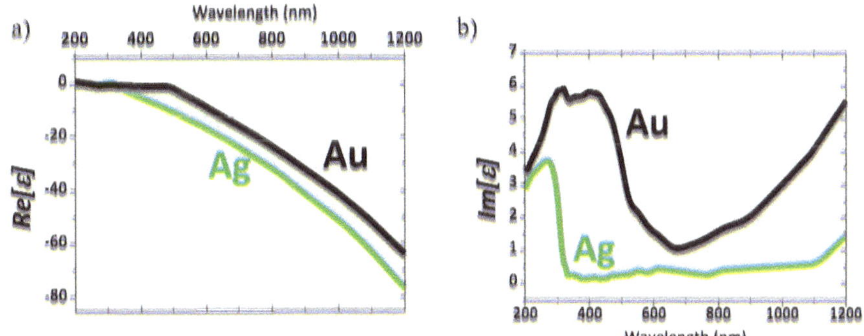

Fig. 18.2 Plot of **a** $Re[\varepsilon]$ and **b** $Im[\varepsilon]$ as a function of varying incident light wavelength for Ag and Au (green and black solid lines respectively). Reprinted with permission from [17]

18.2.3 Electromagnetic (EM) Wave Absorption and Scattering in Plasmonics

The absorption and scattering in plasmonic NPs can be studied and modelled using the Mie scattering theory, a concept put forth by Gustav Mie in the early 1900s. Mie theory allows calculation of the optical cross sections and hence quantify absorption, and scattering and extinction coefficients of a particle treated with EM radiation. The optical coefficients explain the number of photons absorbed or scattered per unit distance. The theory takes into account the concepts for spherical particles and the general expression for scattering (σ_{sca}), and extinction coefficients (σ_{ext}) are given as [18]

$$\sigma_{sca} = \frac{2\pi}{k^2} \sum (2n+1)(|a_n|^2 + |b_n|^2) \tag{18.6}$$

$$\sigma_{ext} = \frac{2\pi}{k^2} \sum (2n+1) Re(a_n + b_n) \tag{18.7}$$

where a_n and b_n are Mie coefficients which in turn are a function of ε_m, ε_p, λ and shape and size of the scatterer. The optical properties of plasmonic NPs are a function of the particle extinction cross section, C_{ext}, which comprises both absorption (C_{abs}) and scattering (C_{sca}) cross-sectional components [18]:

$$C_{ext} = C_{abs} + C_{sca} \tag{18.8}$$

Therefore, from Eqs. 18.4, 18.5, 18.6 and 18.7, and 18.8, we can conclude that.

(a) Properties of plasmonic NPs of any material are defined using its wavelength-dependent complex dielectric constant.
(b) The optical properties (σ_{sca}, σ_{abs}, and σ_{ext}) of the MNP is dependent on the shape, size, and composition of the nanoparticle.

When the incident light frequency matches the LSPR frequency, both absorption and scattering occur. The absorption and scattering properties of plasmonic NPs can be tuned (Fig. 18.3) and used in many biomedical applications, which are discussed in detail in the following sections. According to Eqs. 18.3 and 18.4, the resonance frequency depends upon various controllable parameters such as wavelength of incident light, size, shape, material, and dielectric properties of both the MNP and surrounding media. Therefore, it can be said that LSPR is a tuneable phenomenon. The tunability of LSPR coupled with its ability to amplify light scattering and absorption makes it an excellent candidate for biomedical applications including phototherapy, cancer therapy, bioimaging, and biosensing technologies [19–31].

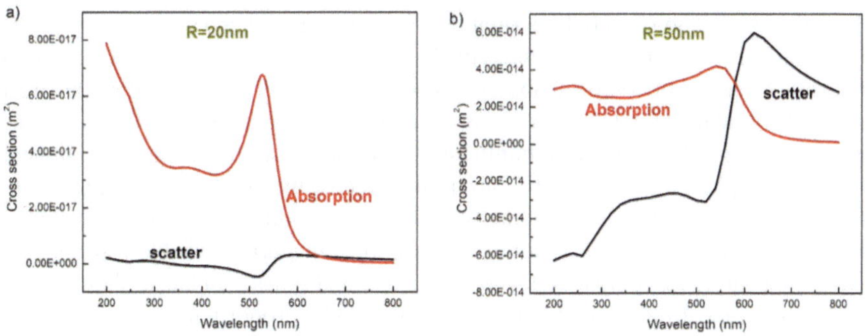

Fig. 18.3 COMSOL simulation results showing absorption and scattering of plasmonic NPs with different radii (R): **a** $R = 20$ nm; **b** $R = 50$ nm

18.3 Pulsed Laser-Induced Phototherapy Mechanisms

As previously stated, with the demand for an alternate approach of combating pathogen-causing ailments and diseases, research towards using EM radiation for pathogen inactivation has accelerated. Tsen and colleagues reported a unique method for virus eradication that combines using an ultrashort pulsed (USP) laser with plasmonic NPs. USP lasers, as the name implies, are pulsed lasers with pulse widths in the picosecond (ps) or femtosecond (fs) range. The near infrared (NIR) region, which is also known as the "therapeutic window" or "tissue optical window", is the spectral range (650–1350 nm) of the EM radiation employed for pathogen inactivation. Outside of this spectral range, light is strongly absorbed by tissue pigments such as haemoglobin and melanin, as well as the water molecules present in cells and tissues, causing damage to healthy mammalian cells. Additionally, because of the low absorption in the NIR range, the amount of light necessary for pathogen inactivation would be minimal, and healthy tissues and cells would remain undamaged.

The effect of NIR radiation on biological tissues, cells, and biomolecules is low when continuous wave (cw) lasers are used, since their power output is consistent throughout the irradiation process. However, the power of the irradiation beam changes with time in the case of pulsed lasers, creating shorter pulses of energy with a higher peak power as compared to cw lasers, making pulsed lasers notably more effective. Furthermore, due to higher penetration depth, pulsed lasers are preferred over cw lasers. The use of USP lasers over long pulse lasers (such as nanosecond (ns) lasers) is essential because pulses longer than 1 ns trigger photochemical reactions in the target and those greater than 1 ps result in thermodynamic processes such as thermal diffusion. In the case of a pulsed laser in the ns regime, there is enough time for thermal diffusion to occur in the target, and therefore the heat generated at the interface interacts with the target's surroundings, generating an imbalance between the interior and exterior of the target. As the temperature of the surrounding material rises, thermal shock-related cracks occur, causing changes in the morphology of the surrounding areas. The shorter the energy pulses, the less heat is generated in the target. For example, photothermal tests have revealed that a single pulse of a ns laser can induce cell death more effectively than a cw laser, implying that plasmonic field amplification of nonlinear absorption processes occurs at high localisation of Au nanospheres at the target [32]. In contrast, employing fs lasers, i.e., USP lasers, eliminates heating of biomolecules, minimising thermal effects, and is rather dominated by molecule photoionisation [33]. This occurs because the timeframe of energy transmission in the fs regime is significantly shorter than required for thermal imbalance, being on the same timescale as ion-to-ion phonon transport. At visible wavelengths, from the conducted comparison study of ns, ps, and fs laser-induced phototherapy the square root of the pulse duration was found to be proportional to the threshold energy required for phototherapy. It was also demonstrated that ablation can be done with ps and fs lasers at substantially lower pulse energy than ns lasers [34].

Even though the approach of employing EM radiation for viral inactivation is relatively well established, there is a lack of a comprehensive understanding of the mechanisms involved. Because of this limited understanding, research towards creating and improving a working prototype has been limited. Therefore, in this section we discuss the two photo-induced mechanisms responsible for pathogen inactivation using USP lasers and MNPs: (a) photothermal effect and (b) photoacoustic effect.

18.3.1 Photothermal Mechanisms

The ability of plasmonic NPs to exhibit distinctive properties such as LSPR has made them an ideal candidate for usage in photothermal applications. When plasmons generated in plasmonic MNPs couple with EM waves, the motion of conduction electrons is amplified, increasing the frequency of collisions with lattice atoms. This increase in collision frequency induces Joule heating, which is the fundamental mechanism for *thermoplasmonics.*

A strong light absorption in MNPs at LSPR frequency implies high light to heat conversion rate. Since the absorbed EM energy by plasmonic MNPs is released as heat into the environment, they can serve as efficient light to heat converters. As a result, the optical properties of plasmonic nanostructures have a direct impact on their photothermal properties, suggesting that the heating efficiency of plasmonic MNPs can be regulated by tuning their shape, size, and composition [35]. Hence the plasmonic photothermal effect is utilized as an efficient heat source for controllable and uniform thermal release at a specific excitation wavelength [36]. The total heat generated (Q) via photothermal mechanism is a function of plasmonic NP concentration (N), absorption cross-section, and laser intensity (I):

$$Q = NC_{abs}I \tag{18.9}$$

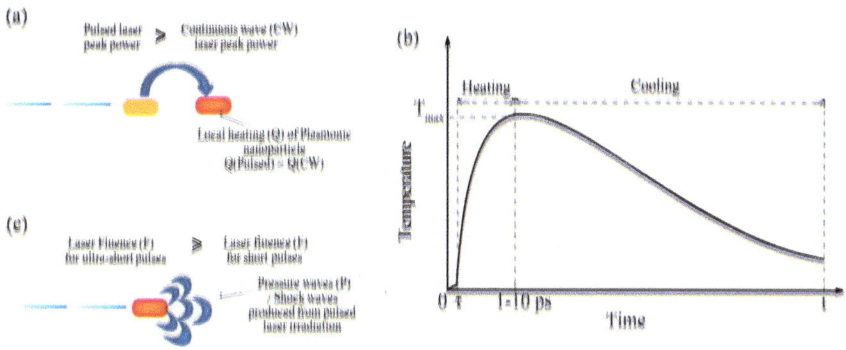

Fig. 18.4 Illustration of **a** Photothermal mechanism. **b** Effect of temperature on the target with respect to time of irradiance of a femtosecond pulsed laser. Reprinted with permission from [37]. **c** Photoacoustic mechanism

The laser intensity is defined as the ratio of the laser peak power and effective focal spot area of laser beam (Eq. 18.10) which proves that the pulsed lasers are more efficient in photothermal generation, as its peak power is high compared to cw lasers (Fig. 18.4a):

$$I = \frac{\text{Laser peak power}}{\text{Effective focal spot area}} \tag{18.10}$$

Heat energy released by plasmonic NPs thermally diffuses from the region of interest to surrounding areas with lower temperatures. The thermal diffusion time (τ_{th}) is given as [38]

$$\tau_{th} = \frac{\delta^2}{4\alpha} \tag{18.11}$$

where α is thermal diffusivity and δ is the optical penetration depth, which is inversely related to the absorption coefficient. When the time of irradiance of a laser pulse (τ) is less than the thermal diffusion time ($\tau < \tau_{th}$), the distribution of laser thermal energy is restricted within the irradiated zone. But when $\tau > \tau_{th}$, heat diffuses outside the irradiated zone and causes damage to nearby areas. The relation between laser irradiance time and temperature rise is shown in Fig. 18.4b as a Gaussian distribution over the spot on the target.

Photothermal conversion occurs when the electrons are excited and follow the electron–electron scattering at sub-ps timescales that lead to the redistribution of hot electrons. Heat is then transmitted to the metal lattice via electron–phonon coupling and dissipated from the lattice to the surrounding medium via phonon–phonon coupling within ns [39]. The energy absorbed by plasmonic NPs can be released by either re-emission of photons, which is known as luminescence, or by the creation of phonons, which generates heat energy. However, prior analyses of plasmonic NPs have shown that the quantum yield of luminescence is less than

1%, suggesting that all of the absorbed energy is converted to heat. This process of converting light into thermal energy is known as photothermal conversion. The strength of light to heat conversion is defined using photothermal efficiency (η). The photothermal efficiency of a plasmonic nanoparticle is quantified using the scattering cross section of the nanoparticle [40]. Scattering should be minimal for efficient light to heat conversion. That simply means that there should be maximum absorption [41]. Therefore, photothermal efficiency depends on the size, shape, dielectric properties of the particle, and composition as well [42]. The efficient photothermal conversion ability of plasmonic MNPs has attracted a lot of applications ranging from solar energy [43], optofluidics, nano-chemistry [44], bioimaging [20], and medical therapy, especially in thermoplasmonic cancer therapy [21, 23] which is discussed in Sect. 18.5 of this chapter.

18.3.2 Photoacoustic Mechanisms

The plasmonically enhanced photoacoustic effect is one of the proposed virus inactivation mechanisms employing USP fs lasers [45]. The photoacoustic effect, also known as *shock wave generation*, is a mechanical phenomenon that does not require direct contact between plasmonic NPs and pathogens. Under the effect of USP laser irradiation, plasma production surrounding plasmonic NPs promotes cavitation and results in the development of shock waves. On absorption of photons, the MNPs convert the energy into heat and result in temperature elevation within the system and its surroundings. In response to heat deposition, the surrounding medium of plasmonic NPs expands thermoelastically generating a pressure wave that a transducer detects as an acoustic or sound signal. The pressure generated is given by

$$P = T_G \sigma_{abs} F \tag{18.12}$$

where F is the laser fluence—optical energy delivered per unit area, and T_G is a Grüneisen parameter, which quantise the vibrational properties [46]. The term laser fluence (F) from Eq. 18.12 is related to the laser pulse energy as

$$F = \frac{\text{Laser pulse energy}}{\text{Effective focal spot area}} \tag{18.13}$$

which again proves the need of pulsed laser in the photoacoustic effect as shown in Fig. 18.3c. On the other hand, F is negligible for cw lasers which makes them inefficient for this effect. This thermoelastic expansion leads to the emission of acoustic waves or shock waves. The shock waves induce vibrations in the virus, which dismantles the protein capsid layer which in turn deactivates the virus. The concept of photoacoustics has been mainly used in bioimaging applications [47–49] given that the acoustic waves scatter much less than the optical waves in tissues which makes it reasonable to employ the photoacoustic technology to generate high-resolution images [16].

A new way for virus inactivation using Raman scattering was hypothesised, namely impulsive stimulated Raman scattering (ISRS) [33, 50]—a mechanism producing low-frequency acoustic vibrations on the protein capsid of the viral particle using a USP laser. The viral capsid acoustic vibrational displacement from the equilibrium position is given by [50]

$$Q_o = \frac{\sqrt{\pi}}{2}\frac{n}{CK\epsilon_0}\alpha_0'\frac{\tau_L}{\omega_0}\cdot I_0\cdot e^{-(\omega_0^2\tau_L^2/4)} \tag{18.14}$$

Here, τ_L is laser temporal pulse width, I_0 is peak laser intensity, ω_0 is the angular frequency of vibrations, $\alpha_0\prime$ is the polarisability derivative, n is the refractive index, c is the speed of light, and $K\epsilon_0$ is the permittivity of the dielectric medium. Therefore, Q_o will be large enough to break the capsid layer, only for very short pulses ($\tau_L \sim fs$), whereas, for cw laser, there is hardly any capsid vibration ($Q_o \sim 0$), as ($\tau_L \sim \infty$) in the term $e^{-(\omega_0^2\tau_L^2/4)}$.

Taking advantage of plasmonic field enhancement that may reach up to four orders of magnitude in certain nanostructures, Raman signals can remarkably be enlarged by 10^8 to even 10^{15}, opening up the possibility in linking ISRS with plasmonic NPs. ISRS may mainly involve two basic mechanisms for the signal enhancement. The most common mechanism is based on the influential EM mechanism caused by the concentrated electric field in the LSPR around plasmonic NPs which can provide enhancement of Raman signal up to 10^{10}. The second mechanism with a much smaller enhancement feature up to 10^2 is attributed to the chemical enhancement caused by an interaction between molecules and NPs [51].

18.3.3 LSPR in Photothermal and Photoacoustic Mechanisms

Au and Ag are the most widely used plasmonic materials for biological applications due to their exceptional biocompatibility. The LSPR band for plasmonic NPs is substantially stronger than other metals and is also present in the visible region of the spectrum (400–700 nm) [52]. Particularly, AuNPs and AgNPs transfer energy into heat with a high efficiency when absorbing light. Through the non-radiative process, nearly all absorbed light is transformed to heat. Because of the enhanced optical cross sections of MNPs as compared to their geometrical cross section, the temperature around and within a single AuNP could be very high. AuNPs, in various forms (spherical, rods, shells, etc.), are the most promising candidates for photothermal effects because they are the strongest nano absorbers. Their absorption coefficient due to profound plasmon resonance is at least three orders of magnitude greater than that of other organic photosensitisers. AuNPs are nontoxic and photostable.

Furthermore, by utilising their form and structure, their maximum absorption can be shifted to the NIR spectral range (700–900 nm), allowing deeper penetration of the laser radiation into biological tissues [36].

For photoacoustic signal amplification, optical to acoustic conversion efficiency should be improved by incorporating various resonating agents. Plasmonic MNPs such as Au and Ag have been proposed as the resonant nanomaterials for enhancing photoacoustic signals [46]. The tuneable LSPR absorption band based on particle size and shape enables them to be extremely efficient energy absorbers at an appropriate wavelength of light, allowing the local thermal effect and energy resonance field to be developed.

18.4 Tuning LSPR by Controlling Size, Shape, and Refractive Index

The optical properties of NPs, such as the optical resonance wavelength, the extinction cross section, and the relative contribution of scattering to extinction, are all highly reliant on the surrounding medium's size, shape, and dielectric properties as evidenced from Eq. 18.4 [35]. Controlling the size and geometry of NPs is one of the most effective methods for harnessing the immense possibilities in the field of plasmonics. However, because of the quantum nature of LSPR in nanostructures, the understanding of size dependence on LSPR is vague. Therefore, it is scientifically vital to realise the physics behind the geometrical dependence of the particle on LSPR to ultimately predict and control the plasmonic responses in nanostructures [53]. Hence, in this part of the book chapter we will discuss the tunability of plasmonic properties by tuning the shape, size, and material of the plasmonic MNP.

18.4.1 Size Dependence of MNPs on LSPR

The optical features of plasmons, such as the LSPR peak, its intensity, and optical efficiency, can be easily controlled by changing the size of plasmonic MNPs (including absorption, scattering, and extinction). As demonstrated in Fig. 18.5, absorption accounts for nearly all of the total extinction for a 20 nm AuNP. As the size increases from 20 to 40 nm, the role of scattering becomes more apparent. At 80 nm, absorption and scattering both almost equally contribute to extinction [16].

Smaller NPs are therefore suitable for photothermal applications, particularly PTT, because light is primarily absorbed by the particles, allowing for effective conversion of light energy into heat for pathogen inactivation or tumour elimination. The particle size not only affects the NP's efficiency but also influences the SPR band. Figure 18.5a shows the LSPR peak of AuNPs with diameter varying from

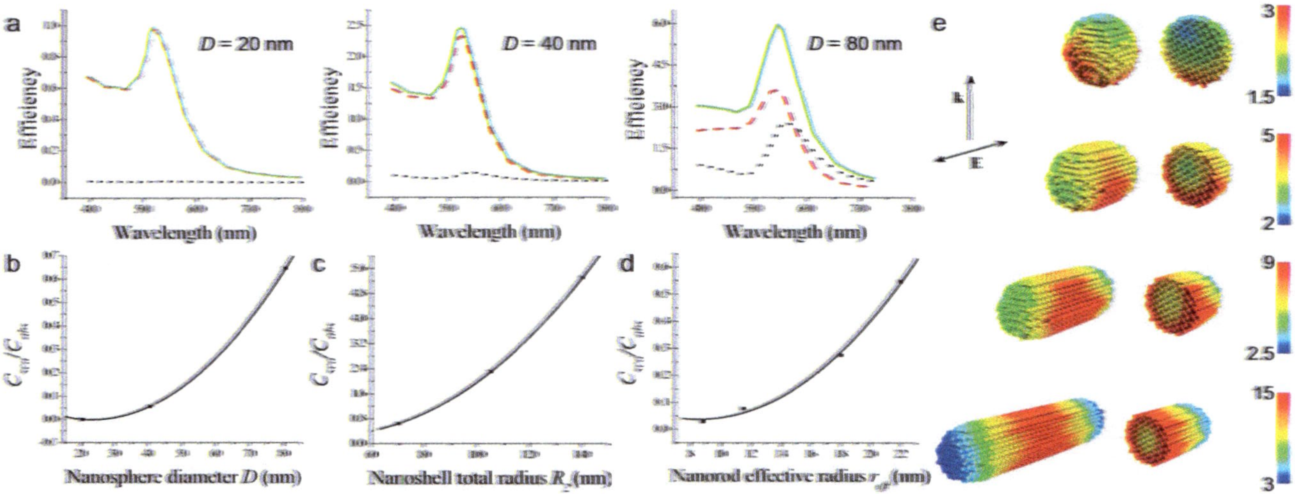

Fig. 18.5 Size- and shape-dependent Mie theory calculations. **a** Optical efficiency spectra for Au nanospheres with diameters (D) = 20, 40, and 80 nm wherein, absorption, scattering, and extinction efficiencies are indicated using red-dashed, black-dotted, and green solid lines, respectively. **b–d** Size-dependent ratio of scattering and absorption cross sections (C_{sca}/C_{abs}) of **b** Au nanosphere, **c** Au nanorod, and **d** silica-Au nanoshell. **e** 3D heat map of power density of Au nanosphere and nanorod at their respective LSPR conditions. Reprinted with permission from [16]

20 to 80 nm. The LSPR peak is roughly 520 nm, in the visible region [52] and the wavelength of the LSPR redshifts as particle size increases.

18.4.2 Effect of Shape of Plasmonic MNPs on LSPR

Researchers have been attracted by the capacity of AuNPs to tune optical properties by synthetic alteration of particle shape, composition, and structure for decades. The LSPR peak splits into two: transverse and longitudinal bands, when the shape of AuNPs changes from spheres to rods, as shown in Fig. 18.6b. A strong band in the NIR region corresponding to electron oscillations along the long axis is known as the longitudinal band, whereas the weak band in the visible region with a wavelength similar to that of Au nanospheres is known as the transverse band [52]. While the transverse band is unaffected by changes in size, the longitudinal band redshifts considerably from the visible to NIR region when the aspect ratio (length/width) of plasmonic MNPs increases.

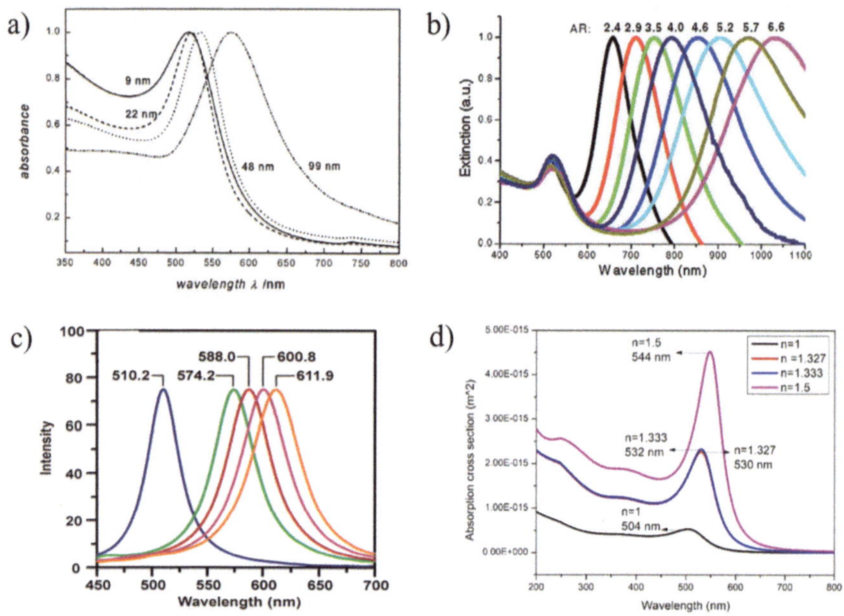

Fig. 18.6 Illustration of **a** LSPR wavelength of Au nanospheres of different sizes. Reprinted with permission from [16]. **b** LSPR wavelength of Au nanorods of different aspect ratios. Reprinted with permission from [52]. **c** Experimental data of single AgNP resonant scattering spectrum in various solvent environments of refractive index (n) = 1.205, 1.33, 1.387, 1.44, 1.49 (left to right). Reprinted with permission from [35]. **d** COMSOL simulation of LSPR wavelength for AuNPs in solvents for n varying from 1 to 1.5

18.4.3 Dependence of Dielectric Constants on LSPR

Slight differences in the refractive index or dielectric contrast between the plasmonic MNP and its surroundings will affect the peak position of the particle's plasmon resonance. The displacement of the plasmon resonance spectral peak as a function of change in the local refractive index may differ for NPs of different shapes. The shift in the plasmon peak was experimentally demonstrated utilising AgNPs in various solvents to vary the dielectric constant of the surrounding medium, hence affecting the refractive index contrast (Fig. 18.6c). It was identified that the plasmon resonance of each individual AgNP is extremely sensitive to tiny changes in the dielectric characteristics of the surrounding media [35, 55]. As the solvent's refractive index rises, the LSPR peak maximum-shifts to longer wavelengths. Similarly, the absorption cross section is observed to fluctuate with the refractive index contrast variation, as shown in Fig. 18.6d, and it is apparent that the LSPR peak and absorbance are reliant on the refractive index contrast between the plasmonic MNPs and the surrounding media.

18.5 Applications of Pulsed Laser-Induced Therapy Using Plasmonics

Biomedical applications of plasmonic nanomaterials irradiated by pulsed laser include biosensing [56], bioimaging [57], drug delivery [58], therapy [59], as well as non-biomedical applications like energy [60], chemical separation [61], nanofluidics [62], and even chemical synthesis [44], all of which require localised heat energy or acoustic vibrations. In this section of the chapter, we will go over a couple of the above-mentioned applications.

(i) Bio-sensing

The early identification of diseases allows more effective treatment and better outcomes. As a result of this, the development and utilisation of quick point-of-care measurement tools capable of biomolecular detection have been increased. Commercially, the most straightforward, affordable, and extensively utilised medical diagnostic procedures for detecting target analytes are lateral flow assays (LFA) [63]. When there is a low level of antigen, LFAs are not really useful for detecting diseases. Therefore, LFAs are coupled with antibody-coated AuNPs to transfer within a nitrocellulose membrane [64]. AuNPs have long been utilised for bio-sensing because their size allows them to easily pass through membrane pores and they can easily be coated with antibodies. Additionally, AuNPs have a high molar absorptivity of light, resulting in a colour that is easily visible as well as optical stimulation by pulsed laser causes MNPs to generate heat [65]. Inclusion of AuNPs increased the sensing efficiency by 10000 folds. Thus, inserting nanostructures with higher light-to-heat conversion efficiency, such as AuNPs, could boost up the sensitivity.

(ii) Photoacoustic imaging

A pulsed laser light beam is used to irradiate the target in photoacoustic imaging. The light absorbed must be turned into an outgoing thermoacoustic wave, which may be detected and used to reconstruct images using an ultrasound transducer. The magnitude of the photoacoustic signal from a heterogeneous medium is very low and plasmonic NPs are used for enhancement [66]. The conventional method to create plasmonic photoacoustic contrast agents is to control the size and shape of NPs to maximise their optical absorption cross section and to determine the peak position. Au nanorods [67] and nanocages [68] with high and tuneable optical absorption cross sections have been effectively used in photoacoustic imaging. Encasing Au nanorods for photoacoustic imaging in a silica shell that is optically non-absorbing has also been developed [69]. To amplify the photoacoustic signal, the heat transfer mechanism from the AuNPs to the signal-generating medium is done by the involvement of the silica shell. The presence of plasmons in photoacoustic imaging results in significant contrast enhancement, proving that these hybrid particles acting as "photoacoustic nanoamplifiers" are high-efficiency contrast agents for photoacoustic imaging [70].

(iii) Thermoplasmonic cancer therapy

The ability to externally control the temperature of plasmonic MNPs has stimulated research in using thermoplasmonics in cancer therapy. Thermoplasmonic cancer therapy involves irradiating plasmonic NPs injected near tumour sites using a laser, and the resulting heating of MNPs kills tumour cells [10, 22, 23, 28, 31, 71]. Tumours in the lungs are extremely difficult to diagnose and treat due to their inaccessibility and limited therapeutic choices. However, the photothermal properties of plasmonic MNPs such as Au nanorods, as well as the ability to fine-tune them simply by modifying structural parameters, have opened up a new route for lung cancer targeted treatment [23]. AuNps absorb the laser light and increase the local temperature, which helps in selective hyperthermia that damages the cancerous area [72]. Continuous wave lasers are generally employed to induce bulk heating but have many potential side effects such as causing damage to surrounding healthy mammalian cells and tissues. However, in another study demonstrated by Knights et al., Au nanorods having a SPR of 850 nm were functionalised with antibodies to evaluate the use of pulsed lasers for PTT [23]. The results obtained demonstrated a reduction of 93% in the cell viability compared to controlled exposures which demonstrated a possible application for minimally invasive therapies for lung cancer using pulsed lasers. Furthermore, multifunctional Au nanorods have also been utilised for plasmonic PTT for destroying pancreatic cancer cells using USP NIR laser radiation [29].

18.6 Conclusion and Outlook

We have extensively reviewed the fundamentals of plasmonics and USP laser-mediated photo-induced mechanisms in this chapter to provide the groundwork for developing an alternative cost-effective and non-toxic technique for diagnostics and therapeutics of diseases caused by pathogens. The upsurge in research on the use of plasmonic MNPs has resulted in a range of plasmonic nanomaterials of various shapes, sizes, and aspect ratios, allowing examination of possible biomedical applications based on photo-induced mechanisms.

Despite several studies establishing the properties of newly synthesised plasmonic NPs and confirming their applicability in biomedical science, their use in practical applications is limited. The limitation stems from the fact that the fundamental problem for employing plasmonic MNPs in biomedical applications is their deposition at the target site. As a result, while there are ongoing experiments that involve injecting MNPs directly into the bloodstream or at the tumour site, additional research is needed for efficient specific NP accumulation and targeting to the treatment site. As micro-optics, miniaturised optomechanical components, light sources, and interconnecting technology with smart devices continue to advance, designing and realising cost-effective, reliable, automated, and portable plasmonic nanomaterials will expand and provide new opportunities over a vast array of applications in all fields including the biomedical industry.

References

1. Medina, E., Pieper, D.H.: Tackling threats and future problems of multidrug-resistant bacteria. Curr. Top. Microbiol. Immunol. **398**, 3–33 (2016)
2. Wong, T., Hsu, L., Liao, W.: Phototherapy in psoriasis: a review of mechanisms of action. J. Cutan. Med. Surg. **17**, 6–12 (2013)
3. Jarrett, P., Scragg, R.: A short history of phototherapy, Vitamin D and skin disease. Photochem. Photobiol. Sci. **16**, 283–290 (2017)
4. Alpert, J.S.: The Jeremiah Metzger lecture: Jeremiah Metzger and the era of heliotherapy. Trans. Am. Clin. Climatol. Assoc. **126**, 219–226 (2015)
5. Zheng, Y.B., Kiraly, B., Weiss, P.S., Huang, T.J.: Molecular plasmonics for biology and nanomedicine. Nanomedicine **7**, 751–770 (2012)
6. Lee, S.E., Lee, L.P.: Biomolecular plasmonics for quantitative biology and nanomedicine. Curr. Opin. Biotechnol. **21**, 489–497 (2010)
7. Qian, C.-G., et al.: Conjugated polymer nanomaterials for theranostics. Acta Pharmacol. Sin. **38**, 764–781 (2017)
8. Cortés, E., Govorov, A. O., Misawa, H., Willets, K. A.: Special topic on emerging directions in plasmonics. J. Chem. Phys. **153**, (2020)
9. Barbillon, G.: Plasmonics and its applications. Materials (Basel) **12**, 10–13 (2019)
10. Yao, C., et al.: Gold nanoparticle mediated phototherapy for cancer. J. Nanomater. **2016**, 1–29 (2016)
11. Yougbaré, S., et al.: Nanomaterials for the photothermal killing of bacteria. Nanomaterials **10**, 1–23 (2020)
12. Maier, S.A.: Localized surface plasmons. In: Plasmonics: Fundamentals and Applications, pp. 65–88 (2007).https://doi.org/10.1007/0-387-37825-1_5

13. Li, M., Cushing, S.K., Wu, N.: Plasmon-enhanced optical sensors: a review. Analyst **140**, 386–406 (2015)
14. Amendola, V., Pilot, R., Frasconi, M., Maragò, O.M., Iatì, M.A.: Surface plasmon resonance in gold nanoparticles: a review. J. Phys. Condens. Matter **29**, 203002 (2017)
15. Laskar, J.M., Mohanta, D., Choudhury, A.: Observation of discrete localized surface plasmon (LSP) states in isolated nanoscaled Ag-particles synthesized via photochemical reduction route. Eur. Phys. J. Appl. Phys. **46** (2009)
16. Kim, M., Lee, J.H., Nam, J.M.: Plasmonic photothermal nanoparticles for biomedical applications. Adv. Sci. **6** (2019)
17. Amendola, V., Pilot, R., Frasconi, M., Maragò, O.M., Iatì, M.A.: Surface plasmon resonance in gold nanoparticles: a review. J. Phys. Condens. Matter **29** (2017)
18. Bohren, C.F., Huffman, D.R., Verlag GmbH, W.-V., KGaA, C.: Absorption and Scattering of Light by Small Particles. http://dnb.ddb.de
19. Liao, H., Nehl, C.L., Hafner, J.H.: Biomedical applications of plasmon resonant metal nanoparticles. Nanomedicine **1**, 201–208 (2006)
20. Sotiriou, G.A.: Biomedical applications of multifunctional plasmonic nanoparticles. WIREs Nanomed. Nanobiotechnol. **5**, 19–30 (2013)
21. Eversole, D., et al.: Femtosecond plasmonic laser nanosurgery (fs-PLN) mediated by molecularly targeted gold nanospheres at ultra-low pulse fluences. Sci. Rep. **10**, 12387 (2020)
22. Turcheniuk, K., et al.: Plasmonic photothermal cancer therapy with gold nanorods/reduced graphene oxide core/shell nanocomposites. RSC Adv. **6**, 1600–1610 (2016)
23. Knights, O., Freear, S., McLaughlan, J.R.: Improving plasmonic photothermal therapy of lung cancer cells with anti-EGFR targeted gold nanorods. Nanomater. (Basel, Switzerland) **10**, 1307 (2020)
24. Kumar, A., Kim, S., Nam, J.-M.: Plasmonically engineered nanoprobes for biomedical applications. J. Am. Chem. Soc. **138**, 14509–14525 (2016)
25. Borghei, Y.-S., Hosseinkhani, S., Ganjali, M.R.: "Plasmonic Nanomaterials": an emerging avenue in biomedical and biomedical engineering opportunities. J. Adv. Res. (2021). https://doi.org/10.1016/j.jare.2021.11.006
26. Wachsmann-Hogiu, S., Smith, Z.J., Kahraman, M.: Editorial: plasmonic technologies for bioanalytical applications. Front. Chem. **7** (2019)
27. McNamara, K., Tofail, S.A.M.: Nanoparticles in biomedical applications. Adv. Phys. X **2**, 54–88 (2017)
28. Khursheed, R., et al.: Biomedical applications of metallic nanoparticles in cancer: current status and future perspectives. Biomed. Pharmacother. **150**, 112951 (2022)
29. Patino, T., et al.: Multifunctional gold nanorods for selective plasmonic photothermal therapy in pancreatic cancer cells using ultra-short pulse near-infrared laser irradiation. Nanoscale **7**, 5328–5337 (2015)
30. Vines, J.B., Yoon, J.-H., Ryu, N.-E., Lim, D.-J., Park, H.: Gold nanoparticles for photothermal cancer therapy. Front. Chem. **7** (2019)
31. He, Y., et al.: Effects and side effects of plasmonic photothermal therapy in brain tissue. Cancer Nanotechnol. **10**, 8 (2019)
32. Huang, X., et al.: Comparative study of photothermolysis of cancer cells with nuclear-targeted or cytoplasm-targeted gold nanospheres: continuous wave or pulsed lasers. J. Biomed. Opt. **15**, 1–7 (2010)
33. Tsen, K.T., et al.: Inactivation of viruses by coherent excitations with a low power visible femtosecond laser. Virol. J. **4**, 50 (2007)
34. Stern, D., et al.: Corneal ablation by nanosecond, picosecond, and femtosecond lasers at 532 and 625 nm. Arch. Ophthalmol. (Chicago, Ill. 1960) **107**, 587–592 (1989)
35. McFarland, A.D., Van Duyne, R.P.: Single silver nanoparticles as real-time optical sensors with zeptomole sensitivity. Nano Lett. **3**, 1057–1062 (2003)
36. Zharov, V.P., Mercer, K.E., Galitovskaya, E.N., Smeltzer, M.S.: Photothermal nanotherapeutics and nanodiagnostics for selective killing of bacteria targeted with gold nanoparticles. Biophys. J. **90**, 619–627 (2006)

37. Yakovlev, E., Shandybina, G., Shamova, A.: Modelling of the heat accumulation process during short and ultrashort pulsed laser irradiation of bone tissue. Biomed. Opt. Express **10**, 3030–3040 (2019)
38. Ganguly, M.: analyzing thermal and mechanical effects of pulsed laser irradiation on tissues. Florida Institute of Technology (2014)
39. Chen, J., Ye, Z., Yang, F., Yin, Y.: Plasmonic nanostructures for photothermal conversion. Small Sci. **1**, 2000055 (2021)
40. Jauffred, L., Samadi, A., Klingberg, H., Bendix, P.M., Oddershede, L.B.: Plasmonic heating of nanostructures. Chem. Rev. **119**, 8087–8130 (2019)
41. Adhikari, S., et al.: Photothermal microscopy: imaging the optical absorption of single nanoparticles and single molecules. ACS Nano **14**, 16414–16445 (2020)
42. Chen, H., et al.: Understanding the photothermal conversion efficiency of gold nanocrystals. Small **6**, 2272–2280 (2010)
43. Yang, B., Li, C., Wang, Z., Dai, Q.: Thermoplasmonics in solar energy conversion: materials, nanostructured designs, and applications. Adv. Mater. **n/a**, 2107351 (2022)
44. Boyd, D.A., Greengard, L., Brongersma, M., El-Naggar, M.Y., Goodwin, D.G.: Plasmon-assisted chemical vapor deposition. Nano Lett. **6**, 2592–2597 (2006)
45. Gellini, C., Feis, A.: Optothermal properties of plasmonic inorganic nanoparticles for photoacoustic applications. Photoacoustics **23**, 100281 (2021)
46. Shen, T.-W., Ou, T.-K., Lin, B.-Y., Chien, Y.-H.: Plasmonic gold nanomaterials as photoacoustic signal resonant enhancers for cysteine detection. Nanomaterials **11** (2021)
47. Ou, X., Liu, Y., Zhang, M., Hua, L., Zhan, S.: Plasmonic gold nanostructures for biosensing and bioimaging. Microchim. Acta **188**, 304 (2021)
48. Das, A., Gomes, A.S.L.: Photoacoustic microscopy and tomography for biomedical imaging with plasmonic TiN nanoparticles. in *Proc. SPIE* vol. PC11990 (2022)
49. Xia, J., Yao, J., Wang, L.V.: Photoacoustic tomography: principles and advances. Prog. Electromagn. Res. **147**, 1–22 (2014)
50. Tsen, S.-W.D., Wu, T.C., Kiang, J.G., Tsen, K.-T.: Prospects for a novel ultrashort pulsed laser technology for pathogen inactivation. J. Biomed. Sci. **19**, 62 (2012)
51. Wang, L., Hasanzadeh Kafshgari, M., Meunier, M.: Optical properties and applications of plasmonic metal nanoparticles. Adv. Funct. Mater. **30**, 2005400 (2020)
52. Huang, X., El-Sayed, M.A.: Gold nanoparticles: optical properties and implementations in cancer diagnosis and photothermal therapy. J. Adv. Res. **1**, 13–28 (2010)
53. Laskar, J.M., Philip, J., Raj, B.: Light scattering in a magnetically polarizable nanoparticle suspension. Phys. Rev. E **78**, 31404 (2008)
54. Kim, M., Lee, J.-H., Nam, J.-M.: Plasmonic photothermal nanoparticles for biomedical applications. Adv. Sci. **6**, 1900471 (2019)
55. Mock, J.J., Smith, D.R., Schultz, S.: Local refractive index dependence of plasmon resonance spectra from individual nanoparticles. Nano Lett. **3**, 485–491 (2003)
56. Zijlstra, P., Paulo, P.M.R., Orrit, M.: Optical detection of single non-absorbing molecules using the surface plasmon resonance of a gold nanorod. Nat. Nanotechnol. **7**, 379–382 (2012)
57. Berciaud, S., Lasne, D., Blab, G.A., Cognet, L., Lounis, B.: Photothermal heterodyne imaging of individual metallic nanoparticles: theory versus experiment. Phys. Rev. B **73**, 45424 (2006)
58. Molina, M., et al.: Stimuli-responsive nanogel composites and their application in nanomedicine. Chem. Soc. Rev. **44**, 6161–6186 (2015)
59. Jaque, D., et al.: Nanoparticles for photothermal therapies. Nanoscale**6**, 9494–9530 (2014)
60. Neumann, O., et al.: Solar vapor generation enabled by nanoparticles. ACS Nano **7**, 42–49 (2013)
61. Boyd, D.A., Adleman, J.R., Goodwin, D.G., Psaltis, D.: Chemical separations by bubble-assisted interphase mass-transfer. Anal. Chem. **80**, 2452–2456 (2008)
62. Liu, G.L., Kim, J., Lu, Y., Lee, L.P.: Optofluidic control using photothermal nanoparticles. Nat. Mater. **5**, 27–32 (2006)
63. Parolo, C., Merkoçi, A.: Paper-based nanobiosensors for diagnostics. Chem. Soc. Rev. **42**, 450–457 (2013)

64. Qin, Z., et al.: Significantly improved analytical sensitivity of lateral flow immunoassays by using thermal contrast. Angew. Chem. Int. Ed. Engl. **51**, 4358–4361 (2012)
65. Anker, J.N., et al.: Biosensing with plasmonic nanosensors. Nat. Mater. **7**, 442–453 (2008)
66. Mantri, Y., Jokerst, J.: Engineering plasmonic nanoparticles for enhanced photoacoustic imaging. ACS Nano **XXXX** (2020)
67. Agarwal, A., et al.: Dual-mode imaging with radiolabeled gold nanorods. J. Biomed. Opt. **16**, 51307 (2011)
68. Skrabalak, S.E., et al.: Gold nanocages: synthesis, properties, and applications. Acc. Chem. Res. **41**, 1587–1595 (2008)
69. Chen, Y.-S., et al.: Silica-coated gold nanorods as photoacoustic signal nanoamplifiers. Nano Lett. **11**, 348–354 (2011)
70. Knights, O.B., Cowell, D.M.J., Carpenter, T.M., Freear, S., McLaughlan, J.R.: Plasmonic gold nanoparticles for combined photoacoustic imaging and plasmonic photothermal therapy using a pulsed laser. In: 2018 IEEE International Ultrasonics Symposium (IUS), pp. 1–4 (2018). https://doi.org/10.1109/ULTSYM.2018.8579902
71. Nomura, S., et al.: Highly reliable, targeted photothermal cancer therapy combined with thermal dosimetry using a near-infrared absorbent. Sci. Rep. **10**, 9765 (2020)
72. Bagley, A.F., Hill, S., Rogers, G.S., Bhatia, S.N.: Plasmonic photothermal heating of intraperitoneal tumors through the use of an implanted near-infrared source. ACS Nano **7**, 8089–8097 (2013)

Chapter 19
Plasmonic Nanostructures for the Detection of Foodborne Pathogens

Htet Htet Kyaw, Myo Tay Zar Myint, and Salim H. Al-Harthi

Abstract The advance of nanomaterials offers a wide range of applications in the food industry. Nanomaterials are used in different applications such as the detection of foodborne pathogens, advanced food packaging, and removal of harmful materials. Nanomaterial has unique physical and chemical properties to enhance sensitivity, selectivity, and rapid detection in various sensing applications. Plasmonic nanostructures have been synthesizing and using optical, electronic, and chemical-based sensors for the detection of target matters. Simple, cost-effective, highly selective, and sensitive portable sensors are the most needed tools to serve our society as a lively environment. This chapter discusses the fundamental properties and the fabrication of plasmonic nanomaterials, their applications, and different types of sensors used in foodborne pathogens.

Keywords Nanomaterial · Pathogen · Sensor · Plasmonic

Abbreviations

EM	Electromagnetic
LSPR	Localized surface plasmon resonance
PSPR	Propagate surface plasmon resonance
LOD	Limit of detection

H. H. Kyaw (✉)
Nanotechnology Research Center, Sultan Qaboos University, 123 Al-Khoud, P.O Box 33, Muscat, Oman
e-mail: htet@squ.edu.om

M. T. Z. Myint · S. H. Al-Harthi
Department of Physics, College of Science, Sultan Qaboos University, 123 Al-Khoud, P.O Box 36, Muscat, Oman
e-mail: myomyint@squ.edu.om

S. H. Al-Harthi
e-mail: salim1@squ.edu.om

D. Mohanta and P. Chakraborty (eds.), *Nanoscale Matter and Principles for Sensing and Labeling Applications*, Advanced Structured Materials 206,
https://doi.org/10.1007/978-981-99-7848-9_19

SERS Surface-enhanced Raman spectroscopy
EEM Electromagnetic enhancement mechanism
CEM Chemical enhancement mechanism
GC Glassy carbon
SPE Screen-printed electrodes

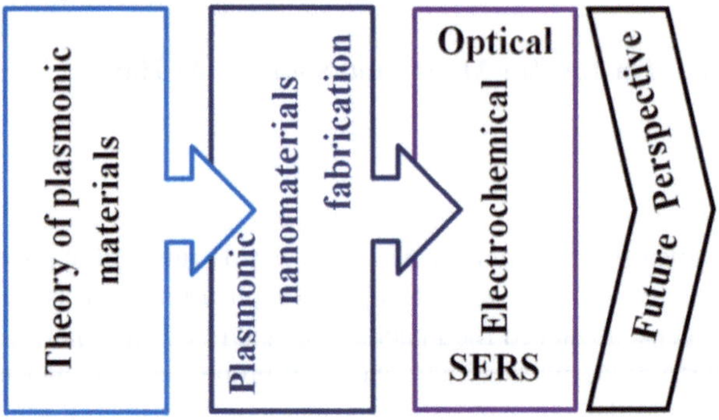

19.1 Introduction

Present-day, food and biosafety are one of the important issues for the society. Life-threatening living organisms and substances exist everywhere in our daily routine. By WHO report, 1 in 10 people falls sick due to the food-related issues annually [1]. These foodborne illnesses are originated from the harmful chemicals (for instance pesticide), pathogenic bacteria, viruses, fungi, and unwanted minerals like heavy metals, etc. Amongst, bacteria-associated sickness is the most concern and caused by most commonly known pathogenic bacteria such as *Escherichia coli* [2], *Staphylococcus aureus* [3], *Campylobacter* and *non-typhoidal Salmonella* (NTS) [4], *Salmonella enterica* and *Listeria monocytogenes* [5], etc. These foodborne pathogenic bacteria are originated from the poultry products, raw dairy products, vegetables, fruits, contaminated water, and polluted environment [6]. According to the World Bank report, the reduction of productivity due to the foodborne sickness and cost to cure the diseases is estimated to be 110 billion dollar annually [7]. Therefore, rapid and accurate early detection of foodborne pathogenic bacteria is an essential tool for modern-day society. As early as 1980, researchers have been using surface plasmon resonance techniques to detect human immunoglobin as immunosensor [8]. Previously, most of the detection system or testing processes were carried out in medical laboratory, which consumed time and human resources, thus making them expensive. Nowadays, inexpensive, high sensitivity and selectivity, user-friendly, rapid, and point-of-care testing have been developed and applied in diverse fields [6, 9–13].

Biosensor can be defined as a type of device, which comprises bio-recognition elements such as enzymes, DNA, cell receptors, antibodies, etc., which can be attached to the specific organism. In general, biosensors are employed in environmental surveillance, quality control for consumer products, health care monitoring system, industrial automation system, pharmaceutical industries and security. Types of biosensor can be categorized into optical, electrochemical, mass and piezoelectric, magnetic, thermal and acoustic and nanoelectromechanical, etc. [14–20].

19.2 The Fundamental Properties of Plasmonic Nanostructures

The meticulous scientific study of the preparation and properties of gold colloids was firstly introduced by Michael Faraday in 1857. He observed and documented the related optical properties between metal colloids and their surrounding media [21, 22]. Many church windows from the period of the middle ages used different types of metallic nanoparticles (of course, those were not known to be as nanomaterials) for decorating glasses without knowing the astonishing optical properties of metal nanoparticles embedded in stained glass [23]. In the early twentieth century, Gustav Mie proposed a mathematical approach for understanding the interaction of electromagnetic wave with small spherical particles suspended in a dielectric medium by applying the classical Maxwell's equation. Mie's work made the optical properties of metallic nanoparticles clear and the applications of surface plasmon attracted a considerable attention in many areas at the end of twentieth century [21]. Metallic particles within the size of 1–100 nm are known as nanoparticles and their properties are different from those of individual atoms or bulk materials [24]. Noble metal nanoparticles such as Au, Ag, and Cu are the most used plasmonic elements in many applications [21]. In terms of electron configurations, Au, Ag, and Cu have completely filled d-subshells (3d, 4d, and 5d) and the metallic properties of these materials are achieved from the lone valence electron in the half-filled s-subshells (4s, 5s, and 6s). They exhibit an intense color due to their strong absorption in the visible region [25]. Other metal nanoparticles such as Pb, Cd, Sn, In, etc. do not show strong color effects and their absorption are in the ultraviolet (UV) region. These metals easily oxidize and are not appropriate to use as plasmonic materials [21].

The characteristic absorption in UV and visible region is related to the collective oscillation of free conduction electrons in the metal nanoparticle surface resonating with the incident electromagnetic (EM) wave [26]. During resonance, the negative charges are accumulated on one side and the positive charges are at the other side creating an electric dipole on the nanoparticle surface. The built-in electric field opposes to that of the applied EM field which forces the electric dipole to the equilibrium position. The electrons oscillate with the resonance frequency, a phenomenon known as localized surface plasmon resonance (LSPR) [25]. Figure 19.1b shows that LSPR occurs in an isotropic shape of metallic nanoparticle with a radius a located in the oscillating electric field. The surface plasmon resonance can also happen in a thin metal film (thickness ~ 50 nm) called 'propagating surface plasmon resonance' (PSPR). Unlike LSPR, the surface plasmon on a thin metal film cannot be excited directly by EM wave due to the mismatch between plasmon dispersion frequency and that of the exciter [27]. Thus, it can be excited by coupling with light through a prism. PSPR occurs where surface wave propagates along a metal film-dielectric interface and decays in the direction normal to the interface as shown in Fig. 19.1a [28].

Other possible electronic excitations also are feasible for plasmonic nanostructures. In metallic materials, the valence and conduction bands are overlapping and build up a continuum of available energy states [23]. If enough energy is received, the valence electrons are triggered from lower-lying *d*-bands to more energetic SP (surface plasmon) conduction bands, called the interband transitions (Fig. 19.1c). In other case, the electrons are excited to higher energies remaining in the same band known as the intraband transitions (Fig. 19.1c). The intraband transitions can be accomplished with lower photon energies; they are not quantized and only happen in metals. The surface plasmon resonance is the process of the intraband transitions [23, 29]. Figure 19.1d and e show the interband and intraband transitions that appear in the optical absorption spectrum of Ag nanoparticles and Au nanoparticles. For Ag nanoparticles, both surface plasmon band and interband transition edges are noticeable. In the case of Au nanoparticles, both the interband and the surface plasmon band overlap [23, 30].

The parameters such as the size, shape, composition, interparticle interactions, and surrounding dielectric medium alter the optical properties of metallic nanoparticles [31]. When the size of metallic nanoparticle is much smaller than the wavelength of excited EM wave, dipolar oscillating is dominant and produces a single sharp peak in SPR absorption spectrum. For bigger particle with size comparable to the wavelength of EM wave, the electron oscillation is not uniform and phase retardation effect emerges. This phase retardation creates higher multipolar resonances such as quadrupole or octupole as an additional peak occurs in the absorption spectrum [30]. Only the absorption process takes place in smaller nanoparticle, however, both absorption and scattering processes exhibit in the bigger nanoparticle. Multipolar resonances also observe in non-spherical metallic nanostructures such as aggregates, nanorods, nanowires, nanocages, nanoprisms, and so on [26]. Various shape of plasmonic nanostructures can tune the absorption at different wavelengths thus desired color can be expressed in the visible region of the EM spectrum [22, 32]. When the metallic nanoparticles are in close proximity with the distance five times smaller than that of the radius ($d < 5a$, d is the center-to-center distance), the electric field from individual particle disturbs the neighboring particle during the excitation of LSPR. This process induces the asymmetric charge distribution around nanoparticle surface and thus more multipolar interactions appear. The multipolar contribution due to the interparticle interactions between closely spaced nanoparticles alters the line shape and the wavelength shift of the LSPR absorption peak (i.e. broaden the LSPR spectrum and blue or red shift depends on the polarization of the external field) [21, 33].

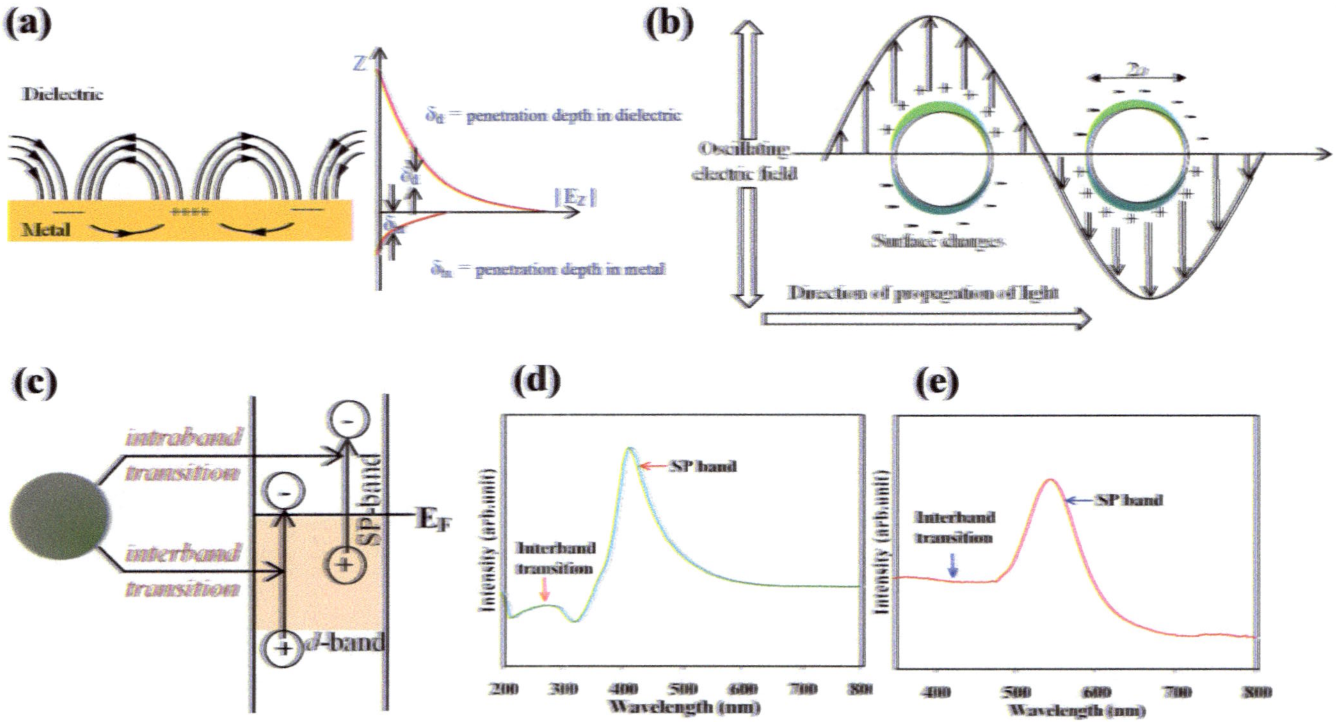

Fig. 19.1 The graphical representation of **a** propagate surface plasmon resonance, **b** localized surface plasmon resonance, **c** the interband and intraband (SP band) transition of metallic nanoparticle and the optical absorption spectrum of **d** Ag nanoparticles and **e** Au nanoparticles deposited on quartz substrate

LSPR absorption intensity and the position of the wavelength also depend on the surrounding dielectric index. By changing the surrounding dielectric medium such as coating of metallic nanoparticles with metal-oxide and polymers or embedding nanoparticles into the polymer metrics can be able to tune the LSPR absorption peak to shorter or longer wavelength (blue and red shift) [30, 34]. Hence, LSPR peak shift via changing the dielectric matrix is widely used in colorimetric, chemical, and biological sensing applications. Another way to change the surrounding environment is to deposit on different substrates such as glass, quartz, fiber optics, silicon, and so on [35–38]. Interestingly, these types of system have multiple dielectric indices: for instance, one side is the metallic nanoparticles interfacing with the substrate which is glass and the other does with multiple dielectric environments such as air, liquid, and solid. This establishes the unusual effects on the plasmonic properties of nanoparticles, which cannot be observed in a colloidal system. Therefore, in terms of reusability and stability, substrate bound metallic nanostructures have great potential for different applications. Besides, combining two or more metals to make alloy nanocomposite structures can further enhance the optical properties of plasmonic materials. For instance, alloying Au (chemical stability) and Ag (highest plasmonic sensitivity) with tuning the composition of each individual metals can have the outstanding plasmonic properties [39, 40]. Many applications of plasmonic metal nanoparticles have emerged in recent years and some up-to-date applications of plasmonic nanostructures utilized in foodborne pathogen detection will be discussed in the next section.

19.3 Fabrication of Plasmonic Nanostructures

Different fabrication techniques have been employed to achieve plasmonic nanomaterials. Generally, top-down and bottom-up approaches are the commonly known method to orchestrate the nanomaterials of mono or multi-elements. Amongst, wet chemical synthesis is the most used technique to synthesize nanomaterials due to being inexpensive, simple and does not require a complex system. Many researchers have been reported the colloidal-based different structures and morphologies of the plasmonic nanomaterials: nanoparticles, nanoclusters, nanorods and bi-metal (gold-silver) heterogeneous nanorods, anisotropic structure (gold-silver-platinum) nanomaterials [41–46]. Likewise, physical deposition method can be used to fabricate nanomaterial with high precision and purity. Evaporation, sputtering, and inert gas condensation techniques are the well-known approaches to make desired nanostructures [47–49]. Apart from the chemical and physical method, lithography is also one of the popular techniques to make nanomaterials especially for large area applications. Recently, the electron beam lithography is getting popular due to its well-defined dimensions, feasible to make large scales and can be used to fabricate complex nanostructures [50–53]. Some of the complex nanostructures are made using a combined method of two different approaches [54, 55].

19.4 Types of (Bio) Sensors

19.4.1 Surface Plasmon Resonance Optical Sensor

Surface plasmon resonance (SPR) optical biosensors have been used for label-free detection of food pathogens, toxins, drug residues, chemical contaminants, pesticides, and so on over the last decade [56–58]. SPR biosensor is an optical detection technique that composes of a transducing medium that can bridge the optical and chemical domain (sensing area) and an electronic system for detecting the SPR signal. In the basic SPR sensing system, a thin gold layer (~50 nm) is coated on the prism to excite the surface plasmon wave. Bio-recognition element is functionalized on the gold film. Upon attachment of the analyte element to the bio-recognition element induces the change of the refractive index at the functionalized gold film (Fig. 19.2a). Then, SPR transducer converts the biological binding event into the detectable signal as the shift in angular spectrum of the reflected light. In SPR system, the detectable signal can be categorized into angular, wavelength, intensity, or phase modulation [28, 56].

The development of SPR sensor for the detection of *Campylobacter jejuni (C. jejuni)* in chicken meat was reported by Masdor et al. [60]. Three different detection methods were developed as (1) polyclonal antibody was directly attached on SPR chip, (2) polyclonal was used as sandwich structure for both capturing as well as secondary antibodies, and (3) sandwich structure combined with antibody conjugated Au nanoparticles for signal amplification. The most sensitive detection method for the *C. jejuni* was achieved using sandwich structure assays with the limit of detection (LOD) of 4×10^4 CFU·mL^{-1}. Another study showed the ultra-low fouling surface using poly (carboxybetaine acrylamide) (pCBAA) brushes functionalized SPR sensor chip for multi-step detection of *E. coli* O157:H7 and *Salmonella* sp. in crude hamburger and cucumber [59]. In their study, a three-step detection assay were fabricated as firstly, antibodies 1 immobilized to the pCBAA surface to capture the

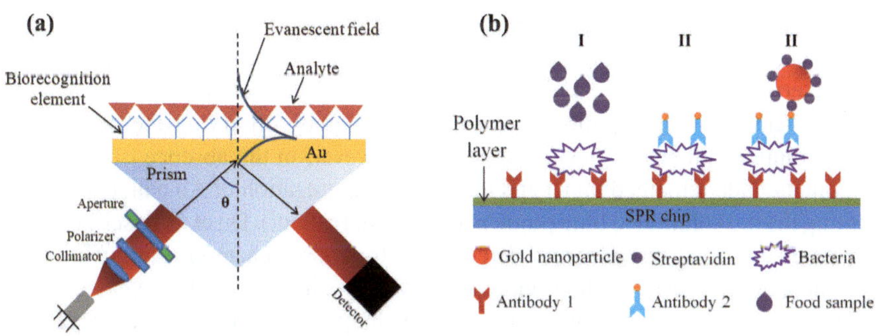

Fig. 19.2 **a** The working principle of prism-based PSPR and **b** The schematic diagram. Adapted from the Reference [59] of a three-step bacteria detection assay in food sample

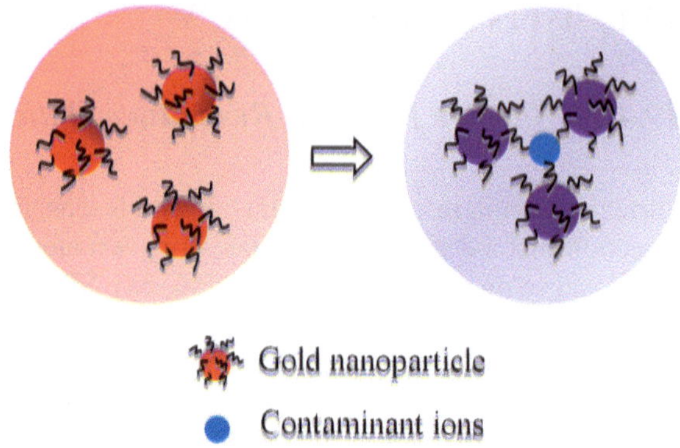

Fig. 19.3 The basic working principle of colorimetric sensor in colloidal form

bacteria. Secondly, the captured bacteria surface was bound with secondary biotinylated antibody 2 followed by the attachment of streptavidin-coated Au nanoparticles onto the biotinylated antibody 2 as shown in Fig. 19.2b. The minimum detection limits using a three-step assay were obtained to be 57 CFU/mL and 17 CFU/mL for *E. coli* and 7.4×10^3 CFU/mL and 11.7×10^3 CFU/mL for *Salmonella* sp. in cucumber and hamburger extracts.

Despite of the utilization of plasmonic nanoparticles as signal amplification in the PSPR system, metallic nanoparticles can also be applied as the LSPR sensor for detection of foodborne pathogens [61, 62]. The metallic nanoparticle has high sensitivity to detect the chemical interaction on their surface (an analyte bound on the functionalized nanoparticles surface) due to LSPR effect. In colloidal-based metallic nanoparticles, any changes or disturbances near their local environment can trigger a visual color change and can be applied as colorimetric sensor in many applications [63]. This change in color is related to the change in refractive index of their surrounding medium, which can be detected as wavelength shift using spectrometer. The schematic representation of the basic colorimetric sensor for the detection of desired pathogens is depicted in Fig. 19.3.

Based on the unique response of LSPR, Sanromán-Igleasias et al. reported a label-free colorimetric method for the detection of bacterial nucleases produced by living bacteria [64]. Authors have disclosed that the nucleuses that were produced by bacteria could recognize and degrade the specific region of oligonucleotide probe and prevented the aggregation of Au nanoparticles by observing the original red color solution (known as positive readout). However, in the absence of nucleuses, oligonucleotide remained intact and induced the aggregation of Au nanoparticles by turning the red solution to blue (known as negative readout). Du et al. described a low pH-based colorimetric sensor for detection of foodborne bacteria using unmodified Au nanoparticles [65]. In their approach, low pH value at 3.2, the negative charge of pathogenic bacteria changed into positive charge and attached to the negatively charge Au nanoparticles by electrostatic interaction. The interaction can be visually detected by color change from red to purple or blue or colorless in accordance with the bacteria concentration (see Fig. 19.3 as a reference).

19.4.2 Surface-Enhanced Raman Spectroscopy Sensor

Surface-enhanced Raman spectroscopy (SERS) technique has been used as the detection of chemical and microbiological hazards since 1990s. Some years later, SERS has been built up as a fingerprint recognition feature for the detection of foodborne microorganisms [66]. SERS technique is simple, non-destructive and delivers direct information of an analyte at the molecular level with high sensitivity and specificity. There are two main established mechanisms for SERS as physical enhancement mechanism also known as electromagnetic enhancement mechanism (EEM) and chemical enhancement mechanism (CEM) [67]. In EEM, noble metal such as gold, silver, or copper is used to fabricate nanostructured SERS substrates to interact with EM wave (Monochromatic light is commonly used). During interaction of light with metal nanostructure, LSPR is formed and induces strong EM fields at metal surface. These strong EM fields are known to be hotspots for SERS to enhance the Raman signal of the analyte molecule, which is adsorbed to the SERS substrates [67, 68]. In chemical enhancement mechanism, the charge transfer process is involved as the analyte molecule chemically bonded to the metal surface. The charge transfer process between the metal surface and the molecule provokes the polarizability of the analyzed molecule, which stimulates the enhancement of the Raman signal. All in all, the degree of enhancement for CEM is much lower than EEM. The basic working principle of SERS sensor is depicted in Fig. 19.4.

High-performance SERS substrates are fabricated by either solution-based or substrate bound nanostructures. The metallic (Au, Ag, Pt, Pd, Fe, etc.), carbon-based (graphene, carbon nanotube, etc.) and semiconductors (CuO, ZnS, CdTe, etc.) nanomaterials are used to fabricate active SERS substrate in recent years [69]. Wet chemical synthesis is the most used technique and colloidal gold, silver and copper nanoparticles with the size between 10 and 200 nm are commonly used for the preparation of solution-based SERS substrate. The tuning of size and shape of the nanoparticles are the important parameters to achieve high intensity and sensitivity of the SERS substrate. Besides, Au nanoparticle has high biocompatibility, chemical stability, and easy to modify the surface. On the other hand, Ag nanoparticle has higher plasmonic activity, stronger SERS activity, and greater EM enhancement in the visible range. A combination of gold and silver as Ag shell Au core (Au@Ag core–shell) nanoparticles [70], Ag-core Au-shell (Ag@Au) nanoparticles [71], Au@Ag

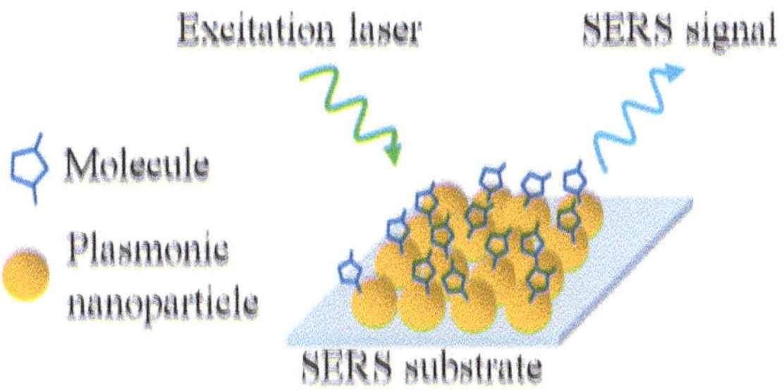

Fig. 19.4 Schematic representation of SERS sensor (Analyte molecule adsorbed on plasmonic nanoparticle which further enhanced the SERS signal)

core–shell nanocubes and nanocuboids [72], Au@Ag core–shell spiky nanoparticles [73] have been synthesized as SERS-active nanostructures to achieve the SERS enhancement.

However, colloidal-based SERS substrates have some drawbacks such as particles instability, hard to reproduce the results and difficult to control the hotspots [68, 74]. In order to overcome such drawbacks, colloidal nanostructures are immobilized onto solid substrates by self-assembly technique [75, 76]. Other deposition techniques such as physical deposition, radiation reduction and laser ablation techniques, nanolithography, microsphere lithography techniques are also used for the fabrication of substrate bound nanostructure in SERS applications [77–81]. Zhu et al. developed a SERS substrate by immobilizing thiol-terminal aptamer on Au nanoparticles deposited PDMS film to serve as a capture substrate. Then, a signal probe made up of 4-mercaptobenzoic acid and amino-terminal aptamers modified hollow gold-silver core–shell nanoflowers were attached on the prior prepared capture substrate. The aptamers on both the capture and signal nanoprobes simultaneously bound to *S. aureus* and the enhancement in Raman signal was achieved with LOD of 13 CFU/mL [75].

19.4.3 Electrochemical (Bio) Sensors

An electrochemical sensor is a class of chemical sensor that provides information about its environment. Electrochemical sensing technique has been used in different domains due to its high sensitivity, selectivity, and low cost. A typical electrochemical sensor consists of three components (Fig. 19.5): a reference (usually made from silver/silver chloride), a counter (mostly platinum), and a working or sensing electrode (made up of conducting material preferably gold and carbon-based materials) [82].

The essential component of an electrochemical sensor is the sensing electrode, which comprises with two main parts as a recognition element that provides a specific

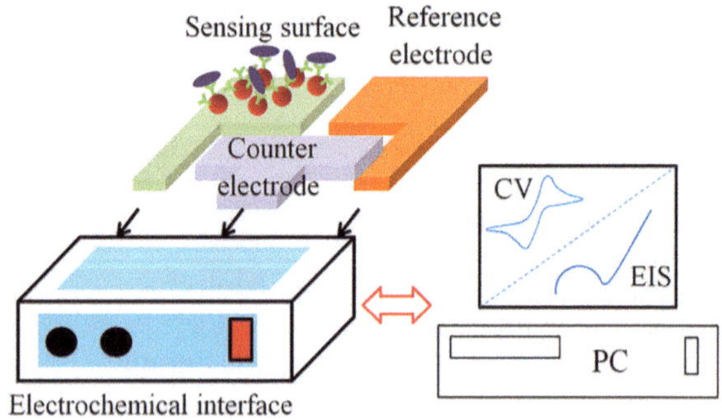

Fig. 19.5 The basic setup for electrochemical sensor

response related to the biological or chemical changes of its environment and a transducer, which converts the biological or chemical response into a detectable signal using modern electronic instruments [83]. The transfer process of the biological or chemical response to an electrical signal can be performed by potentiometry, amperometry, voltammetry, and conductometry methods [84]. Among the various types of biosensors, electrochemical biosensors are widely used in the detection of food analysis [85].

Nowadays, biosensors are developed as label-free detection for simple and rapid detection of water and foodborne pathogens. Different types of nanostructured materials are applied for the fabrication of sensing electrode to improve the sensing ability owing to their unique properties of large surface to volume ratio, high specific surface area, and excellent electrical and optical properties. Moreover, label-free biosensors have the ability to miniaturize and have potential for in situ detection and point-of-care analysis [86]. Plasmonic nanomaterials are the most used in biosensing platforms due to their unique optical properties.

Zarie et al. proposed a type of biosensor made up of Au nanoparticles deposited glassy carbon (GC) electrode followed by thiolated aptamer functionalization [87]. The sensor was applied for the detection of foodborne pathogen *Shigella dysenteriae* (*S. dysenteriae*) by means of electrochemical impedance spectroscopy technique with hexacyanoferrate as an electrochemical probe. The sensor exhibited a linear range between 10^1 and 10^6 CFU/mL with a LOD of 1 CFU/mL. The bio-nanocomposite, chitosan capped Pt nanoparticle on GCE electrode, sensor was used for simultaneous detection of phenylurea herbicides diuron and isoproturon in river water samples using differential pulse adsorptive stripping voltammetry technique [88]. The Pt nanoparticles/chitosan/GCE electrode achieved the LOD of 7 μg/L for isoproturon and 20 μg/L for diuron, respectively.

Apart from the conventional sensing electrode (mostly GC and gold disk), screen-printed electrodes (SPE) are widely used as portable electrodes in electrochemical cells for the on-site analysis purpose [90]. Label-free detection of *Escherichia coli (E. coli)* O157 was achieved by depositing Au nanoparticles on the carbon SPE electrode followed by the functionalization of 3-mercaptopropyl trimethoxysilane and *N*-γ-Maleimidobutyryloxy succinimide and immobilized anti-*E. coli* O157 antibody on the functionalized surface [89]. This sensor was able to detect *E. coli* O157 in the range of 10–10^6 CFU/mL with LOD of 15 CFU/mL. The overall process of fabricating and testing carbon SPE for detection of *E. coli O157* electrochemical biosensor is displayed in Fig. 19.6.

19.5 Conclusions and Future Perspective

Plasmonic nanostructures have been serving our community from the decoration to the health and environmental monitoring applications. Amongst them, tunable or tailor-made plasmonic nanostructure plays an important role in developing simple, inexpensive, highly sensitive, and selective biosensors. Several synthesis techniques

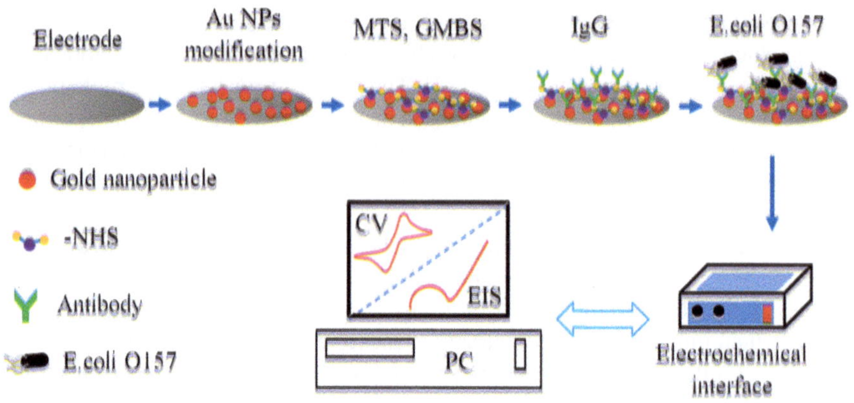

Fig. 19.6 The electrochemical biosensor based on Au nanoparticle-modified carbon SPE for label-free detection of *E. coli O157* bacteria. Adapted from the Reference [89]

have been developed over the decades in order to fulfill the scientific demand for the state-of-the-art sensor technology. Simple and cost-effective LSPR-based sensors, highly sensitive SERS substrates for the detection of molecules and highly selective and sensitive electrochemical sensors are the most promising types of sensors for biological applications. At present, especially in this pandemic era, the availability of the biosensors is an essential tool in our daily routine. Thus, in the near future, development and innovation of highly sensitive and selective as well as inexpensive sensors for commercial product is a must for our society. A personal gadget like a mobile phone with built-in biosensors could be an essential tool for the individuals to lead a safe and sound lifestyle.

References

1. Shang, Y., Xiang, X., Ye, Q., Wu, Q., Zhang, J., Lin, J.-M.: Advances in nanomaterial-based microfluidic platforms for on-site detection of foodborne bacteria. TrAC, Trends Anal. Chem. **147**, 116509 (2022). https://doi.org/10.1016/j.trac.2021.116509
2. Beeman, M.G., Nze, U.C., Sant, H.J., Malik, H., Mohanty, S., Gale, B.K., Carlson, K.: Electrochemical detection of *E. coli* O157:H7 in water after electrocatalytic and ultraviolet treatments using a polyguanine-labeled secondary bead sensor. Sensors **18**(5), 1497 (2018)
3. Guo, Y., Zheng, Y., Liu, Y., Feng, X., Dong, Q., Li, J., Wang, J., Zhao, C.: A concise detection strategy of *Staphylococcus aureus* using N-Succinyl-Chitosan-dopped bacteria-imprinted composite film and AIE fluorescence sensor. J. Hazard. Mater. **423**, 126934 (2022). https://doi.org/10.1016/j.jhazmat.2021.126934
4. Thomas, K.M., de Glanville, W.A., Barker, G.C., Benschop, J., Buza, J.J., Cleaveland, S., Davis, M.A., French, N.P., Mmbaga, B.T., Prinsen, G., Swai, E.S., Zadoks, R.N., Crump, J.A.: Prevalence of *Campylobacter* and *Salmonella* in African food animals and meat: a systematic review and meta-analysis. Int. J. Food Microbiol. **315**, 108382 (2020). https://doi.org/10.1016/j.ijfoodmicro.2019.108382

5. Boukharouba, A., González, A., García-Ferrús, M., Ferrús, M.A., Botella, S.: Simultaneous detection of four main foodborne pathogens in ready-to-eat food by using a simple and rapid multiplex PCR (mPCR) assay. Int. J. Environ. Res. Public Health **19**(3), 1031 (2022). https://doi.org/10.3390/ijerph19031031
6. Weng, X., Zhang, C., Jiang, H.: Advances in microfluidic nanobiosensors for the detection of foodborne pathogens. LWT **151**, 112172 (2021). https://doi.org/10.1016/j.lwt.2021.112172
7. Food-borne Illnesses Cost US$ 110 Billion Per Year in Low- and Middle-Income Countries. World Bank Press Release, 108382 (2018)
8. Liedberg, B., Nylander, C., Lundström, I.: Biosensing with surface plasmon resonance — how it all started. Biosens. Bioelectron. **10**(8), i–ix (1995). https://doi.org/10.1016/0956-5663(95)96965-2
9. Lee, W.-I., Park, Y., Shrivastava, S., Jung, T., Meeseepong, M., Lee, J., Jeon, B., Yang, S., Lee, N.-E.: A fully integrated bacterial pathogen detection system based on count-on-a-cartridge platform for rapid, ultrasensitive, highly accurate and culture-free assay. Biosens. Bioelectron. **152**, 112007 (2020). https://doi.org/10.1016/j.bios.2020.112007
10. Hu, J., Tang, F., Jiang, Y.-Z., Liu, C.: Rapid screening and quantitative detection of *Salmonella* using a quantum dot nanobead-based biosensor. Analyst **145**(6), 2184–2190 (2020). https://doi.org/10.1039/d0an00035c
11. Slavík, R., Homola, J., Brynda, E.: A miniature fiber optic surface plasmon resonance sensor for fast detection of staphylococcal enterotoxin B. Biosens. Bioelectron. **17**(6), 591–595 (2002). https://doi.org/10.1016/S0956-5663(02)00013-1
12. Wu, W., Zhao, S., Mao, Y., Fang, Z., Lu, X., Zeng, L.: A sensitive lateral flow biosensor for *Escherichia coli* O157:H7 detection based on aptamer mediated strand displacement amplification. Anal. Chim. Acta **861**, 62–68 (2015). https://doi.org/10.1016/j.aca.2014.12.041
13. Yamada, K., Choi, W., Lee, I., Cho, B.-K., Jun, S.: Rapid detection of multiple foodborne pathogens using a nanoparticle-functionalized multi-junction biosensor. Biosens. Bioelectron. **77**, 137–143 (2016). https://doi.org/10.1016/j.bios.2015.09.030
14. Wang, Q., Ren, Z.-H., Zhao, W.-M., Wang, L., Yan, X., Zhu, A.-s, Qiu, F.-m, Zhang, K.-K.: Research advances on surface plasmon resonance biosensors. Nanoscale **14**(3), 564–591 (2022). https://doi.org/10.1039/d1nr05400g
15. Qaanei, M., Taheri, R.A., Eskandari, K.: Electrochemical aptasensor for *Escherichia coli* O157:H7 bacteria detection using a nanocomposite of reduced graphene oxide, gold nanoparticles and polyvinyl alcohol. Anal. Methods **13**(27), 3101–3109 (2021). https://doi.org/10.1039/d1ay00563d
16. Lim, J.Y., Lee, S.S.: Sensitive detection of microRNA using QCM biosensors: sandwich hybridization and signal amplification by TiO_2 nanoparticles. Anal. Methods **12**(42), 5103–5109 (2020). https://doi.org/10.1039/d0ay01481h
17. Ravindranath, S.P., Mauer, L.J., Deb-Roy, C., Irudayaraj, J.: Biofunctionalized magnetic nanoparticle integrated mid-infrared pathogen sensor for food matrixes. Anal. Chem. **81**(8), 2840–2846 (2009). https://doi.org/10.1021/ac802158y
18. Ji, J., Pang, Y., Li, D., Huang, Z., Zhang, Z., Xue, N., Xu, Y., Mu, X.: An aptamer-based shear horizontal surface acoustic wave biosensor with a CVD-grown single-layered graphene film for high-sensitivity detection of a label-free endotoxin. Microsyst. Nanoeng. **6**(1), 4 (2020). https://doi.org/10.1038/s41378-019-0118-6
19. Cheng, S., Tu, Z., Zheng, S., Cheng, X., Han, H., Wang, C., Xiao, R., Gu, B.: An efficient SERS platform for the ultrasensitive detection of *Staphylococcus aureus* and *Listeria monocytogenes* via wheat germ agglutinin-modified magnetic SERS substrate and streptavidin/aptamer co-functionalized SERS tags. Anal. Chim. Acta **1187**, 339155 (2021). https://doi.org/10.1016/j.aca.2021.339155
20. Fedyanin, D.Y., Stebunov, Y.V.: All-nanophotonic NEMS biosensor on a chip. Sci. Rep. **5**(1), 10968 (2015). https://doi.org/10.1038/srep10968
21. Ghosh, S.K., Pal, T.: Interparticle coupling effect on the surface plasmon resonance of gold nanoparticles: from theory to applications. Chem. Rev. **107**(11), 4797–4862 (2007). https://doi.org/10.1021/cr0680282

22. Hu, M., Chen, J., Li, Z.Y., Au, L., Hartland, G.V., Li, X., Marquez, M., Xia, Y.: Gold nanostructures: engineering their plasmonic properties for biomedical applications. Chem. Soc. Rev. **35**(11), 1084–1094 (2006). https://doi.org/10.1039/b517615h
23. Garcia, M.A.: Surface plasmons in metallic nanoparticles: fundamentals and applications. J. Phys. D Appl. Phys. **44**(28), 283001 (2011). https://doi.org/10.1088/0022-3727/44/28/283001
24. Eustis, S., El-Sayed, M.A.: Why gold nanoparticles are more precious than pretty gold: noble metal surface plasmon resonance and its enhancement of the radiative and nonradiative properties of nanocrystals of different shapes. Chem. Soc. Rev. **35**(3), 209–217 (2006). https://doi.org/10.1039/b514191e
25. De Sio, L., Placido, T., Comparelli, R., Lucia Curri, M., Striccoli, M., Tabiryan, N., Bunning, T.J.: Next-generation thermo-plasmonic technologies and plasmonic nanoparticles in optoelectronics. Prog. Quantum Electron. **41**, 23–70 (2015). https://doi.org/10.1016/j.pquantelec.2015.03.001
26. Zhang, J.Z., Noguez, C.: Plasmonic optical properties and applications of metal nanostructures. Plasmonics **3**(4), 127–150 (2008). https://doi.org/10.1007/s11468-008-9066-y
27. Liang, A., Liu, Q., Wen, G., Jiang, Z.: The surface-plasmon-resonance effect of nanogold/silver and its analytical applications. TrAc, Trends Anal. Chem. **37**, 32–47 (2012). https://doi.org/10.1016/j.trac.2012.03.015
28. Homola, J., Yee, S.S., Gauglitz, G.: Surface plasmon resonance sensors: review. Sens. Actuators B Chem. **54**(1), 3–15 (1999). https://doi.org/10.1016/S0925-4005(98)00321-9
29. Maier, S.A.: Plasmonics: fundamentals and applications. Springer (2007). https://doi.org/10.1007/0-387-37825-1
30. Evanoff, D.D., Chumanov, G.: Synthesis and optical properties of silver nanoparticles and arrays. ChemPhysChem **6**(7), 1221–1231 (2005). https://doi.org/10.1002/cphc.200500113
31. Su, K.H., Wei, Q.H., Zhang, X., Mock, J.J., Smith, D.R., Schultz, S.: Interparticle coupling effects on plasmon resonances of nanogold particles. Nano Lett. **3**(8), 1087–1090 (2003). https://doi.org/10.1021/nl034197f
32. Foss, C.A., Hornyak, G.L., Stockert, J.A., Martin, C.R.: Template-synthesized nanoscopic gold particles: optical spectra and the effects of particle size and shape. J. Phys. Chem. **98**(11), 2963–2971 (1994). https://doi.org/10.1021/j100062a037
33. Link, S., El-Sayed, M.A.: Shape and size dependence of radiative, non-radiative and photothermal properties of gold nanocrystals. Int. Rev. Phys. Chem. **19**(3), 409–453 (2000). https://doi.org/10.1080/01442350050034180
34. Prakash, J., Pivin, J.C., Swart, H.C.: Noble metal nanoparticles embedding into polymeric materials: from fundamentals to applications. Adv. Colloid Interface Sci. **226**, 187–202 (2015). https://doi.org/10.1016/j.cis.2015.10.010
35. Kyaw, H.H., Myint, M.T.Z., Al-Harthi, S.H., Maekawa, T., Yanagisawa, K., Sellai, A., Dutta, J.: The influence of initial gold nanoparticles layer on migration of silver nanoparticles in silver/glass matrix. Thin Solid Films **685**, 216–224 (2019). https://doi.org/10.1016/j.tsf.2019.06.025
36. Unal, S.: Ion implanted Au nanoparticles in surface plasmon temperature sensing. Mater. Lett. **305**, 130793 (2021). https://doi.org/10.1016/j.matlet.2021.130793
37. Zhou, C., Zou, H., Li, M., Sun, C., Ren, D., Li, Y.: Fiber optic surface plasmon resonance sensor for detection of *E. coli* O157:H7 based on antimicrobial peptides and AgNPs-rGO. Biosens. Bioelectron. **117**, 347–353 (2018). https://doi.org/10.1016/j.bios.2018.06.005
38. Le-The, H., Berenschot, E., Tiggelaar, R.M., Tas, N.R., van den Berg, A., Eijkel, J.C.T.: Large-scale fabrication of highly ordered sub-20 nm noble metal nanoparticles on silica substrates without metallic adhesion layers. Microsyst. Nanoeng. **4**(1), 4 (2018). https://doi.org/10.1038/s41378-017-0001-2
39. Connor, E.E., Mwamuka, J., Gole, A., Murphy, C.J., Wyatt, M.D.: Gold nanoparticles are taken up by human cells but do not cause acute cytotoxicity. Small **1**(3), 325–327 (2005). https://doi.org/10.1002/smll.200400093
40. Gao, C., Hu, Y., Wang, M., Chi, M., Yin, Y.: Fully alloyed Ag/Au nanospheres: combining the plasmonic property of Ag with the stability of Au. J. Am. Chem. Soc. **136**(20), 7474–7479 (2014). https://doi.org/10.1021/ja502890c

41. Pazos-Perez, N., Wagner, C.S., Romo-Herrera, J.M., Liz-Marzán, L.M., García de Abajo, F.J., Wittemann, A., Fery, A., Alvarez-Puebla, R.A.: Organized plasmonic clusters with high coordination number and extraordinary enhancement in surface-enhanced Raman scattering (SERS). Angew. Chem. Int. Ed. **51**(51), 12688–12693 (2012). https://doi.org/10.1002/anie.201207019
42. Zhang, Q., Deng, T.-S., Wei, M.-Z., Chen, X., Cheng, Z., Li, S., Gu, Y.-J.: Symmetric and asymmetric overgrowth of a Ag shell onto gold nanorods assisted by Pt pre-deposition. RSC Adv. **11**(55), 34516–34524 (2021). https://doi.org/10.1039/d1ra07415f
43. Su, G., Yang, C., Zhu, J.-J.: Fabrication of gold nanorods with tunable longitudinal surface plasmon resonance peaks by reductive dopamine. Langmuir **31**(2), 817–823 (2015). https://doi.org/10.1021/la504041f
44. Han, L., Li, C., Zhang, T., Lang, Q., Liu, A.: Au@Ag heterogeneous nanorods as nanozyme interfaces with peroxidase-like activity and their application for one-pot analysis of glucose at nearly neutral pH. ACS Appl. Mater. Interfaces **7**(26), 14463–14470 (2015). https://doi.org/10.1021/acsami.5b03591
45. Zhuo, X., Zhu, X., Li, Q., Yang, Z., Wang, J.: Gold nanobipyramid-directed growth of length-variable silver nanorods with multipolar plasmon resonances. ACS Nano **9**(7), 7523–7535 (2015). https://doi.org/10.1021/acsnano.5b02622
46. Zhu, X.-M., Wan, H.-Y., Jia, H., Liu, L., Wang, J.: Porous Pt nanoparticles with high near-infrared photothermal conversion efficiencies for photothermal therapy. Adv. Healthcare Mater. **5**(24), 3165–3172 (2016). https://doi.org/10.1002/adhm.201601058
47. Chen, G., Zhang, K., Luo, B., Hong, W., Chen, J., Chen, X.: Plasmonic-3D photonic crystals microchip for surface enhanced Raman spectroscopy. Biosens. Bioelectron. **143**, 111596 (2019). https://doi.org/10.1016/j.bios.2019.111596
48. Kim, J.-H., Benelmekki, M.: Interfacial transformation of an amorphous carbon nanofilm upon Fe@Ag@Si nanoparticle landing and its colloidal nanoscrolls: enhanced nanocompositing-based performance for bioapplications. ACS Appl. Mater. Interfaces **8**(48), 33121–33130 (2016). https://doi.org/10.1021/acsami.6b12993
49. Todorov, R., Hristova-Vasileva, T., Katrova, V., Atanasova, A., Milushev, G.: Electronic structure and plasmonic activity in co-evaporated Ag-In bimetallic alloys. J. Alloys Compd. **897**, 163253 (2022). https://doi.org/10.1016/j.jallcom.2021.163253
50. Zakharko, Y., Graf, A., Zaumseil, J.: Plasmonic crystals for strong light-matter coupling in carbon nanotubes. Nano Lett. **16**(10), 6504–6510 (2016). https://doi.org/10.1021/acs.nanolett.6b03086
51. Zaffino, R.L., Mir, M., Samitier, J.: Label-free detection of DNA hybridization and single point mutations in a nano-gap biosensor. Nanotechnology **25**(10), 105501 (2014). https://doi.org/10.1088/0957-4484/25/10/105501
52. Schaal, P.A., Besmehn, A., Maynicke, E., Noyong, M., Beschoten, B., Simon, U.: Electrically conducting nanopatterns formed by chemical e-beam lithography via gold nanoparticle seeds. Langmuir **28**(5), 2448–2454 (2012). https://doi.org/10.1021/la204393h
53. Huang, W., Qian, W., Jain, P.K., El-Sayed, M.A.: The effect of plasmon field on the coherent lattice phonon oscillation in electron-beam fabricated gold nanoparticle Pairs. Nano Lett. **7**(10), 3227–3234 (2007). https://doi.org/10.1021/nl071813p
54. Chowdhury, D., Giordano, M.C., Manzato, G., Chittofrati, R., Mennucci, C., Buatier de Mongeot, F.: Large-area microfluidic sensors based on flat-optics Au nanostripe metasurfaces. J. Phys. Chem. C **124**(31), 17183–17190 (2020). https://doi.org/10.1021/acs.jpcc.0c03023
55. Hwang, C.S.H., Ahn, M.-s, Jeong, K.-H.: Extraordinary sensitivity enhancement of Ag-Au alloy nanohole arrays for label-free detection of *Escherichia Coli*. Biomed. Opt. Express **12**(5), 2734–2743 (2021). https://doi.org/10.1364/boe.420828
56. Homola, J., Hegnerová, K., Vala, M.: Surface Plasmon Resonance Biosensors for Detection of Foodborne Pathogens and Toxins, vol. 7167. SPIE, SPIE BiOS (2009)
57. Altintas, Z.: Surface plasmon resonance based sensor for the detection of glycopeptide antibiotics in milk using rationally designed nanoMIPs. Sci. Rep. **8**(1), 11222 (2018). https://doi.org/10.1038/s41598-018-29585-2

58. Hirakawa, Y., Yamasaki, T., Harada, A., Iwasa, S., Narita, H., Miyake, S.: Development of an immunosensor based on surface plasmon resonance for simultaneous residue analysis of three pesticides—boscalid, clothianidin, and nitenpyram— in vegetables. Anal. Sci. **34**(5), 533–539 (2018). https://doi.org/10.2116/analsci.17P487
59. Vaisocherová-Lísalová, H., Víšová, I., Ermini, M.L., Špringer, T., Song, X.C., Mrázek, J., Lamačová, J., Scott Lynn, N., Šedivák, P., Homola, J.: Low-fouling surface plasmon resonance biosensor for multi-step detection of foodborne bacterial pathogens in complex food samples. Biosens. Bioelectron. **80**, 84–90 (2016). https://doi.org/10.1016/j.bios.2016.01.040
60. Masdor, N.A., Altintas, Z., Tothill, I.E.: Surface plasmon resonance immunosensor for the detection of *Campylobacter jejuni*. Chemosensors **5**(2), 16 (2017). https://doi.org/10.3390/chemosensors5020016
61. Oh, S.Y., Heo, N.S., Shukla, S., Cho, H.-J., Vilian, A.T.E., Kim, J., Lee, S.Y., Han, Y.-K., Yoo, S.M., Huh, Y.S.: Development of gold nanoparticle-aptamer-based LSPR sensing chips for the rapid detection of *Salmonella typhimurium* in pork meat. Sci. Rep. **7**(1), 10130 (2017). https://doi.org/10.1038/s41598-017-10188-2
62. Yoo, S.M., Kim, D.-K., Lee, S.Y.: Aptamer-functionalized localized surface plasmon resonance sensor for the multiplexed detection of different bacterial species. Talanta **132**, 112–117 (2015). https://doi.org/10.1016/j.talanta.2014.09.003
63. He, W., Han, X., Jia, H., Cai, J., Zhou, Y., Zheng, Z.: AuPt alloy nanostructures with tunable composition and enzyme-like activities for colorimetric detection of bisulfide. Sci. Rep. **7**(1), 40103 (2017). https://doi.org/10.1038/srep40103
64. Sanromán-Iglesias, M., Garrido, V., Gil-Ramírez, Y., Aizpurua, J., Grzelczak, M., Grilló, M.-J.: Plasmon-assisted fast colorimetric detection of bacterial nucleases in food samples. Sens. Actuators B Chem. **349**, 130780 (2021). https://doi.org/10.1016/j.snb.2021.130780
65. Du, J., Yu, Z., Hu, Z., Chen, J., Zhao, J., Bai, Y.: A low pH-based rapid and direct colorimetric sensing of bacteria using unmodified gold nanoparticles. J. Microbiol. Methods **180**, 106110 (2021). https://doi.org/10.1016/j.mimet.2020.106110
66. Zhao, X., Li, M., Xu, Z.: Detection of foodborne pathogens by surface enhanced Raman spectroscopy. Front. Microbiol. **9** (2018). https://doi.org/10.3389/fmicb.2018.01236
67. Witkowska, E., Niciński, K., Korsak, D., Dominiak, B., Waluk, J., Kamińska, A.: Nanoplasmonic sensor for foodborne pathogens detection. Towards development of ISO-SERS methodology for taxonomic affiliation of *Campylobacter* spp. J. Biophotonics **13**(5), e201960227 (2020). https://doi.org/10.1002/jbio.201960227
68. Wang, C., Meloni, M.M., Wu, X., Zhuo, M., He, T., Wang, J., Wang, C., Dong, P.: Magnetic plasmonic particles for SERS-based bacteria sensing: a review. AIP Adv. **9**(1), 010701 (2019). https://doi.org/10.1063/1.5050858
69. Yang, B., Jin, S., Guo, S., Park, Y., Chen, L., Zhao, B., Jung, Y.M.: Recent development of SERS technology: semiconductor-based study. ACS Omega **4**(23), 20101–20108 (2019). https://doi.org/10.1021/acsomega.9b03154
70. Samal, A.K., Polavarapu, L., Rodal-Cedeira, S., Liz-Marzán, L.M., Pérez-Juste, J., Pastoriza-Santos, I.: Size tunable Au@Ag core-shell nanoparticles: synthesis and surface-enhanced Raman scattering properties. Langmuir **29**(48), 15076–15082 (2013). https://doi.org/10.1021/la403707j
71. Bu, Y., Lee, S.-W.: The characteristic Ag(core)Au(shell) nanoparticles as SERS substrates in detecting dopamine molecules at various pH ranges. Int. J. Nanomed. **10** Spec Iss (Spec Iss), 47–54 (2015). https://doi.org/10.2147/ijn.s88308
72. Guo, P., Sikdar, D., Huang, X., Si, K.J., Xiong, W., Gong, S., Yap, L.W., Premaratne, M., Cheng, W.: Plasmonic core–shell nanoparticles for SERS detection of the pesticide thiram: size- and shape-dependent Raman enhancement. Nanoscale **7**(7), 2862–2868 (2015). https://doi.org/10.1039/c4nr06429a
73. Huang, Z., Meng, G., Hu, X., Pan, Q., Huo, D., Zhou, H., Ke, Y., Wu, N.: Plasmon-tunable Au@Ag core-shell spiky nanoparticles for surface-enhanced Raman scattering. Nano Res. **12**(2), 449–455 (2019). https://doi.org/10.1007/s12274-018-2238-y

74. Zhou, H., Li, X., Wang, L., Liang, Y., Jialading, A., Wang, Z., Zhang, J.: Application of SERS quantitative analysis method in food safety detection. Rev. Anal. Chem. **40**(1), 173–186 (2021). https://doi.org/10.1515/revac-2021-0132
75. Zhu, A., Ali, S., Xu, Y., Ouyang, Q., Chen, Q.: A SERS aptasensor based on AuNPs functionalized PDMS film for selective and sensitive detection of *Staphylococcus aureus*. Biosens. Bioelectron. **172**, 112806 (2021). https://doi.org/10.1016/j.bios.2020.112806
76. He, S., Chua, J., Tan, E.K.M., Kah, J.C.Y.: Optimizing the SERS enhancement of a facile gold nanostar immobilized paper-based SERS substrate. RSC Adv. **7**(27), 16264–16272 (2017). https://doi.org/10.1039/c6ra28450g
77. Boginskaya, I., Sedova, M., Baburin, A., Afanas'ev, K., Zverev, A., Echeistov, V., Ryzhkov, V., Rodionov, I., Tonanaiskii, B., Ryzhikov, I., Lagarkov, A.: SERS-active substrates nanoengineering based on e-beam evaporated self-assembled silver films. Appl. Sci. **9**(19), 3988 (2019). https://doi.org/10.3390/app9193988
78. Yue, X., Li, H., Lv, X., Tang, J.: Porous silicon photonic crystal/silver composite produced by microwave-assisted reduction: applications to surface-enhanced Raman scattering. Opt. Mater. X **2**, 100027 (2019). https://doi.org/10.1016/j.omx.2019.100027
79. Chu, F., Yan, S., Zheng, J., Zhang, L., Zhang, H., Yu, K., Sun, X., Liu, A., Huang, Y.: A Simple laser ablation-assisted method for fabrication of superhydrophobic SERS substrate on teflon film. Nanoscale Res. Lett. **13**(1), 244 (2018). https://doi.org/10.1186/s11671-018-2658-3
80. Liu, L., Zhang, Q., Lu, Y., Du, W., Li, B., Cui, Y., Yuan, C., Zhan, P., Ge, H., Wang, Z., Chen, Y.: A high-performance and low cost SERS substrate of plasmonic nanopillars on plastic film fabricated by nanoimprint lithography with AAO template. AIP Adv. **7**(6), 065205 (2017). https://doi.org/10.1063/1.4985270
81. Zhu, Z., Li, Q., Bai, B., Fan, S.: Reusable three-dimensional nanostructured substrates for surface-enhanced Raman scattering. Nanoscale Res. Lett. **9**(1), 25 (2014). https://doi.org/10.1186/1556-276x-9-25
82. Mishra, G.K., Sharma, V., Mishra, R.K.: Electrochemical aptasensors for food and environmental safeguarding: a review. Biosensors **8**(2), 28 (2018)
83. Faridbod, F., Gupta, V.K., Zamani, H.A.: Electrochemical sensors and biosensors. Int. J. Electrochem. **2011**, 352546 (2011). https://doi.org/10.4061/2011/352546
84. Shen, Y., Xu, L., Li, Y.: Biosensors for rapid detection of *Salmonella* in food: a review. Compr. Rev. Food Sci. Food Saf. **20**(1), 149–197 (2021). https://doi.org/10.1111/1541-4337.12662
85. Cesewski, E., Johnson, B.N.: Electrochemical biosensors for pathogen detection. Biosens. Bioelectron. **159**, 112214 (2020). https://doi.org/10.1016/j.bios.2020.112214
86. Reta, N., Saint, C.P., Michelmore, A., Prieto-Simon, B., Voelcker, N.H.: Nanostructured electrochemical biosensors for label-free detection of water- and food-borne pathogens. ACS Appl. Mater. Interfaces **10**(7), 6055–6072 (2018). https://doi.org/10.1021/acsami.7b13943
87. Zarei, S.S., Soleimanian-Zad, S., Ensafi, A.A.: An impedimetric aptasensor for *Shigella dysenteriae* using a gold nanoparticle-modified glassy carbon electrode. Microchim. Acta **185**(12), 538 (2018). https://doi.org/10.1007/s00604-018-3075-0
88. Morawski, Fd.M., Winiarski, J.P., de Campos, C.E.M., Parize, A.L., Jost, C.L.: Sensitive simultaneous voltammetric determination of the herbicides diuron and isoproturon at a platinum/chitosan bio-based sensing platform. Ecotoxicol. Environ. Saf. **206**, 111181 (2020). https://doi.org/10.1016/j.ecoenv.2020.111181
89. Vu, Q.K., Tran, Q.H., Vu, N.P., Anh, T.-L., Dang, T.T.L., Matteo, T., Nguyen, T.H.H.: A label-free electrochemical biosensor based on screen-printed electrodes modified with gold nanoparticles for quick detection of bacterial pathogens. Mater. Today Commun. **26**, 101726 (2021). https://doi.org/10.1016/j.mtcomm.2020.101726
90. Torre, R., Costa-Rama, E., Nouws, H.P.A., Delerue-Matos, C.: Screen-printed electrode-based sensors for food spoilage control: bacteria and biogenic amines detection. Biosensors **10**(10), 139 (2020). https://doi.org/10.3390/bios10100139

Chapter 20
Morphology-Dependent Biosensing of Metallic Nanoparticles

Barnika Chakraborty, Rachana Yadwade, and Balaprasad Ankamwar

Abstract Nanoparticles are particles with dimensions that lie within the nanoscale range, typically measured in the order of 10^{-9} meters. It finds its applications in diversified fields such as drug delivery, biosensing, catalysis, environment and energy. In this chapter, we will focus on the biosensing application and how effective nanoparticles are in this field. Amongst the most important factors that govern such activities are the morphologies of the nanoparticles that in turn decide their applications. We are going to focus some light on such morphology-dependent biosensing activities of metals such as gold, silver, copper, i.e. the most abundant and effective nanoparticles in different forms such as rod, wire, cubes, spheres, cages, chips, films and even at times as hybridized nanoparticles to give best results. Also, the advantages of biosensors over other competitors make the discussion noteworthy.

Keywords Biosensing · Surface plasmon resonance (SPR) · LSPR (localized surface plasmon resonance) · Nanosensors

20.1 Introduction

Plasmonic biosensing is a powerful tool that is being used nowadays for effective determination quantitatively for analytes [1]. It finds its applications in the fields of pharmaceutics, diagnostics, safety control projects, health safeguarding, biodefense, etc. [2–4]. However, such interactions get really expensive for the industrial purpose owing to the use of metals like Au and Ag as plasmonic biosensors. They might also involve complexities with modern analytical devices. Several other complexities include sensitivity to contamination and lack of portability. Herein comes the role

'First two authors contributed equally'.

B. Chakraborty · R. Yadwade · B. Ankamwar (✉)
Bio-Inspired Materials Research Laboratory, Department of Chemistry, Savitribai Phule Pune University (Formerly University of Pune), Ganeshkhind, Pune 411007, India
e-mail: balaprasad.ankamwar@unipune.ac.in

D. Mohanta and P. Chakraborty (eds.), *Nanoscale Matter and Principles for Sensing and Labeling Applications*, Advanced Structured Materials 206,
https://doi.org/10.1007/978-981-99-7848-9_20

of metals such as copper, aluminium and other less expensive alloys. Optical sensors sometimes depend on excitation of surface plasmons, commonly called surface plasmon resonance (SPR) sensors. Similarly, localized surface plasmon resonance (LSPR) can also accumulate and manipulate photons at nanoscale level. The resonant wavelength depends on local dielectric environments and on size, shape, composition, orientation [5–7]. The metallic nanostructures have an outstanding property of surface plasmon (SP) resonance. It consists of a collective oscillation of conduction electrons excited by the electromagnetic field of light. Plasmons are a quantum of plasma oscillation. They form collective oscillations of the free electron gas density. For example, at optical frequencies, plasmons do couple with photons to develop quasiparticle. The study regarding surface plasmons was initiated in early twentieth century. The pioneering work elaborating the optical properties of metallic colloids was published by Gustav Mie [8]. The optical properties of metallic nanoparticles were understood after Mie's work in the early twentieth century whereas the further studies had limitation because of lack of resources, potential for controlled synthesis of nanoparticles. As the field of nanotechnology developed at the end of twentieth century, the applications of plasmons achieved numerous applications in various fields [9, 10].

SPR biosensors have been found to be highly effective against biosensor elements like antibodies, which can act against numerous target molecules. In such kind of interactions, the bio-recognition element, i.e. antibody is immobilized. Such interactions never hamper the biological activity and in turn minimize the non-specific binding on the surface. Recently aptamers (DNA or RNA aptamers) are also coming forward an emerging application of bio-recognition element by SPR [11–13]. This growing field regarding light-metal interactions is called plasmonics.

A material can be called as nanomaterial if it shows optical, electrical properties in range of 50–100 nm. Nanomaterials can also be used in areas of gases, DNA, metal ions, protein markers designed for diseases. Processes involved are practical, sensitive enough, robust and of course selective. In biological systems, nanomaterials are effectively found to compete with polymerase chain reaction (PCR) coupled with molecular fluorophore assays [14, 15]. Nanomaterials have also been used in cellular labelling agents in addition to bioconjugation for the past few years. The most general use of nanoparticles is the use of infrared to achieve deep penetration into soft tissues.

20.2 Plasmonic Properties of Metallic Nanoparticles

Plasmonic biosensors being highly specific towards their reaction are the main advantage of the former. Brilliant optical properties, resistance to oxidation, ease in nanopatterning are few among all the immediate plasmonic properties shown by nanoparticles. Some nanomaterials can act good as probes to diseases because of certain factors such as small size, large surface-to-volume ratio, enhanced physical properties that are also related to size such as biodetection itself, shape and composition [16]. Nanomaterials because of their unusual structural robustness can reach unusual binding sites easily where bulk molecules are less effective. This makes detection of diseases and related unusual growth of protein easier. Thus, diseases like cancer can have early detection thereby increasing the survival rate which otherwise could not be detected unless the protein has reached threshold concentration and the disease an advanced state. Also, deliberately the size, shape and characteristics can be altered according to the need or modifying the chemical properties as per requirement. Different properties contribute towards the specified applications of nanoparticles. Amongst which the most important is the morphology. Other properties such as specific emissivity, absorptive nature, light- scattering properties together make nanoparticles the best fit for multiplexed analyte detection.

20.3 Advantages of Biosensors

The advantages of biosensors lie in real-time monitoring of biological interactions between biomolecules. It effectively reduces the use of excess and toxic chemicals. The process is safe and doesn't involve radioactive particles. Biosensors are highly reusable with a high reproducing rate. Thus is also advantageous with a short response time. The process remains green and time-effective throughout, maintaining its efficiency.

Systems Exhibiting Plasmonic Effects

The plasmonic effects were profoundly noticed for system like metal films, waveguides, interferometers, optical fibres, photonic crystals and so on [17–24]. They also exhibit activities as sensors, therapeutics, medical diagnostics [25–29]. Recently, the nanofabrication methodologies have enhanced a lot [30–32].

Effect of Size of Nanomaterials on SPR

Nanomaterials can exist in different forms such as nanospheres, nanorods, nanostars and soon. Nanomaterials are highly in use for their exact characterization with prominent peaks at surface plasmon resonance (SPR) methodology. For instance, for Au nanosphere having a diameter of 50 nm, the peak appears at 520 nm in case of SPR. These peaks display ruby red colour due to the presence of Au colloids [15]. The initial stable colloidal suspension for gold was first prepared by Michael Faraday in 1857 [33]. The shape and boundary conditions of a nanomaterial also highly modify their plasmonic properties and affect the position of peak in SPR too.

20.4 Plasmonic Metal Nanoparticles and Their Characteristics

20.4.1 Gold (Au)

Au shows the best activity among all available plasmonic biosensors exhibiting optical properties effectively. Multiple factors influence the optical properties of gold such as conditions for depositing, metal film annealing, layer of adhesion, size of grain and thicknesses of film [34]. Nanomaterials can be utilized in various forms such as nanocages, nanorods, nanospheres, etc.

Nanorods

Gold nanorods are having narrow size distributions and uniform shape. The absorption peaks are tuneable in the visible and short near infra-red regions. They are used in drug delivery. Heat is generated during the process, which is effective in decreasing toxic cells. Gold nanorods are prepared by micellar solutions of cationic amphiphiles. They can change colour based on the refractive index of solvents. Swarup K. Maji pictured in his work the use of gold nanorods for the plasmon-enhanced electrochemical biosensing of H_2O_2 from cancer cells. He utilized LSPR (Localized Surface Plasmon Resonance) for the purpose. Here, cetyl trimethyl ammonium bromide-capped gold nanorods (AuNRs) were made immobilized on a glassy carbon (GC) electrode for fabrication of a biosensor that detects the presence of hydrogen peroxide. It gave good sensitivity operated at optimum wavelength, intensity, solution *p*H, etc. [35] (Fig. 20.1).

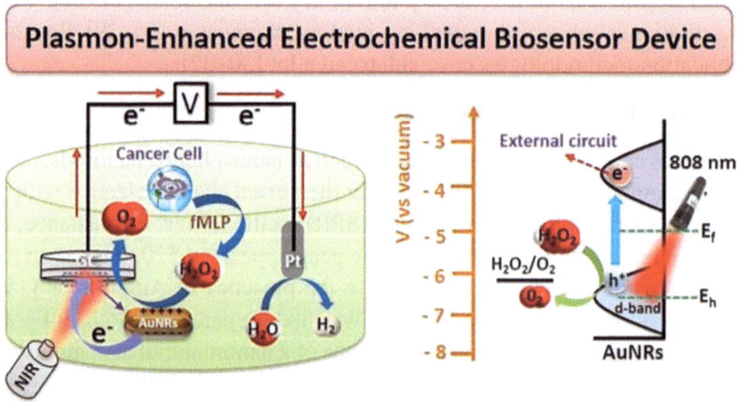

Fig. 20.1 Schematic diagram of electrocatalytic activity of gold nanorods under LSPR excitation Reprinted with permission from Reference [35] Copyright 2019, American Chemical Society

Hybrid gold/graphene oxide nanoparticles

Excess of H_2O_2 concentration can permanently destroy many important biological processes in human body leading to diabetes, nervous disorders, cardiac diseases, Alzheimer's disease, Parkinson's disease, cancer and many others [36, 37]. Therefore, in situ observation of released hydrogen peroxide from live cells is of unlimited consequence for cell functions and clinical diagnosis. H_2O_2 serves as a major compound in infinite cellular activities and is thus a potential marker for living and cancer cells. This detection of secreted H_2O_2 from above cells through an electrochemical technique has many privileges [37].

Nanoporous gold nanoparticle

Gold nanoparticles with tiny pores enable the aggregation of these NPs with the substrate. Erkin Seker et al. worked on nanoporous gold electrode coatings that can increase the efficiency of electrochemical nucleic acid biosensors due to their 3D nanoscale network, facile surface functionalization, high electrical conductivity and biocompatibility. Here, the size of the pores plays a noteworthy role. Small pores make the function more target-oriented, however, large pores increase accessibility. This specific work focuses on the optimization of the condition so that both the above-mentioned issues can be headed upon. This might increase the extent of detection and biofouling resilience towards target DNA [38]. Au nanoparticles and graphene oxide both have high affinity towards DNA adsorption. In this work, they used KCN to dissolve AuNPs and find out the percentage of desorbed DNA. The percentage was found out to be too less, i.e. 5% around. Desorption can be brought about by adding HEPES buffer, NaCl, As (III). The performance as shown by AuNPs is better than that shown by graphene oxide. As in case of the latter the desorption rate is more. Graphene oxide is also non-specific in their desorption activities. They are highly in use [39].

Nanocages

Gold nanocages can be applied as drug carriers and owe other significant applications in drug delivery and release of drug in a controlled manner. AuNPs are bioconjugated with antibodies for selectively targeting the cancer cells. Heat evolved from the heated particles causes defects to selected cancer cells. This process is known as the photothermal effect on metal nanoparticles. The method can be characterized by LSPR (Localized Surface Plasmon Resonance) [27] (Fig. 20.2).

Gold Nanosphere

The properties of gold nanospheres are quite unique and more efficient than other metallic nanospheres having similar diameter. The irregular surface present for gold increases its surface area for absorption thus resulting in enhanced functionalities. Nanospheres can be utilized in certain areas where they give the most prominent colour change. It gives a time-efficient process. The colour change is so clearly visible to the naked eyes as it shifts from red to blue. The process is highly cost-effective and gives high-throughput screening. It can work on concentration of pollutants over food

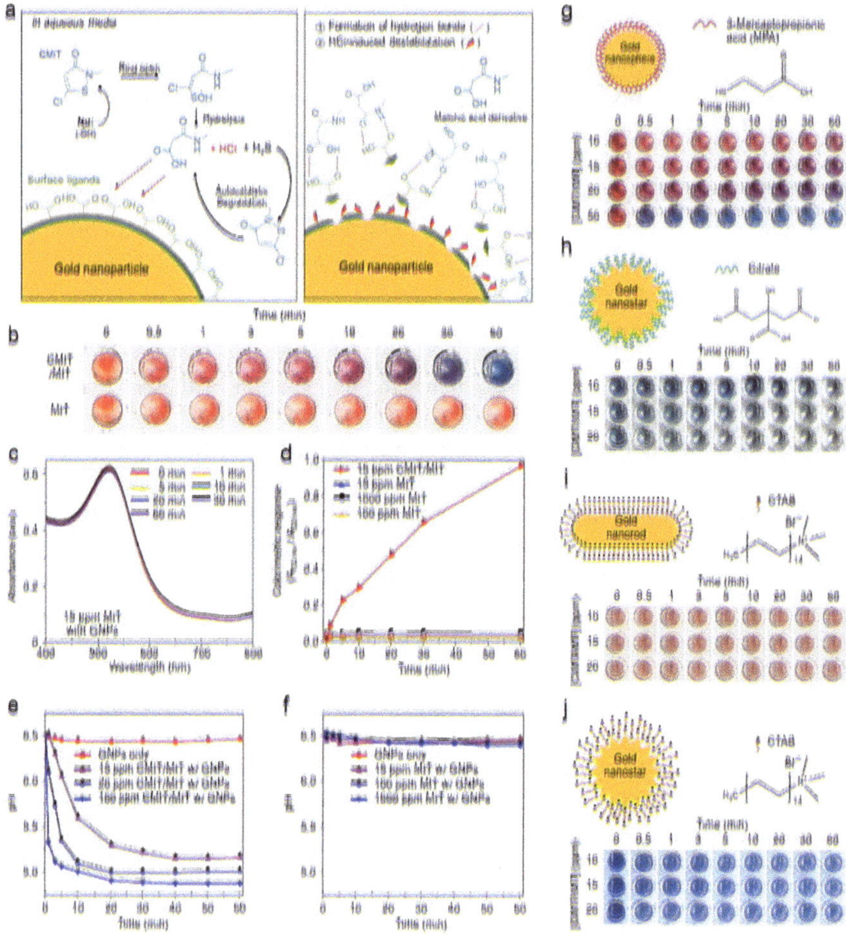

Fig. 20.2 Easy detection based on colour for nanosphere—**a** Aggregation of gold nanoparticles on substrate. **b** Colorimetric detection of CMIT/MIT **c** Plot of absorbance vs wavelength at different time intervals. **d** Plot of colorimetric response versus time at varied concentrations. **e, f** Plot of pH versus time at varied concentrations for aggregated gold nanoparticles. **g** Colorimetric response with respect to gold nanosphere. **h** Colorimetric response with respect to gold nanostar. **i** Colorimetric changes with respect to gold nanorod. **j** Colorimetric changes with gold nanostar. Reprinted with permission from Reference [27] Copyright 2013, American Chemical Society

matters that we consume to decrease health risks to biological bodies. The end results were highly appreciated. The process can be studied by Raman Spectroscopy [26]. When the pollutants and bare gold nanoparticles are present it gives red colouration because of the presence of gold nanoparticles. When the gold nanoparticles get aggregated with the given compound the aggregated AuNPs gave blue colouration that made end-product easily distinguishable. This method can detect CMIT/MIT, identified as a specific pollutant, at a concentration of 15 ppm. Below this threshold, its toxicity may not be as pronounced as commonly perceived. The marking of the concentration specifically and providing a distinguished colour change using nanosphere morphology could participate in biosensing to a good extent. The signal in Raman spectroscopy can be further amplified using these nanoparticles that make detection easier.

The main drawbacks of using gold are quite self-explanatory as it includes high price involvement and is also incompatible with microelectronic technical screening processes.

20.4.2 Silver (Ag)

Ag shows excellent performance mainly because of its low optical losses [28, 29]. Protection of silver by dielectric films stabilizes as well as increases sensitivity [40, 41]. Ag nanoparticles find uses in the food industry, therapeutic uses, drug-delivery system, antitumour activity, dentistry, etc.

Hybrid Ag nanoparticles: Ag/Cu_2O/GO

Ag metal and semiconductors can be utilized to form biosensors that can be used for bioimaging and drug delivery and for detecting H_2O_2, nucleotides and tumour cells. Core–shell nanostructures can be used effectively for their changeable catalytic property. The benefit lies in the fact that we can alter the properties, i.e. shapes and sizes in order to meet our requirements. They also exhibit a wide range of functions. Ag@Cu_2O can detect the amount of dopamine and uric acid in urine and serum samples. If the amount or concentration is more/less the desired amount, it might lead to some diseases. Triangular shaped—they have high superior electrocatalytic property. Properties such as high surface area, increased permeability and catalytic efficiencies influence its functions. These are characterized by Raman spectroscopy, X-Ray diffraction, etc. The products involved are stable enough and the reactions involved are easily reproducible. The core–shell is reinforced with graphene oxide (GO) layer to amplify the activity. In Fig. 20.3—different morphologies are explained. While A shows a prismatic planar structure. Spherical-shaped B shows nearly spherical shape. Cube-shaped C demonstrates uniform cube shaped Ag nanoparticles with sharp edges. The size of each one also varies with nanoprismatic being the smallest. They are synthesized by lithography achieved by higher pressure homogeneously. The method is absolutely green and free from wastes and expenses. The sizes from A, B to C are 38.6 ± 2.3, 48.7 ± 2.3 and 111 ± 2.6 nm, respectively [42]. Nanoparticles show high reactivity in all fields due to various properties. The properties of these morphology-based nanosensors can be altered, exemplified by their electro-oxidation activity towards dopamine, uric acid, guanine, and adenine. The observed activity follows the order of triangular > spherical > cubic shapes. Hence, it is most pronounced for the triangular-shaped hybrid Ag nanoparticles. They can act as catalysts. Pure Ag nanoparticles offer the best sensitivity among all its kinds. The summary table analyses the application of different Ag nanoparticles with varied shapes and sizes. The electrons on metal surface undergo strong interaction with light. Properties like LSPR and SPR can be tuned by simply altering the metallic size and shapes, uniformity, surface coatings, etc. Ag nanoparticles present in different forms and morphologies can enhance biosensing [43, 44] and can be enlisted below:

Nanocubes—Prepared by polyol process and can be utilized for cysteine sensing [45–47].

Nanospheres—Prepared by chemical reduction can be used for plasmonic sensing and anti-microbial activity [48, 49].

Nanotriangles—Prepared by both chemical reduction and nanosphere lithography; can be used for the molecular detection of Alzheimer's disease [49–52].

Nanowires—Prepared by polyol process, it can be utilized for biomolecules plasmonic sensing [45, 52, 53].

Nanorods—Prepared by photochemical, thermal and ORG process can be used for plasmonic biosensing [45, 54].

Nanostars—With a low Ag^+ concentration, it can detect cancer biomarker with a sensitivity of around 10^9 ng/ml in serum. Similar are the activities of other morphologies such as pyramids and flower-like [55]. Ag pyramid morphology gives high yield for the process.

Ag nanoparticles can hold excellent anti-fungal activities too against *Candida albicans*, *Aedes aegyptis*, etc. This works under UV light. They are into anti-parasitic activities and anti-microbial properties also. Ag nanoparticles play a major role in the detection of cancer cells. They act as anti-tumour agent both in vitro and in vivo. Ag nanoparticles induce autophagy in the cells. They can play acceptable roles in drug-delivery systems, dentistry, bone healing, orthopaedic implant, etc.

As a major drawback, surface of silver can get easily oxidized and needs to be protected. They are less stable and less compatible. Hence are less reliable than Au. Hence, combination forms came into play with gold giving rise to Au@Ag or Ag@Au.

Biosensing activities of Au–Ag combined

The size of the nanoparticles depends highly on *p*H of the solution, temperature and time. With the increase of *p*H, the band turns blue thus making the aggregation with the substrate favoured. The Au–Ag alloys have no distinct Au or Ag parts. BSA concentration can be easily detected by their use because of the red shift.

Spherical—The micro-patterns controlled can be utilized for constructing LSPR biochips for high-throughput detection of different concentrations of bovine serum albumin (BSA), an antigen on a glass sample under optimized conditions. The alloys have higher sensitivity to such biosensor applications than pure nanomaterials [56].

Nanoprism—The ordered arrangement of optical geometric parameters increases sensitivity for localized plasmon-based sensors quite efficiently [57].

Nanoplates—It is highly stable against strong etching agents such as H_2O_2 and can be used for detecting biomolecule in a flow channel with Kretchmann configuration. They function in biotinylation. Amidation kind of reaction destroys the coverage of MHA monolayer, and it is at that phase that Ag nanoplates get destabilized [58].

Triangular core–shell—A simple and highly efficient method has been developed for the utilization of plasmonic properties and nanoprobes of nanoparticles in the detection of hydrogen sulphide in living organisms and in vivo applications. It can be detected via SPR peak obtained. The absorption occurs at near infrared region [59].

Ag–Au core–shell nanorods—This phenomenon can be characterized by both surface plasmon resonance (SPR) and Raman Scattering as well detected in UV–vis-NIR absorption area, they display different peaks at different concentrations of Au and Ag. Rhodamine 6G, a molecule, can be detected via this process. It demonstrates the chemisorptive bond effect and morphology-based roles [60]. In certain cases, Au–Ag–Au double shells are employed for the detection of 2-mercapto-1-methyl imidazole in living biological systems [61].

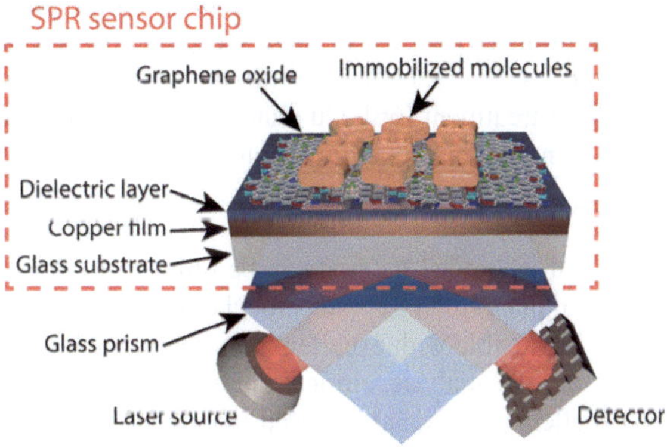

Fig. 20.3 Representation of SPR biosensor on plasmonic copper films protected from further reaction Reprinted with permission from Reference [62] Copyright 2018, American Chemical Society

20.4.3 Copper (Cu)

Advantages of Cu over Au and Ag—Cu is inexpensive as compared to gold and silver. It also shows lower optical losses in visible and NIR ranges. The combined physical and chemical stability of copper and the ease with which it mixes with polymer make it highly advantageous. It is also highly conductive in nature.

Types of morphologies of copper-based nanoparticles with their applications

Copper chips—Stebunov et al. used this basic concept of utilizing copper instead of gold and silver. The major difficulty in this aspect lies in the protection of surfaces from oxidation. This could be brought about by addition of SiO_2 and Al_2O_3 dielectric films which increased sensitivity by 55% and 75%, respectively. SPR technology is utilized for this purpose [62]. It is utilized for selective analysis of oligonucleotide hybridization. Flat copper films—flat films covered with TiO_2 can be used to treat wastewater.

Copper oxide nanoparticle (hybridized)—CuO has a narrow bandgap and can act smartly with good high surface area, electrochemical activity, great stability in solutions, proper redox potential [63, 64]. It can be used to treat cancer stem cells or tumour-initiating cells, especially when it is a pancreatic cancer. They are widely used for non-enzymatic sensing. Copper oxide nanoparticles are cytotoxic against human pancreatic cancer cells. The effects are associated with reactive oxygen species (ROS) levels and reduced mitochondrial potential growth. CuO can inhibit such tumour growth [65].

Mesoporous copper sulphide nanoparticles—It is used to treat localized cancer by inhibiting excessive tumour growth. It serves not only as a drug carrier but also as an imaging to guide chemo-photo therapy. The process involves the utilization of modified hollow mesoporous copper sulfide particles for the effective transfer of the drug. High level of ROS (reactive oxygen species) was generated in near infrared region for photodynamic therapy. This proved to be a promising method in cancer treatment [66]. Cu nanoparticles can be used to study toxicity and detect H_2O_2 generation responding to biosensing strains [67].

ssDNA-templated Cu nanoparticle

An electrochemical method can be utilized for detecting glutathione (GSH) where DNA template CuNPs act selector chemical reporters. This method is highly selective and reproducible with excellent sensitivity. Such applications are highly effective in areas of gene analysis, clinical diagnosis, environmental monitoring [68]. They act as excellent probes and are used in detecting DNA-derived enzyme activities [69]. DNA copper nanoparticle-based biosensors can also be used for detecting reactive quenchers. Poly-thymine template copper nanoparticles—Poly-T template copper nanoparticles can be used for protein detection. It acts as a fluorescent biosensor. Biosensor was sensitive enough, selective and the process was carried out in low cost. The operations involved are also simple ones. Thus, this type of particle finds applications in molecular diagnostics and genomic research [70].

DsDNA-templated Cu nanoparticles

DNA having excellent biocompatibility, base-pairing, convenient synthetic routes can be used for synthesizing nanoparticles. It is studied to have hydrogen peroxide alongside oxidase-based biosensing. As a proof, it could be utilized for detecting several biomolecules. Thus, a vital role is played in biomedical analysis [71]. The rapid oxidation of metal surface still remained as a major drawback.

Detailed morphological dependent biosensing properties of copper oxide (CuO) nanoparticles

Copper oxide has shown great potential in pharmacological activities and biomedical field. Fabrication of the shapes and sizes, tuning them might get difficult but its diversified applications make it worth [72, 73]. Let us have a look in detail considering their biosensing applications, which is also elaborated in Table 20.1, which summarizes applications of different nanoparticles.

CuO-graphene nanocomposites—They can act as glucose sensors [74–77].

CuO nanosheets/CuO nanoflowers/CuO nanocubes—They can act as hydrogen peroxide sensor [78, 79].

CuO nanowires—They can be used as hydrogen peroxide detector and also to detect tumour [80, 81].

Au-CuO nanoflowers—They can be used as detector of alpha fetoprotein [82].

Ag-CuO nanocomposites—Can be used for the detection of carcinoembryonic antigen [83].

Pd-doped CuO nanoparticles can be used for detection of PSA [84].

CuO nanowire@ferrocene—Can be again utilized for the detection of carcinoembryonic antigen [85].

CuO nanoseeds—Utilized for detection of PSA and alpha fetoprotein [86].

Au@AgCuO nanoparticles—Used for the detection of PSA as immunosensor [87].

CuO nanorice/platelet-like CuO/nanohexagon CuO—Used for detecting dopamine [88–91].

CuO nanoparticles—chitosan—It can act as cholesterol sensor [92, 93].

Nickel-oxide CuO, Zinc oxide CuO, Au-CuO—They do act as dopamine sensor [94–96].

CuO nanowires—Can act as DNA sensors [93].

Zn-doped CuO nanoparticles—Can be used to detect kidney cancer [94].

BSA capped CuO—To detect glioma cancer [95].

Leaf-like CuO—For detection of breast cancer [96].

CuO nanoparticle in zeolite—For detection of *E. coli* and similar micro-organisms [95] (Fig. 20.4).

Table 20.1 Preparation and biosensing applications of morphology-based metallic nanoparticles

Shapes of Au		*Characterization method*	*Applications*	*References*
a. Nanorods		LSPR (localized surface plasmon resonance)	Electrochemical biosensing of H_2O_2 from cancer cells	[35]
b. Hybrid Au nanoparticles		SPR (surface plasmon resonance)	Detection of H_2O_2 in live cells, in human urine and prevention of its accumulation	[37]
c. Nanoporous Au particles		SPR (surface plasmon resonance)	Antigen and antibody; protein detection	[38]
d. Nanocages		LSPR	Biomarker to detect kidney injury from synthetic urine	[33]
e. Nanospheres		Raman spectroscopy	Detection of concentration of pollutants over food matters	[34]
Shapes of Ag	*Sizes in nm*	*Method*	*Applications*	*References*
a. Spheres	10–100	Chemical reduction	Plasmonic sensing; catalysis and anti-microbial	[45– 47]
b. Triangles	Width:100; Height: 50	Chemical reduction nanospheres lithography	Molecular detection of Alzheimer's disease	[48–50]
c. Nanocubes	20–45	Polyol process	Cysteine sensing by plasmonic effect	[51]
d. Nanowires	60 (length)	Polyol process	Biomolecule plasmonic sensing	[52, 53]
e. Nanorods	250–300	Photochemical; thermal; ORG (oxidation reduction growth)	Plasmonic biosensing	[54]

(continued)

Table 20.1 (continued)

f. Pyramids		Edge length-50–200	Chemical reduction	Plasmonic biosensing	[41]
g. Flower-like		200–300	Wet-chemical	Catalysis; plasmonic biosensing	[55]
Alloy/core	*Shape*	*Size (nm)*	*Synthesis technique*	*Applications*	*References*
a. Ag–Au alloy	Spherical	20–50	Metal evaporation on glass support	Improved sensitivity in LSPR biosensor application	[56]
b. AgAu alloy	Nanoprisms	150; 50	Nanosphere lithography	Enhanced biosensor applications	[57]
c. Ag@Aucore-shell	Nanoplates	50 (Ag); 0.5 (Au)	Au shell electrode position on Ag nanoplates	SPR biosensor with higher sensitivity developed	[58]
d. Ag@Aucore-shell	Triangular nanoprisms (TNPrs)	60 (Ag core); 1 (Au shell)	Chemical reduction	Easy detection of hydrogen sulphide; highly sensitive	[59]
e. Ag@Au core-shell	Nanorods	35 (Au); 10 (width); 1 (Ag)	Chemical reduction	Chemisorptive bond effect detected by SERS	[60]
Morphology of Cu nanoparticles			*Applications*		*References*
a. Copper chips			Selective analysis of oligonucleotide hybridization		[62]
b. Copper films			Pollutants and wastewater treatment		[64]
c. Copper oxide nanoparticles			Detection of cancer stem cells and tumour at earlier stage		[65]
d. Mesoporous copper sulphide nanoparticles			As a drug carrier; in cancer detection and also guides chemotherapy		[66]
e. Pure copper nanoparticles			To examine hydrogen peroxide generation and create toxicity		[67]
f. ssDNA-templated copper nanoparticles			To detect GSH (glutathione) andreactive quenchers		[68]
g. Poly-T-templated copper nanoparticles			Protein detection		[69, 70]
h. dsDNA-templated copper nanoparticles			To detect hydrogen peroxide and oxidase-based biosensing		[71]

(continued)

Table 20.1 (continued)

Morphology of CuO nanoparticles	*Applications*	*References*
a. NanoAg-CuO/Nafion/CuO-graphene nanocomposites	Acts as a glucose sensor	[74–77]
b. CuO nanosheets inkprinted on Au or Ag/CuO nanoflowers/CuO nanocubes	Acts as a hydrogen peroxide sensor	[78, 79]
c. Nestlike nano-CuO/GCE/CuO nanowires	Hydrogen peroxide sensor, to detect tumour	[80, 81]
d. Au-CuO nanoflowers	Detection of alpha fetoprotein	[82]
e. Ag@CuO nanocomposites	Detection of carcinoembryonic antigen	[83]
f. Pd doped CuO nanoparticles	Detection of PSA	[84]
g. CuO nanowire@ ferrocene	For detection of carcinoembryonic antigen	[85]
h. CuO nanoseeds	For detection of PSA and alpha fetoprotein	[86]
i. Au@AgCuO nanoparticles	For detection of PSA as immunosensor	[87]
j. Carbonpaste/CuO nanorice/platelet-like CuO/ nanohexagon CuO	To detect dopamine	[88–91]
k. CuO-graphene nanocomposite/ultrafine CuO nanoparticles—chitosan	Acts as cholesterol sensor	[92, 93]
l. Nickel-oxide CuO, Zinc oxide CuO, Au-CuO	Acts as dopamine sensor	[94–96]
n. Zn-doped CuO nanoparticles	To detect kidney cancer	[97]
o. BSA-capped CuO	To detect glioma cancer	[98]
p. Leaf-like CuO	To detect breast cancer	[99]
q. CuO nanoparticles in zeolite	To detect *E. coli*	[100]

20.5 Conclusions and Outlook

The applications of metallic nanoparticles in biological field are highly affected by the morphology changes concerning the nanoparticles. The morphology changes can be detected by TEM (Transmission Electron Microscopic Measurements), SEM, fluorescence, Raman Scattering. antigen detection, protein detection, detection of diseases such as cancer, AIDS, heavy metals such as Hg, Pb and smaller entities such as biotoxins, antibiotics, pesticides, narcotics, explosives can be favoured by this process. DNA is targeted too, for such detection too in biological systems [98]. The functionality depends largely on pore size, binding sites and their structures, signal suppression, biofouling resilience—these enhance the detection performance. Nanoparticles with specific spacing, angle and fabrication is configured to enhance plasmonic sensing properties. Zhang *et al.* could determine detection of exosomal surface proteins with aptamer-complexed gold nanoparticles [39]. Recently, hybrid nanopyramid nanoparticles are found to detect intracellular dual microRNA detection. Increasing the portability, absorbance range, accuracy and bioacceptability is

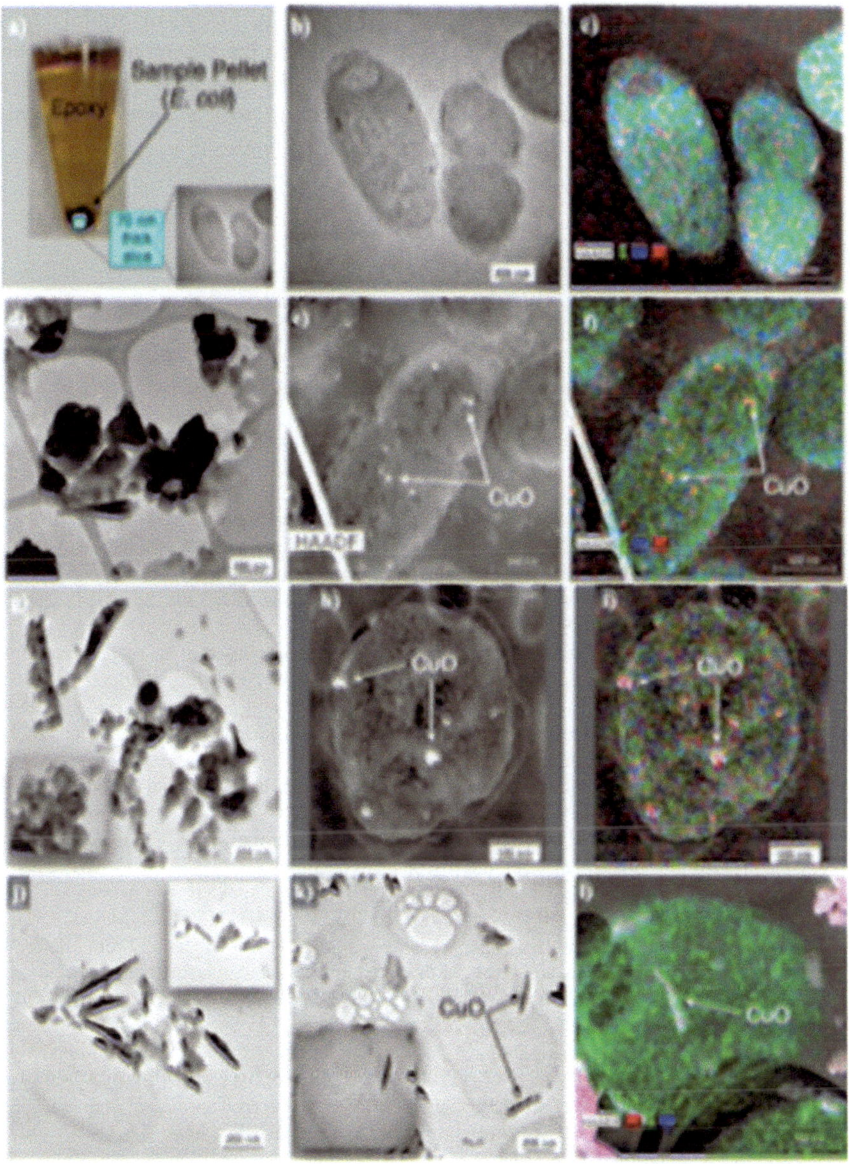

Fig. 20.4 TEM and STEM-EELS images of *Escherichia coli* with **a** experimental sample; **b–c** no exposure of CuONPs; **e–f** bulk of CuO; **h–i** CuOnanopowder; **k–l** nanosheets. TEM images of aggregates of **d** bulk; **g** nanopowder; **j** nanosheets. Reprinted with permission from Reference [72] Copyright 2013, American Chemical Society

highly desired. Devoid of practical applications, plasmonic biosensors find it a long way to make it to the market till date [99, 100]. To summarize, plasmonic biosensors have created a revolution as such opening a huge field of research. However, it is still on the verge of facing several challenges overcoming, which can make it highly effective and a milestone for applications in diversified fields.

References

1. Schasfoort, R.B. (ed.): Handbook of Surface Plasmon Resonance. Royal Society of Chemistry (2017)
2. Cooper, M.A.: Optical biosensors in drug discovery. Nat. Rev. Drug Discov. **1**(7), 515–528 (2002). https://doi.org/10.1038/Nrd838
3. Karlsson, R.: SPR for molecular interaction analysis: a review of emerging application areas. J. Mol. Recognit. **17**(3), 151–161 (2004). https://doi.org/10.1002/Jmr.660
4. Homola, J.: Surface plasmon resonance sensors for detection of chemical and biological species. Chem. Rev. **108**(2), 462–493 (2008). https://doi.org/10.1021/Cr068107d
5. Lee, Y.B., Lee, S.H., Park, S.Y., Park, C.J., Lee, K.S., Kim, J., Joo, J.: Luminescence enhancement by surface plasmon assisted Förster resonance energy transfer in quantum dots and light emitting polymer hybrids with Au nanoparticles. Synth. Met. **187**, 130–135 (2014). https://doi.org/10.1039/C4RA11497C
6. Lin, H.Y., Chen, Y.F.: Giant enhancement of luminescence induced by second-harmonic surface plasmon resonance. Appl. Phys. Lett. **88**(10), 101914 (2006). https://doi.org/10.1063/1.2183819
7. Shahbazyan, T.V.: Theory of plasmon-enhanced metal photoluminescence. Nano Lett. **13**(1), 194–198 (2013). https://doi.org/10.1021/Nl303851z
8. Turkevich, J., Garton, G., Stevenson, P.C.: The color of colloidal gold. J. Colloid Sci. **9**, 26–35 (1954). https://doi.org/10.1016/0095-8522(54)90070-7
9. Barber, D.J., Freestone, I.C.: An investigation of the origin of the colour of the Lycurgus Cup by analytical transmission electron microscopy. Archaeometry **32**(1), 33–45 (1990)
10. Freestone, I., Meeks, N., Sax, M., Higgitt, C.: The Lycurgus cup—a roman nanotechnology. Gold Bull. **40**(4), 270–277 (2007). https://doi.org/10.1007/BF03215599
11. Win, M.N., Klein, J.S., Smolke, C.D.: Codeine-binding RNA aptamers and rapid determination of their binding constants using a direct coupling surface plasmon resonance assay. Nucleic Acids Res. **34**(19), 5670–5682 (2006). https://doi.org/10.1093/Nar/Gkl718
12. Wang, Z., Wilkop, T., Xu, D., Dong, Y., Ma, G., Cheng, Q.: Surface plasmon resonance imaging for affinity analysis of aptamer–protein interactions with PDMS microfluidic chips. Anal. Bioanal. Chem. **389**(3), 819–825 (2007). https://doi.org/10.1007/S00216-007-1510x
13. Jayasena, S.D.: Aptamers: an emerging class of molecules that rival antibodies in diagnostics. Clin. Chem. **45**(9), 1628–1650 (1999). https://doi.org/10.1093/Clinchem/45.9.1628
14. Tyagi, S., Kramer, F.R.: Molecular beacons: probes that fluoresce upon hybridization. Nat. Biotechnol. **14**(3), 303–308 (1996). https://doi.org/10.1038/Nbt0396-303
15. Mulvaney, P.: Surface plasmon spectroscopy of nanosized metal particles. Langmuir **12**(3), 788–800 (1996). https://doi.org/10.1021/La9502711
16. Wong, X.Y., Sena-Torralba, A., Alvarez-Diduk, R., Muthoosamy, K., Merkoçi, A.: Nanomaterials for nanotheranostics: tuning their properties according to disease needs. ACS Nano **14**(3), 2585–2627 (2020). https://doi.org/10.1021/Acsnano.9b08133
17. Anker, J.N., Hall, W.P., Lyandres, O., Shah, N.C., Zhao, J., Van Duyne, R.P.: Biosensing with plasmonicnanosensors. Nanosci. Technol.: Collect. Rev. Nat. J. 308–319 (2010). https://doi.org/10.1038/Nmat2162

18. Stewart, M.E., Anderton, C.R., Thompson, L.B., Maria, J., Gray, S.K., Rogers, J.A., Nuzzo, R.G.: Nanostructured plasmonic sensors. Chem. Rev. **108**(2), 494–521 (2008). https://doi.org/10.1021/Cr068126n
19. Gao, Y., Gan, Q., Xin, Z., Cheng, X., Bartoli, F.J.: Plasmonic Mach-Zehnder interferometer for ultrasensitive on-chip biosensing. ACS Nano **5**(12), 9836–9844 (2011). https://doi.org/10.1021/Nn2034204
20. Shen, Y., Zhou, J., Liu, T., Tao, Y., Jiang, R., Liu, M., Xiao, G., Zhu, J., Zhou, Z.K., Wang, X., Jin, C.: Plasmonic gold mushroom arrays with refractive index sensing figures of merit approaching the theoretical limit. Nat. Commun. **4**(1), 1–9 (2013). https://doi.org/10.1038/Ncomms3381
21. Khodami, M., Berini, P.: Biomolecular kinetics analysis using long-range surface plasmon waveguides. Sens. Actuators, B Chem. **243**, 114–120 (2017). https://doi.org/10.1016/J.Snb.2016.11.120
22. Fan, H., Berini, P.: Bulk sensing using a long-range surface-plasmon dual-output Mach-Zehnder interferometer. J. Lightwave Technol. **34**(11), 2631–2638 (2016)
23. Lee, B., Roh, S., Park, J.: Current status of micro-and nano-structured optical fiber sensors. Opt. Fiber Technol. **15**(3), 209–221 (2009). https://doi.org/10.1016/J.Yofte.2009.02.006
24. Rifat, A.A., Ahmed, R., Yetisen, A.K., Butt, H., Sabouri, A., Mahdiraji, G.A., Yun, S.H., Adikan, F.M.: Photonic crystal fiber based plasmonic sensors. Sens. Actuators, B Chem. **243**, 311–325 (2017). https://doi.org/10.1016/J.Snb.2016.11.113
25. Thanh, N.T.K., Rosenzweig, Z.: Development of an aggregation-based immunoassay for anti-protein A using gold nanoparticles. Anal. Chem. **74**(7), 1624–1628 (2002). https://doi.org/10.1021/Ac011127p
26. Tian, L., Liu, K.K., Morrissey, J.J., Gandra, N., Kharasch, E.D., Singamaneni, S.: Gold nanocages with built-in artificial antibodies for label-free plasmonic biosensing. J. Mater. Chem. B **2**(2), 167–170 (2014). https://doi.org/10.1039/C3TB21551B
27. Kang, T.H., Jin, C.M., Lee, S., Choi, I.: Dual mode rapid plasmonic detections of chemical disinfectants (CMIT/MIT) using target-mediated selective aggregation of gold nanoparticles. Anal. Chem. **92**(6), 4201–4208 (2020). https://doi.org/10.1021/Acs.Analchem.9b04081
28. Park, J.H., Ambwani, P., Manno, M., Lindquist, N.C., Nagpal, P., Oh, S.H., Leighton, C., Norris, D.J.: Single-crystalline silver films for plasmonics. Adv. Mater. **24**(29), 3988–3992 (2012). https://doi.org/10.1002/Adma.201200812
29. Kravets, V.G., Jalil, R., Kim, Y.J., Ansell, D., Aznakayeva, D.E., Thackray, B., Britnell, L., Belle, B.D., Withers, F., Radko, I.P., Han, Z.: Graphene-protected copper and silver plasmonics. Sci. Rep. **4**(1), 1–8 (2014). https://doi.org/10.1038/Srep05517
30. Stewart, M.E., Motala, M.J., Yao, J., Thompson, L.B., Nuzzo, R.G.: Unconventional methods for forming nanopatterns. Proc. Inst. Mech. Eng., Part N: J. Nanoeng. Nanosyst. **220**(3), 81–138 (2006). https://doi.org/10.1243/17403499JNN103
31. Tseng, A.A.: Recent developments in nanofabrication using ion projection lithography. Small **1**(6), 594–608 (2005). https://doi.org/10.1002/Smll.200500113
32. Watt, F., Bettiol, A.A., Van Kan, J.A., Teo, E.J., Breese, M.B.H.: Ion beam lithography and nanofabrication: a review. Int. J. Nanosci. **4**(03), 269–286 (2005). https://doi.org/10.1142/S0219581X05003139
33. Faraday, M.: X. The Bakerian Lecture. Experimental relations of gold (and other metals) to light. Philos. Trans. R. Soc. Lond. **147**, 145–181 (1857). https://doi.org/10.1098/Rstl.1857.0011
34. McPeak, K.M., Jayanti, S.V., Kress, S.J., Meyer, S., Iotti, S., Rossinelli, A., Norris, D.J.: Plasmonic films can easily be better: rules and recipes. ACS Photonics **2**(3), 326–333 (2015). https://doi.org/10.1021/Ph5004237
35. Maji, S.K.: Plasmon-enhanced electrochemical biosensing of hydrogen peroxide from cancer cells by gold nanorods. ACS Appl. Nano Mater. **2**(11), 7162–7169 (2019). https://doi.org/10.1021/Acsanm.9b01675
36. Maruyama, W., Dostert, P., Matsubara, K., Naoi, M.: N-methyl (R) salsolinol produces hydroxyl radicals: involvement to neurotoxicity. Free Radical Biol. Med. **19**(1), 67–75 (1995). https://doi.org/10.1016/0891-5849(95)00013-N

37. Maji, S.K., Sreejith, S., Mandal, A.K., Ma, X., Zhao, Y.: Immobilizing gold nanoparticles in mesoporous silica covered reduced graphene oxide: a hybrid material for cancer cell detection through hydrogen peroxide sensing. ACS Appl. Mater. Interfaces **6**(16), 13648–13656 (2014). https://doi.org/10.1021/Am503110s
38. Matharu, Z., Daggumati, P., Wang, L., Dorofeeva, T.S., Li, Z., Seker, E.: Nanoporous-gold-based electrode morphology libraries for investigating structure–property relationships in nucleic acid based electrochemical biosensors. ACS Appl. Mater. Interfaces **9**(15), 12959–12966 (2017). https://doi.org/10.1021/Acsami.6b15212
39. Zhang, F., Wang, S., Liu, J.: Gold nanoparticles adsorb DNA and aptamer probes too strongly and a comparison with graphene oxide for biosensing. Anal. Chem. **91**(22), 14743–14750 (2019). https://doi.org/10.1021/Acs.Analchem.9b04142
40. Pockrand, I.: Surface plasma oscillations at silver surfaces with thin transparent and absorbing coatings. Surf. Sci. **72**(3), 577–588 (1978). https://doi.org/10.1016/0039-6028(78)90371-0
41. Touahir, L., Jenkins, A.T.A., Boukherroub, R., Gouget-Laemmel, A.C., Chazalviel, J.N., Peretti, J., Ozanam, F., Szunerits, S.: Surface plasmon-enhanced fluorescence spectroscopy on silver based SPR substrates. J. Phys. Chem. C **114**(51), 22582–22589 (2010). https://doi.org/10.1021/Acs.Langmuir.8b00276
42. Gan, T., Wang, Z., Shi, Z., Zheng, D., Sun, J., Liu, Y.: Graphene oxide reinforced core–shell structured Ag@ Cu_2O with tunable hierarchical morphologies and their morphology–dependent electrocatalytic properties for bio-sensing applications. Biosens. Bioelectron. **112**, 23–30 (2018). https://doi.org/10.1016/J.Bios.2018.04.029
43. Farhadi, K., Forough, M., Molaei, R., Hajizadeh, S., Rafipour, A.: Highly selective Hg2+ colorimetric sensor using green synthesized and unmodified silver nanoparticles. Sens. Actuators, B Chem. **161**(1), 880–885 (2012). https://doi.org/10.1016/J.Snb.2011.11.052
44. Loiseau, A., Asila, V., Boitel-Aullen, G., Lam, M., Salmain, M., Boujday, S.: Silver-based plasmonic nanoparticles for and their use in biosensing. Biosensors **9**(2), 78 (2019). https://doi.org/10.3390/Bios9020078
45. Khodashenas, B., Ghorbani, H.R.: Synthesis of silver nanoparticles with different shapes. Arab. J. Chem. **12**(8), 1823–1838 (2019). https://doi.org/10.1016/J.Arabjc.2014.12.014
46. Mahmudin, L., Suharyadi, E., Utomo, A.B.S., Abraha, K.: Optical properties of silver nanoparticles for surface plasmon resonance (SPR)-based biosensor applications. J. Mod. Phys. **6**(08), 1071 (2015). https://doi.org/10.4236/Jmp.2015.68111
47. Cobley, C.M., Skrabalak, S.E., Campbell, D.J., Xia, Y.: Shape-controlled synthesis of silver nanoparticles for plasmonic and sensing applications. Plasmonics **4**(2), 171–179 (2009). https://doi.org/10.1007/S11468-009-9088-0
48. Haes, A.J., Van Duyne, R.P.: A nanoscale optical biosensor: sensitivity and selectivity of an approach based on the localized surface plasmon resonance spectroscopy of triangular silver nanoparticles. J. Am. Chem. Soc. **124**(35), 10596–10604 (2002). https://doi.org/10.1021/Ja020393x
49. Stavytska-Barba, M., Salvador, M., Kulkarni, A., Ginger, D.S., Kelley, A.M.: Plasmonic enhancement of Raman scattering from the organic solar cell material P3HT/PCBM by triangular silver nanoprisms. J. Phys. Chem. C **115**(42), 20788–20794 (2011). https://doi.org/10.1021/Jp206853u
50. Haes, A.J., Chang, L., Klein, W.L., Van Duyne, R.P.: Detection of a biomarker for Alzheimer's disease from synthetic and clinical samples using a nanoscale optical biosensor. J. Am. Chem. Soc. **127**(7), 2264–2271 (2005). https://doi.org/10.1021/Ja044087q
51. Merk, V., Nerz, A., Fredrich, S., Gernert, U., Selve, S., Kneipp, J.: Optical properties of silver nanocube surfaces obtained by silane immobilization. Nanospectroscopy **1**(1) (2015). https://doi.org/10.3390/Bios9020078
52. Cheng, Y., Wang, R., Zhai, H., Sun, J.: Stretchable electronic skin based on silver nanowire composite fiber electrodes for sensing pressure, proximity, and multidirectional strain. Nanoscale **9**(11), 3834–3842 (2017). https://doi.org/10.1039/C7NR00121E
53. Yun, H.J., Kim, S.J., Hwang, J.H., Shim, Y.S., Jung, S.G., Park, Y.W., Ju, B.K.: Silver nanowire-IZO-conducting polymer hybrids for flexible and transparent conductive electrodes

for organic light-emitting diodes. Sci. Rep. **6**(1), 1–12 (2016). https://doi.org/10.1038/Srep34150
54. Hu, Z.S., Hung, F.Y., Chang, S.J., Hsieh, W.K., Chen, K.J.: Align Ag nanorods via oxidation reduction growth using RF-sputtering. J. Nanomater. **2012** (2012). https://doi.org/10.1155/2012/345086
55. Lu, Y., Zhang, C.Y., Zhang, D.J., Hao, R., Hao, Y.W., Liu, Y.Q.: Fabrication of flower-like silver nanoparticles for surface-enhanced Raman scattering. Chin. Chem. Lett. **27**(5), 689-692 (2016). https://doi.org/10.1016/J.Trac.2018
56. Jia, K., Khaywah, M.Y., Li, Y., Bijeon, J.L., Adam, P.M., Déturche, R., Guelorget, B., François, M., Louarn, G., Ionescu, R.E.: Strong improvements of localized surface plasmon resonance sensitivity by using Au/Ag bimetallic nanostructures modified with polydopamine films. ACS Appl. Mater. Interfaces **6**(1), 219–227 (2014). https://doi.org/10.1021/Am403943q
57. Michieli, N., Kalinic, B., Scian, C., Cesca, T., Mattei, G.: Optimal geometric parameters of ordered arrays of nanoprisms for enhanced sensitivity in localized plasmon based sensors. Biosens. Bioelectron. **65**, 346–353 (2015). https://doi.org/10.1016/J.Bios.2014.10.064
58. Gao, C., Lu, Z., Liu, Y., Zhang, Q., Chi, M., Cheng, Q., Yin, Y.: Highly stable silver nanoplates for surface plasmon resonance biosensing. Angew. Chem. Int. Ed. **51**(23), 5629–5633 (2012). https://doi.org/10.1002/Anie.201108971
59. Yang, X., Ren, Y., Gao, Z.: Silver/gold core–shell nanoprism-based plasmonicnanoprobes for highly sensitive and selective detection of hydrogen sulfide. Chem. Eur. J. **21**(3), 988–992 (2015). https://doi.org/10.1002/Chem.201405012
60. Ma, Y., Zhou, J., Zou, W., Jia, Z., Petti, L., Mormile, P.: Localized surface plasmon resonance and surface enhanced Raman scattering responses of Au@ Ag core–shell nanorods with different thickness of Ag shell. J. Nanosci. Nanotechnol. **14**(6), 4245–4250 (2014). https://doi.org/10.1166/Jnn.2014.8202
61. Zhang, L., Chen, P., Loiseau, A., Brouri, D., Casale, S., Salmain, M., Boujday, S., Liedberg, B.: Spatially controlled reduction and growth of silver in hollow gold nanoshell particles. J. Phys. Chem. C **123**(16), 10614–10621 (2019). https://doi.org/10.1021/Acs.Jpcc.8b11864
62. Stebunov, Y.V., Yakubovsky, D.I., Fedyanin, D.Y., Arsenin, A.V., Volkov, V.S.: Superior sensitivity of copper-based plasmonic biosensors. Langmuir **34**(15), 4681–4687 (2018). https://doi.org/10.1021/Acs.Langmuir.8b00276
63. Khan, R., Ahmad, R., Rai, P., Jang, L.W., Yun, J.H., Yu, Y.T., Hahn, Y.B., Lee, I.H.: Glucose-assisted synthesis of Cu_2O shuriken-like nanostructures and their application as nonenzymatic glucose biosensors. Sens. Actuators, B Chem. **203**, 471–476 (2014). https://doi.org/10.1016/J.Snb.2014.06.128
64. Giri, S.D., Sarkar, A.: Electrochemical study of bulk and monolayer copper in alkaline solution. J. Electrochem. Soc. **163**(3), H252 (2016). https://doi.org/10.1149/2.0071605jes
65. Benguigui, M., Weitz, I.S., Timaner, M., Kan, T., Shechter, D., Perlman, O., Sivan, S., Raviv, Z., Azhari, H., Shaked, Y.: Copper oxide nanoparticles inhibit pancreatic tumor growth primarily by targeting tumor initiating cells. Sci. Rep. **9**(1), 1–10 (2019). https://doi.org/10.1038/S41598-019-48959-8
66. Hou, L., Shan, X., Hao, L., Feng, Q., Zhang, Z.: Copper sulfide nanoparticle-based localized drug delivery system as an effective cancer synergistic treatment and theranostic platform. Actabiomaterialia **54**, 307–320 (2017). https://doi.org/10.1016/Jactbio.2017.03.005
67. Li, F., Lei, C., Shen, Q., Li, L., Wang, M., Guo, M., Huang, Y., Nie, Z., Yao, S.: Analysis of copper nanoparticles toxicity based on a stress-responsive bacterial biosensor array. Nanoscale **5**(2), 653–662 (2013). https://doi.org/10.1039/C2nr32156d
68. Wang, Z., Han, P., Mao, X., Yin, Y., Cao, Y.: Sensitive detection of glutathione by using DNA-templated copper nanoparticles as electrochemical reporters. Sens. Actuators, B Chem. **238**, 325–330 (2017). https://doi.org/10.1016/J.Snb.2016.07.078
69. Peng, F., Liu, Z., Li, W., Huang, Y., Nie, Z., Yao, S.: Enzymatically generated long polyT-templated copper nanoparticles for versatile biosensing assay of DNA-related enzyme activity. Anal. Methods **7**(10), 4355–4361 (2015). https://doi.org/10.1039/C5ay00423c

70. Wang, H.B., Zhang, H.D., Chen, Y., Liu, Y.M.: A fluorescent biosensor for protein detection based on poly (thymine)-templated copper nanoparticles and terminal protection of small molecule-linked DNA. Biosens. Bioelectron. **74**, 581–586 (2015). https://doi.org/10.1016/J.Bios.2015.07.021
71. Chen, J., Ji, X., He, Z.: Smart composite reagent composed of double-stranded DNA-templated copper nanoparticle and SYBR green I for hydrogen peroxide related biosensing. Anal. Chem. **89**(7), 3988–3995 (2017). https://doi.org/10.1021/Acs.Analchem.6b04484
72. Verma, N., Kumar, N.: Synthesis and biomedical applications of copper oxide nanoparticles: an expanding horizon. ACS Biomater. Sci. Eng. **5**(3), 1170–1188 (2019). https://doi.org/10.1021/Acsbiomaterials.8b01092
73. Wang, Y., Yang, F., Zhang, H.X., Zi, X.Y., Pan, X.H., Chen, F., Luo, W.D., Li, J.X., Zhu, H.Y., Hu, Y.P.: Cuprous oxide nanoparticles inhibit the growth and metastasis of melanoma by targeting mitochondria. Cell Death Dis. **4**(8), e783–e783 (2013). https://doi.org/10.1038/Cddis.2013.314
74. Chawla, M., Sharma, V., Randhawa, J.K.: Facile one pot synthesis of CuO nanostructures and their effect on nonenzymatic glucose biosensing. Electrocatalysis **8**(1), 27–35 (2017). https://doi.org/10.1007/S12678-016-0337-7
75. Siddiqui, H., Qureshi, M.S., Haque, F.Z.: Effect of copper precursor salts: facile and sustainable synthesis of controlled shaped copper oxide nanoparticles. Optik **127**(11), 4726–4730 (2016). https://doi.org/10.1016/J.Ijleo.2016.01.118
76. Ahmad, R., Vaseem, M., Tripathy, N., Hahn, Y.B.: Wide linear-range detecting nonenzymatic glucose biosensor based on CuO nanoparticles inkjet-printed on electrodes. Anal. Chem. **85**(21), 10448–10454 (2013). https://doi.org/10.1021/Ac402925r
77. Molazemhosseini, A., Magagnin, L., Vena, P., Liu, C.C.: Single-use nonenzymatic glucose biosensor based on CuO nanoparticles ink printed on thin film gold electrode by micro-plotter technology. J. Electroanal. Chem. **789**, 50–57 (2017). https://doi.org/10.1016/J.Jelechem.2017.01.041
78. Yuan, R., Li, H., Yin, X., Zhang, L., Lu, J.: Stable controlled growth of 3D CuO/Cu nanoflowers by surfactant-free method for non-enzymatic hydrogen peroxide detection. J. Mater. Sci. Technol. **34**(9), 1692–1698 (2018). https://doi.org/10.1016/J.Jsmt.2017.11.030
79. Li, D., Meng, L., Xiao, P., Jiang, D., Dang, S., Chen, M.: Enhanced non-enzymatic electrochemical sensing of hydrogen peroxide based on Cu_2O nanocubes/Ag-Au alloy nanoparticles by incorporation of RGO nanosheets. J. Electroanal. Chem. **791**, 23–28 (2017). https://doi.org/10.1016/J.Jelechem.2017.03.010
80. Gao, P., Liu, D.: Hydrothermal preparation of nest-like CuO nanostructures for non-enzymatic amperometric detection of hydrogen peroxide. RSC Adv. **5**(31), 24625–24634 (2015). https://doi.org/10.1039/C5RA00518C
81. Chirizzi, D., Guascito, M.R., Filippo, E., Malitesta, C., Tepore, A.: A novel nonenzymatic amperometric hydrogen peroxide sensor based on CuO@ Cu_2O nanowires embedded into poly (vinyl alcohol). Talanta **147**, 124–131 (2016). https://doi.org/10.1016/j.talanta.2015.09.038
82. Wang, Y., Wu, D., Zhang, Y., Ren, X., Wang, Y., Ma, H., Wei, Q.: Layer-by-layer self-assembly of 2D graphene nanosheets, 3D copper oxide nanoflowers and 0D gold nanoparticles for ultrasensitive electrochemical detection of alpha fetoprotein. RSC Adv. **5**(70), 56583–56589 (2015). https://doi.org/10.1039/C6RA07626B
83. Wang, Y., Fan, D., Wu, D., Zhang, Y., Ma, H., Du, B., Wei, Q.: Simple synthesis of silver nanoparticles functionalized cuprous oxide nanowires nanocomposites and its application in electrochemical immunosensor. Sens. Actuators, B Chem. **236**, 241–248 (2016). https://doi.org/10.1016/J.Snb.2016.06.021
84. Chu, Y., Wang, H., Ma, H., Wu, D., Du, B., Wei, Q.: Sandwich-type electrochemical immunosensor for ultrasensitive detection of prostate-specific antigen using palladium-doped cuprous oxide nanoparticles. RSC Adv. **6**(88), 84698–84704 (2016). https://doi.org/10.1039/C6RA13841A

85. Feng, T., Qiao, X., Wang, H., Sun, Z., Qi, Y., Hong, C.: A porous CuO nanowire-based signal amplification immunosensor for the detection of carcinoembryonic antigens. RSC Adv. **6**(21), 16982–16987 (2016). https://doi.org/10.1039/C5RA26828A
86. Sun, G., Zhang, Y., Kong, Q., Zheng, X., Yu, J., Song, X.: CuO-induced signal amplification strategy for multiplexed photoelectrochemicalimmunosensing using CdS sensitized ZnO nanotubes arrays as photoactive material and AuPd alloy nanoparticles as electron sink. Biosens. Bioelectron. **66**, 565–571 (2015). https://doi.org/10.1016/J.Bios.2014.12.020
87. Yang, Y., Yan, Q., Liu, Q., Li, Y., Liu, H., Wang, P., Chen, L., Zhang, D., Li, Y., Dong, Y.: An ultrasensitive sandwich-type electrochemical immunosensor based on the signal amplification strategy of echinoidea-shaped Au@ Ag-Cu_2O nanoparticles for prostate specific antigen detection. Biosens. Bioelectron. **99**, 450–457 (2018). https://doi.org/10.1016/J.Bios.2017.08.018
88. Ghodsi, J., Rafati, A.A., Shoja, Y., Najafi, M.: Determination of dopamine in the presence of uric acid and folic acid by carbon paste electrode modified with CuO nanoparticles/hemoglobin and multi-walled carbon nanotube. J. Electrochem. Soc. **162**(4), B69 (2015)
89. Krishnamoorthy, K., Sudha, V., Kumar, S.M.S., Thangamuthu, R.: Simultaneous determination of dopamine and uric acid using copper oxide nano-rice modified electrode. J. Alloy. Compd. **748**, 338–347 (2018). https://doi.org/10.1021/Acsbiomaterials.8b01092
90. Sivasubramanian, R., Biji, P.: Preparation of copper (I) oxide nanohexagon decorated reduced graphene oxide nanocomposite and its application in electrochemical sensing of dopamine. Mater. Sci. Eng. B **210**, 10–18 (2016). https://doi.org/10.1016/J.Mseb.2016.04.018
91. Felix, S., Kollu, P., Raghupathy, B.P., Jeong, S.K., Grace, A.N.: Electrocatalytic oxidation of carbohydrates and dopamine in alkaline and neutral medium using CuO nanoplatelets. J. Electroanal. Chem. **739**,1–9 (2015). https://doi.org/10.1016/J.Jelechem.2014.12.006
92. Sharma, V., Mobin, S.M.: Cytocompatible peroxidase mimic CuO: graphene nanosphere composite as colorimetric dual sensor for hydrogen peroxide and cholesterol with its logic gate implementation. Sens. Actuators, B Chem. **240**, 338–348 (2017). https://doi.org/10.1016/J.Snb.2016.08.169
93. Singh, J., Srivastava, M., Roychoudhury, A., Lee, D.W., Lee, S.H., Malhotra, B.D.: Bienzyme-functionalized monodispersed biocompatible cuprous oxide/chitosan nanocomposite platform for biomedical application. J. Phys. Chem. B **117**(1), 141–152 (2013). https://doi.org/10.1021/Jp309639w
94. Liu, B., Ouyang, X., Ding, Y., Luo, L., Xu, D., Ning, Y.: Electrochemical preparation of nickel and copper oxides-decorated graphene composite for simultaneous determination of dopamine, acetaminophen and tryptophan. Talanta **146**, 114–121 (2016). https://doi.org/10.1016/J.Talantta.20155.08.034
95. Khun, K., Ibupoto, Z.H., Liu, X., Mansor, N.A., Turner, A.P.F., Beni, V., Willander, M.: An electrochemical dopamine sensor based on the ZnO/CuOnanohybrid structures. J. Nanosci. Nanotechnol. **14**(9), 6646–6652 (2014). https://doi.org/10.1166/Jnn.2014.9367
96. Aparna, T.K., Sivasubramanian, R., Dar, M.A.: One-pot synthesis of Au-Cu_2O/rGO nanocomposite based electrochemical sensor for selective and simultaneous detection of dopamine and uric acid. J. Alloy. Compd. **741**, 1130–1141 (2018). https://doi.org/10.1016/J.JALLCOM.2018.01.205
97. Xue, Y., Yu, G., Shan, Z., Li, Z.: Phyto-mediated synthesized multifunctional Zn/CuO NPs hybrid nanoparticles for enhanced activity for kidney cancer therapy: a complete physical and biological analysis. J. Photochem. Photobiol. B. **186**, 131–136 (2018). https://doi.org/10.1016/J.Jphotobiol.2018.07.004
98. Joshi, A., Rastedt, W., Faber, K., Schultz, A.G., Bulcke, F., Dringen, R.: Uptake and toxicity of copper oxide nanoparticles in C6 glioma cells. Neurochem. Res. **41**(11), 3004–3019 (2016). https://doi.org/10.1007/S11064-016-2020-Z

99. Kung, M.L., Hsieh, S.L., Wu, C.C., Chu, T.H., Lin, Y.C., Yeh, B.W., Hsieh, S.: Enhanced reactive oxygen species overexpression by CuO nanoparticles in poorly differentiated hepatocellular carcinoma cells. Nanoscale **7**(5), 1820–1829 (2015). https://doi.org/10.1039/C4nr05843g
100. Du, B.D., Phu, D.V., Quoc, L.A., Hien, N.Q.: Synthesis and investigation of antimicrobial activity of Cu_2O nanoparticles/zeolite. J. Nanopart. (2017). https://doi.org/10.1155/2017/7056864

Chapter 21
Structural and Optical Phenomena of Thermally Treated Fullerene-Based Nanocomposites with Metal Nanoparticles for Sensing Applications

Amena Salim, Ritu Vishnoi, Vikesh Chaudhary, Himanshu Dixit, Umesh K. Dwivedi, Pushpendra Kumar, Sunita Bishnoi, Sanjeev Aggarwal, G. D. Sharma, K. Venkataratnam Kamma, and R. Singhal

Abstract The phenomena of structural and optical modifications in metal-fullerene nanocomposite thin films (Cu-fullerene, Ag-fullerene, and Au-fullerene) by thermal annealing at different temperatures have been discussed in this chapter. With the help of thermal co-evaporation unit, the metal-fullerene nanocomposite thin films are grown on glass substrates using different noble metals with fullerene matrix. To enhance the optical and structural properties of nanocomposite especially for tuning of surface plasmon resonance (SPR) wavelength in the visible region, thermal annealing is performed at different temperatures. Both, fullerene C_{60} and C_{70} have been used as matrix to incorporate metal nanoparticles. In each case, low and high metal concentrations are chosen to investigate the SPR tuning with the size of metal

A. Salim · R. Vishnoi · V. Chaudhary · K. V. Kamma · R. Singhal (✉)
Department of Physics, Malaviya National Institute of Technology Jaipur, Malaviya Nagar, Jaipur, Rajasthan 302017, India
e-mail: rsinghal.phy@mnit.ac.in

H. Dixit
Department of Physics, School of Energy Science and Engineering, IIT Kharagpur 721302, India

U. K. Dwivedi
Department of Physics, Amity School of Applied Sciences, Amity University, Jaipur, Rajasthan, India

P. Kumar
Department of Physics, Manipal University Jaipur, Jaipur, Rajasthan 303007, India

S. Bishnoi
Department of Chemistry, Vivekananda Global University, Jaipur 303012, India

S. Aggarwal
Department of Physics, Kurukshetra University, Kurukshetra, Haryana, India

G. D. Sharma
Department of Physics, The LNM Institute of Information Technology, Jamdoli, Jaipur, India

D. Mohanta and P. Chakraborty (eds.), *Nanoscale Matter and Principles for Sensing and Labeling Applications*, Advanced Structured Materials 206,
https://doi.org/10.1007/978-981-99-7848-9_21

nanoparticles. The growth of Cu, Ag, and Au nanoparticles with increasing temperature is investigated with the help of transmission electron microscopy (TEM). Maximum growth of 32.6 nm is obtained in the case of Ag-C_{70} nanocomposite with larger metal concentration; whereas the maximum SPR shift of ~101 nm is obtained for Au nanoparticles in C_{60} matrix at a temperature of 400 °C. In all three metal nanoparticles, a blue shift followed by a redshift is obtained up to 300 °C. The shift towards lower wavelength is mainly due to the transformation of fullerene into the form of amorphous carbon. At higher temperatures, the redshift due to the growth of metal nanoparticles leads over the blue shift which was due to the variation in refractive index of the matrix, and an overall redshift is obtained. The tuning of SPR wavelength of different metal nanocomposites has been demonstrated for their application in sensing-based devices.

Keywords Annealing · Nanocomposites · Fullerene · Optical · Surface plasmon resonance

21.1 Introduction

With advanced optical, electronic, and conducting properties, metal-matrix-based nanocomposites have a vast range of applicability in memory devices, solar cells, biomedical devices, sensors, etc. [1]. For the synthesis of metal nanocomposites, metal nanoparticles (NPs) are used due to their interesting optical, structural, and electrical properties. Nanoparticles of noble metals such as Ag, Au, Cu exhibit many interesting properties due to their size effect as well as high surface-to-volume ratio. They possess the Surface Plasmon Resonance (SPR), which basically involves the collective oscillations of conduction electrons in the presence of an electromagnetic field [2, 3]. The filler matrix for the synthesis of nanocomposites should also be a functional material to enhance the applicability of nanocomposite thin films. Carbon-based matrices such as fullerenes, carbon nanotubes have been an interesting choice as filler for the synthesis of the nanocomposite.

By incorporating metal nanoparticles in the carbon-based matrix (fullerene in the present case), both fullerenes, as well as metal nanoparticles, can be simultaneously used as a bi-functional nanocomposite. For the above-said purpose, fullerene C_{60} and C_{70} have been chosen as a filler (matrix) because they possess exclusive properties such as high refractive index and high electron affinity [4]. These nanocomposites are most convenient for plasmonic integrated devices due to the transparency of fullerene C_{60} and C_{70} in the form of thin-film and their lessened reactivity between noble metals and matrix [5–7]. Among noble metals, Cu nanoparticles have been abundantly used in the field of optics and electronics due to their high conductivity and low cost. On the other hand, Ag nanoparticles that exhibit the strongest SPR band are less reactive with carbon and therefore have been widely used to synthesize. The major concern with Ag and Cu nanoparticles is their instability at ambient conditions due to their oxidation. Ag NPs have a major issue at ambient conditions as their possible oxidation also occurs in different stoichiometry [8, 9]. Due to the oxidation, oxide matrices such as SiO_2, TiO_2, etc. are sometimes not suitable to deposit metal nanoparticles. In this regard, carbon-based matrices are quite appropriate as they can prevent them from oxidation. Au nanoparticles have a minimum possibility to oxidize at ambient condition but its high-cost make is not a very convenient choice for device applications. Another interesting aspect of nanocomposite is that the new novel properties can be generated when metal nanoparticles are embedded in the host matrix, without affecting their original properties. Fullerene C_{60} and $C_{70,}$ both matrices are having very fascinating properties due to its size, its high reactivity and electron affinity, and also biocompatibility [10] and therefore, have been used in the present work to host the nanoparticles of Cu, Ag, and Au in the form of nanocomposite thin films.

The SPR frequency of noble metal nanoparticles is a crucial parameter for their role in many plasmonic-based devices. SPR of noble metal nanoparticles basically depends on the shape, size, and dielectric constant of metal nanoparticles and also on the dielectric properties of the matrix in which nanoparticles are embedded. When the nanoparticles are embedded in a fullerene matrix which itself is absorbing then this complex system can be theoretically understood by Maxwell Garnett Effective Medium (MAG) theory [11]. MAG is a very effective and extensive theory that contains the contribution of different terms related to dielectric properties of particles as well as of matrix in nanocomposites and also calculates the interaction among nanoparticles if they are not very far from each other [12, 13]. According to Clausius–Mossotti relation, simplified by dipolar approximation, the term effective dielectric function of composite medium (ε_{eff}) depends on polarizability α, which contains inclusions (considered to be spherical in shape) and are embedded in host matrix. This dependency can be expressed as:

$$\frac{\varepsilon_{\mathrm{eff}} - \varepsilon_m}{\varepsilon_{\mathrm{eff}} - 2\varepsilon_m} = \frac{f}{\overline{R}^3}\alpha(R) \qquad (21.1)$$

where ε_m = dielectric function of the matrix

$\overline{R}$= mean of NPs radius

f = volume fraction of the inclusions.

The condition for the fulfillment of dipolar approximation holds when the size of inclusion must be smaller than the wavelength and the volume fraction must also be very small [14–16].

The polarizability in the Quasistatic limit can be given as:

$$\alpha_{MG}(R) = R^3 \frac{\varepsilon_{nP} - \varepsilon_m}{\varepsilon_{n_P} + 2\varepsilon_m} \qquad (21.2)$$

ε_{n_P} = dielectric function of inclusion

Substituting Eq. (21.2) in (21.1) gives the Maxwell–Garnett formula

$$\frac{\varepsilon_{\text{eff}}-\varepsilon_m}{\varepsilon_{\text{eff}}+2\varepsilon_m} = f\frac{\varepsilon_1(\omega)-\varepsilon_m}{\varepsilon_1(\omega)+2\varepsilon_m} \tag{21.3}$$

where ε_{eff} = the dielectric function of the effective medium

ε_{np} = real part of the dielectric function of the metal particle

ε_m = dielectric function of the surrounding medium

L = the shape factor (or depolarization factor) of the particle

f = the filling factor (also known as volume concentration), which reflects the extent of the particle-particle interactions employing their dipolar fields.

The real part of the metal nanoparticle dielectric function ε_{np} expresses the optical properties of a single particle, which can be expressed as [17]

$$\varepsilon_{np} = \varepsilon_d' - \frac{\omega_p^2}{\omega^2 + \frac{1}{\tau^{*2}}} \tag{21.4}$$

τ^* is the relaxation time of electron. Here, first term ε_d' denotes the real part of the dielectric function.

This term gives the contribution of core electrons (especially the d electrons). And the second term in Eq. (21.3) is based on Drude free-electron model and gives the contribution from the conduction electrons.

Here, ω_p is the bulk plasmon frequency, which is given as,

$$\omega_p = \left(ne^2/\varepsilon_0 m_{\text{eff}}\right)^{1/2}$$

For the further simplified Eq. (21.3) in the case of free-electron metals, relaxation time can be considered negligible. The equation can then be written as

$$\varepsilon_{np} = \varepsilon_d' - \frac{\omega_p^2}{\omega^2} \tag{21.5}$$

The condition for the occurrence of surface plasmon resonance accordingly with Maxwell–Garnett theory is given by:

$$\varepsilon_{np}(\omega_s)(1-f) + \varepsilon_m(2+f) = 0, \tag{21.6}$$

ω_s denotes the surface plasmon resonance frequency.

The generation and tuning of the SPR frequency of noble metal nanoparticles can be achieved by many processes, such as (a) irradiation by low-energy ion beam, (b) ion-irradiation by high-energy beam, and (c) thermal annealing. The thermal annealing process is chosen as the simplest and cost-effective way for favorable modifications in optical as well as structural properties of synthesized metal-based nanocomposites. By varying metal atomic fraction, the capability of tuning of SPR band can be enhanced effectively by the incorporation of nanoparticles in a matrix having high refractive index, such as fullerenes [18, 19]. Thermal annealing is a process in which a thin film is subjected to high temperatures in a vacuum or inert atmosphere that helps change the properties of the material. It includes the stages namely the recovery recrystallization and grain growth. After annealing, there will be a relief in internal stress, and the consistency of structure will be improved. The stability in fullerenes (C_{60} and C_{70}) with embedded nanoparticles has been studied in literature by various methods, i.e. ion irradiation, thermal annealing, etc.

This chapter discusses the growth of nanoparticles of noble metals, i.e. Cu, Ag, and Au with annealing temperature and also the structural and optical changes occur in metal nanocomposite thin films with various combinations like Cu-C_{60}, Cu-C_{70}, Ag-C_{60}, Ag-C_{70}, Au-C_{60}, and Au-C_{70} occurring due to thermal annealing. For the study of optical properties for plasmonic-based devices, UV–visible absorption spectroscopy has been carried out for all combinations in the spectral region ranging from 400 to 800 nm. In general, the SPR bands for noble metal nanoparticles are found to be at a wavelength of ~400 nm for silver nanoparticles; ~520 nm for gold nanoparticles [20], and ~650 nm for copper nanoparticles [21]. By thermal annealing, both the blue and redshifts in the SPR wavelength have been obtained for these nanocomposite thin films.

From the literature survey, it is found that different metal-matrix combinations of nanocomposite thin films with optical properties with proper tunability could be useful in plasmonic and sensing applications. Many reports are reviewed especially in SPR sensing based on nanocomposite materials, for example, N.A.S. Omar et al. have synthesized the carbon-based nanocomposites using the carbon dots-cellulose nanocrystals (Au/CDs-CNC). In this work, the Kretchmann model was used for the principle of SPR sensors including the SPR experimental setup. He examined the Au/CDs-CNC thin film-based SPR sensor, the efficiency and sensitivity of these sensors were observed for the qualitative detection of Mn^{+2} ions in an aqueous solution. The SPR responses were recorded by allowing the contact of thin films with different concentrations of Mn^{+2} ions solution. The increase in the shift of SPR wavelength reveals the exposure of the sensing layer to Mn^{+2} ion solution, with this the change in refractive index will induce due to the chemical absorption. Hence, the observations due to SPR response with variation in concentration of Mn^{+2} ions showed the proposed sensor has positive response toward low concentration with improved sensitivity [22].

Bharadwaj et al. have synthesized Ag-SnO_2 nanocomposite thin films by using thermal evaporation. The process of annealing was used for the improvement and modifications observed in structural, surface morphological and optical plasmonic properties of nanocomposite thin films with enhancement in CO gas sensing properties. CO gas sensing measurements were done on the fabricated sensors using as deposited and annealed nanocomposite thin films. Before gas sensing, two silver electrodes with dimensions of 10 mm × 0.5 mm and 2 mm on nanocomposite thin films were fabricated and to attain stable resistance these fabricated sensors were kept at temperature 100 °C for 30 min. After that the gas measurements were performed on these sensors [23].

Mishra et al. have considered the fabrication of fiber optics SPR-based probe. Ag and CNT/Cu-nanoparticles were coated over unclad core of fiber for the fabrication of these probes, which are used for the sensing of nitrate. Based on the reduction of nitrate on interaction with Cu nanoparticles and the formation of ammonium ions, sensing was performed. The changes in dielectric properties of CNT/Cu-nanoparticles were found due to adsorption of ammonium ions on CNT, which was measured by shift in SPR wavelength [24].

Saeideh Mohammadzadeh-Asl et al. proved the SPR sensitivity enhancement of gold nanoparticles in his work. For the nanospheres of gold, the LSPR shift was found to be ~522 nm. With the effect of shape and size of Au NPs including nanorods, nanospheres, and nanooctahedrons, maximum SPR shift was found in case of nanorods which was due to the increase in particle size. With this application, a new class of materials were generated by combination of graphene oxide and reduced graphene oxide sheets with gold nanostructures with improved properties and found to be feasible for catalytic biosensing applications [25].

Ozhikandathil et al. have synthesized the gold-Multiwall CNT nanocomposites and used for the work of bio-sensing of recombinant bovine growth hormones (rbST). The demonstration of biosensing applications of gold-MWCNT nanocomposites was done on the basis of detection of LSPR properties for bovine growth hormone. The sensitivity of the nanocomposite is found to superior, which is 1 ng/ml than gold nanoparticles, which was 25 ng/ml [26]. In other literature, Mishra et al. have theoretically proposed fiber optic biosensor based on the property surface plasmonic resonance having configuration of metal/graphene/MoS_2, which was found to be highly sensitive. In this work, a comparable study of sensors framed with different metals like Au, Cu, Al was performed theoretically. The sensor with configuration of Au/graphene/MoS_2 is highly sensitive, which is found to be 5.0 μm/RIU. These results give the verification of using nanocomposites containing gold can be used for sensing purposes [27].

21.2 Synthesis of Metal-Matrix-Based Nanocomposites

Copper-fullerene nanocomposite thin films with different metal concentrations, such as Cu (4.5%)-C_{70} and Cu (8%)-C_{70} are synthesized on glass substrate in a high vacuum, i.e. 10^{-6} mbar using thermal deposition method. Nanocomposite thin films of Ag (4.6%)-C_{60} are synthesized again by thermal co-deposition method on a transparent substrate such as quartz. The Au-fullerene nanocomposites with different Au concentration like Au (2%)-C_{60}, Au (3%)-C_{70} are synthesized by the same technique.

21.2.1 Annealing Studies in Cu-Fullerene Nanocomposite

In this section, the synthesis of Cu-Fullerene nanocomposites and their modifications that occur after thermal annealing have been discussed. The as-deposited Cu-fullerene nanocomposite thin films are annealed using a high tube furnace at different temperatures ranging from 100 to 400 °C in an inert atmosphere. UV–visible spectroscopy has been performed on pristine as well as annealed Cu-fullerene nanocomposite thin films to study their structural and optical modifications. The obtained absorption spectra are shown in Fig. 21.1a and b, where no SPR is seen in the as-deposited samples in both the cases and in the samples annealed up to 200 °C in case of Cu (4.5%)-C_{70} and up to 250 °C in case of Cu (8%)-C_{70}. The absence of SPR in as-deposited films indicates that Cu metal nanoparticles have not formed in matrices properly to excite the plasmons. Under thermal annealing, the SPR peak is expected to occur in both cases at respective wavelengths. A SPR peak is observed at a position of wavelength at 585 nm in the case of Cu (4.5%)-C_{70} (Fig. 21.1a) and at 602 nm in the case of Cu (8%)-C_{70} at an annealed temperature which is about 300 °C (Fig. 21.1b). With a further increment in temperature, the SPR peak is shifted towards a higher wavelength. A redshift of 36 nm is obtained in the case of Cu (4.5%)-C_{70} [28] and ~18 nm in the case of Cu (8%)-C_{70} [29], which is due to the growth of Cu nanoparticles in carbon matrix at higher annealing temperatures. In order to confirm the growth of Cu nanoparticles, the bright-field images of Cu nanoparticles are captured by Transmission Electron Microscopy (TEM). Bright-field images using TEM are given in Fig. 21.1c and d, and the average particle size of Cu in as-deposited films of Cu (4.5%)-C_{70} and of Cu (8%)-C_{70} is found to be 7.2 nm and 4.3 nm, respectively, and are shown in Fig. 21.1c(ii) and d(iii) with the help of Gaussian size distribution curve. With the increase in temperature, a large growth of Cu nanoparticles is seen and the average particle size is found to be 35.68 nm (Fig. 21.1c-iv) in case of Cu (4.5%)-C_{70} sample at 400 °C and 30 nm in the case of Cu (8%)-C_{70} sample at 300 °C (Fig. 21.1d-vi), and duality in particle size distribution is found at 350 °C (Fig. 21.1d-ix).

With an exploration of these results, another investigation is done by replacing the matrix of Cu nanoparticles from fullerene C_{70} to C_{60} and by choosing a higher concentration of Cu metal. By varying temperature ranges from 100 to 400 °C, the thermal annealing effect has been studied in as-deposited Cu (27%)-C_{60} [30].

For the study of Cu-fullerene nanocomposite thin films with high concentration of Cu, Cu (27%)-C_{60} nanocomposites were synthesized using thermal co-deposition technique. From the absorption spectra, no signature of SPR peak is found in an as-deposited sample like the results of Cu-C_{70} nanocomposites with low concentration. At 300 °C, a broad SPR band is found at 300 °C at a wavelength of 694 nm. With further increase in temperature, this band is shifted towards lower wavelength at 400 °C and found to be at 684 nm. The TEM images are recorded for as-deposited, 300 and 400 °C annealed samples and the distribution of Cu nanoparticles is obtained by Image J software. In the as-deposited film, Cu nanoparticles are grown with an average size of 10.6 nm, which increases up to 12.2 nm at a temperature of 300 °C.

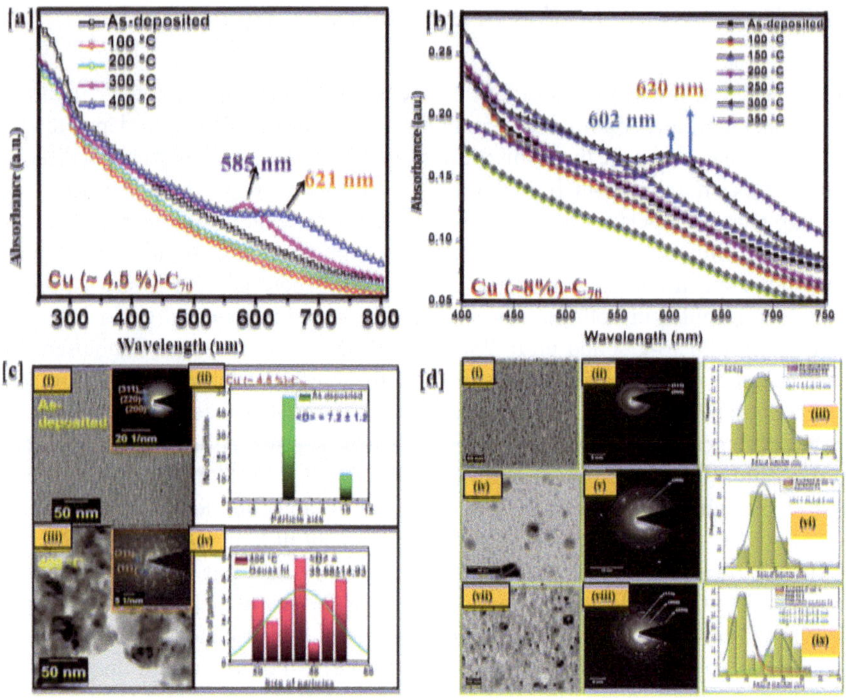

Fig. 21.1 Absorption spectra of pristine and thermally annealed nanocomposite thin films of Cu(4.5%)-C_{70} **a** and Cu(8%)-C_{70} **b** with a low concentration of Cu. Bright-field images of pristine and annealed nanocomposite thin films of Cu(4.5%)-C_{70} with a low concentration of Cu **c**-i and iii. Selected Area Diffraction pattern (SAED) for pristine and annealed thin films is shown inset of each bright field image **c**-i and iii. Particle size distribution of pristine and annealed samples is represented **c**-ii and iv. Bright field images of pristine and annealed nanocomposite thin films of Cu (8%)-C_{70} with low concentration of Cu **d**-i, iv and vii. Selected Area Diffraction Pattern (SAED) for pristine and annealed films shown in **d**-ii, v and viii. Particle size distribution of pristine and annealed samples is represented **d**-iii, vi and ix. **a** and **c** is reprinted from [28], Copyright (2019), with permission from Elsevier. **b** and **d** is reprinted from [29], Copyright (2019), with permission from Elsevier

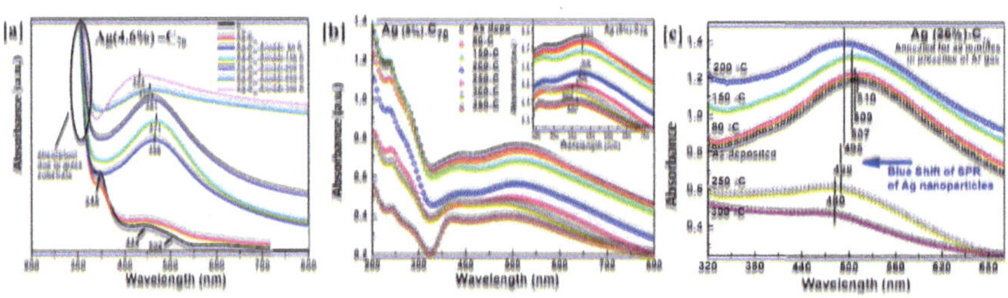

Fig. 21.2 UV–visible absorption spectra of pristine and annealed thin films of nanocomposites of Ag(4.6%)-C_{70} and Ag (5%)-C_{70} with low concentrations of Ag **a**, **b** and Ag(26%)-C_{70} with high concentration of Ag **c**. **a** is reprinted from [31], with the permission of AIP Publishing. **b** is reprinted by permission from [32], [COPYRIGHT] (2017). **c** is reprinted from [33], Copyright (2017), with permission from Elsevier

Much bigger particles are obtained at 400 °C with an average particle size of 27.9 nm and are the main cause of redshift in SPR wavelength, as obtained by UV–visible spectroscopy [30]. The structural changes that occurred in host matrices of fullerene C_{60} and C_{70} with the effect of thermal annealing are revealed by Raman spectroscopy. The intensity of Raman modes in the fullerene matrix decreases with increasing temperature and the formation of D- and G- bands at higher temperatures confirms the complete transformation from fullerene to amorphous carbon in both cases (Cu with C_{60} and C_{70}) [29, 30].

21.2.2 Annealing Studies in Ag-Fullerene Nanocomposite

Among noble metals, Ag nanoparticles give the strongest SPR band in the visible region and Ag nanoparticles have been a topic of scientific studies due to their excellent antibacterial properties. The as-deposited samples of Ag-fullerene are annealed at distinct temperatures ranging from 80 to 300 °C for 30 min at each temperature in an inert atmosphere. No evidence of SPR band is found in pristine thin films which may be due to the very small concentration of the Ag metal in the film. When the pristine sample is annealed at 80 °C for 30 min, the SPR peak is clearly visible at 466 nm. With a further increase in temperature, the SPR peak is shifted to 471 nm at 150 °C. At a temperature of 200 °C, the SPR peak shows a reverse blue shift and is found at 461 nm. Now at 250 °C and 300 °C, the SPR peak is further blue-shifted at 455 and 434 nm, respectively (Fig. 21.2a). In order to understand this behavior (first redshift and then blue shift with temperature), transmission electron microscopy is performed on as-deposited and sample annealed at 300 °C. In bright-field TEM images of an as-deposited sample, a uniform distribution of Ag nanoparticles can

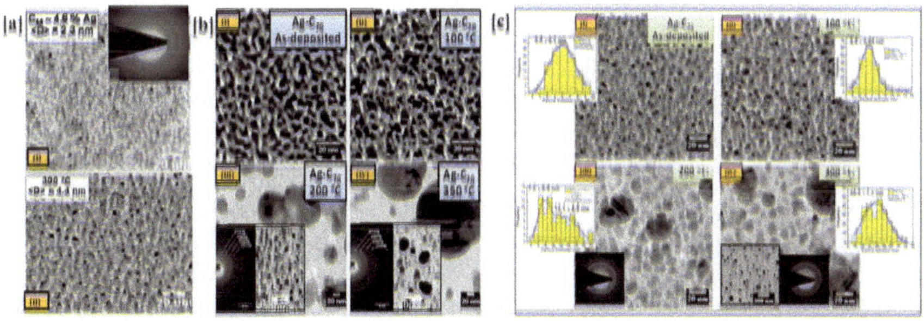

Fig. 21.3 Bright-field images of pristine and annealed nanocomposite thin films of Ag (4.6%)-C_{60} and Ag(5%)-C_{70} with low concentration of Ag **a**-i, ii and **b** i–iv and also Ag(26%)-C_{70} with high concentration of Ag **c** i–iv. The SAED pattern for Ag(4.6%)-C_{60}, Ag(5%)-C_{70} are shown in the inset of **a**-i and **b**-iii, iv. The SAED pattern and particle size distribution of each figure are shown in the inset **b** iii and iv. **a** is reprinted from [31], with the permission of AIP Publishing. **b** is reprinted by permission from [32], [COPYRIGHT] (2017)]. **c** is reprinted from [33], Copyright (2017), with permission from Elsevier

be seen with an average particle size of 2.3 nm (Fig. 21.3 (a-i)). Films annealed at 300 °C, a growth of Ag nanoparticles is observed and average particle size <D> is found to be 4.4 nm ((Fig. 21.3(a-ii)) [31]. The initial redshift in the SPR wavelength is due to the growth of Ag nanoparticles but at a higher temperature, the conversion of fullerene into amorphous carbon gives a SPR shift towards lower wavelength.

In order to study the thermally induced tuning of optical properties in various applications, the fullerene C_{70} matrix is also chosen to host Ag nanoparticles to form an excellent nanocomposite. Fullerene C_{70} due to its high refractive index (~2.3 at Na D line) compared to fullerene C_{60} (~2.1) [4] can effectively enhance the capability of tuning in SPR wavelength. In this case, Ag(5%)-C_{70} nanocomposite is synthesized using thermal co-deposition technique and annealed at different temperature from 80 to 350 °C. A clear but broad SPR peak is observed at 513 nm in an as-deposited sample, as shown in Fig. 21.2b. There is no significant shift in SPR wavelength with an increase in temperature up to 150 °C. From 200 to 300 °C, a blue shift of 33 nm is obtained and the SPR peak position is at 480 nm for the sample annealed at 300 °C. With a further slight increase in annealing temperature (at 350 °C), the SPR peak is shifted towards higher wavelength (redshift of ~16 nm) and the peak position is 496 nm.

In the case of Ag-C_{70} nanocomposite, first, a blue shift and then a redshift is obtained which is contrary to the trend of SPR shift in Ag-C_{60} nanocomposite. To understand this reverse trend, TEM is done on the as-deposited sample and also on the samples, which are annealed at 100, 200, and 350 °C, as shown in Fig. 21.3b. The average particle size in the as-deposited and the sample annealed at 100 °C is found to be 4.5 ± 0.05 and 4.8 ± 0.08 nm, respectively. When the sample is annealed at 200 °C, a bi-modal distribution of Ag nanoparticles is observed with average sizes of 36.4 ± 1.1 and 114.1 ± 2.0 nm. With further increment in temperature (at 350 °C), a tri-modal distribution is observed with average sizes of 27.8 ± 0.6, 155.5 ± 2.3 and 330 ± 4.5 nm. Here, in this case, the redshift is not very large despite the huge growth of Ag nanoparticles, which is mainly due to the bi-modal and tri-modal distribution of the particles, which is affecting the particle–particle interaction. Also at a higher temperature, the particles are deviating from the spherical shape [32].

To further improve the optical properties, the Ag (26%)-C_{70} nanocomposite with enhanced metal concentration in the fullerene C_{70} matrix is synthesized using the co-deposition method. Here also from Fig. 21.2c, there is no significant shift in SPR position up to 150 °C and SPR is more or less around 510 nm. With an increase in temperature, a continuous blue shift of ~30 nm is obtained at 300 °C. The TEM investigations, as shown in Fig. 21.3c (i–iv), exhibit the growth of Ag nanoparticles from 5.9 ± 0.1 to 38.5 ± 1.7 nm at a temperature of 300 °C. Here also, a bi-modal distribution is found at 200 °C with average sizes of 7.5 ± 0.9 and 13.1 ± 2.8 nm (Fig. 21.3 (c-iii)). The continuous blue shift is due to the conversion of fullerene C_{70} into amorphous carbon with the increasing temperature. Ag nanoparticles are finding themselves in a host matrix of much lower refractive index, which reduces the surface polarization and thus a blue shift in SPR wavelength is obtained, which is confirmed by Raman spectroscopy [33].

21.2.3 Annealing Studies in Au-Fullerene Nanocomposites

Noble nanoparticles like Cu, Ag, Au are the most interesting nanoparticulate components with carbon matrix such as fullerene C_{60} and C_{70}. When these nanoparticles are embedded in the carbon matrix to form nanocomposite, they show unique optical properties like SPR band, especially in the range of visible region. One of the advantages of Ag nanoparticles is that it gives the strongest SPR band but the instability of these nanoparticles due to their oxidation is a major problem. Au nanoparticles show the large tuning of SPR with thermal treatment due to the minimum possibility of oxidation at ambient conditions. For annealing studies of Au-fullerene-based nanocomposites, Au nanoparticles are chosen with both C_{60} and C_{70}, separately. Au (2%)-C_{60} and Au (3%)-C_{70} are synthesized by thermal co-deposition and the thermal annealing effect on these nanocomposite-thin films is investigated. In the case of Au (2%)-C_{60} sample, the temperature variation for annealing is in the region from 150 to 300 °C, and in the second case, Au (3%)-C_{60} annealing temperature is varied from 150 to 350 °C. In both cases, no signature of SPR is found in as-deposited nanocomposite thin films, as shown in Fig. 21.4a.

The average particle size of Au nanoparticles calculated with bright-field TEM images is found to be ~1.5 nm in the case of Au-C_{60} (Fig. 21.5 (a-i)) and ~1.8 nm in the case of Au-C_{70} (Fig. 21.5 (b-ii)). With thermal annealing of these samples, no significant change in absorption peak is observed up to a temperature of 200 °C. In the case of the Au-C_{60} sample, a further increase in annealing temperature results in a broad SPR peak at a wavelength of 581 nm at 250 °C, which indicates the formation of small-sized Au nanoparticles at this temperature. At a temperature of 300 °C, a well-defined narrow peak of Au nanoparticles is observed at 554 nm showing a total blue shift of ~27 nm up to 300 °C (Fig. 21.4a). The size of Au nanoparticles at this temperature is found to be 3.2 ± 0.06 nm (Fig. 21.5 (a-iii)) [34]. In the case of the Au (3%)-C_{70} sample, a broad SPR peak at 565 nm is seen at 300 °C (Fig. 21.4 (b-v)). With further increase in annealing temperature, this peak becomes sharp at 350 °C and peak position is reported at 549 nm with a total blueshift of ~16 nm (Fig. 21.4 (b-vi)). The growth of Au nanoparticles is observed in this case, also with the average size of 3.8 ± 0.05 nm at 350 °C, as shown in Fig. 21.5 (b-iv) [35].

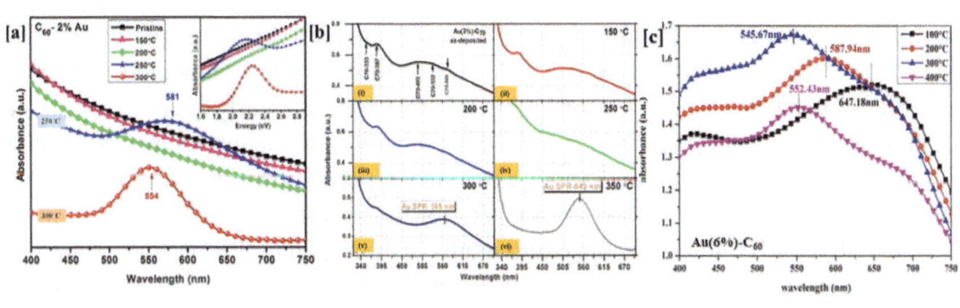

Fig. 21.4 UV–visible absorption spectra of pristine and annealed Au (2%)-C_{60}, Au (3%)-C_{70}, and Au (6%)-C_{60} nanocomposite thin films **a**, **b** and **c**. **a** is reprinted from [34], Copyright (2017), with permission from Elsevier. **b** is reprinted from [35], Copyright (2017), with permission from Elsevier

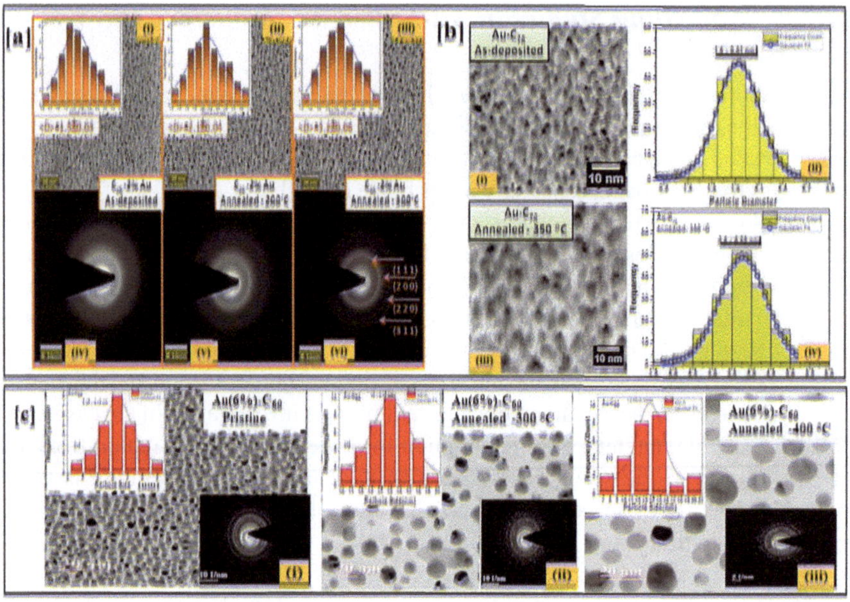

Fig. 21.5 Bright-field TEM images of pristine and annealed thin films of Au (2%)-C_{60} and Au (3%)-C_{70} and Au (6%)-C_{60} shown in **a, b** and **c**. Particle distribution for Au (2%)-C_{60} nanocomposites are shown in the inset **a** i–iii and for Au (3%)-C_{70} nanocomposites **b**-ii and iv. The SAED pattern for Au (2%)-C_{60} is shown in **a** iv–vi. The particle size distribution and SAED pattern for Au (6%)-C_{60} are shown in the inset of **c** i–iii. **a** is reprinted from [34], Copyright (2017), with permission from Elsevier. **b** is reprinted from [35], Copyright (2017), with permission from Elsevier

In both cases, the concentration of Au is very small (2–3 at. %) and the SPR position is shifted towards a lower wavelength with a decrease in FWHM of the peak. The variation in average particle size is also observed with respect to matrix and concentration. For further improvement in structural and optical properties of Au-fullerene nanocomposites and to obtain the desired results, such as improvement in SPR tuning with the growth of Au nanoparticles in fullerene matrix under heat treatment, a higher (6 at. %) metal concentration of gold is chosen. Au-C_{60} nanocomposite thin films are again synthesized by co-deposition and annealed at 100–400 °C. In this case also, owing to quite a small size of nanoparticles, no signature of SPR peak is found in the as-deposited sample. When this sample is annealed at 100 °C, an SPR peak is generated at a wavelength of ~647 nm, as clearly seen in Fig. 21.4c. Both red and blue shifts are observed in this case with increasing temperature. With the increase in temperature at 200 °C, a resonance peak is observed at ~588 nm and 300 °C, the peak is shifted to ~546 nm. Therefore, a large blueshift (towards lower wavelength) of ~101 nm is obtained and finally at 400 °C, the SPR peak is obtained at ~552 nm with a very small redshift. This constant shift in SPR from higher to lower wavelength and again at the high side at 400 °C may also be dependent on the size of Au nanoparticles which is studied in detail in TEM analysis. The particle size in an as-deposited thin film is found to be 6.3 ± 0.2 nm (Fig. 21.5 (c-i)). It is clear from the TEM image that the Au particles are not completely spherical in shape and that's why a broad SPR peak is observed in UV–visible data. The selected area electron diffraction (SAED) pattern for as-deposited Au-C_{60} thin film and annealed samples are also recorded and shown in inset of Fig. 21.5c (i-iii). The rings are observed only due to Au which confirms the poor crystallinity of fullerene in the nanocomposite samples. The rings are seen to get sharper with the rise of temperature, which reveals the increase in size of Au nanoparticles with temperature.

When the temperature of annealing is increased up to 300 °C, the particle size is noted to be 15.0 ± 0.1 nm (Fig. 21.5 (c-ii)). This increment in the size of particles, which is mainly due to the diffusion of small Au nanoparticles into large particles at a higher temperature, leads to a phenomenon known as Ostwald ripening. It is observed that there is a minor decrease in particle size at 400 °C and Au particles are found to be of 13.7 ± 0.3 nm diameter (Fig. 21.5 (c-iii)). From UV–visible absorption spectra, there is a blue shift up to 300 °C and then a very small red-shift in SPR peak at 400 °C.

With increasing temperature, the observed blueshift is mainly due to the disturbance occurring in regularly arranged structure of fullerene C_{60}, which gets converted into amorphous carbon at a particular temperature, as evident from the Raman spectroscopy of as-deposited and annealed samples (Fig. 21.6a). The refractive index of amorphous carbon is much less than that of fullerene C_{60} and therefore, as per the equations of MAG theory, a blueshift is observed in SPR wavelength. The minor redshift is due to the increment in the size of Au nanoparticles due to annealing and at 400 °C, the redshift due to increase in particle size dominates over the blueshift due to the conversion of fullerene C_{60} structure into amorphous carbon. In this particular case, atomic force microscopy (AFM) is also performed to analyze the surface topology and average grain size of pristine and annealed Au-C_{60} thin film. Three-dimensional images of as-deposited and annealed thin films are shown in Fig. 21.6b (i–v). The variation in the roughness of as deposited and annealed thin film at different temperatures is shown in Table 21.1. The roughness increases up to a temperature of 200 °C and then it decreases. The roughness is 3.4 nm at 400 °C.

To understand the chemical state of Au and C, X-ray photoelectron spectroscopy (XPS) is also performed. Figure 21.6 (c-i) shows the survey scan of pristine and annealed thin films at 200 °C and 400 °C, respectively, where the peaks corresponding to their binding energies are demonstrated. This measurement also justifies the presence of C, O, and Au elements on the surface of the nanocomposite thin film. To study the change in the chemical state, detailed scans of Au, C, and O are carried out. Figure 21.6 (c-ii) shows the core-level spectrum of C1s. For the as-deposited and annealed thin films, peaks at 285.47 eV, 284.59 eV, and 284.26 eV are found and are assigned to C–C, C = O, and O–C = O bonding, respectively. Figure 21.6 (c-iii) shows the core-level spectrum of O1s. For the as-deposited thin film, the peak is found at 532.53 eV, which is assigned to as a C–O bond. A small shift in the peaks is observed with increasing annealing temperature, which shows that a small quantity of oxide was present in the as-deposited thin film due to impurity present on the top surface of the sample due to contamination, which was removed after annealing the thin film at temperature 200 °C and 400 °C. Figure 21.6 (c-iv) shows

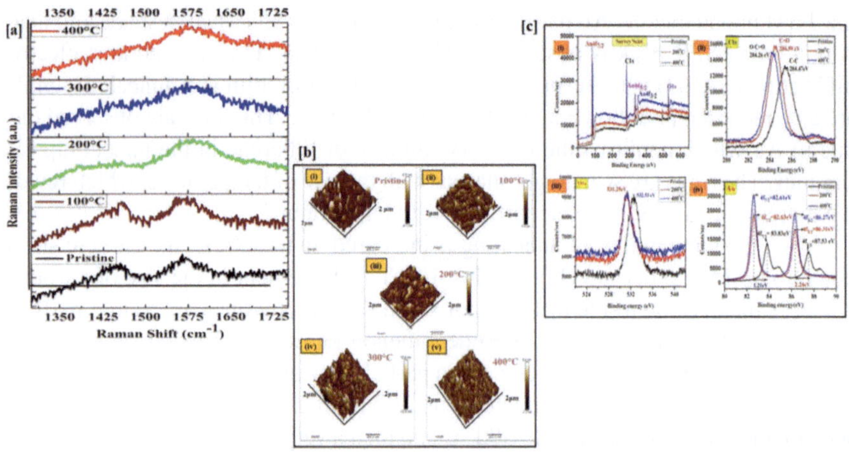

Fig. 21.6 Raman spectra, AFM images and XPS spectra of annealed as well as as-deposited Au-C_{60} nanocomposite thin films, as shown in **a**, **b**, and **c**, respectively

Table 21.1 Variation in roughness with annealing temperature

Temperature	Roughness (nm)
As-deposited	1.6
100 °C	2.2
200 °C	8.4
300 °C	5.6
400 °C	3.4

a core-level spectrum of Au 4f. In as-deposited film, the binding energies observed in $4f_{7/2}$ and $4f_{5/2}$ are at 83.33 eV and 87.53 eV, respectively. The binding energy shift with increasing temperature is found in the Au 4f region, which is attributed to different shapes and sizes of the Au nanoparticles.

The summary of studies on thermal annealing effect on different metal-fullerene nanocomposites with varying metal concentration of noble metals, i.e. Cu, Ag, and Au in carbon matrix is given in Table 21.2.

21.3 Conclusion and Outlook

Metal-matrix nanocomposite thin films, i.e. Cu-fullerene, Ag-fullerene, and Au-Fullerene are successfully synthesized by thermal co-deposition technique. The structural and optical changes occur in different combinations of nanocomposite thin films by the effect of thermal annealing are studied. The main motivation of our

Table 21.2 Comparable studies of thermally treated metal-fullerene nanocomposite thin films

Serial number	Metal matrix nanocomposites	Thermal annealing	SPR position	Average particle size	SPR tuning (nm)	References
(1)	Cu(4.5%)-C_{70} Thickness 50 nm	100–400 °C	No SPR (Pristine) 585 nm (300 °C) 621 nm (400 °C)	7.2 nm (Pristine) 35.68 nm (400 °C)	~36	[22]
(2)	Cu(~8%)-C_{70} Thickness 45 nm	100–350 °C	No SPR (Pristine) 602 nm (300 °C) 620 nm (350 °C)	4.3 nm (Pristine) 30 nm (350 °C)	~18	[23]
(3)	Cu(~27%)-C_{60} Thickness 70 nm	100–400 °C	No SPR (Pristine) 694 nm (300 °C) 684 nm (400 °C)	10.6 nm (Pristine) 27.9 nm (400 °C)	~10	[24]
(4)	Ag (~4.6%)-C_{60} Thickness 45–50 nm	80–300 °C	No SPR (Pristine) 466 nm (80 °C) 471 nm (150 °C) 431 nm (300 °C)	2.3 nm (Pristine) 4.4 nm (300 °C)	~35	[25]
(5)	Ag (~5%)-C_{70} Thickness 50 nm	80–350 °C	513 nm (pristine) 480 nm (300 °C) 496 nm (350 °C)	4.5 nm (Pristine) from 27.8 nm to 330 nm (350 °C)	~16	[27]
(6)	Ag (~26%)-C_{70} Thickness 45 nm	80–300 °C	510 nm (Pristine) 480 nm (300 °C)	5.9 nm (Pristine) 38.5 nm (300 °C)	~20	[28]
(7)	Au (~2%)-C_{60} Thickness 40 nm	150–300 °C	No SPR (Pristine) 581 nm (250 °C) 554 nm (300 °C)	1.5 nm (Pristine) 2.1 nm (200 °C) 3.2 nm (300 °C)	~27	[29]

(continued)

Table 21.2 (continued)

Serial number	Metal matrix nanocomposites	Thermal annealing	SPR position	Average particle size	SPR tuning (nm)	References
(8)	Au (~3%)-C_{70} Thickness 40 nm	150–350 °C	No SPR (Pristine) 565 nm (300 °C) 549 nm (350 °C)	1.8 nm (Pristine) 3.8 nm (350 °C)	~16	[30]
(9)	Au-C_{60}	100–400 °C	No SPR (Pristine) 647 nm (100 °C) 588 nm (200 °C) 546 nm (300 °C) 552 nm (400 °C)	6.34 nm (Pristine) 15 nm (300 °C) 13.69 nm (400 °C)	~101	This work

study is to generate and tuning in optical and structural properties with the effect of a cost-effective thermal annealing technique.

In the case of Cu-fullerene nanocomposite with a lower concentration of Cu (4.5 at.%) embedded in C_{70} matrix, no signature of SPR is found in as-deposited thin films but after annealed at 300 °C, the SPR position is obtained at 585 nm with increasing Cu nanoparticle size from 7.2 nm to 35.6 nm. For Cu (8 at.%) with C_{70}, a growth of Cu nanoparticles from 4.3 to 30.0 nm is obtained. Further increasing annealing temperature peak position is red-shifted with 621 nm in Cu (4.5%)-C_{70} and 620 nm in case of Cu (8%)-C_{70}. In replacing carbon matrix from C_{70} to C_{60} and increasing copper concentration up to 27%, the peak position is obtained at 694 nm at 300 °C annealed temperature and it is blue-shifted at 684 nm at 400 °C due to change in refractive index.

In the case of Ag (5%)-$C_{70,}$ the SPR position is obtained at 513 nm in the as-deposited sample, and with temperature increases up to 300 °C, the SPR peak is shifted towards a lower wavelength up to 480 nm with particle size variation from 27.8 to 330 nm. By replacing the matrix from C_{70} to C_{60}, an overall blue shift of 431 nm is obtained. With higher Ag metal concentration, i.e. 27 at. % in C_{70} matrix, the signature of SPR is found at 510 nm in pristine sample and after annealing at a temperature of 300 °C, a blue shift with a peak position at 480 nm is obtained with this variation in particle size from 5.9 to 35.6 nm.

In the case of Au-C_{60} and Au-C_{70} samples with lower metal concentrations, no signature of SPR is found in as-deposited samples. After annealing at 300 °C, the SPR is generated at a wavelength of 565 nm with a minor increase in particle size from 1.8 nm to 3.8 nm for the Au-C_{60} sample. With a higher concentration of Au, peak position at different temperatures was obtained as 647 nm at 100 °C, 588 nm at 200 °C, 546 nm at 300 °C, and at 400 °C, the SPR peak is red-shifted at 552 nm. The maximum shift of ~101 nm was obtained in the case of Au-C_{60} in comparison with other cases.

With the results of these different cases, an enhancement in the wavelength tuning is observed in visible region due to the plasmonic effect of nanoparticles, which makes these nanocomposites suitable and feasible for LSPR-based sensing in different devices.

It is concluded that this work is based on better tunability in SPR wavelength and proved to be quite feasible for SPR sensing devices. In the present investigation, we have studied the structural and optical modifications based on different metal nanoparticles embedded in fullerene C_{70} and C_{60} matrices having high refractive indices with different metal concentration using thermal co-evaporation method. We have found a large SPR shift of ~ 101 nm and therefore these nanocomposites have better tunability in SPR and improved optical and structural properties. Gold as noble nanoparticles have minimum possibility to oxidize and also Au nanoparticles give the reasonably strong SPR bands and that's why we have chosen this noble metal which results in the improvement in properties required for device applications.

References

1. Okpala, C.C.: The benefits and applications of nanocomposites. Int. J. Adv. Eng. Tech **4**, 12–18 (2014)
2. Yaremchuk, I., Meškinis, Š., Fitio, V., Bobitski, Y., Šlapikas, K., Čiegis, A., et al.: Spectroellipsometric characterization and modeling of plasmonic diamond-like carbon nanocomposite films with embedded Ag nanoparticles. Nanoscale Res. Lett. **10**(1), 1–7 (2015)
3. Yaremchuk, I., Tamulevičienė, A., Tamulevicius, T., Slapikas, K., Balevicius, Z., Tamulevicius, S.: Modeling of the plasmonic properties of DLC-Ag nanocomposite films. physica status solidi(a) **211**(2), 329–335 (2014)
4. Kataura, H., Endo, Y., Achiba, Y., Kikuchi, K., Hanyu, T., Yamaguchi, S.: Dielectric constants of C_{60} and C_{70} thin films. J. Phys. Chem. Solids **58**(11), 1913–1917 (1997)
5. Jensen, A.W., Wilson, S.R., Schuster, D.I.: Biological applications of fullerenes. Bioorg. Med. Chem. **4**(6), 767–779 (1996)
6. Al-Mohamad, A., Allaf, A.W.: Fullerene-60 thin films for electronic applications. Synth. Met. **104**(1), 39–44 (1999)
7. Chen, K.M., Wu, K., Chen, Y., Jia, Y.Q., Jin, S.X., Li, C.Y., et al.: Heterojunctions of solid C70 and crystalline silicon: rectifying properties and barrier heights. App. Phys. Lett. **67**(12), 1683–1685 (1995)
8. Hillenkamp, M., Domenicantonio, G.D., Eugster, O.: Nanotechnology **18**, 015702, 4 (2007)
9. Mcmahon, M.D., Lopez, R., Meyer, V., Feldman, L.C., Haglund, R.F.J.R.: Appl. Phys. B **80**, 915–921 (2005)
10. Scuseria, G.E.: The equilibrium structure of C_{70}. An ab initio Hartree-Fock study. Chem. Phys. Lett. **180**(5), 451–456 (1991)
11. Garnett, J.M.: VII. Colors in metal glasses, in metallic films, and in metallic solutions. II. Phil. Trans. R. Soc. Lond. A **205**(387–401), 237–288 (1906)
12. Cohen, R.W., Cody, G.D., Coutts, M.D., Abeles, B.: Phys. Rev. B **8**, 3689 (1973)
13. Chylek, P., Srivastava, V.: Phys. Rev. B **27**, 5098 (1983)

14. Battie, Y., Resano-Garcia, A., Chaoui, N., Zhang, Y., En Naciri, A.: Extended Maxwell-Garnett-Mie formulation applied to size dispersion of metallic nanoparticles embedded in host liquid matrix. J. Chem. Phys. **140**(4), 044705 (2014)
15. Doyle, W.T.: Phys. Rev. B **39**, 9852 (1989)
16. Wang, J., Lau, W.M., Li, Q.: Effects of particle size and spacing on the optical properties of gold nanocrystals in alumina. J. Appl. Phys. **97**(11), 114303 (2005)
17. Kreibig, U., Vollmer, M.: Theoretical considerations. In: Optical Properties of Metal Clusters, pp. 13–201. Springer, Berlin, Heidelberg (1995)
18. Kreibig, U., Vollmer, M.: Optical Properties of Metal Clusters. Springer, Berlin (1995)
19. Takele, H., Greve, H., Pochstein, C., Zaporojtchenko, V., Faupel, F.: Nanotechnology **17**, 3499 (2006)
20. Kreibig, U., Vollmer, M.: Optical properties of metal clusters, vol. 25. Springer Science & Business Media (2013)
21. Meskinis, S., Ciegis, A., Vasiliauskas, A., Slapikas, K., Tamulevicius, T., Tamuleviciene, A., Tamulevicius, S.: Optical properties of diamond-like carbon films containing copper, grown by high power pulsed magnetron sputtering and direct current magnetron sputtering: structure and composition effects. Thin Solid Films **581**, 48–53 (2015)
22. Omar, N.A.S., Ramli, I., Fen, Y.W., Abdullah, J., Daud, N.F.M., Daniyal, W.M.E.M.M., Mahdi, M.A.: A sensing approach for manganese ion detection by carbon dots nanocomposite thin film-based surface plasmon resonance sensor. Optik, 167435 (2021)
23. Bhardwaj, N., Mohapatra, S.: Structural, optical and gas sensing properties of Ag-SnO_2 plasmonic nanocomposite thin films. Ceram. Int. **42**(15), 17237–17242 (2016)
24. Parveen, S., Pathak, A., Gupta, B.D.: Fiber optic SPR nanosensor based on synergistic effects of CNT/Cu-nanoparticles composite for ultratrace sensing of nitrate. Sens. Actuators, B Chem. **246**, 910–919 (2017)
25. Mohammadzadeh-Asl, S., Keshtkar, A., Dolatabadi, J.E.N., de la Guardia, M.: Nanomaterials and phase sensitive based signal enhancment in surface plasmon resonance. Biosens. Bioelectron. **110**, 118–131 (2018)
26. Ozhikandathil, J., Badilescu, S., Packirisamy, M.: Plasmonic gold decorated MWCNT nanocomposite for localized plasmon resonance sensing. Sci. Rep. **5**(1), 1–11 (2015)
27. Mishra, A.K., Mishra, S.K., Verma, R.K.: Graphene and beyond graphene MoS_2: a new window in surface-plasmon-resonance-based fiber optic sensing. J. Phys. Chem. C **120**(5), 2893–2900 (2016)
28. Vishnoi, R., Sharma, K., Yadav, S., Singhal, R.: Investigation of sequential thermal annealing effect on Cu-C_{70} nanocomposite thin film. Thin Solid Films **680**, 75–80 (2019)
29. Vishnoi, R., Singhal, R., Bhardwaj, J., Inani, H., Kumar, Y., Sharma, A. K., et al.:Thermally-induced plasmonic resonance of Cu nanoparticles in fullerene C_{70} matrix. Vacuum, **159**, 423–429 (2019)
30. Singhal, R., Sharma, K., Vishnoi, R.: Synthesis and characterization of Cu-C_{60} plasmonic nanocomposite. Physica B **550**, 225–234 (2018)
31. Singhal, R., Agarwal, D.C., Mishra, Y.K., Kabiraj, D., Mattei, G., Pivin, J.C., et al.: Synthesis, characterizations, and thermal-induced structural transformation of silver-fullerene C_{60} nanocomposite thin films for applications in optical devices. J. Appl. Phys. **107**(10), 103504 (2010)
32. Singhal, R., Vishnoi, R., Inani, H., Sharma, P., Venkataratnam, K.K., Avasthi, D.K.: Investigations on the thermal stability of fullerene-based (Ag-C_{70}) nanocomposite thin films. Plasmonics **12**(6), 1701–1708 (2017)
33. Singhal, R., Vishnoi, R., Sharma, P., Inani, H., Sharma, G.D., Pivin, J.C.: Thermally-induced tuning of SPR of metal-fullerene Ag (26%)-C_{70} nanocomposite. Surf. Coat. Technol. **324**, 361–367 (2017)
34. Singhal, R., Sharma, P., Vishnoi, R., Avasthi, D.K.: Synthesis and characterizations of Au-C_{60} nanocomposite. J. Alloy. Compd. **696**, 9–15 (2017)
35. Singhal, R., Vishnoi, R., Sharma, P., Sharma, G.D., Chand, S., Kanjilal, D., Pivin, J.C.: Synthesis, characterization, and thermally induced structural transformation of Au-C_{70} nanocomposite thin films. Vacuum **142**, 146–153 (2017)

Chapter 22
Surface Plasmon Resonance (SPR) Biosensors for Antibiotic Residue Detection

Le Thi Thanh Hiep, Khajohnpat Teerasitwaratorn, and Tanujjal Bora

Abstract The presence of pharmaceutical residue especially antibiotic wastes in the environment and food products can cause serious health issues to humans and other organisms. Including controllable risk factors, recognition of their existence in the environment is necessary to ensure a healthy environment. Among various analytical methods to detect pharmaceutical wastes, the development of surface plasmon resonance (SPR) biosensors has been growing very fast due to their simplicity, high sensitivity, fast detection, inexpensiveness, and portability. This book chapter's goal is to give a thorough overview of recent advancements in SPR biosensors for antibiotic residue detection.

Keywords SPR · Nanosystem · Biosensor · Antibiotic

22.1 Introduction

Pharmaceuticals, a revolutionary product in anthropological scientific growth, are being used worldwide to treat humans and animals from diseases. Antibiotics stand out among them as the most prevalent and potent antibacterial agents. Antibiotics are widely used to treat bacterial infections and ward against diseases in people, animals, and plants in the medical, food-producing, livestock, and agricultural areas. They guarantee good health and improve both the quality and yield of the product. According to a recent study [13], the global antibiotic consumption volume given in *defined daily doses* (DDD) has increased from 9.8 billion DDD to 40.2 billion DDD, an increase of more than 300%, during the period 2000–2018. This number is expected to rise further during and in the post-Covid-19 era [68]. The rapid increase in the consumption of pharmaceuticals has simultaneously led to higher leaking of

L. T. T. Hiep · K. Teerasitwaratorn · T. Bora (✉)
Center of Excellence in Nanotechnology, Asian Institute of Technology, PO Box 4, Khlong Luang, Pathum Thani 12120, Thailand
e-mail: tbora@ait.ac.th

D. Mohanta and P. Chakraborty (eds.), *Nanoscale Matter and Principles for Sensing and Labeling Applications*, Advanced Structured Materials 206,
https://doi.org/10.1007/978-981-99-7848-9_22

their residues to the environment posing threat to our life as well as to the health of the ecosystem around, which is an emerging concern today [49].

Pharmaceutical wastes pose major threats to a variety of creatures, including humans, plants, and fauna, and they must be taken seriously. Even at very low amounts, some medication residues can seriously harm the environment [101]. For example, antibiotic resistance has a harmful effect on microorganisms and higher animals, which is mostly induced by antibiotic drug residues in the environment. One of the most widely used and well-known antibiotics for treating bacterial illnesses, usually in animals, is ampicillin. It is a member of the β-lactam class and is eliminated through the urine and feces. As a result, variable concentrations of ampicillin or its metabolites may be found in the environment. Such residues can cause serious effects such as allergic potential or antibiotic resistance in bacterial strains producing an ecological imbalance.

Currently, various chromatographic techniques, such as HPLC (High Performance Liquid Chromatography), LC (Liquid Chromatography), TLC (Thin Layer Chromatography), and GC (Gas Chromatography), are most frequently employed to identify antibiotic residues. Other sparingly used techniques include immunochemical tools (e.g., immunosensors, enzymatic assays, and electrochemical biosensors), which can be used to find drug residues in the environment. Each of these approaches to the detection of pharmaceutical waste offers advantages and challenges. Antibiotic residues can be detected with excellent selectivity and sensitivity using chromatographic techniques. However, they require expensive equipment, intricate sample preparation, and a competent operator. Furthermore, usually, real-time detection is not possible. Consequently, there is an increasing interest in the development of innovative, reliable, and real-time technologies for detecting antibiotic residues in the environment and in food products [31, 55, 92]. In the past few years, detection of chemical residue using optical sensors based on the surface plasmon resonance (SPR) of metallic nanostructures has garnered a great deal of interest due to their remarkable sensitivity and simplicity of detection [43, 64]. The term "*Surface plasmon*" (SP) refers to the free electrons that oscillate coherently at the interface between a metal nanostructure and a dielectric medium when exposed to an appropriate electromagnetic (EM) radiation, e.g., light. A significant amount of the incoming radiation is absorbed by the SPs once a resonance condition is satisfied between the oscillating SPs and the incoming EM radiation, leading in an energy loss in the spectrum of reflected light, which is the basis for the SPR sensor design. SPR sensors have excellent detection sensitivity and portability, making them suitable for faster and on-site detection.

The biggest challenges of SPR sensors today are their complex fabrication processes combining with requirements of other optical instruments, increasing the overall cost of such sensor system. In contrast, a localized surface plasmon resonance (LSPR) sensor potentially reduces the design complexity and, thus, the production cost of sensors. LSPR systems typically use metal nanoparticles instead of thin metallic films, and hence enables an easy sensor design since metal nanoparticles can be produced using simple solution-phase methods. In addition, their surfaces can be functionalized further enabling selective detection of specific chemical and biological species. Due to the binding of analytes to solid surfaces in LSPR sensors, the detection instrument is made portable and simultaneous by the LSPR peak shift [8, 78, 88]. Combining these techniques with the biosensor concepts using bioreceptors has potential soon to provide rapid, accurate and mobile detection of the residual pharmaceutical wastes in food and environment. This book chapter provides an overview of pharmaceutical wastes (mainly focusing on antibiotic residues), their detection techniques, and the recent developments and potential applications of SPR-based biosensors for the detection of antibiotic residues.

22.2 Pharmaceutical Residues in the Environment

Since most pharmaceutical chemicals are polar molecules, they readily dissolve in water. Today, not only is their presence ubiquitous in the geosphere but also in the biosphere including wastewater, groundwater as well as surface water, such as lakes, rivers, ocean, etc. [39, 57, 95]. As indicated in Fig. 22.1, these pharmaceutical wastes originate principally from three application sources: animals and humans, factories, and hospitals. Micropollutant concentrations in water have been found to be remarkably high in recent investigations. According to the *Measured Environment Concentrations* (MEC) database, antibiotics (such as amoxicillin, ciprofloxacin, doxycycline, and many more) are the most often discovered substances in the environment of 71 countries, followed by analgesics [5]. These residues can furthermore contribute to the contamination of food and the environment. For instance, the use of antibiotics in animal and agricultural production can leave residues in animal products such as eggs and meat, as well as in the environment, such as water and soil. Due to the polar nature, they can easily spread into bodies of water, including lakes and rivers. The intake of these contaminated water or food products containing antibiotics can lead to a variety of health problems, including allergy, mutagenicity, bone marrow toxicity, anaphylactic shock, and cancer, not only in humans but also in other living organisms [6].

The most prevalent drugs found in soil and water are antibiotics, analgesics, and hormones. In general, the quantities of pharmaceuticals in surface water and sewage effluent were found to be higher than in soil or manure [61, 76, 94]. A growing number of published investigations demonstrate that pharmaceutical residues are found in the atmosphere around the world. Typically in waterbodies that collect sewage, pharmaceuticals are present in very low amounts, sometimes even lower than *parts-per-trillion (ppt)* levels [5]. Due to their widespread use, medications are continually leaked to the ecosystem and may exhibit persistence [53]. As a result, soil and groundwater contain a high pharmaceutical concentration of various pharmacological and biocidal groups. Both groundwater-dwelling species and human health may be at risk from the use of water and other consumables containing pharmaceutical waste. In contrast to the ordinary pollutants, medicines are designed to be bioactive even at their low concentrations and to act on specified targets in various

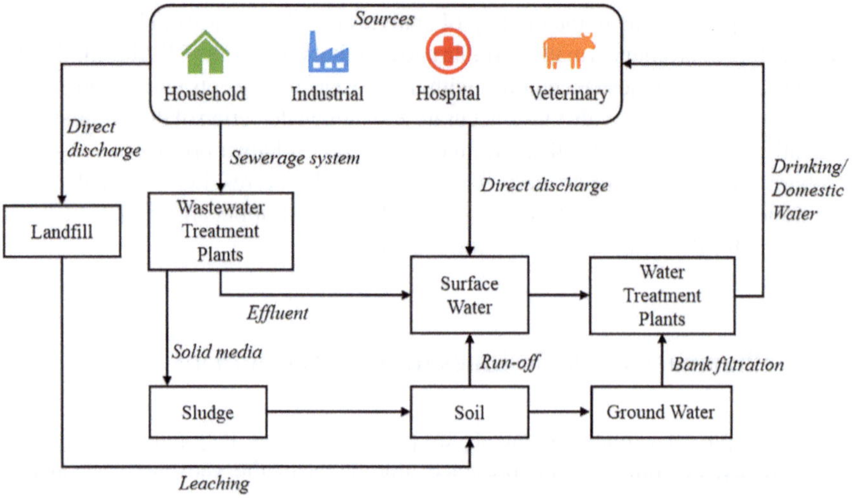

Fig. 22.1 A schematic representation depicting the possible pathways of pharmaceutical residues entering the aquatic environment from their sources

living forms. Therefore, the effects of pharmacological residues on targeted or non-targeted microbial communities should be taken seriously. Studies have demonstrated that some medications can disrupt or alter endocrinology [74], cause the development of antibiotic resistance in microorganisms [4], and have negative effects on natural microbial ecosystems [3, 30]. Microorganisms effectively degrade organic substances and xenobiotics, and deliver nutrition to other organisms, including medications, inside the food chain. Consequently, their healthy population is essential for the ecosystem to function properly.

22.2.1 Effects of Antibiotic Residues on Living Organisms

Antibiotics are intended to treat infections caused by bacteria and help promoting the growth by acting on the various organs and the nervous system [28, 57]. Even though physical and biological processes reduce antibiotic residues in aquatic environments, the quantity of antibiotic residues identified in soil and water is still considerable and might adversely affect any organism that is actively or indirectly exposed to these wastes. Therefore, antibiotic wastes or residues in the environment, even at smaller amounts than parts per trillion (ppt), may exceed the acceptable daily intake (ADI) and have detrimental effects on living organisms [50].

Antibiotic residues have three grave consequences on living organism: (a) allergic effects, (b) antibiotic resistance development, and (c) ecological toxicity. Consumed through food and water, these antibiotic residues can cause allergic effects to humans and animals. The misery of the symptom depends on the amount of residue intake, which in worst case may lead to acute failure of organs or even deaths [6]. Antibiotic resistance, on the other hand, occurs when bacteria change their response to the use of such medicines. This makes antibiotics less effective on the already resistant bacteria and leads to complex medical treatments with increasing mortality [38, 63]. Ecological toxicity is the third concerned effect from antibiotic residues, where these chemicals can be absorbed by plants interfering with their physiological processes. Some negative effects caused by antibiotic residues include disruptions in photosynthesis and mitochondria [93], delay in germinations, biomass reduction and yield drop of agricultural products [66].

22.2.2 Common Antibiotic Residue Detection Methods

Today, a variety of analytical techniques are used to detect and quantify residual antibiotic contaminations in the environment. The most used techniques include chromatography combined with mass spectroscopy: GC/LC, TLC, and HPLC. Chromatography is a separation technique based on the physical qualities of the constituents of a mixture, such as their varying solubilities. The technique uses two phases for separation, a mobile phase composed of a solvent containing the substances to be separated and a stationary phase generally composed of a porous solid medium. During the movement of the mobile phase through the stationary phase placed inside a column, a distinct solute forms as different components of the mobile phase travel at different rates, primarily due to the size and polarity of the components. This procedure is known as *elution*. During sample processing, the elution procedure can be used for separation as well as for purification [12].

The chromatogram produced by the chromatographic procedure depicts the order of elution of the various components. Peaks that appeared on the chromatogram can be then correlated to the components of the separated mixture. TLC and HPLC coupled with MS (Mass spectrometer) or IR (infrared) spectroscopy are the most prevalent chromatographic techniques for analyzing medicines. The primary benefits of chromatography are its great sensitivity, precision, and less sample requirement for analysis. However, they require expert operators, sample pretreatment prior to measurement, and expensive equipment.

TLC is occasionally used for semi-quantitative or quantitative analysis of active substances. It is a technique for separating compounds based on the movement of the mobile phase through a uniform thin layer of solid materials, known as the stationary phase. The thin layer is an adsorbent that is chosen based on the analytical requirements and is adhered to metallic slides or glass. The mobile phase, on the other hand, is a mixture of a single or multiple solvent components in a predetermined ratio. While traversing the thin solid layer, the components of the mobile phase are transported at varying speeds along the flow direction, leading to separation that happens through either adsorption, or distribution, or ion exchange mechanism [82].

Pharmaceutical uses of HPLC are renowned for being the most effective. Similar to TLC, HPLC also uses a solid adsorbent as the stationary phase and a mobile phase comprising aqueous/organic solvents passing through the solid adsorbent. Separation occurs because the components of the separate specimens have distinct dispositions and flow at varying speeds through the stationary phase contained inside a glass or plastic column. For operation, the HPLC method employs a pump to propel the mobile phase along the column and a detector to determine the retention times of individual molecules of the mobile phase. In order to monitor, identify, and quantify antibiotic residues, the processing time in HPLC is affected by the composition of the stationary and mobile phases as well as the complexity of the test sample, and it employs a variety of detectors, including ultraviolet detector (UV), fluorescent (FLD), and diode array (DAD) [42, 91].

22.2.3 Biosensors and Their Types

Emerging technologies, particularly '*biosensors*' have potential applications in a variety of industries, including environmental monitoring, clinical diagnostics, and the food and medicine industry [18, 35]. A biosensor typically consists of three components, viz. a bio-recognition element, a transducer and an electronic system for signal processing. The bio-recognition element, essentially a bioreceptor (e.g., enzymes, antibody, DNA, microorganism, or cells), is allowed to interact with a specific analyte and detects the input physical change due to the binding of the analyte with the bioreceptor. The transducer converts this physical change into appropriate electrical signal, which is then amplified and processed by the electronic system. The primary benefits of biosensors and associated bioanalytical tools over conventional analytical methods include rapid and inexpensive measurements, on-site monitoring, and the ability to assess complicated matrices with minimal sample preparation.

The classification of biosensors depends on the type of bioreceptor or transducer used. Biological receptors or bioreceptors are those that are derived from biological components and react immediately to the chemicals that need to be recognized. Following are some of the commonly used bioreceptors to construct biosensors:

+ Antibodies/Antigen: These receptors exhibit extremely high selectivity and powerful binding to the target analyte. Such biosensor utilizes the very specific binding affinity of antibodies for a specific compound or antigen. The specific nature of the antibody-antigen interaction is analogous to a lock and key fit in that the antigen will only bind to the antibody if it has the correct conformation.

+ Aptamers (nucleic acids): Aptamers are short, single-stranded DNA or RNA molecules that can selectively bind to a specific target. Using these randomly generated sequences, biological recognition can be accomplished.

+ Enzymes: The most dominant kind of receptors are bioreceptors comprised by enzymes due to their binding properties and catalytic activity. Analyte recognition is made feasible by several potential mechanisms: (1) The enzyme that transforms the analyte into a product that can be detected by a sensor; (2) the enzyme that detects inhibition or activation of the enzyme by the analyte; or (3) the enzyme that monitors changes in the characteristics of the enzyme brought on by interaction with the analyte.

+ Molecularly Imprinted Polymers (MIPs): It is a method of leaving cavities in a polymeric matrix that has been imprinted with a particular "template" molecule so that it may be recognized. MIPs are analogs to the natural antibody–antigen systems but prepared synthetically. In addition, they utilize the "lock and key" mechanism to preferentially bind the molecule that served as their synthesis template.

+ Cells: Cells can serve as bioreceptors since they are sensitive to their surroundings and can respond to numerous stimuli. They show superior sensor life and very good reproducibility.

Additionally, the different kinds of signal transducers that are employed affect how a biosensor is classified. Transducers, categorized as *electrochemical*, *optical*, and *piezoelectric* biosensors, are constantly being developed for pharmaceutical detection. Antibiotics can be detected by electro-active signals produced by an electrochemical biosensor's transducer system. When ions bind to the detecting membrane, a difference in membrane potential results, which typically shows a logarithmic relationship with the analyte concentration [35]. Techniques including pulse or cyclic voltammetry, electrochemical impedance spectroscopy, and chrono-amperometry are frequently used to measure this signal [22]. Antibody immobilized on an electrode surface is a common feature of the majority of electrochemical immunosensors discovered in antibiotic research [73], and the mostly they use enzymes as an active biological component to produce signals [16].

Based on specialized transduction techniques, optical biosensors for pharmaceutical detection have been developed. Metal nanostructure-based surface plasmon resonance (SPR)-based biosensors are one of those optical biosensor systems that have already been advanced and are frequently utilized for studies. In a typical SPR platform, the analyte's attachment to its bio-specific, immobilized counterpart on the sensor surface results in shifts in refractive index, which are then analyzed. Analyte binding on the sensor surface results in a change in the refractive index near the surface and the resonant SPR wavelength is altered, which can be correlated to the analyte concentration at the sensor surface [90].

Biosensors based on piezoelectric technology are label-free detecting methods for direct detection. When an analyte binds to a bioreceptor mounted on the surface of a piezoelectric material, say a quartz crystal, the natural piezoelectric frequency of the material shifts which is a function of changes in the material's surface mass. Since the mass changes are proportional to the bioreceptor-analyte complex concentration, the system permits identification and quantification of analyte complexes [100].

Designing a biosensor with bioreceptors combined with a signal transducer has many advantages over conventional detection techniques, including portability, quick detection, high sensitivity, cost-effectiveness, and high specificity with low detection limits. However, the design challenges and performance can differ for each type of biosensor [2, 62]. Optical biosensors (e.g., SPR biosensors) particularly are of great interest as they can provide both label and label-free detection with high sensitivity and selectivity. Especially, label-free designs reduce the complexity of the system preparation offering a huge advantage compared to the label-based detections [72].

22.3 Theory of Surface Plasmons

22.3.1 *Plasmons and Surface Plasmon Resonance (SPR)*

Like how photons are produced when a light wave is quantized, plasma oscillations give rise to quasiparticles called plasmons. A plasmon is, therefore, a collective electromagnetic (EM) oscillation of free electrons, which is quantized in nature, and the Maxwell's EM Equations can be used to define the plasmons' wave nature [47]. Metals and semiconductors' optical characteristics are due to plasmons. For instance, most metals exhibit high plasmon frequencies, which are often found in the UV region and reflect frequencies lower than the plasmon frequency, account for their shining appearance. Through electronic interband transitions, some metals, like gold and copper, absorb visible light energy, giving them their distinctive colors. Using the free electron model, the plasmon energy, E_p, of a material can be determined and is given by,

$$E_p = \hbar\omega_p = \hbar\sqrt{\frac{ne^2}{m\epsilon_0}} \tag{22.1}$$

where $\hbar$ represents the reduced Planck's constant, ω_p represents the frequency of the plasmons, n represents the density of conduction electrons, e represents the charge carried by a single electron with mass m, and ϵ_0 represents the free space permittivity.

The term '*Surface Plasmon*' refers to the surface confined plasmons of a metal structure. When at least one of the dimensions of the metal structure is much smaller than 100 nm, surface confinement occurs and becomes stronger with further reduction in size. Surface plasmons that strongly interact with incoming light of a wavelength corresponding to the size of the metal structure cause polarization. When light strikes the surface of a metallic thin layer interfacing with a dielectric layer and the total internal reflection (TIR) conditions are meet, an electric field is formed due to the positively charged incoming photons and the negatively charged surface plasmons. The attraction and repulsion of the plasmons due to light's alternating electric field then lead to an oscillation. At a specific incident angle, this oscillation becomes maximum giving rise to a resonance condition known as *surface plasmon resonance* (*SPR*), and the incident angle at which resonance occurs is called *SPR angle* [56, 99]. At the intersection of the two mediums (the metal and the dielectric), it creates an evanescent or transient wave of electrons. The resonance conditions change whenever there is a change in the surrounding medium, e.g., the presence of analyte near the metal surface, resulting a shift in the SPR angle. As indicated in Fig. 22.2, this is the primary SPR sensing mechanism, and based on this phenomenon, many types of chemicals and biological species can be identified.

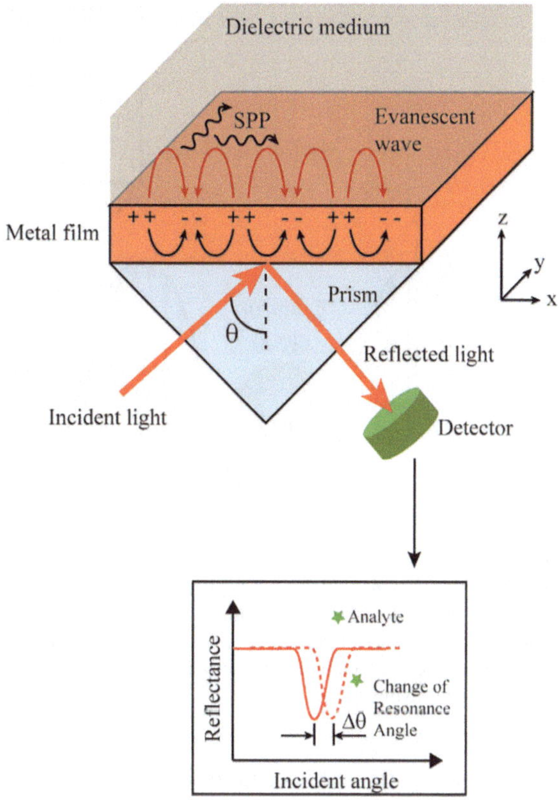

Fig. 22.2 Schematic illustrating the SPR phenomena on a thin metal layer when excited by incident light at the resonance angle (θ). The excited surface plasmons can propagate in the x- and y-directions and are referred to as propagating surface plasmons or surface plasmon polaritons (SPP). The z-direction propagating evanescent wave decays exponentially with distance. Any analyte that binds to the metal surface induces a shift in the resonance angle (θ), which can be exploited for the analyte sensing

22.3.2 Localized Surface Plasmon Resonance (LSPR)

With a few exceptions, localized surface plasmon resonance (LSPR) is almost identical to surface plasmon resonance (SPR). For example, LSPR happens when the metal contains nanoparticles or when metal nanoparticulate thin films have particles that are smaller in size than the wavelength of incident light [87]. In contrast to the SPR mechanism, LSPR generates a very tiny evanescent wave within a very small range of the nanoparticles. The local electric field of metal nanoparticles is augmented at resonance, and this phenomenon has been extensively used for a variety of optical sensing applications, such as surface enhanced Raman scattering (SERS), as shown in Fig. 22.3. It is possible to detect extremely low quantities of chemical and biological components, thanks to the augmented local E-many field's due to the oscillating surface plasmons [11]. Recently, LSPR combined with microfluidics technology for biomedical applications has been reported demonstrating quick, multiplex, and parallel continuous assessments of biomarkers [20].

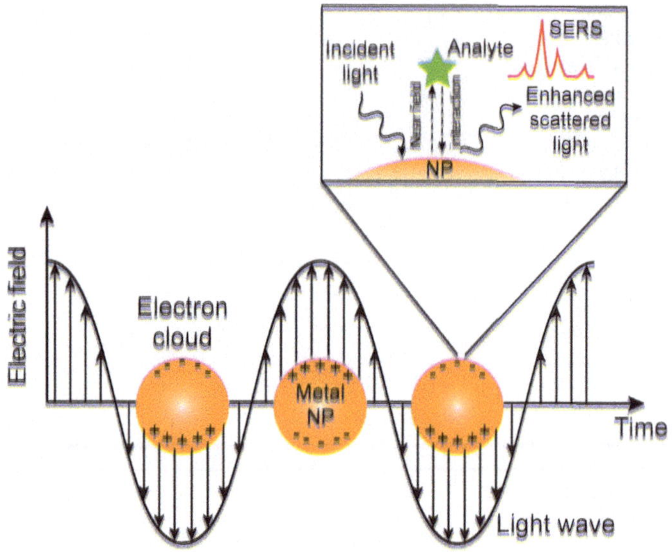

Fig. 22.3 Metal nanoparticles are depicted schematically, with their confined oscillating electron clouds producing localized surface plasmon resonance (LSPR) effect. When resonantly stimulated nanoparticles interact in proximity with an analyte (as shown in the inset). The augmented local E-field from these interactions enhances the light scattering, which can be employed as a probe for SERS-based sensing applications

For high spatial and spectral resolution in LSPR systems, electric field enhancements and light intensity are crucial, which are dependent on the size of the nanoparticles. In the case of spherical metal nanoparticles, the scattering and absorption characteristics of the incident light can be described by the Mie solution to Maxwell's equations, where a relationship for the extinction cross-section (σ_{ext}) which is the combination of the absorption cross-section (σ_{abs}) and the scattering cross-section (σ_{sca}) of the metal nanoparticles is given by:

$$\sigma_{ext} = \frac{9\omega}{c}\varepsilon_m^{3/2} V_0 \frac{\varepsilon_2(\omega)}{[\varepsilon_1(\omega) + 2\varepsilon_m]^2 + \varepsilon_2(\omega)^2} \tag{22.2}$$

where, V_0 represents the volume of the nanoparticle, ω is the angular frequency of the extinction radiation, ε_m is the dielectric constant of the medium surrounding the metal nanoparticles, and ε_1 and ε_2 are the real and imaginary parts of the complex dielectric constant of the metal nanoparticles, respectively. When the term $\varepsilon_2(\omega)$ is weakly dependent on ω or small, resonance conditions occur as the term $\varepsilon_1(\omega)$ in Eq. (22.2) becomes $-2\varepsilon_m$. The surface plasmon resonance peak that gives rise to the specific color of the spherical nanoparticle is observed under this resonance condition at which the EM field is enhanced.

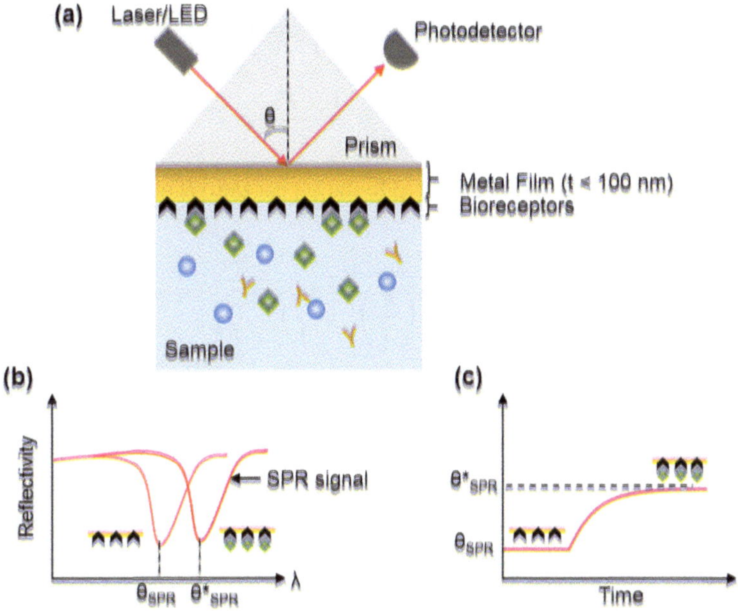

Fig. 22.4 **a** Schematic representation of a SPR-based biosensor showing its various components. The SPR signal in terms of reflectivity is shown in **b**, which has a distinct drop at the SPR angle, θ_{SPR} before the analyte binding. Upon selective analyte binding to the bioreceptors at the sensor surface the SPR angle shifts to a new position denoted by θ^{*}_{SPR} indicating changes in the SPR conditions. The real-time monitoring of the SPR angle during the analyte-bioreceptor interaction is possible as shown in **c** in the form of a sensorgram plot, which can be used to determine the binding kinetic parameters as association/dissociation constants and affinity

For non-spherical particles, extension of Mie theory was used by Richard Gans (known as Gans theory) to predict the scattering characteristics for ellipsoid nanoparticles [32, 33], describing the absorption cross-section for a non-spherical particle as,

$$\sigma_{abs} = \frac{\omega}{3c}\varepsilon_m^{3/2} V \sum_i \frac{\left(1/P_i^2\right)\varepsilon_2}{[\varepsilon_1 + \{(1 - P_i)/P_i\}\varepsilon_m]^2 + \varepsilon_2^2} \quad (22.3)$$

where i represents the three dimensions of the particle and P_i is the depolarization factor which includes P_X, P_Y and P_Z representing each axis of a prolate spheroid particle. The solution to the Eq. (22.3) leads to two absorption peaks representing the transverse and longitudinal plasmon peaks of the metal nanoparticles. Depending on the anisotropy of the particles, the values of ε_1 and ε_2 vary which are governed by the depolarization factors. Mathematically for a prolate particle, the depolarization factors can be expressed as,

$$P_X = \frac{1-e^2}{e^2}\left[\frac{1}{2e}\ln\left(\frac{1+e}{1-e}\right)-1\right] \tag{22.4}$$

$$P_Y = P_Z = \frac{1-P_X}{2} \tag{22.5}$$

where the term *e* in Eq. (22.4) is related to the aspect ratio *R* of the particle, and can be expressed as,

$$e = \sqrt{1-\left(\frac{Y}{X}\right)^2} = \sqrt{1-\frac{1}{R^2}} \tag{22.6}$$

The above mathematical expressions show that both size and shape of the metal nanoparticles are important factors for LSPR sensing. For example, almost sixfold increase in the sensitivity of the longitudinal LSPR signal from gold-nanorods of aspect ratio 3 was estimated by [70] compared to the spherical gold-nanoparticles. LSPR biosensors typically detect the spectral shifts representing changes in the localized refractive index in surrounding dielectric environments caused by the analyte binding to the sensor surface. The dependency of the plasmon frequency (ω_p) on the dielectric function of the surrounding medium can be mathematically expressed, by using the Drude model, as:

$$\varepsilon_1 = 1 - \frac{\omega_p^2}{\omega^2+\gamma^2} \tag{22.7}$$

where γ represents the damping factor of the bulk metal, which can be neglected in the visible and infrared regions as $\gamma \ll \omega_p$ and thus the above equation can be further simplified to

$$\varepsilon_1 = 1 - \frac{\omega_p^2}{\omega^2} \tag{22.8}$$

Under the resonance conditions $(\varepsilon_1 = -2\varepsilon_m)$, and the above equation can be rearranged to

$$\omega_{\max} = \frac{\omega_p}{\sqrt{2\varepsilon_m+1}} \tag{22.9}$$

where $\omega_{\max}$ represents the LSPR peak frequency. For practical applications, the angular frequency, and the dielectric constant terms in Eq. (22.9) are substituted by wavelength $(\lambda = 2\pi c/\omega)$ and refractive index $\left(\varepsilon_m = n^2\right)$ terms, respectively, and hence Eq. (22.9) becomes,

$$\lambda_{\max} = \lambda_p\sqrt{2n_m^2+1} \tag{22.10}$$

where λ_{max} is the LSPR peak wavelength and λ_p represents the wavelength corresponding to the plasma frequency of the bulk metal. Equation (22.10) represents a linear approximation between the LSPR peak wavelength and the refractive index which is vital for LSPR biosensors to detect trace level of analytes with high sensitivity.

22.4 SPR Biosensors for Antibiotic Detection

SPR biosensors encourage the continuous development in the sensing field since they can detect minute changes in the surrounding RI and binding analyte equilibrium. It has been stated that practically every sort of molecular interaction between various types of biological molecules can be accomplished in almost every stage of pharmaceutical research that involves kinetic profiling, target recognition, and/or characterization [69]. Several SPR applications in the realm of pharmaceutical analysis have been published over the past several years. Different pharmaceutical compounds have been investigated using SPR-based sensors to carry out pharmacokinetic drug profiling, such as predictive in-vitro drug, permeability, passive transport through membrane, and characterization of membrane binding affinity [1, 7, 52]. SPR sensors are also used for label-free and raid detection of drug residues in food and clinical samples [10, 69].

Commercialization of SPR biosensors has begun several years ago and one of the most commonly used commercial SPR biosensor is the Biocore® systems (GE Healthcare, Sweden). The Biacore® system with eight channels allows simultaneous detection of eight different analytes or analyte families. Over the years, the Biocore® system is widely used by researchers for the detection of antibiotic residues [9, 15, 23, 24, 36, 37, 45, 86]. Nevertheless, its high cost became a hindrance to its wider applications in the field. An alternative less expensive SPR sensor kit known as Spreeta™ (Texas Instruments, USA) was developed way back in 1990 and released into the market. The system showed less robustness and slow measurements due to its single-channel design. Later the Spreeta™ SPR kit was further modified to have three channels and demonstrated for antibiotic detection in milk [27].

SPR-based biosensors are very sensitive to the refractive index of the surrounding solution and the resonant wavelength or resonant angle changes with the surrounding refractive index. Because of their high sensitivity, high selectivity and label-free characteristics, SPR-based biosensors are widely used in various sensing applications. A SPR biosensor surface contains a layer of bioreceptor, such as antibodies, enzymes, aptamers, cells and molecularly imprinted (MIP) films, to enhance the sensor's specificity and sensitivity. The principle of detection, as depicted in Fig. 22.4, is based on the changes in the refractive index of the sensor surface upon selective binding of an analyte with the immune-recognition molecules or bioreceptors [72].

There has been significant number of applications of the SPR biosensors in recent times, including detection of pharmaceutical residues. For example, the application of

Biocore® systems in this regard has been described earlier. SPR-based immunosensors using antibodies as the bio-recognition layer are explored widely over the past few years. To improve the sensitivity of such systems, competitive inhibition methods were adopted in which the SPR signal response is inversely proportional to the concentration of the target antibiotics enabling the sensors to detect very low concentrations of antibiotics with low limit of detections [25]. Simultaneous detection of multiple antibiotics from a single sample extract is also possible making these SPR-based immunosensors extremely popular for rapid and on-site detection of antibiotics [24, 26]. Protein receptors have been used as a cost-effective alternative to the antibodies to detect antibiotic residues [14, 46]. Enzymatic methods were also developed for highly specific detection of antibiotics residues below the MRL (maximum residue limit) [75].

Aptamers have also gained attentions recently as bio-recognition elements due to their improved stability, robustness, and low cost compared to antibodies and enzymes. Aptamers are single-stranded oligonucleotide sequences (RNA or DNA) with high specificity and affinity towards a variety of target analytes. As aptamers can be synthetically produced in a lab, their mass production is also relatively easy and quick compared to antibodies. For the detection of antibiotics residues, colorimetric SPR-based aptasensors have been developed extensively in the recent time due to their simplicity and high sensitivity [79]. Table 22.1 summarizes some of the colorimetric SPR-based aptasensors developed for the detection of different types of antibiotic residues in food and environment.

Development of LSPR biosensors, where sensing is "highly localized around the nanoparticle" [71], is also rising in recent times. The advantages of LSPR system over the SPR sensor are the simpler and compact design, less sensitive to the environmental factors, possibilities to develop more sensitive and selective methods, and compatibility with multiplexing technology.

Application of LSPR in the field of antibiotic residue detection is relatively new but growing at a faster rate due to the above advantages of this technique over the traditional SPR techniques. Table 22.2 summarizes some of the recent reports on the application of LSPR sensors for the detection of antibiotic in different mediums.

Molecularly imprinted polymers (MIPs) are another budding biosensor element that provides longer sensor life compared to the typically used biorecognition elements [89]. MIPs are obtained by polymerization of a monomer and a cross-linker in the presence of the target molecule, such as an antibiotic molecule, resulting in a synthetic polymer with specific identification function to the target molecule. MIPs have been used to fabricate LSPR gold nanoparticle-based sensors for the detection of neomycin, kanamycin, and streptomycin residues with detection limit in the range of pg/mL [29]. Uniform and controlled thickness of MIP films has been demonstrated recently by using photopolymerization technique and used for the detection of ciprofloxacin showing linear sensor response up to concentrations as low as 10^{-11} M [59].

Table 22.1 Colorimetric SPR-based aptasensors used for the detection of antibiotic residues

Serial number	Antibiotics class	Antibiotics type	Limit of detection	References
1	β-Lactams	Ampicillin	10 ng/mL	[85]
2		Ampicillin	0.1 nM	[80]
3	Aminoglycosides	Kanamycin	3.35 nM	[40]
4		Kanamycin	0.05 nM	[81]
5		Tobramycin	23.3 nM	[60]
6		Streptomycin	25 nM/L	[58]
7		Neomycin	27 nM	[51]
8	Anthracyclines	Daunomycin	17.1 nM	[41]
9		Daunomycin	8.4 μg/L	[96]
10	Chloramphenicol	Chloramphenicol	0.02 ng/mL	[34]
11		Chloramphenicol	0.13 pg/mL	[44]
12		Chloramphenicol	0.03 nM	[48]
13	Fluoroquinolones	Ciprofloxacin	1.3 nM (water) 2.6 nM (human serum) 3.2 (milk)	[54]
14	Tetracyclines	Oxytetracyclines	1 nM	[77]
15		Oxytetracyclines	1 ag/mL	[97]
16		Tetracycline	0.002 ng/mL	[83]
17	Sulfadimethoxine	Sulfadimethoxine	3.41 ng/mL (water) 4.41 ng/mL (fish)	[21]
18		Sulfadimethoxine	10 ng/mL	[98]

22.5 Conclusions and Outlook

Pharmaceutical residues in the environment and food products are dangerous to the living organisms and SPR-based biosensors are promising technology in this regard for the detection of these emerging pollutants. SPR-based biosensors are highly sensitive, selective, and can be portable which makes them highly suitable for on-site detection. Significant amounts of research have been carried out in recent years to develop SPR-based biosensors using biological receptors, such as antibodies, aptamers, enzymes, and MIPs. They show excellent antibiotic residue detection capabilities with extremely low detection limit and very high sensitivity, which are comparable to the conventional lab-based chromatographic techniques. Compared to the traditional SPR biosensors, LSPR-based biosensors provide several advantages, viz. simplicity, compact design, higher stability against environmental factors, and comparable sensitivity. Such systems typically use metal nanoparticles of Au, Ag, or Cu, which have distinct LSPR signal in the visible region making it easy to analyze, and these metal nanoparticles can be produced very easily using cost-effective wet chemical methods. As a result, their applications for the detection

Table 22.2 LSPR sensors for the detection of antibiotic residues

Serial number	Metal nanoparticles	LSPR sensor type	Antibiotic type	Limit of detection	References
1	Silver (Ag) nanoparticles	Colorimetric sensor	Azithromycin	0.053 μM	[19]
2	Gold (Au) nanoparticles and nanorods	Colorimetric sensor	Gentamicin	0.4 ng/mL	[67]
3	Silver (Ag) thin film coated with Ag NPs embedded in MIP-TC polymer film	Fiber optics-based sensor	Tetracycline	2.2 nM	[84]
4	Imprinted boronic acid-functionalized Au nanoparticle	SPR Kretschmann-type spectrometer, NanoSPR 321	Neomycin	2 pM	[29]
			Kanamycin	1 pM	
			Streptomycin	0.2 pM	
5	Au nanoislands	CMOS sensor-based image analysis	Tobramycin	3.4 μM	[17]
6	Au nanoparticles	Colorimetric sensor	Tobramycin	5 nM	[65]

of antibiotic residues in food and environment are increasing in the recent times. However, the design complexity, overall sensor system cost, robustness, and simultaneous measurements are some of the challenges that need to be addressed soon in order for the large-scale commercialization of SPR and LSPR biosensors. This book chapter highlighted some of the key developments happening in this area, exhibiting a wide range of SPR and LSPR-based biosensors reported for antibiotic residue detection.

References

1. Abdiche, Y.N., Myszka, D.G.: Probing the mechanism of drug/lipid membrane interactions using Biacore. Anal. Biochem. **328**, 233–243 (2004)
2. Adrián, J., Fernández, F., Muriano, A., et al.: Biosensors for pharmaceuticals and emerging contaminants based on novel micro and nanotechnology approaches. In: Barceló, D., Hansen, P.-D. (eds.) Biosensors for Environmental Monitoring of Aquatic Systems, pp. 47–68. Springer, Berlin Heidelberg, Berlin, Heidelberg (2009)
3. Alvarino, T., Katsou, E., Malamis, S., et al.: Inhibition of biomass activity in the via nitrite nitrogen removal processes by veterinary pharmaceuticals. Bioresour. Technol. **152**, 477–483 (2014)
4. Andersson, D.I., Hughes, D.: Microbiological effects of sublethal levels of antibiotics. Nat. Rev. Microbiol. **12**, 465–478 (2014)

5. aus der Beek, T., Weber, F.A., Bergmann, A., et al.: Pharmaceuticals in the environment-global occurrences and perspectives. Environ Toxicol. Chem. **35**, 823–835 (2016)
6. Bacanlı, M., Başaran, N.: Importance of antibiotic residues in animal food. Food Chem. Toxicol. **125**, 462–466 (2019)
7. Baird, C.L., Courtenay, E.S., Myszka, D.G.: Surface plasmon resonance characterization of drug/liposome interactions. Anal. Biochem. **310**, 93–99 (2002)
8. Bala, R., Sharma, R.K., Wangoo, N.: Highly sensitive colorimetric detection of ethyl parathion using gold nanoprobes. Sens. Actuators B Chem **210**, 425–430 (2015)
9. Baxter, G.A., Ferguson, J.P., O'Connor, M.C., Elliott, C.T.: Detection of streptomycin residues in whole milk using an optical immunobiosensor. J. Agric. Food Chem. **49**, 3204–3207 (2001)
10. Bettazzi, F., Marrazza, G., Minunni, M., et al.: Biosensors and related bioanalytical tools. In: Palchetti, I., Hansen, P.-D., Barceló, D.B.T.-C.A.C. (eds.) Past, Present and Future Challenges of Biosensors and Bioanalytical Tools in Analytical Chemistry: A Tribute to Professor Marco Mascini, 1st edn, pp. 1–33. Amsterdam, Elsevier (2017)
11. Bora, T.: Recent developments on metal nanoparticles for SERS applications. In: Seehra, M.S., Bristow, A.D. (eds.) Noble and Precious Metals–Properties, Nanoscale Effects and Applications, p. 71573. IntechOpen, London (2018)
12. Braithwaite, A., Smith, F.J.: Chromatographic Methods, 5th edn. Springer, Dordrecht (1995)
13. Browne, A.J., Chipeta, M.G., Haines-Woodhouse, G., et al.: Global antibiotic consumption and usage in humans, 2000–18: a spatial modelling study. Lancet Planet Heal **5**, e893–e904 (2021)
14. Cacciatore, G., Petz, M., Rachid, S., et al.: Development of an optical biosensor assay for detection of β-lactam antibiotics in milk using the penicillin-binding protein 2x*. Anal. Chim. Acta **520**, 105–115 (2004)
15. Caldow, M., Stead, S.L., Day, J., et al.: Development and validation of an optical SPR biosensor assay for tylosin residues in honey. J. Agric. Food Chem. **53**, 7367–7370 (2005)
16. Campaña, A.L., Florez, S.L., Noguera, M.J., et al.: Enzyme based electrochemical biosensors for microfluidic platforms to detect pharmaceutical residues in wastewater. Biosensors **9**, 41 (2019)
17. Cappi, G., Spiga, F.M., Moncada, Y., et al.: Label-Free detection of tobramycin in serum by transmission-localized surface plasmon resonance. Anal. Chem. **87**, 5278–5285 (2015)
18. Chamorro-Garcia, A., Merkoçi, A.: Nanobiosensors in diagnostics. Nanobiomedicine **3**: 1849543516663574 (2016)
19. Chavada, V.D., Bhatt, N.M., Sanyal, M., Shrivastav, P.S.: Surface plasmon resonance based selective and sensitive colorimetric determination of azithromycin using unmodified silver nanoparticles in pharmaceuticals and human plasma. Spectrochim. Acta Part A Mol. Biomol. Spectrosc. **170**, 97–103 (2017)
20. Chen, J.-S., Chen, P.-F., Lin, H.T.-H., Huang, N.-T.: A localized surface plasmon resonance (LSPR) sensor integrated automated microfluidic system for multiplex inflammatory biomarker detection. Analyst **145**, 7654–7661 (2020)
21. Chen, X.X., Lin, Z.Z., Hong, C.Y., et al.: A dichromatic label-free aptasensor for sulfadimethoxine detection in fish and water based on AuNPs color and fluorescent dyeing of double-stranded DNA with SYBR Green I. Food Chem. **309**, 125712 (2020)
22. Chu, X., Wu, L., Liu, X., et al.: Detection of broad-spectrum aminoglycoside antibiotics through fluorescence-labeling aminoglycoside acetyltransferase(6′)-Ii. Anal. Biochem. **376**, 144–150 (2008)
23. Crooks, S.R.H., McCarney, B., Traynor, I.M., et al.: Detection of levamisole residues in bovine liver and milk by immunobiosensor. Anal. Chim. Acta **483**, 181–186 (2003)
24. Dumont, V., Huet, A.-C., Traynor, I., et al.: A surface plasmon resonance biosensor assay for the simultaneous determination of thiamphenicol, florefenicol, florefenicol amine and chloramphenicol residues in shrimps. Anal. Chim. Acta **567**, 179–183 (2006)
25. Ferguson, J.P., Baxter, G.A., McEvoy, J.D.G., et al.: Detection of streptomycin and dihydrostreptomycin residues in milk, honey and meat samples using an optical biosensor. Analyst **127**, 951–956 (2002)

26. Fernández, F., Hegnerová, K., Piliarik, M., et al.: A label-free and portable multichannel surface plasmon resonance immunosensor for on site analysis of antibiotics in milk samples. Biosens. Bioelectron. **26**, 1231–1238 (2010)
27. Fernández, F., Pinacho, D.G., Sánchez-Baeza, F., Marco, M.P.: Portable surface plasmon resonance immunosensor for the detection of fluoroquinolone antibiotic residues in milk. J. Agric. Food Chem. **59**, 5036–5043 (2011)
28. Franco, D., Webb, J., Taylor, C.E.: Antibiotic and sulfonamide residues in meat: implications for human health. J. Food Prot. **53**, 178–185 (1990)
29. Frasconi, M., Tel-Vered, R., Riskin, M., Willner, I.: Surface plasmon resonance analysis of antibiotics using imprinted boronic acid-functionalized au nanoparticle composites. Anal. Chem. **82**, 2512–2519 (2010)
30. Gadipelly, C., Pérez-González, A., Yadav, G.D., et al.: Pharmaceutical industry wastewater: review of the technologies for water treatment and reuse. Ind. Eng. Chem. Res. **53**, 11571–11592 (2014)
31. Gai, P., Gu, C., Hou, T., Li, F.: Ultrasensitive self-powered aptasensor based on enzyme biofuel cell and DNA bioconjugate: a facile and powerful tool for antibiotic residue detection. Anal. Chem. **89**, 2163–2169 (2017)
32. Gans, R.: Über die Form ultramikroskopischer Goldteilchen. Ann. Phys. **342**, 881–900 (1912)
33. Gans, R.: Über die Form ultramikroskopischer Silberteilchen. Ann. Phys. **352**, 270–284 (1915)
34. Gao, H., Gan, N., Pan, D., et al.: A sensitive colorimetric aptasensor for chloramphenicol detection in fish and pork based on the amplification of a nano-peroxidase-polymer. Anal. Methods **7**, 6528–6536 (2015)
35. Gaudin, V.: Advances in biosensor development for the screening of antibiotic residues in food products of animal origin–a comprehensive review. Biosens. Bioelectron. **90**, 363–377 (2017)
36. Gaudin, V., Fontaine, J., Maris, P.: Screening of penicillin residues in milk by a surface plasmon resonance-based biosensor assay: comparison of chemical and enzymatic sample pre-treatment. Anal. Chim. Acta **436**, 191–198 (2001)
37. Gaudin, V., Hédou, C., Sanders, P.: Validation of a biacore method for screening eight sulfonamides in milk and porcine muscle tissues according to European decision 2002/657/EC. J. AOAC Int. **90**, 1706–1715 (2007)
38. Gelband, H., Molly, M.-P., Pant, S., et al.: The state of the world's antibiotics 2015. Wound Heal South Africa **8**, 30–34 (2015)
39. Grenni, P., Ancona, V., Barra Caracciolo, A.: Ecological effects of antibiotics on natural ecosystems: a review. Microchem. J. **136**, 25–39 (2018)
40. Ha, N.R., Jung, I.P., Kim, S.H., et al.: Paper chip-based colorimetric sensing assay for ultra-sensitive detection of residual kanamycin. Process Biochem. **62**, 161–168 (2017)
41. He, L., Zhi, W., Wu, Y., et al.: A highly sensitive resonance scattering based sensor using unmodified gold nanoparticles for daunomycin detection in aqueous solution. Anal. Methods **4**, 2266–2271 (2012)
42. Herzallah, S.M.: Determination of aflatoxins in eggs, milk, meat and meat products using HPLC fluorescent and UV detectors. Food Chem. **114**, 1141–1146 (2009)
43. Homola, J.: Surface plasmon resonance sensors for detection of chemical and biological species. Chem. Rev. **108**, 462–493 (2008)
44. Huang, W., Zhang, H., Lai, G., et al.: Sensitive and rapid aptasensing of chloramphenicol by colorimetric signal transduction with a DNAzyme-functionalized gold nanoprobe. Food Chem. **270**, 287–292 (2019)
45. Huet, A.-C., Charlier, C., Weigel, S., et al.: Validation of an optical surface plasmon resonance biosensor assay for screening (fluoro)quinolones in egg, fish and poultry. Food Addit. Contam. Part A Chem. Anal. Control Expo Risk Assess. **26**, 1341–1347 (2009)
46. Indyk, H.E.: The determination of folic acid in milk and paediatric formulae by optical biosensor assay utilising folate binding protein. Int. Dairy J. **20**, 106–112 (2010)
47. Jackson, J.D.: Classical Electrodynamics, 3rd edn. Wiley, New York (2007)

48. Javidi, M., Housaindokht, M.R., Verdian, A., Razavizadeh, B.M.: Detection of chloramphenicol using a novel apta-sensing platform based on aptamer terminal-lock in milk samples. Anal. Chim. Acta **1039**, 116–123 (2018)
49. Jorgensen, S.E., Halling-Sorensen, B.: Drugs in the environment. Chemosphere **40**, 691–699 (2000)
50. Kadam, A., Patil, S., Patil, S., Tumkur, A.: Pharmaceutical waste management–an overview. Indian J. Pharm. Pract. **9**, 2–8 (2016)
51. Khavani, M., Izadyar, M., Housaindokht, M.R.: Theoretical design and experimental study on the gold nanoparticles based colorimetric aptasensors for detection of neomycin B. Sens. Actuators B Chem. **300**, 126947 (2019)
52. Kim, K., Cho, S., Park, J.H., et al.: Surface plasmon resonance studies of the direct interaction between a drug/intestinal brush border membrane. Pharm. Res. **21**, 1233–1238 (2004)
53. Lapworth, D.J., Baran, N., Stuart, M.E., Ward, R.S.: Emerging organic contaminants in groundwater: a review of sources, fate and occurrence. Environ. Pollut. **163**, 287–303 (2012)
54. Lavaee, P., Danesh, N.M., Ramezani, M., et al.: Colorimetric aptamer based assay for the determination of fluoroquinolones by triggering the reduction-catalyzing activity of gold nanoparticles. Microchim. Acta **184**, 2039–2045 (2017)
55. Li, F., Guo, Y., Wang, X., Sun, X.: Multiplexed aptasensor based on metal ions labels for simultaneous detection of multiple antibiotic residues in milk. Biosens. Bioelectron. **115**, 7–13 (2018)
56. Li, M., Cushing, S.K., Wu, N.: Plasmon-enhanced optical sensors: a review. Analyst **140**, 386–406 (2015)
57. Li, W.C.: Occurrence, sources, and fate of pharmaceuticals in aquatic environment and soil. Environ. Pollut. **187**, 193–201 (2014)
58. Liu, Z., Zhang, Y., Xie, Y., et al.: An aptamer-based colorimetric sensor for streptomycin and its application in food inspection. Chem. Res. Chin. Univ. **33**, 714–720 (2017)
59. Luo, Q., Yu, N., Shi, C., et al.: Surface plasmon resonance sensor for antibiotics detection based on photo-initiated polymerization molecularly imprinted array. Talanta **161**, 797–803 (2016)
60. Ma, Q., Wang, Y., Jia, J., Xiang, Y.: Colorimetric aptasensors for determination of tobramycin in milk and chicken eggs based on DNA and gold nanoparticles. Food Chem. **249**, 98–103 (2018)
61. Maeng, S.K., Sharma, S.K., Lekkerkerker-Teunissen, K., Amy, G.L.: Occurrence and fate of bulk organic matter and pharmaceutically active compounds in managed aquifer recharge: a review. Water Res. **45**, 3015–3033 (2011)
62. Mao, K., Zhang, H., Pan, Y., Yang, Z.: Biosensors for wastewater-based epidemiology for monitoring public health. Water Res. **191**, 116787 (2021)
63. Marshall, B.M., Levy, S.B.: Food animals and antimicrobials: impacts on human health. Clin. Microbiol. Rev. **24**, 718–733 (2011)
64. Masson, J.-F.: Portable and field-deployed surface plasmon resonance and plasmonic sensors. Analyst **145**, 3776–3800 (2020)
65. McKeating, K.S., Couture, M., Dinel, M.-P., et al.: High throughput LSPR and SERS analysis of aminoglycoside antibiotics. Analyst **141**, 5120–5126 (2016)
66. Minden, V., Deloy, A., Volkert, A.M., et al.: Antibiotics impact plant traits, even at small concentrations. AoB Plants **9**, plx010 (2017)
67. Miranda-Andrades, J.R., Pérez-Gramatges, A., Pandoli, O., et al.: Spherical gold nanoparticles and gold nanorods for the determination of gentamicin. Spectrochim. Acta Part A Mol. Biomol. Spectrosc. **172**, 126–134 (2017)
68. Nippes, R.P., Macruz, P.D., da Silva, G.N., Neves Olsen Scaliante, M.H.: A critical review on environmental presence of pharmaceutical drugs tested for the covid-19 treatment. Process. Saf. Environ. Prot. **152**, 568–582 (2021)
69. Olaru, A., Bala, C., Jaffrezic-Renault, N., Aboul-Enein, H.Y.: Surface plasmon resonance (SPR) biosensors in pharmaceutical analysis. Crit. Rev. Anal. Chem. **45**, 97–105 (2015)

70. Pérez-Juste, J., Pastoriza-Santos, I., Liz-Marzán, L.M., Mulvaney, P.: Gold nanorods: synthesis, characterization and applications. Coord. Chem. Rev. **249**, 1870–1901 (2005)
71. Petryayeva, E., Krull, U.J.: Localized surface plasmon resonance: nanostructures, bioassays and biosensing-a review. Anal. Chim. Acta **706**, 8–24 (2011)
72. Piliarik, M., Vaisocherová, H., Homola, J.: Surface plasmon resonance biosensing. In: Rasooly, A., Herold, K.E. (eds.) Biosensors and Biodetection, pp. 65–88. Humana Press, Totowa, NJ (2009)
73. Pollap, A., Kochana, J.: Electrochemical immunosensors for antibiotic detection. Biosensors **9**, 61 (2019)
74. Pomati, F., Castiglioni, S., Zuccato, E., et al.: Effects of a complex mixture of therapeutic drugs at environmental levels on human embryonic cells. Environ. Sci. Technol. **40**, 2442–2447 (2006)
75. Quintanilla-Villanueva, G.E., Luna-Moreno, D., Blanco-Gámez, E.A., et al.: A novel enzyme-based SPR strategy for detection of the antimicrobial agent chlorophene. Biosensors **11**, 43 (2021)
76. Rabiet, M., Togola, A., Brissaud, F., et al.: Consequences of treated water recycling as regards pharmaceuticals and drugs in surface and ground waters of a medium-sized mediterranean catchment. Environ. Sci. Technol. **40**, 5282–5288 (2006)
77. Bin, S.H., Kwon, Y.S., Lee, J., et al.: A novel reflectance-based aptasensor using gold nanoparticles for the detection of oxytetracycline. Analyst **140**, 6671–6675 (2015)
78. Seydack, M.: Nanoparticle labels in immunosensing using optical detection methods. Biosens. Bioelectron. **20**, 2454–2469 (2005)
79. Sharma, R., Ragavan, K.V., Thakur, M.S., Raghavarao, K.S.M.S.: Recent advances in nanoparticle based aptasensors for food contaminants. Biosens. Bioelectron. **74**, 612–627 (2015)
80. Shayesteh, O.H., Ghavami, R.: Two colorimetric ampicillin sensing schemes based on the interaction of aptamers with gold nanoparticles. Microchim. Acta **186**, 485 (2019)
81. Shayesteh, O.H., Khosroshahi, A.G.: A polyA aptamer-based label-free colorimetric biosensor for the detection of kanamycin in human serum. Anal. Methods **12**, 1858–1867 (2020)
82. Shelver, W.L., Larsen, G.L., Huwe, J.K.: Use of an immunoaffinity column for tetrachlorodibenzo-p-dioxin serum sample cleanup. J. Chromatogr. B Biomed. Sci. Appl. **705**, 261–268 (1998)
83. Sheng, Y.-M., Liang, J., Xie, J.: Indirect competitive determination of tetracycline residue in honey using an ultrasensitive gold-nanoparticle-linked aptamer assay. Molecules **25**, 2144 (2020)
84. Shrivastav, A.M., Mishra, S.K., Gupta, B.D.: Localized and propagating surface plasmon resonance based fiber optic sensor for the detection of tetracycline using molecular imprinting. Mater. Res. Express **2**, 35007 (2015)
85. Song, K.-M., Jeong, E., Jeon, W., et al.: Aptasensor for ampicillin using gold nanoparticle based dual fluorescence–colorimetric methods. Anal. Bioanal. Chem. **402**, 2153–2161 (2012)
86. Sternesjö, Å., Mellgren, C., Björck, L.: Determination of sulfamethazine residues in milk by a surface plasmon resonance-based biosensor assay. Anal. Biochem. **226**, 175–181 (1995)
87. Strobbia, P., Languirand, E.R., Cullum, B.M.: Recent advances in plasmonic nanostructures for sensing: a review. Opt. Eng. **54**, 1–21 (2015)
88. Ungureanu, F., Wasserberg, D., Yang, N., et al.: Immunosensing by colorimetric darkfield microscopy of individual gold nanoparticle-conjugates. Sens. Actuators B Chem. **150**, 529–536 (2010)
89. Uygun, Z.O., Uygun, H.D.E., Ermiş, N., Canbay, E.: Molecularly imprinted sensors — new sensing technologies. In: Rinken, T. (ed.) Biosensors - micro and nanoscale applications. IntechOpen, London (2015)
90. Vaisocherová, H., Mrkvová, K., Piliarik, M., et al.: Surface plasmon resonance biosensor for direct detection of antibody against Epstein-Barr virus. Biosens. Bioelectron. **22**, 1020–1026 (2007)

91. Wang, H., Zhou, X.-J., Liu, Y.-Q., et al.: Simultaneous determination of chloramphenicol and aflatoxin M1 residues in milk by triple quadrupole liquid chromatography−tandem mass spectrometry. J. Agric. Food Chem. **59**, 3532–3538 (2011)
92. Wang, X., Dong, S., Gai, P., et al.: Highly sensitive homogeneous electrochemical aptasensor for antibiotic residues detection based on dual recycling amplification strategy. Biosens. Bioelectron. **82**, 49–54 (2016)
93. Wang, X., Ryu, D., Houtkooper, R.H., Auwerx, J.: Antibiotic use and abuse: a threat to mitochondria and chloroplasts with impact on research, health, and environment. BioEssays **37**, 1045–1053 (2015)
94. Watanabe, N., Harter, T.H., Bergamaschi, B.A.: Environmental occurrence and shallow ground water detection of the antibiotic monensin from dairy farms. J. Environ. Qual. **37**, S-78–S-85 (2008)
95. Watkinson, A.J., Murby, E.J., Kolpin, D.W., Costanzo, S.D.: The occurrence of antibiotics in an urban watershed: from wastewater to drinking water. Sci. Total Environ. **407**, 2711–2723 (2009)
96. Wochner, A., Menger, M., Orgel, D., et al.: A DNA aptamer with high affinity and specificity for therapeutic anthracyclines. Anal. Biochem. **373**, 34–42 (2008)
97. Xu, Y., Lu, C., Sun, Y., et al.: A colorimetric aptasensor for the antibiotics oxytetracycline and kanamycin based on the use of magnetic beads and gold nanoparticles. Microchim. Acta **185**, 548 (2018)
98. Yan, J., Huang, Y., Zhang, C., et al.: Aptamer based photometric assay for the antibiotic sulfadimethoxine based on the inhibition and reactivation of the peroxidase-like activity of gold nanoparticles. Microchim. Acta **184**, 59–63 (2017)
99. Yanase, Y., Hiragun, T., Ishii, K., et al.: Surface plasmon resonance for cell-based clinical diagnosis. Sensors **14**, 4948–4959 (2014)
100. Zhou, L., He, X., He, D., et al.: Biosensing technologies for *Mycobacterium tuberculosis* detection: status and new developments. Clin. Dev. Immunol. **2011**, 193963 (2011)
101. Zhou, L., Limbu, S.M., Shen, M., et al.: Environmental concentrations of antibiotics impair zebrafish gut health. Environ. Pollut. **235**, 245–254 (2018)

Chapter 23
Advances in Luminescence-Based Biosensing with Quantum Dots

Debasmita Sinha Ghosh and Abhijit Saha

Abstract The design and development of biosensors have assumed a centre stage for researchers in the present decade owing to the wide range of applications, such as disease diagnosis, drug delivery and monitoring of environmental and food quality. Therefore, it is crucial and required to build effective biosensors that can assess the finer points of biological interactions, even at very small scales, with extraordinary precision and with the highest sensitivities ever. Miniaturisation of the biosensing devices using micro-and nano-fabrication technologies has drawn considerable attention in recent years. With higher sensitivities and smaller detection limits by several orders of magnitude, these nano-objects have clearly improved performances. The fact that nanoparticles have high surface area to volume ratios, which enable the surface to be employed in a greater and far wider range of functional ways, is one advantage shared by all nanomaterials. Additionally, the unique electromechanical, optical, or physicochemical features of these materials are fantastic advantages for the development of biosensors. Furthermore, these nanomaterials can themselves act as transduction elements. This article provides an overview of applications of quantum dots in biosensing by employing luminescence-based methods.

Keywords Quantum dots · Luminescence · Conjugation · Biosensing · Sensitivity

D. Sinha Ghosh
Department of Food and Nutrition, Sarojini Naidu College for Women, West Bengal State University, 30, Jessore Raod, K. B. Sarani, Kolkata 700028, India

A. Saha (✉)
UGC-DAE Consortium for Scientific Research, Kolkata Centre, III/LB-8 Bidhannagar, Kolkata 700106, India
e-mail: abhijit@alpha.iuc.res.in

D. Mohanta and P. Chakraborty (eds.), *Nanoscale Matter and Principles for Sensing and Labeling Applications*, Advanced Structured Materials 206,
https://doi.org/10.1007/978-981-99-7848-9_23

23.1 Introduction

A biosensor is a device that incorporates a biological sensing element that is either closely coupled to or integrated within a transducer and that converts the responses of bioanalyte interactions into an identifiable and repeatable form by using transduction mechanisms [9] to convert a specific biochemical reaction's energy into an electrical form. Nanomaterials, like many other technical advancements, have shown considerable promise in biosensing applications. The choice of substrate for spreading the sensing material determines the sensor performance for the creation of an effective biosensor. A variety of nanomaterials are being steadily included into biosensors, including gold nanoparticles [53], carbon nanotubes (CNTs) [20], magnetic nanoparticles [52] and quantum dots [83].

Numerous variables influence or determine whether a specific type of nanomaterial is appropriate for use in biosensing applications. These elements, coupled with their energy-sensitive and selective reactions, are the main components of their physical and chemical characteristics. Appropriate designing of nanostructured materials is a major requirement before appropriately implementing or utilising the material for the sensing applications. The method of nanofabrication focuses on two crucial processes: the production and design of nanoscale adhesive surfaces using integrated circuit technology and the engineering of nanomaterial surfaces using the technique of micromachining. These four fundamental processes—photolithography, thin film etching/growth, surface etching tactics, and chemical bonding parameters—are used in different ways in this biosensing approach.

The crucial component while using nanomaterials for sensing applications is to monitor and optimise their optical characteristics. Surface Plasmon Resonance (SPR) is an exciting phenomenon when presence of nanoparticles maximises the exact and sharp scale optical response of the sensing materials with the incident light. In surface plasmon resonance, photons at a particular angle of incidence excite the surface electrons of metals, which then propagate parallel to the surface. Thus, SPR signifies collective oscillations of the conduction electrons in metallic nanostructures. Due to their distinctive optical characteristics, which give them a photonic character and good capacity to be utilised as fluorophores, nanoparticles are particularly well suited for this trait, which is known as localised surface plasmon resonance [85]. Since the refractive index of a medium also affects the SPR effect, the detection of analytes is made possible by even slight changes in the refractive index of the sensing medium. Compared to a biosensor, a nanobiosensor is better able to detect the minute interacting phenomena due to SPR, which allows for a considerably higher and more accurate degree of quantification of biological interactions [35].

In this way, regardless of their type, nanoparticles need to be optimised for their performance and impact in accordance with the desired goal before being really implemented for biosensing purposes. For example for effective and sensitive biosensing, metallic NPs like gold and silver nanoparticles, dielectric NPs like SiO_2 or MnO_2, polymeric NPs, and semiconductor NPs or quantum dots have been combined with microstructures [10]. Diverse types of nanoparticles can be ideal for various roles in sensing of different biological systems because they have different electrical properties. For instance, electrical probing of biomolecules typically uses metallic NPs. While semiconductor NPs have been utilised historically as labels or biomarkers, dielectric nanoparticles have been used frequently to immobilise biomolecules.

Gold nanoparticles [40] are the most commonly utilised noble nanoparticle for biosensing applications because of its biocompatibile nature, optical and electrical properties, and ease of manufacture or modification [5]. For the detection of DNA, oligonucleotides, or immunosensors, gold nanoparticles were used to create a variety of effective colorimetric biosensors [81].

Quantum dots (QDs), which are luminescent semiconducting nanocrystals, are another well-known example of nanomaterials used for bioanalytics. Because of unique and fascinating optical as well as electrical characteristics, like broad spectral absorption with size-tunable emission, stronger brilliance, improved photo-stability, QDs are viewed as potential and alluring building blocks in the process of developing effective biosensors with great sensitivity, appreciable selectivity, and brightness. Here, we also offer some insight into potential prospects for QD-based biosensor research.

23.2 Advantages of Quantum Dots

Small semiconductor colloidal particles called quantum dots (QDs) exhibit unique optical and electrical capabilities that are superior to those of larger particles because of quantum confinement. As a result, they are desirable and promising materials with a variety of uses in optoelectronics, cellular imaging, sensing, drug delivery, etc. [51]. They can be produced using atoms from different groups, like II–VI, III–V, or IV–VI of the periodic table [29]. In 1998, the teams led by Alivisatos and Nie first used QDs as innovative luminescent probes for cellular imaging and detection of various biological systems [8, 11]. By only utilising QDs of various sizes, the size-tunable optical properties of QDs, such as their emission spectrum, considerably simplify the detection of various objects [28]. The interest in permitting the optical excitation of numerous QDs with a single frequency of photons has been sparked by the broad excitation and narrow luminescence of quantum dots [47] (Fig. 23.1). Most of the functional constraints of organic fluorophores and fluorescent proteins, which are typically used for biosensor applications, are solved by semiconductor nanoparticles [26]. The applications of quantum dots as the platforms for fabricating sensors of excellent efficiency in detecting a variety of biologically important molecules with very high sensitivity have made great progress [45].

The nanoparticles' surface is altered with the necessary molecules to make them biocompatible and to construct useful QD-labelled probes [5]. QDs have been frequently used as fluorescent markers because of the distinguishing advantages as described above.

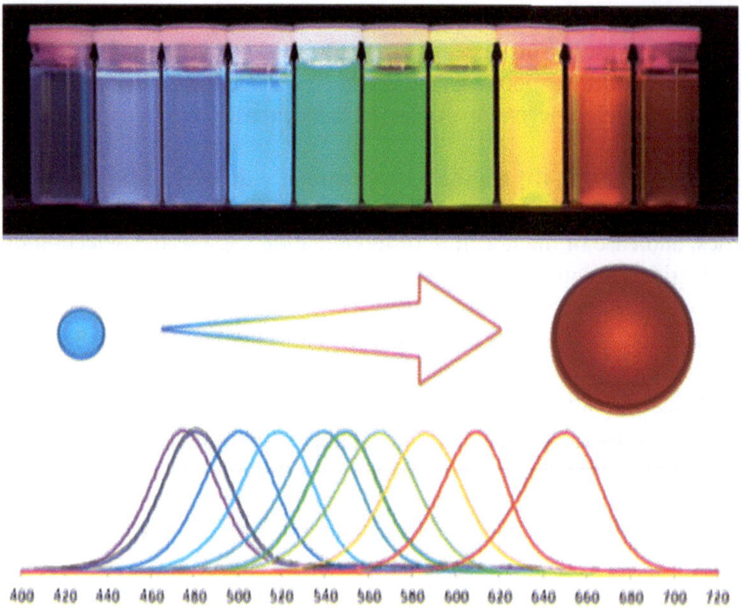

Fig. 23.1 Size-dependent multicolour fluorescence of QDs in aqueous solution [2]

23.3 Quantum Dot-Based Fluorescent Biosensor

Due to their great selectivity and increased sensitivity, fluorometric biosensors have captivated researchers' attention for a considerable amount of time [31]. The optical process known as fluorescence occurs when absorption of photons of shorter wavelength leads to emission of photons of longer wavelength [42]. QDs are being used in fluorescence tagging to replace molecular dyes for biosensing applications due to their superior properties over conventional dyes, like less photobleaching, tunable absorbance and emissions with size, broad excitation and narrow emission spectrum. The broad absorption spectra of the QDs allow emission from different-sized QDs with single wavelength excitation. Among numerous sensors, the most developed fluorescent biosensors are QD-based because of their superior multiplex analysis capability and high sensitivity [77].

The colloidal QDs that have received the greatest attention are primarily derived from chalcogenides (S, Se, Te) of cadmium and these QDs are having spectral features of broad absorption with narrow emission which also vary with size [62]. This phenomenon is caused by the fact that different nanocrystal sizes have varied band gaps in the semiconductor material (the larger the particle, the smaller the band gap) and this results in variable photon emission due to recombination of the electron–hole pair, exciton [78]. Due to wide variety in emission wavelengths of quantum dots of various sizes, multiplexed assay utilising traditional optical transduction is effective [57]. On the other hand, structural defects in crystal lattice might trap the ejected electrons or holes, causing non-radiative relaxation [49]. In order to get around this problem, passivation of these surface imperfections and improvement in quantum efficiency with high photo-stability was achieved by forming a shell of other semiconductor material with a larger band gap range (often ZnS) over the core material referring to development of Core/shell composites [32]. Exceptional photochemical stability of core/shell quantum dots has made these materials suitable replacement of organic fluorophores in a variety of applications, like sensing, imaging, etc. [63].

QDs are currently offered with suitable inert or biocompatible coatings, which may create different functionalities by providing active groups for immobilisation of bioreceptors and also to address issues arising out of any toxicity [6]. Thus, as long as photophysical recombination event of the semiconducting material is not hampered, nearly any type of biomolecule can be joined on the surface of these nanocrystals.

23.3.1 Quantum Dot-Based Sensing Through Fluorescence Quenching Analysis

Information gathered from research on nanoparticle-biomolecule interactions is useful for creating effective NP-based biosensors. QDs have demonstrated their appropriateness to detect analytes as optical probes. The fluorescence characteristics of the QDs, which function as active sensor components, alter in response to the analyte. Fundamental investigations have demonstrated that the surface charges of the QDs can alter as a result of chemical or physical interactions with a particular chemical species (the analyte), which in turn can have a major impact on the photoluminescence emission [21]. The luminescence of the QDs can be quenched as a result of surface state traps being created by the binding of analytes to the nanoparticle surface. On the other hand, binding of analytes can also lead to amplification of the luminescence of the QDs through passivation of surface state traps through filling the electronic defects or bringing them energetically closer to the band boundaries [15]. Such reported methods are quite straightforward, quick to create, and several have demonstrated extremely high sensitivity.

Comprehensive study done by Saha and his group on the interactions between the quantum dots (QDs) and various biomolecules explored the feasibility of probing the biomolecules by a simple and fast quantum dot-based optical method. First attempt was made by them to probe the important diuretic, mannitol with the help of thiol capped QDs by carrying out a detailed study on interactions of mannitol with CdS QDs by optical absorption and photoluminescence (PL) spectroscopy [23]. Mannitol significantly increased the fluorescence of QDs even at micromolar doses. Mannitol's OH group binds to the surface of CdS particles, passivating any surface traps that result from electrical flaws and enhancing luminosity [74]. As seen in Fig. 23.2a, there was a significant size dependence interaction of CdS QDs with mannitol. This is consistent with the pioneering work on tyrosine [59]. The proposed method, which uses cysteine-capped CdS QDs as fluorescence-based biosensors, can be a very quick,

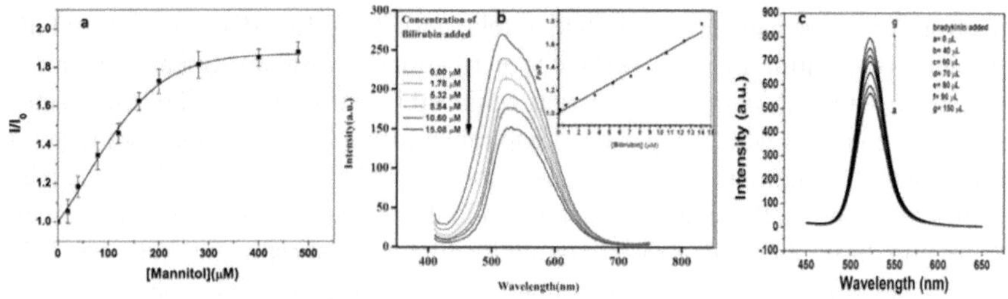

Fig. 23.2 **a** A plot of I/I_0 against mannitol concentration for CdS QDs of different sizes, (▲) 3.0 nm, (■) 2.6 nm, (●) 3.4 nm. Reproduced with permission [22]. Copyright 2010, Elsevier; **b** Photoluminescence quenching of CdSe-BSA QDs in the presence of bilirubin at different concentrations. Stern–Volmer plot of quenching of luminescence of CdSe-BSA QDs (inset), Reproduced with permission [34]. Copyright 2020, American Chemical Society; **c** Luminescence enhancement of CdTe QDs (d_{AV}~4.3 nm) in the presence of varied concentrations of tyr-bradykinin, Reproduced with permission [48] Copyright 2012, American Chemical Society®

easy and potential tool for the monitoring of diuretics. A quantitative correlation between mannitol content and PL enhancement of CdS QDs was obtained.

In a very recent work, it is demonstrated that synthesised bovine serum albumin (BSA)-encapsulated cadmium selenide quantum dots (CdSe-BSA QDs) can strongly interact with bilirubin [34], a marker for jaundice and the same can be employed in the fast sensitive detection of bilirubin in real blood samples. Here, it was demonstrated that the BSA molecule serves to bring BR closer to the quantum dots during interactions of CdSe-BSA QDs with BR in addition to stabilising the quantum dots in an aqueous environment. The photoluminescence spectra of CdSe-BSA QDs are shown in Fig. 23.2b in the presence of various bilirubin concentrations, and the plot of F_0/F versus BR concentration (inset) follows the linear Stern–Volmer relation (Eq. 23.1) [38]. In the presence of other potentially interfering coexisting compounds, it likewise demonstrated a rapid response time and high selectivity towards bilirubin.

$$\frac{F_0}{F} = 1 + K_{SV}[Q] \tag{23.1}$$

where F_0 and F represent the emission intensity of QDs before and after the addition of the quencher (Q), respectively, K_{SV} is the Stern–Volmer constant.

The wrong unfolding of protein molecules may lead to protein-mediated diseases. It is crucial to understand protein conformational changes when nanoparticles are in use as drug-carrier. Chinnathambi et al. studied the interaction mechanism of haemoglobin and thrombin with water-soluble near-infrared silicon quantum dot micelles using fluorescence, circular dichroism spectroscopic techniques and molecular modelling [17]. Analysis of fluorescence spectroscopic results revealed that emission quenching of the proteins by micelles is static at low concentrations and dynamic in nature at higher concentrations. From Circular Dichroism spectral analysis, it was clear that micelle binding over the concentration range could not disturb the secondary structural changes of the plasma proteins. According to the molecular docking data, haemoglobin and thrombin have good binding affinities to Si QD micelles. In addition, molecular dynamics simulations were performed which also showed very good binding affinity of Si QD micelles towards haemoglobin and thrombin protein molecular systems. The observed experimental and simulation results will provide useful information for developing non-toxic NIR QDs as biomarkers for in vivo trials in near future.

A detailed investigation was also reported to probe bradykinin (BK), a vasodilator, employing luminescent CdTe QDs [48]. According to optical spectroscopic investigations, bradykinin's three amino acids may be engaged in interactions with QDs and tyr-BK that alter the protein's structure. With the use of isothermal calorimetry, it was also discovered that the interaction between bradykinin and CdTe QDs is an entropy-driven process. Even in the presence of micromolar concentrations of tyr-BK, the luminescence of QDs dramatically enhanced (Fig. 23.2c), and the enhancement followed the Langmuir binding isotherm. Strong size dependence was seen in the CdTe QDs and the peptide tyr-BK interaction. To illustrate, the binding constant values of tyr-BK with CdTe QDs of sizes of 3.7 nm, 4.1 nm, and 5.0 nm at physiological pH in the excited state were determined to be $95 \times 10^6\ M^{-1}$, $105 \times 10^6\ M^{-1}$, and $124 \times 10^6\ M^{-1}$, respectively [48]. Without requiring any time-consuming processing, the suggested method can be used to detect bradykinin in conditions of illness, such as hypertension, sepsis, and asthma in the range of 0.6–5 μM. Unlike another observation with free tyrosine [59], where more enhancement was observed with smaller size particles, the rise in binding constant with increasing CdTe QD size is not accompanied by an increase in binding constant of similar trend. The interactions between CdTe QDs and glucose oxidase were found to exhibit size dependence, which is also seen in the current situation [58]. Therefore, it is reasonable to suppose that the macromolecular system's structure and substantially larger dimensions make it more compatible for interacting with larger nanoparticles, which leads to stronger luminescence enhancement with the larger sized particles. Colloids of semiconductors that are kinetically stable have a lot of surface defect states. A wide variety of surface energies and structures are covered by these states. At mid band gap energies, a lot of these defect states are present. These states help capture the electron–hole pairs that were initially created. Here, tyr-BK binds to the surface of the QDs, effectively removing these trap states, increasing the light emission in the process.

The interaction between glutathione (GSH)-capped CdTe QDs of 3 nm size and albendazole (ABZ) was investigated using UV–vis absorption, fluorescence spectroscopy, and resonance Rayleigh scattering (RRS) technique [41]. It was shown that the QDs linked to ABZ create novel complexes based on alterations in the UV–vis absorption spectra of the two molecules. In pH 7.1 phosphate buffer solution, it was discovered that ABZ efficiently muted the fluorescence of QDs. The QDs-ABZ complexes, which were greatly stabilised by electrostatic attraction, are primarily responsible for the fluorescence quenching of QDs by ABZ. A notable enhancement in RRS (Resonance Rayleigh scattering), which was proportional to the ABZ concentration within a certain concentration range, was also brought about by the interaction between QDs and ABZ. The ideal reaction conditions, crucial elements, and the impact of coexisting chemicals were investigated. These experimental results will be helpful to develop probe based on quantum dots for the detection of albendazole.

Through similar spectroscopic quenching analysis, new protocols were formulated for the development of QD-based sensors for many other biomolecules, like ascorbic acid [22], folic acid [37], tyrosine [59], nucleic acids [12], flavins [13], alkaline phosphatise [25], glucose oxidase [58], and metal ions like Cu^{2+} [24], [36], etc.

23.3.2 *QD Biosensors Based on FRET Technique*

Target biomolecules can be detected using the fluorescence resonance energy transfer (FRET) technology without the need for washing or separation processes. When the distance between the donor and the acceptor is less than a crucial radius known as the Förster radius [16], energy transfer from the exiting QD (donor) to a quencher (acceptor), such as an organic fluorophore, may occur through a non-radiative route. As a result, the acceptor's emission intensity rises and the donor's emission and excited state lifetime decrease. QDs are regarded as remarkable FRET donor options for biosensing applications when compared to traditional fluorophores. In particular, multiplexed FRET sensing, where donor emission of each channel should be clearly distinct, prefers QDs due to their limited emission bandwidth and broad absorption profile. With greatly improved reproducibility and reliability, multiplexed assays simultaneously detect and quantify numerous biomarkers in a single sample while using fewer sample material, less time on average for analysis, and less sample volumes than separate tests. Depending on the type of analytes, QDs can function as either fluorescence donors [30] or acceptors [76] in QD-based FRET sensors. Additionally, for simple, quick, quantitative, and sensitive detection of target molecules, considerably high signal-to-noise ratio (SNR) and approaches with high sensitivity, such as the single molecule detection (SMD) method, have been applied for developing biosensors on quantum dot platform [44]. The combination of SMD with QDs has enabled in developing numerous sensors supported by a single QD in detecting many biomolecular systems which include deoxyribonucleic acids and its methylated molecules [73] and enzymes [43].

Detection of Nucleic Acid

There are several DNA sensing techniques that have been developed with good selectivity and low amounts of interfering molecules. The vast range of applications for polymerase-based amplification techniques, which are largely employed for DNA testing, are constrained by the need for accurate thermal cycling, intricate primer design, and costly enzymes [86]. In order to facilitate the binding of target DNA, Zhang and co-workers created a DNA test using liposomes encapsulating hundreds of QDs to generate hybrid complexes (Fig. 23.3a) [87]. This method is able to detect single-stranded DNA at attomolar concentrations without the involvement of target amplification. Eminently, this strategy can be employed for simultaneous detection of multiple DNA targets.

FRET has been used to detect DNA homogeneously without the need for washing or separating procedures. This technique has been illustrated by developing a sensor using quantum dot platform for in detecting *Helicobacter pylori*-specific DNA in nanomolar concentration [66]. *H. pylori* is the mostly prevalent pathogen causes ulcers in human [75]. In this assay CdTe quantum dots (QDs) and 5-carboxytetramethylrhodamine (Tamra) were used, respectively to label two sets of oligonucleotide probes. QDs labelled with an amino-modified first oligonucleotide and a Tamra-labelled second oligonucleotide were introduced to the DNA targets, where they hybridised to produce an effective FRET process from QD to Tamra. As no sandwiched hybrid can form if the DNA concerned is not available in the system

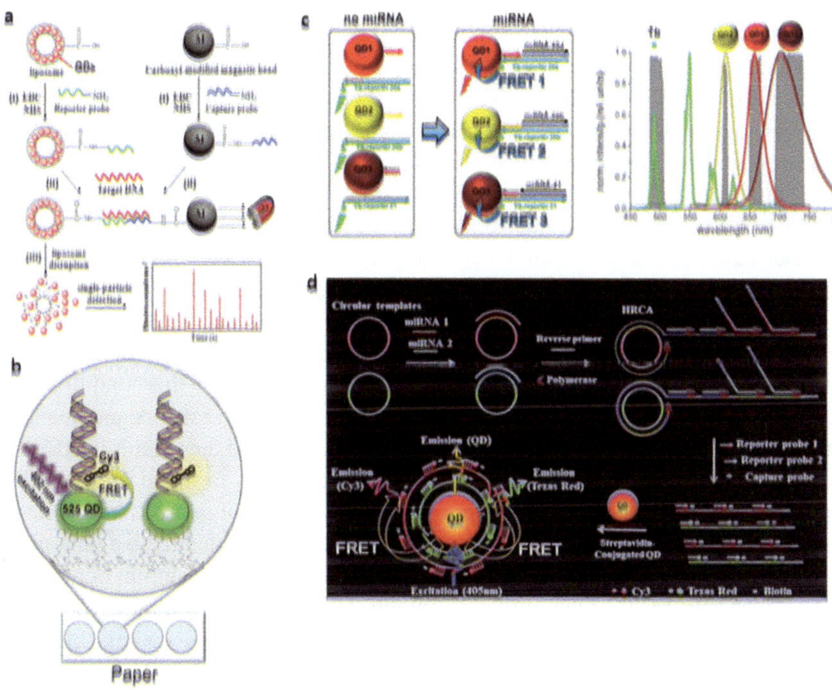

Fig. 23.3 Scheme representing luminescent sensors on QD platform for the detection of DNA and micro-RNA assay **a** The integration of liposomes-QDs complex as single-particle probe of DNA of HIV. Reprinted from [87] 2013 American Chemical Society copyright, **b** Fabrication of a paper-based device with a QD mounted on it for detecting DNA specific for *Helicobacter pylori* [50]. Reproduced with permission of American Chemical Society, **c** Multiplexed microRNA test using a QD-based FRET biosensor, reprinted from [60] American Chemical Society, **d** Integration of HRCA with FRET induced by QDs allows probing of many microRNAs simultaneously. Reprinted from [30] Royal Society of Chemistry®

and thus FRET cannot take place. This quick and easy to perform assay may be used to find deoxyribonucleic acid specified for *H. pylori* at nM level.

Paper-based analytical devices (PADs) hold promise for point-of-care (POC) diagnostics [46] because these are simple to use and inexpensive. Through the use of quantum dots and FRET analysis, Krull and co-workers created a quick point-of-care paper-based diagnostic assay for nucleic acids (Fig. 23.3b) [50]. The detection of Cy3-DNA hybridisation by this device was extremely precise and sensitive (~femtomol concentration). This technique is quicker compared to the older assay method of nucleic acid hybridisation at solid phase because of the fast hybridisation kinetics [1]. Even in the presence of significant volumes of non-complementary DNAs, great sensitivity and excellent specificity are proven without the need of an interface, and it can also be used to identify single nucleotide polymorphisms (SNPs) with ratio of discrimination as much as 19.

Various human diseases like cancers and cardiovascular problems may be caused due to dysregulation of levels of microRNA [19], which may direct expression of the gene through attachment to the $3'$-untranslated segments of mRNA in question [3]. Using multiplexed Förster resonance energy transfer (FRET) between fluorescent complexes of terbium Lumi4-Tb (Tb) and 03 separate semiconductor quantum dots (QDs), Hildebrandt and Qiu detected sensitively three different miRNAs in a single sample of 150 μL with limit of detection as *ca.*1 nM (subpicomol) (Fig. 23.3c) [60]. Short single-stranded DNAs with seven or eight nucleotides are conjugated with three QDs resulting in emission having maxima at 605, 655, and 705 nm. On the other hand, the luminescent Tb complexes are labelled with reporter DNAs of 03 varieties. Even if strands of the reporter DNA partially complement to the short ssDNAs, the temperature (22 °C) of the experiment prevents them from forming stable dsDNA. As the quantum dots remain isolated from terbium complex, energy transfer through fluorescence resonance process cannot happen in this condition. The creation of stable complexes involving terbium, reporter/short ssDNA and quantum dots may be triggered by the target microRNAs, which would then result in effective energy transfer through FRET process from quantum dots to metal terbium. The adopted assay facilitates the simultaneous detection of several microRNAs owing to the exceptionally long life span (order of ms) of Tb coupled with effective time-gated fluorescence energy transfer from the same terbium to distinct quantum dots.

The necessity for 03 QDs different colours to accomplish multiplex detection in the microRNA test is one of its limitations. Zhang and co-workers made an improvement on this work by applying fluorescent resonant energy transfer techniques to demonstrate multiplexed sensing by employing a single quantum-dot-based biosensor (Fig. 23.3d) [30]. It has enormous promise for biomedical research and disease diagnostics and has great potential in the process of detecting microRNAs targeted in cancerous cells.

Wang et al. demonstrated FRET-based sensing of both DNA and microRNA using CdTe/CdS core/shell quantum dots as donor and BHQ2 as quencher [67]. The ligand-exchange approach was used to create the DNA-QDs conjugate. Direct ssDNA attachment to the QD surface results in extremely water-dispersible nanoconjugates that are thiol-labelled. The DNA-QDs conjugate can be utilised as a nanoprobe to create a nanosensor for detecting nucleic acids because it benefits from both the well-controlled recognition properties of DNA and the good optical qualities of QDs. When targeted nucleic acids (DNA and microRNA) are available, sandwich hybrids may form, which bring QD and BHQ2 closer in the system. In consequence, the energy transfers from the QD to BHQ2 leading to efficient fluorescence quenching of the QDs. This test can differentiate between mismatch in single-base and random sequences of nucleic acids and has a high sensitivity and good selectivity.

Detection of Proteins

The diagnosis and treatment of disease depend on the sensitive and specific detection of proteins, like glycoproteins, prostate-specific antigens, and carcinoembryonic antigens [79]. Fluorescence-based detection techniques like fluorescence polarisation (FP), often referred to as fluorescence anisotropy, are frequently used to track molecular interactions in solutions. Here, the fluorophore's light displays uneven intensities along various polarisation axes [14]. Fluorescence polarisation measurement, when compared to other detection techniques, may yield more data, such as the molecular weight and the fluorescence lifespan [33]. The target binding events may be detected by the QDs using fluorescence polarisation, whose signal is extremely sensitive to the surrounding environment [70].

Sang and co-workers fabricated a fluorescent sensor supported on a quantum-dot platform for detecting 10 distinct glycoproteins precisely [64] with detection limit of 0.15 μM. Over 50% of protein-related cellular processes are mediated by glycoproteins, which are viewed as the new biomarkers for disease detection [7]. Fluorescence polarisation (FP), a novel signalling channel, was added into a 'single probe with three signalling channels' sensor array using a novel approach (Fig. 23.4a). The single signalling channel was FP, and the three signalling channels were room temperature phosphorescence, light scattering, and aminophenylboronic-acid-conjugated poly (acrylic acid)-Mn-doped ZnS quantum dots (APBA-QD) as detecting probe. APBA having high affinity for the glycoproteins induces the attachment of targeted glycoproteins onto the surface of quantum dots to produce unique signals of light scattering, phosphorescence, and fluorescence polarisation. This test shows good accuracy in identifying unidentified samples.

Protein detection in a homogeneous format can also be done by QD-based FRET strategy. To diagnose prostate cancer, Hildebrandt and co-workers fabricated immunosensor using quantum dot platform for the very sensitive (1.6 ng/mL) detection of prostate-specific antigens (PSA) (Fig. 23.4b) in samples of 50 L serum [76]. An important biomarker for the early detection of prostate cancer is PSA, which is secreted by the prostate gland's epithelial cells [55]. Here, the PSA-specific primary antibodies are conjugated to the Tb and the QD, respectively. In presence of target PSAs Tb-protein-QD sandwiched nanostructure is created, where effective fluorescence resonance energy transfer from terbium site to the quantum dots happens. This method of presenting antibody epitopes on QD surfaces via self-assembling, modified peptides is easily transferable to various assays of interest. In this assay detection

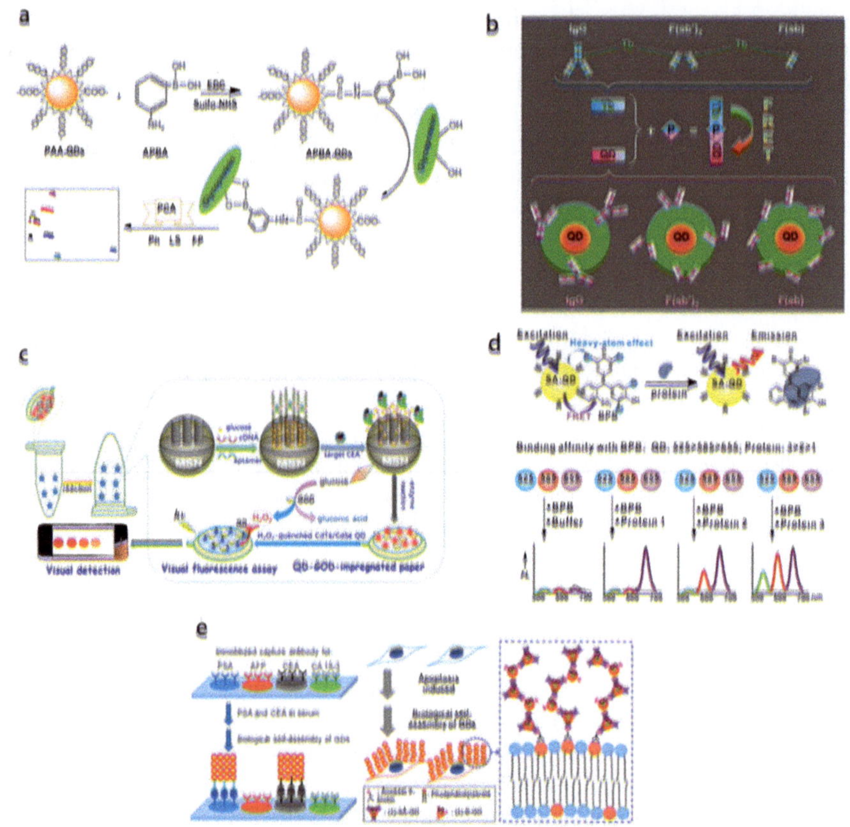

Fig. 23.4 Fluorescence sensors based on QD platform for protein detection are shown schematically. **a** QD-based sensor that has 03 signal pathways in detecting glycoproteins via fluorescence polarisation (Reproduced with permission [64]), **b** Sensor developed on quantum dots to detect prostate-specific antigen through fluorescence resonance energy transfer process [76]; **c** Target-generated H_2O_2 quenches QDs to detect carcinoembryonic antigen [61], **d** Multiplexed protein detection using a QD-based chemical nose [82], **e** Scheme representing an immunosensor developed using QDs for protein detection [54]. All figures were reproduced with permission of American Chemical Society

limit of PSA was 1.6 ng mL^{-1} which is lower compared to that of its clinical cut-off value of 2 ng mL^{-1} [27].

Based on luminescence quenching of quantum dots, the proteins can be detected as well. H_2O_2 can quench the fluorescence of QDs efficiently in as a function of concentrations [80]. Based on this fact, Tang and co-workers introduced a paper-based biosensor to detect carcinoembryonic antigen (CEA) (Fig. 23.4c) [61]. A lot of different types of human cancers can be clinically diagnosed using this protein biomarker (especially colorectal cancer) [4]. In this experiment, through physical adsorption, quantum dots conjugated with glucose oxidase (GOD) was immobilised on paper and mesoporous silica nano-containers (MSNs) loaded with glucose were marked with CEA aptamers to which the targeted CEA could bind, causing MSNs' pores to open leading to the successive liberation of glucose. Large volumes of H_2O_2 are produced as GOD oxidises the released glucose. This H_2O_2 effectively block the fluorescence of QDs bonded to the paper. A relatively inexpensive spectrofluorometer is capable of quickly detecting variations in QD fluorescence. This assay with great sensitivity makes it possible to detect CEA as low as 6.7 pg mL^{-1}.

The aforesaid protein assays still have a single-channel read-out limitation despite their high assay performance. The QDs enable simultaneous multiplexed analysis of various targets at a single excitation because of their size-tunable narrow fluorescence emission. [47]. A multicolour quantum dot (QD)-based multichannel sensing platform was created by Zhang et al. for the quick detection of many proteins in an array-free format (Fig. 23.4d) [82]. In this study, bromophenol blue (BPB), a protein-binding dye, is investigated as a potent reversible quencher of QDs. BPB in combination with multicoloured QDs may produce quenched QD-BPB complexes. Due to the competitive protein-BPB binding caused by the addition of proteins, the QD-BPB complexes will be disrupted, resulting in the separation of BPB from the QDs and the production of distinctive fluorescence patterns. As various proteins possess varying binding affinities for BPB, it is possible to experimentally separate numerous proteins at once. The quickest protein test ever recorded can be performed on this platform in about 1 min, making it extremely promising for quick and high-throughput clinical diagnosis.

For protein detection, the most popular method is enzyme-linked immunosorbent assay (ELISA). However, signal production in ELISA frequently depends on costly and not so stable enzymes (such as, horseradish peroxidase) [18]. Kim and co-workers created an immunosensor using quantum dot platform which can be used with good reliability and to amplify detection of protein without utilising any enzymes (Fig. 23.4e) [54]. The protocol for such sensing involves the modification of surfaces of quantum dots by conjugating with streptavidin (SA) and also with biotin. Procedures used were typical sandwich-type immunoassays, where the capture antibody was immobilised on a plastic substrate and the detection antibody was biotin-labelled. A sandwich comprising capture antibody/protein/detection antibody is formed in the presence of the target protein and the remaining free antibodies are removed by washing with water. Biotinylated detection antibody and streptavidin-QD can bind together, and streptavidin-QD can bind to biotin-QD. The targeted protein binding events were effectively transduced and amplified by the fluorescence of the SA–biotin QD conjugates. This approach offers a platform, which is considered to be a universal one in determination of protein and can also be used to identify additional protein biomarkers, such as antigens specific for prostate and human colon adenocarcinoma.

Detection of Enzyme Activity

Caspases define a class of cysteinyl proteases that are considered to be regulator of apoptosis [69], and these have also been identified as potential targets for diagnosis

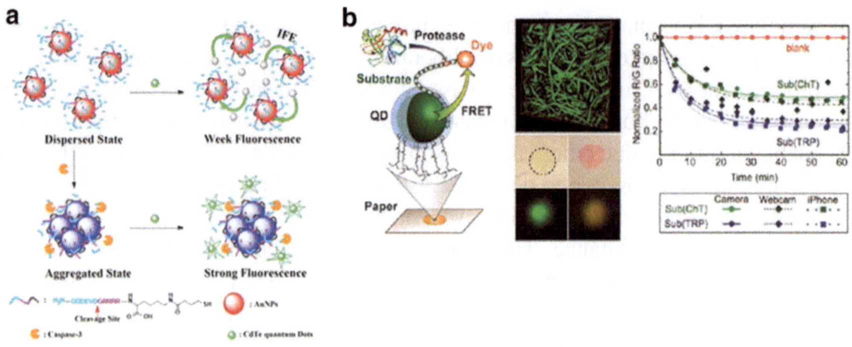

Fig. 23.5 QD-based fluorescence biosensors for detecting enzyme activity are shown schematically. **a** Caspase-3 activity was discovered using QD fluorescence and inner-filter effect of AuNPs. Reproduced with permission [39] Copyright 2013, American Chemical Society, **b** Scheme representing a paper substrate modified with quantum dots for detecting protease. Reprinted from [56] Copyright 2013, American Chemical Society

and treatment for illness [65]. Li et al. devised a straightforward fluorometric test by taking into consideration of inner-filter effect (IFE) of gold nanoparticles (AuNPs) on cadmium telluride quantum dots (CdTe QDs) for the extremely sensitive detection of caspase-3 activity (Fig. 23.5a) [39]. In this assay, by altering AuNPs with formation of Au–S bonds involving the peptide substrate of caspase-3, monodistributed gold nanoparticles–peptide complexes are generated. Because of the substantial overlap of absorption of gold nanoparticles with emission of quantum dots in use, the AuNPs-peptide complexes may effectively suppress the fluorescence of QDs (520 nm). The monodispersed Au-NPs upon labelling with a peptide (peptide-modified AuNPs) possessing the caspase-3 cleavage sequence, showed a tendency to aggregate when exposed to caspase-3 and the absorption band of Au NP shifts from 520 to 680 nm leading to recovery of luminescence of CdTe QDs for caspase-3 sensing. This method of determination can achieve great sensitivity with a limit of detection as 18 pM under ideal experimental circumstances.

The enzyme known as a protease, or proteolytic enzyme, can catalyse the breakdown of proteins into smaller peptides as well as amino acids [68]. FRET biosensor developed employing QDs for protease testing was created by Algar et al. (Fig. 23.5b) [56]. In this assay paper fibres, chemically modified with thiol ligands were used to immobilise CdSeS/ZnS QDs, upon the surface of which the peptide substrates labelled with dye were self-assembled to generate efficient FRET. However, because the targeted protease may break down the peptide substrates when it is present, the dyes get separated from the QDs thereby eliminating FRET. For characterisation, steady-state and fluorescent lifetime imaging microscopy (FLIM) were employed. Quantitative results were obtained within 5–60 min at very low levels of 1–2 nM of protease. It can also be modified to detect other kinds of enzymes.

23.4 QD Biosensors Based on Bioluminescence Resonance Energy Transfer (BRET)

Bioluminescence resonance energy transfer (BRET), which also contributes to the sensor development, is a biophysical technique based on energy transfer mechanism. It takes advantage of the normally occurring non-radiative energy transfer process between a protein that emits light (the donor) and a fluorescent protein that is nearby. While BRET and FRET are similar, the donor energy in BRET comes from the chemical reaction that the donor enzyme catalyses rather than from the absorbed excitation photon. Here, the necessity of an external light source for excitation is eliminated as a light-emitting label of protein itself transfers energy to quantum dots. A quantum-dot-based bioluminescent biosensor, for instance, was used by Shen et al. to demonstrate the sensitive and selective detection of enrofloxacin [84]. This test involves the expression and purification of an anti-FQ single-chain variable fragment-Renilla luciferase (scFv-Rluc) fusion protein and creation of QD-NOR complex by conjugating QD with the antibiotic norfloxacin (NOR). Through the use of BRET, the energy released during the oxidation of the luciferase substrate is transferred to the QDs and excites them, causing them to emit light. Because the targeted enrofloxacin may bind to scFv-Rluc in a competitive manner, the substrate linker that brings luciferase and QDs together was cleaved in the presence of the drug (Fig. 23.6a, b). This assay is effective for determining the activity of matrix metalloproteinases (MMPs) because of high sensitivity and selectivity, having limit of detection as 2.54 ng L^{-1}.

Additionally, utilising quantum-dot biosensors, Tsuboi and Jin created an optical biosensor to detect apoptotic cells [71]. Here, they described the creation of annexin V-functionalised quantum dots (QDs) connected to bioluminescence resonance energy transfer (BRET) and their use in the NIR optical detection of apoptotic cells. The HisRLuc · AnnexinV is used as the bioluminescent label when its histidine moieties directly attach onto the surface of glutathione-encapsulated CdSeTe/CdS quantum dots. Phosphatidylserine (PS) in the plasma membrane of apoptotic cells can be recognised and bound by the resultant annexin-conjugated quantum dots (Annexin VRLuc-QDs), which emit a very strong BRET signal. This allows a precise detection of apoptotic cells by BRET-induced NIR emission. Having high signal-to-noise ratio with a low background signal, this probe has been effectively used to image actynomycin D/cycloheximide (ActD/CHX) treated apoptotic breast cancer cells. Very recently, this group [72] has also reported similar detection and imaging of antibody–drug conjugates (ADC), Kadcyla (a conjugate of Herceptin and emtansine) induced breast tumour apoptosis by employing dual (both Visible and NRI) fluorescing Annexin V−EGFP−QD in the presence of Ca^{2+} ions (Fig. 23.6c).

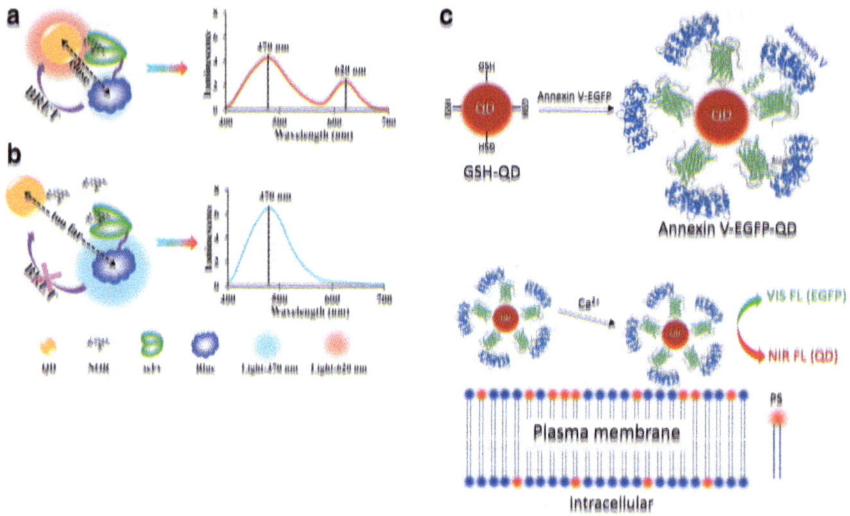

Fig. 23.6 Typical presentation of BRET sensor developed on QD platform in detection of tiny molecules **a** BRET happens when the target is absent. **b** BRET is blocked while the target is present. Reproduced with permission [84] Copyright 2016, American Chemical Society, **c** Sensor for the detection of apoptotic cells is shown schematically. Reproduced with permission [72] Copyright 2022, American Chemical Society®

23.5 Concluding Remarks

Due to their exceptional electronic and optical characteristics, like strong brightness, very good photochemical stability, long photoluminescence lifetime, large Stoke's shift, narrow emission with broad absorption, shape and size-tunable emission, and excellent photo-electrochemical activity, QDs are desirable building blocks to develop efficient biosensors for detecting a variety of biomolecular systems. By combining potent analytical approaches, such as resonance energy transfer through fluorescence, single molecule detection, inner-filter effect, aptamer recognition, and nucleic acid amplification, quantum dots have shown great promise as optical labels in developing fluorescence and bioluminescence-based biosensors as well as electron transducers in fabricating photo-electrochemical sensors. Numerous essential biomolecules, such as DNA, microRNAs, proteins, enzymes, and tiny compounds, can be sensitively, selectively, and multiplexed detected using these QD-based biosensors. They have, therefore, made a significant contribution to fundamental research, early disease diagnosis and therapeutic innovation.

Again, stoichiometry and affinity of QDs to the target biomolecule are of crucial importance for their applications, analysis of binding ions in live cells. The detailed thermodynamic investigation of the behaviour of the 'biomolecule of interest' with QDs and it surface functional group will satisfy these needs for sensing and other biological applications, such as, imaging, drug delivery etc. Hence, calorimetric measurements can be performed between QDs conjugate and 'biomolecule of interest' to develop better guidelines for development of nanosensors. The strategies for conjugation of QDs with specific biomolecule are developing very fast. Slowly, we are gaining new insights into the fundamental processes that are involved in nano-bio interactions. In future, the biofunctionalised QDs can be used in detection of other biological relevant molecules. Further, studies in this field will open up the way to the applications of QDs involving analytical and bioanalytical phenomena. Among the high affinic ligand-receptor interactions, avidin–biotin pair provides a common platform to couple biomolecule onto surface of QDs. Hence, understanding the affinity of the pair when one of them is conjugated to QDs and the fate of fluorescence of QDs during the interactions may help design QDs hybrid with specific affinity for targeted imaging and delivery.

References

1. Algar, W.R., Krull, U.J.: Interfacial transduction of nucleic acid hybridization using immobilized quantum dots as donors in fluorescence resonance energy transfer. Langmuir **25**, 633–638 (2009)
2. Alivisatos, A.P.: Semiconductor clusters, nanocrystals, and quantum dots. Science **271**, 933–937 (1996)
3. Bartel, D.P.: MicroRNAs: genomics, biogenesis, mechanism, and function. Cell **116**, 281–297 (2004)
4. Benchimol, S., Fuks, A., Jothy, S., Beauchemin, N., Shirota, K., Stanners, C.P.: Carcinoembryonic antigen, a human tumor marker, functions as an intercellular adhesion molecule. Cell **57**, 327–334 (1989)
5. Biju, V.: Chemical modifications and bioconjugate reactions of nanomaterials for sensing, imaging, drug delivery and therapy. Chem. Soc. Rev. **43**, 744–764 (2014)
6. Biju, V., Itoh, T., Ishikawa, M.: Delivering quantum-dots to cells: bioconjugated quantum-dots for targeted and nonspecific extracellular and intracellular imaging. Chem. Soc. Rev. **39**, 3031–3056 (2010)
7. Block, T.M., Comunale, M.A., Lowman, M., Steel, L.F., Romano, P.R., Fimmel, C., Tennant, B.C., London, W.T., Evans, A.A., Blumberg, B.S., Dwek, R.A., Mattu, T.S., Mehta, A.S.: Proc. Natl. Acad. Sci. U. S. A. **102**, 779–784 (2005)
8. Bruchez, M., Moronne, M., Gin, P., Weiss, S., Alivisatos, A.P.: Semiconductor nanocrystals as fluorescent biological labels. Science **281**, 2013–2016 (1998)
9. Buch, R.M., Rechnitz, G.A.: Intact chemoreceptor-based biosensors: responses and analytical limits. Biosensors **4**, 215–230 (1989)
10. Chakrabarty, S., Maity, S., Yazhini, D., Ghosh, A.: Surface-directed disparity in self-assembled structures of small-peptide l-glutathione on gold and silver nanoparticles. Langmuir **36**, 11255–11261 (2020)
11. Chan, W.C.W., Nie, S.: Quantum dot bioconjugates for ultrasensitive nonisotopic detection. Science **281**, 2016–2018 (1998)
12. Chatterjee, A., Priyam, A., Bhattacharya, S.C., Saha, A.: PH dependent interaction of biofunctionalized CdS nanoparticles with nucleobases and nucleotides: a fluorimetric study. J. Lumin. **126**, 764–770 (2007)
13. Chatterjee, A., Priyam, A., Ghosh, D., Mondal, S., Bhattacharya, S.C., Saha, A.: Interaction of ZnS nanoparticles with flavins and glucose oxidase: a fluorimetric investigation. J. Lumin. **132**, 545–549 (2012)
14. Chen, R.F., Bowman, R.L.: Fluorescence polarization: measurement with ultraviolet-polarizing filters in a spectrophotofluorometer. Science **147**, 729–732 (1965)

15. Chen, J.L., Zhu, C.Q.: Functionalized cadmium sulfide quantum dots as fluorescence probe for silver ion determination. Anal. Chim. Acta **546**, 147 (2005)
16. Chi, C.W., Lao, Y.H., Li, Y.S., Chen, L.C.: A quantum dot-aptamer beacon using a DNA intercalating dye as the FRET reporter: application to label-free thrombin detection. Biosens. Bioelectron. **26**, 3346 (2011)
17. Chinnathambi, S., Karthikeyan, S., Hanagata, N., Shirahata, N.: Molecular interaction of silicon quantum dot micelles with plasma proteins: hemoglobin and thrombin. RSC Adv. **9**, 14928 (2019)
18. Clark, M.F., Adams, A.N.: Characteristics of the microplate method of enzyme-linked immunosorbent assay for the detection of plant viruses. J. Gen. Virol. **34**, 475–483 (1977)
19. Condorelli, G., Latronico, M.V.G., Cavarretta, E.: MicroRNAs in cardiovascular diseases: current knowledge and the road ahead. J. Am. Coll. Cardiol. **63**, 2177–2187 (2014)
20. Cui, D., Tian, F., Coyer, S.R., Wang, J., Pan, B., Gao, F., He, R., Zhang, Y.: Effects of antisensemyc-conjugated single-walled carbon nanotubes on HL-60 cells. J. Nanosci. Nanotechnol. **7** 1639–1646 (2007)
21. Freeman, R., Girsh, J., Willner, I.: Nucleic acid/quantum dots (QDs) hybrid systems for optical and photoelectrochemical sensing. ACS Appl. Mater. Interfaces **5**, 2815 (2013)
22. Ghosh, S., Bhattacharya, S.C., Saha, A.: Probing of ascorbic acid by CdS/dendrimer nanocomposites: a spectroscopic investigation. Anal. Bioanal. Chem. **397**, 1573–1582 (2010)
23. Ghosh, D., Ghosh, S., Saha, A.: Quantum dot based probing of mannitol: an implication in clinical diagnostics. Anal. Chim. Acta **675**, 165–169 (2010)
24. Ghosh, S., Priyam, A., Bhattacharya, S.C., Saha, A.: Mechanistic aspects of quantum dot based probing of Cu (II) ions: role of dendrimer in sensor efficiency. J. Fluoresc. **19**, 723–731 (2009)
25. Ghosh, D., Roy, C.N., Mondal, S., Kundu, S., Maiti, S., Bag, P.K., Saha, A.: Modulation of catalytic functionality of alkaline phosphatase induced by semiconductor quantum dots: evidence of substrate-mediated protection. RSC Adv. **6**, 5024 (2016)
26. Gill, R., Bahshi, L., Freeman, R., Willner, I.: Optical detection of glucose and acetylcholine esterase inhibitors by H_2O_2-sensitive CdSe/ZnS quantum dots. Angew. Chem., Int. Ed. **120**, 1700 (2008)
27. Greene, K.L., Albertsen, P.C., Babaian, R.J., Carter, H.B., Gann, P.H., Han, M., Kuban, D.A., Sartor, A.O., Stanford, J.L., Zietman, A., Carroll, P.: Prostate specific antigen best practice statement: 2009 update. J. Urol. **189**, S2–S11 (2013)
28. Han, M., Gao, X., Su, J.Z., Nie, S.: Quantum-dot-tagged microbeads for multiplexed optical coding of biomolecules. Nat. Biotechnol. **19**, 631–635 (2001)
29. Hildebrandt, N., Spillmann, C.M., Algar, W.R., Pons, T., Stewart, M.H., Oh, E., Susumu, K., Dı´az, S.A., Delehanty, J.B., Medintz, I.L.: Energy transfer with semiconductor quantum dot bioconjugates: a versatile platform for biosensing, energy harvesting, and other developing applications, Chem. Rev. **117**, 536–711 (2017)
30. Hu, J., Liu, M.-H., Zhang, C.-Y.: Integration of isothermal amplification with quantum dot-based fluorescence resonance energy transfer for simultaneous detection of multiple microRNAs. Chem. Sci. **9**, 4258–4267 (2018)
31. Huber, A.H., Zhu, B., Kwan, T., Kampf, J.P., Hegyi, T., Kleinfeld, A.M.: Fluorescence sensor for the quantification of unbound bilirubin concentrations. Clin. Chem. **58**, 869−876 (2012)
32. Jaiswal, J.K., Mattoussi, H., Mauro, J.M., Simon, S.M.: Long-term multiple color imaging of live cells using quantum-dot bioconjugates. Nat. Biotechnol. **21**, 47–51 (2003)
33. Jameson, D.M., Ross, J.A.: Fluorescence polarization/anisotropy in diagnostics and imaging. Chem. Rev. **110**, 2685–2708 (2010)
34. Karmakar, S., Das, T.K., Kundu, S., Maiti, S., Saha, A.: Physicochemical understanding of protein-bound quantum dot based sensitive probing of bilirubin: validation with real samples and implications of protein conformation in sensing. ACS Appl. Bio Mater. **3**, 8820–8829 (2020)
35. Kelly, K.L., Coronado, E., Zhao, L.L., Schatz, G.C.: The optical properties of metal nanoparticles: the influence of size, shape, and dielectric environment. J. Phys. Chem. B **107**, 668–677 (2003)

36. Kundu, S., Maiti, S., Ghosh, D., Roy, C.N., Saha, A.: Aqueous synthesis of protein-encapsulated ZnSe quantum dots and physical significance of semiconductor-induced Cu(II)ion sensing. ChemPhysChem **18**, 2533–2540 (2017)
37. Kundu, S., Maiti, S., Das, T.K., Ghosh, D., Roy, C.N., Saha, A.: Exploiting the biomimetic and luminescence properties of multivalent dendrimer–semiconductor nanohybrid materials in the ultra-low level determination of folic acid. Analyst **142**, 2491 (2017)
38. Lakowicz, J.R.: Effects of solvents on fluorescence emission spectra. In: Lakowicz, J.R. (ed.) Principles of Fluorescence Spectroscopy, pp. 187–215. Springer US, Boston, MA (1983)
39. Li, J., Li, X., Shi, X., He, X., Wei, W., Ma, N., Chen, H.: Highly sensitive detection of caspase-3 activities *via* a nonconjugated gold nanoparticle–quantum dot pair mediated by an inner-filter effect. ACS Appl. Mater. Interfaces **5**, 9798–9802 (2013)
40. Li, Y., Schluesener, H., Xu, S.: Gold nanoparticle-based biosensors. Gold Bull. **43**, 29–41 (2010)
41. Li, Q., Tan, X., Fu, L., Liu, Q., Tang, W.: A novel fluorescence and resonance Rayleigh scattering probe based on quantum dots for the detection of albendazole (2015)
42. Lichtman, J.W., Conchello, J.-A.: Fluorescence microscopy. Nat. Methods **2**, 910–919 (2005)
43. Ma, F., Liu, W.-J., Tang, B., Zhang, C.-Y.: A single quantum dot-based nanosensor for the signal-on detection of DNA methyltransferase. Chem. Commun. **53**, 6868–6871 (2017)
44. Ma, F., Liu, M., Wang, Z.-Y., Zhang, C.-Y.: Multiplex detection of histone-modifying enzymes by total internal reflection fluorescence-based single-molecule detection. Chem. Commun. **52**, 1218–1221 (2016)
45. Ma, Q., Su, X.: Recent advances and applications in QDs-based sensors. Analyst **136**, 4883–4893 (2011)
46. Martinez, A.W., Phillips, S.T., Whitesides, G.M., Carrilho, E.: Diagnostics for the developing world: microfluidic paper-based analytical devices. Anal. Chem. **82**, 3–10 (2010)
47. Medintz, I.L., Uyeda, H.T., Goldman, E.R., Mattoussi, H.: Quantum dot bioconjugates for imaging, labelling and sensing. Nat. Mater. **4**, 435–446 (2005)
48. Mondal, S., Ghosh, S., Ghosh, D., Saha, A.: Physico-chemical aspects of quantum dot−vasodialator interaction: implications in nanodiagnostics. J. Phys. Chem. C **116**, 9774–9782 (2012)
49. Murphy, C.J.: Peer reviewed: optical sensing with quantum-dots. Anal. Chem. **74**, 520A-526A (2002)
50. Noor, M.O., Shahmuradyan, A., Krull, U.J.: Paper-based solid-phase nucleic acid hybridization assay using immobilized quantum dots as donors in fluorescence resonance energy transfer. Anal. Chem. **85**, 1860–1867 (2013)
51. Page, L.E., Zhang, X., Tyrakowski, C.M., Ho, C.-T., Snee, P.T.: Synthesis and characterization of DNA quantum dot conjugates for the fluorescence ratiometric detection of unlabelled DNA. Analyst **141**, 6251–6258 (2016)
52. Pan, B., Cui, D., Sheng, Y., Ozkan, C., Gao, F., He, R., Li, Q., Xu, P., Huang, T.: Dendrimer-modified magnetic nanoparticles enhance efficiency of gene delivery system. Cancer Res. **67**, 8156–8163 (2007)
53. Pan, B., Cui, D., Xu, P., Li, Q., Huang, T., He, R., Gao, F.: Study on interaction between gold nanorod and bovine serum albumin. Colloids Surface A **295**, 217–222 (2007)
54. Park, J., Park, Y., Kim, S.: Signal amplification *via* biological self-assembly of surface-engineered quantum dots for multiplexed subattomolar immunoassays and apoptosis imaging. ACS Nano **7**, 9416–9427 (2013)
55. Partin, A.W., Kattan, M.W., Subong, E.P., et al.: Combination of prostate-specific antigen, clinical stage, and gleason score to predict pathological stage of localized prostate cancer. JAMA **277**, 1445–1451 (1997)
56. Petryayeva, E., Algar, W.R.: Proteolytic assays on quantum-dot-modified paper substrates using simple optical readout platforms. Anal. Chem. **85**, 8817–8825 (2013)
57. Petryayeva, E., Algar, W.R.: Multiplexed homogeneous assays of proteolytic activity using a smart phone and quantum dots. Anal. Chem. **86**, 3195–3202 (2014)

58. Priyam, A., Chatterjee, A., Bhattacharya, S.C., Saha, A.: Conformation and activity dependent interaction of glucose oxidase with CdTe quantum dots: towards developing a nanoparticle based enzymatic assay. Photochem. Photobiol. Sci. **8**, 362–370 (2009)
59. Priyam, A., Chatterjee, A., Das, S.K., Saha, A.: Size dependent interaction of biofunctionalized CdS nanoparticles with tyrosine at different pH. Chem. Commun. 4122–4124 (2005)
60. Qiu, X., Hildebrandt, N.: Rapid and multiplexed microRNA diagnostic assay using quantum dot-based förster resonance energy transfer. ACS Nano **9**, 8449–8457 (2015)
61. Qiu, Z., Shu, J., Tang, D.: Bioresponsive release system for visual fluorescence detection of carcinoembryonic antigen from mesoporous silica nanocontainers mediated optical color on quantum dot-enzyme-impregnated paper. Anal. Chem. **89**, 5152–5160 (2017)
62. Reiss, P., Protière, M., Li, L.: Core/shell semiconductor nanocrystals. Small **5**, 154–168 (2009)
63. Resch-Genger, U., Grabolle, M., Cavaliere-Jaricot, S., Nitschke, R., Nann, T.: Quantum-dots versus organic dyes as fluorescent labels. Nat. Methods **5**, 763–775 (2008)
64. Sang, L.-J., Wang, H.-F.: Aminophenylboronic-acid-conjugated polyacrylic acid-Mn-doped ZnS quantum dot for highly sensitive discrimination of glycoproteins. Anal. Chem. **86**, 5706–5712 (2014)
65. Schulz, J.B., Weller, M., Moskowitz, M.A.: Caspases as treatment targets in stroke and neurodegenerative diseases. Ann. Neurol. **45**, 421–429 (1999)
66. Shanehsaz, M., Mohsenifar, A., Hasannia, S., Pirooznia, N., Samaei, Y., Shamsipur, M.: Detection of *Helicobacter pylori* with a nanobiosensor based on fluorescence resonance energy transfer using CdTe quantum dots. Microchim. Acta **180**, 195–202 (2013)
67. Su, S., Fan, J., Xue, B., Yuwen, L., Liu, X., Pan, D., Fan, C., Wang, L.: DNA-conjugated quantum dot nanoprobe for high-sensitivity fluorescent detection of DNA and micro-RNA. ACS Appl. Mater. Interfaces **6**, 1152–1157 (2014)
68. Sun, C., Su, K.-H., Valentine, J., Rosa-Bauza, Y.T., Ellman, J.A., Elboudwarej, O., Mukherjee, B., Craik, C.S., Shuman, M.A., Chen, F.F., Zhang, X.: Time-resolved single-step protease activity quantification using nanoplasmonic resonator sensors. ACS Nano **4**, 978–984 (2010)
69. Thornberry, N.A., Lazebnik, Y.: Caspases: enemies within. Science **281**, 1312–1316 (1998)
70. Tian, J., Zhou, L., Zhao, Y., Wang, Y., Peng, Y., Zhao, S.: Multiplexed detection of tumor markers with multicolor quantum dots based on fluorescence polarization immunoassay. Talanta **92**, 72–77 (2012).
71. Tsuboi, S., Jin, T.: Bioluminescence resonance energy transfer (bret)-coupled annexin v-functionalized quantum dots for near-infrared optical detection of apoptotic cells. ChemBioChem **18**, 2231–2235 (2017)
72. Tsuboi, S., Jin, T.: In vitro and in vivo fluorescence imaging of antibody−drug conjugate-induced tumor apoptosis using Annexin V−EGFP conjugated quantum dots. ACS Omega **7**, 2105–2113 (2022)
73. Wang, Z.-Y., Wang, L.-J., Zhang, Q., Tang, B., Zhang, C.-Y.: Single quantum dot-based nanosensor for sensitive detection of 5-methylcytosine at both CpG and non-CpG sites. Chem. Sci. **9**, 1330–1338 (2018)
74. Wang, L.-Y., Zhou, Y.-Y., Wang, L., Zhu, C.-Q., Li, Y.-X., Gao, F.: Synchronous fluorescence determination of protein with functionalized CdS nanoparticles as a fluorescence probe. Anal. Chim. Acta **466**, 87–92 (2002)
75. Watanabe, T., Tada, M., Nagai, H., Sasaki, S., Nakao, M.: *Helicobacter pylori* infection induces gastric cancer in mongolian gerbils. Gastroenterology **115**, 642–648 (1998)
76. Wegner, K.D., Jin, Z., Linde´n, S., Jennings, T.L., Hildebrandt, N.: Quantum-dot-based Forster resonance energy transfer immunoassay for sensitive clinical diagnostics of low-volume serum samples. ACS Nano **7**, 7411–7419 (2013)
77. Wegner, K.D., Hildebrandt, N.: Quantum dots: bright and versatile *in vitro* and in vivo fluorescence imaging biosensors. Chem. Soc. Rev. **44**, 4792–4834 (2015)
78. Weller, H.: Colloidal semiconductor q-particles: chemistry in the transition region between solid state and molecules. Angew. Chem. Int. Ed. Engl. **32**, 41–53 (1993)
79. Wu, L., Qu, X.: Cancer biomarker detection: recent achievements and challenges. Chem. Soc. Rev. **44**, 2963–2997 (2015)

80. Xiangyi, H., Liang, L., Huifeng, Q., Chaoqing, D., Jicun, R.: A resonance energy transfer between chemiluminescent donors and luminescent quantum-dots as acceptors (CRET). Angew. Chem. Int. Ed. **45**, 5140–5143 (2006)
81. Xu, W., Xue, X., Li, T., Zeng, H., Liu, X.: Ultrasensitive and selective colorimetric DNA detection by nicking endonuclease assisted nanoparticle amplification. Angew. Chem. Int. Ed. **48**, 6849–6852 (2009)
82. Xu, Q., Zhang, Y., Tang, B., Zhang, C.-Y.: Multicolorquantum dot-based chemical nose for rapid and array-free differentiation of multiple proteins. Anal. Chem. **88**, 2051–2058 (2016)
83. You, X., He, R., Gao, F., Shao, J., Pan, B., Cui, D.: Hydrophilic high-luminescent magnetic nanocomposites. Nanotechnology **18**, 035701 (2007)
84. Yu, X., Wen, K., Wang, Z., Zhang, X., Li, C., Zhang, S., Shen, J.: General bioluminescence resonance energy transfer homogeneous immunoassay for small molecules based on quantum dots. Anal. Chem. **88**, 3512–3520 (2016)
85. Zeng, S., Yong, K.-T., Roy, I., Dinh, X.-Q., Yu, X., Luan, F.: A review on functionalized gold nanoparticles for biosensing applications. Plasmonics **6**, 491–506 (2011)
86. Zhao, Y., Chen, F., Li, Q., Wang, L., Fan, C.: Isothermal amplification of nucleic acids. Chem. Rev. **115**, 12491–12545 (2015)
87. Zhou, J., Wang, Q.-X., Zhang, C.-Y.: Liposome–quantum dot complexes enable multiplexed detection of attomolar DNAs without target amplification. J. Am. Chem. Soc. **135**, 2056–2059 (2013)

[illegible] [illegible] 2017 [illegible]

[illegible] [illegible]

[illegible] 113, [illegible] (2015)

[illegible] [illegible]

Chapter 24
Recent Advances in Rare-Earth Based Persistent Luminescent Probes

Aftab Ansari and Dambarudhar Mohanta

Abstract Over the past few decades, lanthanide-based nanoparticles (NPs) have found extensive industrial applications due to their excellent photoluminescence as well as phosphorescent behaviour. Their unique photostability owes its origin to the shielding effect of *f*-levels. It is known that while some rare-earth (RE) elements display downconversion (DC) luminescence, others exhibit exquisite upconversion (UC) luminescence. To name a few, RENPs find widely ranged commercial usage in screens, lights, glasses, laser hosts etc. DC-RENPs also have utility as fluorescent agents in various immunoassays, providing better sensitivity as compared to the conventional techniques. However, due to better penetrability in biological systems, low energy near-infrared excited UC RENPs have also drawn the attention of the researchers. Interestingly, both UC and DC-RENPs can be utilized simultaneously in LED devices as multicolour components based on their emission responses. Lately, lanthanide resonance energy transfer has been exploited to augment the luminescent property of RENPs utilized as labels in biosensing devices. In this context, RENPs exhibiting persistent luminescence (PerL) need special mention as it has a great potential in biosensing because of their delayed photoemission responses (up to several days in some cases). This chapter highlights the mechanisms involved in the various emission processes describing the importance of RE-based PerL nanoprobes, specifically in the field of biosensing. A comprehensive discussion on the recent progress in this domain as well as in the combined use of the technique with other existing methods overcoming the limitations and increasing the efficacy has been made.

Keywords Rare-earth · Luminescence · Upconversion · Nanosystem

A. Ansari (✉)
Department of Physics, University of Science and Technology, Meghalaya, India
e-mail: aftyfi9@gmail.com

D. Mohanta
Department of Physics, Tezpur University, PO: Napaam, Tezpur 784028, Assam, India
e-mail: best@tezu.ernet.in

D. Mohanta and P. Chakraborty (eds.), *Nanoscale Matter and Principles for Sensing and Labeling Applications*, Advanced Structured Materials 206,
https://doi.org/10.1007/978-981-99-7848-9_24

24.1 Introduction

The detection of biomolecules is of significant importance for the development of existing diagnostic tools in pharmaceutical therapy, as it offers plenty of scope to understand cellular and molecular biology. As such, the development for advanced, smart probes with enhanced efficiency has been an attractive field of research in materials science as well as biomedical engineering. In this context, a significant advancement of DNA microarrays and/or immunoassays has been achieved over the last few decades resulting in improved detection capability, selectivity, and efficient analysis. Commercially available conventional probes, such as green fluorescent protein (GFP) and other organic dyes have a limited range of applications and they exhibit weak photostability as well as unwanted auto-fluorescence background. Quantum-dots (QDs) have been extensively employed in the place of organic dyes for various assays, including single-protein labelling in order to overcome the limitations in the case of organic fluorophores [2]. However, the semiconductor QDs exhibit "blinking effects" and their inherent cytotoxic behaviour along with the factors, like colour inhomogeneity and thermal quenching also limit their applicability as efficient biosensors and biomarkers.

Lanthanide-based nanoscale systems display excellent optical features and are well-known materials for photophysical applications. Rare-earth nanoparticles (RENPs) possess relatively high resistance to photobleaching and display large Stoke shifts; optical properties encouraging biological applications [4–6]. Due to a large number of emitters present in each RENP, the blinking effect is completely absent [7–9]. Lanthanides owe their luminescence to forbidden transitions among different *f*-electron configurations of the RE ions. Only Ce^{3+} and Eu^{2+} are the known RE ions which involve *d*-level electronic transitions. The *f*-levels are, in general, low-lying levels as a consequence of the shielding effect displayed exclusively by the RE elements. Thus, the *f*-levels are well isolated from the surroundings. The shielding effect by the filled $5p^6 6s^2$ subshells leads to intra*f*-configuration transitions which are parity-forbidden, thereby displaying narrow-line emission. Therefore, the ionic transitions are least influenced by the variation in the environment making the nanosystems highly stable. The *f*-levels consequently lead to sharper absorption and emission profiles as compared to both organic dyes and QDs [11, 12]. These transitions (electronic redistribution) within the *4f* sub-shell are forbidden by the "electric dipole selection rules". However, the coupling with vibrational states relaxes the selection rule. Thus, even molecular vibration could briefly modify the symmetry around the RE ions and lead to transitions. Furthermore, *J* mixing and mixing with opposite-parity wavefunctions could also relax the selection rules. Usually, the ligand orbitals, or charge transfer states are responsible for such a mixing. Considering the shielding effect, mixing remains insignificant explaining why the *f*–*f* transitions have fairly weak oscillator strength. The RENPs can be excited using a transition of the matrix or direct excitation of the RE ion, either in UV or NIR depending on the required Stoke's shift displayed by the ion. Also, subsequent energy transfers to the RE ion, as in the case of co-doping, could also be employed.

Most lanthanide (Ln^{3+}) ions are stable at the trivalent state and the electronic transitions following Laporte's rule [14–16] can be broadly categorized as three different mechanisms:

(i) Charge transfer (CT) transitions: The ligand-to-metal charge transfer (LMCT) and metal-to-ligand charge transfer (MLCT) transitions occur mainly in the UV range, lower than 220 nm.
(ii) *4f-5d* transitions: *4f* electrons promoted to *5d* subshells also have oscillator strengths comparable to CT transitions and thus, can be observed in the UV region as well. In RE, however, *5d* subshells shield the *4f* shells, thereby making these transitions highly dependent on the external environment.
(iii) Intra *f* transitions: These transitions are parity-forbidden resulting in weak intensities but are insensitive to the external environment, allowing them to be of excellent use as optical probes.

24.2 Background

Since its discovery, RE-based materials have been found interesting in bio-organic chemistry as an ideal probe in determining protein binding sites. As early as 1910 itself, the inhibitory effect of lanthanide ions on the beating of perfused frog hearts has been studied [15]. By 1960s, RE has been successfully used to detect the position of Ca^{2+} in trypsin present in pancreatic juice, which is conventionally difficult, given the inert behaviour of a Ca^{2+}. RE has an ionic radius comparable to Ca^{2+} along with similar coordination number and water exchange rates and hence, the substitution of RE ion in Ca^{2+} sites facilitates its detection through luminescent studies. It is the Ca^{2+} transport channels in the cell through which basic lanthanide compounds effortlessly interact with its membrane and can perturb the passage of metal ions. Exploiting this behaviour, in combination with the excellent long lifetimes of RE, time-resolved luminescence immunoassays could be realized as early as the 1980s [17]. Ln^{3+} also find application as temperature probes within a cell with acute sensitivity [18]. These features of lanthanides have labelled them as a proficient alternative of the QDs and dyes acting as biosensing probes, enabling newer opportunities to researchers active in this field.

Ln^{3+}chelates are already commercially available as an improvement in enzyme linked immunosorbent assay (ELISA) based detection techniques [19]. Asati et al. employed RE oxide NPs, CeO, etc. to develop colourimetric immunoassays and found several advantages due to the modification as compared to that with conventional ELISA assay [20]. The team attached folic acid to polyacrylic acid coated CeO NPs and as folic acid can specifically recognize certain tumour cells, it turned out to be more effective than the antifolate receptor antibody generally used in ELISA (Fig. 24.1). In addition to greater stability, the modified technique could also oxidize the substrate, even without H_2O_2 generally required in traditional ELISA technique.

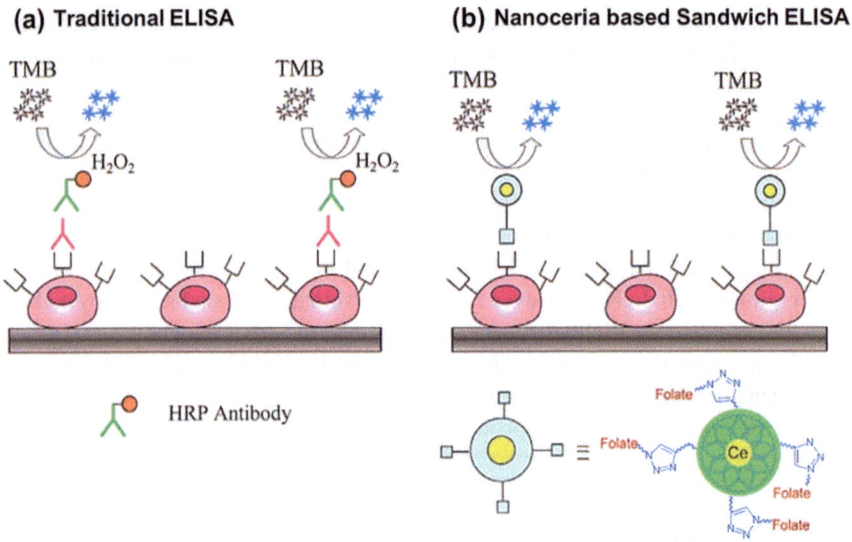

Fig. 24.1 A comparison of traditional ELISA and nanoceria-based ELISA is shown in **a** and **b** respectively. **a** Horseradish peroxidase (HRP) antibody is utilized as secondary antibody which facilitates oxidation of 3,3′,5,5′-tetramethylbenzidine (TMB) upon hydrogen peroxide treatment, resulting in the colour development. In **b**, the oxidase-like activity of nanoceria facilitates the direct oxidation of TMB without the need of HRP or hydrogen peroxide ["with the permission from Asati, A., Santra, S., Kaittanis, C., Nath, S., and Perez, J.M., Oxidase-like activity of polymer-coated cerium oxide nanoparticles. Angewandte Chemie, **121**(13) 2344–2348, (2009). Copyright (2009) American Chemical Society"]

Moreover, the entire process was reportedly cost-effective, as the CeO NPs are relatively cheaper as compared to the traditional ELISA based operations [21]. Previously, many fluorescent chelates have been shown to provide over 100 to 1000-fold better sensitivities, as compared to standard fluorescence based immunoassay techniques [22].

Optical sensing of temperature between 0 and 70 °C could be determined using Eu(III) complexes, the luminescence decay of which are highly reactive gradient in temperature [23, 24]. Temperature sensing using europium(III) β-diketonate complexes, popularly known as $Eu(TTA)_3$, have excellent sensitivity following a luminescence decrement of ~2% for every drop in temperature by 1 °C [16]. The sensitivity is even further augmented for $Eu(TTA)_3$ complexes in poly(vinyl methyl ketone) wherein the decay time drops 2-fold for a rise in temperature from ~1 to 66 °C [24]. Ln^{3+} based luminescent thermometry promises great possibilities in the near future as biosensing devices. Apart from thermometric sensing, the RE complexes such as Polymeric Tb(III) benzene-1,3,5-tricarboxylate are also used in the sensing of anions [25].

Although highly effective RE-based chelates possess several disadvantages, primarily photobleaching, different researchers have tried to nullify the drawback using surface coating techniques, essentially with hydrophobic polymers. The hydrophobic micro-environment thus helps in removing excess water molecules around the chelate, and hence the fluorescent quenching by the water molecules is reduced by a manifold. Ultrasensitive recognition of prostate-specific antigen was reported following similar surface modification of Eu-based nanoparticles [26]. Apart from photobleaching, chelates also have various limitations which include expensive synthesis, a tendency of quick agglomeration and dye leaking which foils their usage in the biological field. Researchers are henceforth bound to look for other beneficial options which involve the usage of RE in nanostructured form. In addition, the long lifetimes of RENPs can be utilized with slower detection arrangements, thereby reducing the undesired signals through direct acceptor excitation in Förster resonant energy transfer (FRET) based experiments [27, 28]. FRET involves a proficient energy transfer between fluorophores i.e. donor to acceptor, both being labelled to individual proteins. Lanthanides prove to be excellent FRET-based agents as large Stokes shift excitation can be attained at a wavelength much shorter than the acceptor absorption. Thus, only the donor gets excited while the acceptor remains in the ground state. RENPs exhibit the narrow emission lines and hence, the acceptor and donor emission lines rarely overlap.

Rare-earth nanoparticle-based biosensors

Most RENPs generally display large Stoke's shift. Based on the type of optical nature displayed by the system, RENPs as biosensing probes can be broadly categorized into three sections of applicability: (i) RENPs displaying downshift in Stoke's lines i.e. down-conversion, or quantum cutting, (ii) RENPs with anti-Stoke shifts or up-conversion, and (iii) RENPs that exhibit persistent luminescence. These three independent processes are detailed below.

24.2.1 Down Conversion RENPs (DC-RENPS)

In general, RENPs display Stoke's shift far greater than that of organic fluorophores. The elimination of background fluorescence can hence be easily isolated following longer lifetimes of the RE system. Eu^{3+} ions give highly intense *d-f* transitions which have been successfully employed for single-particle detection, sodium channel targeting etc. [7]. Nanostructures, being derived from inorganic systems are intrinsically photostable and provide an edge over chelates in immunoassays. Application of novel nanostructures, in order to improve immuno-sensing response, has drawn significant attention over the years. ELISA technique is a well-known immunoassay detection technique and recent research involving lanthanide-based immunoassays prove the two methods to be comparable in terms of limit of detection (LOD). Eu_2O_3 nanoparticles could be used as fluorescent label immunoassay, otherwise known as fluoroimmunoassay, with separation suiting magnetic field for herbicide atrazine [29]. In general, in a fluoroimmunoassay, a RE ion is used as a label to be coupled with an antibody. Since RE possesses longer lifetimes, the short-lived background autofluorescence could be eliminated and the detection becomes easier. RE nanosystem doped with other RE elements allows efficient tuning of the luminescent intensities. Gd_2O_3 doped with Eu nanoparticles are adeptly used to detect protein micropatterns [30]. The spectral ranges of the emission profiles of Eu^{3+}, Tb^{3+}, and Dy^{3+} vary on the basis of which multiple immunoassays could be developed [31, 32]. Moreover, luminescent quenching of the lanthanide emission profiles is extremely applicable in the determination of DNA and/or DNA fragments. Low concentrations down to 2 μg/mL with a limit of detection (LOD) ~0.24, could be detected by measuring the fluorescence of Tb^{3+} ions in LaF_3:Ce^{3+}, Tb^{3+} nanoparticles (~50 nm in radius), thereby proving it to be at par with prevailing techniques of measurement [33]. The calibration curves thus obtained were found to be linear up to 35 μg/mL for ctDNA. Amongst the coexisting substances, the influence of Fe(II), Cu(II) and Citric acid was ascertained even for very low concentrations, thereby leading to a change in the fluorescence intensity. Considering the concentrations of coexisting substances being lower than tolerated concentrations in biological fluids, the method still holds to be applicable for DNA determination. Furthermore, core–shell type structures viz. Fe_3O_4 as core and Gd_2O_3 as shell has been successfully used to detect single nucleotide polymorphism (SNPs); the efficiency of which is comparable to time consuming conventional quantitative RT-PCR techniques [34]. The detection of SNPs is essentially important in the diagnosis of kidney diseases. SNP discrimination over PKD (exon29) over a range of target DNA ~1, 10, 50, 500 pmol/L could be achieved. The team was able to recover 50 μg of gDNA from 5 mL, following an optimized assay using whole blood with a detection range over 200 ngm/L.

DC-RENPs are also extensively used as oxidant sensors, typically, as H_2O_2 sensors. H_2O_2 is associated in beckoning several cellular physiological processes including cellular migration, cell proliferation, and apoptosis. Oxidants also play an imperative role in pathophysiological, neurodegenerative disorders as well as malignant diseases. DC-RENPs allow quantitative and active measurement of H_2O_2 concentration in living cells [35]. Casanova, D and his group demonstrated the use of YVO_4:Eu^{3+} emission to measure the temporal pattern of reactive oxygen species (ROS) produced within the cells. The most intense $^5D_0$-$^7F_2$ transition of dopant Eu^{3+} ions are usually centred at, ~614 nm. Trivalent Eu ions can be reduced to a divalent state through a photoinduced mechanism leading to a suppressed emission intensity feature. The relative photoreduction against the oxidation determines the detectable oxidant concentration with low intensities allowing detection of H_2O_2 ~1 mM (few nanoparticles) and ~2 mM (single particle). Following in vitro oxidation of Eu^{2+} due to H_2O_2 interaction, recovery of the fluorescence could be consequently achieved [35]. Again, the approach also involves the repeated usage and can be applied in the detection of other oxidant species as well.

Eu^{3+}:YVO_4 NPs have been used to track ε-toxin receptors in the membrane of *Madin Darby canine kidney* (MDCK) cells [36]. In this case, the non-blinking feature of RENP was used to observe the confinement potential of bacterial pore-forming ε-toxin receptors. Single-molecule tracking data can also be obtained with organic-dye as protein labels. Use of such RENPs allows to further recreate the force field of toxin receptor micro-domain following a novel tactic based on Bayesian inferences, thereby enhancing its utility and making its usage more beneficial over conventional methods. They studied the receptor motion during 150–300 s and further verified that the system is not limited by the standard image acquisition time of 21.4 or 51.4 ms following the diffusion coefficient and the domain size [36].

DC-RENPs have also demonstrated effectiveness towards its use as FRET donors. Cysteine (Cy5) acceptors adsorbed onto YVO_4:Eu nanoparticles used as donors in FRET-based experiments revealed by a significant variation of both the nanoparticle and the Cy5 fluorescence decay [27]. The FRET efficiency was determined by measuring the lifetimes in and without the presence of the dye and was found to be 0.21, which increases gradually as more dye molecules attach to the nanoparticle, facilitating its utility as a powerful nanosensor [27]. In the absence of any energy transfer, the decay time of the dye ranges from ~208.4 μs while FRET facilitates the decay time to alter up to ~119 μs for range implying FRET efficiency to rise up to ~0.81. Hence a temporal discrimination between the two processes is observed. Such difference allows the detection techniques to be independent of concentrations of acceptor ions and to be free from short-lived direct acceptor excitation signal.

Usually, down-conversion nanoparticles employ UV excitation, which is in general not ideal for applications in living cells as they are known to induce cell damage, which limits their widespread practical applications. This has led the researchers to develop an alternative strategy with possible application under NIR excitation. In fact, the display of anti-Stoke's shift of RENP resolves the drawback suitably.

24.2.2 UP Conversion RENPs (UC-RENPS)

Among the different classes of lanthanides, up-conversion RE nanoparticles (UC-RENPs) display features that are ideal to be used in biosensing. The UC-RENPs emit photons of higher energy upon absorption of two or more photons [37]. There are several processes responsible for the up-conversion which result in extensively varied efficiencies [38]. The emission is therefore, anti-Stoke's shifted; almost certainly in the IR region [39]. The NIR region lacks endogenous fluorescence of biomolecules and thus background fluorescence is minimal. Furthermore, excitation in NIR regime allows deeper penetration into a biological system. Up-conversion processes involve metastable states with longer lifetimes up to a few microseconds and hence, continuous-wave lasers can be used; whereas multiphoton excitation typically requires the use of expensive ultrashort pulsed lasers [40]. With the use of UC-RENP, extremely selective nanoprobes of targets such as Ochratoxin A, Trypsin, and Glucose HIV targets with highly specific detection could be achieved [41–45].

UCNPs as nanoscale thermometers exhibit excellent efficiency in the determination of intracellular temperature. Sevilla and group developed a method which enables thermal sensing within a cell using a single UCNP particle [46]. To generate heat within the cell, the cytoplasm was pre-incubated with 15 × 45 nm gold nanorods followed by laser excitation to raise the intracellular temperature. Such UCNPs prove to be efficient contactless sensing probes responsive to external stimuli only. Further research on such localized sensing techniques enabled to apprehend exact cellular mechanisms in extravagant detailing. A group from Leiden University proposed a well-designed technology with the development of sensitive detection of antigens using UC RENPs [47]. RE microcrystals consisting of yttrium oxysulfide, (fluoride, silicate, or gallate) doped with other UC lanthanides of surface labelled with avidin or antibodies, exhibiting durable emission in the visible range. These surface labelled UC RENPs can bind specifically to antigens on undamaged cells. UC RENPs promise to have a stimulating potential as donors in FRET applications [48]. Anti-Stokes shift further facilitates FRET with a higher excitation wavelength than that of acceptor emission. Consequently, a clean detection of specific FRET signals with minimal or no overlap is observed. Vetrone, F demonstrated FRET by UCNPs to Fluorescent Phycobiliprotein [49]. FRET by UC RENPs for the detection of estradiol (E2) by homogeneous immunoassay in whole blood using $La_2O_2S:Er^{3+}$, Yb^{3+} nanoparticles were reported [50]. The NPs were coated with E2 antibody and subsequently mixed with E2-coupled fluorophore (E2-AF680). The transfer of energy to E2-AF680 acting as an acceptor creates a FRET signal. Upon addition to a blood sample, inborn E2 (contained by the blood) interacts with the acceptor and hence the FRET signal due to labelled E2 reduces, thereby allowing the detection of concentrations in the nanomolar range.

Further improvement in their optical behaviour could also be achieved following already available novel designing strategies which has made this an active field of research in the past [51–54]. Yi and Chow have reported on a significant enhancement of upconversion luminescence (UCL) (up to ~29.6 times), using hydrophobic core/shell (C/S) nanoparticles [51] (Figs. 24.2 and 24.3).

The hydrophobicity arose from oleylamine on the NP surfaces. Interestingly, the C/S were further coated with amphiphilic poly(acrylic acid) (PAA) in order to render hydrophilic properties upon these nanosystems. Such coating of PAA was made possible following hydrophobic interactions of the octyl and isopropyl groups present in PAA with the octadecyl group of oleylamine groups on the nanoparticle surface. The hydrophilic carboxyl group of PAA thus extends outward rendering nanoparticles to be water soluble, allowing a further attachment of biomolecules making them to be potential as bio-probes with high fluorescence efficiency of ~30%.

With a motive to further improve the UCL, Shen et al., determined a general strategy to optimize the doping ratio of the core/shell nanomaterials [52]. With an increase in the concentration of RENP to ~98 mol% they achieved ~15 times of UCL enhancement. It is known that a correlative rise in energy migration between RENPs leading to self-quenching process resembles a rise in self-trapping effects. Therefore, the energy migration i.e. energy transfer (ET) to quenchers, lowers the luminescence intensity. The effect is aggravated in nanosystems with higher surface areas providing additional surface quenching sites. Strikingly, Shen et al., demonstrated that for novel core/shell nanostructures, the ET towards quenching-sites (non-radiative) are distance-dependent, and the introduction of a layer of high quality, inactive shells can in effect overpower energy loss. Alternatively, S. Fischer et al.,

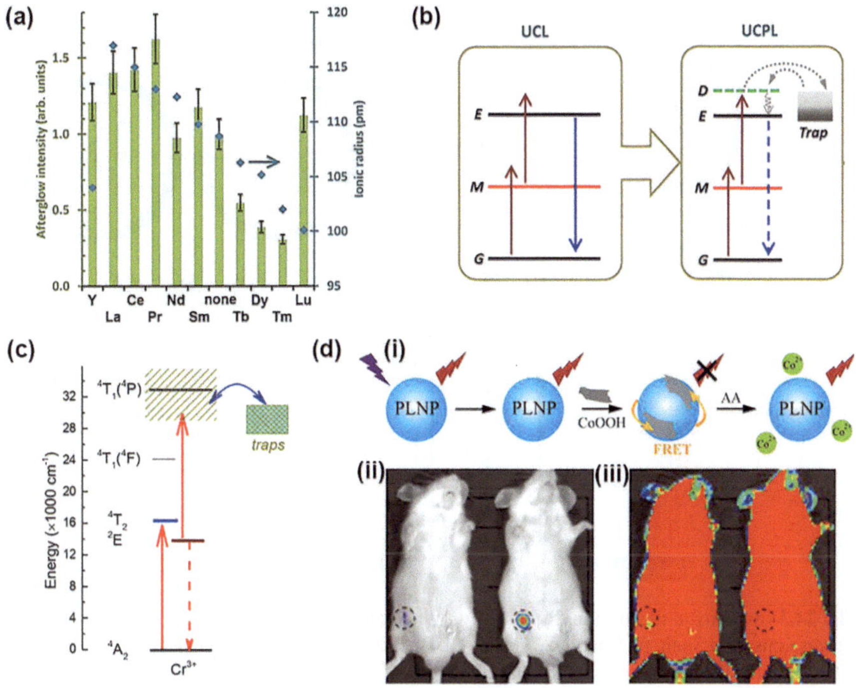

Fig. 24.2 **a** Influence of co-dopants on the afterglow intensity. The diamonds show the ionic radius in octahedral surrounding [1]. ["With permissions from Smet P. F. et al., Persistent luminescence in nitride and oxynitride phosphors: a review. Optical Materials, **36**(11), 1912 (2014). Copyright (2014), Elsevier". **b** Schematic diagrams of up-conversion luminescence (UCL), up-converted persistent luminescence (UCPL), with hypothetical energy level schemes. G, M, E and D represent the ground state, metastable state, emission state, and delocalized state, respectively [3] ["With the permissions from Liu, F., Liang, Y., Pan, Z. Detection of up-converted persistent luminescence in the near infrared emitted by the $Zn_3Ga_2GeO_8$:Cr^{3+}, Yb^{3+}, Er^{3+} phosphor, Phys. Rev. Lett. **113**, 177401 (2014). Copyright (2014), American Physical Society"]. **c** Schematic diagram of two-photon UCC process in Cr^{3+} activated persistent luminescence system [10] ["With the permission from Chen, Y. et al. New up-conversion charging concept for effectively charging persistent phosphors using low-energy visible-light laser diodes. Journal of Material Chemistry C **6**(30), 8003 (2018). Copyright (2019), Royal Society of Chemistry."] **d** Schematic illustration of ascorbic acid (AA) detection using CoOOH-modified $Sr_2MgSi_2O_7$:1% Eu, 2% Dy PLNPs. PL imaging of AA in vivo (ii) PL image of the mouse treated with normal saline and the mouse treated with AA and the nanoprobe (right) without excitation. (iii) PL image of the mice using 420 nm excitation. The centre of the black dashed circle is the injection-site. The colour change from purple to red for the intensity bar represents the increase in the luminescence intensity [13]. "With the permissions from Li, Na et al., A highly selective and instantaneous nanoprobe for detection and imaging of ascorbic acid in living cells and in vivo, Analytical Chemistry **86**(8), 3924 (2014). Copyright (2015), American Chemical Society"

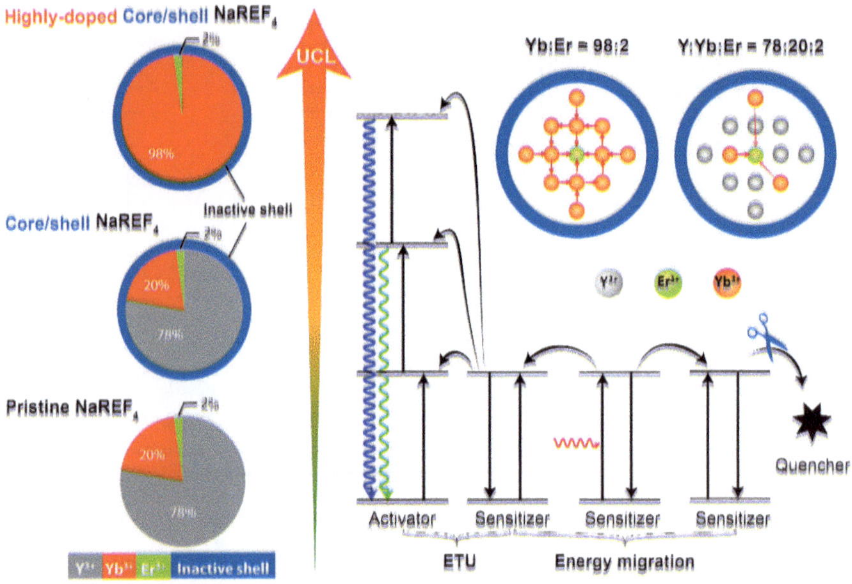

Fig. 24.3 Schematic illustration displaying design strategy of doped core/shell UC-RENPs "With permissions from Shen, B., Cheng, S., Gu, Y., Ni, D., Gao, Y., Su, Q., Feng, W., and Li, F., Nanoscale, **9**(5), 1964 (2016). Copyright (2016), Royal Society of Chemistry"

demonstrated a mechanism which involves tuning the shell thickness for the exploitation of surface quenching processes in order to improve the UCL core–shell nanosystems [53]. Surface quenchers are of the prime reason behind energy loss, prompting the researchers to choose methods that avoid the formation of quenching-sites.

Captivatingly the team has established that surface quenching can be of great advantage, as the non-radiative processes were used to increase an energy level's population, which in turn led to brighter luminescence. The schematic diagram in Fig. 24.4 describes a simplified 3-level rate equation model to populate energy level n3 ($^4I_{11/2}$) and b) energy level n2 ($^4I_{13/2}$). Surface quenching processes are known to be varying with the thickness of the shell. Henceforth, SQAD is treated as a competition involving "surface quenching rates". Thinner shells display surface quenching rates Γ_{SQ}, and Γ_{SQ}, large in magnitude, thereby dominating the entire system. Increment in shell thickness thus becomes non-relative to a higher population of $^4I_{13/2,}$ therefore resulting in higher QY. In addition, it leads to lower Γ_{SQ}, because of which, the population of the $^4I_{13/2}$ decreases. In case of such thickening of shells (as compared to the core), the surface quenching process with the higher energy levels can no longer facilitate the direct population of the $^4I_{13/2}$ energy level. Instead, a more substantial process like ET processes between neighbouring Er^{3+}, and multi-phonon relaxation from $^4I_{11/2}$ mediated spontaneous emission from the green ($^4S_{3/2}$, $^2H_{11/2}$), as well as the blue ($^4G_{11/2}$, $^2G_{9/2}$) Er^{3+} energy levels, is responsible for the population in the $^4I_{13/2}$ energy.

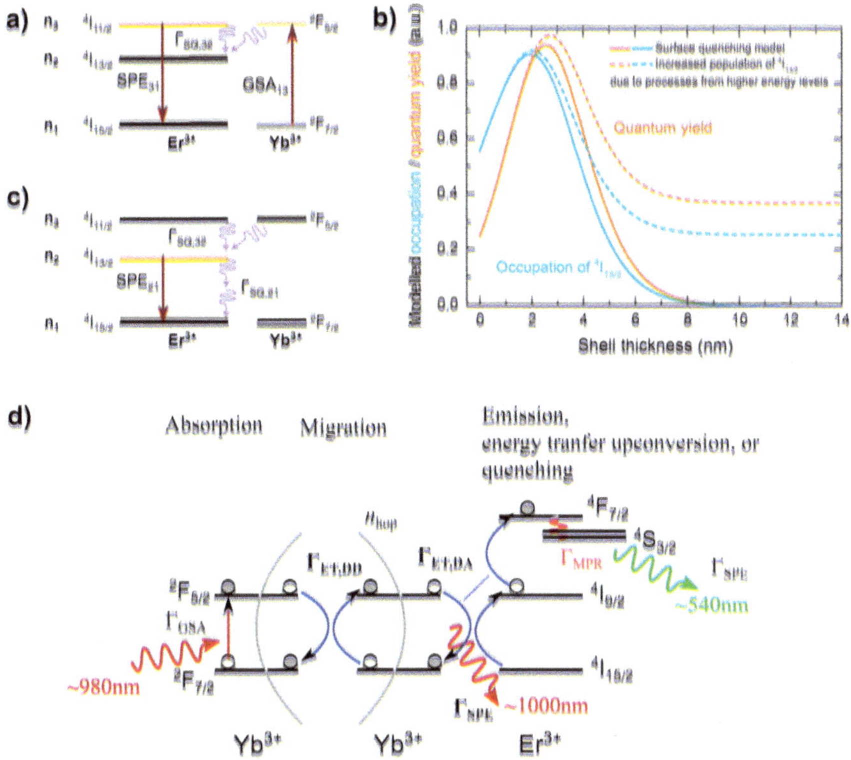

Fig. 24.4 **a, b, c** Schematic model for surface quenching assisted downshifting (SQAD) with emission ~1520 nm for ~980 nm excitation and **d** illustrative description of the related energy migration model. "With permissions from Fischer, S., Bronstein, N.D., Swabeck, J.K., Chan, E.M. and Alivisatos, A.P., Precise tuning of surface quenching for luminescence enhancement in core–shell lanthanide-doped nanocrystals. Nano Letters, **16**(11), 7241 (2016). Copyright (2016), American Chemical Society

In fact, S. Fischer et al., determined the contributions of 17.7, 15.7, and 15.3% of the total spontaneous emission from Er^{3+} ($^4S_{3/2}$, $^2H_{11/2}$) into the $^4I_{13/2}$ energy level for the core nanocrystals and in the samples with a 2.4 nm and 13.0 nm thick shell, respectively [53]. The excitation energy in such systems migrates in consequence of dipole–dipole ET from the rare-earth ion. The quantum-yield value of the NIR → NIR downshifted the luminescence owing to surface quenching-assisted downshifting (SQAD), which was determined to be significantly higher in thinner shells. In other words, surface quenching is largely curbed for very thick shells. The SQAD model is based on an analytical approach assuming an average distance between the lanthanide ions and a 3-energy level system to another. This excitation energy jump in the crystal lattice appears to be continuous until the energy is either transferred to an RE ion or the energy relaxes to the ground state via possible spontaneous emission.

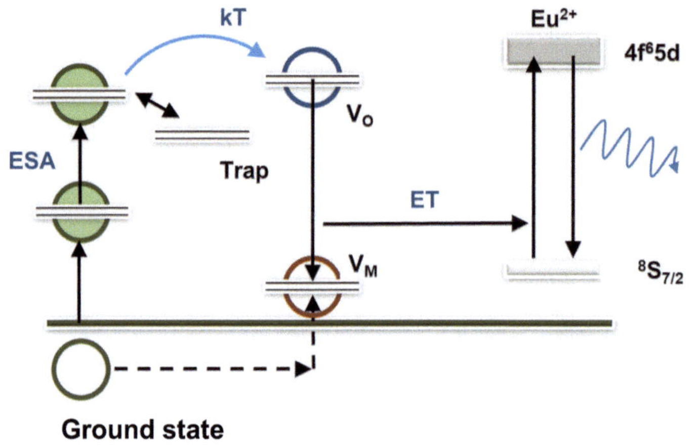

Fig. 24.5 Proposed mechanism of the persistent luminescence, employing Eu^{2+} phosphor. ESA and ET signify excited state absorption (ESA) and Energy transfer, respectively

The UC RENPs, although have extensive utility, suffer from two intrinsic drawbacks: (i) a small energy gap between the excited and ground-state levels favours the efficient non-radiative processes resulting in the low sensitization of the luminescence centre and (ii) relatively shorter lifetimes (ns to ms), which limit the efficiency of time-resolved detection (iii) negative aspects of upconversion, viz. saturation effects. This is because upconversion suggests an upward transition from an excited state (Fig. 24.5).

24.2.3 Persistence Luminescence RENPs (PerL RENPs)

Persistence luminescence is another property of certain RE systems that are used widely in sensing operations, especially in the point for care diagnostic mechanisms [68, 69]. Persistent Luminescence nanoparticles (PerLNPs) are phosphorescent materials which produce a long afterglow and hence can be excited before the system is used in the assay and hence reduce the auto background fluorescence. Such excitation can be achieved via optical excitation (UV, VIS or NIR light), or electronic beam or high-energy irradiation. The excited carriers are decisively deposited or captured on the trapping states which later escape. Even after the removal of the excitation source, the emission from PLNPs is obtained from filled energy traps. Such traps release the energy to luminescent centres following thermal activation and the emission from the luminescent-centres can sometimes last up to several days [69]. The trap depth and the release rate of carriers can be stimulated and predisposed with thermal, optical, or mechanical disorder, thereby enabling the tuning of duration. Alternatively, PerL can also result in the centre-to-centre transitions forming localized centres/trapped electrons and/or holes located as pairs/clusters [70]. A possible mechanism for PerL is shown in Fig. 24.6 [70]. PerL RENPs can have widespread applications in varied fields, viz. in traffic signalling, decorative items, and light emitting 'glow in the dark' sources, etc. [70].

The principle of FRET can also be applied in the preparation of PLNP-based FRET probes. Such a probe was designed by Wu et al. [67]. They combined PLNPs (PEI PLNPs) with antibody and gold nanoparticle conjugates used for the detection of R-fetoprotein (AFP). Following a high molar adsorption coefficient and easy surface modification, the AuNPs were used as a quencher to enhance their FRET efficiency. The conjugate is weakened upon interaction of AFP, following a strong affinity towards the antibody. The inhibition probe displays a very high selectivity to AFP. PEI-PLNPs have zeta potential (ζ) = + 14.25 mV, due to positively-charged $-NH^{3+}$ groups present in the NPs. Their electrostatic interaction with antibodies (negatively charged) on the surfaces of the AuNPs with ζ potential = −28.95 mV (in PBS buffer at pH 7.4) is considered to be responsible for the formation of the FRET inhibition probe PEI-PLNPs/Ab-AuNPs with ζ potential = −0.8 mV.

Understanding of trapping and de-trapping of **stored electrons** is a major challenge in research of afterglow mechanisms requiring to be tackled. Usage of RE as persistent luminescence materials can be strategized as lanthanide-lanthanide co-doping such as $SrAl_2O_4$:Eu^{2+}, Dy^{3+}, and $Sr_4Al_{14}O_{25}$:Eu, Dy [71, 72] or lanthanide-transition metal co-doping such as Ce^{3+} viz. Cr^{3+} co-doped yttrium aluminium gallium garnet and cerium doped garnets [73, 74]. Matsuzawa et al. explained the observed phosphorescence in terms of photoconductivity due to holes, and due to the trapping and thermal release of the holes by Dy^{3+} ions in the system forming a highly dense trapping level situated at a suitable depth with respect to the thermal release rate at room temperature, thus leading to the generation of bright and long phosphorescence [71]. Ueda et al., reported a novel vacuum-referred binding energy of characteristic 3d transition metal ions for PerL Phosphors of Cerium-doped Garnets [74]. They demonstrated that the number of d-electrons, nuclear charge and crystal field splitting largely affect the estimated vacuum-referred binding energy from the trap depth, displaying a zig-zag curve originating from various 3d electron energies. It is this very zig-zag curve which was utilized by the team to successfully develop PLNPs. PersL imaging is still in its early developmental phase and needs to be checked thoroughly for its limitations:

(i) NIR light allows deep tissue penetration and thus in vivo/in situ recharging is challenging.
(ii) NIR light limits the selectivity ranges of elements providing emission located within 650–950 nm.

The combination of UC and persistent luminescence—UCPL can overcome the drawbacks in the bio-field from both the UCL i.e. heating effects due to NIR excitation, and absorption of newly generated up-converted short-wavelength and the PL, such as short-time window of few hours [3, 75, 76]. Feng's group recently proposed a new UCPL concept which combines the key concepts of UC and long-persistent PL processes. They fabricated the first UCPL phosphor ZGGO:Cr, Yb, Er^{3+}, incorporated with upconverting Yb^{3+}/Er^{3+} pair into NIR LPL phosphor

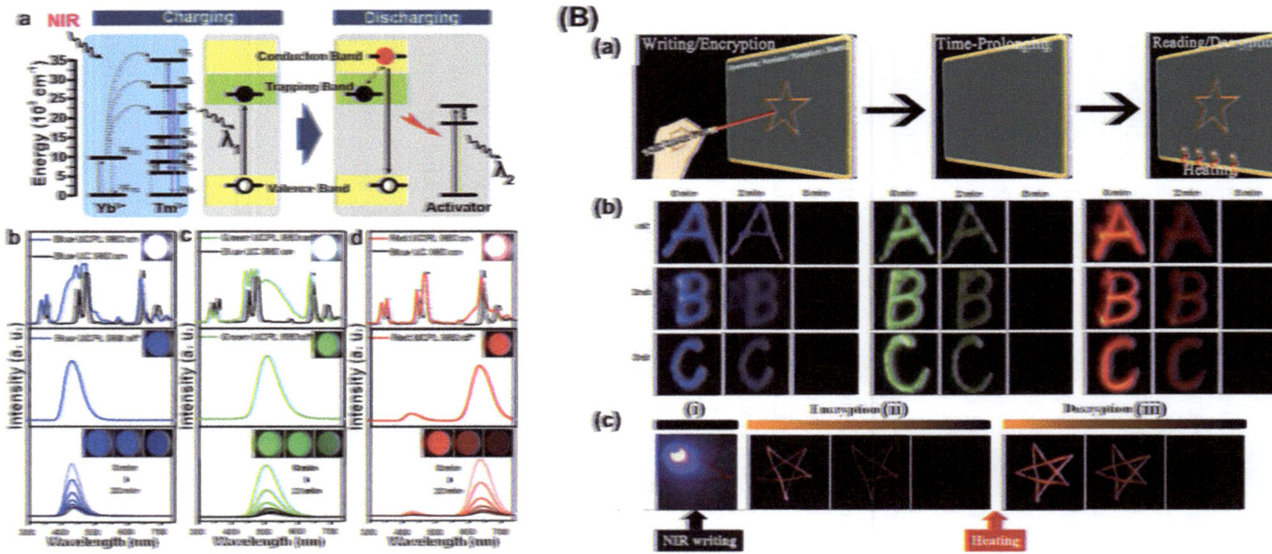

Fig. 24.6 **A a** The energy transfer mechanism of UCPL materials with **b** blue **c** green **d** red emissions obtained when the 980 nm NIR is on (top) or off, i.e. middle, bottom **B a** Schematic representation of data recording with **b** NIR laser pen written characters. **c** NIR writing encryption process and the heat induced decryption (With the permissions from Hu, L., Fan, Y., Liu, L., Li, X., Zhao, B., Wang, R., Wang, P., El-Toni, A.M. and Zhang, F., Orthogonal multiplexed luminescence encoding with near-infrared rechargeable upconverting persistent luminescence composites. Advanced Optical Materials, **5**(22), 1700680 (2017). Copyright (2017), John Wiley and Sons)

$Zn_3Ga_2GeO_8$:1%Cr^{3+} [3]. Upon a 980 nm excitation, the as-prepared phosphor system emitted long-lasting (beyond 24 h) NIR-persistent emission (~700 nm). This UC-PerL concept prospectively seems to have vital implications in biomedical imaging which, in general, requires higher sensitivity, longer observation window, deeper tissue penetrability, and in vivo excitability.

The aforementioned limitations can also be eliminated to a great extent, following exploitations of the phenomenon of defect-mediated increment in glow time of the PerL RENPs. Since, in most cases, the afterglow is due to lattice defect, this aspect needs a thorough examination. The clustering of cation vacancy and the Ln^{3+} ion can alter the hole trap depth and can introduce newer, slightly deeper traps, responsible for the enhancement of PersL intensity [77].

Meanwhile, Hu et al., have reported a novel orthogonal luminescence encoding technique, based on near-infrared (NIR) rechargeable UCPL composites [75]. Their study revealed that with orthogonal encoding signals with tunable multi-colour long-persistent phosphorescence, the UC emission can be functionalized independently both in the presence as well as in the absence of 980 nm excitation with remarkable increment in the coding capacity, as compared to traditional luminescence encoding strategies. Therefore, the relevant strategies offer a momentously increased coding capacity with lesser error rate following a decrement in chromatic aberration of similar colours. The attained NIR rechargeable UCPL materials exhibit great applications for anti-counterfeiting, re-writeable data encryption/decryption, zero-background bio-imaging, non-invasive photo-bio-stimulation, etc.

24.3 Conclusion and Outlook

The contemporary challenge in the development of multimodal nano bio-probes involves efficient multiphoton excitation, while preventing the de-activation. In this context, over the last few decades, lanthanide ions have proven to provide a wide and all-inclusive range of choices of varied and extensive optical properties. Along with the extension of structures, configurations, and manipulative assemblies, following surface designing and fabrication techniques, an eccentric wide ranging yet specific domain of research for scientists and engineers working in the field of bio-sensing has been uncovered, as detailed in Table 24.1. Recent developments and observations of the present scenario advocate multidimensional improvement in the conventional biological monitoring devices which shall be crucial in helping and offering a superior standard of living. Lanthanide-doped hybrid nanostructures find applications in the areas of research covering various sub-domains of biotechnology and clinical diagnostics. To be specific, advances on biosensors centred on time-resolved fluoroimmunoassays, DNA hybridization assays, and fluorescence imaging microscopy, with near-infrared luminescence combined with persistent luminescence promises to be of great utility in the near future.

Table 24.1 Comparative efficiency of various luminescent probes against ranging targets along with the techniques and luminescence processes involved thereof

Process	Probe used	Designated target	Performance (LOD)	Technique employed	References
DC	$LaF_3:Ce^{3+}$, Tb^3 particles	DNA	2 μg/mL	Luminescence spectra	[33]
	Eu_2O_3 NPs	Atrazine	0.5 ng mL^{-1}	Fluoroimmunoassay	[29]
	Europium ion chelate protein-labelling reagent	DNA	6 pM	Time-resolved delayed luminescence imaging microscopy	[55]
	$Eu:Gd_2O_3$ NPs	Phenoxybenzoic acid	1.4 μg L^{-1} PBA	Microarray immunoassay	[56]
	Streptavidin-coated europium (III) nanoparticle label	Biotinylated PSA	0.38 ng/L,	Time-resolved fluoroimmunoassays	[26]
	YVO_4:Eu NPs	Bungarotoxin protein	3 proteins/NP	Microarray	[57]
	$Y_{0.6}Eu_{0.4}VO_4$ NPs	H_2O_2	1–45 μM	Time-resolved monitoring	[35]
UC	$Mn^2:NaYF_4:Yb^{3+}$, Er^{3+} NPs	Ochratoxin A	0.098 ng mL^{-1}	Single-step LRET aptasensor	[41]
	$Y_2O_2S:Yb^{3+}$, Er^{3+}@SiO2	DNA	0.1 pg	Immunochromatography (LF) assay	[58]
	β-$NaYF_4$:Yb, Er@NaYF	Ochratoxin A	0.001 ng mL^{-1}	Ultrasensitive LRET platform	[43]
	$Y_2O_2S:Yb^{3+}$, Er^{3+}	DNA	1 ng/μl	Microarray fluorescence	[59]
	$NaYF_4$:Yb, Er-tetramethylrhodamine	Protein kinases A	0.00005 U μL^{-1}	LRET based assay	[60]
	$Na(Y_{1.5}Na_{0.5})F_6:Yb^{3+}$, Er^{3+}	Avidin	0.5 nM	LRET based assay	[61]
	$NaYF_4$:Yb, Er	DNA	R = 0.995	Luminescence spectra	[62]
	b-NaYF4:Yb^{3+}, Er^{3+}	Adenosine triphosphate	80 nM	Upconversion fluorescence spectra	[63]
	$Y_2O_2S:Yb^{3+}$, Er^{3+}	Human chorionic gonadotropin	10 pg	Immunochromatographic assay	[64]
	$Y_2O_2S:Yb^{3+}$, Er^{3+}	DNA	1 ng ml^{-1}	Upconversion fluorescence spectra	[59]
	$Er^{3+}:NaYF_4$ particles	DNA	1.3 nM	Upconversion fluorescence spectra	[65]
	$NaYF_4$:Yb, Er	Ammonia	400 μM	Upconversion fluorescence spectra	[66]

(continued)

Table 24.1 (continued)

Process	Probe used	Designated target	Performance (LOD)	Technique employed	References
	$NaYF_4$:Yb, Er	Protein kinase activity	0.00005 U/μL	Upconversion fluorescence spectra	[60]
PerL	$Sr_2MgSi_2O_7$:1% Eu, 2% Dy	Ascorbic acid	0.59 μM	Persistent luminescence imaging	[13]
	Eu^{2+} and Dy^{3+} doped $Ca_{1.86}Mg_{0.14}ZnSi_2O_7$	α-Fetoprotein	0.41 μg L^{-1}	LRET based inhibition assay	[67]

References

1. Smet, P.F., et al.: Persistent luminescence in nitride and oxynitride phosphors: a review. Opt. Mater. **36**(11), 1913–1919 (2014)
2. Bouzigues, C., Dahan, M.: Transient directed motions of GABAA receptors in growth cones detected by a speed correlation index. Biophys. J. **92**(2), 654–660 (2007)
3. Liu, F., Liang, Y., Pan, Z.: Detection of up-converted persistent luminescence in the near infrared emitted by the $Zn_3Ga_2GeO_8$:Cr^{3+}, Y^{b3}+, ^{Er}3+ phosphor. Phys. Rev. Lett. **113**(17), 177401 (2014)
4. Shen, J., Sun, L.-D., Yan, C.-H.: Luminescent rare earth nanomaterials for bioprobe applications. Dalton Trans. **42**, 5687–5697 (2008)
5. Vetrone, F., Capobianco, J.A.: Lanthanide-doped fluoride nanoparticles: luminescence, upconversion, and biological applications. Int. J. Nanotechnol. **5**(9–12), 1306–1339 (2008)
6. Escribano, P., et al.: Photonic and nanobiophotonic properties of luminescent lanthanide-doped hybrid organic–inorganic materials. J. Mater. Chem. **18**(1), 23–40 (2008)
7. Beaurepaire, E., et al.: Functionalized fluorescent oxide nanoparticles: artificial toxins for sodium channel targeting and imaging at the single-molecule level. Nano Lett. **4**(11), 2079–2083 (2004)
8. Park, Y.I., et al.: Nonblinking and nonbleaching upconverting nanoparticles as an optical imaging nanoprobe and T1 magnetic resonance imaging contrast agent. Adv. Mater. **21**(44), 4467–4471 (2009)
9. Mialon, G., et al.: New insights into size effects in luminescent oxide nanocrystals. J. Phys. Chem. C **113**(43), 18699–18706 (2009)
10. Chen, Y., et al.: A new up-conversion charging concept for effectively charging persistent phosphors using low-energy visible-light laser diodes. J. Mater. Chem. C **6**(30), 8003–8010 (2018)
11. Huignard, A., Gacoin, T., Boilot, J.-P.: Synthesis and luminescence properties of colloidal YVO_4: Eu phosphors. Chem. Mater. **12**(4), 1090–1094 (2000)
12. Meyssamy, H., et al.: Wet-chemical synthesis of doped colloidal nanomaterials: particles and fibers of $LaPO_4$: Eu, $LaPO_4$: Ce, and $LaPO_4$: Ce, Tb. Adv. Mater. **11**(10), 840–844 (1999)
13. Li, N., et al.: A highly selective and instantaneous nanoprobe for detection and imaging of ascorbic acid in living cells and in vivo. Anal. Chem. **86**(8), 3924–3930 (2014)
14. Laporte, O., Meggers, W.F.: Some rules of spectral structure. JOSA **11**(5), 459–463 (1925)
15. Evans, C.: Chemical properties of biochemical relevance. In: Biochemistry of the Lanthanides, pp. 9–46. Springer (1990)
16. Binnemans, K.: Lanthanide-based luminescent hybrid materials. Chem. Rev. **109**(9), 4283–4374 (2009)
17. Hemmilá, I., Mukkala, V.-M.: Time-resolution in fluorometry technologies, labels, and applications in bioanalytical assays. Crit. Rev. Clin. Lab. Sci. **38**(6), 441–519 (2001)
18. Brites, C., Millán, A., Carlos, L.: Lanthanides in luminescent thermometry. In: Handbook on the Physics and Chemistry of Rare Earths, pp. 339–427. Elsevier (2016)
19. Matsumoto, K., Yuan, J.: Lanthanide chelates as fluorescence labels for diagnostics and biotechnology. Met. Ions Biol. Syst. **40**, 191–232 (2003)
20. Asati, A., et al.: Oxidase-like activity of polymer-coated cerium oxide nanoparticles. Angew. Chem. **121**(13), 2344–2348 (2009)
21. Xu, C., Qu, X.: Cerium oxide nanoparticle: a remarkably versatile rare earth nanomaterial for biological applications. NPG Asia Mater. **6**(3), e90–e90 (2014)
22. Diamandis, E.P., Christopoulos, T.K.: Europium chelate labels in time-resolved fluorescence immunoassays and DNA hybridization assays. Anal. Chem. **62**(22), 1149A-1157A (1990)
23. Brites, C.D., et al.: A luminescent molecular thermometer for long-term absolute temperature measurements at the nanoscale. Adv. Mater. **22**(40), 4499–4504 (2010)
24. Borisov, S.M., Wolfbeis, O.S.: Temperature-sensitive europium (III) probes and their use for simultaneous luminescent sensing of temperature and oxygen. Anal. Chem. **78**(14), 5094–5101 (2006)

25. Chen, B., et al.: A luminescent microporous metal−organic framework for the recognition and sensing of anions. J. Am. Chem. Soc. **130**(21), 6718–6719 (2008)
26. Harma, H., Soukka, T., Lovgren, T.: Europium nanoparticles and time-resolved fluorescence for ultrasensitive detection of prostate-specific antigen. Clin. Chem. **47**(3), 561–568 (2001)
27. Casanova, D., et al.: Single lanthanide-doped oxide nanoparticles as donors in fluorescence resonance energy transfer experiments. J. Phys. Chem. B **110**(39), 19264–19270 (2006)
28. Sapsford, K.E., Berti, L., Medintz, I.L.: Materials for fluorescence resonance energy transfer analysis: beyond traditional donor–acceptor combinations. Angew. Chem. Int. Ed. **45**(28), 4562–4589 (2006)
29. Feng, J., et al.: Functionalized europium oxide nanoparticles used as a fluorescent label in an immunoassay for atrazine. Anal. Chem. **75**(19), 5282–5286 (2003)
30. Dosev, D.K., et al.: Application of luminescent Eu: Gd_2O_3 nanoparticles to the visualization of protein micropatterns. J. Biomed. Opt. **10**(6), 064006 (2005)
31. Hemmilä, I., Ståhlberg, T., Mottram, P.: Bioanalytical Applications of Labelling Technologies: A Review of Trends and Opportunities in Biospecific Assay, Based on the Product Offering of Wallac, An EG&G Company. Wallac, an EG&G Company (1994)
32. Petoud, S., et al.: Stable lanthanide luminescence agents highly emissive in aqueous solution: multidentate 2-hydroxyisophthalamide complexes of Sm^{3+}, Eu^{3+}, Tb^{3+}, Dy^{3+}. J. Am. Chem. Soc. **125**(44), 13324–13325 (2003)
33. Wang, L., Li, P., Wang, L.: Luminescent and hydrophilic LaF_3–polymer nanocomposite for DNA detection. Luminescence **24**(1), 39–44 (2009)
34. Son, A., et al.: Rapid and quantitative DNA analysis of genetic mutations for polycystic kidney disease (PKD) using magnetic/luminescent nanoparticles. Anal. Bioanal. Chem. **390**(7), 1829–1835 (2008)
35. Casanova, D., et al.: Single europium-doped nanoparticles measure temporal pattern of reactive oxygen species production inside cells. Nat. Nanotechnol. **4**(9), 581–585 (2009)
36. Masson, J.-B., et al.: Inferring maps of forces inside cell membrane microdomains. Phys. Rev. Lett. **102**(4), 048103 (2009)
37. Wright, J.: Up-conversion and excited state energy transfer in rare-earth doped materials. In: Radiationless Processes in Molecules and Condensed Phases, pp. 239–295. Springer (1976)
38. Heer, S., et al.: Blue, green, and red upconversion emission from lanthanide-doped $LuPO_4$ and $YbPO_4$ nanocrystals in a transparent colloidal solution. Angew. Chem. Int. Ed. **42**(27), 3179–3182 (2003)
39. Auzel, F., et al.: Rare-Earth Spectroscopy. World Scientific, Singapore (1985)
40. Bouzigues, C., Gacoin, T., Alexandrou, A.: Biological applications of rare-earth based nanoparticles. ACS Nano **5**(11), 8488–8505 (2011)
41. Jo, E.-J., et al.: Single-step LRET aptasensor for rapid mycotoxin detection. Anal. Chem. **90**(1), 716–722 (2018)
42. Wu, M., et al.: An ultrasensitive fluorescent nanosensor for trypsin based on upconversion nanoparticles. Talanta **174**, 797–802 (2017)
43. Dai, S., et al.: An ultrasensitive aptasensor for Ochratoxin A using hexagonal core/shell upconversion nanoparticles as luminophores. Biosens. Bioelectron. **91**, 538–544 (2017)
44. Liu, Y., et al.: A strategy for accurate detection of glucose in human serum and whole blood based on an upconversion nanoparticles-polydopamine nanosystem. Nano Res. **11**(6), 3164–3174 (2018)
45. Geng, D., et al.: Photonic structuring improves the colour purity of rare-earth nanophosphors. Mater. Horiz. **5**(4), 661–667 (2018)
46. Rodríguez-Sevilla, P., et al.: Thermal scanning at the cellular level by an optically trapped upconverting fluorescent particle. Adv. Mater. **28**(12), 2421–2426 (2016)
47. Zijlmans, H., et al.: Detection of cell and tissue surface antigens using up-converting phosphors: a new reporter technology. Anal. Biochem. **267**(1), 30–36 (1999)
48. Kuningas, K., et al.: Homogeneous assay technology based on upconverting phosphors. Anal. Chem. **77**(22), 7348–7355 (2005)

49. Vetrone, F., et al.: Intracellular imaging of HeLa cells by non-functionalized $NaYF_4:Er^{3+}$, Yb^{3+} upconverting nanoparticles. Nanoscale **2**(4), 495–498 (2010)
50. Kuningas, K., et al.: Upconversion fluorescence enables homogeneous immunoassay in whole blood. Clin. Chem. **53**(1), 145–146 (2007)
51. Yi, G.-S., Chow, G.-M.: Water-soluble $NaYF_4$:Yb, Er (Tm)/$NaYF_4$/polymer core/shell/shell nanoparticles with significant enhancement of upconversion fluorescence. Chem. Mater. **19**(3), 341–343 (2007)
52. Shen, B., et al.: Revisiting the optimized doping ratio in core/shell nanostructured upconversion particles. Nanoscale **9**(5), 1964–1971 (2017)
53. Fischer, S., et al.: Precise tuning of surface quenching for luminescence enhancement in core–shell lanthanide-doped nanocrystals. Nano Lett. **16**(11), 7241–7247 (2016)
54. Li, Y., et al.: Cell and nanoparticle transport in tumour microvasculature: the role of size, shape and surface functionality of nanoparticles. Interface Focus **6**(1), 20150086 (2016)
55. Marriott, G., et al.: Time-resolved delayed luminescence image microscopy using an europium ion chelate complex. Biophys. J. **67**(3), 957–965 (1994)
56. Nichkova, M., et al.: Microarray immunoassay for phenoxybenzoic acid using polymer encapsulated Eu:Gd_2O_3 nanoparticles as fluorescent labels. Anal. Chem. **77**(21), 6864–6873 (2005)
57. Casanova, D., et al.: Counting the number of proteins coupled to single nanoparticles. J. Am. Chem. Soc. **129**(42), 12592–12593 (2007)
58. Corstjens, P., et al.: Infrared up-converting phosphors for bioassays. In: IEE Proceedings-Nanobiotechnology. IET (2005)
59. Van De Rijke, F., et al.: Up-converting phosphor reporters for nucleic acid microarrays. Nat. Biotechnol. **19**(3), 273–276 (2001)
60. Liu, C., et al.: Upconversion nanophosphor: an efficient phosphopeptides-recognizing matrix and luminescence resonance energy transfer donor for robust detection of protein kinase activity. Anal. Chem. **86**(12), 6095–6102 (2014)
61. Wang, L., et al.: Fluorescence resonant energy transfer biosensor based on upconversion-luminescent nanoparticles. Angew. Chem. Int. Ed. **44**(37), 6054–6057 (2005)
62. Chen, Z., et al.: Versatile synthesis strategy for carboxylic acid− functionalized upconverting nanophosphors as biological labels. J. Am. Chem. Soc. **130**(10), 3023–3029 (2008)
63. Liu, C., et al.: Efficient fluorescence resonance energy transfer between upconversion nanophosphors and graphene oxide: a highly sensitive biosensing platform. Chem. Commun. **47**(16), 4661–4663 (2011)
64. Hampl, J., et al.: Upconverting phosphor reporters in immunochromatographic assays. Anal. Biochem. **288**(2), 176–187 (2001)
65. Zhang, P., et al.: Design of a highly sensitive and specific nucleotide sensor based on photon upconverting particles. J. Am. Chem. Soc. **128**(38), 12410–12411 (2006)
66. Mader, H.S., Wolfbeis, O.S.: Optical ammonia sensor based on upconverting luminescent nanoparticles. Anal. Chem. **82**(12), 5002–5004 (2010)
67. Wu, B.-Y., et al.: Fluorescence resonance energy transfer inhibition assay for α-fetoprotein excreted during cancer cell growth using functionalized persistent luminescence nanoparticles. J. Am. Chem. Soc. **133**(4), 686–688 (2011)
68. Aitasalo, T., et al.: Persistent luminescence phenomena in materials doped with rare earth ions. J. Solid State Chem. **171**(1–2), 114–122 (2003)
69. Paterson, A.S., et al.: A low-cost smartphone-based platform for highly sensitive point-of-care testing with persistent luminescent phosphors. Lab Chip **17**(6), 1051–1059 (2017)
70. Aitasalo, T., et al.: Mechanisms of persistent luminescence in Eu^{2+}, RE^{3+} doped alkaline earth aluminates. J. Lumin. **94**, 59–63 (2001)
71. Matsuzawa, T., et al.: A new long phosphorescent phosphor with high brightness, $SrAl_2O_4:Eu^{2+}$, Dy^{3+}. J. Electrochem. Soc. **143**(8), 2670 (1996)
72. Lin, Y., et al.: Anomalous luminescence in $Sr_4Al_{14}O_{25}$:Eu, Dy phosphors. Appl. Phys. Lett. **81**(6), 996–998 (2002)

73. Xu, J., et al.: Near-infrared multi-wavelengths long persistent luminescence of Nd3+ ion through persistent energy transfer in Ce^{3+}, Cr^{3+} co-doped $Y_3Al_2Ga_3O_{12}$ for the first and second bio-imaging windows. Appl. Phys. Lett. **107**(8), 081903 (2015)
74. Ueda, J., et al.: Vacuum referred binding energy of 3d transition metal ions for persistent and photostimulated luminescence phosphors of cerium-doped garnets. J. Lumin. **192**, 371–375 (2017)
75. Hu, L., et al.: Orthogonal multiplexed luminescence encoding with near-infrared rechargeable upconverting persistent luminescence composites. Adv. Opt. Mater. **5**(22), 1700680 (2017)
76. Cheng, Y., Sun, K., Ge, P.: Yb^{3+} and Er^{3+} co-doped $ZnGa_2O_4$: Cr^{3+} powder phosphors: combining green up-conversion emission and red persistent luminescence. Opt. Mater. **83**, 13–18 (2018)
77. Hölsä, J., et al.: Role of defect states in persistent luminescence materials. J. Alloy. Compd. **374**(1–2), 56–59 (2004)

Chapter 25
Nanomaterial-Based Sensors for Macrolide Sensing

Noha Hasaneen, Pratishtha Khurana, Rama Pulicharla, Pouya Rezai, and Satinder Kaur Brar

Abstract Recent developments in nanotechnology have empowered researchers to significantly advance all biomes of research, including sensing or detection of emerging contaminants, such as antibiotics, in both simple and complex matrices. This chapter is specifically dedicated to the discussion of the underlying role of various nanomaterials in the detection of macrolide antibiotics and their sensing mechanisms to enable their detection in biological samples, food products, and environmental matrices with confidence and incite further research to plug the existing gaps. In this sense, this chapter includes characteristics of nanosystems, followed by a detailed discussion of optical and electrochemical sensors with nanoparticles for macrolides detection. Finally, shortcomings in existing techniques are identified, and recommendations are provided.

Keywords Antibiotics · Electrochemical sensors · Macrolides · Nanomaterials · Optical sensors · Sensing mechanisms

25.1 Introduction

Size does matter, but nano is even better!

Nano-sized particles and systems, collectively known as nanomaterials, have considerably improved, even revolutionized, science and research. Today, custom-tailored nanomaterials with specific size, shape and surface functionalization offer new opportunities in nearly every field, including and not limited to aviation and

N. Hasaneen · P. Khurana · R. Pulicharla · S. K. Brar (✉)
Department of Civil Engineering, Lassonde School of Engineering, York University, North York, Toronto, ON M3J 1P3, Canada
e-mail: satinder.brar@lassonde.yorku.ca

P. Rezai
Department of Mechanical Engineering, Lassonde School of Engineering, York University, North York, Toronto, ON M3J 1P3, Canada

D. Mohanta and P. Chakraborty (eds.), *Nanoscale Matter and Principles for Sensing and Labeling Applications*, Advanced Structured Materials 206,
https://doi.org/10.1007/978-981-99-7848-9_25

space, automotive engineering, electronics, construction industry, optics, textiles, cosmetics, medicinal science, and sensors. Nanomaterials are now being extensively used in the detection of a wide range of compounds, including antibiotics, especially in complex environmental matrices like soil, wastewater, blood, urine, and food products. Unregulated usage of antibiotics- for both human and veterinary applications has resulted in continuous occurrences of antibiotics in food and environmental compartments, such as wastewaters, soil, and sediments. In this regard, this chapter, particularly, delineates the application of nanomaterials for sensing macrolide antibiotics.

Macrolides are a class of antibiotics used in the management and treatment of various bacterial infections, such as community-acquired severe respiratory tract infections, such as campylobacteriosis, chancroid, tonsilitis, bronchitis, sinusitis, pneumonia, cholera, cystic-fibrosis, toxoplasmosis, among many other. They are categorized as the highest priority, critically important antibiotics by the WHO (World Health Organization) due to their ability to treat severe, life-threatening, rare diseases both in humans and cattle [39]. Moreover, recent injectable macrolides have a long half-life, which allows for once-only dosing regimens [32]. They are primarily bacteriostatic and prevent bacterial multiplication by binding to the 50S subunit of the ribosome and inhibiting bacterial protein synthesis.

Chemically, they possess a macrocyclic lactone ring containing between 12 and 17 atoms, attached to one or more deoxy sugars via glycosidic bonds, and are classified based on the size of the macrocyclic ring, as shown in Fig. 25.1. For instance, a 14-membered ring (e.g., erythromycin and clarithromycin consist of two sugar moieties attached to a 14-atom lactone ring), 15- (e.g., azithromycin), and 16- (e.g., tylosin and josamycin) membered macrolide antibiotics are the most common.

The large lactone ring is substituted by hydroxyl, alkyl or acyl groups, contains one ketone at C7 for 12-membered macrolides, and at C9 for 14-member analogues, and one aldehyde functionality in 16-membered rings. Azithromycin contains a 15-member ring with a tertiary amino-group methyl-substituted nitrogen) as shown in Fig. 25.1. The substitution of hydroxyl groups by neutral or basic sugars renders basic character to the macrolides.

A study has revealed up to 7000, 1300, and 270 ng/L of macrolide antibiotics, including azithromycin, clarithromycin, and erythromycin, in effluents from six wastewater treatment plants (WWTPs) in Canada, with their concentrations ranging from 850-81, 580-4.6, 92-1.4 ng/g dw biosolids, respectively [8]. Similarly, their widespread applications to prevent and treat infections in poultry and livestock have resulted in their occurrence in food samples, such as honey, milk, eggs, and meat products. This continuous exposure to macrolides through water/food sources can trigger the development and dissemination of antibiotic resistance at a faster rate, which is alarming. Therefore, many countries have established maximal residual limits (MRL) of the most used macrolides in different types of food products to monitor their spread. For example, the MRL of erythromycin in Canadian animal food products is 0.05 ppm in cattle milk, 0.2 ppm in sheep muscles, 0.125 ppm in chicken muscles, and 0.2 ppm for tylosin in different food products [28]. In this sense, it becomes imperative to develop sensitive and robust on-site detection techniques

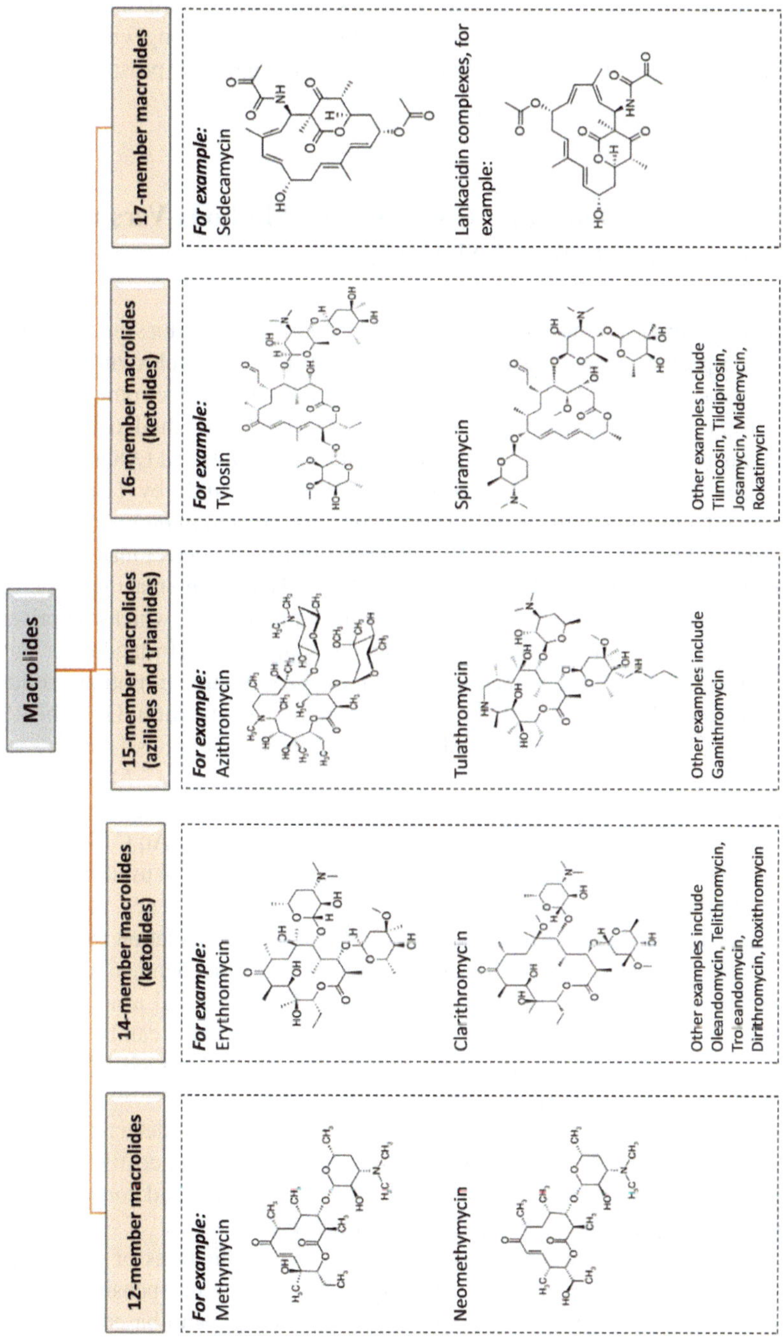

Fig. 25.1 Classification of macrolide antibiotics

in varying samples, such as milk, plasma, water, wastewater, soil, and sediments, where nanomaterials come to the rescue. This chapter, in this regard, discusses the applications of nanomaterials for sensing macrolide antibiotics, along with different sensing mechanisms and associated challenges, and highlights the prospects to stir research forward.

25.2 Methods of Detection: Conventional Methods Versus Nanomaterials-Based Methods

Conventional methods for the detection of macrolide antibiotic residues are primarily based on high-performance liquid chromatography (HPLC), liquid chromatography-mass spectrometry (LC–MS), and gas chromatography-mass spectrometry (GC–MS). For instance, [1] developed a sensitive HPLC-Q-TOF MS method to accurately detect and quantify 12 macrolides in food materials, with LODs and LOQs as 0.01–0.50 ng/g and 0.04–2.0 ng/g, respectively [1]. Similarly, a study by Thangadurai [38] developed a method with GC–MS to analyze azithromycin in biological fluids, with a LOD of 2 ug/ml [38]. LC–MS has also been utilized to detect macrolides in various matrices, including food, biological samples, and environmental matrices. To improve the detection of these trace pollutants, sample preparation, which comprises extraction and concentration of the analyte from our sample, is performed first. This is often done by conventional extraction (dissolution, precipitation, filtration), solid-phase extraction (SPE) and its variants, electro-membrane extraction, QuEChERS (Quick, Easy, Cheap, Effective, Rugged and Safe), ultrasonication-assisted extraction (UAE), etc. [15]. Sample treatment is typically followed by LC–MS analysis, which has been a premier analytical tool for quantification and qualitative analysis of analytes. Generally, reverse phase columns, such as Zorbax SB-Aq C18 column, Hypersil gold C18, Eclipse XDB-C18, and C8 columns, are the most used stationary phases due to their good hydrophobicity, selectivity and high efficiency, along with acetonitrile or methanol as the mobile phase [2, 27, 36]. Soft ionization techniques such as electrospray ionization (ESI) and atmospheric pressure chemical ionization (APCI), matrix-assisted laser desorption/ionization (MALDI), and a newer technique called surface-assisted laser desorption/ionization (SALDI), can be used for macrolides. All these methods have been comprehensively discussed in a recent review by Hu et al. (2020) [15].

These methods are very accurate and quantified- with a limit of detection (LOD) ranging from ng/L to ug/L, however, require skilled personnel and costly, sophisticated equipment. Also, extensive sample preparation- extraction and concentration make the processing time-consuming and labor-intensive.

Various attempts have been made to overcome these limitations of traditional methods and for the development and standardization of rapid, inexpensive, robust, and sensitive detection technologies by the inclusion of nano-based systems.

Different groups of nanomaterials, such as carbon-based and metal-based nanomaterials, quantum dots (QDs) and magnetic nanoparticles, have been extensively included in the detection analysis to reach lower detection limits and improve the sensitivity of and the selectivity of the method. For instance, [5] prepared and utilized core-satellite polydopamine-coated Fe_3O_4 nanoparticle-hollow porous molecularly imprinted polymer (Fe_3O_4@PDA-HPMIPs), combined with conventional HPLC–MS/MS, for quantification of macrolide antibiotics, including spiramycin, tilmicosin, tylosin, josamycin, azithromycin, clarithromycin, and roxithromycin. The linear range was observed to be 0.4–40 ug/kg and the inclusion of nanoparticles allowed to reach lower LODs, ranging from 0.003 ug/kg for spiramycin to 0.075 for tilmicosin, respectively [5]. The use of nanosystems, however, is not limited to detection. They are also used for extraction, which will be discussed in detail in this chapter.

With these reports, we now know that owing to the astounding optical, electronic, and magnetic features of custom-tailored nanomaterials, they have enabled sensing antibiotics, including macrolides for trace concentrations, with confidence. For the detection of macrolides in different matrices, nanomaterials play a significant role owing to their outstanding properties. The properties of nanomaterials are significantly distinct from the bulk material, allowing their widespread applications, including sensing. Some of these key properties include:

(i) ***High surface area to volume ratio***: Nanomaterials have a significantly higher surface area to volume ratio than their bulk analogues, and this property enables them to stand out, especially for surface phenomena, like adsorption and catalysis.
(ii) ***Thermal and electrical conductivity***: Nanoparticles have optical and electronic properties that are distinct from their bulk counterparts. For instance, noble metal nanoparticles exhibit a strong UV–visible extinction band attributed to localized surface plasma resonance (LSPR), is absent in the bulk metal [19].
(iii) ***High thermal and electrical conductivity***: Nanoparticles exhibit extraordinary thermal and electrical conductivity compared to the bulk material, for example: graphene, as compared to graphite [20].
(iv) ***Quantum effects***: Defined by quantum mechanics, quantum effects are unique to nanoparticles and play an important role in the sensing mechanisms. These are discussed in detail in Sect. 25.3 of this chapter.
(v) ***Magnetic properties***: Magnetic characteristics, too, change at the nanoscale. Non-magnetic materials can have magnetic properties at the nanoscale level, which makes them distinctive [34].
(vi) ***Mechanical strength***: Size, surface coating, and coagulation affect the mechanical properties, such as elastic modulus, hardness, friction, stress and strain and adhesion, and define the nanomaterial properties [9].

25.3 Applications for Macrolides Sensing Using Nanomaterials

Now, after discussing the unique nanoparticle properties, let us explore their applications for sensing macrolide antibiotic residues using both optical and electrochemical sensors.

25.3.1 Optical Sensors with Nanomaterials for Macrolides Detection

Optical sensors are known for their simplicity, accuracy, fast response, low cost, and sensitivity. Different types of optical sensors have been used for antibiotic detection in various sample types including fluorescence, colorimetric, chemiluminescence, surface-enhanced Raman scattering (SERS), and surface plasmon resonance (SPR) based sensors. Nanoparticles have been incorporated in some of these techniques for the detection of macrolide antibiotics either as labels or signal transducers. In this chapter, nanoparticles coupled sensing techniques for macrolides detection are discussed. A summary of optical sensors using nanomaterials for macrolides detection is provided in Table 25.1.

Lateral Flow Immunoassay Colorimetric Sensors

Devices using lateral flow immunoassay (LFIA) have been widely used as point-of-care testing (POCT) for food and water quality monitoring, blood and urine analysis, and environmental monitoring, owing to their fast detection, portability, feasibility, high sensitivity, and low-cost analysis. Moreover, LFIA overcomes the washing steps, time-consuming pipetting, and long incubation time of the conventional immunoassay techniques such as ELISA (Enzyme-linked immunosorbent assay). Herein, nanoparticles play an important role as labeling materials for the recognition elements which could be antibodies or aptamers (Aptamer is a synthetic single-strand DNA that has emerged recently as an alternative to antibodies as a recognition element for many types of molecules including antibiotics). Different types of nanoparticles can be used in LFIA such as magnetic nanoparticles, colored nanoparticles (e.g., gold and selenium nanoparticles), luminescent nanoparticles (e.g., quantum dots and dye-doped nanoparticles), and up-converting nanoparticles. The most widely used nanoparticles in antibiotics detection using LFIA are colloidal gold nanoparticles (AuNPs) owing to their stability, low toxicity, facile conjugation, and visual bright color which allows both qualitative detection of the analyte by the naked eye and quantitative detection by using an external strip scan reader. For LFIA sensors, we will discuss device fabrication, analysis mechanism, synthesis of AuNPs, and applications for macrolides detection in different types of samples.

Table 25.1 Summary of nanomaterial-based different sensing techniques for macrolides in various matrices

Optical sensor type	Probe		Antibiotic	Matrix	Detection limit (ng/mL)		References
	Nanomaterial	Recognition element			LOD	Cut off	
LFIA (colorimetric)	Colloidal GNPs	Anti-CLA MAb	Clarithromycin Erythromycin Roxithromycin Azithromycin	Milk	0.095 0.085 0.055 0.175	2.5 5 5 10	[44]
	Colloidal GNPs	Aptamer	Erythromycin	Tap water	3 pM	Not mentioned	[3]
	Colloidal GNPs	Anti-KIT MAb	Kitasamycin Josamycin	Milk	KIT: 1.51 JOS: 1.91	KIT: 25 JOS: 5	[10]
				Egg	KIT: 3.0 ng/g JOS: 2.73 ng/g	KIT: 75 ng/g JOS: 150 ng/g	
	Colloidal GNPs	Anti-TYL MAb	Tylosin Tilmicosin	Milk	Not mentioned	TYL: 8 TIM: 15	[25]
				Pork		TYL: 15 TIM: 30	
	Colloidal GNPs	Anti-TYL MAb	Tylosin	Milk Honey Eggs	0.09	2	[14]
	Colloidal GNPs	Anti-TYL MAb	Tylosin Tilmicosin	Egg	Not mentioned	TYL: 6 ng/g TIM: 14 ng/g	[29]
	Colloidal GNPs	Anti-TYL MAb	Tylosin Tilmicosin	Honey	Not mentioned	TYL: 10 TIM: 20	[37]

(continued)

Table 25.1 (continued)

Optical sensor type	Probe		Antibiotic	Matrix	Detection limit (ng/mL)		References
	Nanomaterial	Recognition element			LOD	Cut off	
	Colloidal GNPs	Anti-DES MAb	Tylosin Tilmicosin Desmycosin	Muscle Liver Fish Eggs	TYL: 10 ng/g TIM: 20 ng/g DES: 5 ng/g	TYL: 40 ng/g TIM: 60 ng/g DES: 10 ng/g	[22]
	Colloidal GNPs	Anti-ERY MAb Anti-SPI MAb Anti-TIM PAb	Erythromycin Spiramycin Tilmicosin Tylosin	Milk	Not mentioned	ERY: 5 SPI: 5 TIM: 10 TYL: 20	[24]
	Time resolved fluorescent microsphere (TRFM)	MAb	Tylosin Tilmicosin	Milk	TYL: 1.72 TIM: 1.39	TYL: 31.25 TIM: 62.50	[42]
Indirect competitive fluorescence-linked immunosorbent assay (ic-FLISA)	QDs	Anti-TYL MAb	Tylosin Tilmicosin	Porcine tissues	TYL: 0.02 ng/g TIM: 0.04 ng/g		[23]
Fluorescence spectrophotometry	Fluorescent gold nanoclusters	MIP	Erythromycin	Urine Saliva	12 nM		[46]

(continued)

Table 25.1 (continued)

Optical sensor type	Probe		Antibiotic	Matrix	Detection limit (ng/mL)		References
	Nanomaterial	Recognition element			LOD	Cut off	
	$CsPbBr_3$ nanoparticles	MIP nanogels	Roxithromycin	Milk Pork Egg	1.7×10^{-2}		[13]
	N,S-CDs	–	Azithromycin	Urine Dispersible tablets	0.76 μM		[11]
Surface plasmon resonance (SPR)	Gold chip	MIP nanoparticles	Erythromycin	Water	290		[35]
Electrochemical sensor technique	*Working electrode*	*Recognition element*	*Antibiotic*	*Matrix*	*Detection limit (μM)*		*References*
DPV	CoN NWs/CC	MIP	Tylosin	Surface water Soil	5.5×10^{-6}		[6]
	$MgCr_2O_4$-MWCNTs/GCE	–	Azithromycin	Plasma Urine Capsules	0.07		[4]
	AB/GCE	–	Erythromycin	Honey	8×10^{-2}		[16]
	GNU/GO/GCE	MIP	Azithromycin	Human serum	0.0001		[18]
SWV	AuNPs/CPE	–	Azithromycin Roxithromycin	Pharmaceutical formulation	0.045 μg/ml 0.15 μg/ml		[40]
LSV	GO-MWCNTs/GCE	–	Azithromycin	Capsules	0.07		[45]

(continued)

Table 25.1 (continued)

Optical sensor type	Probe		Antibiotic	Matrix	Detection limit (ng/mL)		References
	Nanomaterial	Recognition element			LOD	Cut off	
AMP	GR–AuNPs/CS–PtNPs/ gold	MIP	Erythromycin	Milk Honey	2.3×10^{-2}		[26]
BIA-AMP	SPCB-CT/GCE	–	Clarithromycin Erythromycin Roxithromycin Azithromycin	Environmental (water) and drugs	0.16 0.19 0.18 0.15		[41]

(MIP) molecular imprinted polymer, (CoN NWs/CC) cobalt nitride nanowires/carbon cloth, (CS–PtNPs) chitosan–platinum nanoparticles, (GR–AuNPs) graphene–gold nanoparticles, (SDS-SPCEs) screen-printed carbon electrode modified with sodium dodecyl sulfate, (SPCB-CT/GCE) Super P carbon black film-chitosan/glassy carbon electrode, (BIA-AMP) batch injection analysis amperometry, (CPE) carbon paste electrode, (GNU) gold nanourchines, (SWV) square wave voltammetry, (GO-MWCNTs) graphene oxide-multi-walled carbon nanotubes, (AB) acetylene black nanoparticles, (LSV) linear sweep voltammetry

(i) ***LFIA device fabrication and analysis technique***: Lateral flow immunoassay test strips mainly consist of a sample pad, conjugate or probe pad, a detection membrane (glass fiber or nitrocellulose membrane with different pore sizes are usually used), and an absorbent pad attached to polyvinyl chloride (PVC) sheet. Detection membrane characteristics are very important for optimizing the flow rate of the sample by the capillary action. On the membrane, there is a test line with a specific bioreceptor for capturing the labeled recognition element and a control line with goat anti-mouse IgG to ensure that the assay is working properly. However, the appearance of the color on the control line ensures the applicability of the test, the color of the test line can mean positive or negative results based on the used immunoassay technique. There are two types of immunoassay techniques: (i) competitive immunoassay which is divided into the direct and indirect assay, and (ii) sandwich immunoassay. For the direct competitive immunoassay, both the analyte and the bioreceptor at the test line compete for binding to the labeled recognition element (primary antibody), in the presence of the analyte, it will bind to the labeled recognition element, and the binding at the test line will decrease, results in decreasing in the color intensity gradually by increasing the analyte concentration until certain concentration (cut off value), at which the color disappears. In the indirect competitive assay, there are two types of antibodies, primary and secondary antibodies. In the presence of the target, it will bind to the primary antibody which then binds to the secondary labeled antibody and causes no capturing by the coating antigen on the test line, and less color intensity as shown in Fig. 25.2. In the sandwich immunoassay, the analyte is captured between the labeled recognition element and the bioreceptor at the test line, resulting in the appearance of color, and the intensity increases with increasing the analyte's concentration. For antibiotics detection, a competitive assay is usually used as it is suitable for analytes with low molecular size (<1 KDa) which cannot bind to two bioreceptors at the same time [31].

(ii) ***Preparation and characterization of colloidal gold nanoparticles for LFIA***: Colloidal gold nanoparticles are usually prepared by the trisodium citrate reduction method in which trisodium citrate is mixed with boiling chloroauric acid and stirred until the solution color changed from light yellow to wine red and finally stored at 4 °C to be conjugated with the recognition element (Monoclonal antibody, polyclonal antibody, or aptamer). The change in the amounts of the reactants can result in a change in the diameter size of the nanoparticles which can be characterized by the transmission electron microscope. Based on the method used by [24], 2 ml 2% chloroauric acid (w/v) was added to 100 ml boiling double-distilled water, when boiling, 4.5 ml 1% trisodium citrate (w/v) was fast-added and continue boiling for 10 min then cooling the solution and complete to the initial volume by double-distilled water and store at 4 °C. By using the same method but increasing the amount of trisodium citrate from 4.5 ml to 6.2, or 8 ml, you can get a smaller diameter from 40 nm to 30, or 20 nm, respectively [24, 44].

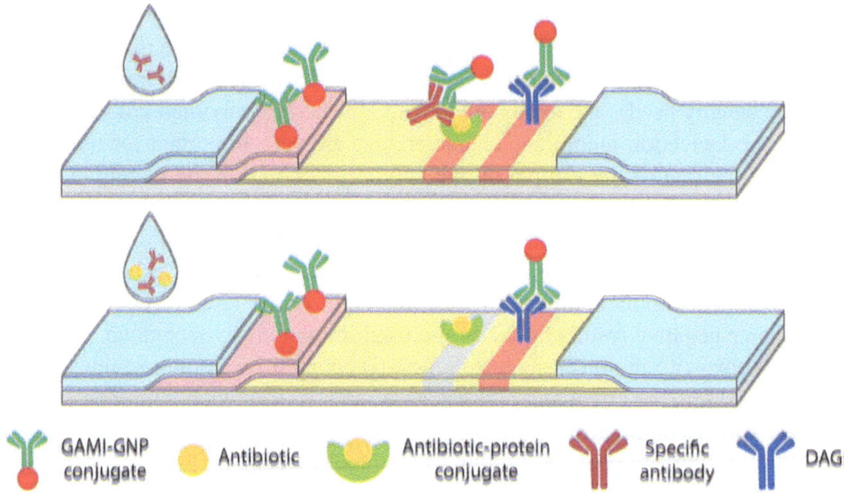

Fig. 25.2 Scheme of the indirect ICA format. (DAGI) donkey antibodies against goat immunoglobulins; (GAMI-GNPs) goat antibodies against mouse immunoglobulins- gold nanoparticles. Adapted from published work [14]

(iii) ***LFIA devices using gold nanoparticles for macrolides detection***: Colloidal gold nanoparticles *(AuNPs)* have been used as colored label material in LFIA test strips for the detection of trace macrolide antibiotics in different types of matrices such as Tap water, food products (Honey, milk, eggs, chicken muscles, pork, liver, and fish). These nanoparticles have a wine-red color which appears on the test strip lines after capturing. The color intensity can be affected by the size of the nanoparticles. Thus, choosing the best size of AuNPs is very important for visualizing the color of the test lines. Increasing the *AuNPs* diameter size from 20 to 40 nm results in an increase in the color intensity from bright red to purple on the strip but does not affect the test sensitivity as investigated by [24] for the detection of four macrolides [24]. Therefore, *AuNPs* with a 40 nm diameter are most used in LFIA for better visualization. Using LFIA allows the detection of several macrolides in ng/ml in just a few minutes with strip costs less than $1 as listed in Table 25.1.

The main disadvantage here is the pretreatment process based on the sample type, which sometimes is time-consuming and could affect the results if not done properly. Herein, erythromycin was detected in spiked tap water samples without pretreatment [3]. Also, erythromycin, spiramycin, tylosin, tilmicosin, kitasamycin, and josamycin were detected in spiked raw milk samples using LFIA strips without any pretreatment [10, 24]. However, sometimes researchers prefer simple pretreatment for milk samples to eliminate any background interference. Therefore, raw milk pretreatment was done for the detection of four macrolides (Erythromycin, clarithromycin, azithromycin, and roxithromycin) by centrifugation of the spiked sample at 6000 g for 10 min, then dilution of the middle emulsion layer 8-folds with phosphate buffered saline (PBS) [44]. Furthermore, kitasamycin and josamycin were detected in spiked egg samples after homogenization and dilution with PBS [10]. In spiked honey samples, tylosin and tilmicosin were detected using test strips just after a 4-folds dilution of the spiked samples [22]. However, pretreatment of tylosin spiked honey can be done by dilution with water and then mixing with ethyl acetate for extraction of tylosin for further purification. Followed by centrifugation at 5000 g for 10 min and finally, ethyl acetate supernatant is evaporated, and the residue is dissolved in PBS [14]. When it comes to solid and more complex samples, extraction would be essential. Thus, different macrolides can be extracted from the homogenized, buffered, centrifuged, and diluted solid samples by ethyl acetate or 50% acetonitrile in 0.02 M PB. Extraction is followed by centrifugation and evaporation of the supernatant, and then dissolving of the residue in buffer and used for analysis such as in the case of tylosin and tilmicosin detection in egg samples [29], pork [25], liver, fish, and muscles [22]. Another disadvantage in LFIA sensing is that it is difficult to differentiate between structurally similar macrolide antibiotics if they are present in the same sample which means low detection selectivity within the same group.

Fluorescence-Based Sensors

Fluorescence sensing is one of the most sensitive, accurate, and easily used techniques for qualitative and quantitative detection of many types of antibiotics. Herein, Fluorescent nanoparticles play an important role and have been widely used as sensing probes for contaminants detection including antibiotic residues in food products and biological samples. Many types of fluorescent nanoparticles have recently emerged for sensing including carbon dots, quantum dots, fluorescent microspheres, and noble metals nanoclusters whose size range from 1 to 100 nm. Compared to conventional organic dyes, fluorescent nanoparticles are more stable and sensitive. Based on the fluorescence intensity and the maximum emission wavelength, the analyte's concentration can be measured. Fluorescence intensity of nanoparticles, after mixing the sample with the probe, can increase (turn on) or decrease (turn off or quenching), or the maximum emission wavelength may decrease (hypsochromic shift) or increase (bathochromic shift) based on the characteristics of the analyte itself and its interactions with the fluorescent probe. The most used technique for the detection of antibiotics using fluorescent nanoparticles is the fluorescence quenching technique. Fluorescence quenching can occur through Forester resonance energy transfer (FERT), inner filter effect (IFE), oxidation, or phase transformation of nanoparticles [13]. Herein, different types of nanoparticles have been used for macrolide antibiotic detection.

(i) ***Gold nanoclusters***: They are bright, stable, less toxic, and sensitive fluorescent nanoparticles. Moreover, gold nanoclusters can be easily functionalized by different functional groups. Coated gold nanoclusters functionalized by carboxylic groups (–COOH) through the mixing of $HAuCl_4$ with GSH in ultrapure water at high temperatures in the dark were used for the detection of erythromycin in urine and saliva [46]. Herein, carboxyl-functionalized gold nanoclusters were bound to amino-functionalized silica (SiO_2) by amide bonding and forming a coat around the nanoclusters. For specific detection of erythromycin, a layer of molecular imprinted polymer (MIP) was coated over the AuNCs-SiO_2 forming a fluorescent probe that can selectively detect erythromycin. In the presence of erythromycin, it would fit in the imprinted cavities of the coating polymer and quench the fluorescence of the AuNCs. Therefore, increasing erythromycin concentration results in decreasing the AuNCs' fluorescence intensity which was detected by spectrofluorometer.

(ii) ***Quantum dots***: One of the most important and massively used fluorescent nanomaterials for sensing are quantum dots which are nanodots of chalcogenides (sulfides, tellurides, selenides) of metals (lead, cadmium, zinc). These semiconducting nanoparticles with high fluorescence quantum yield, high photostability, and well-defined narrow emission bands can be used alone as sensing probes or conjugated with other biomolecules such as protein, antibody, or aptamer. Moreover, they can be embedded inside a shell for increasing sensitivity, stability, or recognition specificity. Quantum dots were used as labels for secondary antibodies in indirect competitive fluorescence-linked immunosorbent assay for the detection of tylosin and tilmicosin in animal tissues. In this, goat anti-mouse secondary antibody was labeled by quantum dots. In the presence of the target analytes, no capturing happens with the coating antigen and no fluorescence appears. However, in the absence of the target, capturing with the coating antigen occurs with a measured fluorescence [23]. Another way to enhance the detection selectivity is using MIP which can be coupled as a recognition element with quantum dots as a fluorescent transducer either by coating quantum dots with a MIP shell or by embedding quantum dots within the MIP nano gel cavities. In this regard, fluorescent lead halide perovskite ($CsPbBr_3$) nanoparticles were loaded in MIP nanogels for selectively detecting roxithromycin in food products based on the fluorescence quenching technique which occurred through phase transformation of the perovskite in the presence of roxithromycin which has oxidative oxime function group [13].

(iii) ***Carbon dots (CDs)***: They are the youngest nanomaterials for sensing with a diameter less than 10 nm, they have emerged as an alternative to quantum dots since 2004 when they were first discovered. CDs mainly consist of a carbon skeleton which can be easily functionalized with different functional groups on their surface such as hydroxyl, carboxyl, amino, epoxy, and amido groups which facilitate the binding with different types of molecules. Moreover, CDs can be doped with heteroatoms to enhance their photophysical characteristics such as sulfur, phosphorus, oxygen, nitrogen, and boron which alter the electron density and result in higher optical quantum yield. Different sizes, doping elements' sources, carbon precursors, and function groups produce different CDs with different chemical and optical characteristics [7]. Compared to QDs, CDs are more stable and hydrophilic with good water solubility which provides a compatible sensing probe for water analysis, are less toxic, and can be synthesized from low-cost resources and by low-cost synthesis techniques. Recently, there have been various successful approaches for synthesizing CDs from environmentally friendly precursors such as food waste, sludge, and bioorganic molecules which result in high quality and less toxic doped CDs. There are several fabrication methods for CDs including microwave-assisted, hydrothermal carbonization, laser ablation, electrochemical, thermolysis, chemical oxidation, and ultrasonic methods [21]. However, the most used technique is the hydrothermal method as it can be done in just one step through a chemical reaction in solvent using high sealing pressure in stainless steel autoclave. As an eco-friendly and low-cost method with high quality and uniform product, the hydrothermal technique was used to produce bright green, fluorescent N, S-CDs for the detection of azithromycin based on the fluorescence quenching mechanism. Herein, they used 3-amino thiophenol as N, S, and carbon source in ethanol, and by hydrothermal carbonization, in stainless steel autoclave, they got a chocolate solution. After evaporating ethanol, they redissolve the residue in ultra-pure water to do centrifugation and filtration to remove large impurities. Then the supernatant was dialyzed against ultra-pure water and lyophilized to finally get the pure orange-yellow powder of N, S-CDs which were used for azithromycin detection in urine and dispersible tablets. Although there is no overlap between azithromycin absorption weak peak and the emission peak of N, S-CDs which means no FERT or IFE occurred, azithromycin can quench their fluorescence by electrostatic interaction with N, S-CDs. Electrostatic interaction was proved by measuring the zeta potential of both N, S-CDs, and azithromycin to find that azithromycin has a negative zeta potential of −4.2 mV, while that of N, S-CDs is positive at 16.2 mV [11].

(iv) ***Time-resolved fluorescent microspheres (TRFM)***: Microspheres dopped with time-resolved fluorescent molecules are distinct from those dopped with conventional fluorescent dyes. These microspheres with different diameters have large fluorescence quantum yield, large stokes shift which is the difference between the wavelengths of maximum excitation and maximum absorption, large signal to background ratio, sharp emission peak, and long fluorescence lifetime. These specific fluorescence molecules are called lanthanide chelate labels and they are used as time-resolved fluorescent molecules such as europium (III), terbium (III), samarium (III), and dysprosium (III) [17]. TRFM could be conjugated with different molecules such as biomolecules through their surface functional groups. One of these microspheres is Europium -chelated fluorescent microsphere with 100 nm diameter and carboxyl groups on the surface. These microspheres were used as labels for monoclonal antibodies to detect tylosin and tilmicosin in milk samples after dilution with phosphate buffer. Before conjugation with antibody, carboxyl groups must be activated first by mixing microspheres with 1-(3-Dimethylaminopropyl)-3-ethyl carbodiimide hydrochloride (EDC) and N-hydroxysuccinimide (NHS). Herein, carboxyl groups react with NHS in the presence of EDC to form semi-stable NHS esters. Then, NHS esters interact with amino groups in biomolecules such as those in lysine amino acids to form stable amide bonds. These TRFM labeled monoclonal antibodies were used for lateral flow immunochromatographic assay of tylosin and tilmicosin based on fluorescence detection using a fluorescence reader under UV [42].

Surface Plasmon Resonance (SPR) Based Sensors

SPR sensor as a powerful and sensitive type of optical sensor relies on the propagation of surface plasmon at the interface of a metal and dielectric medium by excitation with electromagnetic radiation. Herein, a metal film with an optimum thickness of 50 nm of gold or silver is usually used. Although silver shows higher sensitivity, gold is more preferred owing to its higher stability. SPR is considered a label-free and real-time analysis technique for target detection. It mainly determines the analyte by changes in the refractive index which is caused by the binding of target molecules to the recognition element on the metal film surface. To selectively capture the target residues, the recognition element should be immobilized on the SPR surface [33]. In these, molecularly imprinted polymer nanoparticles as synthetic receptors have been used for selective detection of erythromycin. Nanoparticles of molecular imprinted polymer (MIP) with a particle size of 42.6 nm were synthesized by two phases of mini-emulsion polymerization. These MIP nanoparticles were immobilized on the gold film of the SPR chip to produce a selective nanosensor for erythromycin detection in an aqueous solution [35].

Nanomaterials based optical sensors show good sensitivity for the detection of macrolides in different matrices, including and not limited to milk, honey, serum, muscles, tap water, and urine. They are used as colorimetric labels for antibodies in LFIA, fluorescence substrates in fluorescence-based sensors, or as nanofilm in surface plasmon resonance-based sensors. Despite this, the techniques require extensive sample pre-treatment and have a narrow range of linearity, which brings us to the next type of sensors- electrochemical sensors, minimizing the need for sample preparation and are applicable to complex matrices as well. Let us understand these sensors in more detail.

25.3.2 Electrochemical Sensors with Nanomaterials for Macrolides Detection

Electrochemical sensors have received extensive attention in the antibiotic analysis field owing to their distinct low limit of detection compared to most optical sensors. Electrochemical sensing mechanism relies on ions or electrons transfer through a reduction–oxidation reaction *(redox reaction)*. Nanoparticles are applied in electrochemical sensing either by acting as working electrodes or by modifying the electrode's surface or the recognition elements for target residues using different types of sensors such as potentiometric, voltammetric, conductometric and impedimetric sensors [43]. Most highly consumed macrolide antibiotics such as azithromycin, clarithromycin, and others lack good optical properties which makes detection using optical sensors challenging. Therefore, electrochemical sensing for macrolide detection at an extremely low limit of detection is a subject of interest.

Voltammetric sensors are one of the most used and highly sensitive analytical techniques for various types of analytes including antibiotics. In this, electrical current is measured after applying potential, and based on the change in current, target analytes can be determined using current as a function of potential. Different voltammetric techniques have been used for antibiotic detection including amperometry, differential pulse voltammetry (DPV), cyclic voltammetry (CV), square wave voltammetry (SWV), and linear sweep voltammetry (LSV) [12]. Herein, we focus on techniques that include nanomaterials for the detection of macrolide antibiotics. Macrolides can be detected electrochemically as they contain amino groups which act as electron donors. A summary of electrochemical sensing techniques that take advantage of nanomaterials for macrolide detection is provided in Table 25.1.

(i) ***Differential pulse voltammetry (DPV)***: This technique is extremely used for sensitive and quantitative analysis of target traces, where the potential pulse is applied and the difference in current signal just before and at the end of each pulse is recorded. [6] designed a novel Electrochemical sensor using DPV for tylosin detection in surface water and soil samples using cobalt nitride *(CoN) nanoarray* to enhance the sensing performance by offering a large functional surface area and excellent electroconductivity. They grew CoN nanowires on carbon cloth and then coated these nanowires with MIP film for specific tylosin capturing. MIP coating on nanowires has specific cavities for tylosin in which it can be captured by forming non-covalent bonds as shown in Fig. 25.3. After binding of tylosin to these cavities, the measured current was decreased by increasing tylosin concentrations. This can be justified by decreasing the conductivity (electron transfer) after the adsorption of tylosin on MIP cavities [6]. MIP technique is also used for selective detection of azithromycin using a modified glassy carbon electrode (GCE). Herein, gold nanourchines *(GNU)* which have a larger surface area compared to regular AuNPs were coupled with graphene oxide (*GO*) for GCE modification and dramatically improved the current peak [18]. Moreover, nanoparticles such as magnesium chromate ($MgCr_2O_4$) nanoparticles with wide voltage windows and porous structures which provide excellent ions and electrons transfer can improve the performance of the electrode [30]. These nanoparticles are coupled with multi-walled carbon nanotubes (*MWCNTs*), which have remarkable electrical properties with thermal and chemical stability, for modification of GCE to detect azithromycin using DPV [4]. Another approach to enhance the anodic peak currents and sensitivity of GCE is by depositing acetylene black (*AB*) nanoparticles on the electrode surface to form a thin film for detection of the erythromycin [16].

(ii) ***Square wave voltammetry (SWV)***: It can be considered a specific type of DPV as both measure current by applying potential pulses but in SWV the current is measured at the end of the forward pulse and the reverse pulse and the

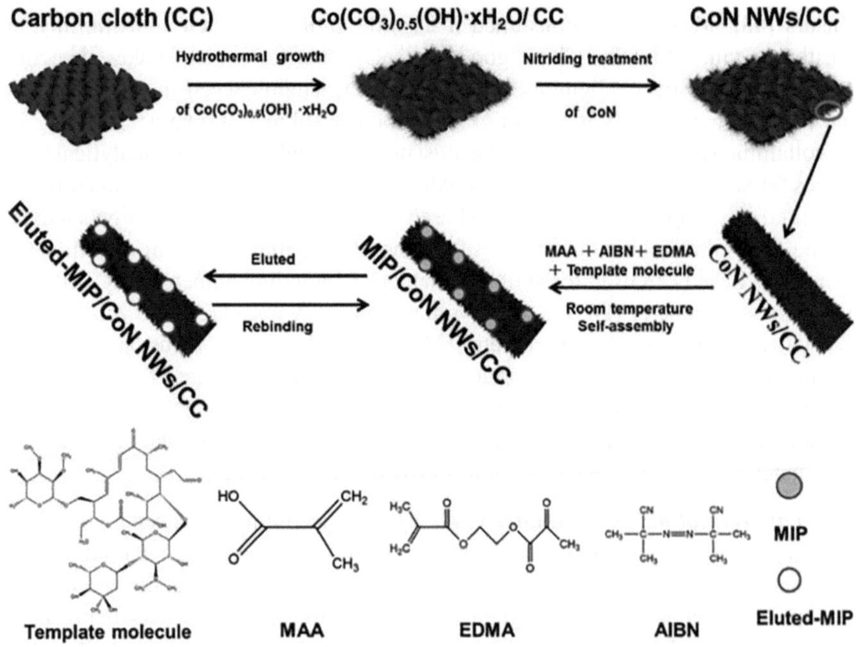

Fig. 25.3 Schematic illustration of preparation of MIP electrochemical sensor on nanowires. Adapted from published work [6]

difference between them is recorded. Two macrolides were detected by SWV using a carbon paste electrode modified with gold nanoparticles (AuNPs/CPE). Gold nanoparticles are known for their unique electrocatalytic properties, large surface area, and biocompatibility. CPE has a wide potential range, and excellent electroconductivity, and can be easily modified with noble metal nanoparticles. When CPE was modified by *AuNPs*, it showed a remarkable increase in the electrode sensitivity and provided a wider linear range for the azithromycin and roxithromycin detection [40].

(iii) ***Cyclic voltammetry (CV)***: Another voltammetric technique based on applying a cycle of potential from positive to negative voltage until reaching the predetermined switching potential and then reversing the scan back to the positive voltage is CV. Compared to DPV, CV can provide a rapid assessment of the electron transfer reaction. CV has been used for studying the electrochemical behavior of different macrolides such as erythromycin [26], tylosin [6], azithromycin [45], and others using nanomaterial modified working electrodes which are discussed in this chapter. By comparing the cyclic voltammogram of the nanomaterial modified electrode with the unmodified one, they found that nanomaterials such as AuNPs, MWCNTs, and PtNPs can result in enhancing the performance of the electrode as they provide higher effective surface area and a remarkable increase in the current signal. CV is often the first technique to be applied during an electrochemical study for a fast and better understanding of the redox process of the analyte.

(iv) ***Linear sweep voltammetry (LSV)***: This technique is similar to CV but here the potential is swept in only one direction from low to high value. Azithromycin was detected in pharmaceutical formulations with LSV using a modified glassy carbon electrode. This modification of the electrode surface with *graphene oxide* and *MWCNTs* plays an important role in increasing the surface area of sensing and enhancing the electrical conductivity. Moreover, the oxidation peak currents with the modified electrode tend to be higher than those with the unmodified one which is a result of incorporating nanoparticles on the electrode surface [45].

(v) ***Amperometry (AMP)***: In this technique, a fixed potential is applied to the working electrode and the current is measured as a function of time. Four macrolides (Azithromycin, erythromycin, clarithromycin, and roxithromycin) were detected in drug and water samples using batch injection analysis amperometry. For this, super P carbon black nanoparticles *(SPCB)* and chitosan were used for enhancing the performance of GCE. Carbon black as carbon nanomaterial has excellent chemical and thermal stability, provides a large surface area, has a good electroconductivity, and can form a stable dispersion. Chitosan was used here as a binder for the nanoparticles on the electrode owing to a large number of functional groups in its structure that help in the immobilization of CB [41]. Moreover, [26] fabricated a gold electrode with chitosan–platinum nanoparticles *(CS–PtNPs)* and graphene–gold nanoparticles *(GR–AuNPs)* on the surface to enhance the electroconductivity of the gold electrode to detect erythromycin. Chitosan is well known for its non-toxicity, biocompatibility, mechanical strength, and the large number of functional groups that can be used in the immobilization of other molecules. Graphene, Au and Pt nanoparticles have been used for their large surface area-to-volume ratio, small particle size, biocompatibility, and good conductivity besides the ability of gold nanoparticles to form covalent bonding with different functional groups such as $-NH_2$, –SH, and –CN groups. Moreover, to enhance analysis selectivity, molecular imprinted polymer bound to gold nanoparticles *(MIP-AuNPs)* was prepared through electropolymerization of the mixture of erythromycin (template), 2-mercapto nicotinic acid (monomer), and $HAuCl_4$ (AuNPs precursor to enhance the conductivity of the MIP). Hence MIP layer on the surface of the nanomaterials' modified electrode was successfully used for erythromycin detection in milk and honey samples using amperometry [26].

25.4 Challenges and Prospects

Sensing macrolide antibiotics in different matrices, especially food products and environmental matrices, is a great challenge due to their trace concentrations that reach a few 'ppb'. In addition, selective detection, that is, differentiating between structurally related macrolide antibiotics within the same group, is a challenge. To overcome this, we require a very sensitive and selective analytical technique to reach this low detection limit with confidence using minimal sample pretreatment. There is an ardent need to develop rapid, sensitive, and selective techniques, with no to minimal sample preparation, for on-site detection of macrolides. Nanomaterial-aided techniques enable lower detection limits, but have some limitations, such as narrow linear range for LFIA, and matrix interferences for the electrochemical sensors. Therefore, future studies should focus on resolving these limitations and develop a platform for selective detection of each macrolide antibiotic without interferences. Developing 'sense and shoot' devices which can not only detect but also eliminate these contaminants, by transforming them into non-toxic by-products, from the environment is an environmentally friendly, one-stop solution to antibiotic contamination.

25.5 Conclusions

In this chapter, we mainly focused on the applications of nanomaterials in sensing of macrolides using both optical and electrochemical sensors. However, macrolides could be detected with conventional methods such as HPLC and LC–MS at low detection limits. These methods require professionally trained personnel and expensive equipment that we cannot rely on for on-site detection. Nanomaterials such as magnetic nanoparticles have been utilized in conventional methods to facilitate the extraction process before detection. Moreover, nanoparticles have been used as labels for antibodies in immunoassay technique for optical sensing of macrolides in several samples. However, LFIA based sensors could detect macrolides in less than 1 ng/ml using colloidal AuNPs, they still have low cut-off values above which we couldn't measure the exact concentration of the target analyte. For instance, roxithromycin was detected in milk with a detection limit of 0.055 ng/ml and cut-off value of 5 ng/ml using LFIA. On the other hand, fluorescence-based sensors can be more sensitive with a wider linear range where roxithromycin was detected in milk using fluorescent $CsPbBr_3$-loaded MIP nanogels with a linear range of 8.4×10^{-2} to 8.4×10^{2} ng/ml and detection limit of 1.7×10^{-2} ng/ml [13]. Moving into electrochemical sensors, nanomaterial modified electrodes have been used for the detection of different macrolide antibiotics with high sensitivity and selectivity using different types of nanomaterials and MIP (receptor mimic) technique. One advantage of the mentioned electrochemical sensors over optical sensors is that working electrodes could be reproducible. All these types of sensors could help in on-site monitoring of these antibiotics in different matrices.

Acknowledgements The authors thank the Natural Sciences and Engineering Research Council (NSERC) of Canada Collaborative Research and Training Experience program (CREATE, CREAT/543238–2020), Training in Applied Biotechnology for Environmental Sustainability (TABES), Discovery Grants (RGPIN-2020-06067; RGPIN-2020-06140 and 554777-2021), and Alliance

Mission (ALLRP 570473-2021), as well as James and Joanne Love Chair in Environmental Engineering for the financial support of this work.

References

1. Amelin, V.G., Bol'shakov, D.S.: Sample preparation, identification, and determination of twelve macrolides in raw food materials and food products using high-resolution mass spectrometry. Mosc. Univ. Chem. Bull. **74**(2), 63–69 (2019). Springer. https://doi.org/10.3103/S0027131419020032
2. Choemunng, A., Na-Bangchang, K.: An alternative liquid chromatography-mass spectrometric method for the determination of azithromycin in human plasma and its application to pharmacokinetic study. **33**(16), 1516–1528 (2010). Taylor & Francis Group. https://doi.org/10.1080/10826076.2010.489009
3. Du, Y., et al.: Preparation of DNA aptamer and development of lateral flow aptasensor combining recombinase polymerase amplification for detection of erythromycin. Biosens. Bioelectron. **181**, 113157 (2021). Elsevier. https://doi.org/10.1016/J.BIOS.2021.113157
4. Ensafi, A.A., Allafchian, A.R., Rezaei, B.: A sensitive and selective voltammetric sensor based on multiwall carbon nanotubes decorated with $MgCr_2O_4$ for the determination of azithromycin. Colloids Surf. B: Biointerfaces **103**, 468–474 (2013). Elsevier. https://doi.org/10.1016/J.COLSURFB.2012.11.021
5. Fan, W., et al.: Application of core–satellite polydopamine-coated Fe_3O_4 nanoparticles–hollow porous molecularly imprinted polymer combined with HPLC-MS/MS for the quantification of macrolide antibiotics. Anal. Methods **13**(11), 1412–1421 (2021). Royal Society of Chemistry. https://doi.org/10.1039/D0AY02025G
6. Zhang, Y., et al.: Directly assembled electrochemical sensor by combining self-supported CoN nanoarray platform grown on carbon cloth with molecularly imprinted polymers for the detection of Tylosin. J. Hazard. Mater. **398**, 122778 (2020). Elsevier. https://doi.org/10.1016/J.JHAZMAT.2020.122778
7. Gayen, B., Palchoudhury, S., Chowdhury, J.: Carbon dots: A mystic star in the world of nanoscience. J Nanomater (2019). https://doi.org/10.1155/2019/3451307
8. Guerra, P., et al.: Occurrence and fate of antibiotic, analgesic/anti-inflammatory, and antifungal compounds in five wastewater treatment processes. Sci. Total Environ. **473–474**, 235–243 (2014). Elsevier. https://doi.org/10.1016/J.SCITOTENV.2013.12.008
9. Guo, D., Xie, G., Luo, J.: Mechanical properties of nanoparticles: basics and applications. J. Phys. D: Appl. Phys. **47**(1), 013001 (2013). IOP Publishing. https://doi.org/10.1088/0022-3727/47/1/013001
10. Guo, L., et al.: Gold immunochromatographic assay for kitasamycin and josamycin residues screening in milk and egg samples. **30**(1), 1189–1201 (2019). Taylor & Francis. https://doi.org/10.1080/09540105.2019.1677567
11. Guo, X., et al.: Azithromycin detection in cells and tablets by N,S co-doped carbon quantum dots. Spectrochim. Acta Part A Mol. Biomol. Spectrosc. **252**, 119506 (2021). Elsevier. https://doi.org/10.1016/j.saa.2021.119506
12. Gupta, V.K. et al.: Voltammetric techniques for the assay of pharmaceuticals—a review. Anal. Biochem. **408**(2), 179–196 (2011). Academic Press. https://doi.org/10.1016/J.AB.2010.09.027
13. Han, J., et al.: Water-stable perovskite-loaded nanogels containing antioxidant property for highly sensitive and selective detection of roxithromycin in animal-derived food products. Sci. Rep. **12**(1), 1–11 (2022). Nature Publishing Group UK. https://doi.org/10.1038/s41598-022-07030-9
14. Hendrickson, O.D., et al.: Development of a double immunochromatographic test system for simultaneous determination of lincomycin and tylosin antibiotics in foodstuffs. Food Chem. **318**, 126510 (2020). Elsevier. https://doi.org/10.1016/J.FOODCHEM.2020.126510

15. Hu, C., et al.: A review of pretreatment and analysis of macrolides in food (Update Since 2010). J. Chromatogr. A **1634**, 461662 (2020). Elsevier. https://doi.org/10.1016/J.CHROMA.2020.461662
16. Hu, X., et al.: Enhanced electrochemical detection of erythromycin based on acetylene black nanoparticles. Colloids Surf. B Biointerfaces **81**(1), 27–31 (2010). Elsevier. https://doi.org/10.1016/J.COLSURFB.2010.06.018
17. Huhtinen, P., et al.: Synthesis, characterization, and application of Eu(III), Tb(III), Sm(III), and Dy(III) lanthanide chelate nanoparticle labels. Anal. Chem. **77**(8), 2643–2648 (2005). American Chemical Society. https://doi.org/10.1021/AC048360I/ASSET/IMAGES/LARGE/AC048360IF00005.JPEG
18. Jafari, S., et al.: An azithromycin electrochemical sensor based on an aniline MIP film electropolymerized on a gold nano urchins/graphene oxide modified glassy carbon electrode. J. Electroanal. Chem. **829**, 27–34 (2018). Elsevier. https://doi.org/10.1016/J.JELECHEM.2018.09.053
19. Khan, I., Saeed, K., Khan, I.: Nanoparticles: properties, applications and toxicities. Arab. J. Chem. **12**(7), 908–931 (2019). Elsevier. https://doi.org/10.1016/J.ARABJC.2017.05.011
20. Krishnan, S.K., et al.: A review on graphene-based nanocomposites for electrochemical and fluorescent biosensors. RSC Adv. **9**(16), 8778–8881 (2019). The Royal Society of Chemistry. https://doi.org/10.1039/C8RA09577A
21. Kurian, M., Paul, A.: Recent trends in the use of green sources for carbon dot synthesis–a short review. Carbon Trends**3** (2021). https://doi.org/10.1016/j.cartre.2021.100032
22. Le, T., et al.: Development of an immunochromatographic assay for detection of tylosin and tilmicosin in muscle, liver, fish and eggs. **24**(4), 467–480 (2013). Taylor & Francis. https://doi.org/10.1080/09540105.2012.716025
23. Le, T., Zhu, L., Yang, X.: A quantum dot-based immunoassay for screening of tylosin and tilmicosin in edible animal tissues. Food Addit. Contam. Part A Chem. Anal. Control Expo. Risk Assess. **32**(5), 719–724 (2015). https://doi.org/10.1080/19440049.2015.1020529
24. Li, X., et al.: Multiplex immunogold chromatographic assay for simultaneous determination of macrolide antibiotics in raw milk. Food Anal. Methods **8**(9), 2368–2375 (2015). Springer New York LLC. https://doi.org/10.1007/S12161-015-0130-X/FIGURES/8
25. Li, X., et al.: Three lateral flow immunochromatographic assays based on different nanoparticle probes for on-site detection of tylosin and tilmicosin in milk and pork. Sens. Actuators B: Chem. **301**, 127059 (2019). Elsevier. https://doi.org/10.1016/J.SNB.2019.127059
26. Lian, W., et al.: Electrochemical sensor based on gold nanoparticles fabricated molecularly imprinted polymer film at chitosan–platinum nanoparticles/graphene–gold nanoparticles double nanocomposites modified electrode for detection of erythromycin. Biosens. Bioelectron. **38**(1), 163–169 (2012). Elsevier. https://doi.org/10.1016/J.BIOS.2012.05.017
27. Liao, Q.G., et al.: Ultrasonic-assisted dispersive liquid–liquid microextraction based on a simple and green deep eutectic solvent for preconcentration of macrolides from swine urine samples. Sep. Sci. Plus **3**(1–2), 22–27 (2020). John Wiley & Sons, Ltd. https://doi.org/10.1002/SSCP.201900064
28. List of Maximum Residue Limits (MRLs) for Veterinary Drugs in Foods—Canada.ca (no date). Available at: https://www.canada.ca/en/health-canada/services/drugs-health-products/veterinary-drugs/maximum-residue-limits-mrls/list-maximum-residue-limits-mrls-veterinary-drugs-foods.html. Accessed 26 Nov 2021
29. Liu, Z., et al.: Immunochromatographic assays based on three kinds of nanoparticles for the rapid and highly sensitive detection of tylosin and tilmicosin in eggs. Microchim. Acta **189**(1), 1–11 (2022). Springer. https://doi.org/10.1007/S00604-021-05151-Y/TABLES/3
30. Maitra, S., Mitra, R., Nath, T.K.: 'Sol-gel derived $MgCr_2O_4$ nanoparticles for aqueous supercapacitor and alkaline OER and HER bi-functional electrocatalyst applications. J. Alloys Compd. **858**, 157679 (2021). Elsevier. https://doi.org/10.1016/J.JALLCOM.2020.157679
31. Parolo, C., et al.: Tutorial: design and fabrication of nanoparticle-based lateral-flow immunoassays. Nat. Protoc. **15**(12), 3788–3816 (2020). Nature Publishing Group. https://doi.org/10.1038/s41596-020-0357-x

32. Pyörälä, S., et al.: Macrolides and lincosamides in cattle and pigs: Use and development of antimicrobial resistance. Vet. J. **200**(2), 230–239 (2014). W.B. Saunders. https://doi.org/10.1016/J.TVJL.2014.02.028
33. Ravindran, N., et al.: Recent advances in surface plasmon resonance (SPR) biosensors for food analysis: a review. (2021). Taylor & Francis. https://doi.org/10.1080/10408398.2021.1958745
34. Roduner, E.: Size matters: why nanomaterials are different. Chem. Soc. Rev. **35**(7), 583–592 (2006). The Royal Society of Chemistry. https://doi.org/10.1039/B502142C
35. Sari, E., et al.: Fabrication of surface plasmon resonance nanosensor for the selective determination of erythromycin via molecular imprinted nanoparticles. Talanta **150**, 607–614 (2016). Elsevier. https://doi.org/10.1016/J.TALANTA.2015.12.043
36. Song, X., et al.: Determination of macrolide antibiotics residues in pork using molecularly imprinted dispersive solid-phase extraction coupled with LC–MS/MS. J. Sep. Sci. **41**(5), 1138–1148 (2018). John Wiley & Sons, Ltd. https://doi.org/10.1002/JSSC.201700973
37. Song, Y., et al.: Simultaneous detection of tylosin and tilmicosin in honey using a novel immunoassay and immunochromatographic strip based on an innovative hapten. **27**(3), 314–328 (2015). Taylor & Francis. https://doi.org/10.1080/09540105.2015.1089843
38. Thangadurai, S.,: Gas chromatographic–mass spectrometric determination of azithromycin in biological fluids. J. Anal. Sci. Technol. **6**(1), 1–6 (2015). Springer. https://doi.org/10.1186/S40543-015-0059-0/FIGURES/4
39. Trott, D.J., et al.: Comparative macrolide use in humans and animals: should macrolides be moved off the World Health Organisation's critically important antimicrobial list? J. Antimicrob. Chemother. **76**(8), 1955–1961 (2021). Oxford Academic. https://doi.org/10.1093/JAC/DKAB120
40. Vajdle, O., et al.: Use of carbon paste electrode and modified by gold nanoparticles for selected macrolide antibiotics determination as standard and in pharmaceutical preparations. J. Electroanal. Chem. **873**, 114324 (2020). Elsevier. https://doi.org/10.1016/J.JELECHEM.2020.114324
41. Veloso, W.B., et al.: Rapid and sensitivity determination of macrolides antibiotics using disposable electrochemical sensor based on Super P carbon black and chitosan composite. Microchem. J. **172**, 106939 (2022). Elsevier. https://doi.org/10.1016/J.MICROC.2021.106939
42. Wang, Y., et al.: Development of time-resolved fluorescence immunochromatographic assays for simultaneously detecting tylosin and tilmicosin in milk in group-screening manner. Foods **10**(8), 1838 (2021). Multidisciplinary Digital Publishing Institute. https://doi.org/10.3390/FOODS10081838
43. Wongkaew, N., et al.: Functional nanomaterials and nanostructures enhancing electrochemical biosensors and lab-on-a-chip performances: recent progress, applications, and future perspective. Chem. Rev. **119**(1), 120–194 (2018). American Chemical Society. https://doi.org/10.1021/ACS.CHEMREV.8B00172
44. Zeng, L., et al.: A paper-based colorimetric assay for rapid detection of four macrolides in milk. Mater. Chem. Front. **3**(10), 2175–2183 (2019). Royal Society of Chemistry. https://doi.org/10.1039/C9QM00429G
45. Zhang, K., et al.: Facile synthesis of the necklace-like graphene oxide-multi-walled carbon nanotube nanohybrid and its application in electrochemical sensing of Azithromycin. Anal. Chim. Acta **787**, 50–56 (2013). Elsevier. https://doi.org/10.1016/J.ACA.2013.05.037
46. Zhang, Y., et al.: SiO_2-MIP core-shell nanoparticles containing gold nanoclusters for sensitive fluorescence detection of the antibiotic erythromycin. Microchim. Acta **184**(7), 2241–2248 (2017). Springer-Verlag Wien. https://doi.org/10.1007/S00604-017-2216-1/TABLES/2

Chapter 26
The Physics of Biofunctionality in Nanoconfined Systems

Alokmay Datta

Abstract Confinement creates an environment where degrees of freedom are reduced drastically. The major consequence of this reduction is a decrease in entropy and, in a closed system, the concomitant increase in enthalpy or, in other words, potential energy. This enhanced potential energy, on one hand, creates ordering in disordered media and on the other increases reactivity quite considerably. We discuss here some specific instances of enhanced biofunctions in confined systems.

Keywords Nanoscale · Confinement · Dispersive · Biofunctionality

26.1 Physics of Confinement

Confinement is apparently a simple and straightforward concept. Any system is said to be confined when its spatial extent is defined within a range of values. The system maybe a particle, either classical or quantum, either a real particle or a quasi-particle, a multi-particle body such as an atom or a molecule, a quasi-continuous bulk such as a solid, liquid or gas, a wave or a field. Confinement can be brought about by the use of an external force, such as by using electrical, magnetic, optical or thermal barriers, or by erecting rigid, material barriers. We shall call such confinement "external" or "artificial" confinement.

Confinement can also be caused by internal forces acting among the constituents of a system: the strong forces confine nucleons within the nucleus, electromagnetic forces confine electrons and nuclei within the atoms, atoms within molecules and condense gases into liquids and liquids into solids [1]. This type of confinement can be designated as "internal" or "natural" confinement. A major task of physical sciences

A. Datta (✉)
Department of Physics, University of Calcutta, Kolkata, India
e-mail: alokmaydatta@gmail.com

CSIR-Central Glass and Ceramic Research Institute, Kolkata, India

Saha Institute of Nuclear Physics, Kolkata, India

D. Mohanta and P. Chakraborty (eds.), *Nanoscale Matter and Principles for Sensing and Labeling Applications*, Advanced Structured Materials 206,
https://doi.org/10.1007/978-981-99-7848-9_26

has been, and still is, understanding this natural confinement and the hierarchy of structures and dynamics caused by it starting from elementary particles all the way to condensed matter and complex matter including living systems.

In comparison, the study of artificially confined systems has been remarkably sparse till recently. Though boundary conditions were ever present in studying any physical problem, their major effects were assumed to be at the boundaries only. Since any system studied was very large compared to the boundary region, this large size set the boundaries at separations big enough to kill any "boundary effect" within the sample. Again, the boundary conditions themselves were essentially "put by hand", especially in the case of rigid material barriers, and this could not really be analyzed any further. Only for infinite and continuously varying systems, such as for fields and waves (in either fields or fluids), did the physics of confinement develop properly and we shall now dwell into some of the salient features of this development.

Boundary conditions for such continuous systems can be summarized as whether the system (or at least a fraction of it) can pass through the boundary or not, i.e., whether the barrier is finite or infinite. To deal with this situation, the physical concepts of reflection and transmission were developed and were then generalized to change in phase at the boundary [2]. The basic idea behind these concepts was a model of the continuous system as a continuous function, i.e., a function $f(\vec{r}, t)$ of space $\vec{r}$ and time t which itself and its derivatives with respect to space and time are defined everywhere and at all times. This function can represent any physically measurable quantity such as the electric and magnetic fields, temperature, pressure, density, velocity, and probability.

According to Fourier theorem, all such functions can be expressed as a linear superposition or a sum of periodic functions such as sine or cosine functions with suitable constant factors. Any periodic function will have a cycle such that its magnitudes at the beginning and end of the cycle are identical and thus the cycle is identical to completing a circle. Any point on the cycle is then represented by its position on this circle, i.e., by an angle between 0 and 2π radian and is called the *phase* φ of the function. The magnitude of the function at any point in space–time is given from its maximum magnitude f_0 and phase as

$$f = af_0\sin\varphi + bf_0\cos\varphi \tag{26.1a}$$

where a and b are real numbers defined by the system. They may stand for electric and magnetic fields for electromagnetic waves, pressure for acoustic waves, pressure/density/velocity for water waves, probability density for a quantum particle, etc. Phase is a sum of two parts, a spatial part given by

$$\varphi_r = \vec{k} \cdot \vec{r} \tag{26.1b}$$

and a temporal part given by

$$\varphi_t = \frac{2\pi}{T} t = \omega t \tag{26.1c}$$

As the space vector is given by $\vec{r} = \hat{i}x + \hat{j}y + \hat{k}z$, the reciprocal *space vector* or *wavevector* is given by $\vec{k} = \hat{i}k_x + \hat{j}k_y + \hat{k}k_z$ with its components in the three orthogonal directions being $k_x = \frac{2\pi}{\lambda_1}$, $k_y = \frac{2\pi}{\lambda_2}$, $k_z = \frac{2\pi}{\lambda_3}$. The lengths λ_i are called the *wavelengths* in the three directions, the distances after which the spatial phase is repeated along these directions, while the time T is the *period* after which the temporal phase is repeated and ω is the *frequency*. In most cases, the wavevector and the space vector are in the same direction, hence

$$\varphi_r = xk_x + yk_y + zk_z \tag{26.1d}$$

The essential effect of a barrier is to change the phase of the continuous system by an amount decided by the nature and strength of forces between the system and the barrier. In fact, this serves to probe this interaction. In particular, the spatial phase is changed by any amount from 0 to π radian. If the change is 0 the barrier is totally transmitting or perfectly transparent, effectively absent. The transmittance of the barrier is 1 and the reflectance is 0. On the other extreme, a change of π radians means a totally reflecting barrier with transmittance 0 and reflectance 1. This is the case for an infinite barrier.

If a barrier introduces a spatial phase change of value between 0 and $\pi/2$, it is said to disperse the system forward, i.e., in transmission while between $\pi/2$ and π the barrier disperses it backward, i.e., in reflection. Thus, other than the two extreme cases, an extended continuous system within barriers has a transmitted and a reflected part, as happens in all finite barriers.

For continuous systems the *momentum* $\vec{p}$ is given as

$$\vec{p} = \frac{E}{\omega} \vec{k} \tag{26.2}$$

with E = energy and $\vec{v}_p = \frac{\omega}{k}\hat{e}$ is the *phase velocity*, where $\hat{e} = \frac{\vec{k}}{|\vec{k}|}$ is the unit vector along the wave vector. Relation (26.2) is valid for electromagnetic and pressure fields such as acoustic waves, linear and nonlinear oscillator systems and fluids in classical physics and "particles", which are in fact "probability fields", in quantum physics. It is to be noted that the phase velocity is not the velocity with which energy is carried in the system but the rate at which a particular phase appears as we move along time.

On the other hand, energy in the system is carried by the *group velocity* $\vec{v}_g = \frac{d\omega}{dk}\hat{e}$ or the *change* in frequency due to a *change* in the wave vector. If barriers do not cause any change in the temporal phase, then their interactions with the systems are called *purely dispersive.* They do not change the frequency and thus do not cause any change in the energy. These interactions change only the momentum and since for all systems energy is some polynomial function of the momentum, they cannot change the magnitude of momentum either. Hence, the only change is in the direction of the momentum. This change is given by the angle θ between the original and final momenta as $\theta = \cos^{-1}\frac{\left(\vec{p}+\Delta\vec{p}\right)\cdot\vec{p}}{p} = \cos^{-1}\frac{\left(\vec{k}+\Delta\vec{k}\right)\cdot\vec{k}}{k}$, with $p = \left|\vec{p}\right|$ and $k = \left|\vec{k}\right|$.

Random phase changes at barriers cause *dissipation.* This term essentially means emergence of new momenta with random magnitude and direction in each of which a part of the total energy is used even though the total energy remains constant. Since these are different microstates, the total macrostate of the system will then have an increase in microstates through interaction with the barriers. The entropy of the system, given by $= k_B \log \Omega$, where Ω is the number of microstates in a macrostate and k_B is Boltzmann's constant, then increases.

Entropy is increased, however, only in transmission through barriers, as the emergent momenta add to the original momentum and are not opposed to it. On the other hand, upon reflection, they oppose the original momentum. In particular, when the system is between two reflecting barriers, not only the emergent random momenta mostly cancel out, the original momentum is also cancelled out. The system becomes essentially stationary and there is a drastic decrease in the number of microstates and thus the entropy. *Confinement thus gives rise to emergence of order in the system.*

In the above discussion, we have considered redistribution of energy through dissipation but not any net *loss* in the total energy. For massless systems like fields this happens when the barrier changes the temporal phase, thus the frequency and the energy. The basic mechanism of this change is through the excitation of internal motions in the barrier that take away energy from the system. This process of *absorption* is ubiquitous and for massive systems like fluids, internal modes are also excited continuously from microscopic interactions *within the system* and not only at the barrier [3].

The main result of this absorption is that some of the momenta are suppressed at each of such absorption events. For each reflection at the barriers some reverse momenta are not generated, hence the cancellation of the random momenta becomes incomplete and the entropy of the system tends to grow with the number of reflections, in other words with time. For massive systems, the additional absorption within the body hugely enhances the rate of entropy increase.

When a reflected wave is superposed on the original wave without any loss due to absorption the result is a wave given by

$$f_s = A\cos\vec{k}\cdot\vec{r}\cos\omega t. \tag{26.3}$$

This is a wave where the spatial and temporal phases are independent. It behaves like a simple harmonic oscillator whose amplitude is a periodic pattern in space with a period λ. At any instant of time, the crests, troughs, and every point in the pattern remain at the same position and that is what is observed. This is called a *stationary or standing wave.* The basic effect of confinement on a continuous system is to create a standing wave, i.e., a periodic pattern.

The introduction of absorption in the system causes deviation from this perfect superposition and retention of waves with random phases. This results finally in the loss of energy from the system and expresses itself as a decay (with arbitrary constant σ) in the amplitude of the standing wave with time, as

$$f_s^{\text{real}} = Ae^{-\sigma t} \cos \vec{k} \cdot \vec{r} \cos \omega t \tag{26.4}$$

Thus all real standing waves decay with time, the decay being dependent on the absorption at the boundaries and in the body of the system. The decay is fast in massive systems and such patterns do not sustain for long in these bodies, as in sound waves in material media (Fig. 26.1a) and liquid waves at the surface (Fig. 26.1b). To sustain them energy has to be fed to them constantly. On the other hand, stationary electromagnetic waves can be sustained much longer without energy input (Fig. 26.1c).

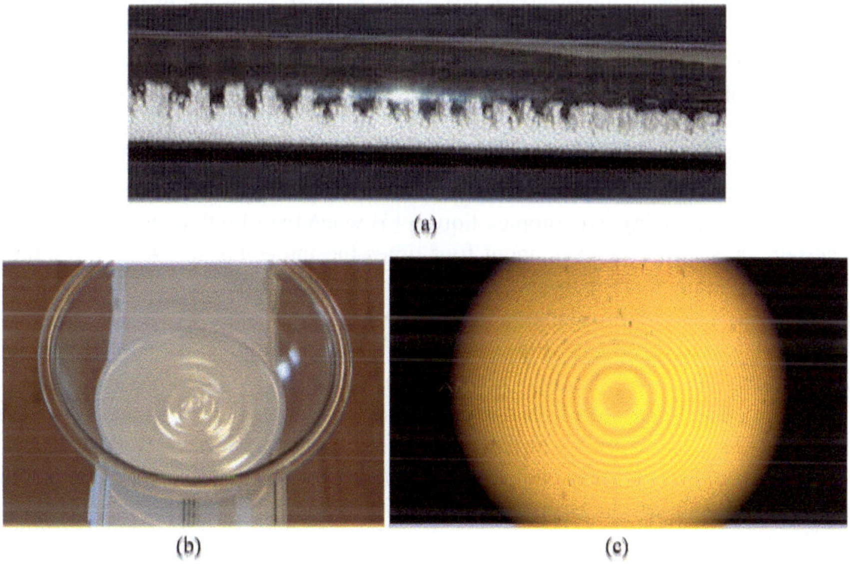

Fig. 26.1 Standing waves in **a** acoustics (Kundt's tube), **b** water surface (ultrasonic vibrator), **c** light (Newton's rings). Credits: **a** https://portal.ttsd.k12.or.us/ (TuHS Physics Homepage); **b** https://courses.lumenlearning.com/suny-osuniversityphysics/chapter/16-6-standing-waves-and-resonance/; **c** http://www.amrita.edu/create

26.2 Physics Under Nanoconfinement

Density waves or variations in material media have wavelengths spanning a range from millimeters to kilometers. Matter, as is well known, is continuous over length scales of micrometers and above but consists of discrete units of nominally constant density at the level of nanometers. In fluids and soft matter, these units are the molecules whereas in crystalline solids they are the unit cells. The size of such microscopic units is thus the indivisible unit in physical phenomena such as wave propagation and confinement, particularly in the transfer of momenta and energy, i.e., momenta and energy of the wave or of any disturbance in the material are carried by these constant density units.

It is obvious that even in micrometer-sized waves an immense number of such units are taking part at every point of the wave. Hence at any point there will be a large number of exchanges of momenta and energy among these molecules and a considerable fraction of this will be at random. This will cause entropy increase and energy loss leading to the decay of any standing wave pattern that maybe created through confinement.

When a material is, however, confined to length scales that are within an order of magnitude of the characteristic unit size, these exchanges are minimized to the point of being negligible. Then the standing wave patterns in density with period equal to that characteristic length are observed to form and are sustained. The simplest example of this spontaneous ordering is the formation of molecular layers in films of simple [4] (Fig. 26.2a) and complex liquids [5] when the film thickness goes below a certain value. This value is about four times the molecular diameter for simple liquids and about four times the radius of the gyration sphere, i.e., the spherical shape a polymer molecule attains in the non-entangled state, which is equivalent to twice the molecular size (Fig. 26.2b). These are examples of one-dimensional confinement with the direction of confinement normal to the film thickness.

In nanometer-sized pores, two-dimensional confinement takes place. Here the best example of standing waves in density is the spontaneous formation of rings of Xenon trapped in Aluminum nanopores with layer periodicity of the atomic diameter of Xenon [6]. Similarly, the trapping of liquids in nanobubbles or as nanodroplets and the consequent molecular layers of concentric spheres are examples of spontaneous order emerging under three-dimensional nanoconfinement.

The nanoconfinement discussed above can be termed as "geometrical". The confining potential maybe represented by an infinite well, i.e., having walls with zero-range repulsive forces of infinite magnitude. Such potentials can be provided, in general, by design through an external agency. In other words, they are man-made or artificial. For example, strong geometrical confinement of fluids in one dimension, achieved through the thinning of fluids to films with thickness ~molecular size, places the fluid between two practically infinite potential barriers, the substrate, and air. However, there is also the more interesting situation where such molecular layers are found between the wall of the vessel of any viscous liquid and the liquid itself.

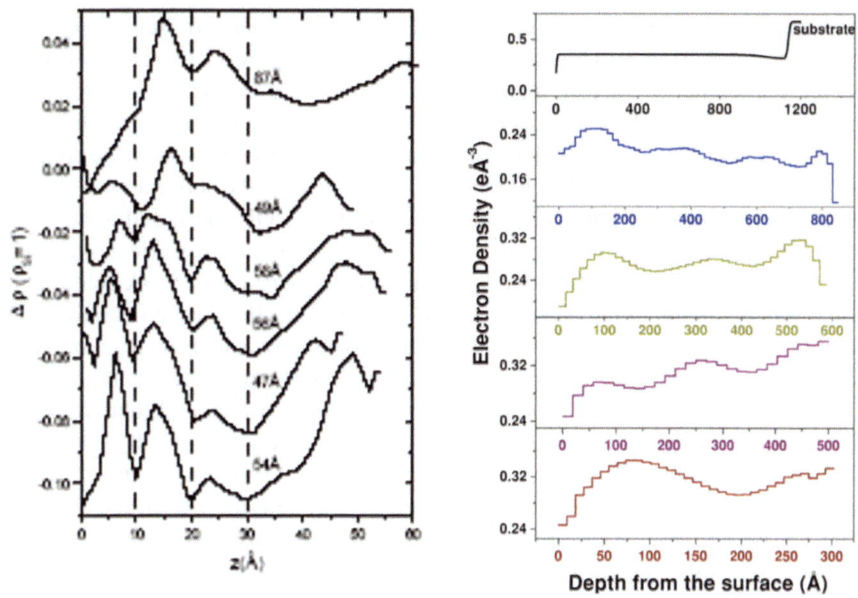

(a) (b)

Fig. 26.2 Layer formation under nanoconfinement **a** in simple liquid TEHOS on Si(001) and **b** in polystyrene on fused quartz. In both, the x-axis is the depth from the film surface while the y-axis is the electron density (number of electrons/$Å^3$) variation from the average electron density of the film in **a** and the electron density itself in **b**. In **a**, the film thickness is shown beside each electron density profile (edp) and the y-coordinates are shifted up by arbitrary amounts for clarity. Density peaks have a period of around 10 Å, coinciding with the molecular size. The last peak is due to the native silica layer on Si surface. In **b**, the film thicknesses from the top are ~6R_g, ~4R_g, ~3R_g, ~2.5R_g, and ~1.5R_g and the period is ~R_g, the radius of gyration of the polymer, indicating the compression of the gyration spheres into ellipsoids with the minor axes along the normal to the film surface (the direction of confinement). The last peak is due to silica. Reference **a** C.-J. Yu et al., *Phys. Rev. Lett.* **82**, 2326 (1999); **b** S. Chattopadhyay and A. Datta, *Physical Review B* **72**, 099,539 (2005)

Here the bulk liquid and the vessel wall form the two potential barriers [7]. This maybe called "natural" geometrical nanoconfinement.

There are other "natural" instances of nanoconfinement, coming from specific interactions. A simple example is a Langmuir monolayer formed by a class of molecules called *amphiphiles* with two parts, a hydrophilic headgroup that is either dissociable or polar and is attracted to water and a hydrocarbon tail that is attracted to other hydrocarbons through van der Waals forces but is repelled from water. If such molecules are spread on water surface in air or at an interface of oil and water, the heads will be allowed to go into the water while the tails can stay in the air on in oil. Thus a monomolecular layer will be formed spontaneously with a nanometer-sized thickness—an instance of ordering under nanoconfinement [8].

There are a number of effects of major physical significance of nanoconfinement. The obvious one is that of emergence of order in disordered media, such as fluids and polymers with no bulk crystalline phases. This signifies a large decrease in entropy which, in equilibrium systems with conserved free energy, leads to a commensurate increase in enthalpy from the bulk phase. This enthalpy excess can be utilized to carry out changes in the materials that are forbidden or difficult to achieve in bulk.

An example of the highly enhanced chemical reactivity coming from the confinement in Langmuir monolayer will be useful to understand the above aspect. Stearic acid is an amphiphilic fatty acid with a carboxylic (COOH) head and a tail of 16 methylene (CH_2) groups ending in a methyl (CH_3) group. It is insoluble in water due to the hydrophobic repulsion coming from the tail and is a very weak acid due to the low dissociation of the head into COO^- and H^+ ions. If we want to convert this acid to cadmium stearate we can never do it through any acid–base reaction since Cd is lower than H in the electrochemical series. The standard procedure is to first convert the acid to sodium stearate, which is soluble in water, and then exchange the Na ion with the Cd ion through a reaction of Na-stearate with $CdCl_2$, which is also soluble in water. These reactions, especially the first one, are not very efficient, i.e., a considerable fraction of acid and Na-stearate remains unreacted, and on top of that, since Cd-stearate is insoluble in most organic solvent as well as water, it is very difficult to recover and purify.

On the other hand, in a Langmuir monolayer of Stearic acid, as the heads dissociate into COO^- and H^+ ions, the cations go into the water but the anions being attached to the tails remain at the interface. If there is even a micromole of $CdCl_2$ in that water, on dissociation into Cd^{2+} and Cl^- ions the cations immediately go to the interface and bond with two COO^- heads and consequently with two tails, forming Cd-stearate at ease and with almost full efficiency. This Cd-stearate can be recovered in near pure form by passing a substrate repeatedly through the interface, whereupon a monolayer of the compound is deposited with each up and down stroke. The resulting multilayer is called Langmuir–Blodgett multilayer [9].

As the ordering emerges spontaneously from confining the system to nanoscales, the system loses momentum. This amounts to the emergence of a short-range repulsive force at the molecular level, which may be termed the *confinement force.* The confinement force can cause inter- and intramolecular rearrangements such as the realignments of the benzene rings in polystyrene [10]. In particular, studies on nanoconfined polymers show that layering entails a reduction of intermolecular forces and an increase in intramolecular bonding (Fig. 26.3a). It can also cause separation of solutes from solvents and can reorganize the solute into ordered structures, such as separating DNA [11] from a solution of polyethyleneglycol and water (Fig. 26.3b), and organizing the DNA into a polarized structure (Fig. 26.3c).

Fluids in highly confined spaces have properties that can differ markedly from that of the bulk fluid. Even a bulk fluid has interfaces, and if the properties of the fluid are strongly altered in the nanometer region of the interface, then this could strongly alter the interfacial transfer kinetics or the fluids' success as a lubricant or adhesive. Properties commonly believed to be altered at the interfaces of a fluid are the viscosity, molecular density, and ordering. The major issue in confined liquids is whether their properties are changed (a) uniformly across the entire confined liquid or (b) vary on a molecular scale as we approach the interface.

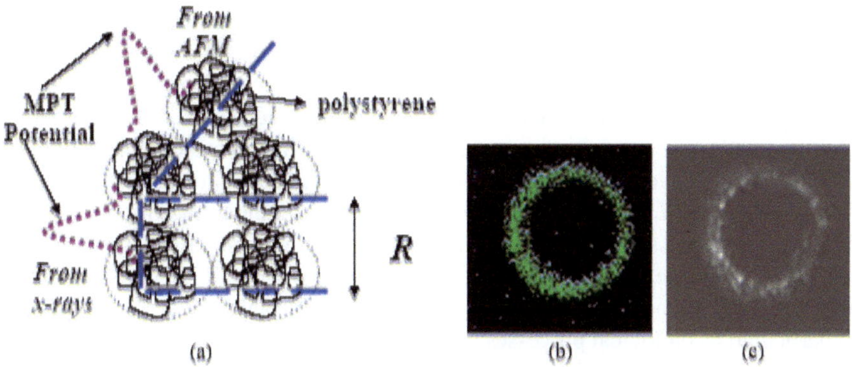

Fig. 26.3 Confinement force and its consequence. **a** The confinement force in a polystyrene film can be modeled by a repulsive modified Pöschl–Teller potential acting between the stacked gyration ellipsoids. **b–c** Phase separation in a DNA–polyethyleneglycol–water solution trapped as nanobubbles in oil. DNA forms a rim around the bubble surface as seen in fluorescence microscopy **b**. It is also polarized normal to bubble surface as seen in polarization microscopy **c**. Reference **a** S. Chattopadhyay et al., *Macromolecules* **40**, 9190 (2007); **b** N. Biswas et al., *Chemical Physics Letters* **539–540**, 157–162 (2012)

Forces emerging from entropy variations or quantum fluctuations are, in general, non-conservative. Since they are also very short range they are independent of the structure of the embedding space–time or, of the embedding space at the smaller ranges of condensed matter. On the other hand, forces emerging from fields are conservative and dependent on the embedding space. Nanoconfinement reduces the dimension d of the embedding space to 2, 1, or 0, depending on whether the confinement is one dimensional, two dimensional, or three dimensional, respectively. Intermolecular forces are long range if they derive from a potential $V(r) \sim r^{-n}$ where $n < d$. Hence nanoconfinement, in particular, geometrical nanoconfinement changes the physics of the fluid by changing the nature of intermolecular forces that are long range, such as dipolar forces, in the bulk to short range in the confined condition. This renders a very large number of fluids into systems dominated by short-range forces. A fundamental consequence of this transition is the near-identical physical behavior of simple and complex fluids (including polymers) under strong geometrical confinement.

The evolution of spontaneous one-dimensional order over two or three molecular layers leading to a nanoconfined state is driven by the confinement-induced novel short-range forces, where the fluid behaves like a fluid or a solid depending on the timescale of study of its dynamics. It is clear that this layering is an order–disorder transition, where the density variation across the layer plays the role of the "order parameter" and "confinement", expressed as the fluid film thickness, that of an "external field" under which simple and complex fluids show essentially similar behavior. This result has intriguing consequences for polymers where the nanoconfined state shows well-defined variations in surface energy that act as generalized growth centers for monodisperse nanoparticles, whose size can be controlled by polymer film thickness. In addition, the polymer shows exciton localization over its gyration radius (R_g) on nanoconfinement, which has potential for quantum localization over mesoscopic length scales. The existence of a critical polymer molecular weight, i.e., critical intermolecular entanglement has been indicated above which there cannot be any layering. A correlation between structure and dynamics of polymers around the layering transition point, as this critical entanglement is approached from below, would go a long way to clarify this fundamental question in complex fluidics in general and polymer dynamics in particular. This elucidation assumes special significance as micro-fluidics is going over to nano-fluidics and is finding applications in areas such as biology, where the major players are complex fluids. In particular, as biological processes involve constant spontaneous reorganization of macromolecules and amphiphilic molecules under confinement, the above ideas and observations are expected to provide crucial understanding.

26.3 Biofunctionality

Biofunctionality of a material can be understood or measured by the effect that material causes on a biological system or process. The central body and unit of all biological structures and activities is the cell or a cell-like structure. Hence biofunctionality has to be defined by the effect of a material or a structure on the cellular structure and/or function, be it the intra- or extra-cellular medium or the cell membrane separating and communicating between them.

From a physical point of view, the biological system is the archetype of a complex system. This means that there are hierarchies of spatial and temporal dimensions in any biological system, starting from small molecules and monomers having sizes of 0.1 to 1 nm and acting within pico- to nanoseconds, going next to individual polymers and macromolecules having sizes ranging from 10 to 100 nm, acting in the nano- to microsecond timescales, then to subcellular structures like organelles and membranes within 1 μm and working at around micro- to milliseconds, then to individual cells carrying their intra- and extra-cellular environments in the 1 to 100 μm lengthscales and the sub-second timescales. This list can continue to tissue and organ scales but the effects of nanoconfined systems on them and, more important, the physical mechanisms are too little understood to merit discussion. What is important, however, that the above-mentioned hierarchies are closely coupled, which is the hallmark of a complex system. Nevertheless, for the sake of a broad physical picture of the mechanisms involved, we are led to separate the hierarchies and study them starting from the lowest rungs.

In organic molecules, the lowest energy internal motions are the large amplitude motions—internal rotation, inversion, ring puckering, and skeletal vibrations [12] (Fig. 26.4). The first happens when a part of the molecule, generally with three-fold or two-fold symmetry, such as the methyl or the methylene group, respectively, rotates relative to the rest of the molecular scaffold. Obviously this rotation cannot be a completely free classical rotation as the bond of this part with the rest of the molecule

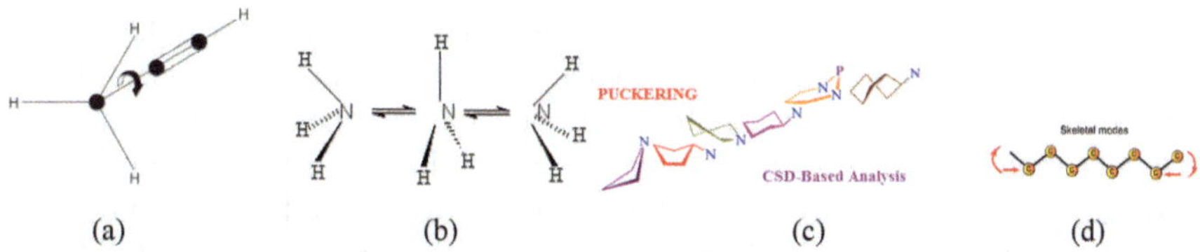

Fig. 26.4 Large amplitude motions in molecules. **a** Internal rotation, **b** Inversion, **c** Ring puckering, **d** Skeletal vibrations. Reference: **a** M. Gilbert, Brydson's Plastics Materials (Eighth Edition), 2017, 59 **b** *Pure Appl. Chem.*, **71**(10), 1919–1981 (1999). **c** B.V. Alviri et al., *Tetrahedron* **74**(1), 28–41 (2018). **d** D. Giordano, *Remote Sensing* **12**(5), 805 (2020)

presents a high three-fold or two-fold sinusoidal barrier to the rotation. However, the rotating part can access its configurations along the rotational trajectory through quantum mechanical tunneling through the barrier and achieve linearly superposed configuration.

Inversion happens in any part of a molecule having one nitrogen and one, two, or three hydrogen atoms. The bonding in this part is such that the N atom is placed normal to the plane containing the H atom(s). The configurations with the N atom on the two sides of this plane are identical, i.e., the molecule has a two-fold symmetry about this plane. Again, the N atom cannot be inverted freely due to the bonds holding it in place and the two configurations are mixed through the corresponding barrier. Similarly, any ring, especially made up of C–C single bonds, can have several identical puckered conformations, depending upon the number of C atoms in the ring. Tunneling through the corresponding barriers creates a superposition of the puckered states, which is the ground energy state.

Skeletal vibrations take place primarily along the hydrocarbon backbones of organic molecules. Any organic molecule can be considered as a derivative of a basic hydrocarbon and these vibrations are the collective motions of the C–C bonds. Somewhat similar to the phonons in crystalline solids, these can be classified into different modes, in particular, transverse and longitudinal, each mode can be analyzed as an uncoupled linear harmonic oscillator, quantized and the energy be assumed to be carried by a quasi-particle. The only difference of these vibrations with other molecular vibrations is that they involve a large number of bonds and consequently have very low energies.

All these large amplitude motions take place at meV energies, at the equivalent temperatures of a few kelvins. Thus they are also the fastest motions in any biological system. In small organic molecules, such as amino acids, they constantly change the molecular configurations thus changing the molecular structure and their functions. However, their more crucial role is in the monomers of all the macromolecules. The internal rotations in the methylene groups composing the long chains and the skeletal vibrations change the configurations of the chains, while the internal rotations in the terminal or pendulous methyl and other groups, as well as the inversion of the nitrogen-containing groups, stabilize the chains and the macromolecules through interactions with neighboring molecules.

The primary existence of an isolated macromolecule or polymer is as the gyration sphere, created through the complete orientational disorder in its monomers. This is achieved if all the monomers are allowed to undergo internal rotations, inversions, and puckering, and all skeletal vibration modes are allowed. The macromolecule or polymer is then in its state of maximum internal entropy. Similarly, a particular set of large amplitude motions will give rise to the double-helix structure of DNAs, the α-helix of proteins, and the lipid raft or membrane. Thus the different spatio-temporal hierarchies are connected mainly through an increase or decrease of conformational entropy in the small molecules or, most importantly, monomers which translates to structural reorganization at the macromolecular levels, and then to assemblies of these macromolecules to specific biological structures.

26.4 Biofunctionality Under Nanoconfinement

Nanoconfinement of any type reduces drastically the number of conformational states by reducing the large amplitude motion degrees of freedom. In particular, it affects very strongly the internal rotations and the transverse modes of skeletal vibrations, the two motions that cause the majority of conformational transitions. This large reduction in the number of conformational states of the monomers puts severe restrictions on the allowed structures of the macromolecules and their assemblies, and also changes the structures of most of them.

For example, restrictions on the internal rotations of monomers in polymer molecules attached as a monolayer to surfaces do not allow the molecules to fold up into gyration spheres even when they are not interacting. They are forced to remain elongated normal to the substrate surface and form polymer brushes. While geometrical nanoconfinement of polymers in films allows the gyration spheres to form but the stacking of these spheres in layers does not allow the full spectrum of internal rotational states of the monomers. This causes an elongation of the "sphere" into an "ellipsoid" creating anisotropy in the molecular structure. The ellipsoidal axes in the layers are also ordered in some orientation relative to the film surface. Thus a spatially as well as orientationally ordered structure emerges from nanoconfinement in both cases, creating a huge decrease in entropy.

As discussed above, this decrease in entropy is followed by an equally large increase in enthalpy in the system, endowing it with a potential to cause interactions with its surroundings with magnitudes not observed in bulk. In biological systems, such interactions of the macromolecular assemblies are expressed as enhanced biofunctionality of the system. This, in fact, is the basis of biofunctionality of nanoparticles, especially of metals and metal oxides.

Metallic and metal oxide nanoparticles do not interact directly with biomolecules and their assemblies in general. They require an organic layer coating or capping them through which they interact and the basis of the interaction lies in the nanoconfinement of the organic molecules. In these layers, the organic molecules lose most of their allowed conformational states and become ordered through interactions with

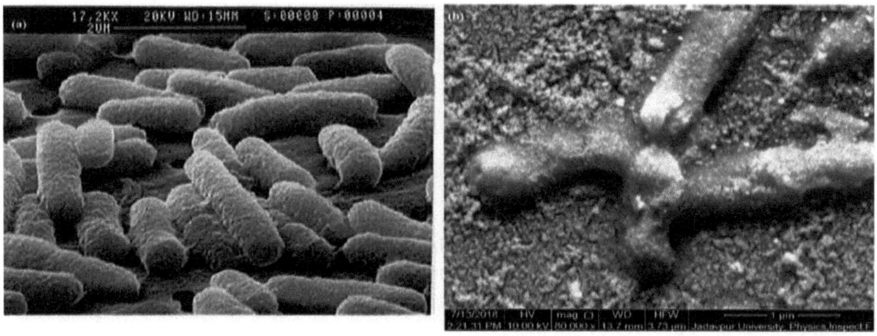

Fig. 26.5 Bactericidal effect of polyvinyl alcohol-coated Ag nanoparticles. **a** Pristine *Escherichia coli* (*E. coli*) and **b** *E. coli* after treatment with polyvinyl alcohol-coated Ag nanoparticles—the nanoparticles can be seen on the surfaces of the destroyed bacteria. Reference: R. Parveen et al., *J. Surf. Sci. Technol.* **36**, 137 (2020)

the surface states of the nanoparticles. Thus the huge surface energy excess of the nanoparticles is utilized to create ordering in the capping layer. The consequent large decrease in internal entropy goes to enhance the enthalpic potential of the organic monolayer to interact with the molecules of the biological system, causing drastic changes in the structures and compositions of the system. In very broad and general terms, thus the excess surface free energy of nanoparticles is converted to potential energy of the nanoconfined capping layer, creating some novel biofunctionality of the nanoparticle.

As an example, Ag nanoparticles are known to possess bactericidal properties. However, the nanoparticles themselves interact weakly with biological molecules through partial ionization. Their efficacy is seen to be dramatically enhanced when capped with polyvinyl alcohol and acted on bacterial culture (Fig. 26.5). What is more interesting is the fact that the bactericidal efficacy is maximum precisely for the nanoparticles which has a monolayer of the polymer covering its entire surface, particles with more than one monolayer coverage or incomplete coverage are remarkably lower in bioefficacy [13]. This underscores the importance of enthalpy enhancement consequent to entropy reduction due to nanoconfinement of the polymer in the capping layer.

26.5 Concluding Remarks

The focus in this chapter is on the essential physics of confinement, nanoconfinement, biofunctionality, and the effect of nanoconfinement on biofunctionality rather than on any review of the instances in which nanoconfined systems affect biofunctionality. While such reviews provide extensive coverage on the topic and are doubtlessly important, the detailed physics of the different aspects are sometimes missed in that coverage, and a fresh look at them may be helpful in understanding these aspects with clarity. It is hoped that it will thus empower the student and the researcher in reaching out to new areas of biofunctionality of nanosystems.

References

1. Manini, N.: Introduction to the Physics of Matter. Springer (2020)
2. Pippard, A.B.: The Physics of Vibration. Cambridge University Press (2007)
3. Besant, W.H.: A Treatise on Hydrodynamics and Hydrostatics. Nabu Press (2010)
4. Yu, C.-J., et al.: Phys. Rev. Lett. **82**, 2326 (1999)
5. Sanyal, M.K., et al.: Europhys. Lett. **36**, 265 (1996); Chattopadhyay, S., Datta, A.: Phys. Rev. B **72**, 099539 (2005); Cortes, L.B.G., et al.: J. Phys.: Condens. Matter **29**, 064003 (2017)
6. Donnelly, S.E., et al.: Science **296**, 507 (2002)
7. Yu, C.-J., et al.: Europhys. Lett. **50**, 487 (2000)
8. Kaganer, V.M., Möhwald, H., Dutta, P.: Rev. Mod. Phys. **71**, 779 (1999)
9. Sanyal, M.K., et al.: Phys. Rev. B **65**, 033409 (2002)
10. Chattopadhyay, S., et al.: Macromolecules **40**, 9190 (2007)
11. Biswas, N., et al.: Chem. Phys. Lett. **539–540**, 157–162 (2012)
12. Lister, D.G., Macdonald, J.N., Owen, N.L.: Internal Rotation and Inversion: Introduction to Large Amplitude Motion in Molecules. Academic Press (1978)
13. Parveen, R., et al.: J. Surf. Sci. Technol. **36**, 137 (2020)

Chapter 27
Development of Rapidly Quenched Amorphous-Nanostructured Materials for Sensor Applications

Somnath Das, Rajat Kumar Roy, Dev Kumar Mahato, and Ashis Kumar Panda

Abstract The rapidly quenched soft magnetic amorphous-nanostructured materials have been used according to their properties targeted toward their potential applications. In rapid solidification process (RSP), different techniques, such as melt spinning, in-water quenching for the production of soft magnetic materials in the form of ribbons/foils, and microwires, have been discussed. The FeSiB-based soft magnetic alloy has been prepared in the form of amorphous ribbon and used as a sensing element in magnetostrictive-sensor (MsS) device for the generation and detection of ultrasonic guided wave for pipeline inspection. CoFe-based prepared microwires show interesting giant magneto-impedance (GMI) behavior.

Keywords Rapid solidification process (RSP) · Magnetic · Amorphous · Nanostructured · Long-range ultrasonic testing (LRUT) · Magnetostrictive sensor (MsS) · Giant magneto-impedance (GMI)

The original version of the chapter has been revised: Reference list entries have been renumbered. A correction to this chapter can be found at
https://doi.org/10.1007/978-981-99-7848-9_29

S. Das · R. K. Roy · A. K. Panda (✉)
CSIR-National Metallurgical Laboratory, Jamshedpur 831007, India
e-mail: akpanda@nmlindia.org

D. K. Mahato
Department of Physics, National Institute of Technology Patna, Patna 800005, India
e-mail: devk@nitp.ac.in

D. Mohanta and P. Chakraborty (eds.), *Nanoscale Matter and Principles for Sensing and Labeling Applications*, Advanced Structured Materials 206,
https://doi.org/10.1007/978-981-99-7848-9_27

27.1 Introduction

The development of sensor materials has witnessed a major breakthrough with the advent of metastable magnetic alloys. Such alloys are prepared through non-equilibrium processing techniques. In such techniques, a typical cooling rate of 10^5–10^6 K/s [1] is achievable. Such cooling rate is achievable through rapid solidification route like melt spinning to get ribbon/foils [2] or in-water quenching for microwire [3] preparation. The other technique followed are Cu-mould casting, sputtering, and gas atomization to get materials in the form of rod, thin films, and powder, respectively.

In addition to rapid quenching technique to get a metastable state, alloy modification plays a vital role to get amorphous or nano-composite phase. The optimal content of metalloids (Silicon, Boron) is critically important for getting amorphous-nanocomposites of Fe- or Fe-Co-based materials. The metastable magnetic alloys can be either positively, negatively, or zero magnetostrictive in nature. The high Fe-containing alloys are usually positively magnetostrictive and can be used in application deploying time-varying strain. In a typical example, high magnetostrictive material is used for the generation of mechanical waves in plates or pipes for their structural health monitoring (SHM). Such an application involves a Magnetostrictive Sensor (MsS) technology.

In many industrial applications, low or vanishing magnetostriction has been used in transformers, magnetic field, or a magneto-impedance sensor. Such materials reveal Giant Magneto-impedance (GMI) property in their glassy state. The GMI-based sensors find applications in biomedical, stress, torque, and very low magnetic field sensors. There has been a lot of advances in Giant Magneto-impedance (GMI) and Magnetostrictive Sensors (MsS) technologies with promising features toward the development of smart sensors in industrial sites.

The present article reviews the fundamental concepts pertaining to MsS [4–6] and GMI [7–9] sensors in reference to their scope of applications.

27.2 Magnetostrictive Sensor (MsS) Technology

The Magnetostrictive Sensor (MsS) is basically a transducer-sensor-based technology. The magnetostriction-based methodology primarily uses this technique for inspection of pipes. Prior to the advent of MsS technique, the elastic waves were generated through piezo-electric transducers (PZT) [4]. However, the inherent bulky nature of PZT system limits portability to the inspection site. This limitation was overcome with the advent of MsS system [10], wherein the sensor element was crystalline CoFe-based alloy sheets.

Recent developments in MsS technology have fathomed into application of amorphous/nanostructured magnetostrictive-sensor element for generation of ultrasonic guided waves. At CSIR-National Metallurgical Laboratory, Jamshedpur, India, such materials have been prepared and used in MsS sensor for pipe inspection.

Fig. 27.1 **a** Photograph of melt spinning system, **b** melt ejection during spinning, **c** melt spun ribbon

27.2.1 Preparation of MsS Sensor Material

Pure elements Fe, Si, B were melted in vacuum arc melting furnace purged with Argon gas. To ensure effective homogenization, the alloy ingot was remelted several times. The master alloy was taken in quartz crucible and induction melted in Argon atmosphere to get cylindrical ingots. Subsequently, the cylindrical ingot was taken in quartz crucible with a slit-type nozzle at the bottom. This crucible was fixed inside a melt spinning system (Fig. 27.1a), where it was placed over a rotating copper wheel. The alloy ingot was induction melted and ejected over the rotating wheel [11] using Argon gas (Fig. 27.1b). The melt is rapidly quenched in the form of continuous ribbon of FeSiB alloy having thickness of 25–30 μm (shown in Fig. 27.1c). Ribbon of desired width can be produced by changing the crucible slit length.

27.2.2 Magnetic Property Considerations

Conventionally, the crystalline materials included either FeCo or Fe-Co-V-based crystalline alloys were used for MsS transducer application. Although these alloys exhibit a high saturation magnetostriction of $\lambda_s = +80$ ppm, they are achievable at a high magnetizing field of 50 Oe (4000 A/m) [12]. Thus, FeCo need strong and bulky magnets to induce magnetostriction which limits the portability of the MsS System. In contrast, the amorphous materials used in the present investigation, due to their isotropic nature of the material, lead to absence of magnetocrystalline anisotropy with a low magnetizing field required for magnetostrictive actuation. Moreover, these materials in the absence of crystallinity are ductile enough to be pasted easily on the pipes. The saturation magnetostriction of the FeSiB ribbon is typically around +30 ppm obtainable at a field below 0.6 Oe [13].

27.2.3 MsS Transducer Methodology

The application of magnetostrictive-sensor transducer methodology is based on the synergistic phenomenon of "Joule Effect" [14] and "Villari Effect" [15]. In the Joule effect, when an external magnetic field is applied on the ferromagnetic material, it leads to a small dimensional change (magnetostriction) in the ferromagnetic material along the direction of applied field. While Villari effect (inverse magnetostriction) refers to a change in permeability of the material due to induced mechanical stress (or strain).

27.2.4 Configuration of Transducer-Sensing System

The configuration of MsS transducer-sensor system comprises arrangements of transmitter and sensing cables for the generation and detection of ultrasonic pulsed signal. The transmitter excites the signal while the receiver comprising of sensing cable receives the signal in the form of electrical voltage. Basically, there are two ways for the arrangement of such transmitter and sensing cables. One of the arrangements is based on pulse-echo technique in which the transmitter on the pipe generates a mechanical wave. This wave traverses through the pipe and is reflected from the pipe end. The reflected signal is picked up by a receiver, placed adjacent to the transmitter. While in another sensor configuration involving pitch-catch technique, the transmitter and the receiver are placed at different locations of the pipe. The receiver picks up the direct excitation signal from the transmitter. The transmitter excites the mechanical wave and, at the other end, sensing cable picking up the signal traversing through the pipe in the form of electrical voltage.

The present investigation focuses on the single transducer-based pulse-echo method because the identification of any defect and its position is easier than that of pitch-catch technique. Another advantage is to easily adjust the transmitter and sensing cables, as they are placed on the same position other than in pitch-catch technique.

27.2.5 MsS Measurement System and Application

The stress relieving annealed ribbon pasted circumferentially over the test pipe. The MsS system comprising of three coil/cable system, namely, transmitter (T_c), biasing (B_c), and sensing/receiving (R_c) placed over the pasted ribbon [16]. The programmable LabView software generates the amplified pulsed sinusoidal wave through DAQ (Data Acquisition) unit, which is then fed to transmitter cable, thus resulting in the time-varying magnetic field to vibrate the ribbon. Due to the vibration, the mechanical wave, which is basically the ultrasonic guided wave, is generated and that traverses through the pipe.

Simultaneously, the DC biasing current sent to the biasing cable enhances the transduction efficiency of the travelling signal. The reflected signal picked up by the secondary sensing/receiving cable and fed to LabView for the further analysis. The laboratory test setup of MsS measurement system is shown in Fig. 27.2. In the present setup, the system has the facility to precisely control the DC magnetizing field as against the conventionally used permanent magnets.

The MsS sensor transducer found an important application in the area of structural health monitoring of pipeline system using melt spun ribbon as sensing core element. The present investigation carried out with the industrial-grade aluminium pipe which is 3 m in length and having outer diameter is 48 mm with 8.5 mm wall thickness. For a typical aluminium pipe, some of the typical operating MsS parameters were (a) excitation frequency (68 kHz), (b) AC driving current (32 mA), and (c) DC bias current (200 mA). The optimized parameters for MsS System used for aluminium pipe are shown in Table 27.1. The driving frequency is chosen from the DISPERSE Software, which gives the plot of frequency versus group velocity. The driving frequency must lie on the non-dispersive region, i.e., the velocity remains constant with the changing frequency so that pure modes can be excited. The present investigation focuses on

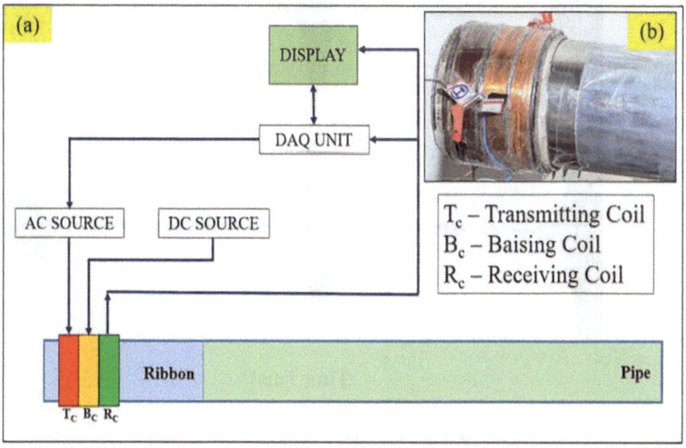

Fig. 27.2 **a** Laboratory setup of MsS measurement system, **b** Magnetostrictive sensor

Table 27.1 MsS optimized parameters for Al pipe

MsS parameters	Value
Excitation frequency	68 kHz
AC driving current	32 mA
DC bias current	200 mA
Group velocity	4.87922 m/ms

the generation of axially symmetric L(0, 2) mode which is highly sensitive to internal and external defects. The typical group velocity is 4.87922 m/ms at 68 kHz driving frequency, obtained from DISPERSE software. An artificially simulated defect in the form of hole at the middle (1.5 m from the pipe end) of the pipe has been created and the post-analysis of the signal gives the information about the presence of the defect on the pipe. The layer plot of as-received and defect pipe is illustrated in Fig. 27.3. The BW is the backwall signals which are basically the end wall reflected signals from the pipe end. As in the first layer, there is no defect on the pipe so it will only show BW signals which is the pipe-end wall reflections but in the second layered signal, an another peak is coming in between the primary and first backwall BW signal which indicates the presence of defect on the pipe. From the graph, the amplitude of first BW of defect free pipe is 0.894 V, but after simulated defect (3 mm diameter hole), the amplitude of first BW signal reduces to 0.887 V simultaneously the amplitude of defect signal increases to 0.059 V. After measuring the time and velocity of the MsS output signal, one can detect the location of the defect present on the pipe.

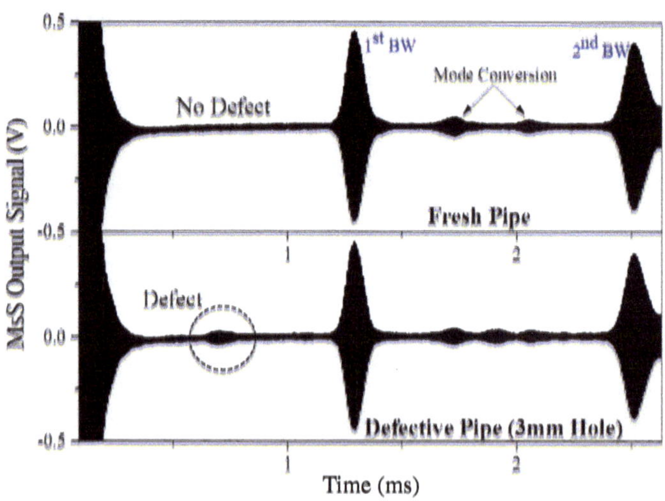

Fig. 27.3 Layer plot of as-received and defective (Hole: 3 mm) Al pipe

27.3 Giant Magneto-Impedance (GMI)-Based Technology

The giant magneto-impedance sensors rely on the GMI effect. When a soft magnetic amorphous/nanostructured ferromagnetic conductor is subjected to a small alternating current (AC), then there is a large change in the AC complex impedance which occurs with the application of varying DC magnetic field. This phenomenon is termed as GMI effect. It has an electromagnetic origin which is strictly related to the frequency-dependent penetration depth of the magnetic conductor [17].

The ratio of impedance variation under external applied DC magnetic field at a fixed frequency is used to represent the GMI effect.

$$\frac{\Delta Z}{Z}\% = \left[\frac{|Z|(H) - |Z|(H_{\max})}{|Z|(H_{\max})}\right] \times 100 \tag{27.1}$$

where $|Z|(H)$ = Impedance of the material in the absence of DC field;
$|Z|(H_{\max})$ = Impedance of the material at maximum DC field applied.

The GMI-effect-based magnetic sensors offer several advantages over conventional magnetic sensors. The soft magnetic amorphous materials cast in the form of wires having diameter in the order of 100 μm are used as sensing core element of magnetic sensors. These microwires have superior electrical, mechanical, and magnetic properties due to the short-range order. The formation of amorphous state depends on the alloy composition as well as the processing condition. All these novel properties make soft magnetic amorphous/nanostructured microwires very attractive for unique technological applications and also provide opportunities for fundamental micro-magnetic studies mainly due to their unique magnetic domain structure. The sensitivity of GMI sensor is found to be very high and improved characteristics among the micro-magnetic sensor family [18].

27.3.1 *Preparation of Microwires*

The master alloy and ingot is prepared in similar fashion as explained in Sect. 27.2.1. At CSIR-National Metallurgical Laboratory, Jamshedpur, India, the amorphous wires have been prepared by in-water quenching technique. The amorphous microwires can be prepared by ejecting the melted alloy in flow of water, which rotates at a high speed of about 250–300 rpm. The water is introduced in a hollow drum which is spinning to keep water inside due to centrifugal force. The rapid transfer of heat freezes the melt when it hits the water. To obtain an amorphous structure, the required cooling rate becomes 10^4–10^6 K/s. Before performing experiment, the following parameters must be optimized to get continuous amorphous microwires: (a) the distance and angle between the surface of the water and the end of the quartz tube; (b) size of the nozzle; (c) temperature of the cooling water; and (d) melt temperature, argon pressure, and ejection speed of the melt. The amorphous structure is usually observed when its

Fig. 27.4 **a** Photograph of rotating drum wire caster, **b** melt ejection, **c** microwire

diameter varies from 80 to 250 μm. The solidified wire is collected by centrifugal force on the inner side of the rotating drum. The cooling of the alloy takes place from the outer surface of the alloy, which acquires a cylindrical shape when falls into the water. During the process of cooling, the outer surface of the wire solidifies first and subsequently the inner surface solidifies and shrinks. The illustration of microwire formation from the in-water quenching rotating drum wire caster is shown in Fig. 27.4.

27.3.2 Measurement of GMI Behavior

The GMI sensor system for the magneto-impedance measurement consists of different arrangements of the probes with the amorphous microwires. Generally, there are the two methods to arrange the probe and microwire to measure magneto-impedance of the material. The first one is the fluxgate technique which is based on four-probe method. In this measurement, probe of one pair is used to send AC current or AC driving field and other pair is used for impedance measurement. The two current probes connected to the two ends of the wire sample while the two voltage probes connected to the wire sample in between the two probes. Another one is the "fluxmetric method", which is based on two-probe method. In this measurement technique, single probe is used to send the current as well as measured the magneto-impedance of the wire sample. The two probes are connected to the two ends of the wire sample used to send the AC driving field and the same probe measures the magneto-impedance of the wire. The present investigation focuses on the four-probe-based fluxgate method for the magneto-impedance measurement of $(Co_{0.94}Fe_{0.06})_{72.5}$ $Si_{12.5}$ B_{15} microwire sample. In both the techniques, the size of the wire sample used for magneto-impedance measurement is of 10 cm in length. In both the technique, the convention to calculate the magneto-impedance ratio is similar as mentioned in Eq. (27.1).

The present investigation focuses on the annealed $(Co_{0.94}Fe_{0.06})_{72.5}Si_{12.5}B_{15}$-based soft magnetic microwires of 125 μm in diameter prepared through in-water quenching. The four-probe technique is used for the present experimentation. The microwire of length 10 cm is placed along the axis of the Helmholtz coil, which is perpendicular to the earth's magnetic field (to minimize field's effect on sample). The use of Helmholtz coil is that it provides a uniform DC magnetic field parallel to the axis of the microwire. The optimized frequency and AC driving current for the maximum GMI ratio are 1 MHz and 2 mA, respectively. The magneto-impedance of prepared microwires was measured using impedance analyzer (Agilent 4294A). The GMI ratio of the first harmonic signal can be calculated using Eq. (27.1). The property of microwires is evaluated in terms of percentage change in impedance with respect to a maximum applied DC magnetizing field. The impedance and GMI plot of amorphous microwire is shown in Fig. 27.5a and b. The percentage change in GMI (Fig. 27.5b) obtained using Eq. (27.1) gives information regarding the magneto-impedance response of the amorphous wire in an external magnetic field. This field can be dependent on magnetic transition in a test material or an industrial component.

The rescaled plot of Fig. 27.5b is shown in Fig. 27.6 to show the dual peak (DP) behavior. This characteristic dual peak is observed in a typical zero/near-zero magnetostrictive amorphous wires [19]. The peak GMI value obtained was 202%. In such

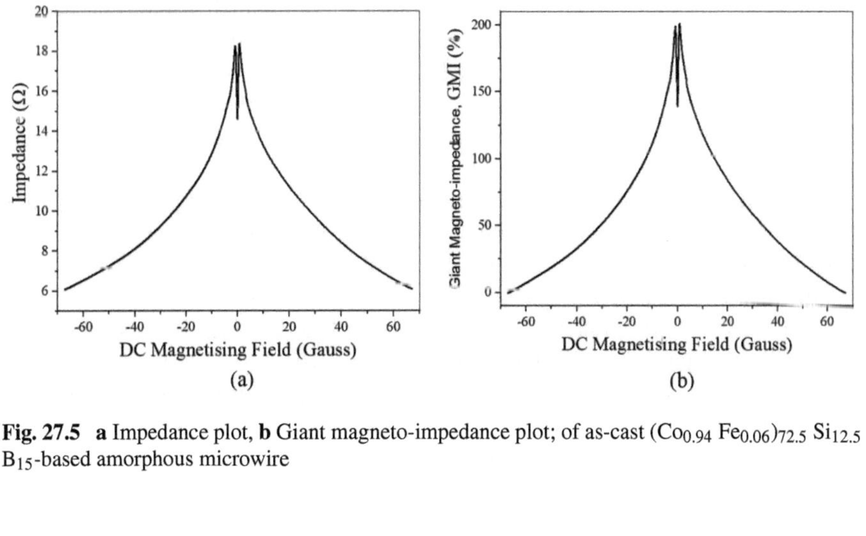

Fig. 27.5 **a** Impedance plot, **b** Giant magneto-impedance plot; of as-cast $(Co_{0.94}$ $Fe_{0.06})_{72.5}$ $Si_{12.5}$ B_{15}-based amorphous microwire

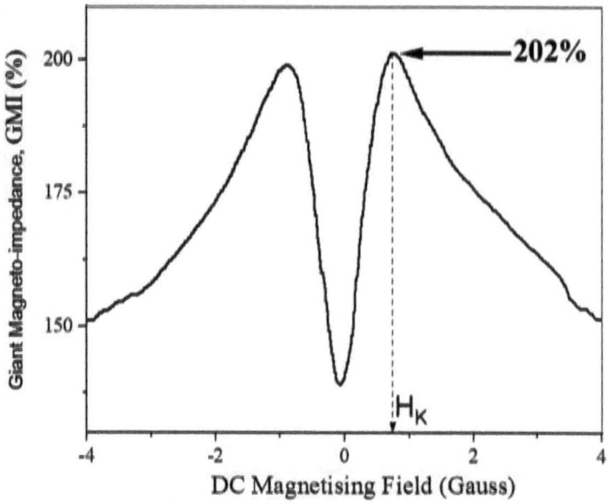

Fig. 27.6 Rescaled plot of GMI dual peak for $(Co_{0.94}Fe_{0.06})_{72.5}Si_{12.5}B_{15}$-based amorphous microwire

wires, the lateral stress in the as-quenched wires induces circumferential magnetic domains which enhance the circumferential permeability of the material. Consequently, an anisotropy field (H_k) is observed as shown in Fig. 27.6. The anisotropy field "H_k" value obtained is as low as 0.75 Gauss which is useful from the perspective of sensor application.

27.3.3 Scope of GMI Sensor

The GMI effect has attracted considerable scientific and technological interest especially because of its applicability in sensing very feeble magnetic field. The detection of different levels of magnetic field using magnetic sensor can be broadly categorized into three regimes [20], viz, (a) the detection of small-to-smaller magnetic field by SQUID (superconducting quantum interference device) with a sensitivity of 10^{-10} Gauss, (b) the detection of small-to-medium magnetic field by magneto-impedance and fluxgate sensor with a sensitivity of 10^{-6} Gauss, and (c) the detection of medium-to-large magnetic field by magnetostrictive devices with the sensitivity of 10 Gauss and 1 Gauss, respectively.

The GMI microwires are used in biosensors [21]. With small power dissipation per unit area, the microwires are highly sensitive at very low magnetic field. The GMI-based biosensors are targeted toward the detection of bio-molecules. The GMI microwires are also used in magnetic devices used for the detection of parked vehicles [22]. The sensor works in such a manner that the parked vehicle is detected by sensing the mechanical tension generated by the vehicle's wheel fixed at its one end. The device counts the number of parked vehicles as well as their parked locations. The GMI microwires are used in coding devices. The group of microwires are annealed and characterized with different values of remanence and switching field. The output information stored in a tag and can be identified by different peak voltages from each wire [23]. GMI microwire plays an important role in the design of stress sensor. Current "I" passes through a single conducting wire and the resulting circular field is generated and magnetize the soft magnetic amorphous microwire, inducing sharp voltage pulses at the coil. The above phenomenon is useful for pulse generator devices and magnetic switches with compact size and very sharp pulses.

27.4 Conclusion and Future Outlook

The soft magnetic amorphous/nanostructured ribbons and microwires are prepared through rapid solidification process (RSP) using melt spinning and in-water quenching techniques, respectively. The rapidly quenched FeSiB-based ribbon prepared through melt spinning technique can be used as MsS sensor element for pipeline inspection. The melt spun ribbon being amorphous and soft magnetic in nature, needs low magnetizing field to induce magnetostriction. Low amount of magnetizing field AC I_{rms} = 5–10 Oe and DC bias field = 10–25 Oe are required. The MsS sensor was used to generate ultrasonic guided wave on aluminium pipe. A defect in the form of a hole created in the pipe is detected in the form of small MsS signals along with the backwall signals. In contrast to the high magnetostrictive FeSiB-based amorphous ribbons, the CoFe-based alloy microwires reveal giant magneto-impedance (GMI) property. The rapidly quenched microwires exhibit a large change in GMI behavior of about 202% in their as-prepared state. With both fundamental and technological interests, it could provide ample scope in the field of nanoelectronics, magnetoelectrics, and nano-bio-interface engineering.

References

1. Nderu, J.N., Shinokawa, Y., Yamasaki, J., Humphrey, F.B., Ogasawara, I.: IEEE Trans. Magn. **32**, 751 (1996)
2. Christofaro, N.: Amorphous metals in electric power distribution application. MRS Bull. **23**, 50–56 (1998)
3. Sarkar, P., Roy, R.K., Panda, A.K., Mitra, A.: Optimization of process parameters for developing FeCoSiB amorphous microwires through in-rotating-water quenching technique, results for Appl. Phys A: Mater. Sci. Prices **111**, 575–580 (2013). https://doi.org/10.1007/s00339-012-7260-4
4. Panda, A.K., Sharan, P.K., Roy, R.K., Murthy, G.V.S., Mitra, A.: Generation and detection of guided Waves in a defective pipe using rapidly quenched magnetostrictive ribbons. Smart Mater. Struct. **21**, 045015 (2012). https://doi.org/10.1088/0964-1726/21/4/045015
5. Ling, E.H., Rahim, R.A.: A review on ultrasonic guided wave technology. Austr. J. Mech. Eng. (2017). Taylor & Francis. https://doi.org/10.1080/14484846.2017.1373385
6. Vinogradov, S., Cobb, A., Light, G.: Review of magnetostrictive transducers (MsT) utilizing reversed widemann effect. AIP Conf. Proc. **1806**, 020008 (2017)
7. Zhukova, V., Ipatov, M., Zhukova, A.: Thin magnetically soft wires for magnetic microsensors. Sensors **9**, 9216–9240 (2009). https://doi.org/10.3390/s91109216

8. Dzhumazoda, A., Panina, L.V., Nematov, M.G., Ukhasov, A.A., Yudanov, N.A., Morchenko, A.T., Qin, F.X.: Temperature-stable magnetoimpedance (MI) of current-annealed Co-based amorphous microwires. J. Magn. Mag. Mater. (2018). https://doi.org/10.1016/j.jmmm.2018.10.111
9. Khan, M.A., Sun, J., Li, B., Przybysz, A., Kosel, J.: Magnetic sensors–a review and recent technologies. Eng. Res. Express **3**, 022005. IOP Publishing (2021). https://doi.org/10.1088/2631-8695/ac0838
10. Wu, J., Tang, Z., Yang, K., Lv, F.: Signal strength enhancement of magnetostrictive patch transducers for guided wave inspection by magnetic circuit optimization. Appl. Sci. **1477**(9) (2019). https://doi.org/10.3390/app9071477
11. Panda, A.K., Roy, S., Singh, S.R., Rao, V., Pramanik, S., Chattoraj, I., Mitra, A., Ramachandrarao, P.: Effect of quenching rate on the properties of melt-spun FeNbCuSIB ribbons. Mater. Sci. Eng. A **304–306**, 457–461 (2001). https://doi.org/10.1016/S0921-5093(00)01490-8
12. Lafont, T., Gimeno, L., Delamare, J., Labedev, G.A., Geoffroy, O.: Magnetostrictive-piezoelectric composite structures for energy harvesting. J. Micromech. Microeng. **22**, 094009, 6 (2012). https://doi.org/10.1088/0960-1317/22/9/094009
13. Naoe, M., Yamamoto, H., Yamanaka, S.: Magnetic properties of amorphous FeSiB films prepared by co-sputtering. J. App. Phys. **50**, 7606 (1979). https://doi.org/10.1063/1.326859
14. Joule, J.P.: On the effects of magnetism upon the dimensions of iron and steel bars. Phil. Mag. **30**, 76–87 (1847). https://doi.org/10.1080/14786444708645656
15. Villari, E.: Change of magnetization by tension and by electric current. Ann. Phys. Chem. Lpz. **126**, 87–122 (1865)
16. Panda, A.K., James, A.P., Mohan, A., Dey, S., Roy, R.K., Mitra, A.: Investigation on transverse defect and cavitation depth in ferrous pipe using melt spun FeSiB ribbon as magnetostrictive sensing element. NDT E Int. **80**, 15–22 (2016). https://doi.org/10.1016/j.ndteint.2016.02.002
17. Panina, L.V., Mohri, K., Bushida, K., Noda, M.D.: J. Appl. Phys. **76**, 6198 (1994)
18. Mandal, K., Puerta, S., Vazquez, M., Hernando, A.: IEEE Trans. Magn. **36**, 3257 (2000)
19. Pal, S.K., Manik, N.B., Mitra, A.: Dependence of frequency and amplitude of the ac current on the GMI properties of Co-based amorphous wire. Mater. Sci. Eng. A **415**(2006), 195–2001 (2006)
20. Tannous, C., Gieraltowski, J.: Giant magneto-impedance and its applications. J. Mater. Sci.: Mater. Electron. **15**, 125–133 (2004)
21. Chiriac, H., Herea, D.D., Corodeanu, S.: Microwire array for giant magneto-impedance detection of magnetic particles for biosensor prototype. J. Magn. Magn. Mater. **311**(2007), 425–428 (2007). https://doi.org/10.1016/j.jmmm.2006.11.207
22. Marin, P., Hernando, A.: Application of amorphous and nanocrystalline magnetic materials. J. Magn. Magn. Mater. **215–216**, 729–734 (2000)
23. Hernando, A., Vazquez, M., Zhukov, A.P., Larin, V.S., Thorkunov, A.V., Antoneneko, A.N.: Patent P9601993 ES

Chapter 28
Magnetic Nanostructures for Transport Control and Sensing Applications

Dipti R. Sahu

Abstract Magnetic nanostructures in the form of nanomaterials, thin films, multilayers, and junctions were studied for sensing application. Different nanostructures are synthesized using chemical and physical method. Magnetic nanostructure showing interesting magnetic, transport and sensing properties are reported here. This chapter gives an overview of different magnetic nanostructures showing difference in their transport behaviour and sensing characteristics. This field opens new possibilities for development of magnetic sensors for the next-generation applications.

Keywords Magnetic · Nanostructure · Thin film · Sensing

28.1 Introduction

In recent years, nanostructures have been the focus of intense research due to their extraordinary physical, chemical, optical, mechanical, electrical and magnetic properties [1–4]. Different nanostructures such as metal, carbon nanotube, oxides, graphene, and composite nanostructures of various materials have been extensively explored for diverse applications in various fields including sensing [4–7]. The nanostructure material provides several benefits for sensing application owing to their nanometer size and high surface-to-volume ratios compared to that of traditional bulk and thin films. The classifications of nanosensors are based on their functional characteristics. Among all nanostructures, the magnetic nanostructures are a promising class of sensing materials due to their unique magnetic properties, interesting physical phenomena, chemical stability, and potential applications [5, 8, 9]. Magnetic nanostructures are synthesized in a wide range of geometries which have different electronic structure, domain structure, magnetic orientation, magnetic

D. R. Sahu (✉)
Department of Natural and Applied Sciences, Namibia University of Science and Technology, Windhoek, Namibia
e-mail: dsahu@nust.na

D. Mohanta and P. Chakraborty (eds.), *Nanoscale Matter and Principles for Sensing and Labeling Applications*, Advanced Structured Materials 206,
https://doi.org/10.1007/978-981-99-7848-9_28

relaxation, and intrinsic properties at nanoscale. However, magnetic nanostructures face experimental challenges at nanoscale due to the magnetic fluctuations [10, 11].

This problem opened a new challenge to the researchers in the synthesis of more complex nanomagnetic materials for technological applications. Exotic magnetic nanostructures in the form of thin films, multilayers, skyrmions and core–shell structure are tailored using novel approaches in the modelling and synthesis process [12, 13]. Novel spin textures due to the interactions of neighbouring spins and coupling are observed. The nano-magnetism in the nanostructures is an exciting phenomenon for future technologies towards the development of innovative devices with new applications, such as sensing. Magnetic gas sensors, electrochemical sensors, magnetic biosensors for pharmaceutical and biomedical applications are developed. It is reported that sensing behaviour is measured by the variation in magnetic properties and controlled by the size, shape and structures of the active materials. The exotic spin configurations, vortex state in nanostructures, interface and surface of the nanostructures coated with other substances also produced interesting sensor devices for medical applications in diagnosis and magnetic resonance imaging [14].

Nowadays, skyrmions based on magnetic multilayer nanostructure systems are developed which facilitate the production of new type of memory and sensors devices with better properties [15–17]. Further developments are made for high-resolution spin-sensitive imaging technique with the manipulation of individual skyrmions [18, 19]. Skyrmion is also sensed using a simple magnetic tunnel junction (MTJ). New nanostructures are designed for hosting small size skyrmion sensor for communication, transmission of spin information and low-power current sensor for global interconnect [20, 21]. The tunnel magnetoresistance (TMR) skyrmion is also reported by Chen et al. in a multilayer skyrmionic racetrack with ferromagnetic electrode separated by a thin tunnel barrier [17].

Based on the importance of magnetic nanostructure, special attention must be taken to control the size and shape of nanomaterials during synthesizing process; the microstructure and morphology of sensing materials; improvement on the specificity and stability of functional molecules; correlation between composition, structure, and surface reactivity of nanomaterials; and improving the sensitivity and stability of the nanosensors [2, 22, 23]. The sensor's behaviour and properties can be controlled and monitored by the fabrication of hybrid structure and tuning the nanostructure properties of materials with device structure and on the construction of specially designed structures [23]. In magnetic sensors, the magnetic and transport behaviour of the active materials is important. The active material will be characterized using magnetization, magnetoresistance, ferromagnetic resonance, magneto-optical Kerr effect and magnetostatic wave oscillation effect [24, 25]. The low-dimensional nanosensors with increased sensitivity and extended capabilities operating at wide range of temperature and low magnetic field including at room temperatures have many advantages and uses [8, 23, 26, 27]. The present chapter presents an overview of some magnetic nanostructures used for transport-control sensing applications.

28.2 Magneto-Transport Phenomena

In general, the conventional electronic materials or devices are controlled by charge and functionalities of these devices are mainly due to electric current. Therefore, charge transport is the main phenomena in the electronic devices. However, in magnetic materials, electron spin plays an important role. Spin-dependent properties are associated with the motion of electrons and interaction between the charges. In a magnetic system, the transport behaviour is always related to spin transport associated with an electrical current. Therefore, a unique transport phenomenon is observed in nanomagnetic system due to interplay between spin and charge of the electrons.

Various magneto-transport phenomena occur due to the anisotropic magnetoresistance (AMR), giant magnetoresistance (GMR), colossal magnetoresistance (CMR), tunnelling magnetoresistance (TMR), spin torque transfer (STT) and quantum anomalous Holzer effect (QAHE) [28]. However, in most of the phenomenon dealing with only nanostructure of elemental metal and their alloys where the spin transport only persists over distances of the order of a few nanometers. In case of a layered structure, a few nanometers of individual layer thicknesses play important role for spin-transport effects. The interfaces between neighbouring layers also effectively contribute to the transport behaviour of the structure in the way of spin-dependent reflection, transmission and scattering. Therefore, description of transport phenomenon and physical picture of the basic mechanisms are needed for proper understanding.

28.2.1 *Giant Magnetoresistance (GMR)*

GMR phenomenon is the change in electrical resistance of the material due to reorientation of the magnetization by the application of magnetic field. The GMR is observed in magnetic materials ranging from heterogenous bulk system over magnetic nanomaterials to multi-layered thin films and permanent magnets [9, 29]. Different geometries such as current in plane and current in perpendicular are used for the study of GMR. In the in-plane GMR, transport is treated as nonlocal and current in perpendicular. The local approximation is often adequate. The basic geometry of multilayer which shows GMR effect is shown in Fig. 28.1.

The relative alignment of the magnetizations of adjacent ferromagnetic layer determines the change in electrical resistance, which gives value of giant magnetoresistance effect. If R_P is the resistance for parallel alignment of adjacent ferromagnetic films and R_{AP} the resistance for antiparallel alignment, then GMR is defined as

$$\Delta R/R_P = R_{AP} - R_P/R_P \tag{28.1}$$

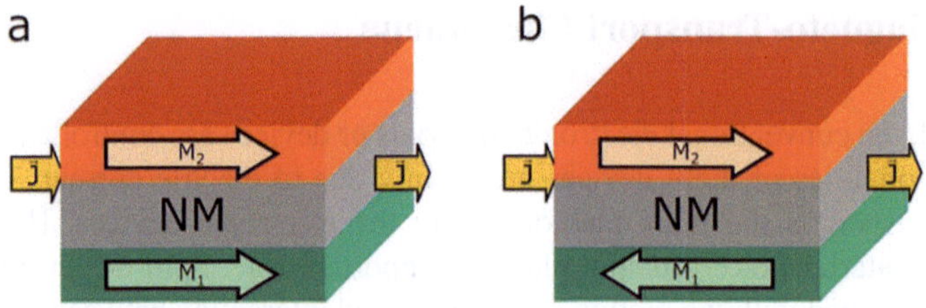

Fig. 28.1 Representation of GMR effect in layered magnetic system in the current in-plane direction. **a** Parallel configuration; **b** antiparallel configuration [9]

Normally, the resistance is highest for antiparallel alignment yielding a positive $\Delta R/R_P$. This situation corresponds to normal GMR effect. In tri-layers consisting of only two ferromagnetic layers separated by a non-ferromagnetic spacer layer, the GMR ratio $\Delta R/R_P$ can reach values of up to 36% at RT, whereas multilayers with many repetitions of ferromagnetic/non-ferromagnetic bilayers yield GMR ratios of up to 80%. In magnetic nanostructures, the scattering of electrons based on the configuration of the magnetic layer determines the change in resistance. It is known that the scattering of electrons takes place differently depending on the spin orientation of the magnetic layers. When magnetization of the two magnetic layers is in a parallel configuration, there is almost no scattering thus offering low resistance. However, in the antiparallel state, the spin-up and spin-down electrons are scatted equally giving rise to high resistance [29, 30]. This scenario is shown in Fig. 28.2, with the equivalent circuit.

Further, the GMR angular dependence can also be expressed as

$$R = R_P + \Delta R\left(1 - \cos^2\left(\frac{\theta}{2}\right)\right), \tag{28.2}$$

where θ is the angle between the magnetization of the two FM layers and $\Delta R = R_{AP} - R_P$.

The GMR sensors are sensitive to angular variation and determined based on the cosine function. Only the component of magnetic field along the direction of its sensitivity axis can be detected by the sensor. In this case, spin-dependent scattering of conduction electrons and spin-dependent potential barriers between successive layers are taken into consideration for the description of angular variation of GMR.

28.2.2 Tunnelling Magnetoresistance (TMR)

TMR phenomenon is like GMR where non-magnetic layers are insulating. However, TMR occurs when an electron tunnels from one ferromagnetic layer to another through a thin insulating layer based on the relative orientation of the magnetizations of the ferromagnetic electrodes [32, 33]. The TMR ratio is generally much

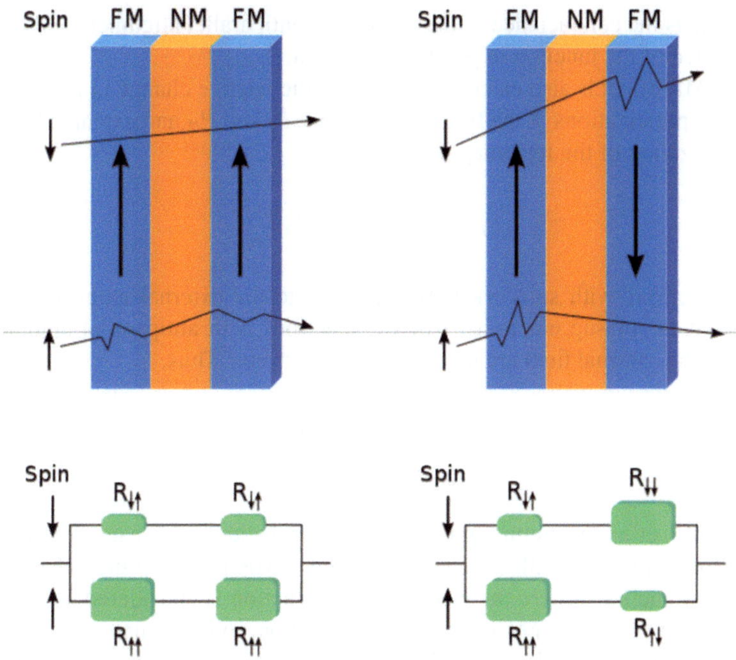

Fig. 28.2 Illustration of the GMR effect and the equivalent circuit diagram [31]

larger than that of GMR. The GMR effect is about 5–15%, whereas TMR effect is 20–70%. Different concepts, such as Landauer approach, ballistic electron, chemical potential, fermi energy of the electrodes, and spin channels, are taken into account for the description and discussion of the TMR effect in magnetic nanostructures. The TMR was explained by the tunnelling Hamiltonian theory [34] and a scheme of this effect is depicted in Fig. 28.3a. The TMR can be written as

$$\mathrm{TMR} = \mathrm{G_{AP}} - \mathrm{G_P}/\mathrm{G_{AP}} = \mathrm{R_{AP}} - \mathrm{R_P}/\mathrm{R_{AP}} = 2P_I P_{II}/1 - P_I P_{\mathrm{II}}, \tag{28.3}$$

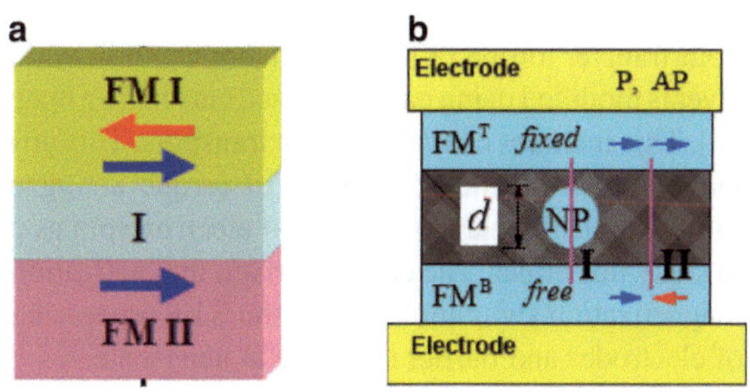

Fig. 28.3 **a** TMR multilayers structure [35], **b** TMR nanostructure with nanoparticles (NPs) [36]

where G_{AP} is the conductance of the junction in antiparallel alignment and G_p is the conductance of the junction in parallel alignment.

The TMR effect is also calculated based on the relative change in resistance due to the spin polarizations of the two ferromagnets, P_I and P_{II} and explained using the spin polarization of the ferromagnetic electrodes.

$$P_I = \frac{n_I^{\uparrow}-n_I^{\downarrow}}{n_I^{\uparrow}+n_I^{\downarrow}}\ P_{II} = \frac{n_{II}^{\uparrow}-n_{II}^{\downarrow}}{n_{II}^{\uparrow}+n_{II}^{\downarrow}} \tag{28.4}$$

The electrons with spin orientation parallel to the external magnetic field are considered as spin-up electrons, whereas electrons with antiparallel alignment of spin with the external field are the spin-down electrons. Thus,

$$\begin{aligned} G_P \alpha n_I^{\uparrow} n_{II}^{\uparrow} + n_I^{\downarrow} n_{II}^{\downarrow} \\ G_{AP} \alpha n_I^{\uparrow} n_{II}^{\downarrow} + n_I^{\downarrow} n_{II}^{\uparrow} \end{aligned} \tag{28.5}$$

The TMR of nanostructure tunnel junctions embedded with nanoparticles (NPs) can be found in Fig. 28.3b. In a single crystal, TMR is due to the coherent spin-polarized tunnelling where the electron wave function is considered as symmetric function. However, in the nanoparticle-based junction, coherent tunnelling along with resonance tunnelling of electrons takes place, which is like a double barrier magnetic tunnel junction [37, 38]. It is also reported that the electron momentum states of the nanoparticle located inside the insulating layer affect the TMR magnitudes and TMR sign at small voltages [36]. In a nanostructure-based junction, the role of the lowest quantum well states and small values of the wave number is crucial to observe quantum conductance regime for TMR. The deposition of middle layer which can form different nanostructures due to the variation of thickness of the layer or dispersion of nanoparticle indicates peak-like TMR in some cases of tunnel junction [39]. The peak-like TMR response or amplitude is generally explained by the result of the large spin transfer torque (STT) effect or resonance tunnelling [40–42].

28.2.3 Spin-Torque Transfer (STT)

Spin-transfer torque (STT) is a phenomenon in which the orientation of a magnetic layer is modified using pain polarized current in a magnetic junction or spin valve [43–45]. This includes charge and spin transport. This effect is one type of magnetization manipulation by the application of torque acting on the ferromagnetic electrode [46]. A noncollinear description of electron spin as presented in Fig. 28.4 is needed to study this effect. The current-driven magnetization dynamics and current-induced magnetization switching explain the STT process based on the magnetic behaviour of electrodes and barrier structure at nanoscale.

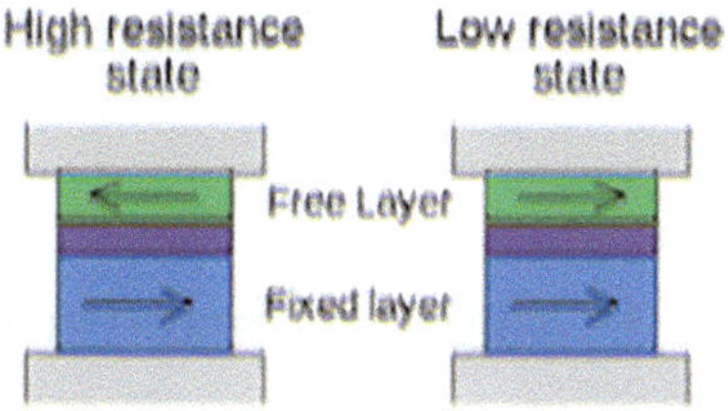

Fig. 28.4 The noncollinear description of electron spin

Spin-transfer torque phenomenon shown in Fig. 28.5 explains that the electron with specific spin orientation which is filtered by pinned layer transfers magnetization direction to the free layer which favours parallel state. The electron with spin orientation antiparallel to the pinned layer magnetization cannot pass as the bands are occupied. They will accumulate in the free layer, which favours antiparallel state [29]. In the case of the large STT effect, non-ideal magnetization alignments take place in the layers and nanostructures, and STT will grow up rapidly with applied voltage. The STT can average out the magnetic moments of the nanostructures at some threshold voltage. Thus, the TMR value can drop down to the one of non-magnetic nanostructure cases, while the TMR value can be greater in the initial magnetic state for the nanostructures. In the case magnetic nanostructure, magnetic moment coupling between nanoparticle and ferromagnetic free layer happened. The coupling corresponds to co-aligned and anti-aligned magnetic moment orientations of the nanoparticles with the magnetization alignment of the free magnetic layer in parallel and antiparallel state. The coupling along with the size dispersion of nanoparticles shows TME effect in resonant oscillation state [48].

Nowadays, ferromagnetic resonance (FMR) combined with magnetoresistance (GMR or TMR) is used for the study of the magnetization dynamics of nanoscale materials [48, 49]. The spin transfer torque-driven FMR (STT-FMR) is widely exploited for FMR studies and sensing behaviour of various nanostructures, multi-layers, magnetic tunnel junctions [50–53], metallic nanotubes, nanopillars or nanowires, and devices [54–57]. Other phenomenon, such as Coulomb blockade, is used for understanding the transport through quantum dots. Recently, spin blockade

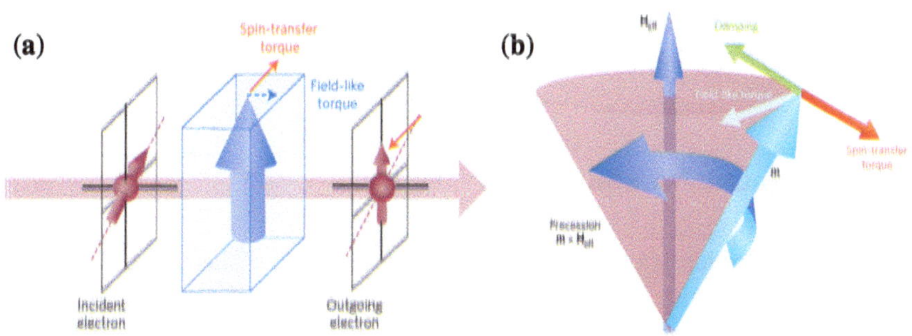

Fig. 28.5 Spin-transfer torque phenomenon: **a** Current-induced torque, **b** Dynamics of STT [47]

mechanism is applied to describe the transport phenomena in double-quantum-dots [58].

28.3 Magnetic Nanostructures and Materials

Magnetic structures of a material are determined based on the ordered arrangement of magnetic spins, which are also related to the crystal structures and the unit cells. However, the magnetic structures of materials change when the size of the particles approach the nanometer range. The magnetic moments of the nanostructures are arranged spontaneously under certain principles without external fields due to various coupling interactions between magnetic atoms. Again, in addition to the exchange interaction, the magnetic anisotropy and dipole interaction play important role for the formation of noncollinear magnetic structures [59]. Various physical parameters affect the nanostructure dimensions and develop potentially diverse nanostructure geometries with different properties and novel applications which are explained by new concepts. The sensing properties of the material depend on the microstructure and morphology of materials, modification of matrix materials by doping and the formation of new nanostructure of the materials [20]. Therefore, good magnetic nanostructure is very much important for transport and sensing applications.

It is widely accepted that magnetic materials show interesting properties and behaviour for different applications. Moreover, the science of nanoscale magnetic materials has contributed a lot for the fundamental understanding of magnetism and turned to be significantly more attractive with enhancing behaviour for numerous advanced technologies. From application point of view, as hard magnetic materials, these materials are used in magnetic separation, magneto-biology, and electrical machinery and magnetic medicine and as soft magnetic materials in stator or rotator parts of generators and motors [60]. As a functional material its application includes giant magnetic resistors, magnetic fluids, skyrmion-based functional devices, magneto-strictive materials and magnetic refrigeration materials. There is also a systematic change in the magnetic properties and exchange coupling happens, when these materials are exposed to environment and magnetic field, which is a good indication for sensing applications. In this direction, different types of nanomagnetic materials and nanostructures are presented.

28.4 Magnetic Nanomaterials

Magnetic nanoparticle is a topic of growing interest because of their versatile applications. Nanosized particles have application in structural, optical, magnetic, electronic and specially in medical field for treatment of in cancer, cardiovascular diseases and neurological disorders. The synthesis of nanoparticles by different methods, functionalization and characterization of nanoparticles demonstrates strong transport and sensing properties at room temperature [61, 62]. Especially magnetic nanoparticles such as iron oxide are studied extensively for different applications [63, 64]. However, its applications are limited in sensing and detection due to low magnetization. Therefore, there is a need to study other magnetic materials in pure form, different structure or coated with some other materials for further application [65, 66].

28.4.1 FeCo Nanoparticles

FeCo is a soft magnetic nanoparticle which shows high saturation magnetization (Ms) and Curie temperature, high permeability, low coercivity and low anisotropy [67, 68]. Sau et al. reported that cubic geometry FeCo nanoparticles are good for magnetic-based biosensing devices due to larger contact area which create more robust binding to a sensor platform [69, 70]. Furthermore, Treguer-Delapierre et al. mentioned that the shape and size of magnetic nanoparticles strongly affects the properties of magnetic nanoparticles [71]. Selvan et al. confirmed that nanoparticles must be coated with some protective layer such as a thin layer of silica which protects the magnetic cores from degradation and gives uniform magnetic properties [72, 73] In addition, coating increases the interfacial contact, permits the facile surface functionalization, and decreases the void fraction for enhanced sensitivity of magnetic nanoparticles [70, 74]. Baselt et al. also reported that for sensing application which depends on molecular recognition (e.g. ligand or antibody binding), magnetic nanoparticles are coated with a robust, biocompatible and readily modifiable protective layer [75, 76]. Recently, Li et al. developed FeCo electrochemical sensor where FeCo nanoparticles are used to modify the magnetic composites for detection of synthetic metals related to human health [77]. From literature, it is observed that most of the studies are focussed on spherical or cubic FeCo nanoparticles coated with protective layer or embedding FeCo in a silica matrix [78–82] for stable magnetic properties needed for sensor applications. The magnetization data on FeCo nanoparticle of different shapes and sizes are still lacking for diverse sensing applications. FeCo nanoparticles of about 20 nm synthesized using coprecipitation method and some coated with silica using sol–gel process are presented for magnetic biosensing applications. The magnetic hysteresis curve of FeCo nanoparticles is depicted in Fig. 28.6a. The FeCo nanoparticles synthesized at lower temperature show a higher magnetic saturation value in comparison with the room-temperature magnetization value of silica-coated FeCo.

This magnetization result indicates that the synthesized FeCo nanoparticles have good magnetic sensing characteristics which can be useful for biomedical applications. Further, the synthesized FeCo nanoparticles can also be detected by the change in magnetoresistance which will bind to the sensor by direct or sandwich way for applications [83].

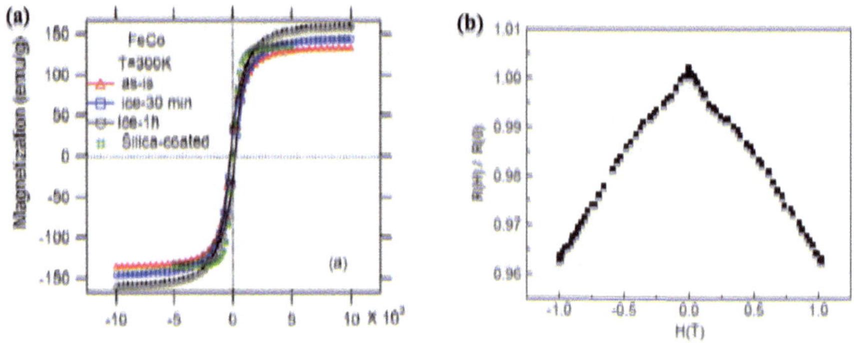

Fig. 28.6 **a** The magnetic hysteresis of FeCo nanoparticles [81], **b** Room temperature MR as a function of H [90]

28.4.2 CMR Nanomaterials

Colossal magnetoresistance (CMR) materials have novel physical and transport properties. It is reported that CMR effect is promising for high field application and sensor development technologies [84–86]. However, the main problem for development of CMR sensor technologies to higher readiness level arises due to the low temperature sensitivity to the magnetic field and high sensitivity to ambient temperature dependence of the material [87]. The nanostructured CMR material provides new magnetoresistance (MR) features comparing with that of other materials [88, 89]. The CMR materials such as $LaSrMnO_3$ (LSMO) and $LaCaMnO_3$ (LCMO) prepared using chemical versatile technique show interesting transport behaviour [90]. Size and shape of nanoparticles and nanostructures control the transport properties exhibiting metal–insulator transition at different temperature. Typically, metal-to-insulator (T_{MI}) transition shown by LCMO CMR samples is around 265 K with large MR of 15.4%. This magnetoresistance sensing characteristics is associated with spin-polarized tunnelling or spin-dependent scattering [91]. This boundary-induced MR always favours low temperatures and depends on the grain size. However, LSMO sample shows T_{MI} at about 340 K with MR of about 5.7% up to the field of 1 T.

The magnetoresistance value changes with change of magnetic field and MR curves become liner at higher external magnetic field as shown in Fig. 28.6b. The microstructure of the grain domain plays important role where long-range magnetic order is increased by increasing the external applied field. The relationship between MR and H indicates good transport characteristics of the materials. This CMR manganite nanostructure can act as a low magnetic field sensor. Here the sensitivity of the signal will be measured with respect to the change in magnetic field. The magnetoresistance of the polycrystalline sample exhibiting the CMR effect generally does not saturate up to high fields. The sensitivity of CMR nanostructure can be described as the normalized value of the voltage change across the sensor's resistance. In GMR sensor, the magnetoresistance saturates at a certain magnetic low field, but CMR sensor does not saturate up to high field. A CMR sensor can also respond to the low fields emanating from the recording media.

28.4.3 Co/Pt Multilayer

Co/Pt multilayers are potential sensing magneto-optic recording materials with perpendicular magnetic anisotropy as alternatives to other magnetic materials and multilayers. Co/Pt ultrathin multilayers are also used for detecting very low field at room temperature and high-sensitivity Hall sensor [92]. Muehlenhoff et al. reported spin canting effect of Co/Pt multilayers and both single and double structures of Co/Pt as GMR sensors [93]. Co/Pt multilayers always show magnetic anisotropies in out-of-plane fields. Gabureaca et al. showed that a CoPt ultrathin multilayer can be used as a high-sensitivity Hall sensor at room temperature for detecting very low fields [94].

In the development of Co/Pt-based GMR sensor in wide range of magnetic field, the magnetization results of the multilayer specially the free layer systems are important [94, 95]. Growth of the textured multilayer with proper phase is necessary. A precise proportion of Co to Pt must be maintained for phase stability, and hence lower magnetic transition temperature and associated properties can be achieved for sensing application. The highly fcc-textured Co/Pt multilayer films [Pt (2 nm)/[Co (0.4 nm)/Pt (2 nm)]5/Sapphire (0001)] synthesized by the pulsed laser deposition (PLD) method show remarkable enhancement in the coercivity field. The temperature dependence of the field-cooled (FC) magnetization curves of Co/Pt multi-layered film shown in Fig. 28.7a over a temperature range of 5–410 K with the field applied parallel to the film's surface indicates the onset of the ferromagnetic transition just above 400 K, one of the record-high magnetic transitions in such films without any antiferromagnetic capping layer. This remarkable enhancement in both magnetic transition and the coercivity field is a good sign for magneto Hall sensor, which is attributed due to the coupling of ferromagnetic nanoparticles in the multilayers. This is generally observed where ferromagnetic nanoparticles are embedded in an antiferromagnetic matrix [96]. Furthermore, the temperature dependence of the ferromagnetic resonance (FMR) linewidths presented in Fig. 28.7b clearly demonstrates that the film would show the ferromagnetic response up to 401 K. To emphasize the ferromagnetic behaviour above 400 K, the ferromagnetic resonance curve has been shown in the inset of Fig. 28.4 at $T = 401$ K, which is almost the limiting temperature in our machine. A small peak above 3 kg is the response from the cavity. The shape of the resonance curve clearly illustrates that the film is still in the ferromagnetic state at 401 K. This result demonstrates that the quality fabricated by the PLD technique is superior enough for real applications compared with other techniques.

Featuring such a peak is the fingerprint of the FMR effect controlling the magnetization dynamics and the magneto-impedance variations. This FMR variation results indicate that multilayer system can be used in the integrated system as sensor

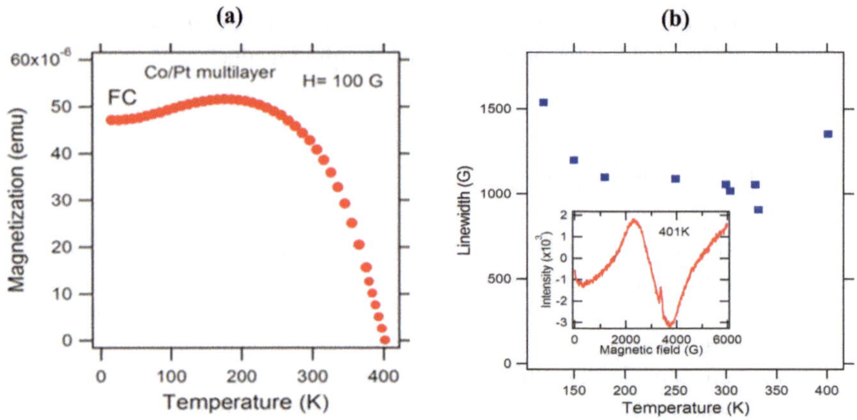

Fig. 28.7 **a** The temperature-dependent field-cooled (FC) magnetization curves, **b** The ferromagnetic resonance (FMR) line width variation with temperature

elements. Co/Pt multilayer showing higher sensitivity at high fields brings the possibility to design single sensor at the distinct field ranges. It is reported that, high magneto-impedance sensitivity material can be integrated with ferrimagnetic insulators for future technological applications, as ultra-fast sensors. These results provide good results for design and development of magneto-resistive sensor and sensor materials in devices.

28.4.4 CoFe/Bi/Co Multilayers

Magnetic junctions based on magnetic layer and interlayer such as insulators, semiconductors, or superconductors have different applications with interesting magnetic and transport properties. Properties such as giant magnetoresistance (GMR) and spin-polarized tunnelling have potential applications in devices. Litvinov et al. reported Co/Cu/Co magnetoresistance-based sensor coated with a thin layer of alumina or silica for biosensing application with functionalization of a molecular recognition element [97]. The tri-layer structure CoFe/Bi/Co, with Bi as a semimetal, is deposited by dc-sputtering. This structure shows excellent magnetic behaviour with the existence of ferromagnetic coupling or weak AF coupling [98]. The oscillatory type of interlayer coupling is shown in Fig. 28.8 by the tri-layer with period of oscillation of about 18 nm. The same period of oscillation has occurred in the tri-layer showing an increase of interlayer thickness and a decay of amplitude of oscillation. The strength of the interlayer coupling also changes with the change of temperature [99]. This type of coupling at room temperature is good for magneto-engineering and advanced sensor applications.

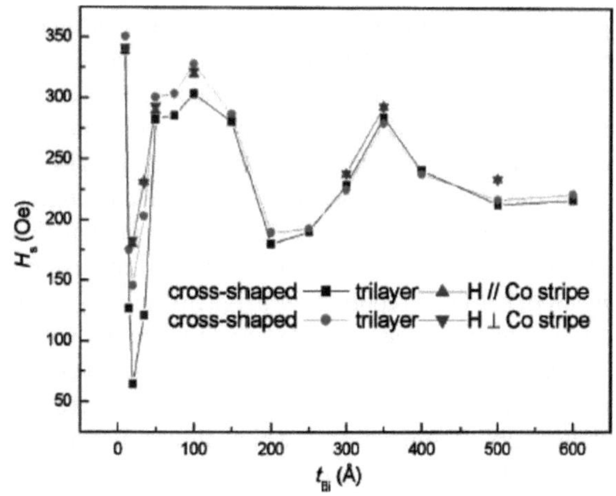

Fig. 28.8 Oscillatory trend of H_S variation with the thickness of bilayer [98]

Recently, it is also reported that with the emerging advanced magnetic 2D materials, magnetic gas sensors can also be designed and developed based on the performance of different materials [100]. Different magnetic multilayer systems such as above can be designed with the use of 2D magnetic materials, and the magnetic properties of the active material system can be measured for the development of magnetic gas sensors.

28.4.5 *Magnetic Tunnel Junctions (MTJ)*

Nowadays magnetic tunnel junctions (MTJ) are considered for many magnetic sensor applications due to small size, low cost, high sensitivity, low power consumption, and high potential for future magnetic sensing devices [101]. Freitas et al. reported high field sensitivity TMR sensor based on MTJ, which can detect small changes in magnetic fields [102, 103]. A very high value of MR ratio for tunnelling magnetoresistance and tunnelling electro-resistance effects is observed. Fujiwara et al. also suggest that high-sensitivity TMR sensors in combination with large magnetic nanostructures or nanoparticles have potential for biomarker detection applications. In the MTJ, the ultrathin barrier layer in addition to ferromagnetic layer plays an important role [104, 105]. However, there are many parameters involved for the design of MTJ sensors which needs to be optimized for specific applications. As obvious, the measurement of magnetoresistance and its optimization are crucial for the design of magnetic sensor. A linear MTJ can be designed based on the three-dimensional magnetic field measurements [106].

It is also known that tunnel magnetoresistance behaviour is generally shown by the MTJs where two ferromagnetic electrodes are separated by a thin insulating barrier. Wei et al. reported that with ferroelectric barrier, MTJ sensors have great potential for multifunctional devices and electronics [107]. Zurauskiene et al. have mentioned that CMR-based nanostructured lanthanum manganite films show a room-temperature enhancement of low-field magnetoresistance which can act as ferromagnetic electrode in MTJ for magnetic sensor applications [108, 109]. The growth and optimization of manganite film thickness in MTJ low-field sensors is necessary to know the signal responses to magnetic field changes.

MTJ with two LaSrMnO (LSMO) ferromagnetic electrodes with a barrier layer of $PbZrTiO_3$ (PZT), grown on STO substrate using pulsed laser deposition (PLD) method, has been reported for magneto-sensing applications [110]. The MTJ shows a ferromagnetic hysteresis at room temperature with magnetic onset at about 360 K.

TMR behaviour of LSMO/PZT/LSMO/STO MTJ at various temperatures is presented in Fig. 28.9. At very low temperature about 5 K, zero-filed MR curve shows very prominent peak which almost diminishes at room temperature of about 300 K. MR value of 5% is observed at room temperature compared to 16% at 5 K. The effective barrier layer is very much important to MTJ characteristics [111]. The ferroelectric polarization of PZT barrier layer influences the tunnelling behaviour. The magnetic behaviour of ferromagnetic layers, formation of local current domain and ferroelectric character of barrier layer also change the MR behaviour of MTJ [112, 113]. The magnetic properties of LSMO layer effectively influence MR more if the MTJ is fabricated at higher temperature as reduced magnetization is reduced if it is fabricated above room temperature. This type of MTJ showing TMR characteristics is very good for sensor applications. In a simple magnetic tunnel junction (MTJ) like above, skyrmion-based sensing can be used for energy-efficient signal conversion, magneto-impedance as a simple and medical imaging applications.

28.5 Conclusion and Future Prospective

Magnetic nanostructures such as nanoparticles, thin film, multilayer and junctions are grown successively for magneto-sensor application. The developed nanostructure has shown interesting transport and sensing characteristics. However, from the scientific point of view, more comprehensive researches in the field of nanosensors are necessary for futuristic device applications, such as improving the performance of existing materials, design and development of new nanostructures, design of device architecture, etc. More investigations are needed for the development of skyrmion-based MTJ and for efficient magnetic technologies for sensing pico-tesla signals using spintronic-based magneto-sensing devices. This field will also open new possibilities for the development of MR sensors for the next generation of magnetomyography study.

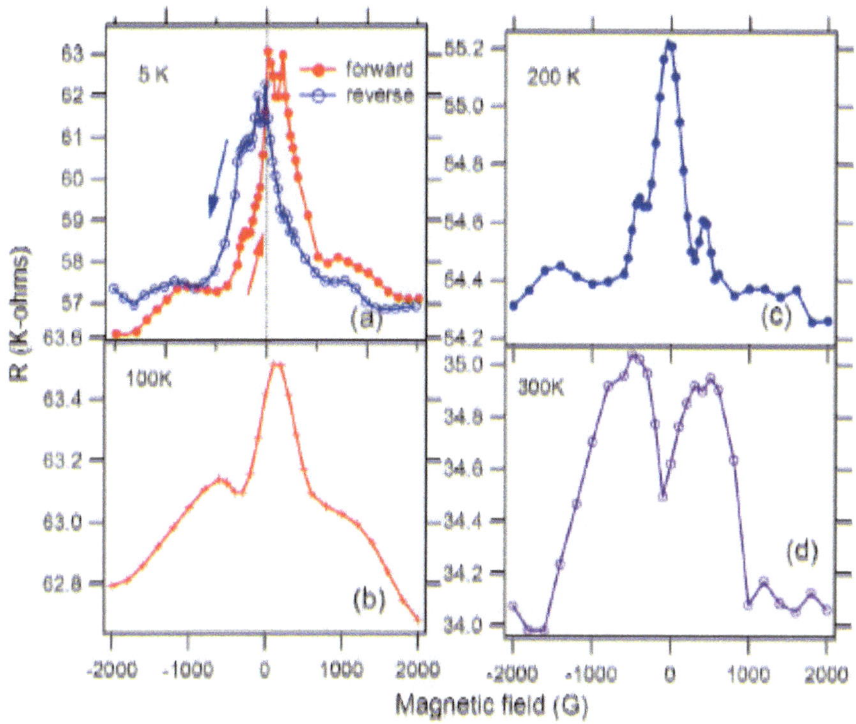

Fig. 28.9 The TMR curves of MTJ at low and high temperatures [107]

Acknowledgements Author acknowledges and thanks the collaborator for contribution to this work.

References

1. Jeevanandam, J., Barhoum, A., et al.: Review on nanoparticles and nanostructured materials: history, sources, toxicity and regulations. Beilstein J. Nanotechnol. **9**, 1050–1074 (2018)
2. Pradhan, S.C., Phadikar, J.K.: Nonlocal elasticity theory for vibration of nanoplates. J. Sound Vib. **325**, 206–223 (2009)
3. Polytechnique Montréal: A giant step toward miniaturization: nanotechnology transforms molecular beams into functional nano-devices. ScienceDaily (2013). www.sciencedaily.com/releases/2013/04/130403154422.htm
4. Aykaç, A., Gergeroglu, H., et al.: An overview on recent progress of metal oxide/graphene/CNTs-based nanobiosensors. Nanoscale Res. Lett. **16**, 65–70 (2021)
5. Grundler, D., Hengstmann, T., et al.: Magnetic nanostructures for lateral spin-transport devices. Braz. J. Phys. **34**, 2B (2004)
6. Li, Y., Cheng, H., et al.: The progress of magnetic sensor applied in biomedicine: a review of non-invasive techniques and sensors. J. Chin. Chem. Soc. **68**(2), 216–227 (2021)
7. Pandey, R., Chusuei, C.: Carbon nanotubes, graphene, and carbon dots as electrochemical biosensing composites. Molecules **26**(21), 6674 (2021)

8. Zuo, S., Heidari, H., et al.: Miniaturized magnetic sensors for implantable magnetomyography. Adv. Mater. Technol. **5**, 1–15 (2020)
9. Ennen, I., Kappe, D., et al.: Giant magnetoresistance: basic concepts, microstructure, magnetic interactions and applications. Sensors **16**, 904 (2016)
10. Fisher, P., Sanz-Hernandez, et al.: Launching a new dimension with 3D magnetic nanostructures. APL Mater. **8**, 010701 (2020)
11. Vavassori, P., Gobbi, M., et al.: Magnetic Nanostructures for the manipulation of individual nanoscale particles in liquid environments. J. Appl. Phys. **107**, 09B301–09B3015 (2010)
12. Murthy, S., Maria, P., et al.: Metal oxide nanoparticles and their applications in nanotechnology. SN Appl. Sci. **1**, 607 (2019)
13. Djamal, M., Ramli, R.: Giant magnetoresistance sensors based on ferrite material and its applications (2020). https://doi.org/10.5772/intechopen.70548
14. Peixoto, L., Et, M., al,: Magnetic nanostructures for emerging biomedical applications. Appl. Phys. Rev. **7**, 011310 (2020)
15. Azim, Z.A., Chen, M.C.: Skyrmion sensor-based low-power global interconnects. IEEE Trans. Magn. **53**, 1 (2017)
16. Hanneken, C., Otte, F., et al.: Electrical detection of magnetic skyrmion by tunnelling non collinear magneto resistance. Nat. Nanotechnol. **10**, 1039–1042 (2005)
17. Chen, H., Bouckaert, W., Majetich, S.A.: Tunnel magnetoresistance detection of skyrmions. J. Magn. Magnet. Mater. **541**, 168552 (2022)
18. Wiesendanger, R.: Nano-scale magnetic skyrmions in metallic films and multilayers: a new twist for spintronics. Nat. Rev. Mater. **1**, 16044 (2016)
19. Chu, Y., Li, L.L., et al.: Generation and manipulation of skyrmions and other topological spin structures with rare metal. Rare Met. **41**, 200–2216 (2022)
20. Brede, J., et al.: Long-range magnetic coupling between nanoscale organicmetal hybrids mediated by a nanoskyrmion lattice. Nat. Nanotechnol. **9**, 1018–1023 (2014)
21. Cinchetti, M.: Topology communicates. Nat. Nanotechnol. **9**, 965–966 (2014)
22. Su, D., Wu, K., et al.: Detection of influenza a virus in swine nasal swab samples with a wash-free magnetic bioassay and a handheld giant magnetoresistance sensing system. Front. Microbiol. **10**, 1077 (2019)
23. Krishna, V.D., Wu, K., et al.: Nanotechnology: review of concepts and potential application of sensing platforms in food safety. Food Microbiol. **75**, 47 (2018)
24. Barman, A., Mondal, S., et al.: Magnetization dynamics of nanoscale magnetic materials: a perspective. J. Appl. Phys. **12**, 170901 (2020)
25. Shinde, P.V., Rout C.S.: Magnetic gas sensing: working principles and recent developments Nanoscale Adv. **3**, 1551 (2021)
26. Maksymov, I.S., Kostylev, M.: Magneto-electronic hydrogen gas sensors: a critical review. Chemosensors 10, 49 1–40 (2022)
27. Lueng, C., Lupo, P., et al.: Sensitivity of ferromagnetic resonance in PdCo alloyed films to hydrogen gas. Int. J. Hydrogen Energy **44**, 7715–7724 (2019)
28. Zhang, W., Somg, Y.: Multilayered Magnetic Thin Films for Electron Transport Control and Signal Sensing: From GMR, CMR, TMR to Quantum Anomalous Holzer Effect, Chapter 3 (2021). https://doi.org/10.1002/9783527344987.ch3
29. Vetro, F.A.: Effect of Heat current on magnetization dynamics in magnetic insulators and nanostructures. Doctoral thesis (2015). École Poly technique Federale de lasuane 1–88
30. Baibich, M.N., et al.: Giant magnetoresistance of (001) Fe/(001)Cr magnetic superlattices. Phys. Rev. Lett. **61**, 2472–2475 (1988)
31. Alex-enraver. Spin-valveGMR.svg.URL http://uploadwikimedia.org/wikipedia/commons/9/93/Spin-valve_GMR.svg
32. Zhang, X., Butler, W.: Theory of giant magnetoresistance and tunneling magnetoresistance. In: Handbook of Spintronics. Springer Science and Business Media LLC 1–56 (2015)
33. Fermon, C., Van de Voorde, M.: Nanomagnetism: Applications and Prospectives. Wiley Publications (2017) ISBN 978-3-27-33985-3
34. Itoh, H., Inoue, J.C.: Theory of tunnel magnetoresistance. J. Magn. Soc. Jpn. **30**, 1–37 (2006)

35. Julliere, M.: (1975) Tunneling between ferromagnetic films. Phys. Lett. A **54**, 225–226
36. Useinov, N., Useinov, K.H., et al.: Tunnel magnetoresistance in magnetic tunnel junctions with embedded nanoparticles. In: IEEE International Magnetics Conference (INTERMAG), 1–1 (2015)
37. Useinov, A., Kosel, J., et al.: Resonant tunnel magnetoresistance in double-barrier planar magnetic tunnel junctions. Phys. Rev. B **84**, 085424 (2011)
38. Useinov, A., Kosel, J.: Spin asymmetry calculations of the TMR-V curves in single and double-barrier magnetic tunnel junctions. IEEE Trans. Magn. **47**(10), 2724–2727 (2011)
39. Yang, H., Yang, S.H., Parkin, S.S.P.: Crossover from Kondo-assisted suppression to co-tunneling enhancement of tunneling magnetoresistance via ferromagnetic nanodots in MgO tunnel barriers. Nano Lett. **8**(1), 340–344 (2008)
40. Slonczewski, J.C.: Current-driven excitation of magnetic multilayers. J. Magn. Magn. Mater. **159**, L1 (1996)
41. Useinov, A., Chshiev, M., Manchon, A.: Controlling the spin-torque efficiency with ferroelectric barriers. Phys. Rev. B **91**, 064412 (2015)
42. Useinov, A., Kalitsov, A., Velev, J., Kioussis, N.: Ferroelectric control of spin-transfer torque in multiferroic tunnel junctions. Phys. Rev. B **91**, 094408 (2015)
43. Kaviraj, B., Sinha, J.: Relativistic torques induced by currents in magnetic materials: physics and experiments. RSC Adv. **8**, 25079–25093 (2018)
44. Baraduc, C., Cshiev, M., et al.: Introduction to spin transfer torque. In: Nanomagnetism and Spintronics, pp. 173–192 (2010)
45. Pathak, S., Youm, C., Hong, J.: Impact of spin-orbit torque on spin-transfer torque switching in magnetic tunnel junctions. Sci. Rep. **10**, 2799 (2020)
46. Ralph, D.C., Stiles, M.D.: Spin transfer torques. J. Magn. Magn. Mater. **320**, 1190–1216 (2008)
47. Brataas, A., Kent, A.D., et al.: Current-induced torques in magnetic materials. Nat. Mater. **11**, 372–381 (2012)
48. Useinov, A., Kosel, J., Useinov, N.K., Tagirov, L.R.: Resonant tunnel magnetoresistance in double-barrier planar magnetic tunnel junctions. Phys. Rev. B **84**, 085424 (2011)
49. Wang, C., Seinige, H., Tsoi, M.: Ferromagnetic resonance driven by an ac current: a brief review. Low Temp. Phys. **39**, 247–251 (2013)
50. Tulapurkar, A.A., et al.: Spin-torque diode effect in magnetic tunnel junctions. Nature **438**, 339–342 (2005)
51. Sankey, J.C., et al.: Measurement of the spin-transfer-torque vector in magnetic tunnel junctions. Nat. Phys. **4**, 67–71 (2007)
52. Wang, C., et al.: Angular dependence of spin-transfer torque in magnetic tunnel junctions. Phys. Rev. B **79**, 224416 (2009)
53. Kubota, H.etal,: Quantitative measurement of voltage dependence of spin-transfer torque in MgO-based magnetic tunnel junctions. Nat. Phys. **4**, 37–41 (2008)
54. Sankey, J.C., et al.: Spin-transfer-driven ferromagnetic resonance of individual nanomagnets. Phys. Rev. Lett. **96**, 227601 (2006)
55. Chen, W., Beaujour, J.M.L., De Loubens, G., Kent, A.D., Sun, J.Z.: Spin-torque driven ferromagnetic resonance of CoNi synthetic layers in spin valves. Appl. Phys. Lett. **92**, 012507 (2008)
56. Mourachkine, A., Yazyev, O.V., Ducati, C., Ansermet, J.P.: Template nanowires for spintronics applications: nanomagnet microwave resonators functioning in zero applied magnetic field. Nano Lett. **8**, 3683–3687 (2008)
57. Biziere, N., Murè, E., Ansermet, J.-P.: Microwave spin-torque excitation in a template-synthesized nanomagnet. Phys. Rev. B **79**, 012404 (2009)
58. Wu, K., Saha, R., et al.: Magnetic nanosensor-based virus and pathogen detection strategies before and during COVID-19. ACS Appl. Nano Mater. **3**(10), 9560–9580 (2020)
59. Luo, S., Uou, L.: Skyrmion devices for memory and logic applications. APL Mater. **9**, 050901 (2021)

60. Joa, Y., Yang, C.: A review on electrospun magnetic nanomaterials: methods, properties and applications. J. Mater Chem C **9**, 9042–9092 (2021)
61. Rizvi, M., Gerengi, H. et al.: Functionalization of nanomaterial: synthesis and characterization. ACS Symp. Ser. **1418**(1), 1-–26 (2022)
62. Zou, D., Jin, L., et al.: Rapid detection of salmonella in milk by biofunctionalised magnetic nanoparticle cluster sensor based on nuclear magnetic resonance. Int. Dairy J. **91**, 82–88 (2019)
63. Pastucha, M., Farka, Z., et al.: Magnetic nanoparticles for smart electrochemical immunoassays: a review on recent developments. Microchim. Acta. **186**(5), 312 (2019)
64. Tran, H.V., Ngo, N.M., et al.: Multifunctional iron oxide magnetic nanoparticles for biomedical applications: a review. Materials **15**, 503 (2022)
65. Liu, S., Yu, B., et al.: Preparation, surface functionalization and application of Fe_3O_4 magnetic nanoparticles. Adv. Colloid Interface Sci. **281**, 102165 (2020)
66. Katz E.: Synthesis, properties and applications of magnetic nanoparticles and nanowires—a brief introduction. Magnetochemistry **5**, 61 1–15 (2019)
67. Kodama, D., Shinoda, K., et al.: Synthesis of Fe-Co alloys particle by modified polyol process. IEEE Trans. Magn. **42**(10), 2796–2798 (2006)
68. Zhang, Z., Zhnag, J., et al.: (2018) Structure and magnetism evolution from FeCo nanoparticles to hollow nanostructure conversion for magnetic applications. ACS Appl. Nano Mater. **1**(10), 5837–5842 (2018)
69. Sau, T., Rogach, A.L.: Non spherical noble metal nanoparticles: colloid-chemical synthesis and morphology control. Adv. Mater. **22**, 1781–1804 (2010)
70. Dovgolevsky, E., Haick, H., et al.: Direct observation of the transition point between quasi-spherical and cubic nanoparticles in a two-step seed-mediated growth method. Small **2**, 2059–2066 (2008)
71. Treguer-Delapierre, M., Majimel, J., Mornet, S., Duguet, E., Ravaine, S.: Utilization of gold nanostructures in biomedical applications. Gold Bull. **41**, 195–207 (2008)
72. Guerrero-Martinez, A., Perez-Juste, J., Liz-Marzan, L.M.: Recent progress on silica coating of nanoparticles and related nanomaterials. Adv. Mater. **22**, 1182–1195 (2010)
73. Selvan, S.T.: Silica-coated quantum dots and magnetic nanoparticles for bioimaging applications (mini-review). Biointerphases **5**, FA10−FA15 (2010)
74. Weichold, O., Tigges, B., Bertmer, M., Moller, M.A.: Comparative study on the dispersion stability of aminofunctionalised silica nanoparticles made from sodium silicate. J. Colloid Interface Sci. **324**, 105–109 (2008)
75. Baselt, D.R., Lee, G.U., Natesan, M., Metzger, S.T., Sheehan, P.E., Colton, R.J.: A biosensor based on magnetoresiatnce technology. Biosens. Bioelectron. **13**, 731–739 (1998)
76. Koh, I., Josephson, L.: Magnetic nanoprticle sensor. Sensors **9**, 8130–8145 (2009)
77. Li, L., Zheng, B., Tang, M., Feng, J., Huang, K., Liu, Y.: Research on electrochemical molecular imprinting sensor of P-FeCo nanoparticles. In: IOP Conf. Series: Materials Science and Engineering 914, 012006 (2020)
78. Thirumal, E., Prabhu, D., Chattopadhyay, K., Ravichandran, V.: Magnetic, electric and dielectric properties of FeCo alloy nanoparticles dispersed in amorphous matrix. Phys. Status Solidi A **207**, 2505–2510 (2010)
79. Carta, D., Casula, M.F., Bullita, S., Falqui, A., Corrias, A.: Cubic silica-coated and amine-functionalized FeCo nanoparticles with high saturation magnetization. J. Nanopart. Res. **13**, 3489–3501 (2011)
80. Falqui, A., Corrias, A., Gass, M., Mountjoy, G.A.: Transmission electron microscopy study of Fe-Co alloy nanoparticles in silica aerogel matrix using HREM, EDX, and EELS. Microsc. Microanal. **15**, 114–124 (2009)
81. Zhang, K., Holloway, T., Bahoura M., Pradhan, A., Prabakaran, R., Pradhan, J., Smith, S., Hall, J., Ramesh, G., Sahu, D., Huang, J.L.: Europium doped Gd_2O_3 and FeCo nanoparticles for biomedical applications. In: Proc. SPIE 7291, 729104-1−729104-10 (2009)
82. Harris, M.T., Brunson, R.R., Bryers, C.H.J.: The base-catalyzed hydrolysis and condensation reactions of dilute and concentrated TEOS solutions. Non-Cryst. Solids **121**, 397–403 (1990)

83. Kolhatkar, A.G., Nekrashevich, I., et al.: Cubic silica-coated and amine-functionalized FeCo nanoparticles with high saturation magnetization. Chem. Mater. **25**, 1092–1097 (2013)
84. Pla, D., Jimenez, C., Burriel, M.: Engineering of functional manganites grown by MOCVD for miniaturized devices. Adv. Mater. Interfaces **4**, 1600974 (2017)
85. Balevicius, S., Zurauskiene, N., et al.: Nanostructured thin manganite films in megagauss magnetic field. Appl. Phys. Lett. **101**, 092407 (2012)
86. Pekała, M., Pękała, K., Drozd, V.: Magnetotransport study of nanocrystalline and polycrystalline manganites La0.8Sr0.2MnO3 in high magnetic fields. J. Appl. Phys. **117**, 175902 (2015)
87. Israel, C., Calderón, M.J., Mathur, N.D.: The current spin on manganites. Mater. Today **10**, 24–32 (2007)
88. Wang, G., Wang, Z.D.: Synthesis and magnetocaloric effect of $(La_{0.47}Gd_{0.2})\ Sr_{0.33}MnO_3$ polycrystalline nanoparticles. Mater. Sci. Eng. B **116**, 183–188 (2006)
89. Naguyen, L.H., Hung, L.X., et al.: Composites (1–x) La0.7Ca0.3MnO3/ xLa0.7Sr0.2Ca0.1MnO3: electrical transport properties and enhancing of low-field-magnetoresistance and colossal magnetoresistance. J. Alloy. Compd. **849**, 1566071–1566110 (2020)
90. Sahu, D., Roul, B.K., et al.: Synthesis of $La_{0.7}Sr_{0.3}MnO_3$ materials by versatile chemical technique. Phys. B **369**, 209–214 (2005)
91. Damay, F., Driscoll, J.L., et al.: Magnetotransport properties of $La_{0.7}Ca_{0.3}MnO_3$ thick films prepared by spray pyrolysis. J. Mater. Res. **16**, 784–789 (2001)
92. Zhang, S.I., Teng, I., et al.: Large enhancement of the anomalous Hall effect in Co/Pt multilayers sandwiched by MgO layers. Appl. Phys. Lett. **97**, 222504 (2010)
93. Muehlenhoff, C., Vogler, C., Raberg, W.: Spin-canting effects in GMR sensors with wide dynamic field range. IEEE Sensor **21**, 12 (2021)
94. Gabureaca, M.S., Alvarezb, I.S.E., Marrows, C.H.: Co/Pt hall sensors for low field detection. Procedia Chemistry **1**, 851–854 (2009)
95. González, J.A., Andrés, J.P., López Antón, R.: Applied trends in magnetic rare earth/transition metal alloys and multilayers. Sensors **21**, 5615 (2021)
96. Skumryev, V., Stoyanov, S., et al.: Beating the superparamagnetic limit with exchange bias. Nature (London) **423**, 850–853 (2003)
97. Litvinov, D., Willson, R.: U.S. Patent Application No. 20100188075 (2010)
98. Hsu, J., Sahu, D.R.: Interlayer coupling in a trilayer junction having bismuth as spacer layer. Appl. Phys. Lett. **86**, 192501–192503 (2005)
99. Hsu, J., Sahu, D.: A transition of interlayer coupling from antiferromagntic to ferromagnetic observed at low temperature in CoFe/Bi/Co trilayers. J. Appl. Phys. **101**, 09D114 (2007)
100. Shinde, P.V., Rout, C.S.: Magnetic gas sensing: working principles and recent developments. Nanoscale Adv. **3**, 1551–1568 (2021)
101. Jin, Z., Koo, T.M., et al.: Highly sensitive magnetic sensor for detecting magnetic nanoparticles based on magnetic tunnel junctions at a low static field. AIP Adv. **11**, 015046 (2021)
102. Freitas, P.P., Ferreira, R., Cardoso, S., Cardoso, F.: Magnetoresistive sensors. J. Phys.: Condens. Matter **19**, 165221 (2007)
103. Fujiwara, K., Oogane, M., et al.: Magnetocardiography and magnetoencephalography measurements at room temperature using tunnel magneto-resistance sensors. Appl. Phys. Express **11**, 023001 (2018)
104. Garcia, V., Bibes, M., et al.: Ferroelectric control of spin polarization. Science **327**, 1106 (2010)
105. Pantel, D., Goetze, S., et al.: Reversible electrical switching of spin polarization in multiferroic tunnel junctions. Nat. Mater. **11**, 289 (2012)
106. Nguyen, H., Persson, A.: Tailoring the properties of a magnetic tunnel junction to be used as a magnetic field sensor. In: Conf. on Solid State Physics and Materials VII (SPMS 2011), Ho Chi Minh, 7–9 November 2011 (2011)
107. Wei, Y., Matzen, S., et al (2019) Magnetic tunnel junctions based on ferroelectric $Hf_{0.5}Zr_{0.5}O_2$ tunnel barriers. Phys. Rev. Appl. **12**, 031001

108. Zurauskiene, N., Stankevich, V., et al.: Enhancement of room-temperature low-field magnetoresistance in nanostructured lanthanum manganite films for magnetic sensor applications. Sensors **22**, 4004 (2022)
109. Vertelis, V., Balevicˇius, S., et al.: The application of a CMR-B-scalar sensor for the investigation of the electromagnetic acceleration of type II superconductors. Sensors **21**, 1293 (2021)
110. Mundle, R., Konda, B., et al.: Maganite based magnetic tunnel junction with piezoelectric barrier. J. Appl. Phys. **105**, 07C907 (2009)
111. Meijer, R.C., I, et al.: Electrical current distribution across a metal–insulator–metal structure during bistable switching. J. Appl. Phys. **90**, 2892 (2001)
112. Szot, K., Speier, W., et al,: Switching the electrical resistance of individual dislocations in single-crystalline $SrTiO_3$. Nat. Mater. **5**, 312–320 (2006)
113. Pradhan A.K., Hunter. D., Sahu. D.R., et al.: Magnetic properties of $La_{0.6}Sr_{0.4}MnO_3$ thin films on $SrTiO_3$ and buffered Si substrates with varying thickness. J. Appl. Phys. **103**, 023914 (2008)

Correction to: Development of Rapidly Quenched Amorphous-Nanostructured Materials for Sensor Applications

Somnath Das, Rajat Kumar Roy, Dev Kumar Mahato, and Ashis Kumar Panda

Correction to: Chapter 27 in: D. Mohanta and P. Chakraborty (eds.), *Nanoscale Matter and Principles for Sensing and Labeling Applications*, Advanced Structured Materials 206, https://doi.org/10.1007/978-981-99-7848-9_27

The original version of this chapter has been revised by renumbering the entries for the reference list throughout. The correction to this book have been updated with the changes.

The updated version of this chapter can be found at
https://doi.org/10.1007/978-981-99-7848-9_27

D. Mohanta and P. Chakraborty (eds.), *Nanoscale Matter and Principles for Sensing and Labeling Applications*, Advanced Structured Materials 206,
https://doi.org/10.1007/978-981-99-7848-9_29

MIX
Papier aus verantwortungsvollen Quellen
Paper from responsible sources
FSC® C105338

If you have any concerns about our products,
you can contact us on
ProductSafety@springernature.com

In case Publisher is established outside the EU,
the EU authorized representative is:
Springer Nature Customer Service Center GmbH
Europaplatz 3, 69115 Heidelberg, Germany

Printed by Libri Plureos GmbH
in Hamburg, Germany